3

AF758353

Realizing Africa's Rice Promise

FSC
www.fsc.org
MIX
Paper from
responsible sources
FSC® C013604

The Africa Rice Center (AfricaRice) is a leading pan-African research organization working to contribute to poverty alleviation and food security in Africa through research, development and partnership activities. It is one of the 15 international agricultural research Centers that are members of the CGIAR Consortium. It is also an intergovernmental association of African member countries.

The Center was created in 1971 by 11 African countries. Today its membership comprises 24 countries, covering West, Central, East and North African regions, namely Benin, Burkina Faso, Cameroon, Central African Republic, Chad, Côte d'Ivoire, Democratic Republic of Congo, Egypt, Gabon, The Gambia, Ghana, Guinea, Guinea-Bissau, Liberia, Madagascar, Mali, Mauritania, Niger, Nigeria, Republic of Congo, Senegal, Sierra Leone, Togo and Uganda.

AfricaRice temporary headquarters is based in Cotonou, Benin. Research staff are also based in Senegal, Nigeria, Tanzania, Côte d'Ivoire, Liberia and Sierra Leone.

For more information, visit www.AfricaRice.org

Realizing Africa's Rice Promise

Edited by

Marco C.S. Wopereis

Africa Rice Center

David E. Johnson

International Rice Research Institute

Nourollah Ahmadi

Centre de coopération internationale en recherche agronomique pour le développement

Eric Tollens

Katholieke Universiteit Leuven

and

Abdulai Jalloh

West and Central African Council for Agricultural Research and Development

www.cabi.org

AfricaRice

CABI is a trading name of CAB International

CABI
Nosworthy Way
Wallingford
Oxfordshire OX10 8DE
UK

Tel: +44 (0)1491 832111
Fax: +44 (0)1491 833508
E-mail: info@cabi.org
Website: www.cabi.org

CABI
38 Chauncey Street
Suite 1002
Boston, MA 02111
USA

Tel: +1 800 552 3083 (toll free)
Tel: +1 (0)617 395 4051
E-mail: cabi-nao@cabi.org

© CAB International 2013. All rights reserved. No part of this publication may be reproduced in any form or by any means, electronically, mechanically, by photocopying, recording or otherwise, without the prior permission of the copyright owners.

A catalogue record for this book is available from the British Library, London, UK.

Library of Congress Cataloging-in-Publication Data

Realizing Africa's rice promise / editors: Marco C.S. Wopereis ... [et al.].
p. cm.
Includes bibliographical references.
ISBN 978-1-84593-812-3 (hbk : alk. paper)
1. Rice--Africa. 2. Rice farming--Africa. 3. Sustainable agriculture--Africa. I. Wopereis, M.

SB191.R5R42 2013
633.1'8--dc23

2013017093

ISBN-13: 978 1 84593 812 3

Commissioning editor: Nigel Farrar
Editorial assistant: Emma McCann
Production editor: Tracy Head
Copy-editor: Guy Manners (freelance and latterly Green Ink, www.greenink.co.uk)
Cover photo: Jonne Rodenburg/AfricaRice

Typeset by SPi, Pondicherry, India.
Printed and bound in the UK by CPI Group (UK) Ltd, Croydon, CR0 4YY.

Contents

Contributors ix

Foreword xiii
Papa Abdoulaye Seck

Acknowledgements xv

Introduction xix
Marco C.S. Wopereis

SECTION 1 RICE IN AFRICA: AN OVERVIEW

1 **From WARDA to AfricaRice: an Overview of Rice Research for Development Activities Conducted in Partnership in Africa** 1
Eric Tollens, Matty Demont, Moussa Sié, Aliou Diagne, Kazuki Saito and Marco C.S. Wopereis

2 **Africa's Rice Economy Before and After the 2008 Rice Crisis** 24
Papa Abdoulaye Seck, Ali A. Touré, Jeanne Y. Coulibaly, Aliou Diagne and Marco C.S. Wopereis

3 **Estimation of Cultivated Area, Number of Farming Households and Yield for Major Rice-growing Environments in Africa** 35
Aliou Diagne, Eyram Amovin-Assagba, Koichi Futakuchi and Marco C.S. Wopereis

4 **Farmer Perceptions of the Biophysical Constraints to Rice Production in Sub-Saharan Africa, and Potential Impact of Research** 46
Aliou Diagne, Didier Y. Alia, Eyram Amovin-Assagba, Marco C.S. Wopereis, Kazuki Saito and Tebila Nakelse

SECTION 2 RICE GENETIC DIVERSITY AND IMPROVEMENT

5 **A Continent-wide, Product-oriented Approach to Rice Breeding in Africa** 69
Takashi Kumashiro, Koichi Futakuchi, Moussa Sié, Marie-Noëlle Ndjiondjop and Marco C.S. Wopereis

6 **Rice Varietal Release Systems in Africa** 79
Kayode Abiola Sanni, Ali A. Touré, Aliou Diagne, Fatimata Bachabi, Rosemary Murori, Rakesh Kumar Singh and Moussa Sié

7 **Diversity of Rice and Related Wild Species in Africa** 87
Kayode Abiola Sanni, Daniel D. Tia, David K. Ojo, Ayoni S. Ogunbayo, Mouritala Sikirou and N. Ruaraidh Sackville Hamilton

8 **Gene Flow in African Rice Farmers' Fields** 95
Edwin Nuijten and Paul Richards

9 **Making Rice Genomics Work for Africa** 108
Susan McCouch, Rod A. Wing, Mandè Semon, Ramaiah Venuprasad, Gary Atlin, Mark E. Sorrells and Jean-Luc Jannink

10 **Unlocking the *Oryza glaberrima* Treasure for Rice Breeding in Africa** 130
Mathias Lorieux, Andrea Garavito, Julie Bouniol, Andres Gutiérrez, Marie-Noëlle Ndjiondjop, Romain Guyot, César Pompilio Martinez, Joe Tohme and Alain Ghesquière

11 **Rice Genetic Improvement for Abiotic Stress Tolerance in Africa** 144
Khady N. Dramé, Baboucarr Manneh and Abdelbagi M. Ismail

12 **Integration of Molecular Markers in Rice Improvement: A Case Study on Resistance to *Rice yellow mottle virus*** 161
Marie-Noëlle Ndjiondjop, Laurence Albar, Mounirou Sow, Nasser Yao, Gustave Djedatin, Deless Thiemélé and Alain Ghesquière

13 **Hybrid Rice in Africa: Challenges and Prospects** 173
Raafat A. El-Namaky and Matty Demont

14 **Development of an Integrated Rice Seed Sector in Sub-Saharan Africa: Meeting the Needs of Farmers** 179
Amadou M. Bèye, Thomas Remington, Marco C.S. Wopereis and Aliou Diagne

SECTION 3 SUSTAINABLE PRODUCTIVITY ENHANCEMENT

15 **Towards a Better Understanding of Biophysical Determinants of Yield Gaps and the Potential for Expansion of the Rice Area in Africa** 188
Kazuki Saito, Andrew Nelson, Sander J. Zwart, Abibou Niang, Abdoulaye Sow, Hiroe Yoshida and Marco C.S. Wopereis

16 **Managing Weeds of Rice in Africa** 204
Jonne Rodenburg and David E. Johnson

17 **Managing the Major Diseases of Rice in Africa** 213
Yacouba Séré, Denis Fargette, Myimaorga Emanuel Abo, Kerstin Wydra, Mohamed Bimerew, Amos Onasanya and Salomon Kofi Akator

18 **Managing Insect Pests of Rice in Africa** 229
Francis E. Nwilene, Souleymane Nacro, Manuele Tamò, Philippe Menozzi, Elvis A. Heinrichs, Abdoulaye Hamadoun, Dona Dakouo, Cyrille Adda and Abou Togola

19 **Bird Damage to Rice in Africa: Evidence and Control** 241
Yann de Mey and Matty Demont

20 Increasing Rice Productivity through Improved Nutrient Use in Africa 250
Stephan M. Haefele, Kazuki Saito, Kabirou M. N'Diaye, Frank Mussgnug, Andrew Nelson and Marco C.S. Wopereis

21 Assessing and Improving Water Productivity of Irrigated Rice Systems in Africa 265
Sander J. Zwart

22 Inland Valleys: Africa's Future Food Baskets 276
Jonne Rodenburg

SECTION 4 RICE VALUE CHAIN DEVELOPMENT

23 Consumer Preferences for Rice in Africa 294
Pieter Rutsaert, Matty Demont and Wim Verbeke

24 Tailoring African Rice Value Chains to Consumers 303
Matty Demont and David Neven

25 Improving Grain Quality of Locally Produced Rice in Africa 311
Koichi Futakuchi, John Manful and Takeshi Sakurai

26 Developing Competitive Rice Value Chains 324
J. Dirck Stryker

27 Mechanizing Africa's Rice Sector 332
Joseph Rickman, Jean Moreira, Martin Gummert and Marco C.S. Wopereis

SECTION 5 WORKING WITH RICE COMMUNITIES

28 Integrating Gender Considerations in Rice Research for Development in Africa 343
Afiavi Agboh-Noameshie, Abdoulaye Kabore and Michael Misiko

29 Towards a New Approach for Understanding Interactions of Technology with Environment and Society in Small-scale Rice Farming 355
Edwin Nuijten, Marina Temudo, Paul Richards, Florent Okry, Béla Teeken, Alfred Mokuwa and Paul C. Struik

30 Innovative and Effective Ways to Enhance Rural Learning in Africa 367
Paul Van Mele, Jonas Wanvoeke, Josephine Rodgers and Blythe McKay

31 Raising Rice Yields and Beyond: An Experience of Collective Learning and Innovation in Lowland Rice Systems in Madagascar 379
Toon Defoer and Marco C.S. Wopereis

SECTION 6 RICE IN AFRICA: LOOKING AHEAD

32 Impact of Rice Research on Income, Poverty and Food Security in Africa: An *Ex-ante* Analysis 390
Aliou Diagne, Didier Y. Alia, Marco C.S. Wopereis, Kazuki Saito, Tebila Nakelse and Papa Abdoulaye Seck

33 Realizing Africa's Rice Promise: Priorities for Action 424
Marco C.S. Wopereis, Aliou Diagne, David E. Johnson and Papa Abdoulaye Seck

Index 437

Contributors

Abo, Myimaorga Emanuel, National Cereals Research Institute (NCRI) Badeggi, PMB 8 Bida, Niger State, Nigeria; E-mail: meabo2003@yahoo.com

Adda, Cyrille, AfricaRice, 01 BP 2031, Cotonou, Benin; E-mail: c.adda@cgiar.org

Agboh-Noameshie, Afiavi, AfricaRice, 01 BP 2031, Cotonou, Benin; E-mail: a.agboh-noameshie@cgiar.org

Ahmadi, Nourollah, Centre de coopération internationale en recherche agronomique pour le développement (CIRAD), Avenue Agropolis, 34398 Montpellier cedex 5, France; E-mail: Nourollah.ahmadi@cirad.fr

Akator, Salomon Kofi, AfricaRice, 01 BP 2031, Cotonou, Benin; E-mail: k.akator@cgiar.org

Albar, Laurence, Institut de recherche pour le développement (IRD), UMR Diversité, Adaptation et Développement des Plantes (DIADE), IRD-CIRAD-UM2, BP 5045, 34394 Montpellier cedex 5, France; E-mail: Laurence.albar@ird.fr

Alia, Didier Y., AfricaRice, 01 BP 2031, Cotonou, Benin; E-mail: d.alia@cgiar.org

Amovin-Assagba, Eyram, AfricaRice, 01 BP 2031, Cotonou, Benin; E-mail: e.amovin-assagba@cgiar.org

Atlin, Gary, International Maize and Wheat Improvement Center (CIMMYT), Cornell University, 310 Bradfield Hall, Ithaca, NY 14853, USA; E-mail: g.atlin@cgiar.org

Bachabi, Fatimata, AfricaRice, 01 BP 2031, Cotonou, Benin; E-mail: f.bachabi@cgiar.org

Bèye, Amadou M., AfricaRice, 01 BP 4029, Abidjan 01, Côte d'Ivoire; E-mail: a.beye@cgiar.org

Bimerew, Mohamed, Centre for Tropical and Subtropical Agriculture and Forestry (CeTSAF), Georg-August Universität Goettingen, Buesgenweg 1, D-37077 Goettingen, Germany

Bouniol, Julie, Institut de recherche pour le développement (IRD), BP 64501, 34394 Montpellier cedex 5, France; E-mail: julie.bouniol@ird.fr

Coulibaly, Jeanne Y., AfricaRice, 01 BP 2031, Cotonou, Benin, E-mail: j.coulibaly@cgiar.org

Dakouo, Dona, Institut de l'environnement et des recherches agricoles (INERA) Station de Farako-Bâ, 01 BP 910 Bobo-Dioulasso, Burkina Faso; E-mail: dakouo@hotmail.com

de Mey, Yann, Centre for Environmental Studies, Hasselt University, Agoralaan Gebouw D, BE-3590, Diepenbeek, Belgium; E-mail: yann.demey@uhasselt.be

Defoer, Toon, International Centre for development oriented Research in Agriculture (ICRA), PO Box 88, 6700 AB Wageningen, Netherlands; E-mail: toondefoer1@gmail.com

Demont, Matty, AfricaRice Sahel station, *Current address:* International Rice Research Institute (IRRI), DAPO Box 7777, Metro Manila, Philippines; E-mail: m.demont@irri.org

Diagne, Aliou, AfricaRice, 01 BP 2031, Cotonou, Benin; E-mail: a.diagne@cgiar.org

Djedatin, Gustave, AfricaRice, 01 BP 2031, Cotonou, Benin; E-mail: g.djedatin@cgiar.org

Dramé, Khady N., AfricaRice Tanzania station, Mikocheni B/Kawe, Avocado Street, PO Box 33581, Dar es Salaam, Tanzania; E-mail: k.drame@cgiar.org

El-Namaky, Raafat, AfricaRice Sahel station, BP 96, Saint-Louis, Senegal; E-mail: r.elnamaky@cgiar.org

Fargette, Denis, Institut de recherche pour le développement (IRD), UMR RPB, BP 64501, 34394 Montpellier cedex 5, France; E-mail: denis.fargette@ird.fr

Futakuchi, Koichi, AfricaRice, 01 BP 2031, Cotonou, Benin; E-mail: k.futakuchi@cgiar.org

Garavito, Andrea, Institut de recherche pour le développement (IRD), BP 64505, 34394 Montpellier cedex 5, France; E-mail: neagef@gmail.com

Ghesquière, Alain, Institut de recherche pour le développement (IRD), BP 64505, 34394 Montpellier cedex 5, France; E-mail: alain.ghesquiere@ird.fr

Gummert, Martin, International Rice Research Institute (IRRI), DAPO Box 7777, Metro Manila 1301, Philippines; E-mail: m.gummert@irri.org

Gutiérrez, A., International Center for Tropical Agriculture (CIAT), Apartado Aéreo 6713, Cali, Colombia; E-mail: a.gutierrez@cgiar.org

Guyot, R., Institut de recherche pour le développement (IRD), BP 64501, 34394 Montpellier cedex 5, France; E-mail: Romain.Guyot@ird.fr

Haefele, Stephan, formerly International Rice Research Institute (IRRI), Present address: Australian Centre for Plant Functional Genomics, School of Agriculture, Food and Wine, Plant Genomics Centre, University of Adelaide, Waite Campus, Hartley Grove, Urrbrae, SA 5064, Australia; E-mail: stephan.haefele@acpfg.com.au

Hamadoun, Abdoulaye, Institut d'économie rurale Centre régional de recherche agricole (IER/CRRA), BP 262 Sotuba, Bamako, Mali; E-mail: abdoulayehamadoun@yahoo.fr

Heinrichs, Elvis A., University of Nebraska, Lincoln, NE 68583-0748, USA; E-mail: eheinrichs2@unl.edu

Ismail, Abdelbagi M., International Rice Research Institute (IRRI), DAPO Box 7777, Metro Manila 1301, Philippines; E-mail: abdelbagi.ismail@cgiar.org

Jalloh, Abdulai, West and Central African Council for Agricultural Research and Development (CORAF/WECARD), BP 48, Dakar RP, Senegal; E-mail: abdulai.jalloh@coraf.org

Jannink, Jean-Luc, United States Department of Agriculture, Agriculture Research Service, Cornell University, 407 Bradfield Hall, Ithaca, NY 14853, USA; E-mail: jeanluc.work@gmail.com

Johnson, David, International Rice Research Institute (IRRI), DAPO Box 7777, Metro Manila 1301, Philippines; E-mail: d.johnson@irri.org

Kabore, Abdoulaye, AfricaRice, 01 BP 2031, Cotonou, Benin; E-mail: ab.kabore@cgiar.org

Kumashiro, Takashi, AfricaRice, 01 BP 2031, Cotonou, Benin; E-mail: t.kumashiro@cgiar.org

Lorieux, Mathias, Institut de recherche pour le développement (IRD), BP 64501, 34394 Montpellier Cedex 5, France; E-mail: Mathias.lorieux@ird.fr

McCouch, Susan, Cornell University, Ithaca, NY 14853-1901, USA; E-mail: srm4@cornell.edu

McKay, Blythe, Farm Radio International, 1404 Scott Street, Ottawa, K1Y 4M8, Canada; E-mail: bmckay@farmradio.org

Manful, John L., AfricaRice, 01 BP 2031, Cotonou, Benin; E-mail: j.manful@cgiar.org

Manneh, Baboucarr, AfricaRice Sahel station, BP 96, Saint-Louis, Senegal; E-mail: b.manneh@cgiar.org

Martinez, Cesar Pompilio, International Center for Tropical Agriculture (CIAT), Apartado Aéreo 6713, Cali, Colombia; E-mail: c.p.martinez@cgiar.org

Menozzi, Philippe, Centre de coopération internationale en recherche agronomique pour le développement (CIRAD), AfricaRice, 01 BP 2031, Cotonou, Benin; E-mail: p.menozzi@cgiar.org

Misiko, Michael, International Maize and Wheat Improvement Center (CIMMYT), PO Box 5689, Addis Ababa, Ethiopia; E-mail: m.misiko@cgiar.org

Mokuwa, Alfred, Wageningen University, Hollandseweg 1, 6706 KN Wageningen, Netherlands; E-mail: alfred_mokus@yahoo.com

Moreira, Jean, AfricaRice, 01 BP 2031, Cotonou, Benin; E-mail: j.moreira@cgiar.org

Murori, Rosemary, International Rice Research Institute Regional Office, Mikochen B Kawe, Dar es Salaam, Tanzania; E-mail: r.mutegi-murori@irri.org

Mussgnug, Frank, AfricaRice, *Current address:* International Rice Research Institute – South Asia Rice Breeding Hub, c/o ICRISAT, Patancheru – 502324, Andhra Pradesh, India; E-mail: f.mussgnug@irri.org

Nacro, Souleymane, Institut de l'environnement et des recherches agricoles (INERA) Station de Farako-Bâ, 01 BP 910, Bobo-Dioulasso, Burkina Faso; E-mail: snacro2006@yahoo.fr

Nakelse, Tebila, AfricaRice, 01 BP 2033, Cotonou, Benin; E-mail: t.nakelse@cgiar.org

N'Diaye, Kabirou M., AfricaRice Sahel station, BP 96, Saint-Louis, Senegal; E-mail: k.ndiaye@cgiar.org

Ndjiondjop, Marie-Noëlle, AfricaRice, 01 BP 2031, Cotonou, Benin; E-mail: m.ndjiondjop@cgiar.org

Nelson, Andrew, International Rice Research Institute (IRRI), DAPO Box 7777, Metro Manila 1301, Philippines; E-mail: a.nelson@cgiar.org

Neven, David, Food and Agriculture Organization of the United Nations (FAO), Viale delle Terme di Caracalla, 00153 Rome, Italy; E-mail: David.Neven@fao.org

Niang, Abibou, AfricaRice, 01 BP 2031, Cotonou, Benin; E-mail: abibou.niang@cgiar.org

Nuijten, Edwin, Wageningen University, Hollandseweg 1, 6706 KN Wageningen, Netherlands; E-mail: Edwin.Nuijten@wur.nl

Nwilene, Francis E., AfricaRice Nigeria station, PMB 5320, Oyo Road, Ibadan, Nigeria; E-mail: f.nwilene@cgiar.org

Ogunbayo, A.S., AfricaRice, 01 BP 2031, Cotonou, Benin; E-mail: a.ogunbayo@cgiar.org

Ojo, David Kolawole, Department of Plant Breeding and Seed Technology, Federal University of Agriculture, Abeokuta, Nigeria; E-mail: dkojo58@gmail.com

Okry, Florent, Wageningen University, Hollandseweg 1, 6706 KN Wageningen, Netherlands; E-mail: okryflorent@gmail.com

Onasanya, Amos, AfricaRice Nigeria station, PMB 5320, Oyo Road, Ibadan, Nigeria; E-mail: a.onasanya@cgiar.org

Remington, Tom, Catholic Relief Services, BP 665, Avenue Muyinga #5, Bujumbura, Burundi; E-mail: Tom.Remington@crs.org

Richards, Paul, Wageningen University, Hollandseweg 1, 6706 KN, Wageningen, Netherlands; E-mail: paul.richards@wur.nl

Rickman, Joseph, International Rice Research Institute (IRRI), 41 rua da Gorongosa, Maputo, Mozambique; E-mail: j.rickman@cgiar.org

Rodenburg, Jonne, AfricaRice Tanzania station, Mikocheni B/Kawe, Avocado Street PO Box 33581, Dar es Salaam, Tanzania; E-mail: j.rodenburg@cgiar.org

Rodgers, Josephine, Countrywise Communication Ltd, 103 Main Road, Wilby, Northants, NN8 2UB, UK; E-mail: jo@countrywise.com

Rutsaert, Pieter, Ghent University, Coupure links 653, B-9000 Ghent, Belgium; E-mail: pieter.rutsaert@ugent.be

Sackville Hamilton, N. Ruaraidh, International Rice Research Institute (IRRI), DAPO Box 7777, Metro Manila 1301, Philippines; E-mail: r.hamilton@irri.org

Saito, Kazuki, AfricaRice, 01 BP 2031, Cotonou, Benin; E-mail: k.saito@cgiar.org

Sakurai, Takeshi, Hitotsubashi University, 2-1 Naka, Kunitachi, Tokyo, Japan; E-mail: sakurai@ier.hit-u.ac.jp

Sanni, Kayode A., AfricaRice, 01 BP 2031, Cotonou, Benin; E-mail: k.sanni@cgiar.org

Seck, Papa Abdoulaye, AfricaRice, 01 BP 2031, Cotonou, Benin; E-mail: p.seck@cgiar.org

Semon, Mandè, AfricaRice Nigeria station, PMB 5320, Oyo Road, Ibadan, Nigeria; E-mail: m.semon@cgiar.org

Séré, Yacouba, AfricaRice Tanzania station, Mikocheni B/Kawe, Avocado Street, PO Box 33581, Dar es Salaam, Tanzania; E-mail: y.sere@cgiar.org

Sié, Moussa, AfricaRice, 01 BP 2031, Cotonou, Benin; E-mail: m.sie@cgiar.org

Sikirou, Mouritala, Institut national des recherches agricoles du Bénin (INRAB); E-mail: alanioss@yahoo.fr

Singh, Rakesh Kumar, International Rice Research Institute Regional Office, Mikochen B Kawe, Dar es Salaam, Tanzania; E-mail: rksingh10@yahoo.com

Sorrells, Mark E., Cornell University, 240 Emerson Hall, Ithaca, NY 14853-1902, USA; E-mail: mes20@cornell.edu

Sow, Abdoulaye, AfricaRice Sahel station, BP 96, Saint-Louis, Senegal; E-mail: a.sow@cgiar.org

Sow, Mounirou, AfricaRice, 01 BP 2031, Cotonou, Benin; E-mail: m.sow@cgiar.org

Struik, Paul C., Centre for Crop Systems Analysis, Wageningen University, PO Box 430, 6700 AK Wageningen, Netherlands; E-mail: Paul.Struik@wur.nl

Stryker, J. Dirck, Associates for International Resources and Development (AIRD), 455 Bartlett Avenue, Arlington, MA 02476, USA; E-mail: dstryker@aird.com

Tamo, Manuele, International Institute of Tropical Agriculture (IITA), 08 BP 0932, Tri Postal, Cotonou, Benin; E-mail: m.tamo@cgiar.org

Teeken, Béla, Wageningen University, Hollandseweg 1, 6706 KN Wageningen, Netherlands; E-mail: Bela.Teeken@wur.nl

Temudo, Marina, Tropical Research Institute (IICT), Apartado 3014, 1301 Lisboa, Portugal; E-mail: marinatemudo@gmail.com

Thiemélé, Deless, Centre National de la Recherche Agronomique (CNRA), Station de Recherche Bimbresso, Adiopodoume, Côte d'Ivoire; E-mail: delessthiemele@yahoo.fr

Tia, Daniel D., AfricaRice, 01 BP 2031, Cotonou, Benin; E-mail: d.tia@cgiar.org

Togola, Abou, AfricaRice, 01 BP 2031, Cotonou, Benin; E-mail: a.togola@cgiar.org

Tohme, Joe, International Center for Tropical Agriculture (CIAT), Apartado Aéreo 6713, Cali, Colombia; E-mail: j.tohme@cgiar.org

Tollens, Eric, Centre for Agricultural and Food Economics, Katholieke Universiteit Leuven, Celestijnenlaan 200E box 2411, BE-3100 Leuven (Heverlee), Belgium; E-mail: eric.tollens@biw.kuleuven.be

Touré, Ali A., AfricaRice, 01 BP 2031, Cotonou, Benin; E-mail: a.toure@cgiar.org

Van Mele, Paul, AfricaRice, *Current address:* Agro-Insight, Ghent, Belgium; E-mail: paul@agroinsight.com; paulvanmele@gmail.com

Venuprasad, Ramaiah, AfricaRice Nigeria station, PMB 5320, Oyo Road, Ibadan, Nigeria; E-mail: r.venuprasad@cgiar.org

Verbeke, Wim, Ghent University, Coupure links 653, B-9000 Ghent, Belgium; E-mail: wim.verbeke@ugent.be

Wanvoeke, Jonas, AfricaRice, *Current address:* Wageningen University, PO Box 47, 6700 AA Wageningen, Netherlands; E-mail: jonas.wanvoeke@wur.nl

Wing, Rod A., University of Arizona, 1657 E. Helen Street, Tucson, AZ 85721, USA; E-mail: rwing@ag.arizona.edu

Wopereis, Marco C.S., AfricaRice, 01 BP 2031, Cotonou, Benin; E-mail: m.wopereis@cgiar.org

Wydra, Kerstin, Centre for Tropical and Subtropical Agriculture and Forestry (CeTSAF) – Tropenzentrum, Georg-August Universität Goettingen, Buesgenweg 1, D-37077 Goettingen, Germany; E-mail: kwydra@uni-goettingen.de

Yao, Nasser, AfricaRice, 01 BP 2013, Cotonou, Benin; E-mail: n.yao@cgiar.org

Yoshida, Hiroe, National Agriculture and Food Research Organization, Kannondai 3-1-1, Tsukuba, Ibaraki 305-8517, Japan; E-mail: hiroey@affrc.go.jp

Zwart, Sander J., AfricaRice, 01 BP 2031, Cotonou, Benin; E-mail: s.zwart@cgiar.org

- **Nigeria:** *Focal points:* Vivian Ojehomon (National Cereals Research Institute), Olorunfemi Ogundele (Nigerian Institute for Social and Economic Research) and Samuel B. Adebayo (National Bureau of Statistics, NBS); *Team members:* Victor Okoruwa (University of Ibadan), Olupomi Ajayi (AfricaRice) and Bandele Ogunlana (NBS)
- **Rwanda:** *Focal points:* Chantal Ingabire (Rwanda Agricultural Board, RAB) and Justine Mukayiranga (RAB); *Team members:* Innocent Ndikumana (RAB), E. René Gasoré (RAB) and Joseph Mutware (RAB)
- **Senegal:** *Focal points:* Alioune B. Dieng (Institut sénégalais de recherches agricoles, ISRA) and Mamadou Sagna (Directrice de l'Analyse, de la Prévision et des Statistiques, DAPS); *Team members:* Amadou Abdoulaye Fall (ISRA), Maïmouna Lô Gueye (DAPS), Abdoulaye Mbengue (DAPS), Mbengue Mékha Babou (DAPS), Fara Dione (DAPS), Gorgui Djibril Diallo (DAPS), Ousmane Diagne (DAPS), Oumy Seck Lô (DAPS), Sylvie Dasylva (DAPS), Balde Some (DAPS) and Moustapha Niang (DAPS)
- **Sierra Leone:** *Focal points:* Mahmood Ahmed Nazir (Sierra Leone Agricultural Research Institute) and Mohamed Ajuba Sheriff (Planning, Evaluation, Monitoring and Statistics Division)
- **Tanzania:** *Focal points:* Elisha Martine Mkandya (Ilonga Agricultural Research Institute) and Malemi Nyanda (Policy and Planning, Statistics Unit, Ministry of Agriculture, Food and Cooperatives)
- **Togo:** *Focal points:* Tsatsu K. Domenyo (Institut togolais de recherche agronomique) and Ezi Tsogli (Direction des Statistiques Agricoles, de l'Information et de la Documentation); *Team members:* Egnonto M. Koffi-Tessio (Laboratoire pour la Réduction de la Pauvreté et la Sécurité Alimentaire Durable, Université de Lomé, LARPSAD), Lébéné Kougnenya (LARPSAD), Kodjo Dokodjo (LARPSAD) and Kodjo Labare (LARPSAD)
- **Uganda:** *Focal points:* Stella Adur (National Crops Resources Research Institute, National Agricultural Research Organisation, NaCRRI), Benard Odur (NaCRRI) and Mary Frances Nakakawa (NaCRRI); *Team member:* Godfrey Asea (NaCRRI).

We are very grateful to Guy Manners (formerly freelance and now with Green Ink) for editorial assistance and most of all for his patience and perseverance.

We also wish to thank Emma McCann, Editorial Assistant at CABI, and her predecessor Chris Shire, who answered all our technical queries, following up CABI production staff as necessary.

The Editors

Introduction

Marco C.S. Wopereis*
Africa Rice Center (AfricaRice), Cotonou, Benin

In many countries in sub-Saharan Africa (SSA), rice used to be consumed only during special events, perhaps once or twice a year. From a luxury crop it has now become the most rapidly growing food source in SSA. This is driven by many factors, including urbanization, changes in employment patterns, rising income levels, shifts in consumer preferences, and rapid population growth. Rice is the leading provider of food calories in West Africa and in Madagascar, and it is now the third largest source of food energy in Africa as a whole. Because of its ease of preparation, storage and cooking, low preparation costs, low price and steady supplies (often through imports), rice has become a staple food for the poorest sectors of urban populations. The proportion of Africans living in urban areas is expected to continue to increase rapidly from its current (2012) level of 38% to reach 48% by 2030. The increasing trend in rice consumption across the continent is, therefore, likely to continue for the foreseeable future.

The increasing demand for rice means that in 2009 nearly 37% of the rice consumed in SSA was imported. This translated into imports of 9.8 million tonnes (Mt), worth more than US$5 billion. This reliance on imports is a very risky, expensive and unsustainable strategy. The risks became painfully clear in 2008 during food riots in major African capitals in protest against high rice prices caused by traditional exporting countries like Vietnam, India and Egypt closing their borders and banning exports. Rice is one of the most protected commodities in the world and only about 7% of global production is traded on the international market.

Average rice prices in 2012 were about 2.5 times the price levels in 2000 and the rice price on the world market has seen tremendous fluctuations. Prices are predicted to remain high and volatile because of declining production capacity in major rice-producing countries in Asia as a result of increasing pressure on land and water resources and growing demand. Increasing rice prices will adversely affect poor and low-income households who spend a larger proportion of their revenue on staple food relative to high-income households. In a region where more than 40% of the population lives below the extreme poverty line of $1 a day, coping with high rice prices will mean poorer households taking measures which could include a reduction in their intake of essential nutrients, especially in urban areas, and long episodes of food deprivation and malnutrition. Moreover, in countries such as Guinea-Bissau and Sierra Leone, where annual per-capita rice consumption is above 100 kg, the incidence of higher rice prices has the potential to trigger political disturbances.

* Corresponding author: m.wopereis@cgiar.org

Seck *et al.* (2012) estimate that between 2010 and 2035 an additional 116 Mt of milled rice will be needed worldwide to meet increasing demand. About a quarter of this (30 Mt) will be needed in Africa (an increase of 130% compared to 2010 consumption levels), and one-third of that volume (10 Mt) will be needed in Nigeria alone, illustrating the increasing importance of Africa in terms of rice consumption worldwide.

This book is published at a time when there is greatly increased awareness that rice has become a strategic commodity in Africa, potentially fuelling economic growth and contributing to hunger and poverty reduction across the continent. Many African countries have embarked on ambitious programmes to boost their rice production capacity, most of them as a response to the 2008 rice crisis. Africa Rice Center (AfricaRice) has provided technical support to 21 countries within the framework of the Coalition for African Rice Development (CARD)[1] and to other AfricaRice member states that are not members of CARD – assisting them with the development of national rice development strategies (NRDS). These strategies now need to be turned into concrete action plans to boost the rice sector. The critical challenge facing the African rice sector is to enhance performance in production, processing and marketing to respond to a major concern that needs to be turned into an opportunity: the growing demand for rice as a preferred staple.

As this book shows, Africa has sufficient land and water resources to produce enough rice to feed its own population and, in the long term, generate export revenues – this is 'Africa's rice promise'. The purpose of this book is to provide a comprehensive overview of Africa's rice sector and ongoing rice research and development activities, and indicate priorities for action on how to realize this promise.

Section 1 ('Rice in Africa: An Overview') contains four chapters that set the stage for this book. We first (Chapter 1) look back at over 20 years of rice research for development in Africa, highlighting some of the major institutional and technological challenges encountered and achievements obtained since about 1990. Next (Chapter 2), we zoom in on Africa's rice economy and discuss trends in rice demand and supply before and after the 2008 rice crisis. New estimates on cultivated area, number of farming households and yields for major rice-growing environments in Africa are provided in Chapter 3. Farmer perceptions of biophysical constraints to rice production are presented in Chapter 4.

Rice varieties, designed for farmers' growth environments and market demand, are the backbone of rice sector development and are central in Section 2 ('Rice Genetic Diversity and Improvement', ten chapters). First (Chapter 5), a new systematic, continent-wide and product-oriented approach to rice breeding in Africa is presented to enhance farmer access to new varieties specifically designed for his or her rice-growing environment and market demand. This approach has been in use by AfricaRice, national and international partners since 2010. Next, we review the status of varietal release systems in Africa and the associated challenges and opportunities (Chapter 6), followed by an overview of the diversity of rice and related wild species in Africa and opportunities for more effective use of these genetic resources for rice varietal development (Chapter 7). Chapter 8 provides insight into gene-flow processes in farmers' fields between cultivated and wild species, the associated emergence of interspecific progenies and the role of selection of new plant types by populations of farmers over long periods of time, adding to *in-situ* genetic diversity. Chapter 9 zooms in on the rapidly changing world of rice genomics and how this field can accelerate the development of new rice varieties in Africa. Biological, logistical and institutional challenges to the development of functioning breeding pipelines making effective use of these tools in Africa are discussed. Chapter 10 reviews recent advances in the understanding of interspecific sterility in rice with emphasis on the reproductive barriers between the two cultivated species. Chapter 11 provides a state-of-the-art overview of genetic improvement of rice for abiotic-stress tolerance (i.e. drought, salinity, excess water, phosphorus deficiency, iron toxicity and extreme temperatures) using both modern and more conventional genetic approaches. Chapter 12 discusses sources of resistance to one particularly devastating rice disease in Africa (*Rice yellow mottle virus*, RYMV). Chapter 13 zooms in on the prospects for hybrid rice by presenting AfricaRice's hybrid-rice breeding programme and results of tests conducted with Chinese rice hybrids in Africa. Chapter 14 presents data on rice farmers' seed sources and discusses opportunities for the development of a viable and integrated rice seed sector in Africa.

Section 3 ('Sustainable Productivity Enhancement', eight chapters) discusses Africa's highly diverse production environments and opportunities to enhance rice productivity in a sustainable manner, tackling the major yield- and productivity-reducing and -limiting factors. Chapter 15 provides a general overview of rice environments in Africa, and provides the reader with a better understanding of biophysical determinants of yield gaps (differences between actual farmers' yields and what would be possible with improved management) and potential expansion of rice area in Africa. Chapter 16 zooms in on the weed flora of Africa's rice systems, the most frequent and widespread biotic constraint to rice productivity in Africa. The authors review problem weed species, and weed management strategies across ecosystems. Future weed management issues related to crop intensification, labour and water shortages, and changing environmental conditions (including climate change and the evolution of herbicide-resistant weed ecotypes) are discussed. Chapter 17 provides an overview of Africa's 'big three' diseases: RYMV, blast and bacterial leaf blight. Progress made with respect to genetic control and field management of these diseases is presented and outstanding challenges and opportunities for integrated management of rice diseases are discussed. Chapter 18 provides an overview of insect pest damage to rice in Africa, both pre- and postharvest. Opportunities for field-level, landscape-level and postharvest management of these pests and future challenges related to climate change are discussed. Chapter 19 deals with an important pest that is rarely quantified in Africa, bird damage. The chapter reviews pest bird species and literature on physical crop losses suggesting an average loss of 15–20%, and proposes several measures to protect farmers against avian risk. Chapter 20 reviews the current state of knowledge on farmer practices, challenges and opportunities with respect to nutrient use in the three major rice ecosystems in Africa. Chapter 21 presents an overview of water productivity of irrigated rice at field and scheme levels in Africa based on a literature survey and in the context of increasing pressure on water due to expansion and intensification of irrigated rice systems and climate change. Chapter 22 reviews the diversity of inland-valley systems across West, East and Southern Africa, and opportunities and challenges for their sustainable development for rice-based systems.

Section 4 ('Rice Value Chain Development', five chapters) takes the reader through the various opportunities and challenges related to the development of sustainable and profitable rice value chains in Africa. Chapter 23 provides an overview of consumer preferences in 17 countries in SSA based on literature, expert knowledge and AfricaRice surveys. Chapter 24 moves on to discuss how to tailor rice value-chain development to consumer preferences. The authors identify key systemic constraints and challenges to the competitiveness of African rice and explore opportunities for upgrading African rice value chains and tailoring them to consumers in end-markets. Chapter 25 discusses the crucial issue of grain quality, an important prerequisite to making rice production competitive in Africa. Ways to improve grain quality are discussed, ranging from the choice of variety to management factors before and after harvest. The chapter concludes with recommendations for making African produced rice more competitive *vis-à-vis* imports. Chapter 26 compares small rice-processing units with larger mills that deliver vastly different qualities of rice for different markets in Rwanda and provides recommendations for improving the competitiveness of locally produced rice through the phased introduction of modern processing facilities. Chapter 27 reviews the status of agricultural mechanization in SSA and presents mechanization options that could make a difference along the rice value chain.

Section 5 ('Working with Rice Communities') contains four chapters. Chapter 28 discusses gender roles in rice farming in SSA, emphasizing the importance of women in rice farming. An approach to ensure that gender is mainstreamed in the research agenda is proposed and outstanding challenges are discussed. Chapter 29 looks in detail at how rice farmers innovate, select and match technologies to local environmental conditions, looking for optimal interactions of technologies with agroecological, socio-economic and cultural factors. Chapter 30 focuses on dimensions of rural learning in Africa, with a special focus on rice farmers. The potentially powerful role of videos in rural learning is illustrated. Chapter 31 presents a case study on improving rice productivity in rainfed lowland systems in northern Madagascar using a participatory learning and action-research approach (PLAR). Opportunities to use the methodology elsewhere are discussed.

Section 6 ('Rice in Africa: Looking Ahead') has two chapters. In Chapter 32, we look ahead with respect to rice research for development and estimate the potential impact of rice research on income and poverty in SSA over the next 10 years. Estimates are made within the context of the Global Rice Science Partnership, the CGIAR Research Programme on rice, led globally by the International Rice Research Institute and in Africa by AfricaRice. The concluding chapter (33) presents a list of priorities for action to boost Africa's rice sector in a sustainable and equitable manner, based on the information contained in this book, the authors' own experience from field-work in Africa, and discussions with many actors and facilitators in Africa's multiple rice value chains.

It has taken several years to compile the information contained in this book. I sincerely hope that it will contribute to **realizing Africa's rice promise**.

Note

[1] The Coalition for African Rice Development (CARD) is an initiative of the Japan International Cooperation Agency (JICA) and the Alliance for a Green Revolution in Africa (AGRA). It aims to double rice production between 2008 and 2018. CARD Steering committee members are African Development Bank, AfricaRice, AGRA, Food and Agriculture Organization of the United Nations, Forum for Agricultural Research in Africa, International Fund for Agricultural Development, International Rice Research Institute, JICA, Japan International Research Center for Agricultural Sciences, NEPAD Planning and Coordinating Agency and the World Bank.

Reference

Seck, P.A., Diagne, A., Mohanty, S. and Wopereis, M.C.S. (2012) Crops that feed the world 7: rice. *Food Security* 4(1), 7–24.

1 From WARDA to AfricaRice: an Overview of Rice Research for Development Activities Conducted in Partnership in Africa

Eric Tollens,[1] Matty Demont,[2] Moussa Sié,[3] Aliou Diagne,[3] Kazuki Saito[3] and Marco C.S. Wopereis[3]*

[1]Centre for Agricultural and Food Economics, Katholieke Universiteit Leuven, Leuven, Belgium; [2]Africa Rice Center (AfricaRice), Saint-Louis, Senegal; [3]Africa Rice Center (AfricaRice), Cotonou, Benin

Introduction

The West Africa Rice Development Association (WARDA) was created in 1971 by 11 West African states (Burkina Faso, Côte d'Ivoire, The Gambia, Ghana, Liberia, Mali, Mauritania, Niger, Senegal, Sierra Leone and Togo) – with the assistance of the United Nations Development Programme (UNDP), the Food and Agriculture Organization of the United Nations (FAO) and the Economic Commission for Africa (ECA) – as an autonomous research organization and an intergovernmental association of member states (WARDA, 2001a). The highest governing body of the Center is the Council of Ministers of Agriculture of member states, with statutory meetings being held once every 2 years. In 1975, WARDA joined the growing contingent of international agricultural research centres affiliated through the CGIAR (now the CGIAR Consortium) (CGIAR, 2013). Over the following two decades, six additional countries (Benin, Cameroon, Chad, Guinea, Guinea-Bissau and Nigeria) joined the Association to bring the membership to 17. When it was realized that WARDA's products were gaining ground in countries beyond its traditional mandate region of West and Central Africa, the name Africa Rice Center was adopted in January 2003 (WARDA, 2004). In 2007, a new vision was formulated to encourage such countries to become full members of the Association. This prompted five more countries to join in the midst of the global food crisis of 2007–2008 (Central African Republic, the Democratic Republic of Congo, the Republic of Congo, Egypt and Uganda). These were followed by Gabon (September 2009) and Madagascar (February 2010). Thus, today (June 2013) the Association's 24 member states represent West, Central, East and North African regions. When Africa Rice Center became the official legal name of the Center and the Association in 2009, the abbreviation AfricaRice was adopted in both English and French and was to be applied retroactively.[1]

Like other CGIAR-supported centres, AfricaRice has a Board of Trustees composed of nominees from member states and from

* Corresponding author: m.wopereis@cgiar.org

© CAB International 2013. *Realizing Africa's Rice Promise* (eds M.C.S. Wopereis *et al.*)

non-member states. They work together to ensure that AfricaRice management conforms to the resolutions of the Council of Ministers and to the CGIAR guidelines on governance and management in implementing the Center's approved 2011–2020 Strategic Plan (AfricaRice, 2011b).

In this chapter, we look back over rice research for development in Africa, with emphasis on the past two decades. We focus on AfricaRice, but contributions of many other institutions and partners are acknowledged. This is not a comprehensive historical overview of rice research and development in Africa. Rather, we highlight some of the major institutional and technological challenges encountered and achievements obtained since about 1990. Much more information on a wide range of research-for-development activities conducted in partnership in Africa can be found in subsequent chapters of this book.

The chapter starts with a brief overview of the Task Force mechanism, which has been the main vehicle used by AfricaRice to conduct research in partnership with the national agricultural research systems (NARS) of its member states and other African nations and an overview of some other important partnerships. The remainder of the chapter is structured in three thematic areas: (i) genetic diversity and improvement; (ii) crop and natural-resources management; and (iii) policy, impact assessment and rice value-chain development.

Task Forces and Other Partnership Mechanisms

The Task Force mechanism

The Task Force mechanism was introduced as a novel approach to building partnerships between the Center and the NARS in 1991. The mechanism responded to concerns highlighted by two working groups of NARS representatives (WARDA, 1999). Consequently, each Task Force had four primary objectives:

- to coordinate regional research activities, thereby reducing duplication and identifying the most complementary forms of collaboration among rice research programmes;
- to provide national scientists with more complete and rapid access to information and results from regional research;
- to test and transfer technologies in a targeted and systematic manner; and
- to target technical, material and financial assistance to national programmes in such a way as to strengthen the regional rice research system as a whole (Fakorede and Yoboué, 2001).

Between 1991 and 1995, ten Task Forces were established (although, because of re-organization and mergers, a maximum of nine operated at any one time), covering mangrove swamp, upland rice breeding, lowland rice breeding, irrigated rice breeding, Sahel resource management, integrated pest management (IPM), problem soils, cropping systems, rice economics and technology transfer.

Specifically designed to address constraints to rice production identified by the NARS, the Task Forces operated through (annual) meetings, joint research activities, monitoring tours, visiting fellowships, and training (WARDA, 1999). The Task Forces provided for sharing of information and resources so that – on a regional scale – there would be no duplication of effort, isolated national researchers could interact with their peers from other countries, and no one NARS was overstretched in achieving research objectives. The Task Forces attempted to ensure that each national partner and AfricaRice conducted activities according to their institutional comparative advantages, thereby achieving greater overall impact on a regional scale. Finally, the Task Forces also served to help AfricaRice prioritize its work with direct guidance from national partners. A 218-page summary of the first 7 years of AfricaRice–NARS Task Forces activities was published in 2001 (Fakorede and Yoboué, 2001).

In 1998, the decision was taken to merge the AfricaRice–NARS Task Forces with the Rice Network of the West and Central African Council for Agricultural Research and Development (WECARD/CORAF). This was prompted by a review of networks funded by the United States Agency for International Development (USAID) and in light of the fact that the two networks involved the same NARS scientists. Moreover, the composition of the newly created (January 1998) AfricaRice National Experts Committee was almost identical to that of the WECARD/CORAF Executive Committee (WARDA, 1999). Consequently, the Rice Research and Development Network for West and Central Africa (Réseau Ouest et Centre Africain du Riz, ROCARIZ) was created in 1999 (WARDA, 2000). ROCARIZ followed the Task Force mechanism, but replaced the annual meetings of individual Task Forces with a biennial Regional Rice Research Review (4Rs) (e.g. Sanyang *et al.*, 2003; Narteh *et al.*, 2006), which brought together as many of the NARS scientists as possible, with members of the individual Task Forces meeting as 'breakout groups' during the Review. Unfortunately, funding for ROCARIZ dried up and the network ceased to function in 2006 (AfricaRice, 2012).

However, AfricaRice did manage to continue some of the networking activities of some of the Task Forces (e.g. Breeding and Economics) through projects (AfricaRice, 2012; A.A. Touré, Cotonou, Benin, 2013, personal communication).

In 2010, the Second Africa Rice Congress urged African governments to renew commitment to rice research and development, and supported AfricaRice in its proposal to revive the task force mechanism (AfricaRice, 2012). Consequently, by March 2013, there were Africa-wide Rice Task Forces covering breeding, agronomy, processing and value addition, policy and gender, with the final one (mechanization) scheduled for launch later that year (GRiSP, 2013).

Other partnerships

The Consortium for the Sustainable Use of Inland Valley Agro-Ecosystems in Sub-Saharan Africa (**Inland Valley Consortium**, IVC), convened by AfricaRice, is composed of 12 West African NARS and a number of international (IITA, ILRI, IWMI,[2] FAO, WorldFish and WECARD/CORAF) and advanced research institutes (CIRAD,[3] Wageningen University). It was founded in 1993 with the objective to develop, in concerted and coordinated action, technologies and operational support systems for intensified but sustainable use of inland valleys in sub-Saharan Africa. Extensive biophysical and socio-economic characterization work was done during the first phase in 18 key sites in 1994–1999 (e.g. Andriesse *et al.*, 1994; Windmeijer *et al.*, 1998). A second phase (2000–2006) focused on technology development for inland-valley systems. The participatory learning and action-research (PLAR) approach was developed and diffused during this time (see Defoer and Wopereis, Chapter 31, this volume). Lack of funding after 2006 severely restricted IVC operations. Since 2012, IVC has continued as the Inland Valley Community of practice, which better reflects its new modus operandi.

The **Human Health Consortium** (1994–2000) brought together six multidisciplinary West African research institutions to evaluate the health and social impacts of various degrees of wetland water management and irrigation in the humid rain-forest, savannah and Sahel zones of Côte d'Ivoire and Mali. The research concluded that most types of water management in the region have minimal impact on the occurrence of the two main water-associated diseases, malaria and schistosomiasis (WARDA, 2001a).

Various research and development partnerships grew up around and to support the work that led to the development of the NERICA varieties and subsequently promoted them across the continent. First, the **Interspecific Hybridization Project** itself, launched in 1996 – just 3 years after the first successful development of fertile progeny from crosses between African cultivated rice (*Oryza glaberrima*) and Asian cultivated rice (*Oryza sativa*) – brought together three CGIAR centres (AfricaRice, IRRI and CIAT),[4] Cornell University (USA), Institut de recherche pour le développement (IRD, France), Tokyo University (Japan) and Yunnan Academy of Agricultural Sciences (China). The Project sought to develop and utilize the breeding techniques and their products (WARDA, 2001a).

The mechanism devised to disseminate the NERICA varieties and to help resource-poor upland-rice farmers identify varieties best suited to their particular agroecological and socio-economic contexts – participatory varietal selection (PVS) – itself became the focus of a research and development network, the **Participatory Rice Improvement and Gender/User Analysis**, which ran for several years from about 1999 (WARDA, 2001a). Then in 2001, the **African Rice Initiative (ARI)** was established to promote the widespread and rapid diffusion of the NERICA and complementary technologies throughout the rice-growing areas of Africa. Early work targeted upland rice in five pilot countries in West Africa, while enabling PVS and community-based seed systems (CBSS) activities to start in East and Southern Africa (WARDA, 2002a; Bèye *et al.*, 2011).

The African component of the **International Network for Genetic Evaluation of Rice (INGER-Africa)** started life as part of the International Rice Testing Program, a systematic global programme for the collection, distribution and testing of rice genetic materials convened by IRRI. Initially (from its inception in 1985) coordinated by IITA, INGER-Africa was relocated to AfricaRice in 1997.

Since 2008, AfricaRice has been a member of the steering committee of the Coalition of African Rice Development (CARD), led by the Japan International Cooperation Agency (JICA) and the Alliance for a Green Revolution in Africa (AGRA) (see www.riceforafrica.org).

Since 2011, AfricaRice has been participating in CGIAR Research Programmes, in particular the Global Rice Science Partnership (GRiSP), led globally by the International Rice Research Institute (IRRI). AfricaRice is leading the implementation of GRiSP in Africa.

Genetic Diversity and Improvement

Rice breeding

Introducing, testing and disseminating varieties

In the early days (i.e. in the 1970s and 1980s), the AfricaRice philosophy was that rice varieties developed by other organizations could simply be introduced to the region, so the Center's early work comprised coordinated variety trials for the major rice-growing environments (irrigated, mangrove, rainfed lowland and upland). The principal sources of these varieties were IRRI, IITA and Institut de Recherche Agronomique Tropicale (IRAT).[5] This led to the adoption of well-known varieties in West Africa such as BG90-2, Bouaké189, IR1529-680-3, C74 and Jaya for lowland environments and Dourado Precoce, IRAT 144, IRAT 13, ITA 257, ITA 150 and ITA 235 for upland environments.[6]

At AfricaRice, breeding programmes for the main rice-growing environments started in earnest when the Center moved from Liberia to Côte d'Ivoire in 1988, with upland breeding based at M'bé, Côte d'Ivoire, lowland breeding through an AfricaRice breeder based at IITA, Ibadan, Nigeria (taking over from the IITA rice breeder in 1990), breeding for irrigated systems based at Saint-Louis, Senegal, and breeding for the mangrove-swamp environment based at the Rokupr research station, Sierra Leone, alongside the Sierra Leone national programme. Breeding lines were named WAB in Côte d'Ivoire, WAT in Nigeria, WAS in Senegal and WAR in Sierra Leone. In 2010, this naming system was revised to ARS in Senegal, ART in Nigeria, ARB in Côte d'Ivoire and ARC for Cotonou, Benin, reflecting the name change from WARDA to AfricaRice.

The most promising lines (bred or introduced) for the rainfed lowland and irrigated ecosystems south of the Sahel zone were included in the WITA series, the most well-known being WITA 4 (from line TOX 3100-44-1-2-3-3). For the irrigated Sahelian environment, the Sahel series was started, which includes the short-duration Sahel 108 (IR 13240-108-2-2-3, an IRRI line), and the medium-duration varieties Sahel 201 (BW 293-2) and Sahel 202 (ITA 306) now widely grown in Senegal and Mauritania. New aromatic Sahel varieties (Sahel 177, Sahel 328 and Sahel 329) are currently (2012) making rapid headway in Senegal. For the upland environment the most popular varieties were WAB 56-50, WAB 56-104 and WAB 56-125 (before the arrival of the NERICA varieties, see below). Breeding for the mangrove ecosystem was devolved to the Sierra Leone national programme in 1986.

In the 1990s, AfricaRice started developing wide crosses between Asian rice *O. sativa* (improved tropical *japonica*) and African rice (*O. glaberrima*). The idea was to combine specific assets of *O. glaberrima* (e.g. weed competitiveness and tolerance to diseases) with the yield potential of tropical *japonica* parents (Jones *et al.*, 1997; Dingkuhn *et al.*, 1999b). The most promising seven lines were subsequently named 'New Rices for Africa' (NERICA 1 to NERICA 7). The term 'NERICA' is used to indicate interspecific hybrid progeny derived from crosses between

O. glaberrima and *O. sativa*. The NERICA varieties had their first taste of success in Guinea in the late 1990s, primarily through PVS programmes (WARDA, 2000, 2001b). The first NERICA varieties to be officially released were in Côte d'Ivoire in December 2000 (NERICA 1 and NERICA 2). NERICA is now a household name in Africa and earned AfricaRice the King Baudouin award in 2000 (WARDA, 2001c) and, in the person of Dr Monty P. Jones, the World Food Prize in 2004.

WARDA (2002b) lists the varieties released or generally adopted in West and Central Africa up to 2002. Dalton and Guei (2003) estimated that a total of 197 improved varieties were released in seven West African countries between 1960 and 2000, of which 40% (80 out of 197) were developed by national programmes using genetic material that originated within national programmes without any direct or indirect involvement of the CGIAR, but about half of them (38) with support from IRAT/CIRAD. A further 13 varieties were produced by the AfricaRice mangrove rice programme. A total of 27% (54 out of 197) were varieties developed by the CGIAR from genetic material developed by the CGIAR. The third most important source of released varieties, purified landraces, made up 14% of accessions (27 out of 197). The remainder (16%) were developed collaboratively by national and international agricultural research centres, including the CGIAR. These figures do not include the NERICA varieties which were just entering farmers' fields in and around 2000.

Despite the difficult conditions under which AfricaRice worked between 2002 and 2005 (forced relocation from Côte d'Ivoire to Mali and eventually Benin as a result of the civil war in Côte d'Ivoire) AfricaRice's breeding work resulted in an additional 11 varieties for upland conditions (NERICA 8 to NERICA 18). Based on work that started in the late 1990s, AfricaRice and partners also developed a family of NERICA varieties adapted to the lowlands. This new generation of NERICA varieties, derived from crosses between *O. sativa* subsp. *indica* and *O. glaberrima*, currently consists of 60 genotypes (NERICA-L 1 to NERICA-L 60), the most widely grown being NERICA-L 19. The main breeding objectives were yield potential, grain quality, broad adaptation to diverse lowlands in the region, and tolerance to *Rice yellow mottle virus* and African rice gall midge (Sié *et al.*, 2008a). This work earned AfricaRice, in the person of Dr Moussa Sié, the Fukui International Koshihikari Rice Prize from Japan in 2006.

AfricaRice breeders and colleagues from NARS are making increasing use of marker-assisted selection and other genomics techniques (see McCouch *et al.*, Chapter 9, this volume) to 'upgrade' existing varieties with one or more genes that will give them an edge in farmers' fields or in market segments. For example, in partnership with IRD, AfricaRice has developed varieties that are resistant to *Rice yellow mottle virus* through incorporation of the *rymv1-2* gene (see Ndjiondjop *et al.*, Chapter 12, this volume) and these varieties are currently (2012) being tested in farmers' fields in West Africa. Through the newly established Africa-wide Rice Breeding Task Force and as part of the Global Rice Science Partnership (GRiSP), AfricaRice and partners are conducting a systematic, continent-wide and product-oriented breeding approach that is expected to accelerate varietal development. Specific varietal-development pipelines are being defined for different growth environments and market-segments (see Kumashiro *et al.*, Chapter 5, this volume).

Box 1.1. Rice breeding in Egypt

Egypt's rice breeding programme started in 1917. The breeding programme has developed an impressive number of medium-duration (120–130 days) high-yielding varieties with good grain quality (Giza and Sakha series) suited to the Egyptian market (short grain, low amylose), mostly resistant to blast. Promising hybrid rice varieties have recently been developed as well. These varieties have contributed to the very high rice yields that can be obtained in the Egyptian climate (10–12 t/ha). The challenge for Egypt is to develop short-duration varieties that can be grown with less water, are resistant to blast and tolerate heat and moderate salinity levels while maintaining yields.

Physiology research

Detailed physiological work on varietal responses of irrigated lowland rice in the Sahel to temperature and photoperiod was done in the early 1990s (Dingkuhn and Miezan, 1995; Dingkuhn *et al.*, 1995). A total of 49 varieties were characterized for their photothermal responses through 'rice garden' trials (sown every month over a 1.5 year period). The study identified varieties best suited for different growing seasons in the Sahel. A relationship between minimum air temperature at booting stage and spikelet sterility was established (Dingkuhn *et al.*, 1995). This work led to the development of the RIDEV (RIce DEVelopment) simulation model (see next section, 'Crop and Natural Resources Management').

Work by Asch *et al.* (1997, 1999) and Asch and Wopereis (2001) revealed varietal and seasonal differences in salinity tolerance for the irrigated Sahelian environment, leading to a physiological model of sodium uptake in the rice plant and screening tools for breeding for salinity tolerance.

Studies conducted by Sahrawat (2000, 2004) and Audebert and Sahrawat (2000) identified the role of iron-tolerant rice genotypes and other plant nutrients in reducing iron toxicity in lowland rice and related physiological mechanisms. Screening methodologies for tolerance to iron toxicity are discussed by Asch *et al.* (2005).

A large number of studies on varietal adaptation to different water regimes have been undertaken since the mid-1990s (e.g. Dingkuhn *et al.*, 1999b; Sie *et al.*, 2008b). Saito *et al.* (2010c) evaluated 14 rice varieties in seven experiments to investigate genotype × environment (G×E) interaction for yield, and to identify high-yielding varieties and plant characteristics associated with high yield. Three environment groups were identified from a pattern analysis on yield. Grouping was related to water availability, distinguishing: (i) an aerobic environment, with rice grown under aerobic conditions with supplementary irrigation; (ii) a hydromorphic environment, with rice grown under rainfed conditions with drought spells at the vegetative stage; and (iii) a permanently flooded environment. These results indicate that a systematic effort is needed to screen a wide range of varieties across hydrology gradients to identify varieties that perform well across or within a specific target population of environments. In 2011, a high-throughput phenotyping facility for drought resistance was established at AfricaRice, Cotonou in collaboration with CIRAD.

Weeds are serious constraints to rice production in Africa across all rice environments. An illustrated guide to weeds of rice in West Africa was produced by Johnson (1997). Weed competitiveness of rice plants is considered highly desirable, especially in low-input systems. In the 1990s, AfricaRice developed a plant-type concept for high-yielding, weed-competitive cultivars adapted to low-input management conditions through analysing data from field experiments using a simulation model (Jones *et al.*, 1997; Dingkuhn *et al.*, 1998). The idea was to combine specific assets of *O. glaberrima* such as weed competitiveness and the yield potential of *O. sativa* tropical *japonica* parents. Since the mid-1990s, a number of studies have been conducted to identify varieties with strong weed competitiveness, plant traits related to weed competitiveness, and to develop screening tools (e.g. Dingkuhn *et al.*, 1999a; Rodenburg *et al.*, 2009; Saito *et al.*, 2010b). Saito *et al.* (2010b) indicate that grain yield, plant height at maturity and visual growth vigour at 42–63 days after sowing **under weed-free conditions** are useful selection criteria for developing superior weed-competitive rice genotypes.

The 18 upland NERICA varieties were tested for pre- and post-attachment resistance to *Striga hermonthica*, an obligate hemiparasitic weed that causes severe yield losses in cereals, including rice. Some NERICA varieties showed good pre- and post-attachment resistance to this very damaging weed (Cissoko *et al.*, 2011; Jamil *et al.*, 2011) and research is ongoing to determine the genes behind these different types of resistance. Varieties combining pre- and post-attachment resistances are expected to provide a breakthrough in terms of *Striga* control in rice-based systems in sub-Saharan Africa.

Saito and Futakuchi (2009) and Saito *et al.* (2012) proposed *aerobic* rice and *aus* varieties[7] with high yielding ability, strong weed competitiveness, and superior adaptation to low soil fertility for inclusion in upland breeding programmes in Africa.

Crop and Natural Resources Management

Sahel irrigated systems

In the early 1990s, AfricaRice focused on the development of best-fits between varietal choice and cropping calendars in irrigated lowland systems in the Sahel. From this detailed physiological work (Dingkuhn and Miezan, 1995; Dingkuhn *et al.*, 1995), the RIDEV decision-support tool was derived and physiological knowledge was also captured in the crop simulation model ORYZA_S (Dingkuhn, 1995). ORYZA_S allowed regional analysis of climatic risk to irrigated rice cropping in the Sahel and zonation of regions with potential for double cropping. RIDEV provided advice on best-bet sowing dates and varietal choice, and timing of N fertilizer application, drainage before harvest, and harvest for any site under irrigation in the Sahel.

In the mid- and late 1990s, the research focus shifted to analysis of farmer decision making, their major constraints and opportunities, and determinants of rice productivity (e.g. Adesina, 1996; Adesina and Djato, 1996). Donovan *et al.* (1999) and Wopereis *et al.* (1999) reported on such work in Senegal (Sahel), Mali (Sudan savannah), Mauritania (Sahel) and Burkina Faso (northern Guinea savannah). Results showed that average farmers' yields varied between 3.8 t/ha and 7.2 t/ha, resulting in an overall average of 4.5 t/ha. Yields of individual farmers were highly variable, ranging from 0.3 t/ha to 8.7 t/ha. Maximum yields reached by farmers were only at 40–60% of 10-year-averages of ORYZA_S simulated potential yield (limited by temperature and solar radiation). The yield gap between farmers' average yields and farmers' highest yield was between 0.7 t/ha and 4.1 t/ha, with an average of 2.6 t/ha, indicating considerable scope for improving yields. Surveys on soil degradation in irrigated systems in the Sahel revealed the importance of double cropping and drainage to combat salinity in the Senegal River delta (Ceuppens and Wopereis, 1999). Numerous studies were conducted to quantify and reduce the risk of alkalinization and sodification (e.g. Boivin *et al.*, 2002; Van Asten *et al.*, 2003a,b).

Box 1.2. Irrigated systems outside of the Sahel

AfricaRice has conducted little research on irrigated systems outside of the Sahel. However, Becker and Johnson (1999a) conducted surveys in irrigated systems of the forest zone of Côte d'Ivoire. Yields varied between 0.2 t/ha and 7.3 t/ha, with average yields of 3.2 t/ha under partial irrigation and 4.2 t/ha in fully irrigated systems. Age of seedlings at transplanting, timeliness of operations and application of P fertilizer explained 60% of observed variability.

Haefele *et al.* (2000, 2001) observed the following main agronomic constraints in the Senegal River valley: (i) mismatches between timing of N fertilizer application and critical N-demanding growth stages of the rice plant; (ii) non-use of P fertilizer on P-deficient soils; (iii) largely neglected or inefficient weed management; and (iv) delayed harvesting. Consequently, technology-specific coalitions were established in the Senegal River valley focused on three key issues: soil fertility management, weed management, and timely harvest and postharvest practices. These coalitions developed action plans and worked with farmers and agricultural machinery manufacturers to evaluate and adapt technologies at key sites, often through test plots and regular field visits. To address the harvest and postharvest problems, a thresher–cleaner and a stripper–harvester were imported from the Philippines and a consortium of research and development partners and small-machinery manufacturers was formed to develop Senegalese prototypes of both machines (Donovan *et al.*, 1998; Wopereis *et al.*, 1998).

Gradually a basket of 'integrated rice management' (IRM) options was developed for the Senegal River valley (Box 1.3). These options provided guidance to extension agents and farmers in terms of good agricultural principles and practices. During farmer visits to test plots and field tests of the thresher–cleaner, various issues related to rice cropping were debated, including best age to transplant rice seedlings, control of pests and diseases, water management, access to fertilizers, credit, and certified seed. Over time, AfricaRice staff developed a powerful learning tool to facilitate these debates: a cropping calendar depicting timing of key management interventions (i.e. sowing, transplanting, weeding, fertilizer application, harvesting) as a function of rice development stage (Wopereis *et al.*, 2003).

These cropping calendars could be easily adjusted to any choice of sowing date × site × cultivar combination along the Senegal River valley using the RIDEV decision-support tool (Dingkuhn, 1995; Wopereis *et al.*, 2003). Another direct consequence of the debates in farmers' fields was the development of a manual with technical references on irrigated rice cropping in the Senegal River valley (ADRAO and SAED, 2000), as a support for research and

Box 1.3. Integrated rice management (IRM) options for the Senegal River valley (as formulated in the 1990s by AfricaRice and partners)

1. Land preparation: cultivate on soil suitable for irrigated rice (i.e. heavy clay soils; local soil series terminology: Holladé and Faux-Holladé soils), make sure the field is properly tilled and levelled.
2. Varietal choice: use pre-germinated certified seeds; for the dry season: Sahel 108 (good grain quality, but salinity sensitive) or I Kong Pao (low grain quality, salinity tolerant); and for the wet season: Sahel 108, Jaya, Sahel 201 or Sahel 202.
3. Sowing date: guided by RIDEV to avoid spikelet sterility due to cold or heat.
4. Seeding rates: use certified seed and 100 kg/ha and 40 kg/ha, respectively, for direct seeding and transplanting.
5. Maximum recommended fertilizer rates: 100 kg triple super phosphate (TSP, 20% P) or diammonium phosphate (DAP, 20% P, 18% N) per hectare and 250–300 kg urea (46% N) per hectare, depending on location along the Senegal River. TSP is applied as a base fertilizer, while urea is applied in three splits. The first dose of 40% is applied at the start of tillering, and another dose of 40% at panicle initiation. A final dose of 20% is applied at the booting stage of the crop. Timing is guided by RIDEV.
6. Weed management: a mixture of propanil (6 l/ha) and 2,4-D (1.5 l/ha) applied a few days before first urea application (at 2–3 leaf stage of the weeds), complemented with one manual weeding before the second urea application.
7. Water management: directed at maximizing the efficiency of fertilizers and herbicides, consists of applying herbicides in completely drained fields and reducing water levels in the field to 3 cm for about 4–5 days at each fertilizer application. The rice field is completely drained 15 days after flowering to promote uniform ripening of the grains, but primarily to allow for a timely harvest.
8. Harvest and postharvest: harvesting at maturity, i.e. if about 80% of the panicles are yellow. Threshing within 7 days after timely harvest, preferably with the 'ASI' thresher–cleaner prototype developed for Sahelian conditions.

extension staff. Similar work was conducted for irrigation schemes in Burkina Faso (Segda *et al.*, 2004, 2005). Scaling-out of the IRM options was done through training of extension staff and promotional campaigns. A number of these modules have been converted into farmer training videos (entitled 'Rice Advice'), translated into more than 30 local African languages (Van Mele *et al.*, 2010).

Kebbeh and Miézan (2003) confirmed the potential of IRM to raise rice productivity. They observed that technologies that are of greatest direct interest to farmers and that are within their reach are adopted first, such as improved soil fertility and weed management. Yield increases were positively correlated to the number of IRM options farmers were able to adopt.

The Senegalese version of the thresher–cleaner was officially released by the minister of agriculture of Senegal in 1997 and obtained the 'Grand Prix du Président de la République du Sénégal pour les Sciences' in 2003. There are now hundreds of these machines in Senegal, Mali and Mauritania. The project to develop a local version of the stripper–harvester was abandoned. During field tests, farmers clearly indicated that they did not appreciate the fact that the machine left rice straw standing in the field. Follow-up work led to the development of a reaper to deal with this problem. More recent work on mechanization (2011–2012 cropping seasons) focused on testing mini-combine harvesters imported from the Philippines in Mali and Senegal, and subsequent local adaptation, fabrication and maintenance.

Rainfed lowland systems

Becker and Johnson (2001b) studied the effects of improved water control and crop management on lowland rice productivity in West Africa. Bunding significantly increased rice yield across sites by about 40% and controlled weeds, with approximately 25% less weed biomass in bunded than in open plots. Mineral fertilizer N application significantly increased rice yields (18% on average across sites) in bunded fields only. Levelling together with bunding facilitated water management and decreased weed competition – as many weed species are not well adapted to permanently flooded conditions (e.g. Kent and Johnson, 2001) – and generally increased nutrient use efficiencies, in particular in well-drained fields. Water management and regular drainage will also avoid problems with iron toxicity in inland-valley lowlands. Direct and indirect effects of iron toxicity in inland valleys can lead to 40–45% rice yield reductions in lowlands, depending on the extent of the problem, water, soil and crop management (e.g. cultivar choice) as well as on the availability of other soil nutrients (Becker and Asch, 2005; Audebert and Fofana, 2009). A synthesis of 15 years of multi-disciplinary research on iron toxicity in West Africa by AfricaRice and NARS partners was provided by Audebert *et al.* (2006).

Sakurai (2006) reported on a landmark study conducted in Côte d'Ivoire to explore factors that influence the expansion and intensification of rice production in rainfed lowland sites. Results showed that: (i) expansion of lowland rice cultivation was driven by population pressure and accessibility of the market; and (ii) the adoption of water control technologies is enhanced by the presence of immigrants and accessibility of the market. Investment in water supply canals was strongly influenced by land tenure security. Improved water control greatly enhanced rice productivity.

Erenstein (2006) and Erenstein *et al.* (2006) also studied increased inland-valley use for rice (and vegetable) production around large urban centres in West Africa. They concluded (in line with the conclusions of Sakurai, 2006) that market access was a necessary, but not sufficient factor for technological intensification of lowland use. They also concluded that better targeting of development efforts in terms of enabling lowland intensification is needed. Remote sensing or remote-sensing-derived products have been used to map inland valleys in the past (e.g. Thenkabail and Nolte, 1996). AfricaRice developed an automated mapping procedure based solely on a digital elevation model, which is globally available at a spatial resolution of 30 m (AfricaRice *et al.*, 2012). This standardized methodology is currently being implemented and validated for the entire West African region (S.J. Zwart and C.A. Linsoussi, Cotonou, Benin, 2012, personal communication).

Given the diversity and dynamics of farmer reality and growing conditions in rainfed lowland systems, AfricaRice developed a locally adapted and integrated approach to increase rice productivity in inland-valley production systems in Africa (Wopereis and Defoer, 2007). The approach is based on participatory learning and action-research (PLAR) and IRM. It is essentially a farmer-participatory, step-wise approach to put inland valleys under rice production using good agricultural principles and practices (Defoer *et al.*, 2004; Wopereis *et al.*, 2007).

Discussing these different approaches to technology development based on diagnostic and yield gap surveys, Wopereis and Defoer (2007) note that the need to use a 'PLAR approach' increases as one moves from high- to low-precision systems and from relatively uniform to more diverse production systems. Technology development can be more advanced in Sahelian irrigated lowland systems before evaluation by farmers; farmers in inland valleys need flexible technologies that can be adjusted relatively easily to local settings. Farmers in both low- and high-precision systems can benefit tremendously from decision-support tools and improved knowledge of agroecological and socio-economic principles. More on this approach is given in Defoer and Wopereis (Chapter 31, this volume).

Upland rice-based systems

Becker and Johnson (2001a) analysed cropping intensity effects on upland rice yield and sustainability in four agroecological zones in Côte d'Ivoire. Increased cropping intensity and reduced fallow duration were associated with yield reduction. Intensification-induced yield loss was about 25% (a drop from an average 1.5 t/ha to 1.1 t/ha) and was mainly related to increased weed infestation and declining soil quality.

Following the above study, cropping system alternatives using cover legumes as short-duration fallow crops were tested for reducing weed pressure and sustaining rice yield (Becker and Johnson 1998, 1999b; Akanvou *et al.*, 2000). Furthermore, to increase benefits from the systems, the timing of legume establishment in relation to rice and the effects of removing, burning, mulching or incorporating fallow residues prior to the rice crop on rice and weed growth were determined. Legume fallows appeared to offer good potential to sustain rice yields. Timing of legume establishment into upland rice depended on choice of legume, choice of rice variety, and their crop densities as well as environmental conditions (Saito *et al.*, 2010a). Incorporating or mulching of fallow residues provided no significant yield advantage compared to burning. On-farm participatory legume evaluation was also carried out and selected legumes were grown by farmers. Through this work, AfricaRice researchers learned that, in order to be successful, solutions such as improved fallows must consider the biophysical and socio-economic specifics of prevailing systems.

Box 1.4. Madagascar

Work in Madagascar by the national research institute (Foibem-pirenena Fikarohana ampiharina amin'ny Fampandrosoana ny Ambanivohitra, FOFIFA), Université d'Antananarivo and CIRAD on conservation agriculture in upland rice-based systems has led to expansion of improved fallow systems combined with mulching as residue management in no-tillage system in mid-west Madagascar (about 1800 ha). One of the major reasons for the expansion is the use of *Stylosanthes guianesis* as a fallow crop which seems to reduce *Striga* infestation (Michellon *et al.*, 2011).

Phosphorus deficiency is a major limiting factor in upland rice systems in the humid forest zone (Somado *et al.*, 2011). Responses of upland rice to P were studied by Sahrawat *et al.* (1995, 1997). Numerous other studies have been conducted to test the response of upland rice yield to P fertilizer application, including locally available rock P in West Africa in the 1990s and 2000s (e.g. Sahrawat *et al.*, 2003). Somado *et al.* (2011) reviewed previous studies related to agronomic management options for alleviating P deficiency, and suggested that integrated approaches combining agronomic management options with genetically tolerant or high P-use efficient varieties are essential.

A number of fertilizer management and IPM options were developed for upland NERICA varieties (e.g. Nwilene *et al.*, 2008a; Oikeh *et al.*, 2008, 2009; Sokei *et al.*, 2010; Ogha *et al.*, 2011; Touré *et al.*, 2011). However, these options have not yet been tested sufficiently in farmers' fields to enable out-scaling.

Across systems

Systematic evaluation of rice germplasm for African rice gall midge (AfRGM, *Orseolia oryzivora*) resulted in identification of over 50 primary sources of resistance among *O. glaberrima* and traditional *O. sativa* varieties (Nwilene *et al.*, 2002). AfRGM is attacked by the parasitoids *Platygaster diplosisae* and *Aprostocetus procerae* in rice-production systems (Nwilene *et al.*, 2008b). However, the level of parasitization is low because the parasitoid populations build up too late in the season to prevent heavy AfRGM infestation. Meanwhile, a related gall midge (paspalum gall midge, PGM, *Orseolia bonzii*) that forms galls on *Paspalum* grass is an important alternative host for the two parasitoids of AfRGM. Hence, habitat manipulation to increase the carry-over of parasitoids from PGM (which does not attack rice) to AfRGM, such as dry-season cultivation to encourage *Paspalum scrobiculatum* abundance early in the wet season, could improve the natural biological control of AfRGM.

Coyne *et al.* (1999) studied the prevalence of plant-parasitic nematodes associated with rice in Ghana and Côte d'Ivoire in rainfed upland, hydromorphic and lowland rice fields. Thirty days after the introduction of rice, nematode species diversity across ecosystems was reduced by 57% to 17 species. At harvest, species diversity was 55% lower than in adjacent forest and vegetation. With progression of the season, a small number of nematode genera became numerically dominant, while most nematode genera were present at low mean density. Lowland rice communities were characterized by low nematode intensity and low species diversity. An overview of nematode research in rice in West Africa is provided by Coyne and Plowright (2004).

Sy and Séré (1996) discuss the three major pathogens of rice in Africa: blast fungus

(*Magnaporthe oryzae*), *Rice yellow mottle virus* (RYMV) and the bacterium responsible for leaf blight (*Xanthomonas oryzae* pv. *oryzae*). Since the early 1990s, a large number of studies have been conducted on the spatial variability of isolates of these diseases and corresponding resistance genes in rice (e.g. Séré *et al.*, 2007; see also Séré *et al.*, Chapter 17, this volume).

Policy, Impact Assessment and Rice Value-Chain Development

Policy

During the 1980s and 1990s, AfricaRice and its member countries focused their attention on the effects of structural adjustment programmes on local rice production in West Africa. In particular, use was made of the Policy Analysis Matrix (PAM) approach to assess regional comparative advantage of rice across rice environments and markets, and the roles of policy in influencing competitiveness. Economists from national systems were trained in this approach through the Rice Policy Task Force (WARDA, 1997). At the turn of the millennium, AfricaRice began to take a more holistic, value-chain approach to rice policy. A 2-year rice-sector study in Nigeria culminated in 2003 with a strategy for the development of the national rice economy. Recommendations emphasized the need for:

- a consistent agriculture and trade policy, including price protection for local rice;
- upgrading quality management along the value chain;
- increasing value-chain efficiency; and
- enhancing policy dialogue among stakeholders and with government, including the establishment of a national rice stakeholders' forum (Erenstein *et al.*, 2003).

The Nigeria Rice Alliance was established and open debate of policy reform ensued. New policies were implemented in line with the new strategy, including 100% import-tariff increase in 2003 and 150% increase in 2005. Meanwhile, the government maintained subsidies on fertilizers and other agro-inputs. In terms of funding, US$400 million was allocated to boost agricultural production, including rice (AfricaRice, 2011a).

Between 2000 and 2006, AfricaRice conducted a broad review of policy in West Africa, using the Policy Analysis Matrix (AfricaRice, 2011a). The Center maintained the Africa Rice Statistics Databank and produced periodic *Africa Rice Trends* (e.g. WARDA, 2008). AfricaRice also conducted research on rice market efficiency and rice-sector competitiveness (AfricaRice, 2011a).

Policy comes to the fore as the food crisis looms

In June 2007, AfricaRice organized a 2-day workshop of the Africa Policy Research and Advocacy Group, at which it predicted the rice crisis which effectively happened in 2008. That message was delivered by AfricaRice's Director General at the AfricaRice Council of Ministers meeting in Abuja, Nigeria in September 2007, forewarning the ministers of the looming crisis, and encouraging them to turn the impending crisis into an opportunity. Specific recommendations delivered were to:

- establish seed legislation and encourage the involvement of the private sector in seed supply and trade;
- reduce the import tax on small-scale farm and processing machinery that can increase rice farmers' labour efficiency and improve grain quality;
- work together to reduce fertilizer prices (fertilizers in Africa are two to six times more expensive than in Asia and Europe);
- improve capacity at research, extension, processing and marketing levels;
- promote large-scale use of upland and lowland NERICA rice varieties; and
- significantly increase the share of high-yielding irrigated and lowland rice farming (Seck, 2007; AfricaRice, 2011a).

This work resulted in an Emergency Rice Initiative, funded by several donors, that helped over 110,000 farmers gain access to subsidized seed of improved varieties, fertilizer and improved crop-management methods during the 2008 food crisis. AfricaRice also contributed to the development of national rice development strategies for 21 African countries under the Coalition for African Rice Development (AfricaRice, 2011a).

Through its Council of Ministers and National Experts Committee (composed of the directors of the NARS) and engagement with the political authorities (ECOWAS, UEMOA, UNECA West Africa, CILSS,[8] AGRHYMET Regional Centre, etc.) of the region, AfricaRice has guided a gradual change in the policy environment from unfavourable to more favourable (aided by the food crisis) through buy-in from political leaders in the diagnosis of the problems and possible solutions. Frequent spikes in international rice prices – mainly the result of export restrictions in major Asian exporting countries, low stocks and speculation – have been a sobering lesson for most African rice-importing countries. They now realize that they **must** develop their local production. In countries where there are big vested interests in the importing of rice through large-scale, often diversified, powerful companies with political connections, changing the policy framework in favour of smallholder rice production and milling operations will be particularly hard.

Impact studies

Matlon *et al.* (1998) reviewed the adoption and impact of 'modern varieties' in West Africa. Hard data were scarce, but they report 95% adoption of modern varieties Bouaké 189 and IR8 in irrigated systems in the humid zone in Côte d'Ivoire. However, only 40% of the upland-rice area in the same country was planted with improved varieties (original data from AfricaRice–NARS Task Force). Meanwhile, an estimated 68% of the region's rainfed lowland areas was planted with modern varieties, with farmer adoption rates ranging from a low of 5–10% in western Côte d'Ivoire, through 20% in eastern Côte d'Ivoire and 35% in Sierra Leone, to 95% in Niger State, Nigeria. AfricaRice activities had led to high adoption rates in Sahelian irrigated systems in Mali (90%), the Senegal River valley (Mauritania and Senegal) (90%), Niger and Burkina Faso. An *ex-ante* impact study indicated an internal rate of return for the then AfricaRice Sahel irrigated-rice breeding programme in the range 40–55% (Fisher *et al.*, 1995, cited by Matlon *et al.*, 1998). These high internal rates of return were confirmed by Fall (2005). Matlon *et al.* (1998) also report various adoption rates for modern varieties in the mangrove-swamp rice environment of between 10% (for Guinea-Bissau) and 100% (Casamance, Senegal and Sierra Leone). They report additional rate-of-return studies, which are included in our own summary (see Table 1.1).

In 2001–2002, AfricaRice conducted a major study on the impact of improved rice varieties, from both national and international research centres in all West African rice-growing environments. The study estimated that genetic enhancement and transfer had increased the value of rice production by $93/ha (Dalton and Guei, 2003). The study also confirmed that while irrigated and rainfed lowland ecosystems have largely benefited from varietal improvements, the upland

Table 1.1. Meta-analysis of internal rates of return to rice genetic improvement R&D in West and Central Africa.

Country	Period	Internal rate of return (%)	Source
Sierra Leone	1976–1986	18–21	Tré (1995)
Guinea-Bissau	1980–1994	26	Seidi (1997)
Sierra Leone	1993	34	Edwin and Masters (1998)
Burkina Faso	1973–1999	43	Ouedraogo and Ouedraogo (2003)
Niger	1982–2000	66	Ibro *et al.* (2001)
Mauritania	1985–2000	68	Bah (2001); Fall (2005)
Mali	1979–1998	75	Kergna (2003)
Senegal	1985–2000	75–78	Fall (2005)
The Gambia	1988–2000	81	Bittaye *et al.* (2002)
Senegal	1995–2004	121	Fisher *et al.* (2001)
Chad	1983–2000	182	Dastre *et al.* (2003)

rice ecosystem lagged behind due to much lower adoption rates and limited yield gains.

AfricaRice (2011b) estimated that over 700,000 ha of land in Africa was under NERICA varieties in 2010 in 35 countries, with 243,000 ha in Nigeria, 140,000 ha in Guinea, 60,000 ha in The Gambia, 90,000 ha in Mali and 60,000 ha in Uganda. Compared with earlier estimates (Adegbola *et al.*, 2006; Diagne *et al.*, 2006), this shows a considerable increase in the area under NERICA varieties in Africa (see also Wopereis *et al.*, 2008; Diagne *et al.*, 2011). The probability of adoption and amount of area planted were enhanced in Uganda by membership of a farmers' group, formal education of the head of the farming household and the number of farming household members (Kijima *et al.*, 2006, 2008, 2011). There were significant positive effects on rice yields through NERICA adoption in Benin and The Gambia (Dibba *et al.*, 2012). However, no significant effects on rice yields were observed in Côte d'Ivoire and Guinea. Women profited generally more than men in almost all countries (Agboh-Noameshie *et al.*, 2006; Diagne *et al.*, 2011). In Nigeria, uptake of NERICA varieties was reportedly high in Kaduna and Ekiti states, with up to 30% of farmers in Ekiti and 42% of farmers in PVS villages in Kaduna cultivating NERICA varieties in 2005 (Spencer *et al.*, 2006). In Uganda, NERICA varieties have the potential to increase per-capita annual income by $20 (12% of actual per-capita income) and to decrease the poverty incidence by 5 percentage points, measured by the head count ratio (from 54.3% to 49.1%; Kijima *et al.*, 2006).

The global rice sector has been particularly responsive to agricultural R&D investment: the median annual rate of return amounting to 51–60% (Alston *et al.*, 2000; Evenson, 2001). Some 47% of the global R&D payoffs of the CGIAR, for instance, have been attributed to rice breeding (Raitzer and Kelley, 2008) and, in turn, 29% of the returns from rice genetic enhancements in West Africa, in particular, were attributable to the CGIAR (Dalton and Guei, 2003). Alene *et al.* (2007) show that rice R&D yields high returns in different ecosystems, but particularly in the moist savannah. However, only a handful of studies have actually attempted to estimate the rate of return to R&D in the sub-Saharan African rice sector. Our meta-analysis (presented in Table 1.1) suggests a high median rate of return of 68% to rice genetic improvement R&D in West and Central Africa. These findings concur with Renkow and Byerlee (2010), who conclude that the most profound documented positive impacts of the CGIAR are recorded for crop genetic improvement research. It is thus surprising that more is not invested in agricultural R&D, particularly in rice R&D.

Rice value-chain development

The importance of a value-chain approach to rice-sector development and the importance of rice quality came to the forefront with the Nigerian Rice Sector Policy Study of 2002–2003 (see above) and was highlighted by Lançon *et al.* (2004). It was clear that local rice in Nigeria was being discounted by around 30% compared to imports, mainly because of a lack of cleanness and presentation. By contrast, in Guinea and Mali certain varieties of local rice may receive a price premium (USAID, 2009). In order to improve the quality of local rice, institutional innovations are needed that make producers more responsive to end-users' requirements, and attach much more importance to milling and cleaning, and identity preservation (no mixing of different rice varieties).

The growing awareness of grain quality and marketing issues came to a head in the Center's 2006 External Programme and Management Review, which recommended adoption of a value-chain focus.

In 2008, AfricaRice and its national partners started a series of experimental auctions in Benin, Burkina Faso, Cameroon, The Gambia, Mauritania, Senegal and Uganda in order to assess consumers' willingness to pay for rice quality and marketing attributes (Demont *et al.*, 2012, 2013a,b). The main finding was that local rice is or can become competitive with imported rice after the necessary quality improvements (see Demont and Neven, Chapter 24, this volume).

Since 2011, AfricaRice and partners have adopted the concept of 'Rice Sector Development Hubs', referred to by Seck (2012) as 'fronts de développement agricole et rural'. These Hubs represent key rice-growing environments and different market opportunities across African countries and are linked to major national or regional rice-development efforts to facilitate broader uptake of rice knowledge and technologies. The geographic positioning of each Hub is determined in national workshops, convened by the NARS. Activities in the Hubs focus on producing sufficient quantities of the right quality of rice and rice-based products of interest to the national or regional markets in a sustainable manner. Hubs are regions strategic for rice development, where local innovations and research products and services are tested, adapted and integrated in 'baskets of good agricultural practices' (GAP, i.e. IRM options) with feedback provided to researchers on technology performance. Hubs also work to improve value chains by investing in institutional innovations and market development. Care is taken that women and youth are not marginalized, but rather strengthened in the process of rice value-chain development. By January 2013, some 59 Hubs had been identified in 20 countries in sub-Saharan Africa.

Surveys are currently (2012–2013) being conducted in the Hubs to understand farmers' practices and to identify constraints to rice production, processing and marketing. Gaps between actual yields obtained by farmers and what would be possible through improved management are being quantified (rice yield and productivity gaps). The surveys will also help to assess farmers' needs, gender dimensions, institutional and political arrangements, and linkages among value-chain actors. The results from these surveys will be used to set research priorities and for developing 'GAP baskets' for each Hub, analogous to what was done in the 1990s for irrigated systems in the Sahel (see Box 1.3), but with greater precision, across rice-growing environments and moving beyond production to postharvest issues and rice processing.

Hubs are built around large groups of farmers and involve other value-chain actors, such as rice millers, input dealers and rice marketers. Change agents from research, NGOs and extension agencies work with these actors to evaluate technological and institutional innovations, facilitate diffusion of knowledge and establish linkages along the rice value chain. These types of interactions are stimulated through the establishment of multi-stakeholder platforms (MSPs).

MSPs refer to a process which aims to bring stakeholders together in a new form of communication, 'decision finding' and, possibly, decision making on a particular issue (Hemmati, 2002). MSPs are also platforms for interaction among different stakeholders who share a common resource and interact to improve mutual understanding, create trust, define roles and engage in joint action (Thiele *et al.*, 2011). MSP participants can include those directly involved in agricultural production, processing and marketing, and also public- and private-sector partners with technical knowledge about extension and agriculture.

In Benin, MSP participants noted that the most significant outcomes of multi-stakeholder involvement were capacity development and increased rice yields in an inland-valley setting (Dossouhoui and Kinha, 2012). In relation to capacity development, participants noted how the MSP process helped actors who had otherwise worked individually to come together to complete certain activities – notably cultivation, processing and wholesaling. Actors felt that they had gained more confidence and technical experience through the MSP process, which helped the collective but also the actors/producers themselves. One of the MSPs in Benin noted that their yields had almost doubled, which was explained by access to improved rice varieties (i.e. IR841, NERICA-L 14, NERICA-L 19 and NERICA-L 20; Dossouhoui and Kinha, 2012).

In Mali, MSP participants noted similar benefits related to collective management of natural resources; however, their benefits focused more on improving governance (Bengaly *et al.*, 2012). Members felt that by participating in the MSP process, they better understood others' needs and resource-management objectives. They felt more cohesion among various resource-user groups such as fishers and potato farmers, who may not have been working in a collaborative capacity prior to MSP establishment. In addition, MSP actors noted that involvement in multi-actor processes provided access to rice technology (introduction of improved rice varieties). Women parboilers who were MSP members had become better organized with respect to rice parboiling and processing to sell their product. In terms of governance, the MSP facilitated negotiations among various sub-groups (e.g. fishers, potato farmers) for the use and management of irrigation systems within the inland valley. In addition, institutionally the MSP became the conduit to develop and implement the 5-year community plan for land and water development within the inland valley.

Conclusions

Agricultural R&D can be classified into two main categories: supply-shifting and demand-lifting. Supply-shifting R&D can be further sub-divided into: (i) **yield-increasing R&D** (e.g. genetic enhancement, breeding, biotechnology, IRM, crop husbandry); and (ii) **cost-reducing R&D** (e.g. technical change, collective management, efficiency increase, enhanced crop protection, labour-saving technologies). Demand-lifting R&D can be further sub-divided into: (iii) **quality-enhancing R&D** (e.g. grain quality, homogeneity and purity, visual, cooking, sensory and nutritional characteristics); and (iv) **marketing R&D** (e.g. processing, storage, transport, aggregation, distribution, value-chain development, access to markets, branding, advertising and generic promotion). Marketing R&D can also be supply-shifting (e.g. by reducing marketing and transaction costs and increasing efficiency of marketing systems). A final category which may have both supply-shifting and demand-lifting effects is: (v) **policy R&D** (creating a more conducive general policy climate for agribusiness, production, marketing and consumption).

Looking back at over 20 years of rice research for development in Africa, it is clear that most rice R&D has been related to supply-shifting and policy R&D. The outbreak of the civil war in Côte d'Ivoire in 2002 and the forced relocation of AfricaRice to Mali and then Benin in 2005 severely disrupted the research programmes, forcing the Center to essentially focus on rice-breeding efforts and networking activities. Since 2007, the Center has regained stable ground and research activities have increased rapidly, including both supply-shifting and demand-lifting R&D as documented in this book. The annual budget of the Center tripled from about $10 million in 2007 to $32 million in 2013. This book itself is an excellent illustration of the renewed vigour of the Center.

The new 2011–2020 strategic plan of the Center (AfricaRice, 2011b) emphasizes the need to conduct demand-lifting R&D through one of its seven priority areas, 'Creating market opportunities for smallholder farmers and processors by improving the quality and the competitiveness of locally produced rice and rice products'. The creation of Rice Sector Development Hubs, which is central in the 2011–2020 AfricaRice strategy (AfricaRice, 2011b), also points to the importance of combining different types of R&D and testing and developing technologies with partners along the rice value chain. The objective in these Hubs is to produce rice or rice-based products that respond to consumer preferences in urban and rural markets in quantities that are of interest to rice traders, who would usually import such products. Hubs will be strategically selected to allow linkage with major national or regional rice-development efforts to facilitate broader uptake of rice knowledge and technologies.

Supply-shifting R&D and demand-lifting R&D complement each other. Supply-shifting R&D creates value, but part of the value is transmitted to consumers through lower prices. Synchronous demand-lifting R&D redistributes part of this consumer welfare back to producers and strengthens vertical links between production and consumption throughout the value chain. Hence, the ultimate challenge for African policy makers, researchers and donors will be to find the optimal mix of investment between supply-shifting and demand-lifting R&D (Demont and Rizzotto, 2012) and ensure optimal links with public- and private-sector partners involved in Africa's multiple national and regional rice-sector development efforts (Seck *et al.*, 2012). As an association of currently 24 member states (February 2013), recognized by the African Union as the Center of Excellence for Rice Research in Africa, AfricaRice is well placed to coordinate these rice research-for-development efforts across the continent over the decades to come, in close collaboration with partners.

Notes

[1] AfricaRice is used throughout the rest of this chapter and book to refer to the Association and Center from its inception to the present day. One notable exception to this rule is the use of the correct name as it appears on publications in both citations and lists of references.
[2] IITA, International Institute of Tropical Agriculture; ILRI, International Livestock Research Institute; IWMI, International Water Management Institute.
[3] CIRAD, Centre de coopération internationale en recherche agronomique pour le développement.
[4] IRRI, International Rice Research Institute; CIAT, International Center for Tropical Agriculture.
[5] IRAT was responsible for crop genetic improvement in the French colonies and continued its rice breeding activities in most of francophone West Africa after independence. Many varieties were created through this research during the 1960s and 1970s – in particular, upland varieties developed by crossing African *japonica* (63-104, 63-83) and Brazilian *japonica* (Iguape Cateto, Dourado Précoce) with Asian temperate *japonica* (e.g. Lung Sheng 1). These crosses formed the basis of the IRAT series (e.g. IRAT 110, IRAT 112, IRAT 144, IRAT 146, IRAT 147). IRAT 13 was a product of gamma mutation on 63-83 from Senegal. West African countries such as Burkina Faso, Côte d'Ivoire, Mali and Senegal continued to receive bilateral support from IRAT and later CIRAD to strengthen upland rice breeding, in some cases up to the 1990s (Dalton and Guei, 2003). CIRAD continues to support the rice varietal improvement programme for high-altitude upland systems in Madagascar in collaboration with the NARS.
[6] During this period (1970s and 1980s), IITA conducted rice varietal improvement research at its headquarters in Ibadan, Nigeria. Its early programme (1974–1975) was primarily for upland rice environments, because it was believed that for irrigated systems with good water control there were many Asian-bred dwarf cultivars which could yield very well under West African conditions (Miézan and Sié, 1997). This resulted in the ITA series, the most well-known varieties being ITA 212 and ITA 306 released for irrigated systems in 1986, and ITA 150 released for upland cultivation in 1992 in Nigeria.
[7] Aus cultivars are very early maturing, drought-tolerant rice varieties grown under upland conditions in Bangladesh and West Bengal state of India during the so-called *Aus* season (March–June).
[8] ECOWAS, Economic Community of West African States; UEMOA, West African Economic and Monetary Union (Union économique et monétaire ouest-africaine); UNECA, United Nations Economic Commission for Africa; CILSS, Interstate committee for drought control in the Sahel region (Comité permanent Inter-Etats de Lutte contre la Sécheresse dans le Sahel).

References

Adégbola, P.Y., Arouna, A., Diagne, A. and Souléïmane, A.A. (2006) Evaluation de l'impact économique des nouvelles variétés de riz NERICA au Bénin: évidence avec les modèles basés sur l'approche contre factuel. Paper presented at the first Africa Rice Congress, Dar es Salaam, Tanzania, 31 July–4 August.

Adesina, A.A. (1996) Factors affecting the adoption of fertilizers by rice farmers in Côte d'Ivoire. *Nutrient Cycling in Agroecosystems* 46, 29–39.

Adesina, A.A. and Djato, K.K. (1996) Farm size, relative efficiency and agrarian policy in Côte d'Ivoire: Profit function analysis of rice farms. *Agricultural Economics* 14, 93–102.

ADRAO and SAED (2000) *Manuel Pratique pour la Riziculture Irriguée dans la Vallée du Fleuve Sénégal.* Association pour le développement de la riziculture en Afrique de l'Ouest, Saint-Louis, Senegal.

AfricaRice (2011a) *Lessons from the Rice Crisis: Policies for food security in Africa.* Africa Rice Center, Cotonou, Benin.

AfricaRice (2011b) *Boosting Africa's Rice Sector: A research for development strategy 2011–2020.* Africa Rice Center, Cotonou, Benin.

AfricaRice (2011c) Multi-national NERICA rice dissemination. Project report. Africa Rice Center, Cotonou, Benin.

AfricaRice (2012) The new Africa-wide Rice Breeding Task Force. In: *Africa Rice Center (AfricaRice) Annual Report 2011: A new rice research for development strategy for Africa.* Africa Rice Center, Cotonou, Benin, pp. 7–11.

AfricaRice, CBF (Cellule Bas-Fonds), ITRA (Institut Togolais de Recherche Agronomique) and IWMI (International Water Management Institute) (2012) Sawah, Market Access and Rice Technologies for Inland Valleys (SMART-IV) project Annual technical project report 2011–2012. Africa Rice Center, Cotonou, Benin.

Agboh-Noameshie R.A., Kinkingninhoun-Medagbe, F.M., Diagne, A. Simtowe, F.P. and Biaou, G. (2009) Who benefit more from NERICA varieties? Gender impact on production and income in Benin. Paper presented at the Conference on Perspectives on Impact Evaluation, an International Initiative for Impact Evaluation, Cairo, Egypt, 29 March–2 April.

Akanvou, R., Becker, M., Chano, M., Johnson, D.E., Gbaka-Tcheche, H. and Toure, A. (2000) Fallow residue management effects on upland rice in three agroecological zones of West Africa. *Biology and Fertility of Soils* 31, 501–507.

Alene, A.D., Manyong, V.M., Tollens, E.F. and Abele, S. (2007) Targeting agricultural research based on potential impacts on poverty reduction: Strategic program priorities by agro-ecological zone in Nigeria. *Food Policy* 32, 394–412.

Alston, J.M., Chan-Kang, C., Marra, M.C., Pardey, P.G. and Wyatt, T. (2000) A meta-analysis of rates of return to agricultural R&D: *Ex pede herculem*? *Research Report* 113. International Food Policy Research Institute, Washington, DC.

Andriesse, W., van Duivenbooden, N., Fresco, L.O., Windmeijer, P.N. (1994) A multi-scale approach to characterize inland valley agro-ecosystems in West Africa. *Netherlands Journal of Agricultural Science* 42: 159–179.

Asch, F. and Wopereis, M.C.S. (2001) Responses of field-grown irrigated rice cultivars to varying levels of floodwater salinity in a semi-arid environment. *Field Crops Research* 70, 127–137.

Asch, F., Dingkuhn, M., Wopereis, M.S.C., Dörffling, K. and Miezan, K.M. (1997) A conceptual model for sodium uptake and distribution in irrigated rice. In: Kropff, M.J., Teng, P.S., Aggarwal, P.K., Bouma, J., Bouman, B.A.M., Jones, J.W. and van Laar, H.H. (eds) *Applications of Systems Approaches at the Field Level*, vol. 2. Kluwer Academic Publishers, Dordrecht, Netherlands, pp. 201–217.

Asch, F., Dingkuhn, M., Wittstock, C. and Dörffling, K. (1999) Sodium and potassium uptake of rice panicles as affected by salinity and season in relation to yield and yield components. *Plant and Soil* 207, 133–145.

Asch, F., Becker, M. and Kpongor, D.S. (2005) A quick and efficient screen for resistance to iron toxicity in lowland rice. *Journal of Plant Nutrition and Soil Science* 168, 764–773.

Audebert, A. and Fofana, M. (2009) Rice yield gap due to iron toxicity in West Africa. *Journal of Agronomy and Crop Science* 195, 66–76.

Audebert, A. and Sahrawat, K.L. (2000) Mechanisms for iron toxicity tolerance in lowland rice. *Journal of Plant Nutrition* 23, 1877–1885.

Audebert, A., Narteh, L.T., Kiepe, P., Millar, D. and Beks, B. (2006) *Iron Toxicity in Rice-based Systems in West Africa*. Africa Rice Center, Cotonou, Benin.

Bah, M.S. (2001) *Impact économique de la recherche et de la vulgarisation sur le riz irrigué en Mauritanie*. Centre National de Recherche Agronomique et de Développement Agricole (CNRADA), Kaedi, Mauritania.

Becker, M. and Asch, F. (2005) Iron toxicity in rice – conditions and management concepts. *Journal of Plant Nutrition and Soil Science* 168, 558–573.

Becker, M. and Johnson, D.E. (1998) Legumes as dry season fallow in upland rice-based systems of West Africa. *Biology and Fertility of Soils* 27, 358–367.

Becker, M. and Johnson, D.E. (1999a) Rice yield and productivity gaps in irrigated systems of the forest zone of Côte d'Ivoire. *Field Crops Research* 60, 201–208.

Becker, M. and Johnson, D.E. (1999b) The role of legume fallows in intensified upland rice-based systems of West Africa. *Nutrient Cycling in Agroecosystems* 53, 71–81.

Becker, M. and Johnson, D.E. (2001a) Cropping intensity effects on upland rice yield and sustainability in West Africa. *Nutrient Cycling in Agroecosystems* 59, 107–117.

Becker, M. and Johnson, D.E. (2001b) Improved water control and crop management effects on lowland rice productivity in West Africa. *Nutrient Cycling in Agroecosystems* 59, 119–127.

Bengaly, M., Konaté, P., Dembélé, U., Yattara, K. and Senou, O. (2012) *Rapport Annuel: Les plates-formes multi-acteurs (PMA) pour le renforcement des capacités de changement institutionnel et de développement des acteurs au niveau des bas-fonds*. Institut d'Economie Rurale (IER), Bamako, Mali and Africa Rice Center, Cotonou, Benin.

Bèye, A.M., Jones, M. and Simpson, B. (2011) *The Community-based Seed System: The case of traditional rice farming system. The Technician's Manual*. AfricaRice Training Course Collection. Africa Rice Center, Cotonou, Benin.

Bittaye, A., Jones, K. and Drammeh, E. (2002) Economic impact assessment of the rice research program in The Gambia. National Agricultural Research Institute (NARI), Serekunda, The Gambia.

Boivin, P., Favre, F., Hammecker, C., Maeght, J.L., Delarivière, J., Poussin, J.C. and Wopereis, M.C.S. (2002) Processes driving soil solution chemistry in a flooded rice-cropped vertisol: analysis of long-time monitoring data. *Geoderma* 110, 87–107.

Ceuppens, J. and Wopereis, M.C.S. (1999) Impact of non-drained irrigated rice cropping on soil salinization in the Senegal River delta. *Geoderma* 92, 125–140.

CGIAR (2013) Centers chronology. www.cgiar.org/who-we-are/history-of-cgiar/centers-chronology/ (accessed 18 March 2013).

Cissoko, M., Boisnard, A., Rodenburg, J., Press, M.C. and Scholes, J.D. (2011) New Rice for Africa (NERICA) cultivars exhibit different levels of post-attachment resistance against the parasitic weeds *Striga hermonthica* and *Striga asiatica*. *New Phytologist* 192, 952–963.

Coyne, D. and Plowright, R.A. (2004) *Nematode Parasites of Rice.* The Africa Rice Center (WARDA), Bouaké, Côte d'Ivoire, and International Rice Research Institute, Los Baños, Philippines.

Coyne, D.L., Thio, B., Plowright, R.A. and Hunt, D.J. (1999) Observations on the community dynamics of plant parasitic nematodes of rice in Côte d'Ivoire. *Nematology* 1(4), 433–441.

Dalton, T.J. and Guei, R.G. (2003) Productivity gains from rice genetic enhancements in West Africa: countries and ecologies. *World Development* 31, 359–374.

Dastre, A.M., Betetmbaye, D., Ngawara, K.N. and Bagda, V. (2003) *Evaluation économique de l'impact de la recherche et de la vulgarisation sur le riz au Tchad.* Institut Tchadien de Recherche Agronomique pour le Développement, N'Djaména, Chad.

Defoer, T., Wopereis, M.C.S., Idinoba, P.A., Kadisha, T.K.L., Diack, S. and Gaye, M. (2004) *Curriculum for Participatory Learning and Action Research (PLAR) for Integrated Rice Management (IRM) in Inland Valleys of Sub-Saharan Africa: Facilitators' manual.* Africa Rice Center (WARDA), Cotonou, Benin, and IFDC, Muscle Shoals, Alabama.

Demont, M. and Rizzotto, A.C. (2012) Policy sequencing and the development of rice value chains in Senegal. *Development Policy Review* 30, 451–472.

Demont, M., Zossou, E., Rutsaert, P., Ndour, M., Van Mele, P. and Verbeke, W. (2012) Consumer valuation of improved rice parboiling technologies in Benin. *Food Quality and Preference* 23, 63–70.

Demont, M., Rutsaert, P., Ndour, M., Verbeke, W., Seck, P.A. and Tollens, E. (2013a) Experimental auctions, collective induction and choice shift: Willingness-to-pay for rice quality in Senegal. *European Review of Agricultural Economics* 40(2), 261–286.

Demont, M., Rutsaert, P., Ndour, M. and Verbeke, W. (2013b) Reversing urban bias in African rice markets: evidence from Senegal. *World Development* 45, 63–74.

Diagne, A., Sogbossi, M.J., Diawara, S. and Camara, A. (2006) L'étendue du succès de la dissémination des variétés NERICA en Guinée: estimation des superficies emblavées. [Mimeo]. Centre du riz pour l'Afrique (ADRAO), Cotonou, Benin.

Diagne, A., Midingoyi, S.-K. G., Wopereis, M. and Akintayo, I. (2011) Increasing rice productivity and strengthening food security through New Rice for Africa (NERICA). In: Punam, C.-P. and A. Manka (eds) *Yes Africa Can: Success stories from a dynamic continent.* The International Bank for Reconstruction and Development/The World Bank, Washington, DC, p. 492.

Dibba, L., Fialor, S.C., Diagne, A. and Nimoh, F. (2012) The impact of NERICA adoption on productivity and poverty of the small-scale rice farmers in The Gambia. Rice. *Food Security: The Science, Sociology and Economics of Food Production and Access to Food* 4(2), 253–265.

Dingkuhn, M. (1995) Climatic determinants of irrigated rice performance in the Sahel. III. Characterizing environments by simulating the crop's photothermal responses. *Agricultural Systems* 48, 435–456.

Dingkuhn, M. and Asch, F. (1999) Phenological responses of *Oryza sativa*, *O. glaberrima* and inter-specific rice cultivars on a toposquence in West Africa. *Euphytica* 110, 109–126.

Dingkuhn, M. and Miezan, K.M. (1995) Climatic determinants of irrigated rice performance in the Sahel. II. Validation of photothermal constants and characterization of genotypes. *Agricultural Systems* 48, 411–434.

Dingkuhn, M., Sow, A., Samb, A., Diack, S. and Asch, F. (1995) Climatic determinants of irrigated rice performance in the Sahel. I. Photothermal and micro-climatic responses of flowering. *Agricultural Systems* 48, 385–410.

Dingkuhn, M., Jones, M.P., Johnson, D.E. and Sow, A. (1998) Growth and yield potential of *Oryza sativa* and *O. glaberrima* upland rice cultivars and their interspecific progenies. *Field Crops Research* 57, 57–69.

Dingkuhn, M., Johnson, D.E., Sow, A. and Audebert, A.Y. (1999a) Relationships between upland rice canopy characteristics and weed competitiveness. *Field Crops Research* 61, 79–95.

Dingkuhn, M., Audebert, A.Y., Jones, M.P., Etienne, K. and Sow, A. (1999b) Control of stomatal conductance and leaf rolling in *O. sativa* and *O. glaberrima* upland rice. *Field Crops Research* 61, 223–236.

Donovan, C., Miézan, K.M., Wopereis, M.C.S., Diack, B.S. and Douthwaite, B. (1998) Technology transfer from Asia to Africa sets the stage for public and private sector collaboration in new technology in Senegal. *International Rice Research Notes* 23(2), 41–42.

Donovan, C., Wopereis, M.C.S., Guindo, D. and Nébié, B. (1999) Soil fertility management in irrigated rice systems in the Sahel and savanna regions of West Africa. Part II. Profitability and risk analysis. *Field Crops Research* 61, 147–162.

Dossouhoui, F. and Kinha, C. (2012) Rapport finale d'atelier de capitalisation des acquis et de relance de la dynamique des PMA de Dogbo et Houeyogbe. Africa Rice Center, Cotonou, Benin.

Edwin, J. and Masters, W.A. (1998) Returns to rice technology development in Sierra Leone. *Etudes et Recherches Sahéliennes/Sahelian Studies and Research* 1, 67–74.

Erenstein, O. (2006) Intensification or extensification? Factors affecting technology use in peri-urban lowlands along an agro-ecological gradient in West Africa. *Agricultural Systems* 90, 132–158.

Erenstein, O., Lançon, F., Akande, S.O., Titilola, S.O., Akpokodje, G. and Ogundele, O.O. (2003) Rice production systems in Nigeria: a survey. West Africa Rice Development Association, Abidjan, Côte d'Ivoire.

Erenstein, O., Oswald, A. and Mahaman, M. (2006) Determinants of lowland use close to urban markets along an agro-ecological gradient in West Africa. *Agriculture, Ecosystems and Environment* 117, 205–217.

Evenson, R.E. (2001) Economic impacts of agricultural research and extension. In: Gardner, B.L. and Rausser, G.C. (eds) *Handbook of Agricultural Economics*, Vol. 1. Elsevier, Amsterdam, Netherlands, pp. 573–628.

Fakorede, M.A.B. and Yoboué, N. (2001) *Summary of WARDA/NARS Task Forces Activities 1991–1997.* West Africa Rice Development Association, Bouaké, Côte d'Ivoire.

Fall, A.A. (2005) Impact économique de la recherche rizicole au Sénégal et en Mauritanie. *Agronomie Africaine* 5, 53–62.

Fisher, M.G., Masters, W.A. and Sidibé, M. (2001) Technical change in Senegal's irrigated rice sector: impact assessment under uncertainty. *Agricultural Economics* 24, 179–197.

GRiSP (2013) Beyond the 'good old times'. In: *GRiSP (Global Rice Science Partnership) Annual Report 2012.* International Rice Research Institute, Los Baños, Philippines, pp. 22–23.

Haefele, S., Johnson, D.E., Diallo, S., Wopereis, M.C.S. and Janin, I. (2000) Improved soil fertility and weed management is profitable for irrigated rice farmers in Sahelian West Africa. *Field Crops Research* 66, 101–113.

Haefele, S., Wopereis, M.C.S., Donovan, C. and Maubuisson, J. (2001) Improving productivity and profitability of irrigated rice production in Mauritania. *European Journal of Agronomy* 14(3), 181–196.

Hemmati, M. (2002) *Multi-stakeholder Processes for Governance and Sustainability: Beyond deadlock and conflict.* Earthscan, London.

Ibro, G., Oumar, A.A., Amir, S.Y., Hamidou, O., Yaovi, S. and Hassane, A.M. (2001) *Analyse économique de l'impact de la recherche sur le riz au Niger.* Institut National de la Recherche Agronomique du Niger, Niamey, Niger.

Jamil, M., Rodenburg, J., Charnikhova, T. and Bouwmeester, H.J. (2011) Pre-attachment *Striga hermonthica* resistance of NERICA cultivars based on low strigolactone production. *New Phytologist* 192, 964–975.

Johnson, D.E. (1997) *Weeds of Rice in West Africa.* West Africa Rice Development Association, Bouaké, Côte d'Ivoire.

Jones, M.P., Dingkuhn, M., Aluko, G.K. and Semon, M. (1997) Interspecific *Oryza sativa* L. × *O. glaberrima* Steud. progenies in upland rice improvement. *Euphytica* 92, 237–246.

Kebbeh, M. and Miézan, K.M. (2003) *Ex-ante* evaluation of integrated crop management options for irrigated rice production in the Senegal River valley. *Field Crops Research* 81, 87–94.

Kent, R.J. and Johnson, D.E. (2001) Influence of flood depth and duration on growth of lowland rice weeds, Côte d'Ivoire. *Crop Protection* 20(8), 691–694.

Kergna, A. (2003) *Evaluation économique de l'impact de la recherche et de la vulgarisation sur le riz au Mali.* Institut d'Economie Rurale, Bamako, Mali.

Kijima, Y., Sserunkuuma, D. and Otsuka, K. (2006) How revolutionary is the 'NERICA Revolution'? Evidence from Uganda. *The Developing Economies* 44(2), 252–267.

Kijima, Y., Otsuka, K. and Sserunkuuma, D. (2008) Assessing the impact of NERICA on income and poverty in central and western Uganda. *Agricultural Economics* 38, 327–337.

Kijima, Y., Otsuka, K. and Sserunkuuma, D. (2011) An inquiry into constraints on a green revolution in sub-Saharan Africa: the case of NERICA rice in Uganda. *World Development* 39(1), 77–86.

Lançon, F., Erenstein O., Touré, A.A. and Akpokodje G. (2004) Qualité et compétitivité des riz locaux et importe sur les marches urbains ouest-africains. *Cahiers d'études et de recherches francophone / Agricultures* 13(1), 110–115.

Matlon, P.J., Randolph, T. and Guei, R. (1998) Impact of rice research in West Africa. In: Pingali, P.L. and Hossain, M. (eds) *Impact of Rice Research*. International Rice Research Institute, Los Baños, Philippines, pp. 383–401.

Michellon, R., Husson, O., Moussa, N., Randrianjafizanaka, M.T., Naudin, K., Letourmy, P., Andrianaivo, A.P., Rakotondramanana, Raveloarijoana, N., Enjalric, F., Penot, E and Séguy, L. (2011) *Striga asiatica*: a driving-force for dissemination of conservation agriculture systems based on *Stylosanthes guianensis* in Madagascar. Paper presented at the 5th World Congress of Conservation Agriculture incorporating 3rd Farming Systems Design Conference, Brisbane, Australia, September 2011.

Miézan, K.M. and Sié, M. (1997) Varietal improvement for irrigated rice in the Sahel. In: Miézan, K.M., Wopereis, M.C.S., Dingkuhn, M., Deckers, J. and Randolph, T.F. (eds) *Irrigated Rice in the Sahel: Prospects for sustainable development*. West Africa Rice Development Association, [Bouaké, Côte d'Ivoire], pp. 443–455.

Narteh, L.T., Millar, D. and Beks, B. (Eds) (2006) *Proceedings of the 3rd Regional Rice Research Review (4Rs 2004)*. Accra, Ghana, 27 September to 1 October 2004. Africa Rice Center (WARDA), Cotonou, Benin.

Nwilene, F.E., Williams, C.T., Ukwungwu, M.N., Dakouo, D., Nacro, S., Hamadoun, A., Kamara, S.I., Okhidievbie, O., Abamu, F.J. and Adam, A. (2002) Reactions of differential rice genotypes to African rice gall midge in West Africa. *International Journal of Pest Management* 48(3), 195–201.

Nwilene, F.E., Agunbiade, T.A., Togola, A., Youm O., Ajayi, O., Oikeh, S.O., Ofodile. S. and Falola, O.O. (2008a) Efficacy of traditional practices and botanicals for the control of termites on rice at Ikenne, southwest Nigeria. *International Journal of Tropical Insect Science* 28, 37–44.

Nwilene, F.E., Togola, A., Agunbiade, T.A., Ogah, E.O., Ukwungwu, M.N., Hamadoun, A., Kamara, S.I. and Dakouo, D. (2008b) Parasitoid biodiversity conservation for sustainable management of the African rice gall midge, *Orseolia oryzivora* (Diptera: Cecidomyiidae) in lowland rice. *Biocontrol Science and Technology* 18(10), 1075–1081.

Ogha, O.O., Omoloye, A.A., Nwilene, F.E. and Nwogbaga, A.C. (2011) Effect of neem seed kernel extracts in the management of rice stem borers in the field in Nigeria. *Nigerian Journal of Biotechnology* 23, 13–21.

Oikeh, S., Touré, A., Sidibé, B., Niang, A., Semon, M., Sokei, Y. and Mariko, M. (2009) Responses of upland NERICA rice varieties to nitrogen and plant density. *Archives of Agronomy and Soil Science* 55, 301–314.

Oikeh, S.O., Houngnandan, P., Abaidoo, R.C., Rahimou, I., Toure, A., Niang, A. and Akintayo, I. (2008) Integrated soil fertility management involving promiscuous dual-purpose soybean and upland NERICA enhanced rice productivity in the savannas. *Nutrient Cycling in Agroecosystems* 88, 29–38.

Ouedraogo, S. and Ouedraogo, M. (2003) *Evaluation économique de l'impact de la recherche et de la vulgarisation sur le riz au Burkina Faso*. Institut de l'Environnement et des Recherches Agricoles, Ouagadougou, Burkina Faso.

Raitzer, D.A. and Kelley, T.G. (2008) Benefit–cost meta-analysis of investment in the international agricultural research centers of the CGIAR. *Agricultural Systems* 96, 108–123.

Renkow, M. and Byerlee, D. (2010) The impacts of CGIAR research: a review of recent evidence. *Food Policy* 35, 391–402.

Rodenburg, J., Saito, K., Kakai, R.G., Toure, A., Mariko, M. and Kiepe, P. (2009) Weed competitiveness of the lowland rice varieties of NERICA in the southern Guinea savanna. *Field Crops Research* 114, 411–418.

Sahrawat, K.L. (2000) Elemental composition of the rice plant as affected by iron toxicity under field conditions. *Communications in Soil Science and Plant Analysis* 31, 2819–2827.

Sahrawat, K.L. (2004) Iron toxicity in wetland rice and the role of other nutrients. *Journal of Plant Nutrition* 27, 1471–1504.

Sahrawat, K.L., Jones, M.P. and Diatta, S. (1995) Responses of upland rice to phosphorus in an Ultisol in the humid forest zone of West Africa. *Fertilizer Research* 41, 11–17.

Sahrawat, K.L., Jones, M.P. and Diatta, S. (1997) Direct and residual phosphorus effects on yield and phosphorus efficiency of upland rice in an Ultisol. *Nutrient Cycling in Agroecosystems* 48, 209–215.

Sahrawat, K.L., Jones, M.P., Diatta, S. and Sika, M. (2003) Long-term phosphorus fertilizer effects on phosphorus uptake, efficiency, and recovery by upland rice on an Ultisol. *Communications in Soil Science and Plant Analysis* 14, 999–1011.

Saito, K. and Futakuchi, K. (2009) Performance of diverse upland rice cultivars in low and high soil fertility conditions in West Africa. *Field Crops Research* 111, 243–250.

Saito, K., Azoma, K. and Oikeh, S. (2010a) Combined effects of *Stylosanthes guianensis* fallow and tillage management on upland rice yield, weeds and soils. *Soil and Tillage Research* 107(2), 57–63.

Saito, K., Azoma, K. and Rodenburg, J. (2010b) Plant characteristics associated with weed competitiveness of rice under upland and lowland conditions in West Africa. *Field Crops Research* 116, 308–317.

Saito, K., Azoma, K. and Sokei, Y. (2010c) Genotypic adaptation of rice to lowland hydrology in West Africa. *Field Crops Research* 119(2–3), 290–298.

Saito, K., Sokei, Y. and Wopereis, M.C.S. (2012) Enhancing rice productivity in West Africa through genetic improvement. *Crop Science* 52, 484–494.

Sakurai, T. (2006) Intensification of rainfed lowland rice production in West Africa: present status and potential green revolution. *The Developing Economies* 44, 232–251.

Sanyang, S., Ajayi, A. and Sy, A.A. (Eds) (2003) *Proceedings of the Second Biennial Regional Rice Research Review (4Rs 2002).* M'bé, Bouaké, Côte d'Ivoire, 9–12 April 2002. Africa Rice Center (WARDA), Bouaké, Côte d'Ivoire.

Seck, P.A. (2007) Rice crisis in Africa, myth or reality? Presentation at the 26th AfricaRice Council of Ministers Session, Abuja, Nigeria, 26 September.

Seck, P.A. (2012) *Eloge de la Recherche: Passion et tension d'un chercheur africain.* Editions Panafrika, Silex – Nouvelles du Sud, Dakar, Senegal.

Seck, P.A., Diagne, A., Mohanty, S. and Wopereis, M.C.S. (2012) Crops that feed the world 7: rice. *Food Security* 4, 7–24.

Segda, Z., Häfele, S.M., Wopereis, M.C.S., Sedogo, M.P. and Guinko, S. (2004) Agro-economic characterization of rice production in a typical irrigation scheme in Burkina Faso. *Agronomy Journal* 96, 1314–1322.

Segda, Z., Häfele, S.M., Wopereis, M.C.S., Sedogo, M.P. and Guinko, S. (2005) Combining field and simulation studies to improve fertilizer recommendations for irrigated rice in Burkina Faso. *Agronomy Journal* 97, 1429–1437.

Seidi, S. (1997) An economic analysis of mangrove rice research, extension and seed production in Guinea-Bissau: preliminary evidence from the Tombali region. Mimeo. Department of Agricultural Economics, Purdue University, West Lafayette, Indiana.

Séré, Y., Onasanya, A., Akator, K., Afolabi, A. and Abo, M.E. (2007) Serological differentiation indices and phylogenetic analysis of *Rice yellow mottle virus* isolates in Côte d'Ivoire. *Journal of Biological Sciences* 7(7), 1147–1154.

Sié, M., Sere, Y., Sanyang, S., Narteh, L.T., Dogbe, S., Coulibaly, M.M., Sido, A., Cisse, F., Drammeh, E., Ogunbayo, S.A., Zadij, L., Ndri, B. and Toulou, B. (2008a) Regional yield evaluation of interspecific hybrids (*O. glaberrima* × *O. sativa*) and intraspecfic (*O. sativa* × *O. sativa*) lowland rice. *Asian Journal of Plant Sciences* 7, 130–139.

Sié, M., Futakuchi, K., Gridley, H., Mande, S., Manneh, B., Ndijondijop, M.N., Efisue, A., Ogunbayo, S.A., Moussa, M., Tsunematsu, H. and Samejima, H. (2008b) Drought research at WARDA: current situation and prospects. In: Serraj, R., Bennett, J. and Hardy, B. (eds) *Drought Frontiers in Rice: Crop improvement for increased rainfed production.* World Scientific Publishing, Singapore, and International Rice Research Institute, Los Baños, Philippines, pp. 61–73.

Sokei, Y., Akintayo, I., Doumbia, Y., Gibba, A., Keita, S. and Assigbe, P. (2010) Growth and yield performance of upland NERICA varieties in West Africa. *Japanese Journal of Crop Science* 79(Extra issue 2), 2–3.

Somado, E.A., Kiepe, P. and Niang, A. (2011) Alleviating phosphorus deficiency in rice-based systems in humid Africa. In: Yanagihara, S. (ed.) Next Challenges in Rice Development for Africa: Workshop for New Collaboration between JIRCAS and AfricaRice. *JIRCAS Working Report* 70. Japan International Research Center for Agricultural Sciences, Tsukuba, Japan, and Africa Rice Center, Cotonou, Benin, pp. 29–35.

Spencer, D.S.C., Dorward, A., Abalu, G., Philip, D. and Ogunbile, D. (2006) Evaluation of adoption of NERICA and other improved upland rice varieties following varietal promotion activities in Nigeria. A study for the Gatsby and Rockefeller Foundations, Final Report.

Sy, A.A. and Séré, Y. (1996) *Manuel de formation en pathologie du riz.* Association pour le Développement de la Riziculture en Afrique de l'Ouest, Bouaké, Côte d'Ivoire.

Thenkabail, P.S. and Nolte, C. (1996) Capabilities of Landsat-5 Thematic Mapper (TM) data in regional mapping and characterization of inland valley agroecosystems in West Africa. *International Journal of Remote Sensing* 17, 1505–1538.

Thiele, G., Devaux, A., Reinoso, I., Pico, H., Montesdeoca, F., Pumisacho, M., Andrade-Piedra, J., Velasco, C., Flores, P., Esprella, R., Thomann, A., Manrique, K. and Horton, D. (2011) Multi-stakeholder platforms for linking small farmers to value chains: evidence from the Andes. *International Journal of Agricultural Sustainability* 9(3), 423–433.

Touré, A., Rodenburg, J., Saito, K., Oikeh, S., Futakuchi, K., Gumedzoe, D. and Huat, J. (2011) Cultivar and weeding effects on weeds and rice yield in a degraded upland environment of the coastal savanna. *Weed Technology* 25, 322–329.

Tré, J.-P. (1995) The rates of return to mangrove rice research in West Africa. MSc thesis, Purdue University, West Lafayette, Indiana.

USAID (2009) *Global Food Security Response: West Africa rice value chain analysis.* United States Agency for International Development, Washington, DC.

Van Asten, P.J.A., Wopereis, M.C.S., Haefele, S., Ould Isselmou, M. and Kropff, M.J. (2003a) Explaining yield gaps in farmer-identified degraded and non-degraded soils in a Sahelian irrigated rice scheme. *Netherlands Journal of Agricultural Science* 50, 277–296.

Van Asten, P.J.A., Barbiero, L., Wopereis, M.C.S. and Maeght, J.L. (2003b) Actual and potential salt-related soil degradation in an irrigated rice scheme in the Sahelian zone of Mauritania. *Agricultural Water Management* 60, 13–32.

Van Mele, P., Wanvoeke, J. and Zossou, E. (2010) Enhancing rural learning, linkages and institutions: the rice videos in Africa. *Development in Practice* 20(3), 414–421.

WARDA (1997) Rice or semiconductors? In: *WARDA Annual Report 1997.* West Africa Rice Development Association, Bouaké, Côte d'Ivoire, pp. 50–55.

WARDA (1999) Task Forces: a successful mode of operation with national partners. In: *WARDA Annual Report 1998.* West Africa Rice Development Association, Bouaké, Côte d'Ivoire, pp. 9–15.

WARDA (2000) *WARDA Annual Report 1999.* West Africa Rice Development Association, Bouaké, Côte d'Ivoire.

WARDA (2001a) *WARDA: 1970–2001.* West Africa Rice Development Association, Bouaké, Côte d'Ivoire.

WARDA (2001b) *NERICAs in Guinea: Towards food security. The history and keys of success.* West Africa Rice Development Association, Bouaké, Côte d'Ivoire.

WARDA (2001c) *Bintu and Her New Rice for Africa: Breaking the shackles of slash-and-burn farming in the world's poorest region.* West Africa Rice Development Association, Bouaké, Côte d'Ivoire.

WARDA (2002a) The African Rice Initiative: taking the NERICAs to sub-Saharan Africa. In: *WARDA Annual Report 2001–2002.* West Africa Rice Development Association, Bouaké, Côte d'Ivoire, pp. 9–14.

WARDA (2002b) *Promising Technologies for Rice Production in West and Central Africa.* West Africa Rice Development Association, Bouaké, Côte d'Ivoire, and Food and Agriculture Organization of the United Nations, Rome.

WARDA (2004) The Africa Rice Center – recognizing WARDA's role in sub-Saharan Africa. In: *WARDA Annual Report 2002–2003.* WARDA – The Africa Rice Center, Bouaké, Côte d'Ivoire, pp. 7–12.

WARDA (2008) *Africa Rice Trends 2007.* Africa Rice Center (WARDA), Cotonou, Benin.

Windmeijer, P.N., Stomph, T.J., Adam, A., Coppur, R., De Ridder, N., Kandeh, M., Mahaman, M. and Van Loon, M. (1998) Transect sampling strategies for semi-detailed characterization of inland valley systems. *Netherlands Journal of Agricultural Science* 46(1), 15–25.

Wopereis, M.C.S. and Defoer, T. (2007) Moving methodologies to enhance agricultural productivity of rice-based lowland systems in sub-Saharan Africa. In: Bationo, A., Waswa, B., Kihara, J. and Kimetu, J. (eds) *Advances in Integrated Soil Fertility Management in Sub-Saharan Africa: Challenges and opportunities.* Springer, Dordrecht, Netherlands, pp. 1077–1091.

Wopereis, M.C.S., Miézan, K.M., Donovan, C., Ndiaye, A.M. and Douthwaite, B. (1998) A new Senegalese thresher / cleaner responds to small-farmer postharvest needs. *International Rice Research Notes* 23(2), 39–41.

Wopereis, M.C.S., Donovan, C., Nebié, B., Guindo, D. and N'Diaye, M.K. (1999) Soil fertility management in irrigated rice systems in the Sahel and savanna regions of West Africa. Part I. Agronomic analysis. *Field Crops Research* 61, 125–145.

Wopereis, M.C.S., Haefele, S.M., Dingkuhn, M. and Sow, A. (2003) Decision support tools for irrigated rice based systems in the Sahel. In: Struif-Bontkes, T.E. and Wopereis, M.C.S. (eds) *Decision Support*

Tools for Smallholder Agriculture in Sub-Saharan Africa: A practical guide. IFDC, Muscle Shoals, Alabama, and ACP–EU Technical Centre for Agricultural and Rural Cooperation (CTA), Wageningen, Netherlands, pp. 114–126.

Wopereis, M.C.S., Defoer, T., Idinoba, M.E., Diack, S. and Dugué, M.J. (2007) *Participatory Learning and Action Research (PLAR) for Integrated Rice Management (IRM) in Inland Valleys of Sub-Saharan Africa: Technical manual.* Africa Rice Center (WARDA), Cotonou, Benin, and IFDC, Muscle Shoals, Alabama.

Wopereis, M.C.S., Diagne, A., Rodenburg, J., Sié, M. and Somado, E.A. (2008) Why NERICA is a successful innovation for African farmers: a response to Orr *et al.* from the Africa Rice Center. *Outlook on Agriculture* 37(3), 169–176.

2 Africa's Rice Economy Before and After the 2008 Rice Crisis

Papa Abdoulaye Seck, Ali A. Touré,* Jeanne Y. Coulibaly, Aliou Diagne and Marco C.S. Wopereis
Africa Rice Center (AfricaRice), Cotonou, Benin

Introduction

Rice is the most rapidly growing food commodity in sub-Saharan Africa (SSA), mainly driven by urbanization. The opportunity costs of women's labour and the ease and rapidity of cooking rice are key factors in urban settings. Urbanization is often accompanied by increased consumption of food away from the home, which has spurred rice demand due to the convenience of rice storage, preparation and cooking. With the proportion of Africans living in urban areas expected to increase from the current 38% to 48% by 2030 (AfricaRice, 2011b), rice consumption in Africa is expected to continue to grow for the foreseeable future. Household surveys reveal that urban consumers on lower incomes tend to spend a greater share of their total budget on rice than higher-income households (AfricaRice, 2011b). These developments mean that rice is no longer a luxury food, but has become the main source of calories for low-income households, particularly in West Africa. Rice is also rapidly gaining in importance in other parts of the continent. Rice is the leading provider of food calories in West Africa and in Madagascar, and it is now the second largest source of food energy in SSA as a whole (Fig. 2.1). The increasing role of rice in the food basket of SSA consumers has made rice a political crop, in the sense that its price and accessibility influence social stability. Africa depends to a large extent on imports, however, and in 2008, African imports accounted for 32% of the rice traded globally, most of it from Asia.

Increasing wealth in Asia has resulted in greater demand for meat, thereby drawing maize increasingly into the animal feed market. This has resulted in steadily rising world cereal prices (since 2003) and the simultaneous emergence of Africa as a major player in the world rice market. Some influences on world cereal prices may be transient – for example, weather-related crop failures, spikes in oil prices, and demand for ethanol from maize – but the long-term outlook for rice production in Africa is bolstered by the signs that Asia's consumption will outstrip its capacity to produce. In 2012, Asia accounted for 86% of rice consumption and 78% of rice exports worldwide, and it is facing increasing local demand with already high pressure on land and water resources.

Rice is often considered one of the most protected commodities in the world and only about 7% of global rice production is traded on the international market. In this distorted market, the major producing countries may close their borders to trade during periods of perceived supply shortage, as happened in 2007 and 2008.

* Corresponding author: a.toure@cgiar.org

© CAB International 2013. *Realizing Africa's Rice Promise* (eds M.C.S. Wopereis *et al.*)

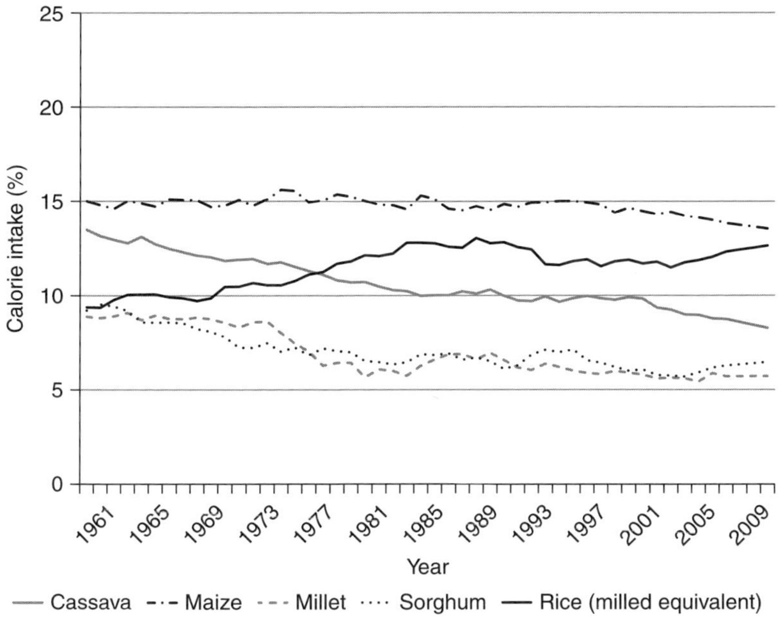

Fig. 2.1. Contribution of staple food crops in diets in sub-Saharan Africa. (Data from FAOSTAT, 2011.)

In 2008, Vietnam, India and Egypt banned rice exports for several months, pushing up rice prices, as predicted by the Africa Rice Center (AfricaRice) in 2007 (AfricaRice, 2011a). Sub-Saharan Africa's reliance on rice imports became painfully visible in 2008 during the food riots in several major capitals. In Africa, these riots were mostly related to high rice prices. Prices are predicted to remain high due to declining production capacity in major rice-producing countries in Asia and growing demand.

With the upward spikes in food prices in Africa, many African governments (assisted by the international donor community) embarked upon ambitious programmes to boost their rice-production capacity, mostly as a response to the 2008 rice crisis. In this chapter, we review trends in rice demand and production across the continent, placing particular emphasis on what happened before and after the rice crisis, and discuss the challenges faced in attempts to boost Africa's rice sector. All data were retrieved from the United States Department of Agriculture (USDA, 2013), unless indicated otherwise. The USDA database was the source of choice, because it contains more recent data than FAOSTAT.

Rice Production

Rice paddy production in SSA increased from 5.6 million tonnes (Mt) in 1980 to 18.2 Mt in 2010, and in Africa as a whole from 8.2 Mt to 24.8 Mt over the same period. The relative contributions of yield increase and harvested area in production for the periods 1980–1990, 1990–2000 and 2000–2010 for both Africa and sub-Saharan Africa are shown in Table 2.1. The relative contribution of enhanced yield increased over time, and the contribution of yield increase and harvested area expansion to production increase in SSA reached approximate parity between 2000 and 2010. The picture is more erratic for Africa as a whole, reflecting important yield gains in Egypt during 1990–2000 followed by stagnating yields and harvested area in 2000–2010.

Figures 2.2 and 2.3 present the trend in average paddy rice yield and harvested area from the 1960s to 2012. A clear shift in gear is visible in average paddy rice yield from 2007–2008 onwards, which is analysed in more detail below.

Total paddy rice production in SSA increased during 2000–2012 by 7.5 Mt from 11.5 Mt to 19.0 Mt. Contributions of yield increase and harvested area expansion before

Table 2.1. Relative contributions of yield increase and harvested area expansion to rice production increase in sub-Saharan Africa (SSA) and Africa for the periods 1980–1990, 1990–2000 and 2000–2010. (Data derived from USDA, 2013.)

	1980–1990			1990–2000			2000–2010		
Region	Production increase (Mt)	Yield contr.[a] (%)	Area contr. (%)	Production increase (Mt)	Yield contr. (%)	Area contr. (%)	Production increase (Mt)	Yield contr. (%)	Area contr. (%)
SSA	4.2	26	74	1.7	33	67	6.7	47	53
Africa	3.8	6	94	5.4	63	37	7.4	37	63

[a]contr., contribution.

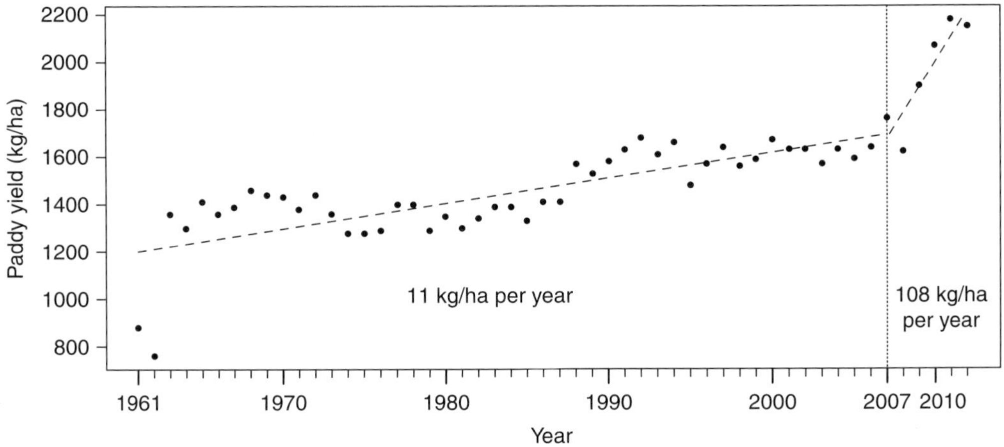

Fig. 2.2. Average rice paddy yield in sub-Saharan Africa showing segmented regression for the pre- and post-crisis periods. (Data from USDA, 2013.)

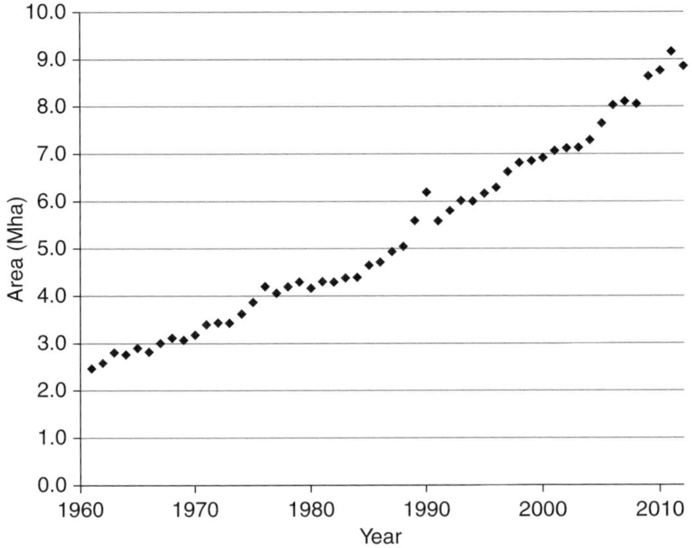

Fig. 2.3. Total harvested area under rice in sub-Saharan Africa. (Data from USDA, 2013.)

and after the rice crisis are indicated in Table 2.2. Whereas about 25% of production increase before the rice crisis can be attributed to yield increase, and 75% to harvested area expansion, after the rice crisis these percentages are more or less reversed, with yield increases contributing 71% and area expansion 29%.

Tables 2.3, 2.4 and 2.5 provide annual growth rates for production, average yield and harvested area, respectively, for each major region, SSA and Africa as a whole. Annual production growth rate (Table 2.3) over the period 2000–2012 was 5.5% per year for SSA, with a clear difference between the periods before the rice crisis (i.e. 2000–2007; 3.2% per year) and after the rice crisis (i.e. 2007–2012; 8.4% per year). Trends in the major regions were similar, except for North Africa, which was heavily influenced by the production decline in Egypt in the period 2007–2012.

Table 2.2. Relative contributions of yield increase and harvested area expansion to rice production increase in SSA and Africa for the periods 2000–2007 (before the rice crisis) and 2007–2012 (after the rice crisis). (Data from USDA, 2013.)

	2000–2007			2007–2012		
Region	Production increase (Mt)	Yield contr.[a] (%)	Area contr. (%)	Production increase (Mt)	Yield contr. (%)	Area contr. (%)
SSA	2.8	24	76	4.7	71	29
Africa	3.7	22	78	4.2	54	46

[a]contr., contribution.

Table 2.3. Rice production (equivalent milled) growth rates (% per year) for major regions in Africa for the period 2000–2012 and sub-periods 2000–2007 (before the rice crisis) and 2007–2012 (after the rice crisis). The period 2010–2012 is added because 2011 and 2012 were affected by poor weather conditions across the continent (drought, floods). (Data from USDA, 2013.)

Region	2000–2012	2000–2007	2007–2012	2010–2012
Central Africa	1.2	−1.3	4.5	2.1
East Africa[a]	5.8	5.0	6.5	−0.1
North Africa	0.9	2.5	−4.6	−3.4
Southern Africa	6.1	−1.8	29.2	12.4
West Africa	5.5	2.5	9.7	3.7
Sub-Saharan Africa	5.5	3.2	8.4	2.3
Africa	4.1	3.0	4.5	0.7

[a]For Rwanda and Uganda, no data available up to 2000; therefore the periods start from 2001 (2001–2012 and 2001–2007).

Table 2.4. Rice paddy yield growth rates (% per year) for major regions in Africa for the period 2000–2012 and sub-periods 2000–2007 (before the rice crisis) and 2007–2012 (after the rice crisis). The period 2010–2012 is added because 2011 and 2012 were affected by poor weather conditions across the continent (drought, floods). (Data from USDA, 2013.)

Region	2000–2012	2000–2007	2007–2012	2010–2012
Central Africa	0.0	−1.4	0.7	−0.6
East Africa[a]	2.8	2.0	2.6	−2.1
North Africa	0.3	1.2	−2.0	−5.6
Southern Africa	1.9	1.2	11.8	10.8
West Africa	2.5	−0.4	8.0	4.3
Sub-Saharan Africa	2.6	0.4	5.8	1.9
Africa	1.4	0.3	2.6	0.2

[a]For Rwanda and Uganda, no data available up to 2000; therefore the periods start from 2001 (2001–2012 and 2001–2007).

Table 2.5. Rice harvested area growth rates (% per year) for major regions in Africa for the period 2000–2012 and sub-periods 2000–2007 (before the rice crisis) and 2007–2012 (after the rice crisis). The period 2010–2012 is added because 2011 and 2012 were affected by poor weather conditions across the continent (drought, floods). (Data from USDA, 2013.)

Region	2000–2012	2000–2007	2007–2012	2010–2012
Central Africa	1.2	–0.1	3.9	3.0
East Africa[a]	2.9	3.0	3.8	2.0
North Africa	–0.1	0.8	–2.6	2.3
Southern Africa	3.9	–3.0	14.6	1.1
West Africa	2.4	2.5	1.6	–0.6
Sub-Saharan Africa	2.5	2.4	2.4	0.5
Africa	2.3	2.3	2.1	0.6

[a]For Rwanda and Uganda, no data available up to 2000; therefore the periods start from 2001 (2001–2012 and 2001–2007).

Annual yield growth rate[1] (Table 2.4) over the period 2000–2007 was 0.4% per year for SSA. However, annual yield growth rate jumped to 5.8% per year for the period 2007–2012, despite a relatively low yield growth rate in the period 2010–2012 (1.9% per year). The slowing rate of yield growth during these last two years can be explained by the fact that rice cropping in SSA is predominantly rainfed (see Diagne *et al.*, Chapter 3, this volume), so production and productivity are strongly influenced by the rainfall regime. Harvests in 2011–2012 were affected by poor weather conditions (drought and floods). In real terms, and despite two relatively bad years, average rice yield in SSA increased in the period 2007–2012 by 78 kg/ha per year. A segmented regression analysis[2] was performed on rice yield in SSA over the period 1961 to 2012 partitioned into two intervals, 1961–2007 and 2007–2012. The analysis shows that rice yield increased by about 11 kg/ha per year from 1960 to 2007 ($R^2 = 0.6$) and by 108 kg/ha per year from 2007 to 2012 ($R^2 = 0.8$) (see Fig. 2.2).

In comparison, rice yield worldwide, driven by the Green Revolution in Asia, increased by 52 kg/ha per year over the period 1960–2010. Cereal growth rates after the Second World War amounted to 78 kg/ha per year in the UK and 50 kg/ha per year in the USA. The rice yield growth rate in SSA – as a response to renewed commitments to boosting Africa's rice sector after the rice crisis in 2008 – is, therefore, similar to growth rates witnessed on other continents after the introduction of technological innovation and institutional change. These trends are visible in West, East and Southern Africa, but not in Central Africa (virtually no change in yield in the period 2007–2012) and North Africa (decline in average yield, driven by Egypt). Major rice-growing countries contributing to the average yield increase after the rice crisis include: Côte d'Ivoire, Ghana, Guinea, Liberia, Madagascar, Mali, Nigeria, Senegal and Sierra Leone.

Harvested area growth rates (Table 2.5) over the periods 2000–2007 and 2007–2012 in SSA were a constant 2.4% per year. In West Africa, the rate of expansion of harvested area declined somewhat after the rice crisis compared to before the rice crisis. Production increases in Central Africa in the period 2007–2012 were driven by expansion of harvested area after the rice crisis.

Rice Consumption

Total rice consumption[3] (Table 2.6) over the period 2000–2012 increased from 16 Mt to 29 Mt in Africa and from 12 Mt to 24 Mt in SSA. Annual growth rates during this period in Africa were 4.3% per year compared to 5% per year for SSA. These percentages represent huge quantities of rice. For 2012, a 5% per year increase in consumption in SSA is roughly equivalent to 1.2 Mt of milled rice per year that needs to be either produced or imported. An upward trend in consumption is particularly clear for West Africa.

Future Trends in Rice Production and Consumption in sub-Saharan Africa

Rice consumption growth rate in SSA was estimated at 5% per year from 2000 to 2012 (Table 2.6). Assuming that this growth rate and all other parameters influencing demand remain equal, rice consumption would increase from 24 Mt of milled rice in 2012 to 36 Mt of milled rice in 2020 (Fig. 2.4). The bulk of this consumption (80%) will occur in West and East Africa.

Rice production projections to 2020 were made using 2012 as a starting point. The 19 Mt of paddy rice production in 2012 was converted to a production of 12 Mt of milled rice, assuming a milling recovery of 65%. Next, projections were made assuming a production

Table 2.6. Rice consumption growth rates (% per year) for major regions in Africa for the period 2000–2012 and sub-periods 2000–2007 (before the rice crisis) and 2007–2012 (after the rice crisis). The period 2010–2012 is added because 2011 and 2012 were affected by poor weather conditions across the continent (drought, floods). (Data from USDA, 2013.)

Region	2000–2012	2000–2007	2007–2012	2010–2012
Central Africa	3.6	4.3	6.0	6.6
East Africa[a]	4.6	4.1	5.7	3.0
North Africa	1.5	2.9	–2.5	2.5
Southern Africa	4.9	9.2	1.7	12.3
West Africa	5.4	4.2	9.7	13.6
Sub-Saharan Africa	5.0	4.4	7.9	10.3
Africa	4.3	4.0	5.9	8.7

[a]For Rwanda and Uganda, no data available up to 2000; therefore the periods start from 2001 (2001–2012 and 2001–2007).

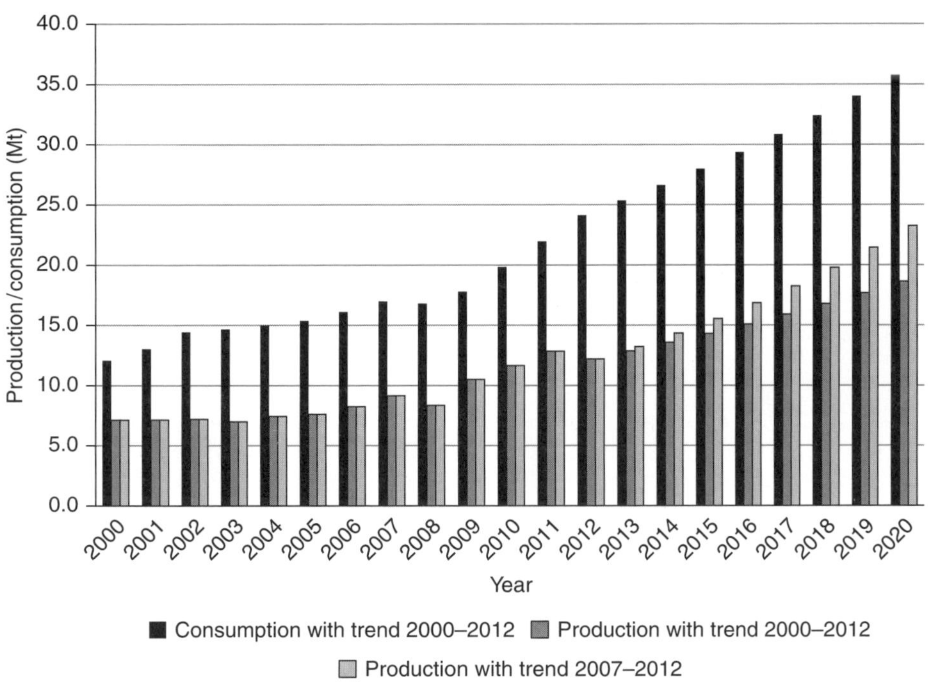

Fig. 2.4. Projection of milled rice production and consumption in sub-Saharan Africa using different assumptions (see text). (Data from USDA, 2013.)

growth rate of 5.5% per year as witnessed for the period 2000–2012 and a growth rate of 8.4% per year as witnessed for the period 2007–2012 (Table 2.3). With the 5.5% per year growth rate, SSA rice production would increase to 19 Mt of milled rice in 2020, creating a demand for imports estimated at 17 Mt of milled rice. With the growth rate seen after the rice crisis (2007–2012), milled rice production in 2020 would be at 23 Mt, closing the gap between demand and supply by 4 Mt compared to the baseline 2000–2012 scenario.

Rice Imports

Rice imports into SSA stabilized at around 7–9 Mt from 2008 to 2011, illustrating that rice production on the continent was able to keep pace with the increase in rice consumption. However, a clear surge in imports occurred in 2012 (Fig. 2.5).

Several factors explain this surge in rice imports. First, many West African countries incurred production setbacks because of severe drought spells in 2011. In addition to rice, production of other staple crops – including maize, millet and sorghum – was also undermined by drought. Consumers substituted other grains with rice, thereby creating a greater dependence on imported rice.

According to the *FAO Rice Market Monitor* (FAO, 2012), the announcement of increased border protection in Nigeria prompted Nigerian importers to complete their procurement before the implementation of the new import duties (tariffs on husked rice were raised from 5% to 30% and for milled and semi-milled rice from 30% to 50%). Second, the flow of imports was facilitated by the recourse to lower import tariffs and/or ceiling reference prices at the retail level in several West and East African countries. In 2012, Côte d'Ivoire adopted a three-month suspension of duty and taxes, which contributed to a growth of imports from 850,000 t to 1.4 Mt, 60% more than in 2011 (USDA, 2013). The same year, Côte d'Ivoire, Mali and Senegal established ceiling reference prices at the retail level to facilitate consumers' access to rice. In East Africa, Burundi, Kenya and Rwanda faced a strong demand for imported rice with the adoption of lower import tariffs in 2012. Lastly, the above-average population growth in SSA and the resulting increase in urbanization have been identified as contributing factors to an increase in demand for food, particularly rice (e.g. Kessides, 2005; Zuberi and Thomas, 2012). The growing demand for rice provides a strong impetus to continue to improve growth and efficiency of local rice production, but also to develop policies to control large imports that can impede the development of the domestic rice sector.

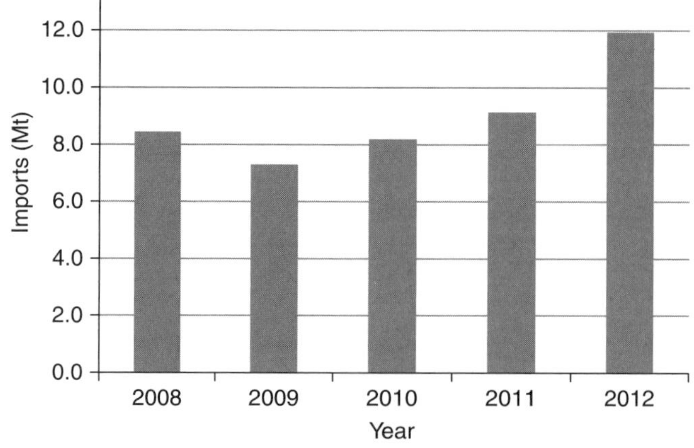

Fig. 2.5. Rice imports into sub-Saharan Africa during the period 2008–2012. (Data from USDA, 2013.)

Competitiveness of Domestic Rice Production

To assess the potential for domestic rice production following the rice crisis, AfricaRice and national (NARS) partners initiated rice competitiveness studies in collaboration with Michigan State University. Selected estimates of domestic resources cost (DRC) related to rice production are presented in Table 2.7: if the value of DRC is below 1, local rice is competitive against imports. Results indicate that local rice production systems are competitive and they make efficient use of domestic resources (see AfricaRice *et al.*, 2011, for more details). It must be noted, however, that domestically produced rice often fetches a lower price in urban markets in Africa because of perceived lower quality. The challenge often lies in postharvest practices, with the end result that African produced rice does not reach the standard of imported rice in the eyes of the African consumer (see Futakuchi *et al.*, Chapter 25, this volume).

General Discussion

Using projections of population from the United Nations and of income from the Food and Agricultural Policy Research Institute (FAPRI), Seck *et al.* (2012) predicted that global rice demand will rise from 439 Mt (milled rice) in 2010 to 496 Mt in 2020 and further increase to 555 Mt in 2035. An additional 116 Mt of rice will, therefore, be needed by 2035 to feed growing populations. In Africa, where rice is the most rapidly growing food source, about 30 Mt more rice will be needed by 2035, representing an increase of 130% from 2010. About one-third of this extra rice in Africa will be needed in Nigeria alone.

Promising developments in the African rice sector and improvements in rice productivity and competitiveness achieved after the 2008 food crisis as presented in this chapter can only be sustained over time if appropriate policies are adopted to incentivize producers and secure investments in the rice value chain. The most debated policy instrument is the Common External Tariff (CET). Small-scale producers' organizations in West Africa are currently (March 2013) pushing hard for an increase in the level of CET from 10% to 35%, in order to secure investments, provide incentives for rice production, and control for massive rice imports. Another argument is that the increase in CET level will be consistent with the Economic Community of West African States' (ECOWAS) regional agricultural policy commitment to promote food sovereignty. The proposed tariff is expected to be flexible, which means that the rate in force will fluctuate around the 35% reference level. The tariff will closely follow the trends of the international and the domestic regional markets: in times of abundant supply and very low prices on the international market that could threaten domestic investments, the CET could be raised above its reference value. Similarly, when international prices are very

Table 2.7. Estimations of domestic resources cost (DRC). (From AfricaRice *et al.*, 2011, with permission from Africa Rice Center.)

Country	National	Rainfed rice	Lowland rice	Irrigated rice
Benin	0.65	0.61	0.64	0.62
Burkina Faso	0.50	0.75	0.38	0.18
Côte d'Ivoire	0.61	0.88	0.73	0.60
Ghana	0.34	1.17	0.40	0.29
Guinea	0.68	n/a	n/a	n/a
Mali	0.51[a]	n/a	n/a	n/a
Nigeria	0.69	0.70	0.58	0.78
Senegal	0.75	0.83	n/a	0.55
Togo	0.41	0.50	0.23	0.68

[a]DRC of Mali is for the rice sector at the Office du Niger.
n/a, data not available.

high or if particularly poor harvests are anticipated within the ECOWAS community, the CET could be set below 35%. The argument for more protection of domestic rice production is rooted in infant industry theory, which stipulates that smaller countries need to protect their emerging industries (Edwards, 1993) in order to develop. Although criticized, empirical evidence in the rice economy in Asia as well as in Africa strongly supports this theory. First of all, the success of major rice producers and exporters was due to a large extent to the protectionist policy implemented by their governments. The majority of rice-producing countries in Asia had a policy of maintaining a very high import tariff on rice. Rice self-sufficiency and protection of the domestic sector from entry of cheaper rice are the reasons often provided to justify such high tariffs. The second empirical evidence is that trade liberalization and the reduction of import tariff, promoted mainly by the World Bank and the International Monetary Fund during the structural adjustment period, encouraged massive rice imports into Africa and hindered the development of the local rice industry. The period before 1980 saw import restrictions, licensing, high tariff rates, and government support to the development of local rice production. Those interventions enabled countries like Côte d'Ivoire to achieve self-sufficiency in rice in 1977. From 1980 to 2000, major policy shifts took place through structural adjustment policies. Countries liberalized their domestic sector, reduced or dismantled their import restrictions and became outward-oriented to cover domestic rice consumption. As a result, self-sufficiency ratios began decreasing while imports grew. After 2000, some countries (Nigeria and Ghana) reverted back to more protection of their local sector, but lack of harmonization of tariff policy across countries[4] in West Africa did not allow a reduction in import volume, thus the large gap between production and imports persisted. These facts demonstrate that more protectionism together with coordination and harmonization of import tariffs are necessary to boost the domestic sector and control large inflows of rice.

Higher import tariffs undoubtedly play an important role in the performance of the domestic rice sector, but they should not be the only policy option and cannot alone solve the structural and marketing problems that hamper the expansion of demand and supply of locally produced rice. Additional policy measures need to be implemented throughout the rice value chain. As Demont and Neven (Chapter 24, this volume) point out, policy makers and stakeholders need to gradually upgrade the domestic rice value chain according to a three-stage sequence: (i) investment in value adding through quality enhancement, certification, branding and labelling; (ii) adoption of supply-shifting strategies, including investments in research, extension and storage infrastructure; and (iii) implementation of demand-lifting actions based on the development of marketing strategies.

More specifically, investment in value addition will raise rice quality: mechanized harvesting and threshing operations to ensure product quality; promotion of private-sector investment in efficient rice processing technologies with built-in capacity for de-stoning, polishing and sorting homogeneous high-quality rice; packaging and labelling. Adoption of supply-shifting strategies will need to be based on sustainable area extension, enhanced access to quality seed of improved high-yielding varieties, promotion of producer incentives including better access to credit, facilitation of access to input subsidies (particularly fertilizer subsidy), and increased private-sector and local-community investment in storage infrastructures. Promotion to raise demand may include advertising, generic promotion and consumer subsidy on local rice. Consumer subsidy will not only provide vulnerable consumers with access to local rice, but also stimulate the demand for rice. Effective targeting of the beneficiaries of the subsidy will be made possible with the use of modern technologies including biometry (Gelb and Decker, 2011). Regional integration and inter-country cooperation within Africa are necessary to more efficiently address some of the cross-country challenges, including tariffs on imported rice and fertilizer subsidy.

Regional cooperation will be needed to develop efficient market regulation policies, including regional rice storage and bulk purchase. Regional storage is considered as a policy instrument to stabilize price variability and ensure food security. Although countries possess national facilities for grain storage, the 2008 food crisis illustrated the importance of developing a stock of rice at regional level. Quantities of national stock released during the crisis were often small and had only limited effect in drawing food prices down (ECOWAS, 2008). It is, apparently, therefore essential to stock rice at regional level to address price instability caused by domestic shocks as well as the international market. Prices will be stabilized through a release of stocks during periods of high domestic prices and accumulation of stock during periods of abundant production or low international prices. Market regulation will be through public or public–private purchasing and selling capacity. The use of regional public inventories to achieve price stabilization has been common in economic community blocs such as Europe, the USA and Asia (Larson *et al.*, 2004; Yao *et al.*, 2005). Using data for the Middle East and North Africa region, Larson *et al.* (2012) were able to demonstrate that when the importing region is sufficiently large, setting a grain reserve is effective to hold domestic price below a specific target and to smooth global prices. Bulk purchase could facilitate increased access and provide affordable prices to consumers, particularly the most vulnerable. This policy will also help countries to address diverse forms of distortions introduced by the main exporters and importers and reduce most of the transaction costs of importing rice. Despite the fact that SSA imports a third of the international rice supply (AfricaRice, 2011b), there is no regional instrument for regulating rice imports into the sub-Saharan regional economic blocs. Yet, by aggregating their procurement of rice from the world rice market, countries could be in a position to exercise some buyer power on the international rice import markets or at least improve their bargaining power *vis-à-vis* the multinational grain trading firms or the state trading agencies on the export side. Support for this approach comes from Fiamohe *et al.* (2012), who show that if West African nations would aggregate their purchase of rice from Thailand, they could exercise a strong importer oligopsony[5] market power and distort prices below their competitive level. But taken individually, only Nigeria could be in a position to exercise market power. The effectiveness of the bulk-purchase strategy will rely on agreement among stakeholders in the rice economy, including exporters, traders, commercial banks and importers.

Conclusions

Africa's rice economy has seen a tremendous boost since the rice crisis occurred in 2008. Over the period 2007–2012, average rice yield in SSA increased in real terms by 108 kg/ha per year, comparable to Green Revolution growth rates in Asia or what was witnessed in the USA and Europe after the Second World War. Given the tremendous growth in rice consumption across the continent, this positive trend must be continued and even enhanced, to ensure that imports can be kept at a manageable level.

Together with a continued emphasis on enhanced production, it will be essential to invest in harvest and postharvest equipment and infrastructure to improve processing and marketing of domestic rice to ensure that rice produced is of the quality standard of imported rice and can find a market. Appropriate policy instruments need to be applied at national and regional levels to encourage the development of Africa's rice sector. Such instruments must be rules-based and predictable, avoiding uncertainty which could discourage investment or undercut the emergence of a dynamic private rice sector. Continued development of Africa's rice sector will take multiple paths and great caution is needed because of the complexity, political sensitivity and context specificity of the land issue within and across countries. African countries need to ensure that investments lead to win–win situations for all involved, not least the resource-poor local farmers.

Notes

[1] 'Growth rate' as used in this chapter is the ordinary least squares growth rate used by the World Bank (http://data.worldbank.org/about/faq/specific-data-series). It represents the average annual growth rate over the entire period and has the advantage of taking into account intermediate values of the series. It does not necessarily match the actual growth rate over any given period.

[2] 'Segmented regression' is a regression analysis method in which the independent variable is partitioned into intervals and a separate line segment is fit to each interval (http://en.wikipedia.org/w/index.php?title=Segmented_regression&oldid=541637502, accessed 7 March 2013).
[3] 'Total consumption' (or domestic consumption) refers to the sum of production and imports of milled rice. This apparent rice consumption is slightly different from real consumption by the value of stock and exports which are not very substantial in most of Africa.
[4] For example, Nigeria has a very high import tariff, whereas its neighbour Benin has a low tariff of 10%. This is also the case with Ghana and Côte d'Ivoire: tariff rate in Ghana ranged between 20% and 37%, whereas Côte d'Ivoire implemented the West African Economic and Monetary Union tariff of 10%. These differences in import tariffs encouraged smuggling across borders and did not help high import tariff countries to reduce their rice imports.
[5] 'Oligopsony': 'a state of the market in which only a small number of buyers exists for a product' (Oxford Dictionaries [online], http://oxforddictionaries.com/definition/english/oligopsony?q=oligopsony, accessed 4 March 2013).

References

AfricaRice (2011a) *Lessons from the Rice Crisis: Policies for food security in Africa.* Africa Rice Center, Cotonou, Benin.

AfricaRice (2011b) *Boosting Africa's Rice Sector: A research for development strategy 2011–2020.* Africa Rice Center, Cotonou, Benin.

AfricaRice, SNRA (Système National de Recherche Agricole) and MSU (Michigan State University) (2011) Analyse de la competitivite de la filiere riz et maïs. Rapport Technique de Synthèse. Sous-contrat No. 61-9812A. Africa Rice Center, Cotonou, Benin.

ECOWAS (2008) Mémorandum relatif à la hausse des prix des denrées alimentaires: situation, perspectives, stratégies et mesures recommandées. Economic Community Of West African States, Abuja, Nigeria.

Edwards, S. (1993) Trade liberalization, and growth in developing countries. *Journal of Economic Literature* 31(3), 1358–1393.

FAO (2012) *FAO Rice Market Monitor* 15(4).

FAOSTAT (2011) Percentage contribution of various staple food crops in diets in SSA. Available at: http://faostat.fao.org (accessed December 2011).

Fiamohe, R., Bamba, I., Seck, P.A. and Diagne, A. (2012) Regional bulk purchase of imported rice initiative by ECOWAS: a feasibility assessment. *OIDA International Journal of Sustainable Development* 5(10), 79–89.

Gelb, A. and Decker, C. (2011) Cash at Your Fingertips: Biometric for transfers in developing and resource rich countries. *CGD Working Paper* 253. Center for Global Development, Washington, DC.

Kessides, C. (2005) The Urban Transition in sub-Saharan Africa: Implications for economic growth and poverty reduction. Africa region. *Working Paper Series* No. 97. The World Bank, Washington, DC.

Larson, D.F., Anderson, J.R. and Varangis, P. (2004) Policies on managing risk in agricultural markets. *The World Bank Research Observer* 19(2), 199–230.

Larson, D.F., Lampietti, J., Garel, C., Cafiera, C. and Roberts, J. (2012) Food Security and Storage in the Middle East and North Africa. *Policy Research Working Paper* 6031. The World Bank, Washington, DC.

Seck, P.A., Diagne, A., Mohanty, S. and Wopereis, M.C.S. (2012) Crops that feed the world 7: rice. *Food Security* 4, 7–24.

USDA (2013) Production, supply and distribution online. Foreign Agricultural Service, United States Department of Agriculture. www.fas.usda.gov/psdonline/psdQuery.aspx (accessed 31 January 2013).

Yao, T.R., Shively, G.E. and Masters, W.A. (2005) How Successful are Government Interventions in Food Markets? Insights from the Philippines rice market. *Working Paper.* Department of Agricultural Economics, Purdue University, West Lafayette, Indiana.

Zuberi, T. and Thomas, K.J.A. (2012) Demographic Projections, the Environment and Food Security in sub-Saharan Africa. *Working Paper* 2012-001. United Nations Development Programme Regional Bureau for Africa, New York, New York.

3 Estimation of Cultivated Area, Number of Farming Households and Yield for Major Rice-growing Environments in Africa

Aliou Diagne,* Eyram Amovin-Assagba, Koichi Futakuchi and Marco C.S. Wopereis
Africa Rice Center (AfricaRice), Cotonou, Benin

Introduction

Rice is an extremely versatile crop which can grow under a range of water regimes (in dry- and wetland conditions) and temperatures (at low and high altitudes and latitudes) (Seck *et al.*, 2012; Saito *et al.*, Chapter 15, this volume). The various rice environments are characterized mainly by the main source of water for the plant – for example, rainfall (direct rainfall and/or inflow), irrigation (water controlled through a system of canals, etc.), water table, uncontrolled flood water, and sea/brackish water. This has led to the distinguishing of rainfed upland, rainfed lowland, irrigated upland, irrigated lowland, mangrove-swamp and deep-water environments. Taking other environmental factors into account, a more detailed classification of rice environments is possible; for instance, 'high-altitude rainfed upland rice' versus 'low-altitude rainfed upland rice' (see Saito *et al.*, Chapter 15, this volume).

Figure 3.1 depicts in a schematic manner the three major rice-growing environments in Africa: upland (excluding hydromorphic fringes, see below), rainfed lowland (inclusive of hydromorphic fringes) and irrigated lowland (WARDA, 2004).

Upland environments are situated at the high end of the toposequence, where rice depends solely on rainfall as the water table is out of the reach of rice roots for much of the growing season. At the lower end of the toposequence, rice plants can reach the water table or profit directly from flood water. Along the toposequence, interactions exist between environments (e.g. water and nutrient flow from upland to lowland). These externalities influence the environmental sustainability of rice farming. Furthermore, interlinkages can blur the boundaries between rice environments, as in the hydromorphic zones, between the upland and lowland areas, which rather than being flooded for long periods have a water table close enough to the surface for rice roots to reach during the growing season. Another unclear transition exists between rainfed and irrigated lowlands, where a continuum of water management exists ranging from the strictly rainfed (no water control or only drainage), which may evolve (via investments in water control measures) to the fully irrigated lowlands. In this chapter, lowlands with partial water control (intensified lowland) are classified as rainfed lowland environments.

* Corresponding author: a.diagne@cgiar.org

© CAB International 2013. *Realizing Africa's Rice Promise* (eds M.C.S. Wopereis *et al.*)

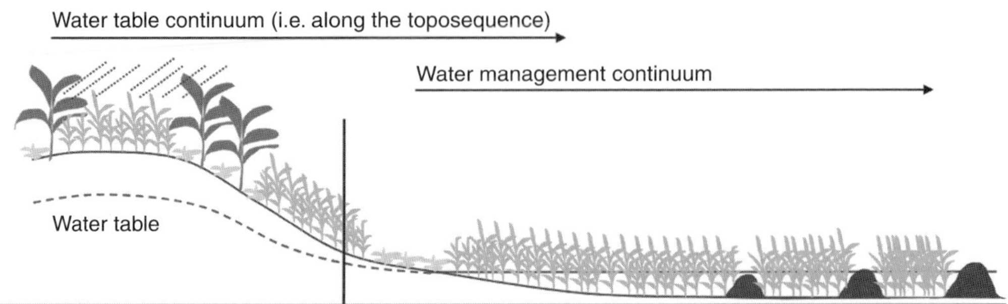

Ecology	Upland	Lowland fringe (hydromorphic)	Lowland	Intensified lowland	Irrigated lowland
Main water supply	Rainfall	Rainfall + water table	Rainfall + water table + unregulated flood water	Regulated flood water	Irrigation
Agroecological zone	Guinea savannah–humid forest zone	Guinea savannah–humid forest zone	Sudan savannah to humid forest zone	Sudan savannah to humid forest zone	Sahel to humid forest zone
Main stresses	Drought Weeds Pests and diseases P and N deficiency Soil erosion Soil acidity	Drought Weeds Pests and diseases P and N deficiency Soil erosion Soil acidity Fe toxicity	Drought/flooding Weeds Pests and diseases P and N deficiency Fe toxicity	Drought/flooding Weeds Pests and diseases P and N deficiency Fe toxicity	Weeds Pests and diseases Salinity Alkalinity P and N deficiency Fe toxicity

Fig. 3.1. Major rice environments and their attributes in Africa. (Adapted from WARDA, 2004, reproduced with permission from Africa Rice Center.)

Within each environment, rice is produced as part of a wide range of production systems (Saito *et al.*, Chapter 15, this volume). The heterogeneity and the close interaction of rice production with other farm and off-farm activities highlight the need for technical components that can be readily used by producers to build more integrated production systems. The search for greater integration is above all a search for greater resource use efficiency through the creation of positive interactions within rice production systems, and between rice and other farm and non-farm activities. Seen in this way, the focus on integrated production systems encompasses many of the concerns associated with natural-resources management.

Country-based information on potential and actual rice cultivation areas in each rice environment is useful to prioritize rice research and development because of the differences in farming practices and productivity among environments. In this chapter, we estimate actual rice cultivation areas and farming population by rice environment in Africa as of 2009, based on survey data and secondary data from country publications and FAOSTAT (2012).

Methodology

The level of detail in the information available on rice cultivation areas in Africa varies among countries. Estimates of total annual rice harvested area are provided by FAOSTAT (2012) for most rice-producing countries in Africa. The United States Department of Agriculture (USDA) provides similar statistics for many African countries. Unfortunately, these total area estimates are not disaggregated by rice-growing environment.[1] Here, we use farm-household survey data from the rice statistics survey conducted by Africa Rice Center (AfricaRice) and its national (NARS) partners in 2009 (AfricaRice, 2010) to estimate the total cultivated rice area by rice-growing environment in Africa.[2] Because of differences in the classification of rice-growing environments between countries, we use four environments in our study: irrigated, (rainfed) upland, (rainfed) lowland and 'other'. All rice-growing environments found in the countries and described using different terminologies which may vary from one country to another were reclassified into these four environments in order to obtain consistent estimates and meaningful comparison across countries (Table 3.1). For the purpose of the estimation, we classified the countries into four groups on the basis of the source of the data used to do the estimation (see Table 3.2 for the country groupings).

In the first two groups of countries (Table 3.2), the estimations of total rice area and number of farming households were based exclusively on data from the rice statistics survey database. For the first group, the data includes the population inflation weights used to obtain the national-level estimates from the sample estimates of the average total household rice area by environment and the distribution of rice-farming households by rice environment (see Appendix for details). These survey population weights[3] are missing from the survey data for the second group of countries. To obtain the national values from the sample estimates for the second group of countries, we used total national rice area from the 2009 rice statistics divided by the average household rice area to derive the total number of rice-farming households in the country. From this estimate and the average rice area by environment, we used the sample distribution of rice-farming households by rice environment to obtain estimates of the total number of rice-farming households and total rice areas by environment per country.

The third group comprises seven country members of the Coalition for African Rice Development (CARD), which were also part of the 2009 rice statistics survey. These countries also had missing survey population extrapolation weights. In addition, these countries did not have estimates of total rice cultivated area in their country reports. Thus, we used the information on rice area and production by rice environment for 2008 available in their National

Table 3.1. Classification of rice-growing environments by country in the rice statistics survey. (Modified from AfricaRice, 2011.)

Countries[a]	Irrigated	Lowland	Other
Benin, Burkina Faso, Cameroon, CAR, Côte d'Ivoire, DRC, Ghana, Guinea, Kenya, Madagascar, Nigeria, Rwanda, Tanzania	Irrigated lowland Upland with supplementary irrigation	Lowland Hydromorphic[b] Upland with groundwater[b]	Mangrove Other
The Gambia	Irrigated Upland with supplementary irrigation	Lowland Upland with groundwater	Mangrove
Mali	Irrigated	Lowland	
Senegal	Irrigated Upland with supplementary irrigation	Lowland Upland with groundwater	Mangrove
Sierra Leone		Inland-valley swamp Boliland[c]	Mangrove Other
Togo	Irrigated Upland with supplementary irrigation	Lowland Upland with groundwater	
Uganda	Irrigated Upland with supplementary irrigation	Lowland Hydromorphic[b] Upland with groundwater[b]	

[a]CAR, Central African Republic; DRC, Democratic Republic of Congo.
[b]Hydromorphic and upland with groundwater share the same definition, but the NARS partners differentiated them in the survey.
[c]Boliland is lowland where flooded conditions continue for 2–4 months in a year.

Table 3.2. Distribution of rice area by rice environment in Africa (2009).

		Distribution of area by rice environment (ha)				
Group[a]	Country	Irrigated	Upland	Lowland	Other	All
Group 1	Guinea	52,526	532,329	381,756	39,211	1,005,822
	Nigeria	86,079	557,256	1,032,935	219,427	1,895,697
	Senegal	94,185	36,178	43,948	0	174,311
	Sierra Leone	0	453,531	117,720	28,266	599,517
Group 2	Benin	4,798	10,407	23,552	0	38,757
	Burkina Faso	13,328	16,022	61,743	0	91,093
	Cameroon	25,212	60,926	19,635	0	105,773
	Kenya	17,521	449	414	0	18,384
	Togo	3,689	4,345	27,876	0	35,910
	Mali	335,269	174,747	134,851	0	644,867
	Madagascar	782,487	78,439	322,688	0	1,183,614
	Côte d'Ivoire	37,110	615,325	314,863	973	968,271
Group 3	CAR	377	18,310	2,731	373	21,791
	DRC	71,910	248,485	28,021	23,378	371,794
	The Gambia	27,916	19,350	25,231	7,013	79,510
	Ghana	24,496	8,126	129,533	2,485	164,640
	Rwanda	14,433	0	0	0	14,433
	Tanzania	254,679	9,382	677,806	571	942,438
	Uganda	361	65,790	72,109	0	138,260
Group 4	Mozambique	72,684	36,182	73,954	0	182,820
	Liberia	7,433	143,662	82,013	0	233,108
	Ethiopia	6,313	3,852	2,811	155	13,131
	Zambia	6,724	4,696	11,775	6	23,201
	Chad	51,579	51,961	37,734	2,075	143,349
	Niger	5,023	4,279	4,727	797	14,826
	Guinea-Bissau	8,479	38,025	47,521	11,042	105,067
	Mauritania	7,261	1,630	4,779	0	13,670
	Egypt	518,520	12,935	12,810	31,203	575,468
	Morocco	6,392	160	172	1,076	7,800
	Algeria	63	43	35	8	149
	Angola	4,023	4,153	6,036	0	14,212
	Burundi	18,585	1,258	7,778	0	27,621
	Comoros	12,001	4,314	6,266	0	22,581
	Republic of Congo	207	1,028	1,008	469	2,712
	Gabon	80	253	258	84	675
	Malawi	17,951	9,678	28,338	52	56,019
	Somalia	1,890	1,142	710	102	3,844
	South Africa	378	114	572	0	1,064
	Sudan[b]	2,529	2,285	1,438	206	6,458
	Zimbabwe	117	55	145	0	317
Total Africa		2,594,608	3,231,102	3,748,292	368,970	9,942,974
Total Africa (%)		26	32	38	4	100

[a]Group 1: countries with all data including sampling weights; Group 2: countries with full information on total rice area and farmers' distribution across rice-growing environments from rice statistics survey data 2009; Group 3: countries with full information on total rice area from NRDS 2008 data, distributions across rice-growing environments and average rice area by environment from rice statistics survey data 2009; Group 4: countries with total rice in 2009 from FAOSTAT (2012), farmers' distribution and average area by rice-growing environment predicted using spatial econometrics.

[b]Including what is now South Sudan.

Rice Development Strategies (NRDS) and followed the same estimation procedure as for the countries in the second group.

Countries in the fourth group are the remaining rice-producing countries in Africa which were not covered by the 2009 rice statistics survey.[4] For this group, we first used the 2009 rice statistics survey data of the first three groups of countries and the GIS coordinates at the third level of administrative sub-division in the country to estimate a spatial forecasting model based on the universal kriging interpolation method (Calder and Cressie, 2009). We then used the estimated model parameters to predict the average total household rice area and proportion of rice-farming households in each rice-growing environment for all the countries in the fourth group. And, finally, we used the 2009 total country rice harvested areas from FAOSTAT (2012), which we multiplied by the predicted averages and proportions to estimate the total rice area and number of rice-farming households in 2009 for each rice environment for each country in this fourth group (see Appendix for more details).

Rice yields per rice environment were estimated for countries covered by the 2009 rice statistics survey (Groups 1, 2 and 3). These yields were obtained by dividing total rice production by the total rice-cropped area of the household.

Results

The estimated rice areas disaggregated by country and by rice environment are shown in Table 3.2. The estimated total number of rice-farming households by rice-growing environment are presented in Table 3.3, but will not be discussed as the trends are the same as for the area estimates. The total rice area harvested in Africa in 2009 is estimated at 9,942,974 ha. Countries with high total rice areas (500,000 ha or more) are Nigeria (1,895,697 ha), Guinea (1,005,822 ha), Madagascar (1,183,614 ha), Côte d'Ivoire (968,271 ha), Tanzania (942,438 ha), Mali (644,867 ha), Sierra Leone (599,517 ha) and Egypt (575,468 ha). The disaggregation by rice environment shows rainfed lowland (3,748,292 ha) and upland (3,231,102 ha) to be the dominant environments with 38% and 32% of the total rice areas, respectively. The total area under irrigation is estimated at 2,594,608 ha, which represents about 26% of the total rice area on the continent. The other rice-growing environments are estimated to occupy only 4% of the total rice area.

The total cultivated rice area estimated for 2009 in this study is 4.5% higher than the estimated 9,514,792 ha for 2008 (Seck *et al.*, 2012). This may be due to differences in the reference years or that Seck *et al.* (2012) used a different methodology to estimate the total area.[5] When the estimates from the two studies are compared country by country, the most important **increases** in the estimated total rice areas are for Côte d'Ivoire (568,271 ha, +142%), The Gambia (45,510 ha, +134%), Cameroon (55,772 ha, +112%), Mozambique (72,820 ha, +66%), Senegal (48,982 ha, +39%), Mali (162,315 ha, +34%) and Tanzania (232,437 ha, +33%). Substantial **decreases** in the estimated total rice area are also recorded in three of the major rice-producing countries in Africa: Sierra Leone (400,483 ha, –40%), Nigeria (486,304 ha, –20%) and Egypt (169,923 ha, –23%).

Table 3.4 presents average paddy rice yield per rice environment as obtained from the survey data of countries in Groups 1, 2 and 3. These yields were obtained at the household level by dividing total household rice production by total household rice-cropped area and then averaged across the country.

Highest country average household rice yields (per season) were in Rwanda, Mali, Cameroon, Senegal and Nigeria. These relatively high yields are due to the high yield in irrigated growing environment, except for Nigeria and Cameroon where lowland yields exceed irrigated ones. The highest national average yields were recorded in Rwanda (4.43 t/ha), Mali (4.01 t/ha) and Senegal (3.90 t/ha) for irrigated environments and in Cameroon (3.20 t/ha) and Nigeria (3.02 t/ha) for lowlands. Countries with lowest rice yields across rice environments include the Central African Republic, the Democratic Republic of Congo and Sierra Leone.

Table 3.5 gives estimates for rice yields in irrigated environments in the Sahel–savannah and sub-humid agroecological zones. Sahel–savannah irrigated zones occur throughout The Gambia and Mali, and in the extreme north of Benin; the Sahel

Table 3.3. Number of rice-farming households by rice environment in Africa (2009).

		Distribution of rice-farming households by rice-growing environment (number)				
Group[a]	Country	Irrigated	Upland	Lowland	Other	All
Group 1	Guinea	39,911	522,917	297,555	26,518	886,901
	Nigeria	16,557	135,186	262,379	42,448	456,570
	Senegal	93,468	30,315	44,960	0	168,743
	Sierra Leone	0	400,906	97,307	24,880	523,093
Group 2	Benin	8,594	20,033	43,765	0	72,392
	Burkina Faso	13,256	25,750	123,251	0	162,257
	Cameroon	21,778	63,229	23,503	0	108,511
	Kenya	7,107	273	195	0	7,575
	Togo	3,631	5,603	29,517	0	38,751
	Mali	162,425	146,654	59,399	0	368,478
	Madagascar	669,254	107,771	389,329	0	1,166,354
	Côte d'Ivoire	27,480	493,906	287,929	1,297	810,612
Group 3	CAR	633	36,470	6,554	633	44,290
	DRC	61,199	261,946	40,774	21,466	385,385
	The Gambia	13,060	5,803	8,708	2,281	29,852
	Ghana	6,580	2,709	32,353	621	42,263
	Rwanda	55,648	0	0	0	55,648
	Tanzania	125,563	3,044	251,506	380	380,493
	Uganda	1,098	109,127	133,852	0	244,077
Group 4	Mozambique	37,986	23,287	42,094	0	103,367
	Liberia	4,644	104,174	56,688	0	165,506
	Ethiopia	3,836	2,207	1,554	251	7,848
	Zambia	3,887	3,001	5,799	534	13,221
	Chad	38,326	38,268	23,366	1,441	101,401
	Niger	1,901	2,205	1,577	316	5,999
	Guinea-Bissau	8,593	34,193	30,393	9,650	82,829
	Mauritania	5,943	1,092	3,087	0	10,122
	Egypt	864,200	9,097	9,097	27,291	909,685
	Morocco	9,541	106	106	848	10,601
	Algeria	50	28	22	6	106
	Angola	2,250	2,802	3,535	0	8,587
	Burundi	20,757	569	4,727	0	26,053
	Comoros	6,457	2,910	4,213	0	13,580
	Congo	138	750	822	382	2,092
	Gabon	49	173	170	55	447
	Malawi	10,276	5,857	14,111	2,708	32,952
	Somalia	1,484	796	493	74	2,847
	South Africa	173	73	299	0	545
	Sudan[b]	2,046	1,607	1,000	151	4,804
	Zimbabwe	60	36	80	0	176
Total Africa		2,349,839	2,604,873	2,336,069	164,231	7,455,012
Total Africa (%)		32	35	31	2	100

[a]Group 1: countries with all data including sampling weights; Group 2: countries with full information on total rice area and farmers' distribution across rice-growing environments from rice statistics survey data 2009; Group 3: countries with full information on total rice area from NRDS 2008 data, distributions across rice-growing environments and average rice area by rice-growing environment from rice statistics survey data 2009; Group 4: countries with total rice area from FAOSTAT (2009), farmers' distribution and average area by rice-growing environment predicted using spatial econometrics.

[b]Including what is now South Sudan.

Table 3.4. Average rice yield (t/ha) by rice environment for countries in Groups 1, 2 and 3 (2009).

	Irrigated			Upland			Lowland			Other		
Country	n	Mean	SD	n	Mean	SD	n	Mean	SD	n	Mean	SD
Benin	41	2.07	0.9	75	1.51	0.8	159	1.83	0.8	0	–	–
Burkina Faso	51	2.21	1.4	99	1.73	1.2	455	1.71	1.1	0	–	–
Cameroon	227	1.94	1.2	659	2.95	1.4	245	3.20	1.4	0	–	–
Côte d'Ivoire	39	1.68	0.7	665	1.22	0.6	389	1.61	0.7	2	1.10	0.1
The Gambia	62	2.16	2.0	25	1.70	1.8	39	1.28	0.6	11	1.30	0.0
Ghana	78	1.60	0.7	32	1.01	0.4	394	1.16	0.6	8	1.00	0.4
Guinea	53	1.24	0.4	601	1.12	0.4	357	1.10	0.4	32	1.09	0.3
Kenya	226	1.53	0.9	13	1.25	0.5	9	1.29	1.1	0	–	–
Madagascar	446	1.69	1.0	70	1.04	0.7	257	1.71	0.9	0	–	–
Mali	618	4.01	1.4	558	2.33	1.0	226	2.85	1.8	0	–	–
Nigeria	244	2.24	1.6	1815	1.44	0.8	3448	3.02	2.1	973	2.55	1.2
CAR	7	0.49	0.2	407	0.57	0.3	71	0.53	0.2	7	0.53	0.2
DRC	57	1.34	0.4	244	1.00	0.4	36	0.88	0.3	20	1.13	0.4
Rwanda	168	4.43	2.2	0	–	–	0	–	–	0	–	–
Senegal	370	3.90	1.7	358	1.12	0.7	868	1.22	0.6	0	–	–
Sierra Leone	0	–	–	780	0.55	0.4	218	0.56	0.4	41	0.77	0.4
Tanzania	305	1.40	0.6	8	0.63	0.3	527	1.13	0.6	1	1.08	0.0
Togo	68	2.06	0.8	104	1.40	0.8	549	1.82	0.9	0	–	–
Uganda	4	1.85	0.8	480	1.72	1.0	588	1.79	1.0	0	–	–
All countries in Groups 1, 2 and 3[a]	3064	2.22	1.0	6993	1.23	0.2	8835	1.89	0.7	1095	2.07	0.5

n, sample size matching the criteria; SD, standard deviation, –, not applicable.
[a]The yield for all countries in Groups 1, 2 and 3 is the average country yield weighted by the estimated area per country.

Table 3.5. Average rice yield (t/ha) by irrigation environment for countries in Groups 1, 2 and 3 (2009).

	Sahel–savannah irrigated systems			Sub-humid irrigated zones		
Country	n	Mean	SD	n	Mean	SD
Benin	13	2.12	1.0	28	2.05	0.9
Burkina Faso	23	2.78	1.0	28	1.74	1.5
Cameroon	48	2.95	1.6	179	1.68	1.0
The Gambia	62	2.16	2.0	0	–	–
Mali	618	4.01	1.4	0	–	–
Nigeria	109	2.23	1.7	135	2.24	1.6
Senegal	365	3.91	1.7	5	1.96	1.2
All	1238	3.65	1.7	375	1.92	1.3

n, sample size matching the criteria; SD, standard deviation; –, not applicable.

agroecological zone occurs in northern Cameroon, the Senegal River valley and Casamance regions in Senegal, the northern states of Nigeria (Borno, Jigawa, Kano, Katsina, Kebbi, Sokoto, Yobe and Zamfara) and Boucle du Mouhoun, Central, North and Sahel regions of Burkina Faso. Sub-humid irrigated zones are irrigated zones located outside the Sahel–savannah.

Notes

[1] Balasubramanian *et al.* (2007) provide estimates of country total rice areas disaggregated by rice-growing environments.
[2] The estimation methodology is detailed in the Appendix.
[3] 'Survey population weights' are coefficients associated with sampled units which are used to extrapolate from the sample to the whole population of farmers. In essence, the coefficient gives the number of farmers represented by each sample farmer.
[4] Liberia and Mozambique, which are included in this group, were covered by the 2009 rice statistics survey, but data analyses were not conducted.
[5] Seck *et al.* (2012) used a simpler procedure to estimate the distribution of total rice area by rice-growing environment for these countries: the percentage distribution of area by environment of a similar neighbouring country for each country of this group in combination with the estimated total rice area in 2008 for that country as given by FAOSTAT (2012).

References

AfricaRice (2010) Rice data systems for sub-Saharan Africa: Contribution to the Japan–AfricaRice Emergency Rice Project Synthesis Report. Africa Rice Center, Cotonou, Benin.
Balasubramanian, V., Sie, M., Hijmans, R.J. and Otsuka, K. (2007) Increasing rice production in sub-Saharan Africa: opportunities and challenges. *Advances in Agronomy* 94, 55–133.
Calder, C. and Cressie, N.A. (2009) Kriging and variogram models. In: Kitchin, R. and Thrift, N. (eds) *International Encyclopedia of Human Geography*, Vol. 1. Elsevier, Oxford, pp. 49–55.
FAOSTAT (2012) http://faostat.fao.org/ (accessed 21 June 2012).
Kastelec, D. and Košmelj, K. (2002) Spatial interpolation of mean yearly precipitation using universal kriging. In: Mrvar, A. and Ferligoj, A. (eds) *Developments in Statistics*. FDV, Ljubljana, Slovenia, pp. 149–162.
Ribeiro, P.J. Jr and Diggle, P.J. (2001) geoR: a package for geostatistical analysis. *R NEWS* 1(2), 15–18.
Seck, P.A., Diagne, A., Mohanty, S. and Wopereis, M.C.S. (2012) Crops that feed the world 7: rice. *Food Security* 4(1), 7–24.
WARDA (2004) *Strategic Plan 2003–2012*. Africa Rice Center (WARDA), Bouaké, Côte d'Ivoire.

Appendix 3.1 Methodology for Estimating Total Rice Cultivated Area by Rice-growing Environment

In this appendix we describe the methodology used to estimate a country's total cultivated rice area and number of rice-farming households in 2009 by rice environment. The estimated total area for the whole continent was obtained by aggregating the country estimates.

For a country with a population of farmers (i.e. farming households) of size N and J rice environments each with N_j number of rice farmers, let a_{ij} denote the total rice area in environment j of a farmer i and d_{ij} be a dummy binary indicator taking the value 1 if the farmer grows rice in environment j and zero otherwise, $i = 1, \ldots, N$ and $j = 1, \ldots, J$. Let also A_j denote the country's total cultivated rice area in environment j, f_j the proportion of farmers in the country growing rice in environment j and $\overline{a}_j$ the average total household cultivated rice area in environment j. Then we have for $j = 1, \ldots, J$, $N_j = \sum_{i=1}^{N} d_{ij}$, $A_j = \sum_{i=1}^{N} a_{ij} = \sum_{i=1}^{N} d_{ij}a_{ij}$, $f_j = \frac{N_j}{N}$ and $\overline{a}_j = \frac{A_j}{N_j}$. The country's total rice area A is thus given by $A = \sum_{j=1}^{J} A_j$.

From these expressions, we can obtain from a (possibly multistage stratified) random sample of rice-farming households consistent estimates of the country's total number of farmers growing rice in environment j and the corresponding total rice area in environment j. More precisely, we have $\widehat{N}_J=\sum_{i=1}^{n} w_i d_{ij}$ and $\widehat{A}_J=\sum_{i=1}^{n} w_i d_{ij} a_{ij}$, where w_i $i=1,..,n$, are the survey population weights such that $\sum_{i=1}^{n} w_i=N$.[1] Similarly, the country's proportion of farmers growing rice in environment j and the average total household rice area in environment j are consistently estimated by $\widehat{f}_J=\frac{1}{\sum_{i=1}^{n} w_i}\sum_{i=1}^{n} w_i\, d_{ij}$ and $\widehat{\overline{a}}_j=\frac{1}{\sum_{i=1}^{n} w_i\, d_{ij}}\sum_{i=1}^{n} w_i\, d_{ij}\, a_{ij}$, respectively.

For the countries in the second and third groups where we did not have the survey population weights, we have estimated $\widehat{A}_j$ and $\widehat{N}_j$ by $\widehat{A}_J=\widehat{N}_J\times\widehat{\overline{a}}_j$, and $\widehat{N}_j=N\times\widehat{f}_j$, respectively, where $\widehat{f}_J=\frac{1}{n}\sum_{i=1}^{n} d_{ij}$ is the sample proportion of farmers growing rice in environment j and $\widehat{\overline{a}}_j=\frac{1}{n\widehat{f}_j}\sum_{i=1}^{n} d_{ij}\, a_{ij}$ is the sample average of the total household rice area under environment j. We estimate N, the total number of rice-farming households in the formulae above, by dividing the total cultivated rice area obtained from the 2009 survey country reports (for the second group) and the NRDS (for the third group) by the sample average of total household rice area across all rice environments.

In the 2009 rice statistics survey, the area information was collected for all of the plots of each household. Unfortunately, the information on the rice environment of each household plot was not collected. Instead, the rice environment information was collected at the household level by asking the household to indicate by order of importance all the environments where it was growing rice. Hence, we cannot know the a_{ij} quantities in the formulae above from the rice statistics survey data. Only the d_{ij} quantities are available from the rice statistics survey. This makes it impossible to estimate A_j from the formulae above. To circumvent the problem of non-availability of the rice-growing environment-specific household cultivated area variables, we use the household-level rice environment information and the total household cultivated rice area $a_i=\sum_{j=1}^{J} a_{ij}$, which can be computed from the rice statistics survey data (since the information on the size of each household rice cultivated plot was collected) and proceed as follows.

Let $j_i^*=argmax\{a_{ij}; j=1,\dots J\}$ designate the household's most important rice-growing environment in terms of area occupied (the first rice-growing environment listed by the household) and let d_{ij}^* be the dummy binary indicator taking the value 1 if $j=j_i^*$ and 0 otherwise. We can then approximate A_j, the country's total rice area in environment j, by the quantity $A_j^*=\sum_{i=1}^{N} d_{ij}^*\, a_i$. It is clear that $A_j^*=A_j$ if every household grows rice in only one environment. Also, A_j^* is close to A_j if the area of a household's minor (secondary) rice-growing environment is small (i.e. if a_{ij} is small for $j\neq j_i^*$). The reality is very close to these two conditions as the vast majority of African rice farmers generally grow rice in small plots located in the same environment (Seck *et al.*, 2012). This turned out to be overwhelmingly the case in the 2009 rice statistics survey data, where less than 10% of farmers grew rice in more than one rice-growing environment. Moreover, this percentage becomes insignificant after grouping the various rice-growing environments into the four major environments (Table 3.A.1). Hence, A_j^* is a very good approximation of A_j when using the data from the 2009 rice statistics survey. To obtain an estimate $\widehat{A}_j^*$ of A_j^* and the coresponding approximate quantities it suffices to replace in the estimation formula above d_{ij} by d_{ij}^* and a_{ij} by a_i.

Spatial interpolation: kriging

Let $\{(Z_s): s\in D\subset R^2\}$ be a spatial process specified as follow: $Z_s=m(s)+X_s$ where $m(s)=E(Z_s)<\infty$ is the trend of the process and $X_s=(Z_s-E(Z_s))$ the fluctuation (Calder and Cressie, 2009).

Let $\{s_i: i=1,\dots,n\}$ be the geostatistical data, i.e. a set of known locations of the process: $s_i=(x_i, y_i)$ with x_i being the longitude and y_i the latitude at location s_i. For this chapter, s_i represents data points at the third sub-division level for Benin, Burkina Faso, Cameroon, CAR, Côte d'Ivoire, Guinea, Kenya, Rwanda, Sierra Leone, Togo and Uganda. The second sub-division was used for DRC, The Gambia, Ghana, Madagascar, Nigeria and Tanzania, whereas the first sub-division level was used for Mali and Senegal according to the availability of data. For countries of the fourth group, we randomly selected within each country 25 locations where further predictions will be made. In total, $n=695$ locations were obtained from the above cited 19 countries on which models are based to predict values at the other

Table 3.A.1. Estimation details for countries of Group 1.

	Estimated area (ha)	Standard error	95% Confidence interval	
Senegal				
Irrigated	94,185	2,734	88,813	99,557
Upland	36,178	1,242	33,736	38,621
Lowland	43,948	1,090	41,809	46,086
Total estimated	174,311	4156	166,160	182,462
Guinea				
Irrigated	52,526	3,584	45,320	59,731
Upland	532,329	13,142	506,522	558,137
Lowland	381,756	9,862	362,362	401,150
Others	39,211	4,061	30,940	47,482
Total estimated	1,005,822	17,618	971,252	1,040,392
Nigeria				
Irrigated	86,079	4,868	76,503	95,655
Upland	557,256	12,063	533,600	580,911
Lowland	1,032,935	21,240	991,292	1,074,577
Others	219,427	12,823	194,267	244,586
Total estimated	1,895,697	28,071	1,840,668	1,950,723
Sierra Leone				
Upland	453,531	13,154	427,713	479,348
Lowland	117,720	5,824	106,255	129,185
Others	28,266	3,602	21,059	35,474
Total estimated	599,517	14,842	570,397	628,636

850 locations where no data are available. The country predicted value is the average value of the predicted values at sampled locations.

The mean part of the model can be specified as a function of given covariates and/or coordinates, but we specified it as a polynomial function of the coordinates: $m(s) = \beta_0^* + \beta_1^* x + \beta_2^* y$,

where $(\beta_0^*, \beta_1^*, \beta_2^*) = \arg\min \sum_{i=1}^{n} (z_{s_i} - \beta_0 - \beta_1 x_i - \beta_2 y_i)^2$ are coefficients of the linear trend.

For the sake of kriging, X_s was assumed to be a random process with an existing Gaussian variogram $2\gamma(h) = V(Z_{s_i} - Z_{s_j})$, $h = d(s_i, s_j)$ being the Euclidian distance between s_i and s_j, with the parameter γ estimated as (see Calder and Cressie, 2009):

$$\hat{g}_h = \begin{cases} 0 & if\,|h| = 0 \\ C_0 + C_1\left(1 - \exp\left(\dfrac{-3h^2}{a^2}\right)\right) & if\,|h| > 0 \end{cases} \text{; where } a = range,\ C_0 = nugget\ effect = V(Z_s),\ C_0 + C_1 = sill$$

The range is the longest distance with correlated values of the process; the nugget effect represents the semivariance of the process on microscale, i.e. very close to 0 (Kastelec and Košmelj, 2002) and the sill is the maximum semivariogram value. The range, nugget effect and sill parameters were estimated by maximum likelihood using the package geoR (Ribeiro and Diggle, 2001). The predicted values of Z are obtained as weighted linear combinations of the available data: $\hat{Z}_{s_0} = \sum_{i=1}^{n} \lambda_i Z_{s_i}$ where s_0 is an unobserved location. The weights λ_i are obtained by minimizing the mean square error: $\min(Z_{s_0} - Z_{s_0})^2$ subject to the condition of an unbiased estimator $\sum_{i=1}^{n} \lambda_i = 1$ (Calder and Cressie, 2009).

We used this spatial interpolation method to predict the proportion of rice-farming households in irrigated, upland and lowland environments (the proportion of farmers in other environments, where applicable, is deduced from the first three) and the average total household area in each of the four environments for each country of the fourth group.

Note

[1] We note that $W_i = \frac{N}{n}$ for a simple random sample. But most countries have used a two-stage stratified random sampling procedure (see AfricaRice, 2010, and the country reports cited therein for details of the countries' sampling procedures and survey populations).

4 Farmer Perceptions of the Biophysical Constraints to Rice Production in Sub-Saharan Africa, and Potential Impact of Research

Aliou Diagne,* Didier Y. Alia, Eyram Amovin-Assagba, Marco C.S. Wopereis, Kazuki Saito and Tebila Nakelse
Africa Rice Center (AfricaRice), Cotonou, Benin

Introduction

Average rice yield in Africa (2.15 t/ha; USDA, 2013)[1] is low compared with other continents; this is to a large extent a result of the fact that rice cropping in sub-Saharan Africa (SSA) is predominantly rainfed (see Diagne *et al.*, Chapter 3, this volume). Important gaps exist, however, between actual farmers' yields in a growing season and what would be possible with improved crop management practices. The maximum or **potential yield** (Yp) in farmers' fields is defined as the yield of a cultivar when grown under the conditions to which it is adapted, where nutrients and water are not limiting, and pests and diseases are effectively controlled (Evans, 1993). Estimates of Yp are usually based on crop simulation modelling studies (see Saito *et al.*, Chapter 15, this volume).

Solar radiation, carbon-dioxide concentration, temperature and crop characteristics are the major **yield-defining** factors that determine Yp. Breeding efforts may help raise Yp – for example, by introducing hybrid rice varieties. **Yield-limiting** factors are related to shortage of water or nutrients (or both) and determine water- or nutrient-limited production levels in a given rice environment. **Yield-reducing** factors induce yield losses by reducing or hampering growth, including abiotic and biotic factors. Biotic factors include weeds, pests and diseases; abiotic factors include salinity, alkalinity and iron toxicity.

Attainable yield (Ya) refers to the yield that can be achieved with best management practices that control yield-limiting and yield-reducing factors in a economically optimal manner. Under irrigated conditions, this is typically about 80% of Yp. The **yield gap** is commonly defined as the difference between Yp or Ya and actual average farmers' yields. A range of socio-economic reasons underpin these yield gaps at harvest and the substantial losses that often occur after harvest (see Rickman *et al.*, Chapter 27, this volume), such as lack of availability of key inputs (labour, fertilizer, etc.) and sub-optimal knowledge of improved management practices.

In this chapter, we quantify farmer perceptions of major biotic and abiotic constraints that limit and reduce rice yields in farmers' fields in SSA. We also estimate the potential impact of research addressing such constraints.

* Corresponding author: a.diagne@cgiar.org

© CAB International 2013. *Realizing Africa's Rice Promise* (eds M.C.S. Wopereis *et al.*)

Methodology

Data used in this chapter are derived from farm-household surveys conducted in 2009–2010 in 21 countries by Africa Rice Center (AfricaRice) in collaboration with partners from national agricultural research systems (NARS) and national agricultural statistics systems. Data were complete and of acceptable quality for 18 countries: Benin, Burkina Faso, Cameroon, Central African Republic (CAR), Côte d'Ivoire, Democratic Republic of Congo (DRC), The Gambia, Ghana, Guinea, Kenya, Madagascar, Nigeria, Rwanda, Senegal, Sierra Leone, Tanzania, Togo and Uganda. These 18 countries accounted for almost 87% of area harvested and 84% of total rice production in SSA in 2010 (FAOSTAT, 2010).

Surveys were designed to produce nationally representative rice production data. A two-stage stratified random sampling methodology was used in most countries (for details see AfricaRice, 2010) with regions, provinces, departments or states used as strata. Within each stratum, rice-producing villages were randomly selected in the first stage and rice-farming households were randomly selected in the second stage. The sample size ranged from 395 (Rwanda) to 10,500 (Nigeria). Detailed descriptions of the sampling methodologies and data analyses can be found in the country reports and the synthesis report (AfricaRice, 2010).

Farmer perceptions of abiotic and biotic constraints

We used the following grouping of abiotic and biotic constraints:

1. Biotic constraints: weeds (considering all weed species, except parasitic weeds), diseases (*Rice yellow mottle virus*, blast and bacterial leaf blight), insects (African rice gall midge, termites and stem borers), birds and rodents.
2. Abiotic constraints (climate-related): drought, flooding and extreme temperatures.
3. Abiotic constraints (soil-related): non-fertile soil (low organic-matter content, poor water-holding capacity), nutrient deficiencies (N, P, K, Zn), Fe toxicity, salinity, alkalinity, acidity and soil erosion.

In each rice environment (upland, rainfed lowland, irrigated and others, including mangrove-swamp and deep-water environments), we assessed the relative prevalence of these biotic and abiotic constraints and the losses they cause as perceived by farmers. The proportion of farmers who experienced one of the constraints was estimated along with the average area affected within their field and the yield loss incurred on a whole-field basis when the constraint occurred.

Data on rice production constraints were collected in each country through a sequence of three surveys, involving: (i) rice experts in the country; (ii) focus groups in sample villages; and (iii) individual farmers. Rice experts' knowledge enabled us to identify the major biotic and abiotic constraints occurring in the country in each rice environment and to develop materials to be used during interviews with farmers (e.g. pictures of symptoms and descriptions in local languages). For this purpose, an exhaustive list of biotic and abiotic constraints was available (complete with local names), along with pictures of accompanying symptoms – a resource developed by AfricaRice scientists and refined over time, reflecting experience in collecting this type of information gained since 2000.

Data collected in the village focus groups enabled enumerators to establish a list of biotic and abiotic constraints for each village (by rice environment) (list one). We also identified the five major biotic or abiotic constraints that negatively affected rice production in each village (list two).

From the lists and symptom illustrations, farmers identified the constraints that they experienced and provided their perceptions of their incidence and their importance in general in terms of scope and severity. For each constraint known in the village, information on awareness of the constraint and its occurrence in the village was elicited from each sample farmer and for each rice environment. Farmers scored or ranked the relative importance of each constraint that they were aware of on a scale of: 0 – constraint does not occur (i.e. in the village); 1 – minor constraint; 2 – intermediate constraint; 3 – major constraint.

Next, for each of the five major constraints, the farmer was asked the following set of questions:

1. Do you know the constraint? If so:
2. Have you experienced the constraint in each of the past three years, i.e. 2008, 2007, 2006? (This chapter uses only results for 2008.)

3. What was the proportion of area affected in your field in 2008, 2007, 2006? (This chapter uses only data for 2008.)
4. What was the percentage reduction in yield for your entire field because of this constraint as compared to a year without it?

There are several points that should be noted on farmer perception of the relative importance of constraints. First, the set of indicators derived from the information collected using the two lists are not independent. Indeed, the ordinal ranking of the perceived relative importance of the constraint obtained from the first list can be taken as a summary measure of the farmer perception of the prevalence of the constraint as measured in (2) and the perception of its scope and severity when it occurs as measured in (3) and (4), respectively. Second, farmers' perceptions of the relative importance of a constraint can be very different from the real importance of the constraint (in terms of scope and severity) if it were measured. In other words, farmers may have the wrong perception about the relative importance of the constraint. However, it is important to keep in mind that it is this subjective perception, right or wrong, that drives farmers' production decisions, including the *ex-ante* preventive and *ex-post* mitigating actions they take with respect to the constraint. Hence, in the end the **real effects** of the constraint on each farmer's productivity outcomes will be the results of both the **real importance** (as it would be measured objectively by an expert) and that farmer's **perception** of its importance (which dictates the *ex-ante* preventive and *ex-post* mitigating actions he or she takes). Thus, in determining the relative importance of a constraint in terms of its negative effects on productivity, both the objective assessments of knowledgeable experts and the subjective assessments of farmers are important and complementary. Furthermore, one can argue that the divergence between the two assessments is smaller with the ordinal ranking based on the first list than with the estimates of proportion of areas affected and yield loss in (3) and (4) above. This is because of its ordinal nature, the first assessment is a measure of relative importance which can lead to the same ordinal ranking of a constraint by a farmer and a knowledgeable expert even if they give different estimates of yield loss and area affected by the constraint.[2]

From the information collected, we developed four alternative measures of relative importance of a biophysical constraint as perceived by a sample of farmers randomly selected from a population of rice farmers.

We use the following notation. First, let $\tilde{S}$ be the set of all biotic and abiotic constraints existing in a country and let n be the sample size. For any constraint $s \in \tilde{S}$ and any farmer i, $i = 1,...,n$, let $r_i^s \in \{0,1,2,3\}$ be the rank given to the constraint s by farmer i when expressing his or her perception of the relative importance of constraint s. Also, for any given rank $r \in \{0,1,2,3\}$, let d_{ir}^s be the dummy binary indicator that takes the value 1 if $r_i^s = r$ and the value 0 if $r_i^s \neq r$. Also for any constraint $s \in \tilde{S}$ and any farmer i, let e_i^s be the dummy binary indicator taking the value 1 if the farmer has experienced the constraint s (in a particular year) and 0 otherwise; let a_i be the total rice area of farmer i, a_i^s the area affected by the constraint s and $\alpha_i^s = \frac{a_i^s}{a_i}$ the proportion of the land affected by the constraint; $\Delta y_i^s = 1 - \frac{y_{i1}^s}{y_{i0}^s}$ the farmer's perceived percentage yield reduction on a whole-field basis when he or she experiences the constraint, where y_{i1}^s and y_{i0}^s are the yield obtained by the farmer (total quantity of harvested rice divided by the total cultivated rice area) with and without experience of the constraint, respectively. Finally, to reflect the restrictive nature of the top-five list of constraints (list two), we will define for any farmer i a sub-set $\tilde{S}_i^m$ of $\tilde{S}$ of five major constraints identified in his or her village and the dummy binary indicator m_i^s that takes the value 1 if $s \in \tilde{S}_i^m$ and 0 otherwise.

From these notations, we defined five measures for assessing the relative importance of each biotic or abiotic constraint s found in the different rice environments as perceived by a sample of size n of rice farmers randomly selected from a rice-farming population:

1. The percentage of sample farmers who perceive the constraint s to have a relative importance of rank r, $r = 0,1,2,3$ (does not exist in the village, minor, medium, major): $f_r^s = \frac{1}{n}\sum_{i=1}^{n} d_{ir}^s$
2. A farmer perception index of the relative importance of the constraint as given by the average

rank given to the constraint by sample farmers, normalized to lie between the values 0 and 1: $\bar{\bar{r}}^s = \frac{1}{3} \times \bar{r}^s$ where $\bar{\bar{r}}^s = \frac{1}{n}\sum_{i=1}^{n} r_i^s = \frac{1}{n}\sum_{i=1}^{n}\sum_{r=0}^{3} rd_{ir}^s$ is the average ranking of the constraint.[3]

3. Percentage of farmers experiencing the constraint in 2008, provided it is among the five 'major' constraints in the village: $E^{s,2008} = \frac{1}{n}\sum_{i=1}^{n} m_i^s e_i^s$.

4. Proportion of area affected for farmers who experienced the constraint in 2008, provided it is among the five 'major' constraints in the village: $A^{s,2008} = \frac{1}{n_m^s}\sum_{i=i}^{n} m_i^s e_i^s \alpha_i^s$.

5. Percentage reduction in yield on a whole-field basis compared to no-constraint conditions for farmers who have experienced the constraint, provided it is among the five 'major' constraints in the village: $\Delta y^{s,2008} = \frac{1}{n_m^s}\sum_{i=1}^{n} m_i^s e_i^s \Delta y_i^s$.

Unfortunately, even if the sample is random, all five constraint indicators derived above are biased estimates of the respective population measures of the relative importance of a constraint. Indicators 1 and 2 are based on list one and suffer from **population non-awareness bias** that arises from the fact that awareness of the existence of a constraint is not universal in the rice farming population. This non-awareness bias, which is a characteristic of the population, is the same type of bias identified by Diagne (2006) in the context of adoption of a new variety the existence of which is not universally known in the population (see also Diagne and Demont, 2007; Diagne, 2009).

Indicators 3, 4 and 5 are based on the second list of constraints (top five) and also suffer from the same population non-awareness bias. They also suffer from a **sample minority-exclusion bias** caused by the restricted nature of the second list, which (by the design of the survey instruments) excluded the 'non-major' constraints. This exclusion is intentional, made in order to reduce the time taken to collect the information related to farmers' estimates of the proportion of area affected and yield loss when they experience a constraint. Hence, in contrast to the non-awareness bias, the exclusion bias is a property of the sample and not a property of the population.

In Appendix 4.1, we formally demonstrate the following facts about these two biases. First, everything else being equal, population non-awareness introduces a **downward** bias in all five constraint indicators if the measure of relative importance of the constraint at the individual farmer level is monotonically and positively related to awareness of the existence of the constraint – for any farmer, the perceived relative importance of the constraint cannot be lower when he or she is aware of the constraint compared to when he or she is not. This positive monotonicity is clearly satisfied by all five measures of relative importance of a constraint.[4] Furthermore, the downward bias introduced by non-awareness disappears completely when all farmers are aware of the existence of the constraint; conversely, and everything else being equal, the bias is more severe for constraints that farmers are less likely to be aware of compared to the ones they are more likely to be aware of. This is particularly the case for most abiotic constraints, which are usually more difficult for farmers to diagnose. Hence, we should expect abiotic constraints in the sample to have generally lower sample estimates for indicators 1 and 2 compared to biotic constraints.

Second, everything else being equal, sample minority-exclusion introduces an **upward** bias in constraint indicators 3, 4 and 5 if the measure of relative importance of the constraint at the individual farmer level is positively correlated with the constraint being major for the farmer – for any farmer, the perceived relative importance of the constraint cannot be lower when the constraint is a major constraint compared to when it is not. Again, positive correlation with the constraint being a major constraint is clearly satisfied by all five measures of relative importance of a constraint. Also, as with non-awareness, the upward bias introduced by sample minority-exclusion disappears completely when all the constraints are major constraints for all farmers.

Third, the two biases introduced by population non-awareness and sample minority-exclusion are additive and operate in opposite directions. Hence, the direction of the overall bias in the three sample indicators of relative importance of a constraint in (3)–(5) is indeterminate and depends on the relative sizes of the two biases. Moreover, the overall bias may vanish if the biases cancel out each other.

Potential Impact of Research Aiming to Reduce Yield Loss Caused by Biotic and Abiotic Constraints

Next, we assessed the potential economic and poverty impact of reducing part of the yield gap caused by the major biotic and abiotic constraints in SSA, directly in the 18 African countries surveyed and indirectly through extrapolation to another 18 rice-producing countries in Africa using secondary data (i.e. 36 countries in total; for more details see Diagne *et al.*, Chapter 32, this volume).

The evaluation of the potential impact of reduction of the yield loss incurred by farmers when they experience a biotic or abiotic constraint is based on an econometric model that links yield loss, farmers' profits and village poverty levels. For this purpose, research to address a biotic or abiotic constraint is assumed to lead to some percentage reduction in yield loss caused by the constraint when it occurs. The impact on income and production is assessed directly at the farmer level, while the impact on poverty is done at village level (see Appendix 4.2).

We use an autoregressive model for each outcome (farmer income and village poverty headcount) with the contemporaneous yield loss of each constraint to assess the impact of research technology addressing the constraint. The general form of the equation estimated is $E(Z_t \mid Z_{t-1}, e^s, \Delta y^s, x) = \alpha Z_{t-1} + \beta e^s \Delta y^s + \gamma e^s + \sigma(x)$, where Z_t is the outcome variable at year t and Z_{t-1} that for the preceding year; Δy^s stands for the yield loss caused by the constraint s; e^s is a binary variable indicating the experience or not of a given constraint ($e^s = 1$ and $e^s = 0$ indicate the experience and the non-experience of the constraint by a farmer, respectively); x is a vector of covariates that encompasses household socio-demographic and economic characteristics (the head of household's age, gender, education level and occupational status; household size, farm size, community infrastructures, etc.); and α, β, γ and σ are the model parameters to be estimated (with the parameter σ being modelled as a function of x). For the purpose of this estimation, we assume the yield loss to be exogenous to the farmer's decision, thus ordinary least squares estimation would yield consistent parameter estimates.

Following Diagne *et al.* (2012), a first-order autoregressive (AR1) model is used to estimate impact parameters and project annual individual impact over time up to 2035, in line with the 25-year vision of success of the Global Rice Science Partnership (GRiSP; IRRI *et al.*, 2010). The estimated impact parameters are combined with secondary data on the number of rice-farming households and the average household size to get aggregated impact at country level under the assumption that the technology adoption follows a logistic diffusion curve (for more details see Diagne *et al.*, Chapter 32, this volume). The attainable yield loss reduction by research for each constraint and each environment was obtained through a consultation process involving rice scientists at AfricaRice. For each constraint, assumptions on initial and peak adoption rates with the respective year they may be reached are also used.

Results

The estimations of the five constraint indicators in the different rice environments as perceived by the rice farmers in our sample are presented in Tables 4.1 (for the biotic constraints), 4.2 (for the soil-related abiotic constraints) and 4.3 (for the climate-related constraints).

Perception, experience and effects of major biotic constraints

About 76% of farmers (see Table 4.1) reported having experienced (in 2008) at least one of the biotic constraints in the list of the five major bio-physical constraints identified in the village. The proportion was 60% for irrigated, 81% for upland and rainfed lowland, and 88% for other rice environments. When these constraints occurred, about 30% of the harvested areas were affected, causing on average 22% of yield loss across all rice environments and all countries. The area affected and the yield losses due to biotic constraints varied slightly across rice environments. The estimated areas affected and yield losses were, respectively, 29% and 21% in irrigated, 30% and 23% in upland, 30% and 21% in rainfed lowland, and 32% and 19% in other environments.

Among biotic constraints, weed infestation was ranked as the most important by far, followed by birds and rodents, the various insects and rice diseases (Table 4.1).

Table 4.1. Farmers' perceptions of the relative importance of biotic constraints across rice environments in 18 countries in sub-Saharan Africa.

Biotic constraints	Percentage of farmers assigning a given rank to the constraint (f_r^s)				Normalized average rank given to the constraint by farmers ($\bar{\bar{r}}^s$) (scale: 0–100)	Percentage of farmers who had experienced the constraint ($E^{s,2008}$)	Average percentage of areas affected by the constraint when experienced ($A^{s,2008}$)	Average percentage yield reduction caused by the constraint when experienced ($\Delta y^{s,2008}$)
	None (r=0)	Minor (r=1)	Intermediate (r=2)	Major (r=3)				
All rice environments								
All biotic constraints					93	76	30	22
Weeds	2	8	20	70	86	53	33	22
Birds and rodents	4	14	24	59	79	45	29	21
Diseases					61	17	25	20
Bacterial leaf blight	36	25	24	15	39	5	22	17
Blast	29	26	24	21	46	7	28	22
Rice yellow mottle virus (RYMV)	23	30	27	20	48	4	23	19
Insects					76	50	27	20
African rice gall midge	34	16	19	30	49	14	27	20
Stem borers	24	32	25	19	46	11	29	23
Termites	18	24	28	30	56	14	31	23
Other insects	6	26	31	37	66	29	26	19
Irrigated								
All biotic constraints					93	60	29	21
Weeds	2	9	27	62	83	39	27	18
Birds and rodents	4	16	22	58	78	31	30	19
Diseases					56	15	26	19
Bacterial leaf blight	34	30	23	13	39	4	21	16
Blast	36	28	24	12	38	5	32	19
RYMV	39	25	22	14	37	4	30	25
Other diseases	41	29	19	11	33	4	23	21
Insects					66	37	30	22
African rice gall midge	57	10	19	14	30	4	36	29
Stem borers	32	31	24	14	40	9	35	25
Other insects	12	24	36	29	61	25	29	21
Upland								
All biotic constraints					96	81	30	23
Weeds	1	6	14	79	90	54	32	22
Birds and rodents	2	8	19	71	87	51	32	24
Diseases					66	20	22	19

Continued

Table 4.1. Continued.

Biotic constraints	Percentage of farmers assigning a given rank to the constraint (f_r^s) None (r=0)	Minor (r=1)	Inter-mediate (r=2)	Major (r=3)	Normalized average rank given to the constraint by farmers ($\bar{\bar{r}}^s$) (scale: 0–100)	Percentage of farmers who had experienced the constraint ($E^{s,2008}$)	Average percentage of areas affected by the constraint when experienced ($A^{s,2008}$)	Average percentage yield reduction caused by the constraint when experienced ($\Delta y^{s,2008}$)
Bacterial leaf blight	43	19	19	19	38	5	16	15
Blast	31	23	22	24	46	10	26	22
Other diseases	42	23	18	17	36	7	17	15
Insects					83	55	27	21
African rice gall midge	32	11	15	42	55	21	25	19
Stem borers	25	33	19	23	46	14	27	24
Termites	11	19	27	43	67	16	30	24
Rainfed lowland								
All biotic constraints					92	81	30	21
Weeds	2	9	20	69	86	61	36	23
Birds and rodents	5	19	28	48	73	46	28	21
Diseases					58	14	23	18
Bacterial leaf blight	38	29	21	12	36	4	17	16
Blast	29	29	21	21	45	4	25	18
RYMV	25	34	23	18	45	4	23	18
Other diseases	34	31	21	15	39	5	27	23
Insects					73	53	26	19
African rice gall midge	34	24	17	25	44	13	26	19
Stem borers	24	35	23	18	45	10	31	22
Termites	17	32	27	24	53	14	28	22
Others								
All biotic constraints					93	88	32	19
Weeds	0	5	41	54	83	56	38	21
Birds and rodents	0	2	29	69	89	64	25	17
Diseases					76	26	41	23
Bacterial leaf blight	2	7	68	23	71	13	51	26
Blast	1	8	68	23	71	14	28	18
Other diseases	1	8	62	29	73	2	34	31
Insects					84	62	34	18
African rice gall midge	0	3	54	43	80	18	37	21
Stem borers	1	3	71	25	73	9	24	13
Other insects	0	6	44	50	81	35	24	14

Table 4.2. Farmers' perceptions of relative importance of soil-related abiotic constraints across rice environments in 18 countries in sub-Saharan Africa.

Soil-related constraints	Percentage of farmers assigning a given rank to the constraint (f_r^s)				Normalized average rank given to the constraint by farmers ($\bar{\bar{r}}^s$) (scale: 0–100)	Percentage of farmers who have experienced the constraint ($E^{s,2008}$)	Average percentage of areas affected by the constraint when experienced ($A^{s,2008}$)	Average percentage yield reduction caused by the constraint when experienced ($\Delta y^{s,2008}$)
	None (r=0)	Minor (r=1)	Medium (r=2)	Major (r=3)				
All rice environments								
All soil-related constraints					77	18	37	27
Non-fertile soil	14	24	33	30	59	6	45	32
Salinity/alkalinity	39	20	22	19	41	1	32	28
Iron toxicity	43	20	19	18	37	2	36	32
Irrigated								
All soil-related constraints					73	19	36	30
Non-fertile soil	12	26	36	27	59	5	43	29
Salinity/alkalinity	53	20	16	11	28	7	34	31
Iron toxicity	47	24	17	13	32	2	30	29
Upland								
All soil-related constraints					80	18	34	25
Non-fertile soil	15	21	28	36	62	7	43	32
Rainfed lowland								
All soil-related constraints					76	17	37	28
Non-fertile soil	15	26	33	27	57	7	45	34
Iron toxicity	42	22	18	18	37	2	38	34
Others								
All soil-related constraints					78	20	50	26
Non-fertile soil	54	34	5	7	11	7	59	27
Salinity/alkalinity	65	23	3	9	9	3	28	15

Weeds

Weeds cause economic losses to agricultural crops, and require some action to reduce their effects on crop production (Zimdahl, 2007). Various categories of rice weeds are common in rice production in Africa, including sedges, broad-leaved species, grasses, parasitic weeds and aquatic weeds (see Rodenburg and Johnson, Chapter 16, this volume). Weeds are the predominant biotic constraint, as perceived by farmers. An estimated 70% of farmers perceive weeds as a major problem across the rice environments, with the highest percentage reported for upland areas (79%). Similar reports were obtained with the perception index. An estimated 53% of rice farmers experienced weed

Table 4.3. Farmers' perceptions of relative importance of climate-related abiotic constraints across rice environments in 18 countries in sub-Saharan Africa.

Climate-related constraints	Percentage of farmers assigning a given rank to the constraint (f_r^s) None (r = 0)	Minor (r = 1)	Intermediate (r = 2)	Major (r = 3)	Normalized average rank given to the constraint by farmers ($\bar{\bar{r}}^s$) (scale: 0–100)	Percentage of farmers who experienced the constraint ($E^{S,2008}$)	Average percentage of areas affected by the constraint when experienced ($A^{S,2008}$)	Average percentage yield reduction caused by the constraint when experienced ($\Delta y^{s,2008}$)
All rice environments								
All climate-related constraints					76	24	36	27
Drought	16	24	30	30	58	10	37	29
Flooding	25	26	24	25	50	5	37	27
Heat	30	33	23	13	40	1	28	16
Cold	35	30	20	16	39	1	33	29
Irrigated								
All climate-related constraints					75	17	35	28
Drought	20	24	30	26	54	7	31	28
Flooding	26	24	26	24	49	6	40	34
Heat	37	34	19	10	35	1	36	17
Cold	27	37	18	18	43	2	24	16
Upland								
All climate-related constraints					77	25	35	28
Drought	17	18	31	35	61	13	38	31
Flooding	29	19	24	27	50	4	34	25
Heat	28	29	27	16	44	1	30	11
Cold	37	24	22	16	39	1	37	28
Rainfed lowland								
All climate-related constraints					75	28	37	26
Drought	15	30	27	28	56	11	38	28
Flooding	25	33	20	22	46	6	38	27
Heat	35	39	17	9	33	2	26	17
Cold	43	33	13	11	31	1	41	41
Others								
All climate-related constraints					80	16	25	18
Drought	1	5	66	27	73	2	29	28
Flooding	2	4	54	41	78	5	16	13
Heat	1	8	63	28	72	0.1	0	0
Cold	1	4	68	26	73	5	0	0

infestation across rice environments, affecting 33% of their rice area, causing 22% of rice yield loss in 2008. The area affected by weed infestation ranged from 27% in irrigated environments to 38% in other environments, while yield losses ranged from 18% in irrigated environments to 22% in uplands.

Comparison across countries (data not shown) indicated that the highest proportions of farmers experiencing weed infestation were observed in Madagascar (82%), Togo (78%), DRC (77%), Uganda (76%), Côte d'Ivoire (74%) and Burkina Faso (72%). The largest proportions of fields affected were reported in Burkina Faso (57%), Côte d'Ivoire (49%), Togo (43%), Kenya (41%) and Benin (40%). Greatest yield losses were observed in Kenya (43%) and Côte d'Ivoire (40%).

Birds and rodents

Birds feed on rice grains before germination, during crop establishment and during grain filling (see also de Mey and Demont, Chapter 19, this volume). Rodents cause damage at all stages of rice cultivation. Bird and rodent attacks were perceived as a major biotic constraint, coming in second after weed infestation. About 59% of farmers ranked bird and rodent attacks as a major constraint across rice environments, 24% of farmers considered the attacks of intermediate importance, and 14% considered them as a minor constraint. About 45% of rice farmers experienced bird and rodent attacks in 2008, affecting 29% of the area and leading to an estimated 21% yield loss.

About 31% of irrigated-rice farmers experienced losses to birds and rodents in 2008 compared to 64% in other environments. The proportion of the areas affected and yield loss caused by birds and rodents varied little across rice environments. There were, however, noticeable differences across countries (data not shown). The proportion of farmers reporting bird and rodent problems in 2008 ranged from 5% in Rwanda to 91% in Nigeria. The percentage area affected ranged from 8% in Guinea to 51% in Kenya, while the yield loss ranged from a low of 9% in DRC and Guinea to up to 44% in Kenya.

Diseases

There are many diseases that affect rice plants (see Séré *et al.*, Chapter 17, this volume). In this analysis, we focused on bacterial leaf blight, blast and *Rice yellow mottle virus* (RYMV). The normalized average score for all diseases across rice environments is estimated to be 61%, which corresponds to more than the intermediate rank. This means that farmers perceive diseases to be of high relative importance. The average normalized score of all diseases is relatively high in upland environments (66%), meaning that they are perceived to be of high relative importance in uplands. About 17% of rice farmers experienced at least one disease in 2008, affecting 25% of the area and leading to an estimated 20% yield loss.

An assessment based on the perception index shows that bacterial leaf blight is the most important disease in irrigated environments and RYMV and blast in rainfed lowlands.

Looking at differences across countries (data not shown), 71% of Guinean farmers experienced diseases as a major constraint in 2008, followed by CAR (63%) and Madagascar (41%). The area affected by diseases ranged from a low 10% in Guinea to a high of 50% in Kenya, and yield loss ranged from 2% (The Gambia) to 47% (Kenya).

Insects

Several insect species attack the rice plant during its growth stage (see Nwilene *et al.*, Chapter 18, this volume). In this analysis, we focused on African rice gall midge, stem borers and termites. In 2008, insect attacks were experienced least in the irrigated environments (37%) and most often in the other rice environments (62%). About half of rice farmers experienced at least one insect attack across rice environments in 2008, affecting 27% of the area and leading to an estimated 20% yield loss.

Differences between countries were relatively large (data not shown). An estimated 84% of farmers from Sierra Leone and Guinea experienced at least one insect attack in 2008, followed by Burkina Faso (72%) and Madagascar (62%). The average area affected by this constraint ranged from a low 11% in Guinea to a high of 49% in Kenya. The yield loss caused ranged from a low of 8% in DRC to 45% in Kenya.

Perception, experience and effects of major abiotic constraints

Abiotic constraints are dealt with in two groups: soil-related abiotic constraints and climate-related abiotic constraints.

Soil-related constraints

Soil-related constraints included in this survey included: low soil fertility (poor soil fertility, deficiencies in soil macro-nutrients [N, P, K] and Zn, and acidity), salinity and alkalinity, Fe toxicity and other soil-related constraints (soil erosion).

Across all rice-growing environments, the normalized average score for soil-related constraints is about 77%, which means that farmers perceive soil-related constraints to be of high relative importance. According to the perception index, soil-related problems are more important in the uplands than in the irrigated or rainfed lowland environments.

About 18% of rice farmers reported having experienced at least one soil-related constraint as a major constraint in 2008, affecting 37% of their rice area, leading to a yield loss of 27%. The proportion of farmers who had experienced soil-related constraints was 19% in irrigated, 18% in upland and 17% in rainfed lowland. The area affected by these constraints by rice environment ranged from 34% in upland to 50% in other environments. The resulting yield losses ranged from 25% in upland to 30% in irrigated environments.

In Burkina Faso, Senegal and Togo, almost 37% of rice farmers reported major soil problems, both in terms of the proportion of farmers perceiving them of major importance and the proportion of farmers having experienced at least one soil constraint in 2008. For the other countries, this proportion ranged from 0.4% in Rwanda to 31% in The Gambia. The highest share of area affected was 56% (observed in Burkina Faso) and the lowest proportion of area affected was 1% (in Rwanda). The minimum yield loss recorded was 6% in CAR and maximum yield loss was 52% in Kenya.

Drought

An estimated 30% of farmers perceived drought as a major problem across rice environments, with the highest percentage reported for upland areas (35%). Across rice environments, an estimated 10% of rice farmers experienced drought affecting 37% of their rice area, causing 29% of rice yield loss in 2008. Upland and rainfed-lowland rice farmers were most affected by drought in 2008 (11% or more).

Comparison of individual countries (data not shown) indicated that the highest proportions of farmers experiencing drought were in Rwanda (45%), followed by Cameroon (30%) and Burkina Faso (28%). The largest proportion of fields affected was reported in Senegal (51%), followed by Burkina Faso and The Gambia (46%), Benin and Côte d'Ivoire (44%). Greatest yield losses were observed in The Gambia (46%), followed by Senegal (45%) and Côte d'Ivoire (41%).

Flooding

An estimated 25% of farmers perceived flooding as a major problem across rice environments. An estimated 5% of rice farmers experienced flooding across rice environments, affecting 37% of their rice area, causing 27% of rice yield loss in 2008.

Comparison of individual countries (data not shown) indicated that the highest proportion of farmers experiencing flooding was in Kenya (17%), followed by Burkina Faso (11%), Benin and Togo (9%). The largest proportion of fields affected was reported in Burkina Faso (56%), followed by Cameroon (53%), Togo (50%) and Côte d'Ivoire (47%). Greatest yield losses were observed in Rwanda (53%), followed by Kenya (45%), Côte d'Ivoire (42%), Burkina Faso (40%) and Benin (38%).

Extreme temperatures

Very few farmers reported extreme temperatures (cold or heat) across rice environments and countries.

Potential Impact of Research Addressing Biophysical Constraints

We assessed the potential impact of rice research on the biophysical constraints facing rice farmers. The assessments assumed that research targeting a particular biotic or abiotic constraint will generate technological options which, if adopted by farmers, will reduce yield loss due to the occurrence of the constraint by a reasonable magnitude. Assumptions on reduction in yield loss that can reasonably be expected were derived from scientific expert opinions taking into account the presently observed average losses (from the survey) and the chances of finding solutions through research (Table 4.4).

The relative magnitude of yield loss reduction expected from research ranged from 20% (for birds and rodents) to 35% (for weeds and soil-related constraints). Technology options to reduce yield loss due to biophysical constraints are expected to be available for farmer use starting in 2014. The adoption of these technologies is assumed to follow a logistic diffusion curve (see Appendix 4.2 for details on the logistic diffusion curve parameters). The assumed starting adoption rate is 1% for all constraints and the peak ranges from 25% for solutions that address insects to 45% for solutions that address diseases. Peak adoption is assumed to be reached after 2025 for most solutions with half of this peak reached around 2020–2023.

Individual impact on farmer's income and village poverty headcount

Appendix Tables 4.A.2.1 and 4.A.2.2 present the estimation of the farmer income and village poverty rate determination models, respectively. Only the estimates corresponding to the parameters α, β and γ are reported to save space. For the farmer income model (Table 4.A.2.1), the estimated values for the β coefficients that measure the effects of yield loss on farmer income range from –2.78 for diseases to –1.06 for birds and rodents,

Table 4.4. Characteristics of scientific options to reduce yield loss due to biophysical constraints and adoption patterns (start of simulation is 2010).

Characteristic	Weeds	Insects	Birds and rodents	Diseases	Soil-related constraints	Climate-related constraints
Yield loss						
Actual yield loss, AL (%)	22	20	21	20	27	27
Relative reduction, R (fraction between 0 and 1)	0.35	0.25	0.20	0.25	0.35	0.25
Absolute reduction, AR=AL×R (%)	8	5	5	5	10	7
Remaining yield loss after technology adoption, RL= AL–AR (%)	14	15	17	15	18	20
Logistic adoption rate curve parameters						
Adoption rate in the first year (%)	1	1	1	1	1	1
No. adopters in the first year of adoption (thousands)	72	68	62	23	24	33
Starting year of adoption	2014	2014	2014	2014	2014	2014
Peak adoption rate	0.35	0.25	0.30	0.45	0.40	0.35
Peak no. adopters (million)	4.97	1.93	4.5	4.29	2.14	1.93
Year of peak adoption rate	2025	2025	2030	2025	2027	2025
Year when adoption rate is half of peak adoption	2020	2023	2020	2020	2022	2022
Logistic growth parameter	0.29	0.18	0.30	0.31	0.14	0.22

and they are all statistically significantly different from zero at the 10% level except for the ones for diseases and climate-related constraints. The estimated lagged income effects as measured by the α coefficients are all significantly different from zero at the 1% level with estimated values ranging from 0.93 for climate-related constraints to 0.81 for birds and rodents. In addition, the total effect of the experience of the constraint is negatively related to the income as expected and statistically significantly different from zero at the 10% level except for diseases and climate-related stresses.

For the village poverty rate determinants models (Table 4.A.2.2), the estimated effects of yield loss on village poverty rate range from 0.17 for soil-related constraints to 0.02 for birds and rodents and insects, with only the coefficient for weeds being statistically significantly different from zero. The estimated lagged village poverty effects are very high and statistically significantly different from zero at the 1% level, with values ranging from 0.95 for weeds to 0.81 for birds and rodents. In addition, the total effect of the experience of the constraint is positively related to the village poverty rate as expected, but not statistically different from zero except for weeds.

The estimated coefficients of the models are used to forecast the individual impact over time as described by the formula in Appendix 4.2. The results represent the change over time in the income and poverty headcount as a result of the adoption of technology options generated by research to help farmers mitigate the effects of the different constraints. Average increases in household total annual income from research to mitigate the effect of the losses are estimated in the starting year to be $37 for weed infestation, $21 for insects, $16 for birds and rodents, $58 for diseases, $54 for soil-related constraints and $33 for climate-related constraints. Average annual increase in household income grows over time to reach at least $25 from technology options that mitigate bird and rodent damage, up to $140 from technology solutions to mitigate diseases (with $110 for soil-related stresses, $91 for climate-related stresses, $80 for weeds and $40 for insect attacks).

In terms of poverty reduction, the village poverty headcount reductions range from 0.33% from bird and rodent technology options to 5.9% from soil-related technology options. This reduction in poverty will grow to reach 0.5% from technology options dealing with bird and rodent attacks, 9.7% for options that alleviate soil-related constraints, 4.5% for options that alleviate weed infestation and 5.5% for options that alleviate climate-related constraints in 2035.

These estimated average household income benefits and village poverty rate reductions were combined with the projection of adoption of the technology options based on the logistic diffusion model to obtain the estimates of aggregate gross impact of adoption of the technology options generated by rice research that addresses the various biophysical constraints (Figs 4.1, 4.2 and 4.3).

Aggregate impact on farmer's income

Figures 4.1, 4.2 and 4.3 present the annual nominal income benefits, the discounted annual income benefits and the discounted cumulated income benefits ($ millions), respectively.

The expected average annual nominal income gain from research addressing all constraints in the 36 rice-producing countries considered in our analysis is $14 million in 2014 and growing to reach $917.3 million in 2035. This corresponds to an average annual aggregate nominal income gain of about $344 million per year for the period 2014–2035. The discounted value of these benefits is $11 million in 2014 and $258 million in 2035, with an average discounted annual benefit for the period 2014–2035 of $127 million. By aggregating the discounted annual income benefits across time we obtain a total aggregated discounted income benefit of $2.8 billion by 2035.

Comparing the different constraints, the highest impact is given by research addressing weed infestation (due to higher number of adopters), with an average annual nominal income benefit of $118 million and a cumulated discounted income benefit over the period 2014–2035 of $947 million. This is followed by research addressing rice diseases, with an average annual nominal income benefit of $83 million ($667 million cumulative discounted benefit over the period 2014–2035). Research that mitigates soil-related constraints comes in third position, with an average nominal income benefit of $41 million ($322 million discounted

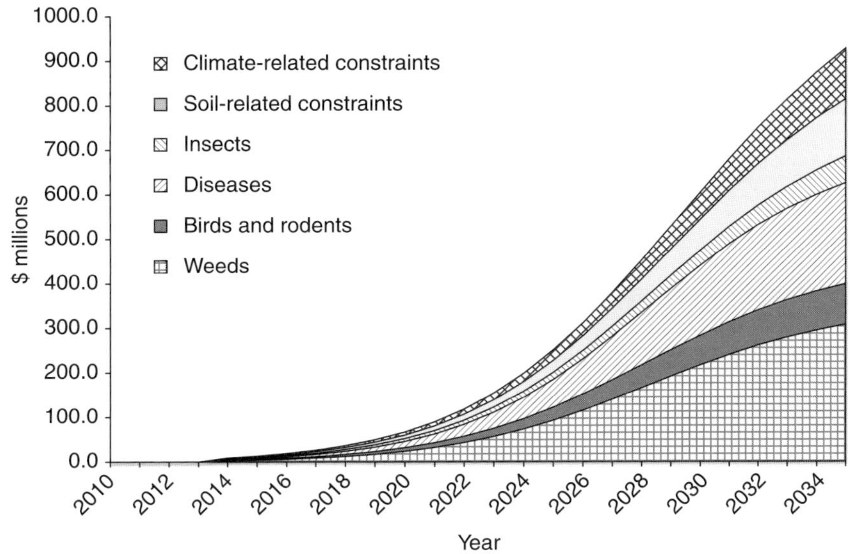

Fig. 4.1. Annual nominal income benefits of research addressing biophysical constraints.

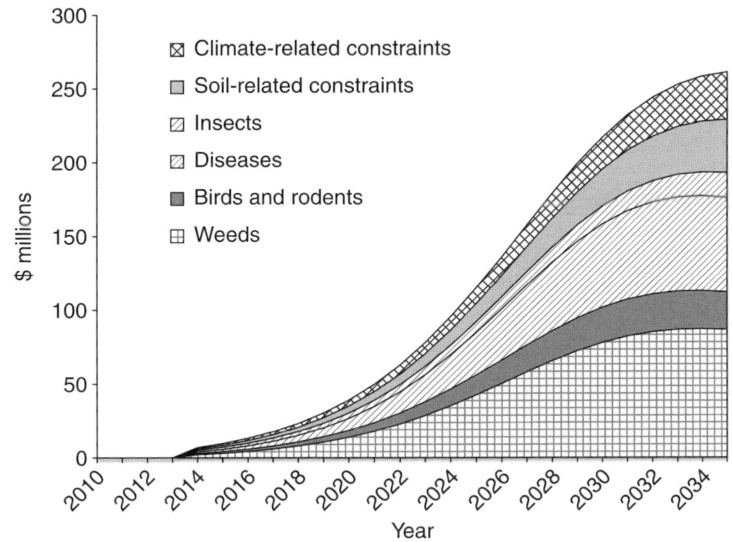

Fig. 4.2. Discounted annual income benefits of research addressing biophysical constraints (5% discounted).

cumulative benefits for the period 2014–2035). Research that mitigates bird and rodent attacks comes in fourth position, with an average nominal income benefit of $36 million and a discounted cumulative value of $292 million for the period 2014–2035, and in fifth position research that alleviates climate-related constraints, with an average annual income gain of $36 million ($281 million cumulative over the period 2014–2035). In comparison, the income benefits derived from research addressing insects will be substantially lower, with $21 million (corresponding to $171 million cumulative benefits over the period 2014–2035).

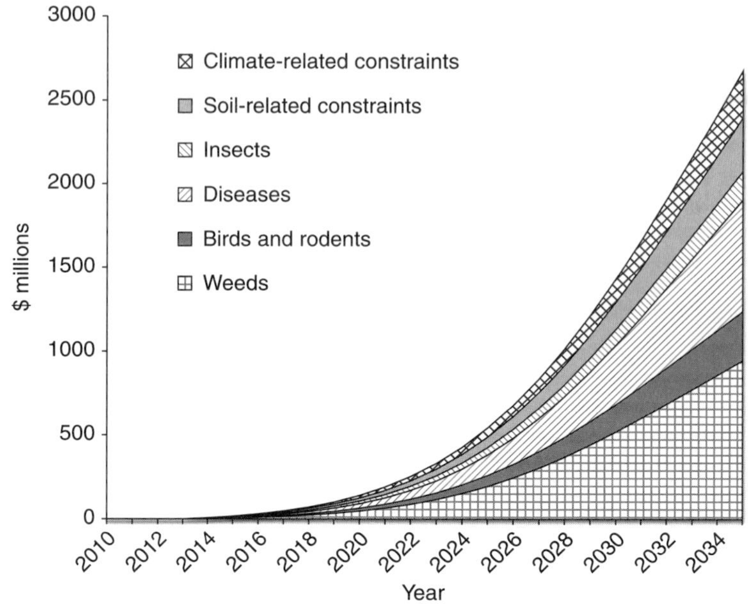

Fig. 4.3. Cumulated discounted income benefits of research addressing biophysical constraints (5% discounted).

Aggregate impact on poverty reduction

The poverty reduction resulting from the increase in rice-farming household incomes as a result of adopting technologies to address biophysical constraints are presented in Fig. 4.4.

It is expected that at least 12,162 people will be lifted above the poverty line in 2014, the starting year of adoption of the technological options resulting from rice research addressing all the biophysical constraints in Africa. This number will grow rapidly to reach 1.5 million in 2035. These numbers represent the maximum reductions in the number of poor farmers each year across all technology options addressing the different biotic and abiotic constraints (taking into account the fact that a person is lifted out of poverty only once even if he or she experiences an income gain from more than one technology option).[5] Although research addressing soil-related constraints will initially lead to the highest reduction in poverty, technology options that mitigate the effects of weed infestation are the ones that will eventually achieve the highest poverty reduction (starting around 2022, because of higher growth in its number of adopters; reaching the maximum 1.5 million lifted out of poverty without the other constraint-mitigating technology options). This is followed by technology addressing soil-related constraints starting in 2028 (and which without the weed technology options would have lifted 0.9 million people out of poverty in 2035), and by those addressing climate-related constraints starting in 2030 (with 0.6 million people lifted out of poverty in 2035 without technologies addressing weed infestation and diseases). Thus, technology options addressing soil-related constraints moved from achieving the highest poverty reduction numbers in 2014 to being second in 2035 because of relatively slower growth in its number of adopters. Research addressing insects and birds and rodents are those that consistently deliver the lowest poverty reduction numbers without the other technologies across the years (leading respectively to 0.18 million and 0.16 million people they could lift out of poverty in 2035 without the other technology options).

Our estimates are slightly higher than the ones reported by IRRI *et al.* (2010), who conducted a similar analysis for Africa using data from a similar but different set of surveys. However, our analysis is based on data from 18 countries and extrapolates the results to 36 rice-producing countries, while the IRRI *et al.* (2010) analysis used data from 12 of the 18 countries

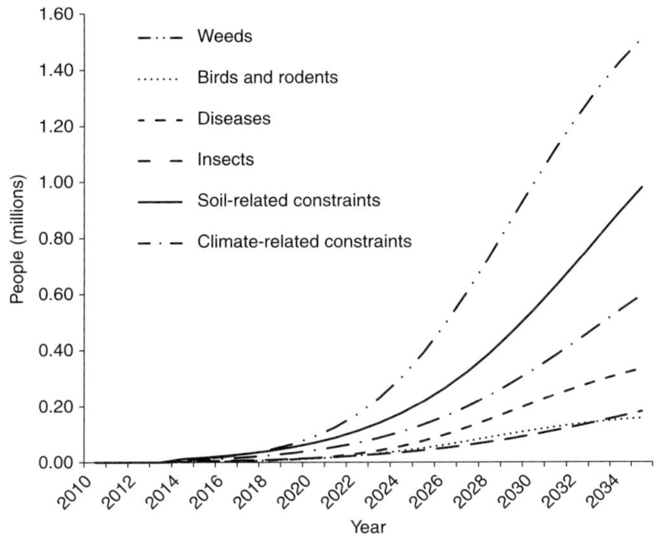

Fig. 4.4. Number of people lifted out poverty (PPP $1.25 poverty line) by research addressing biophysical constraints.

and extrapolated the results to 31 countries only. Nevertheless, the results from the two studies are qualitatively similar. In both studies, alleviation of weed infestations provides the highest impact, followed by alleviation of diseases.

Conclusions

In this chapter, we have analysed the biophysical constraints in rice production in 18 major rice-producing countries in Africa using survey data of farmers' perceptions of the occurrence and relative importance of these constraints collected in 2009. The biophysical constraints were grouped into three broad categories: biotic constraints, soil-related constraints, and climate-related constraints. From the results of the analysis, a large proportion of farmers across all countries and rice environments perceive several biophysical constraints as major production constraints. The average area affected by all biophysical constraints across all rice environments and all countries in 2008 was 30%, with 22% of yield loss. These effects vary significantly across constraints, rice environments and countries. Across all categories of biotic and abiotic constraints, the proportion of farmers giving a high rank to a constraint was consistently highest in Burkina Faso and Madagascar, while the area affected and yield loss incurred was consistently highest in Kenya.

The biotic constraints appear to be the most important biophysical constraints perceived by farmers. Among these biotic constraints, weed infestation is the most important, followed by insects and birds and rodents. It is perhaps not surprising that these constraints are all very 'visible' to the farmers and involve major mitigation efforts. The importance of weeds as major rice production constraints has been reported in several previous studies. Somado *et al.* (2008) also found weeds to be the most important biophysical constraint in SSA, with annual losses estimated at around 2.2 million tonnes (Mt). The losses have been reported to vary from 30% to 100% according to locality (Balasubramanian *et al.*, 2007; Rodenburg and Demont, 2009; Rodenburg and Johnson, Chapter 16, this volume). Damage by birds is also considered to be important (de Mey and Demont, Chapter 19, this volume). Additionally, stem borers, and bacterial and virus diseases are reported to be major biotic constraints that significantly reduce rice productivity (Seck *et al.*, 2012). Among the abiotic constraints, poor soil fertility is the most important soil-related constraint, while drought and flooding are the most important climate-related constraints. This latter finding concurs with Balasubramanian *et al.* (2007).

The *ex-ante* evaluation of the impact of adoption of technology options from research addressing the different biophysical constraints in rice production showed great potential impact in terms of both additional income generated and poverty reduction. These impact studies were based on farmer perceptions of yield-limiting and yield-reducing factors in rice cropping in SSA. It is this subjective perception, right or wrong, that drives a farmer's decision-making. Hence, in the end the **real effects** of the constraint on a farmer's productivity outcomes will be the result of both the **real importance** (as it would be measured objectively by yield-gap surveys as proposed by Saito *et al.*, Chapter 15, this volume) and the farmer's **perception** of that importance (reflected in the *ex-ante* preventive and *ex-post* mitigating actions the farmer takes in relation to the constraint). Thus, in determining the relative importance of a constraint in terms of its negative effects on rice productivity, both the objective assessments of knowledgeable experts and the subjective assessments of farmers are important and complementary.

Notes

[1] 'Average yield' calculated from total annual paddy production divided by total annual harvested area in 2012.

[2] See, however, discussion later in the chapter on the inherent population and sample selection biases in the indicators derived below caused respectively by unawareness of the existence of some constraints by some farmers and the restrictive nature of the second list of constraints.

[3] The $\bar{\bar{r}}^s$ index can also be viewed as a weighted average of the ranking of the constraint by farmers in terms of relative importance with the weight of a rank value corresponding to the proportion of sample farmers giving that rank value to the constraint: $\bar{\bar{r}}^s = \frac{1}{3n} \sum_{i=1}^{n} \sum_{r=0}^{3} r d_{ir}^s = \frac{1}{3} \sum_{r=0}^{3} r f_r^s$. Hence, the $\bar{\bar{r}}^s$ index is an aggregation of two measures of relative importance of a constraint: (i) a measure of its perceived incidence, scope and severity at the individual farmer level as indicated by the rank r; and (ii) a measure of the prevalence of its occurrence (in the village) being perceived by farmers as indicated by the proportion f_r^s of farmers giving it the rank r.

[4] In fact, positive monotonicity is trivially satisfied by the ordered rank measure. To see why, we note that only farmers who are aware of the existence of the constraint s are able to give their ordered ranks for s. Hence, r_i^s can be observed only for farmers who are aware of the existence of constraint s. In other words, r_i^s is missing for s-constraint unaware farmers. For this group of farmers, it is as if the constraint s did not occur in their villages or anywhere else (since for them the constraint does not exist). That is, it is as if the observed rank given to s by a farmer in this s-constraint unaware group is equal to 0, the rank corresponding to the non-occurrence of the constraint in his or her village, which is the lowest possible value for r_i^s. This may not be the case, however, for other measures of relative importance. Indeed, if we take the experience measure of relative importance, in principle a farmer may experience a constraint without being aware of its existence. Although this is unlikely for most constraints, it is still possible for constraints that are difficult for farmers to diagnose. Even in these cases, it is safe to assume that positive monotonicity is satisfied.

[5] In other words, the number of people lifted out of poverty cannot be aggregated across the various technology options (this would be double counting). For the same reason, the poverty reduction numbers are not cumulative across time.

References

AfricaRice (2010) Rice data systems for sub-Saharan Africa: Contribution to the Japan–AfricaRice Emergency Rice Project Synthesis Report. Africa Rice Center, Cotonou, Benin, p. 73.

AfricaRice (2011) *Boosting Africa's Rice Sector: A research for development strategy 2011–2020.* Africa Rice Center, Cotonou, Benin.

Balasubramanian, V., Sie, M., Hijmans, R.J. and Otsuka, K. (2007) Increasing rice production in sub-Saharan Africa: opportunities and challenges. *Advances in Agronomy* 94, 55–133.

Diagne, A. (2006) Diffusion and adoption of NERICA rice varieties in Côte d'Ivoire. *The Developing Economies* 44(2), 208–231.
Diagne, A. (2009) Technological Change in Smallholder Agriculture: Bridging the adoption gap by understanding its source. CEGA Working Paper No. AfD-0905. Center of Evaluation for Global Action, University of California, Berkeley, California.
Diagne, A. and Demont, M. (2007) Taking a new look at empirical models of adoption: average treatment effect estimation of adoption rates and their determinants. *Agricultural Economics* 37, 201–210.
Diagne, A., Alia, D.Y., Wopereis, M.C.S. and Saito, K. (2012) Impact of rice research on income and poverty in Africa: an *ex-ante* analysis. Paper presented at the International Association of Agricultural Economics 28th Triennial Conference, Foz do Iguacu, Brazil, 18–24 August.
Evans L.T. (1993) *Crop Evolution, Adaptation and Yield.* Cambridge University Press, Cambridge.
FAOSTAT (2010) http://faostat.fao.org/site/452/default.aspx (accessed 5 November 2010).
FAOSTAT (2012) http://faostat.fao.org/site/452/default.aspx (accessed 12 February 2012).
Imbens, G.M. and Woodridge, J.M. (2009) Recent developments in the econometrics of program evaluation. *Journal of Economic Literature* 47, 5–86.
IRRI, AfricaRice and CIAT (2010) *Global Rice Science Partnership (GRiSP).* International Rice Research Institute, Los Baños, Philippines; Africa Rice Center, Cotonou, Benin; and International Center for Tropical Agriculture, Cali, Colombia.
Just, R.E, Hueth, D.L. and Schmitz, A. (2004) *The Welfare Economics of Public Policy: A practical approach to project and policy evaluation.* Edward Elgar, London.
Marshall, A. (1930) *Principles of Economics.* Macmillan, London.
Rodenburg, J. and Demont, M. (2009) Potential of herbicide-resistant rice technologies for sub-Saharan Africa. *AgBioforum* 12(3&4), 313–325.
Rubin, D. (1974) Estimating causal effects of treatments in randomized and non-randomized studies. *Journal of Educational Psychology* 66, 688–701.
Seck, P.A., Diagne, A., Mohanty, S. and Wopereis, M.C.S. (2012) Crops that feed the world 7: rice. *Food Security* 4(1), 7–24.
Somado, E.A., Guei, R.G. and Keya, S.O. (2008) *NERICA: The New Rice for Africa – A Compendium.* Africa Rice Center (WARDA), Cotonou, Benin.
USDA (2013) Production, supply and distribution online. Foreign Agricultural Service, United States Department of Agriculture. www.fas.usda.gov/psdonline/psdQuery.aspx (accessed 7 February 2013).
World Bank (2010) *World Development Indicators 2010.* The World Bank, Washington, DC. Available at: http://data.worldbank.org/data-catalog/world-development-indicators (accessed 2 December 2010).
Zimdahl, R.L. (2007) *Fundamentals of Weed Science.* Academic Press, London.

Appendix 4.1. The Population Non-Awareness and Sample Minority-Exclusion Biases in the Farmers' Perceptions of the Relative Importance of Biotic and Abiotic Constraints

In this appendix, we derive the expressions and signs of the non-awareness and sample minority-exclusion biases in the five measures of farmers' perceptions of the relative importance of biotic and abiotic constraints.

Population non-awareness bias

For this purpose, we consider the full population and use expected population values instead of sample averages as in the main text. Also, for any constraint $s \in \tilde{S}$ we will use the random variable x^s to denote generically any of the perception measures of relative importance of the constraint as given by a farmer randomly selected from the population of rice farmers.[1] First, let ω^s be a dummy binary indicator, with $\omega^s = 1$ if the farmer is aware of the existence of the constraint s, $\omega^s = 0$ otherwise and, following the counterfactual outcome framework, define the two potential outcomes x_1^s and x_0^s of x^s associated with the two mutually exclusive states of awareness and non-awareness of the existence of the constraint, respectively by: $x^s = x_1^s$ if $\omega^s = 1$ and $x^s = x_0^s$ if $\omega^s = 0$. The two potential outcomes x_1^s and x_0^s are supposed to exist for any farmer in the population. However, for any given farmer, we can only observe one of the two potential outcomes because a farmer can only be in one of the two mutually exclusive states (being aware or unaware).[2] In particular, the value of the perception indicator x_1^s is missing for farmers who are not aware of the existence of the constraint. Hence, we cannot directly compute $E(x_1^s)$, its expected value for the full population of rice farmers. And yet, it is $E(x_1^s)$, the mean perception of relative importance of the constraint when all farmers are aware of its existence, that truly informs on the whole population of rice farmers' perception of the relative importance of the constraint s (as measured by x^s at the individual farmer level).[3] On the other hand, the sample measures of perception of relative importance of a constraint given in points (3)–(5) in the main text will, as the sample size grows, all converge to their population counterparts corresponding each to $E(x^s)$, which is the population mean of the observed values of x^s. Consequently, the population non-awareness bias (PNAB) is defined as the difference between the expected population value, to which the directly computed sample average of the observed values of x^s converges, and the population mean perception of relative importance of the constraint when all farmers are aware of its existence. That is, $PNAB^s = E(x^s) - E(x_1^s)$.

To show that $PNAB^s$ is negative when the positive monotonicity with respect to non-awareness condition is satisfied, we decompose the unconditional expectation $E(x^s)$ into its conditional parts as:

$$E(x^s) = E(x^s|\omega^s=1)\,P_\omega^s + E(x^s|\omega^s=0)(1-P_\omega^s) = E(x_1^s|\omega^s=1)\,P_\omega^s + E(x_0^s|\omega^s=0)(1-P_\omega^s)$$

where $P_\omega^s = Prob\{\omega^s = 1\}$ is the probability that a farmer randomly selected from the population is aware of the existence of the constraint. Similarly, we decompose the unconditional expectation $E(x_1^s)$ into its conditional parts as:

$$E(x_1^s) = E(x_1^s|\omega^s=1)P_\omega^s + E(x_1^s|\omega^s=0)(1-P_\omega^s)$$

Hence, from the two equations above we obtain the expression of the PNAB as:

$$PNAB^s = E(x^s) - E(x_1^s) = -(1-P_\omega^s)\{E(x_1^s|\omega^s=0) - E(x_0^s|\omega^s=0)\}$$

This shows that PNAB is negative if we assume that the perception of relative importance of a constraint is positively monotonically related to awareness for all farmers, which means that $x_1^s \geq x_0^s$ uniformly (implying that $E(x_1^s\,|\,\omega^s=0) \geq E(x_0^s\,|\,\omega^s=0)$). Also, the above expression of the population non-awareness bias shows that it is a decreasing function of the probability of awareness and hence, everything else being equal, it is more severe for constraints the existence of which farmers are less likely to be aware of compared to those they are more likely to be aware of.

Sample minority-exclusion bias

The sample minority-exclusion bias results from the restriction of the farmer perception of the relative importance of a constraint to those that he or she considers major constraints. For any given constraint s, the mean perception of its observed relative importance under such restriction is given by the conditional mean $E(x^s\,|\,m^s = 1)$, where m^s is a dummy binary indicator with $m^s = 1$ if constraint s is a major constraint for the farmer and $m^s = 0$ otherwise. Hence, the expected sample minority-exclusion bias is defined as the difference between this conditional mean perception of relative importance and its unconditional counterpart, $E(x^s)$. That is, $SMEB^s = E(x^s\,|\,m^s = 1) - E(x^s)$.

As above, we decompose the conditional expectation $E(x^s)$ into its conditional parts, but this time with respect to the constraint being a major constraint or not for the farmer, to obtain:

$$\begin{aligned} SMEB^s &= E(x^s\,|\,m^s = 1) - \{E(x^s\,|\,m^s = 1)P_m^s - E(x^s\,|\,m^s = 0)\,(1 - P_m^s)\} \\ &= (1 - P_m^s)\,\{E(x^s\,|\,m^s = 1) - E(x^s\,|\,m^s = 0)\} \end{aligned}$$

where $P_m^s = Prob\{m^s = 1\}$ is the probability that constraint s is a major constraint for a farmer randomly selected from the population. Hence, the sign of the expected sample minority-exclusion bias is the sign of the expression within the curly bracket. But, this expression is positive if we assume that the expected perceived relative importance of the constraint when the constraint is major for the farmer is not lower compared to when it is not (i.e. if the perception of relative importance of a constraint is positively correlated with the constraint being a major constraint for the farmer). The above expression also shows that the expected sample minority-exclusion bias is more severe for constraints that are less likely to be major constraints for farmers.

The combination of population non-awareness and sample minority exclusion

The expected bias introduced by the combination of the constraint not being universally known in the population and not being a major constraint for all the farmers is given by the quantity $E(x^s|m^s=1) - E(x_1^s)$, which is the difference between the mean perception of the observed relative importance of the constraint for the sub-population of farmers for which the constraint is major and the population mean perception of relative importance of the constraint when all farmers are aware of its existence.

By subtracting and adding $E(x^s)$ to this difference we have:

$$E(x^s|m^s=1) - E(x_1^s) = E(x^s|m^s=1) - E(x^s) + E(x^s) - E(x_1^s) = SMEB^s + PNAB^s$$

This shows that the two biases introduced by population non-awareness and sample minority exclusion are additive and operate in opposite directions.

Notes

[1] More precisely, $x^s \in \{r_i^s, d_i^s e_i^s, \alpha_i^s, \Delta y_i^s\}$. Working at the population level means that we can omit the *i* subscript in our notation and use the mathematical expectation operator instead of sample averages.

[2] We note that observed value x^s of a farmer is linked to the two associated potential outcomes by the following relationship: $x^s = \omega^s x_1^s + (1-\omega^s) x_0^s$.

[3] It is important to note the difference between the mean perception for the full population, $E(x_1^s)$, and the mean perception for the sub-population of farmers who are aware of the existence of the constraint, which is given by the conditional mean $E\left(x_1^s \mid \omega^s = 1\right)$. The latter is usually greater than the full population mean perception of relative importance because of the likely positive correlation between awareness of the existence of a constraint and perception of its relative importance.

Appendix 4.2. Theoretical Framework of Evaluation of Potential Impact of Research Addressing Biophysical Constraints

To show theoretically the relationship between the yield loss and the various producer outcomes, we use the producer quasi-rent function (defined as the excess of gross receipts over total variable costs) as welfare measurement and the counterfactual or potential outcomes framework introduced by Rubin (1974), which has now become the standard framework for impact assessment (Imbens and Wooldridge, 2009).

As explained by Just *et al.* (2004), a change in the producer quasi-rent is a willingness-to-pay measure of the change in producer welfare. In contrast, change in the producer profit (which is equal to the quasi-rent minus total fixed cost) and change in the producer surplus (the area under the supply curve) introduced by Marshall (1930) are not in general willingness-to-pay measures of change in producer welfare. Quasi-rent, profit and producer surplus coincide only when there is no fixed cost and markets are complete (see Just *et al.*, 2004, chapter 4 for more details on the measurement of changes in producer welfare).

Let the producer quasi-rent be expressed as $\pi = PQ - C$, where P is the output unity price, Q is the quantity of output produced and C is the total variable cost. Also, let e^s be the binary variable indicating the experience or not of a given constraint s with $e^s = 1$ indicating experience of the constraint and $e^s = 0$ indicating non-experience of the constraint by a population unit (a farmer or a village). Under the potential outcome framework, each population unit has *ex-ante* two potential quasi-rents: $\pi_1 = P_1Q_1 - C_1$ when he or she experienced the constraint, and $\pi_0 = P_0Q_0 - C_0$ when he or she did not experience the constraint. Thus, the observed quasi-rent is $\pi = \pi_0 + e^s(\pi_1 - \pi_0)$.

We have $\pi_1 - \pi_0 = (P_1Q_1 - C_1) - (P_0Q_0 - C_0) = P_1(Q_1 - Q_0) + P_1Q_0 - P_0Q_0 - (C_1 - C_0)$, hence $\pi_1 - \pi_0 = P_1(y_1a_1 - y_0a_0) + P_1Q_0 - P_0Q_0 - (C_1 - C_0)$, where $y = \frac{Q}{a}$ is the observed yield and a is the observed total area cultivated. Noting that $a_1 = a_0 = a$, we get:

$$\pi_1 - \pi_0 = P_1a_0(y_1 - y_0) + P_1Q_0 - C_1 - (P_0Q_0 - C_0)$$

and

$$\pi = P_0Q_0 - C_0 + e^sP_1a_0(y_1 - y_0) + e^s(P_1Q_0 - C_1) - e^s(P_0Q_0 - C_0)$$

or

$$\pi = e^sP_1a_0(y_1 - y_0) + e^s(P_1Q_0 - C_1) + (1 - e^s)\pi_0.$$

Now, letting $\beta = -P_1\, a_0y_0$, $\Delta y^s = 1 - \dfrac{y_1}{y_0}$, $\gamma = P_1Q_0 - C_1 - \pi_0$ and $\sigma = \pi_0$, we have:

$$\pi = \beta e^s \Delta y^s + \gamma e^s + \sigma$$

By making the coefficients random and dependent on the socio-demographic characteristics X of the producer and assuming additive separability, they can be expressed as $\beta = \beta(X) + \mu_1$, $\gamma = \gamma(X) + \mu_2$, $\sigma = \sigma(X) + \mu_3$, $\mu = \mu_1 + \mu_2 + \mu_3$. Hence, we obtain the following relation:

$$E(\pi \mid X = x, Es = es) = \beta(x)e^s\Delta y^s + \gamma(x)e^s + \sigma(x) + \mu$$

which shows theoretically how the change in the yield loss due to a given production constraint can affect the producer quasi-rent function and consequently his or her total income. This relation also shows that the village poverty headcount is affected by the occurrence of constraints in the village because constraints affect the income of the village members.

Projection over time and aggregation at country level

The different steps followed to project the impact over time and to extrapolate it at country level are described here following Diagne *et al.* (Chapter 32, this volume).

Step 1: Projection of impact over time

The estimation of the models gives the impact at starting year of availability of technology. This year corresponds $t_0 = 2010 + t_d$ with t_d being the estimated number of years to technology delivery (from 2010). Using the AR1 model parameters estimated, we forecast the mean impact starting in a given year t_0 to any subsequent year $t_0 + \rho$ in the future as:

$$E\Delta y_{t_0+\rho} = \beta \sum_{j=0}^{\rho-1} \alpha^j E(\Delta y_{t_0+\rho-j}) = \beta r_e \frac{1-\alpha^{\rho}}{1-\alpha} \quad \text{and}\ \rho = 1,2,\ldots$$

Where r_e stands for the constant reduction in yield loss. This formula gives the ρ – period ahead forecasted value for the outcome y. Finally, the annual nominal income gained is discounted at the rate of 5% and cumulated to get gross benefit at farmer level.

Step 2: Estimation of the number of rice farmers and rice farming population

The extrapolation from farmer and village levels to country level is based on the estimation of the total number of rice farmers in each country. Due to the lack of national estimates of the total number of rice farmers per country, we combined household-survey and secondary data to get these estimates.

The total number of rice farming households N_h in each of the countries included in our analysis was estimated by taking the ratio of the country's total rice harvested area S (obtained from FAOSTAT, 2010) and the average rice area per household s_h (estimated from the farm-household surveys) and projected over time assuming a constant population growth rate of $g = 2.5\%$ (average rural population growth rate in SSA from the World Development Indicators; World Bank, 2010). The formula used is $N_h = \frac{S}{s_h} \times (1+g)^{\rho}$, where ρ stands for time.

Step 3: Determination of the number of adopting farmers

The total number of adopting farmers was derived by using a logistic adoption model, starting with a 1% adoption rate in 2014 and with a peak adoption rate of 20% in 2025.

Assuming a logistic diffusion curve, the number of adopters at each time t is given by the following formula:

$$N_t^a = \frac{N_p^a}{\left(1 + \left(\frac{N_p^a}{N_0^a} - 1\right) e^{-\gamma(t-t_0)}\right)}$$

where N_t^a = Number of adopters at time t; N_p^a = Peak number of adopters; $N_p^a = N\alpha_0(1+\beta_0)^{(t_p-t_0)}$ with α_0 the percentage of rice farmers who experienced a given constraint; β_0 population growth rate and N number of rice farmers; N_0^a = starting number of adopters and

$$\gamma = \frac{1}{\left(t_{\frac{p}{2}} - t_0\right) \log\left(\frac{N_p^a - N_0^a}{N_0^a}\right)}$$

with t_p = the year of peak of adoption, $t_{\frac{p}{2}}$ = year when adoption reaches half of its peak value and t_0 = starting year of adoption.

Adopting farmers will be made aware of the technologies through video, radio and other learning tools, farmer-to-farmer training and out-scaling by development partners from both the public and private sectors.

Step 4: Extrapolation of impact from household and village levels to country level

For each country included in the analysis, the farmer individual impact on income is extrapolated to country level by multiplying the average estimated impact by the estimated total number of direct and indirect beneficiaries in the country. Diagne *et al.* (2012) have shown how this provides a consistent estimate of the total benefit to rice farmers at the national level.

The poverty impact estimated at the village level was multiplied by the total rice-farming population size in the country. Evidence that this provides a consistent estimate of the reduction in the total number of poor rice farmers at the national level is provided by Diagne *et al.* (2012).

Table 4.A.2.1. Econometric model of impact of yield loss on household income.[a]

Dependent variable (household total income in 2008)	Weeds	Insects	Birds and rodents	Diseases	Soil-related constraints	Climate-related constraints
Yield loss due to constraint (β)	−1.20***	−1.07**	−1.06*	−2.78	−1.43***	−1.13
	(−2.71)	(−2.23)	(−1.66)	(−1.02)	(−2.70)	(−0.19)
Whether the farmer experienced the constraint (γ)	−25.19***	−20.09**	−20.97*	−20.37	−30.99***	−3.58
	(−2.71)	(−2.23)	(−1.66)	(−1.02)	(−2.70)	(−0.19)
Household total income in 2007 (α)	0.89***	0.87***	0.81***	0.91***	0.88***	0.93***
	(95.87)	(86.44)	(68.55)	(41.72)	(51.41)	(56.26)
No. observations	6,577	5,243	5,456	1,087	2,240	1,558
R-squared	0.632	0.643	0.536	0.684	0.619	0.716

[a]The estimated coefficients for the demographic variables and the constant term (σ(x))are omitted to save space. t-statistics in parentheses. *** $P < 0.01$, ** $P < 0.05$, * $P < 0.1$.

Table 4.A.2.2. Econometric model of impact of yield loss on village level poverty headcount.[a]

Dependent variable (village poverty headcount in 2008)	Weeds	Insects	Birds and rodents	Diseases	Soil-related constraints	Climate-related constraints
Yield loss due to constraint (β)	0.04**	0.02	0.02	0.03	0.17	0.10
	(2.01)	(0.99)	(0.35)	(0.52)	(0.63)	(0.60)
Whether the farmer experienced the constraint (γ)	0.66**	0.35	0.25	0.27	0.65	0.33
	(2.01)	(0.99)	(0.35)	(0.52)	(0.63)	(0.60)
Village poverty headcount in 2007 (α)	0.95***	0.92***	0.81***	0.93***	0.83***	0.89***
	(132.99)	(105.78)	(56.50)	(62.14)	(34.64)	(56.71)
No. observations	1,869	1,907	1,809	769	562	801
R-squared	0.932	0.900	0.744	0.893	0.790	0.875

[a]The estimated coefficients for the demographic variables and the constant term (σ(*x*)) are omitted to save space. t-statistics in parentheses. *** $P < 0.01$, ** $P < 0.05$, * $P < 0.1$.

5 A Continent-wide, Product-oriented Approach to Rice Breeding in Africa

Takashi Kumashiro,* Koichi Futakuchi, Moussa Sié, Marie-Noëlle Ndjiondjop and Marco C.S. Wopereis
Africa Rice Center (AfricaRice), Cotonou, Benin

Introduction

Urbanization, changes in employment patterns, rising income levels, shifts in consumer preferences, and rapid population growth have contributed significantly to the widening gap between rice supply and demand in sub-Saharan Africa (SSA) (Seck *et al.*, 2012). In 2008, Africa imported about 10 million tonnes (Mt) of milled rice, at a cost of US$3.6 billion. Continued reliance on imports is unsustainable and potentially dangerous, as it may affect food security and civil stability on the continent. Rice is one crop for which SSA can become self-sufficient given the continent's available land and water resources.

Rice is grown across a large diversity of growth environments in Africa, from the salty delta of the Senegal River to the highlands of Madagascar. The Africa Rice Center's (AfricaRice) gene bank alone contains about 17,000 African rice germplasm accessions (Sanni *et al.*, Chapter 7, this volume). Rice environments are generally divided into rainfed upland, rainfed lowland, irrigated lowland, deep-water/floating and mangrove-swamp production systems (for details, see Saito *et al.*, Chapter 15, this volume). In addition to this tremendous diversity of rice-growing environments, there are also large differences in terms of consumer preferences for rice, both between and within countries. For example, surveys in Benin showed large differences between consumer preferences in urban and rural areas (see Futakuchi *et al.*, Chapter 25, this volume).

The challenge to develop new rice varieties for such diversity is huge and compounded by the fact that training and recruitment of rice breeders has been neglected across the continent since the 1990s. There are very few rice breeders left, and those remaining are mostly over 50 years of age, often with additional managerial responsibilities.

In response to this tremendous lack of research capacity, in 1991 AfricaRice established rice-breeding Task Forces (collective research and development efforts by national and AfricaRice breeders, based on the principles of sustainability and build up of critical mass at the national and regional levels) catering to the different rice environments in West and Central Africa. These Task Forces were later merged into a regional rice research and development network in West and Central Africa (ROCARIZ). Funding for ROCARIZ ceased in 2005 and this has severely disrupted AfricaRice's ability to collaborate and plan activities with national agricultural research system (NARS) partners.

AfricaRice established a new, continent-wide Rice Breeding Task Force in 2010, aiming

* Corresponding author: t.kumashiro@cgiar.org

© CAB International 2013. *Realizing Africa's Rice Promise* (eds M.C.S. Wopereis *et al.*)

to accelerate rice varietal development through continent-wide varietal evaluation of nominated elite lines from AfricaRice and its international and NARS partners.

Despite the difficulties, AfricaRice and partners have contributed to the release of many new rice varieties since 1960 (Sanni *et al.*, Chapter 6, this volume). However, many rice farmers still do not have access to new varieties that are better adapted to their farming environment and likely to sell well on the market. This is partly related to cumbersome, non-functional or non-existent varietal release procedures (Sanni *et al.*, Chapter 6, this volume) or non-functional seed systems (Bèye *et al.*, chapter 14, this volume). However, there is also a clear need to develop rice varieties better targeted to rice environments and market segments.

This chapter reviews the continent-wide, product-oriented breeding approach promoted by AfricaRice within the framework of the Global Rice Science Partnership (GRiSP). GRiSP is a CGIAR research programme for rice, led globally by the International Rice Research Institute (IRRI) and for Africa by AfricaRice (IRRI *et al.*, 2010).

A Systematic, Continent-wide Product-oriented Approach to Rice Breeding

The systematic, continent-wide product-oriented approach to rice breeding pursues varietal development in different phases, with feedback loops as schematically represented in Fig. 5.1. These phases feed into 'varietal development pipelines' tailored to different growth environments and market segments. The important phases in this breeding approach are: (i) definition of breeding goals; (ii) pre-breeding activities; (iii) breeding activities; and (iv) multi-environment testing (MET). These phases are discussed in more detail below.

Definition of breeding goals

Rice breeding in Africa aims to develop improved cultivars with high and stable grain yield – varieties that are able to resist environmental stresses that reduce or limit productivity, and that respond well to nutrient inputs. Traditional varieties usually have good yield stability, but often low yield potential. In addition to these two general requirements, grain quality has become important in meeting the demands of consumers for healthy, high-quality food. Grain quality is essential for marketability of locally produced rice in SSA. Long slender grain and aromatic rice is preferred by urban consumers; rice of medium amylose content (20–25%) is widely acceptable, but rice with high amylose content (>25%) is preferred in Nigeria.

Breeding goals must, therefore, address the most important challenges of the target rice-growing environment (e.g. iron-toxicity in some rainfed lowlands, salinity in mangrove-swamp areas), the level of input use (low-, medium- or high-input system) and specific desired traits of the variety (e.g. grain quality, growth duration and plant height). The breeding goals must also respond to the needs of the various stakeholders in the rice value chain (with due consideration of gender issues), including rice farmers, millers, traders and consumers. For example, a varietal development pipeline for rice grown in irrigated Sahelian systems and tailored to the urban market may have as breeding goals: high and stable yield potential, tolerance to salinity, resistance to *Rice yellow mottle virus* (RYMV) and bacterial leaf blight, short duration, medium height, and aromatic, long and slender grains.

Pre-breeding

Activities in the pre-breeding phase aim to accelerate and facilitate the varietal development process in the different varietal development pipelines. This includes: (i) *ex-situ* conservation and characterization of different *Oryza* species germplasm in AfricaRice's gene bank; (ii) development of novel populations through enhanced recombination of cultivated and wild gene pools; (iii) phenotyping for key agronomic traits, resistance to diseases and responses to major stresses, including those of importance to adapt to climate change (e.g. tolerance to drought, high temperatures, salinity); and (iv) identification of genes responsible for these traits through quantitative trait loci (QTLs) mapping or association analysis, then fine mapping.

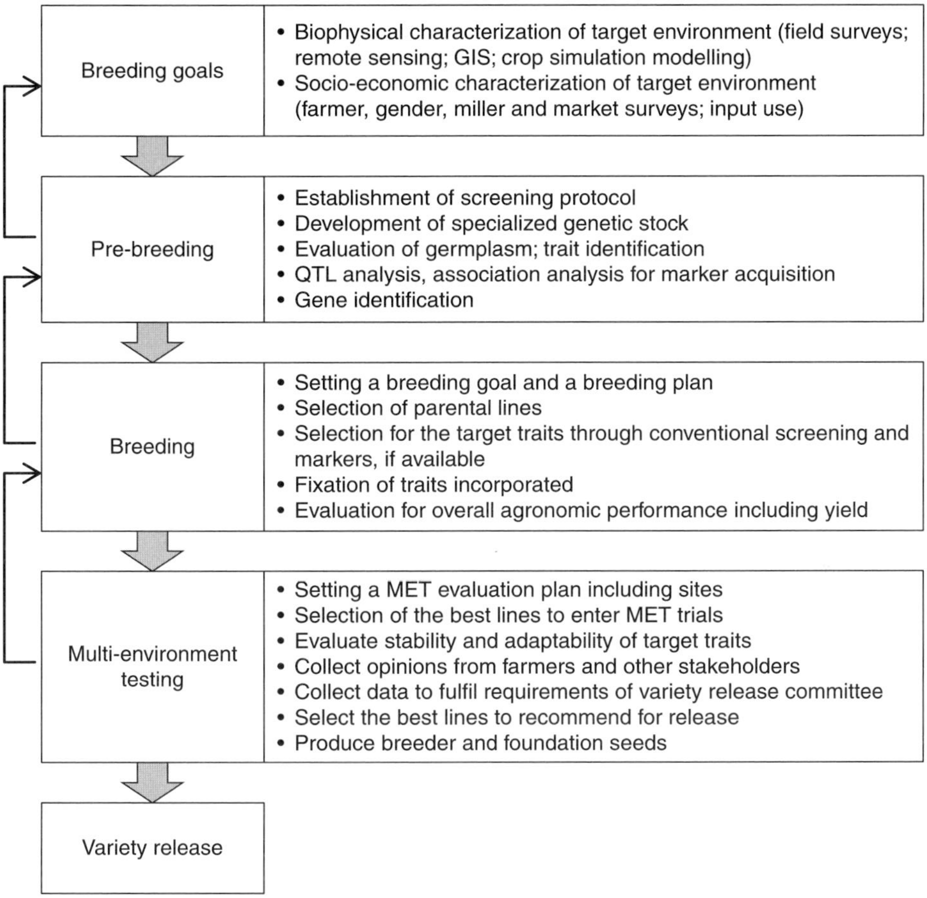

Fig. 5.1. Stages of a varietal development pipeline: definition of breeding goals, pre-breeding activities, breeding activities and multi-environment testing (MET).

AfricaRice and partners have focused in particular on gene discovery related to tolerance to abiotic stresses such as drought, different types of submergence, salinity, cold, heat, iron-toxicity and phosphorus deficiency (Dramé *et al.*, Chapter 11, this volume), and resistance to biotic stresses including blast, bacterial blight and RYMV (Séré *et al.*, Chapter 17, this volume) and African rice gall midge (AfRGM) (Nwilene *et al.*, Chapter 18, this volume; see also Lorieux *et al.*, Chapter 10, this volume). Among these target traits, novel resistant or tolerant germplasm has been successfully identified for salinity (unpublished), cold (unpublished), RYMV (Albar *et al.*, 2006) and AfRGM (Nwilene *et al.*, 2009).

Institut de recherche pour le développement (IRD, France) has been a collaborator of AfricaRice in studies on RYMV, and a new resistance gene (*RYMV2*) has been identified in *O. glaberrima* (Thiémélé *et al.*, 2010). The International Center for Tropical Agriculture (CIAT) and IRD have developed novel populations based on chromosome segment substitution (chromosome segment substitution lines, CSSL) that have *O. glaberrima* segments introgressed in different genetic backgrounds. These populations are used by collaborators for the localization of genes/QTLs for important traits such as drought tolerance, panicle architecture, root development, iron-toxicity tolerance and bacterial leaf blight resistance.

Through the IRD–CIAT–AfricaRice collaboration, efforts have been made to overcome the sterility barrier in the interspecific breeding of the two cultivated rice species to make greater use of useful traits of *O. glaberrima*; one of the project outputs is the development of 'interspecific bridge lines', which possess a high content of the *O. glaberrima* genome and produce fertile hybrids with *O. sativa* (Lorieux *et al.*, Chapter 10, this volume).

Japan International Research Center for Agricultural Sciences (JIRCAS) has worked with AfricaRice on drought research, focusing on deep rooting. Screening of 654 rice accessions identified 17 with deep-rooting capabilities; efforts to identify associated QTLs are under way (Tsunematsu and Samejima, 2011). AfricaRice and JIRCAS are also collaborating on tolerance to phosphorus deficiency and resistance to blast. Diversity analysis of blast isolates in Africa is conducted using 'differential varieties', each of which carries one resistance gene in the same genetic background (these varieties were developed under long-term collaboration between Japan and IRRI). To develop a variety with durable blast resistance, DNA markers linked with field resistance (Fukuoka *et al.*, 2009; Hayashi *et al.*, 2010) identified by the National Institute of Agrobiological Sciences (NIAS), Japan, are being used. For the development of rice cultivars capable of using a greater portion of phosphorus from the soils, markers for the *Pup1* (Heuer *et al.*, 2009) gene were introduced from JIRCAS and are being used to detect donors among upland varieties in Africa.

Breeding

The most suitable parents and breeding methods are selected on the basis of the breeding target. The availability of a reliable phenotyping method to identify desirable individuals or lines giving adequate performance for each trait of interest, especially for quantitative traits such as drought tolerance and yield potential, is crucial for the success of any breeding programme.

In principle, 'hot spots' with regular occurrence of a particular abiotic or biotic constraint offer the breeder a natural screening platform. However, spatial and temporal variability of the constraint may be large. For instance, iron-toxicity in the AfricaRice breeding hot spot in Burkina Faso is unevenly distributed in the field. For diseases and insects, prevailing strains and constraint severity may vary from year to year. Therefore, hot-spot screening should be coupled with screening in a controlled environment, to ensure more reliable evaluation of phenotypic performance.

AfricaRice will benefit from the global phenotyping platform that is being established within GRiSP, focusing mainly on tolerance to abiotic stresses and yield potential. AfricaRice is developing additional facilities for phenotyping for major abiotic stresses (e.g. drought, cold, Fe-toxicity) and the most important diseases (i.e. blast, bacterial leaf blight, RYMV) under controlled conditions, where stress intensity is managed to mimic the target environment and artificial inoculation is carried out using the prevalent strains.

Use of molecular markers

Marker-assisted selection (MAS) improves the efficiency of the breeding process when molecular markers tightly linked to the trait of interest are available and when phenotypic performance for the trait cannot be evaluated reliably or in a timely fashion on a routine basis. Loci or genes responsible for important agronomic traits in rice (e.g. disease resistance, abiotic-stress tolerance, grain quality) have been identified through QTL analyses and association analyses. The Q-TARO ('QTL annotation rice online', http://qtaro.abr.affrc.go.jp) (Yonemaru *et al.*, 2010) and Gramene (http://www.gramene.org/) databases summarize and update important QTLs for rice.

AfricaRice is using MAS on a routine basis to incorporate resistance to biotic and abiotic stresses in different rice varietal development pipelines (Table 5.1). For example, MAS was used to incorporate the resistance allele, *rymv1-2*, of the *RYMV1* gene into a number of major varieties grown in Africa, such as IR64. Near-isogenic lines (NILs) of IR64 bearing the resistance allele were developed and evaluated in Burkina Faso, Côte d'Ivoire, Mali and Nigeria (Ndjiondjop *et al.*, Chapter 12, this volume). Likewise, markers linked to genes/QTLs for tolerance to abiotic stresses are also routinely used in MAS. For instance, the *Sub1* gene (Septiningsih *et al.*, 2009) is being introgressed into major African varieties such as WITA 4 and NERICA-L 19 to increase tolerance to flash floods during the early vegetative stage.

Table 5.1. Ongoing marker-assisted selection (MAS) at AfricaRice, 2012.

Trait	Locus	Reference(s)
RYMV1 resistance	*rymv1-2*	Ndjiondjop *et al.* (1999); Thiémélé *et al.* (2010)
RYMV2 resistance	*rymv2*	Thiémélé *et al.* (2010)
Bacterial blight resistance	*Xa4*	Sun *et al.* (2003)
	xa5	Jiang *et al.* (2006)
	Xa7	Chen *et al.* (2008)
Blast resistance (field)	*Pb1*	Hayashi *et al.* (2010)
	pi21	Fukuoka *et al.* (2009)
Salt tolerance	*saltol*	Thompson *et al.* (2010)
Submergence tolerance	*sub1*	Xu *et al.* (2006)
Drought tolerance (upland)	*DTY12.1*	Bernier *et al.* (2007)
Drought tolerance (lowland)	*DTY 3.1*	Venuprasad *et al.* (2009)
Cold tolerance	*Ctb1, Ctb2*	Saito *et al.* (2010)
	qRCT6b	Negussie (2011 unpublished)

In addition to using MAS to target a single gene or major QTL, AfricaRice is also using a more advanced MAS scheme where QTLs for target traits are detected within the segregating progenies of elite lines crossed for their complementarities. This marker–phenotype information serves to guide the breeding process towards the ideal mosaic of QTLs or favourable chromosomal segments from the two parents. Such a breeding scheme, involving several successive generations of crossing of progenies bearing complementary QTLs, is referred to as marker-assisted recurrent selection (MARS) or genotype construction (Stam, 1995; Peleman and van der Voort, 2003). AfricaRice, Centre de coopération internationale en recherche agronomique pour le développement (CIRAD) and partners are implementing a MARS scheme to increase drought tolerance in lowland rice. Bi-parental populations are being developed for mapping QTLs involved in drought tolerance and yield potential; the breeding scheme (directed recombination between specific individuals carrying the favourable allele of the detected QTLs) is designed to produce or approach the ideal genotype accumulating favourable alleles for as many QTLs as possible.

Breeding for high yield

Most varietal development pipelines in Africa address the incorporation of resistance or tolerance to various stresses, because there are so many constraints that hinder the expression of the full potential of a variety in farmers' fields (see Dramé *et al.*, Chapter 11, this volume). However, more emphasis is now being placed on raising yield potential in different growth environments. This is important because high-yielding varieties that have been bred and do well in environments without major yield-reducing or yield-limiting factors will usually also out-perform in environments where a particular stress interferes, unless the stress becomes too severe (e.g. Bänziger *et al.*, 1997, for soil fertility status in maize; Kumar *et al.*, 2008, for drought tolerance in rice; Saito, 2010, for rice competitiveness against weeds). Several approaches are used to increase yield potential.

The first approach is to validate and incorporate yield-component QTLs detected elsewhere (Ando *et al.*, 2008; Obara *et al.*, 2010; Ookawa *et al.*, 2010) into popular varieties in Africa.

The second approach is to apply recurrent selection, a multi-parental scheme designed to facilitate the accumulation of favourable alleles for complex traits such as yield through successive cycles of recombinations and phenotypic selection (Gallais, 1990; Guimaraes, 2005). Such a scheme was used successfully in Brazil to achieve rapid gains in grain yield for upland rice (e.g. Breseghello *et al.*, 2009, 2011). The third approach is to develop F_1 hybrid varieties adapted to African rice environments. Hybrid rice has been very successful in Asia, especially in China and India. AfricaRice started work on hybrids in 2009, evaluating hybrids from China and developing an in-house hybrid-rice breeding programme for the irrigated environment at the AfricaRice Sahel station in Senegal. Some parental lines, including male sterile lines and restorer lines, were introduced from IRRI, Egypt and China, and are being tested (El-Namaky and Demont, Chapter 13, this volume).

Grain quality

Given the importance of grain quality in enabling locally produced rice to compete with imported rice (Fofana *et al.*, 2010; Futakuchi *et al.*, Chapter 25, this volume), it is important to assess grain quality when choosing the parental lines and then during the breeding process. Since grain quality is easily affected by many factors (e.g. timing of harvest, moisture content, storage condition of harvested grain), repeated evaluation of properly prepared grain samples is necessary. At AfricaRice, all elite lines entering the multi-environment testing phase are evaluated for quality traits, including basic parameters directly related to farmers' income, such as chalkiness, grain colour and milling properties. Attention is also paid to the needs of the various stakeholders. For instance, among the three aromatic varieties released in Senegal in 2009, Sahel 177 has the highest yield potential, Sahel 328 the shortest growth duration and Sahel 329 is the most appreciated for its taste.

Multi-environment Testing (MET) Network, the Africa-wide Rice Breeding Task Force

In 2010, the Africa-wide Rice Breeding Task Force was launched by AfricaRice involving NARS from 25 countries (Plate 1). The objectives of the MET network are to evaluate the stability of traits incorporated in breeding processes and to identify varieties that best fit the growth conditions in target regions and the markets. The Task Force also accumulates data on the performance of new elite lines, thereby facilitating varietal release procedures. Furthermore, by exposing scientists from NARS and farmers to these elite lines during the testing phase, dissemination will be facilitated (Table 5.2). Moreover, the Task Force is also building the capacity of rice breeders, by enabling experienced and younger breeders to work together on evaluation trials.

The MET conducted by the Africa-wide Rice Breeding Task Force consists of a series of three consecutive trials. Promising breeding lines developed by AfricaRice, national or international partners (e.g. IRRI and CIAT) are nominated for evaluation in one or several rice-cultivation environments: lowland, irrigated, upland, high elevation and mangrove (Plate 1). All nominated lines should be fixed and accompanied by supporting data on traits incorporated during the breeding process and with information on yield performance. These characteristics are checked at AfricaRice before incorporation into the MET network.

The first phase (MET-1) consists of an initial evaluation of about 100 lines selected from the nominated lines. Each national partner evaluates these lines at sites in its country. Such sites may be at an experimental station under optimal management to evaluate yield potential, or may be a 'hot spot' to check the performance of the nominations in a stressed growth environment. Trials are replicated three times and include common and local checks. Data collected at all sites are analysed centrally at AfricaRice, including comprehensive genotype-by-environment (G×E) analysis. Based on the results of these analyses and breeders' observations, about 30 lines are advanced to the next phase.

The second phase (MET-2) serves to evaluate and confirm the performance of the selected lines. These lines are cultivated at the same sites as MET-1 using the same experimental design with three replications. An important feature of MET-2 is that farmers and other stakeholders (e.g. millers and traders) are invited to participate in varietal selection and their opinions on the performance of all entries are collected through a participatory varietal selection (PVS) procedure (Sié *et al.*, 2010). Members of national varietal release committees are also invited. Using the data collected, observations by the breeders and the opinions of stakeholder groups, NARS partners select up to ten lines.

In the third phase (MET-3), the ten lines selected during MET-2 are evaluated in at least three sites per country during one or more growing seasons, depending on varietal release requirements. Entries are arranged in an alpha-lattice

Table 5.2. Principles of multi-environment testing (MET) conducted by the Africa-wide Rice Breeding Task Force.

Trial phase	Name of trial	Characteristics of trials	Sites/ country	No. lines in trial	Exp design	No. lines to advance	Evaluator(s)
MET-1	MET (multi-environment trial)	Evaluation of lines nominated by breeders	1	About 100 lines	Alpha-lattice with 3 replications	About 30	NARS breeders
MET-2	PET (participatory evaluation trial)	Evaluation of lines selected in MET	1	About 30	Alpha-lattice with 3 replications	About 10	NARS breeders Farmers VRC members Other stakeholders
MET-3	PAT (participatory advanced trial)	Evaluation of lines selected in PET by NARS breeders	3	About 10 best lines	RCBD with 4 replications	1–3 lines recom-mended for release	NARS breeders Farmers VRC members Other stakeholders
	FAT (farmers' adoption trial)	Evaluation of a few lines selected in PET by farmer	50	3 lines among 10 for each farmer	No replication		

Exp = experimental; NARS = national agricultural research system; RCBD = randomized complete block design; VRC = varietal release committee.

design (Patterson and Williams, 1976) with at least four replications. All stakeholders are again invited to familiarize them with the new lines and voice their opinions to help select lines for further advancement. Among the ten lines, farmers are invited to select three lines and cultivate these in their own fields, together with a common check and their own variety (Sié *et al.*, 2010).

The ultimate decision to nominate a particular variety for release in a country will be made by the country's rice breeder involved in the Task Force, based on evaluation of all data acquired before and during the MET phase.

In 2011, AfricaRice and partners decided to assign a new name to particularly promising breeding lines that result from Task Force activities: 'ARICA', which stands for 'Advanced RICes for Africa'.

ARICA varieties can be considered as the next generation of rice varieties for Africa, after the success of the 'NEw RICes for Africa' (NERICAs) developed in the 1990s and the first decade of this century. For a breeding line to be nominated as an ARICA line it must have a clear advantage over the best check varieties in a region, backed by quality data over at least three seasons. Moreover, at least one country should show interest in nominating the line for varietal release. Unlike the NERICA varieties, they are not restricted to interspecific crosses. Any line that shows promise, regardless of its origin, can become an ARICA line as long as the data collected are convincing. The ultimate decision on naming an ARICA line is taken by AfricaRice, and is based on data gathered in the Task Force trials, and any other data gathered during the breeding process.

During the 2013 Task Force meeting, five nominations for ARICA naming were examined and accepted to become the first ARICA lines. ARICA1, 2 and 3 are suited for the rainfed lowland growth environment and are proposed for varietal release in Mali (ARICA1, ARICA2 and ARICA3) and Nigeria (ARICA2 and ARICA3). ARICA4 and 5 are suited for the upland growth environment and have just been released in Uganda. All these ARICAs out-yielded local checks including NERICA-L 19 in the rainfed lowland environment and NERICA 4 in the upland environment. In addition, ARICA3 has better grain quality, higher milling recovery, lower chalkiness and shorter cooking time than NERICA-L 19.

Data management

Rapid access to information on germplasm, including advanced breeding lines and varieties, is essential to the use and deployment of breeding lines for further research and evaluation. Data management plays a central role starting from germplasm stored in a gene bank up to information on performance of elite breeding lines about to be released – high-quality data needs to be available for all the fixed breeding lines generated.

Data acquired by NARS in the MET trials are collected and sent to AfricaRice for central analyses and archiving, following the standards of the International Rice Information System (McLaren *et al.*, 2005). These data can be used to facilitate varietal release in a particular country or new varietal development activities.

Conclusions

Setting correct breeding goals is essential in any breeding process. Each region, country or market niche will have specific requirements in terms of traits that ideally need to be incorporated in a new variety. However, in general, very little knowledge is available about the target environment, both in terms of biophysical settings and preferences of farmers, millers and rice consumers. Much greater and more precise information with respect to such requirements is needed. AfricaRice is gathering such information through focused field surveys and trials and increasing use of crop simulation modelling to deal with spatial and temporal variability of growth conditions, in collaboration with partners such as CIRAD, IRRI and Wageningen University. AfricaRice and partners are also conducting surveys in key rice-growing regions across Africa, interviewing farmers, millers, traders and consumers. This information will allow more precise definition of breeding goals and a greater variety of varietal development pipelines. It will also allow collaborating NARS to specify in greater detail their requirements for their growth and market conditions, enabling more targeted selection of germplasm entering the MET-1 trials of the Africa-wide Rice Breeding Task Force. Greater breeding precision and a larger diversity of varietal development pipelines will only be possible by strengthening national rice-breeding capacity in Africa.

Acknowledgements

The authors wish to acknowledge the funding provided by the Government of Japan for the Africa-wide Rice Breeding Task Force and also that provided by the CGIAR Generation Challenge Program to the second phase of the Rice-Challenge Initiative, through which the drought-tolerance MARS work is being conducted.

References

Albar, L., Bangratz-Reyser, M., Hébrard, E., Ndjiondjop, M.N., Jones, M. and Ghesquière, A. (2006) Mutations in the eIF(iso)4G translation initiation factor confer high resistance of rice to *Rice yellow mottle virus*. *The Plant Journal* 47, 417–426.

Ando, T., Yamamoto, T., Shimizu, T., Ma, X.F., Shomura, A., Takeuchi, Y. and Yano, M. (2008) Genetic dissection and pyramiding of quantitative traits for panicle architecture by using chromosomal segment substitution lines in rice. *Theoretical and Applied Genetics* 116, 881–890.

Bänziger, M., Betrán, F. and Lafitte, H.R. (1997) Efficiency of high-nitrogen selection environments for improving maize for low-nitrogen target environments. *Crop Science* 37, 1103–1109.

Bernier, J., Kumar, A., Venuprasad, R., Spaner, D. and Atlin, G. (2007) A large-effect QTL for grain yield under reproductive-stage drought stress in upland rice. *Crop Science* 47, 507–518.

Breseghello, F., Morais, O.P., Castro, E.M., Prabhu, A.S., Bassinello, P.Z., Pereira, J.A., Utumi, M.M., Ferreira, M.E. and Soares, A.A (2009) Recurrent selection resulted in rapid genetic gain for upland rice in Brazil. *International Rice Research Notes* 34, 1–4.

Breseghello, F., Peixoto de Morais, O., Pinheiro, P.V., Simões Silva, A.C., Castro, E.M. Guimarães, E.P., Pereira de Castro, A., Pereira, J.A,, Lopes, A.M., Utumi, M.M. and Pereira de Oliveira, J. (2011) Results of 25 years of upland rice breeding in Brazil. *Crop Science* 51, 914–923.

Chen, S., Huang, Z., Zeng, L., Yang, J., Liu, Q. and Zhu, X. (2008) High-resolution mapping and gene prediction of *Xanthomonas oryzae* pv *oryzae* resistance gene *Xa7. Molecular Breeding* 22, 433–441.

Fofana, M., Adhikari, K., Futakuchi, K., Dossou, J., Bleoussi, R.T.M. and Dieng, I. (2010) Urban consumer acceptability of newly introduced rice with local and imported rice in Benin: a comparison. *Indian Food Industry* 29, 37–45.

Fukuoka, S., Saka, N., Koga, H., Ono, K., Shimizu, T., Ebana, K., Hayashi, N., Takahashi, A., Hirochika, H., Okuno, K. and Yano, M. (2009) Loss of function of a proline-containing protein confers durable disease resistance in rice. *Science* 325, 998–1001.

Gallais A. (1990) *Théorie de la Sélection en Amélioration des Plantes.* Masson, Paris, France.

Guimaraes E.P (2005) *Population Improvement: A way of exploiting the rice genetic resources of Latin America.* Food and Agriculture Organization of the United Nations, Rome, Italy.

Hayashi, N., Inoue, H., Kato, T., Funao, T., Shirota, M., Shimizu, T., Kanamori, H., Yamane, H., Hayano-Saito, Y., Matsumoto, T., Yano, M. and Takatsuji, H. (2010) Durable panicle blast-resistance gene *Pb1* encodes an atypical CC-NBS-LRR protein and was generated by acquiring a promoter through local genome duplication. *The Plant Journal* 64, 498–510.

Heuer, S., Lu, X., Chin, J.H., Tanaka, J. P., Kanamori, H., Matsumoto, T., De Leon, T., Ulat, V.J., Ismail, A.M., Yano, M. and Wissuwa M. (2009) Comparative sequence analyses of the major quantitative trait locus phosphorus uptake 1 (*Pup1*) reveal a complex genetic structure. *Plant Biotechnology Journal* 7, 456–457.

IRRI, AfricaRice and CIAT (2010) *Global Rice Science Partnership (GRiSP).* International Rice Research Institute, Los Baños, Philippines; Africa Rice Center, Cotonou, Benin; and International Center for Tropical Agriculture, Cali, Colombia.

Jiang, G.H., Xia, Z.H., Zhou, Y.L., Wang, J., Li, D.Y., Chen, R.S., Zhai, W.X. and Zhu, L.H. (2006) Testifying the rice bacterial blight resistance gene *xa5* by genetic complementation and further analyzing *xa5* (*Xa5*) in comparison with its homologues *TFIIAgamma1. Molecular Genetics and Genomics* 275, 354–366.

Kumar, A., Bernier, J., Verulkar, S., Lafitte, H.R. and Atlin, G.N. (2008) Breeding for drought tolerance: direct selection for yield, response to selection and use of drought tolerant donors in upland and lowland-adapted populations. *Field Crops Research* 107, 221–231.

McLaren, C.G., Bruskiewich, R.M., Portugal, A.M. and Cosico, A.B. (2005) The International Rice Information System. A platform for meta-analysis of rice crop data. *Plant Physiology* 139, 637–642.

Ndjiondjop, M.N., Albar, L., Fargette, D., Fauquet, C. and Ghesquière, A. (1999) The genetic basis of high resistance to *Rice yellow mottle virus* (RYMV) in cultivars of two cultivated rice species. *Plant Disease* 83, 931–935.

Nwilene, F.E., Okhidievbie, O., Agunbiade, T.A., Traore, A.K., Gaston, L.N., Togola, M.A. and Youm, O. (2009) Antixenosis component of rice resistance to African rice gall midge *Orseolia oryzivora. International Rice Research Notes* 34, 1–6.

Obara, M., Tamura, W., Ebitani, T., Yano, M., Sato, T. and Yamaya, T. (2010) Fine-mapping of *qRL6.1*, a major QTL for root length of rice seedlings grown under a wide range of NH_4^+ concentrations in hydroponic condition. *Theoretical and Applied Genetics* 121, 535–547.

Ookawa, T., Hobo, T., Yano, M., Murata, K., Ando, T., Miura, H., Asano, K., Ochiai, Y., Ikeda, M., Ebitani, T., Ozaki, H., Angeles, E.R., Hirosawa, T. and Matsuoka, M. (2010) New approach for rice improvement using a pleiotropic QTL gene for lodging resistance and yield. *Nature Communications* 1, 132. doi:10.1038/ncomms1132.

Patterson, H.D. and Williams, E.R. (1976) A new class of resolvable incomplete block designs. *Biometrika* 63, 83–92.

Peleman, J.D. and van der Voort, J.R. (2003) Breeding by design. *Trends of Plant Science* 8, 330–334.

Saito, K. (2010) Weed pressure level and the correlation between weed competitiveness and rice yield without weed competition: an analysis of empirical data. *Field Crops Research* 117, 1–8.

Saito, K., Hayano-Saito, Y., Kuroki, M. and Sato, Y. (2010) Map-based cloning of the rice cold tolerance gene *Ctb1. Plant Science* 179, 97–102.

Seck, P.A., Diagne, A., Mohanty, S. and Wopereis, M.C.S. (2012) Crops that feed the world 7: rice. *Food Security* 4, 7–24.

Septiningsih, S.M., Pamplona, A.M., Sanchez, D.L., Neeraja, C.N., Vergara, G.V., Heuer, S., Ismail, A.M. and Mackill, D.J. (2009) Development of submergence-tolerant rice cultivars: the *Sub1* locus and beyond. *Annals of Botany* 103, 151–160.

Sié, M., Dogbé, S. and Diatta, M. (2010) *Participatory Varietal Selection of Rice – The technician's manual.* Africa Rice Center, Cotonou, Benin.

Stam, P. (1995) Marker-assisted breeding. In: Van Ooijen, J.W. and Jansen, J. (eds) *Biometrics in Plant Breeding: Applications of molecular markers.* Proceedings of the 9th Meeting of the EUCARPIA Section Biometrics in Plant Breeding. Centre for Plant Breeding and Reproduction Research (CPRO-DLO), Wageningen, Netherlands, pp. 32–44.

Sun, X., Yang, Z., Wang, S. and Zhang, Q. (2003) Identification of a 47 kb DNA fragment containing *Xa4*, a locus for bacterial blight resistance in rice. *Theoretical and Applied Genetics* 106, 683–687.

Thiémélé, D., Boisnard, A., Ndjiondjop, M.N., Chéron, S., Séré, Y., Aké, S., Ghesquière, A. and Albar, L. (2010) Identification of a second major resistance gene to *Rice yellow mottle virus, RYMV2,* in the African cultivated rice species, *O. glaberrima. Theoretical and Applied Genetics* 121, 169–179.

Thomson, M.J., de Ocampo, M., Egdane, J., Rahman, M.A., Sajise, A.G., Adorada, D.L., Tumimbang-Raiz, E., Blumwald, E., Seraj, Z.I., Singh, R.K., Gregorio, G.B. and Ismail, A.M. (2010) Characterizing the *Saltol* quantitative trait locus for salinity tolerance in rice. *Rice* 3, 148–160.

Tsunematsu, H. and Samejima, H. (2011) Rice varieties with deep root system in West Africa for drought tolerance. In: Yanagihara, S. (ed.) Next Challenges in Rice Development for Africa: Workshop for New Collaboration between JIRCAS and AfricaRice. *JIRCAS Working Report* No. 70. Japan International Research Center for Agricultural Sciences, Tsukuba, Japan and Africa Rice Center, Cotonou, Benin, pp. 37–44.

Venuprasad, R., Dalid, C.O., Del Valle, M., Zhao, D., Espiritu, M., Sta Cruz, M.T., Amante, M., Kumar, A. and Atlin, G.N. (2009) Identification and characterization of large-effect quantitative trait loci for grain yield under lowland drought stress in rice using bulk-segregant analysis. *Theoretical and Applied Genetics* 120, 177–190.

Xu, K., Xu, X., Takeshi Fukao, T., Patrick Canlas, P., Maghirang-Rodriguez, R., Heuer, S., Ismail, A.M., Bailey-Serres, J., Ronald, P.C. and Mackill, D.J. (2006) *Sub1A* is an ethylene-response-factor-like gene that confers submergence tolerance to rice. *Nature* 442, 705–708.

Yonemaru, J., Yamamoto, T., Fukuoka, S., Uga, Y., Hori, K. and Yano, M. (2010) Q-TARO: QTL annotation rice online database. *Rice* 3, 194–203.

6 Rice Varietal Release Systems in Africa

Kayode Abiola Sanni,[1]* Ali A. Touré,[1] Aliou Diagne,[1] Fatimata Bachabi,[1] Rosemary Murori,[2] Rakesh Kumar Singh[2] and Moussa Sié[1]
[1]*Africa Rice Center (AfricaRice), Cotonou, Benin;* [2]*International Rice Research Institute, Regional Office, Dar es Salaam, Tanzania*

Introduction

National rice breeding programmes in Africa differ in strength, with some focusing only on testing lines introduced from outside, while others also conduct their own hybridization and line-development programmes targeting specific market segments and rice-growing environments in their respective countries. International partners, such as Africa Rice Center (AfricaRice), International Rice Research Institute (IRRI), International Center for Tropical Agriculture (CIAT), Centre de coopération internationale en recherche agronomique pour le développement (CIRAD) and International Institute of Tropical Agriculture (IITA) have backstopped and contributed to rice varietal development in Africa since 1960 (see Tollens *et al.*, Chapter 1, this volume).

Countries also differ greatly with respect to seed regulation. Seed regulation encompasses two basic areas: seed quality control (certification) and variety regulation. Variety regulation primarily seeks to control the release of new varieties developed by public and private breeding programmes for seed production and marketing. National seed laws establish national seed boards (councils, authorities) that govern varietal release regulations. Common features of varietal release regulations are the establishment of: (i) a mandatory procedure for testing varieties proposed for release; (ii) a national varietal release committee (NVRC), which recommends or rejects release based on test results; and (iii) an official register of released varieties, recording names and main agronomic characteristics of varieties that have successfully passed the tests and have been recommended for release (Louwaars, 2002). An officially released variety is a new variety that has been tested according to the standards of a country and recommended by the NVRC of that same country to be of proven value, registered and made available to the public.

In practice, some African countries keep a register of varieties that are 'adopted' in their country, because they do not operate a varietal release mechanism covering the three features mentioned above. In that case, 'adopted varieties' are those that are widely cultivated and for which the country has deemed it important to include them in the national crop register.

If international standards (UPOV, 1978, 1991) are followed, new varieties can only be registered if they satisfy four criteria: novelty or value for cultivation and use (VCU), distinctness, uniformity and stability. Distinctness, uniformity and stability are often referred to as a group: the 'DUS criteria'.

* Corresponding author: k.sanni@cgiar.org

© CAB International 2013. *Realizing Africa's Rice Promise* (eds M.C.S. Wopereis *et al.*)

Novelty means that the new rice variety performs better than the existing varieties for one or more traits of agronomic or technological importance. The VCU of a new variety is tested and compared to local check varieties using standard protocols measuring key agronomic data such as yield, growth duration, grain quality, and resistance to biotic and abiotic stresses.

Distinctness means that the variety is visually distinguishable from existing registered varieties in one or more morphological (shape, colour, height, leaf length, etc.) and agronomic (disease resistance, growth duration, etc.) traits.

Uniformity or homogeneity means that at any development stage all individual plants are identical for all plant characteristics.

Stability means that the variety remains identical to its initial description in its essential characteristics after repeated cycles of reproduction or propagation.

There are ongoing efforts to harmonize regulations for registration of varieties, for seed trade across borders, and for phytosanitary documents needed for seed movement across Africa, by the Economic Community of West African States (ECOWAS), the East African Community (EAC) and the South African Development Community (SADC). These efforts aim at encouraging seed system development to broaden the choice of varieties and enhance accessibility to and quality of seeds for farmers.

In the ECOWAS region, the harmonization system has started with a regional varietal catalogue published in 2008. Common crop variety release systems, seed certification standards and accreditation of seed producers, quarantine pest lists and seed import–export manuals are being prepared. EAC also maintains a regional catalogue of varieties released in different member countries and is in the process of finalizing common seed certification standards and accreditation of seed producers. In the SADC region, a regional varietal release system has existed since 2009. A common quarantine pest list, seed import–export manuals, seed certification standards and accreditation of seed producers, and seed tests using International Seed Testing Association (ISTA) rules as a common standard have been established. SADC is most advanced in the harmonization process.

This chapter provides an overview of the degree to which varietal release systems are in place in Africa and to what extent they are functional at the national level. It also provides an overview of rice varieties released or adopted (in case of absent or non-functional varietal release systems) in Africa since 1960.

Methodology

Questionnaires were sent out to AfricaRice collaborators from national agricultural research institutes of 30 African countries in 2010, asking for information on the procedures for varietal release in their respective countries and to provide an update on varieties released or adopted since 1960. Participating countries are listed in Tables 6.1 and 6.2. The data from this survey were complemented with information contained in the national and regional catalogues of varieties, published literature and information available from the International Network for Genetic Evaluation of Rice (INGER-Africa) in AfricaRice.

Results and Discussion

Rice varietal release systems in Africa

National seed boards are responsible for convening and chairing the NVRC meetings and generally meet once a year. However, some committees may not meet for several years for a variety of reasons (financial constraints are often mentioned). The functionality of the varietal release system in a country can be assessed by considering whether or not its NVRC meets to assess and recommend varieties for release before those varieties are made available to the public.

We classified countries into four groups based on: (i) the existence of a varietal release system (comprising varietal testing, regular NVRC meetings to judge test results and official varietal registration); (ii) functionality of the varietal release system; and (iii) existence of at least a varietal register in the absence of a varietal release system.

Of the 30 countries surveyed (Table 6.1), 18 have a varietal release system in place (Groups 1 and 2). Of these, 13 have regular NVRC meetings (Group 1; although Gabon has yet to release any rice variety), and 5 countries have non-effective

Table 6.1. Rice varietal release systems in 30 African countries.

Country	Existence of varietal release system	Release system functional?	Registration required for growing commercially	Existence of varietal register	Registration required for certified seed sale	Varietal descriptor part of registration process
Group 1						
Ghana	Yes	Yes	No	Yes	Yes	Yes
Burkina Faso	Yes	Yes	Yes	Yes	Yes	Yes
Côte d'Ivoire	Yes	Yes	Yes	Yes	Yes	Yes
Egypt	Yes	Yes	Yes	Yes	Yes	Yes
Ethiopia	Yes	Yes	Yes	Yes	Yes	Yes
Gabon	Yes	Yes	Yes	Yes	Yes	Yes
Kenya	Yes	Yes	Yes	Yes	Yes	Yes
Mali	Yes	Yes	Yes	Yes	Yes	Yes
Mozambique	Yes	Yes	Yes	Yes	Yes	Yes
Nigeria	Yes	Yes	Yes	Yes	Yes	Yes
Senegal	Yes	Yes	Yes	Yes	Yes	Yes
Tanzania	Yes	Yes	Yes	Yes	Yes	Yes
Uganda	Yes	Yes	Yes	Yes	Yes	Yes
Group 2						
Benin	Yes	No	No	Yes	Yes	Yes
Burundi	Yes	No	Yes	Yes	Yes	Yes
Democratic Republic of Congo	Yes	No	Yes	Yes	Yes	Yes
Madagascar	Yes	No	Yes	Yes	Yes	Yes
Mauritania	Yes	No	Yes	Yes	Yes	Yes
Group 3						
The Gambia	No	n/a	No	Yes	No	Yes
Guinea-Bissau	No	n/a	No	Yes	No	Yes
Liberia	No	n/a	No	Yes	No	Yes
Niger	No	n/a	No	Yes	No	Yes
Republic of Congo	No	n/a	No	Yes	No	Yes
Rwanda	No	n/a	No	Yes	No	Yes
Sierra Leone	No	n/a	No	Yes	No	Yes
Guinea	No	n/a	No	Yes	Yes	Yes
Group 4						
Cameroon	No	n/a	No	No	No	n/a
Central African Republic	No	n/a	No	No	No	n/a
Chad	No	n/a	No	No	No	n/a
Togo	No	n/a	No	No	No	n/a

n/a = not applicable.

NVRCs (the committee exists on paper but is not yet functional; Group 2). Eight countries do not have a varietal release mechanism, but operate a varietal register (Group 3), and 4 countries lack both a varietal release system and a varietal register (Group 4).

In countries without a formal varietal release system, varietal registers are made up of varieties that are already being cultivated and are recognized as such by the authority maintaining the register, e.g. the ministry of agriculture, the national seed board or the national agricultural research institute. Feedback obtained from the national partners indicated that there is considerable variation in consistency and

Table 6.2. Current status of application of DUS and VCU requirements for varietal release and registration in 30 African countries.

Country	Published DUS protocols	DUS required for rice varietal release or registration	Published VCU protocols	VCU required for rice varietal release or registration	Data from other countries allowed	PVS data accepted for release
Group 1						
Ghana	Yes	Yes	Yes	Yes	No	No
Burkina Faso	No	Yes	No	Yes	No	Yes
Côte d'Ivoire	No	Yes	No	Yes	No	No
Egypt	Yes	Yes	Yes	Yes	No	No
Ethiopia	Yes	Yes	Yes	Yes	No	No
Gabon	No	Yes	No	Yes	No	Yes
Kenya	Yes	Yes	Yes	Yes	No	No
Mali	No	Yes	No	Yes	Yes	Yes
Mozambique	No	Yes	Yes	Yes	Yes	Yes
Nigeria	No	Yes	Yes	Yes	No	Yes
Senegal	No	Yes	No	Yes	No	Yes
Tanzania	Yes	Yes	Yes	Yes	No	No
Uganda	Yes	Yes	No	Yes	Yes	No
Group 2						
Benin	No	Yes	No	Yes	Yes	Yes
Burundi	No	Yes	No	Yes	No	No
Democratic Republic of Congo	No	Yes	No	Yes	No	Yes
Madagascar	Yes	Yes	Yes	Yes	No	No
Mauritania	No	Yes	No	Yes	No	No
Group 3						
The Gambia	n/a	No	No	No	No	n/a
Guinea-Bissau	n/a	No	No	No	No	n/a
Liberia	n/a	No	No	No	No	n/a
Niger	n/a	No	No	No	No	n/a
Republic of Congo	n/a	No	No	No	No	n/a
Rwanda	n/a	No	No	No	No	n/a
Sierra Leone	n/a	No	No	No	No	n/a
Guinea	n/a	No	No	No	No	n/a
Group 4						
Cameroon	n/a	n/a	No	n/a	No	n/a
Central African Republic	n/a	n/a	No	n/a	No	n/a
Chad	n/a	n/a	No	n/a	No	n/a
Togo	n/a	n/a	No	n/a	No	n/a

n/a = not applicable.

accuracy among countries in keeping the varietal register up to date. Some countries do not promptly update their varietal register when new varieties are released or adopted.

Official release of a variety is required before it can be grown commercially (grown for sale) in all the countries with a release system, except Benin and Ghana (Table 6.1). Most West African countries have sent their varietal register to be included in the West African catalogue of plant species and varieties. In East Africa, Kenya, Tanzania and Uganda have included their varieties in the EAC list of crop varieties.

The registration of varieties in the national varietal catalogue is required for the sale of certified seeds in all 18 countries with a varietal release system. Since the other countries do not have a varietal release system, this condition is not applicable to them.

Table 6.2 provides an insight into requirements with respect to varietal release in the 30 surveyed countries. In the 18 countries with varietal release system (Groups 1 and 2), DUS and VCU information are needed for a new rice variety to be released and registered, but only seven countries have published protocols for the conduct of DUS testing and just eight countries have published protocols for VCU testing. In countries without protocols, the new varieties are compared with the best existing variety for the targeted traits to decide whether or not they will be released.

The list of traits measured for VCU (grain yield, disease resistance, plant height, etc.) and DUS (shape, colour, height, leaf length, etc.) varies across countries. Tests are repeated for at least two or three seasons. Requirement for the number of locations depends on the environment for which the variety is being recommended.

In most countries, the national seed board is responsible for assembling and conducting national performance trials (NPTs) from which VCU and DUS data are obtained. Once the VCU and DUS data have been recorded, they are then submitted to the NVRC for consideration. In Kenya, Tanzania and Uganda, the NPTs are conducted by the national seed boards for a set fee, and this may hinder some breeders from submitting their varieties, especially in the public sector.

To complement the VCU data from NPTs, independent trials, grown on farmers' fields by the farming community are required. Some countries – such as Benin, Mali, Mozambique and Uganda – accept VCU and DUS data from other countries with similar agro-ecological conditions to complement in-country data.

Agronomic data collected by breeders and socio-economic information from participatory varietal selection (PVS) are acceptable as credible VCU data for varietal release in eight countries.

History of rice varietal release/adoption in Africa

Between 1960 and 2010, over 700 rice varieties were released or adopted in sub-Saharan Africa (Table 6.3). The number of varieties released/adopted varied between countries and years. A total of 250 varieties were released between 1960 and 1990, i.e. about 8 per year. This increased to an average of 16 per year during the next decade (1991–2000), reaching 29 varieties per year in the period 2001–2010, with more countries progressively becoming involved in rice research and development.

The highest number of varieties released between 1960 and 2010 was in Burkina Faso (64), followed by Mali (62), Nigeria (57), Côte d'Ivoire (56) and Senegal (50). Rice cultivation is just taking off in Gabon and the country has not yet officially released any rice variety. The relatively high release rates in Burkina Faso, Mali, Nigeria, Côte d'Ivoire and Senegal are probably related to their long histories of rice cultivation and relatively advanced varietal release systems compared to other countries (they are all Group 1 countries). Another factor could be that some countries, such as Mali, accept agronomic data from other countries with similar growing conditions to complement their own data. Although the countries of East Africa are well advanced in their seed and varietal release systems, they have not yet released large numbers of varieties because rice only recently became a major crop in the region. Egypt is the most advanced in terms of rice production, but not many varieties have been released (only ten), because the country has a system in place that ensures that only a limited number of varieties are released over a period of time.

Varieties were released for four rice-growing agroecosystems: upland, rainfed lowland, irrigated lowland and mangrove. Of the released varieties, the largest number of varieties was released for upland systems (34%), followed by irrigated lowland (31%), rainfed lowland (24%) and mangrove (9%), while the growth environment is not known for 2% of the varieties.

Conclusions

The survey results show that varietal release systems differ greatly among countries in sub-Saharan Africa. Countries could be classified into four groups, based on the existence and functionality of the varietal release system and, in the absence of a varietal release system, whether a varietal register exists. Thirteen of the 30 countries surveyed had a functional varietal release system (Group 1 countries).

Table 6.3. Varietal release/adoption history by country and rice-growing environment in 30 African countries between 1960 and 2010.

Country	Irrigated				Rainfed lowland				Upland				Mangrove				Total	Source of information
	Pre-90	91 to 00	01 to 10	NA	Pre-90	91 to 00	01 to 10	NA	Pre-90	91 to 00	01 to 10	NA	Pre-90	91 to 00	01 to 10	NA		
Varieties released by country																		
Burkina Faso	3	1	0	0	20	9	4	0	18	6	3	0	0	0	0	0	64	ECOWAS (2008); survey through national partner
Côte d'Ivoire	7	5	2	0	8	1	2	0	13	14	4	0	0	0	0	0	56	ECOWAS (2008); MINAGRI (2002); CNRA (2011); survey through national partner
Democratic Republic of Congo	0	1	2	0	0	2	1	0	1	1	11	0	0	0	0	0	19	Survey through national partner
Egypt	2	9	2	0	0	0	0	0	0	0	0	0	0	0	0	0	13	Survey through national partner
Ethiopia	0	0	0	0	0	0	1	0	0	0	4	0	0	0	0	0	5	Survey through national partner
Gabon	0	0	0	0	0	0	0	0	0	0	0	0	0	0	0	0	0	Survey through national partner
Ghana	1	1	11	0	6	1	11	0	0	0	10	0	0	0	0	0	41	ECOWAS (2008); survey through national partner
Kenya	0	0	7	0	0	0	0	0	0	0	5	0	0	0	0	0	12	KEPHIS (2010); survey through national partner
Madagascar	0	0	16	0	0	0	0	0	0	0	6	0	0	0	0	0	22	Survey through national partner
Mali	9	4	11	0	3	0	11	0	1	0	10	6	5	0	2	0	62	ECOWAS (2008); survey through national partner
Mauritania	3	3	6	0	1	0	0	0	0	0	1	0	0	0	0	0	14	ECOWAS (2008); survey through national partner
Mozambique	0	4	0	0	0	4	0	0	0	0	0	0	0	0	0	0	8	Survey through national partner
Nigeria	13	1	1	0	18	0	2	0	18	0	3	0	1	0	0	0	57	Guei and Traoré (2001); NACGRAB (2009)
Senegal	2	5	23	0	5	4	0	0	3	1	2	0	1	3	1	0	50	ECOWAS (2008); survey through national partner
Tanzania	6	0	2	0	0	0	0	0	0	0	6	0	0	0	0	0	14	Survey through national partner
Uganda	0	0	0	0	0	0	0	0	0	4	6	0	0	0	0	0	10	Survey through national partner
Sub-total	46	34	83	0	61	21	32	0	54	26	71	6	7	3	3	0	447	
Varieties adopted by country																		
Benin	3	1	0	0	6	2	2	0	3	5	0	0	0	0	0	0	22	MAEP (2011)
Burundi	0	0	0	0	0	0	0	0	1	5	9	0	0	0	0	0	15	Survey through national partner

Central African Republic[a]	2	1	0	0	0	0	1	4	2	0	0	0	0	1	0	0	11	Survey through national partner
Chad[a]	0	0	0	0	0	0	0	0	0	4	0	0	0	0	0	0	4	Guei and Traoré (2001); survey through national partner
The Gambia[a]	8	4	1	0	2	0	0	0	2	0	5	0	1	4	2	0	29	ECOWAS (2008); survey through national partner
Guinea[a]	0	0	0	0	3	5	6	0	0	4	5	0	7	6	0	0	36	Dalton and Guei (2003); survey through national partner
Guinea-Bissau[a]	0	0	0	0	1	2	0	0	0	5	1	0	2	13	0	0	24	Survey through national partner
Liberia[a]	0	0	0	0	0	0	2	0	7	3	5	0	0	0	0	0	17	Guei and Traoré (2001); survey through national partner
Niger[a]	4	4	4	0	0	0	3	0	0	0	1	0	0	0	0	0	16	ECOWAS (2008); survey through national partner
Republic of Congo[a]	0	0	0	0	0	0	0	0	0	0	3	0	0	0	0	0	3	Survey through national partner
Rwanda[a]	0	0	12	0	0	0	8	0	0	0	0	0	0	0	0	0	20	ISAR (2010)
Sierra Leone[a]	7	0	0	0	6	0	2	0	8	1	6	0	6	2	6	0	44	ECOWAS (2008); survey through national partner
Togo[a]	1	1	3	0	0	0	5	0	0	1	0	4	0	0	0	0	15	ECOWAS (2008); survey through national partner
Sub-total	25	11	23	0	18	9	29	4	23	28	37	4	16	26	8	0	261	
Total	71	45	106	0	79	30	61	4	77	54	108	10	23	29	11	0	708	

[a]Countries without varietal release system; NA = not available.

Varietal testing and release systems in sub-Saharan Africa are fragmented, which makes the system overall costly and prone to duplication – since the same variety is often tested in all countries where it is being targeted for marketing. Also, most national varietal catalogues are not updated regularly, which makes it difficult for seed companies and farmers to keep up to date with every improved variety. It is not surprising, therefore, that seed systems in Africa are underdeveloped and access to good performing (rice) seed is still a major issue and constraint (see Bèye *et al.*, Chapter 14, this volume).

To increase rice production in sub-Saharan Africa, well-coordinated rice breeding efforts (see Kumashiro *et al.*, Chapter 5, this volume), functional national varietal release systems, and regional efforts to facilitate seed trade across borders are essential.

References

CNRA (2011) *Repertoire des acquis de recherche du CNRA*, Tome 2: *Cultures vivrieres.* Centre National de Recherche Agronomique, Abidjan, Côte d'Ivoire.

Dalton, T.J. and Guei, R.G. (2003) Productivity gains from rice genetic enhancements in West Africa: countries and ecologies. *World Development* 31(2), 359–374.

ECOWAS (2008) Rice [*Oryza sativa* (L.) and *Oryza sativa* (L.) × *Oryza glaberrima* Steud.]. In: *West African Catalogue of Plant Species and Varieties.* Food and Agriculture Organization of the United Nations, Rome, Italy.

Guei, R.G. and Traoré, K. (2001) New approach to germplasm exchange for a sustainable increase of rice biodiversity and production in Africa. *International Rice Commission Newsletter* 50, 49–58.

ISAR (2010) *Description of New Rice Varieties Introduced in Rwanda.* Rwanda Agricultural Research Institute, Kigali, Rwanda.

KEPHIS (2010) National Crop Variety List – Kenya. Kenya Plant Health Inspectorate Service, Nairobi, Kenya.

Louwaars, N.P. (2002) Variety controls. In: Louwaars, N.P. (ed.) *Seed Policy, Legislation and Law: Widening a narrow focus.* The Haworth Press, New York, pp. 131–142.

MAEP (2011) Riz [*Oryza sativa* (L.) et *Oryza glaberrima (Steud.)*]. In: *Catalogue Béninois des Espèces et Variétés Végétales (CaBEV).* Imprimerie ATG, Cotonou, Benin.

MINAGRI (2002) *Catalogue officiel des variétés de riz.* Ministère de l'Agriculture, Abidjan, Côte d'Ivoire.

NACGRAB (2009) Rice (*Oryza sativa*). In: *Crop Varieties Released and Registered in Nigeria*, No. 2. Xavier Communications, Ibadan, Nigeria.

UPOV (1978) *Act of 1978: International Convention for the Protection of New Varieties of Plants of December 2, 1961, as Revised at Geneva on November 10, 1972, and on October 23, 1978.* Available at: www.upov.int/export/sites/upov/upovlex/en/conventions/1978/pdf/act1978.pdf (accessed 4 February 2013).

UPOV (1991) *Act of 1991: International Convention for the Protection of New Varieties of Plants of December 2, 1961 as Revised at Geneva on November 10, 1972, on October 23, 1978, and on March 19, 1991.* www.upov.int/export/sites/upov/upovlex/en/conventions/1991/pdf/act1991.pdf (accessed 4 February 2013).

7 Diversity of Rice and Related Wild Species in Africa

Kayode Abiola Sanni,[1]* Daniel D. Tia,[1] David K. Ojo,[2] Ayoni S. Ogunbayo,[1] Mouritala Sikirou[3] and N. Ruaraidh Sackville Hamilton[4]

[1]Africa Rice Center (AfricaRice), Cotonou, Benin; [2]Department of Plant Breeding and Seed Technology, Federal University of Agriculture, Abeokuta, Nigeria; [3]Department of Crop Production, Faculty of Agricultural Sciences, University of Abomey-Calavi, Benin; [4]International Rice Research Institute (IRRI), Los Baños, Philippines

Introduction

In-depth understanding and wise management of genetic resources and genetic diversity is of great importance to progress in rice breeding and sustainable production (Guimaraes, 2002). Africa possesses an enormous wealth of rice genetic resources which can contribute to the broadening of the genetic base of rice varieties grown in farmers' fields to ensure that they become better adapted to the multiple biotic and abiotic stresses in Africa. This diversity is under threat from both environmental and socio-economic factors, such as urbanization and climate change (Kiambi *et al.*, 2008).

Rice belongs to the genus *Oryza*, classified under the tribe *Oryzeae*, sub-family *Oryzoideae* (*Ehrhartoideae*) of the grass family *Poaceae* (*Gramineae*). The *Oryza* genus was named by Linnaeus (1753). The taxonomy of the genus has undergone numerous revisions. Vaughan *et al.* (2003) recognized 23 species. GRIN Taxonomy (USDA, ARS, National Genetic Resources Program, 2012), a standard widely adopted by the genetic-resources community, now recognizes 26 species. Of these, two are cultivated in Africa, *O. sativa* and *O. glaberrima*, both of which are diploid (genome AA, 2n = 24).

Oryza sativa was introduced to East Africa, probably in the first century AD, via traders from India (Harlan and Stemler, 1976; Ng *et al.*, 1991) and subsequently to West Africa by the Portuguese in about AD 1500 (Portères, 1962).

According to Portères (1962, 1976), *O. glaberrima* was first domesticated from the wild ancestor *O. barthii* (formerly known as *O. breviligulata*) by people living in the inland delta of the upper Niger River about 3500 years BP. The species spread to two secondary centres of domestication, one along the coast of The Gambia, Casamance (Senegal) and Guinea-Bissau and the second in the Guinea forest between Sierra Leone and the western part of Côte d'Ivoire. However, Harlan and Stemler (1976) suggest that *O. glaberrima* was domesticated from *O. barthii* in several different localities within the vast forest and savannah areas. Many studies have confirmed that *O. barthii* is the progenitor of *O. glaberrima* and both are restricted to West Africa, while *O. longistaminata* (from which *O. barthii* is derived) is widely distributed in Africa (Brar and Khush, 2003).

* Corresponding author: k.sanni@cgiar.org

© CAB International 2013. *Realizing Africa's Rice Promise* (eds M.C.S. Wopereis *et al.*)

Oryza also contains 24 wild species distributed throughout the tropics and sub-tropics and representing 10 genome types (AA, BB, CC, BBCC, CCDD, EE, FF, GG, HHJJ and HHKK) (Vaughan *et al.*, 2003). Africa is home to representatives of five of these genome types: AA (*O. longistaminata*, *O. barthii*, *O. glaberrima* and *O. sativa*), BB (*O. punctata*), BBCC (*O. schweinfurthiana*), CC (*O. eichingeri*) and FF (*O. brachyantha*) (Vaughan *et al.*, 2003).

The collection, conservation and utilization of genetic diversity of these rice resources and related wild species in Africa were entrusted to the Africa Rice Center (AfricaRice) and resources are kept in trust for humanity under the auspices of the Food and Agriculture Organization of the United Nations (FAO). This chapter provides an overview of efforts by AfricaRice and partners since 1970 to build up an *ex-situ* collection. The current status of knowledge with respect to these accessions is presented. The chapter concludes with a discussion of the challenges remaining for expansion, characterization and more effective use of the collection.

Collection Efforts and Conservation

Between 1973 and 1977, AfricaRice received collections from national and international research institutes working on rice in Africa, including the Office de la Recherche Scientifique et Technique d'Outre-Mer (ORSTOM), Institut de Recherches Agronomiques Tropicales (IRAT) and the International Institute for Tropical Agriculture (IITA) (Abifarin, 1988). Since 1978, it has continued to receive materials from various collaborators and has also organized its own missions to collect landraces from all rice-growing environments, mostly in West Africa. The mangrove-swamp rice station in Rokupr (Sierra Leone) collected landraces from the mangrove-swamp environment; the then irrigated rice station based in Saint-Louis (Senegal) collected lowland rice; the station based in Mopti (Mali) collected deep-water/floating rice and the upland station based in M'bé/Bouaké (Côte d'Ivoire) collected upland landraces. Between 2000 and 2008, collections were made through various missions by AfricaRice scientists and partners in Burkina Faso, Côte d'Ivoire, Niger, Sierra Leone and Togo.

About 20,000 accessions of rice are being conserved in the gene bank of AfricaRice, made up of both cultivated (*O. glaberrima* and *O. sativa*) and wild species (*O. barthii*, *O. longistaminata*, *O. punctata*, *O. brachyantha* and *O. eichingeri*) from all across the world, with over 80% of the germplasm accessions from Africa (Table 7.1). AfricaRice, the International Rice Research Institute (IRRI) and national partners are now collaborating to expand the germplasm collection to include entries from North, Central, East and Southern Africa (Table 7.1).

The accessions conserved in the AfricaRice gene bank are from about 85 countries and are kept in trust for humanity under the Multi-Lateral System (MLS) of access and benefit sharing within the purview of the International Treaty on Plant Genetic Resources for Food and Agriculture of FAO as part of the global *ex-situ* collections. These accessions are kept in a medium-term storage at 5°C at the temporary headquarters of AfricaRice in Cotonou (Benin) and in a long-term conservation facility at −18°C at IITA in Ibadan (Nigeria). Two safety duplications are also maintained at the National Center for Genetic Resources Preservation (NCGRP), Fort Collins, USA and Svalbard Global Seed Vault (SGSV), Norway in collaboration with Global Crop Diversity Trust as part of the global effort for crop diversity security.

Seeds of cultivated species are multiplied during the dry season when good-quality seeds (free from disease and pest attack) can be produced. Wild species are grown in pots in a quarantined screen house. After harvest, seeds are partially dried, threshed and hand cleaned. Seeds are then brought to a cool drying room maintained at 18°C and 20% relative humidity to attain a moisture content of 6% before medium- and long-term conservation. The temperature and humidity of the storage rooms are monitored daily, and the viability of the seeds in medium-term storage is monitored every 5 years for each accession.

AfricaRice has established a database for the accessions stored in its gene bank, which is being bar-coded. Data associated with each accession consist of passport information and phenotypic and genotypic data.

Table 7.1. Number of rice germplasm accessions per region in the AfricaRice gene bank as of April 2012.

Origin	*O. glaberrima*	*O. sativa*	Wild species	Total
Central Africa	83	190	101	374
East Africa	14	743	13	770
North Africa	0	53	0	53
West Africa	2,400	12,025	311	14,736
Southern Africa	0	869	53	922
Total Africa	2,497	17,033	482	20,012
Others	0	3,153	4	3,157

Current Status of Knowledge

Molecular diversity analysis

The population structure of *O. glaberrima* was analysed using 93 simple sequence repeat (SSR) markers on 198 accessions by Semon *et al.* (2004). The accessions were selected among 1136 accessions in the AfricaRice gene bank to maximize geographical and morphological diversity. Five genetically distinct groups were identified. Two of those groups (groups 4 and 5) shared ancestry with the two subspecies of *O. sativa* (*indica* and *japonica*, respectively). It was suggested that the three other groups corresponded to the three ecotypes reported by Portères (1970). Group 1 contained plants that were relatively tall and was found throughout West Africa. This group may represent the ancestral floating type first domesticated in the inland delta of the upper Niger River. Group 2, mostly from Nigeria, consisted of late-maturing entries. This group represented the non-floating (lowland) rice type that migrated to Nigeria along the Niger River. The third group originated from Liberia and was characterized by a larger number of tillers per plant and longer grain, corresponding phenotypically to the upland rice type.

Barry *et al.* (2007) used 11 SSR markers to characterize 26 *O. glaberrima* and 144 *O. sativa* accessions collected in maritime Guinea. They revealed the existence of two major ecotypes – floating and erect – among the *O. glaberrima* accessions. Moreover, they detected an original genetic compartment, highlighting the occurrence of *O. glaberrima* × *O. sativa* hybridization.

Dramé *et al.* (2011) used a highly polymorphic set of 30 SSR markers to characterize 74 accessions of *O. glaberrima* collected from various ecosystems in eight countries. They differentiated the two major ecotypes – a floating photosensitive type grown in deep water, including coastal mangrove areas, and an early erect ecotype grown in upland or moderately inundated lowlands (Ghesquière *et al.*, 1997) – and an intermediate group between *O. sativa* and *O. glaberrima*.

Morphological characterization

Guei *et al.* (2005) assessed the genetic diversity of 434 *O. sativa* landraces from Côte d'Ivoire using morphological traits such as plant height, leaf length, number of days to heading and maturity, tillering ability, panicle length and grain size (weight, length and width) as the principal discriminatory characteristics. The accessions were classified into seven groups, each with specific traits. This information could be used by breeders in making parental selection based on their traits of interest.

Sanni *et al.* (2008) analysed the geographical patterns of phenotypic diversity of 880 *O. sativa* landraces of upland rice collected from the four main rice-production regions in Côte d'Ivoire: Gagnoa (west-central), Touba (north-western), Boundiali (northern) and Danané (western). Variation was observed across the geographical zones and ecological regions with respect to the 13 traits measured. The majority of accessions from the north and north-west were early maturing, a sign of adaptation to the short rainfall period in these zones. Meanwhile, most accessions from the west and west-central regions were late maturing.

Jones *et al.* (1999) studied 1300 *O. glaberrima* accessions and reported considerable variation in seedling vigour. More than 90% of the accessions showed very high seedling vigour, and 15% also had very rapid vegetative growth and produced a large number of tillers and droopy lower leaves within 20 to 30 days after sowing, resulting in rapid ground cover and good weed suppression.

AfricaRice has initiated a comprehensive morphological characterization of the whole *O. glaberrima* collection (about 2500 accessions) in order to understand its diversity, as well as to make the information available for breeding activities.

Trait Identification in *O. glaberrima*

Oryza glaberrima constitutes a rich reservoir of adaptive traits essential for new rice varieties for Africa. It has sources of resistance to African rice gall midge (Nwilene *et al.*, 2002), for weed competitiveness (Rodenburg *et al.*, 2009), for tolerance to iron toxicity (Sahrawat and Sika, 2002), and for drought tolerance (Ndjiondjop *et al.*, 2010; Bimpong, 2011; Bocco, 2012).

New screening, inspired by identification of a source of resistance to *Rice yellow mottle virus* (RYMV) among *O. sativa*, *O. glaberrima* and *O. barthii* (Thottappilly and Rossel, 1993), led to the identification of new resistant accessions. Among these accessions, TOG5681 and TOG5672 exhibited two new resistance alleles (*rymv1-3* and *rymv1-4*, respectively), different from the resistance allele *rymv1-2* first identified in an *O. sativa* accession Gigante (Albar *et al.*, 2006). Furthermore, using 337 *O. glaberrima* accessions, covering a wider geographic area in Africa, a new gene named *RYMV2*, different from *RYMV1*, was identified (Thiémélé *et al.*, 2010).

Screening a set of 107 *O. glaberrima* accessions, selected on the basis of their geographical origin, for resistance to African strains of *Xanthomonas oryzae*, Djedatin *et al.* (2011) identified 20 accessions showing resistance to strains from Mali.

Challenges for More Effective Use of Genetic Resources

Enhancing use

Arguably the greatest challenge facing gene bank managers is how to facilitate effective use of the germplasm. Germplasm without information about its major characteristics has little value, hindering its wide use. AfricaRice has worked with Bioversity International and IRRI to define standards for characterizing rice (Bioversity International *et al.*, 2007). However, it is neither practically nor economically feasible to evaluate all accessions in a gene bank for agriculturally important traits. Even if it were feasible, much of the genetic variation present in germplasm collections is hidden by factors such as epistasis and epigenetics – valuable traits may be expressed in the progeny of crosses between accessions in ways that cannot be predicted simply by evaluating the accessions. Traditionally, gene bank managers have characterized their accessions for highly heritable traits that are simple enough to record for all accessions; however, most such traits have little agronomic relevance.

To overcome this problem, the concept of 'core collections' (van Hintum *et al.*, 2000) has become established as an effective 'entry point' to start exploring the potential value of collections. A core collection consists of a sub-set of the germplasm believed to provide a reasonable representation of the genetic diversity of a particular species, and small enough to be evaluated in detail. In most cases, however, a core set has been derived from a sub-set of the germplasm, but not the entire collection. Furthermore, traits evaluated have usually been limited. A core set should be derived from an entire collection of a particular species, and be evaluated for the majority of agriculturally important traits.

Thanks to rapid advances in genomics, molecular markers such as SSRs and single nucleotide polymorphisms (SNPs) can be used to characterize the diversity of rice germplasm at reasonable cost, and the information can be used to define more representative core collections for detailed evaluation. Development of core sets of wild species and landraces is an urgent task in Africa. Realizing the importance of this, AfricaRice – in collaboration with Institut de recherche pour le développement (IRD) – has started to develop core sets of *O. glaberrima* and *O. barthii* based on SNP markers and will subsequently evaluate them for major agronomic traits including disease resistance.

However, the core collection concept itself suffers limitations. First, although intended as an entry point to explore the collection, mechanisms have not been developed to go beyond the core to explore the rest of the collection. Second, since core collections are constructed to represent the greatest possible diversity in a reasonably small sub-set regardless of the trait(s) of interest, they are by definition a generic entry point; for the user with a specific breeding or research objective, it is often possible and suitable to construct a sub-set better tailored to their specific needs.

One proven alternative is the ecogeographic approach to select use-specific sub-sets (Sackville Hamilton *et al.*, 2002; Mackay and Street, 2004). Assuming that accessions are adapted to the environment they came from, we may use knowledge of the distribution of accession origins to predict probable patterns of variation. However, creating such a sub-set requires intensive analysis. There is a need to mainstream this approach so that gene bank curators or even users can more easily select tailor-made sub-sets designed to better meet the specific needs of each user.

A new approach that brings the promise of radically improving the effectiveness of using germplasm collections, is the genomics approach (McCouch *et al.*, 2012; McCouch *et al.*, Chapter 9, this volume). In the 3 years from 2008, the cost per raw megabase of DNA sequence decreased from approximately US$500 to less than $0.10. We can now envision that in the future the genome of every accession present in the collection will be sequenced. Having sequence information of every accession, discovery of new alleles for agronomically important traits can be done by association analysis based on available phenotypic data (McCouch *et al.*, 2012).

The keys to this approach will be to tie all phenotyping to genotyping or sequencing, and to involve gene banks more in gene discovery. For that, gene bank curators should no longer characterize independently of other phenotyping efforts; for rice, this work should be integrated into the phenotyping network of the Global Rice Science Partnership (GRiSP).

Collection of germplasm from under-represented regions

Since 1978, AfricaRice has organized collection missions with national (NARS) partners to collect landraces and wild species from all rice-growing environments, mostly in West Africa (Burkina Faso, Côte d'Ivoire, Niger, Sierra Leone and Togo). NARS scientists have also collected in Côte d'Ivoire, Mali, Senegal and Sierra Leone. Collection efforts are under way with NARS partners and IRRI in East Africa.

However, there may still be areas and countries that are under-represented in the AfricaRice gene bank. We intend to check the gene bank database for under-represented areas and, when needed, establish collection programmes with NARS partners in compliance with their national priorities, policies and regulations on collecting germplasm and traditional knowledge. Complete information associated with the collected germplasm (e.g. geographical origin, including latitude, longitude and altitude) will be recorded.

It will be essential to plan the further collection of African rice in the context of the global rice strategy (Hay and Sackville Hamilton, 2010). Compared with many other crops, the rice gene pool is already relatively well conserved *ex situ*, with over 500,000 accessions in gene banks worldwide. Annual investment in conservation is high. The additional cost of adding a new accession to a collection is much higher than the recurrent annual cost of conserving the accession (Shands *et al.*, 2011). Thus, it is becoming increasingly important to ensure that efforts to collect additional germplasm genuinely add additional diversity.

However, no methodology exists to address this need systematically, or to estimate when a collection can be considered 'complete'. As genotyping and sequencing become ever more affordable, it is now possible – as well as increasingly important – to develop such a methodology: to determine how much genetic novelty is added to a collection when new accessions are added; to quantify the diminishing rate of return on sequentially adding more accessions; and to devise the cost–benefit relationships to guide how much more collecting is needed. We will rely on the geographic information system- (GIS-) based global gap analysis of the Global Crop Diversity Trust (Hijmans *et al.*, 2001; Ramírez-Villegas *et al.*, 2010) to establish priorities for collection of the wild relatives of rice.

Up-to-date databases

A wide range of information about rice accessions is recorded by various stakeholders. This includes information on their genealogy and origins, conservation status, availability and use, genetic and phenotypic characteristics, agronomic adaptation, performance and quality, and the performance of their progeny. Much of this information is generated through scientific research by those who receive germplasm from gene banks, and much is in the domain of traditional knowledge held by the indigenous communities that originally grew landraces now conserved in the gene bank. For effective conservation and use of accessions, all this information should ideally be integrated and made available. However, the task of collating and publishing it is beyond the capacity of any gene bank or any other laboratory. One effective strategy to help fill a critical component of the information gap is based on recognizing that the most important scientific knowledge generated by recipients of accessions is typically published in scientific journals. Information in a published journal can be incorporated into the germplasm database. With a large proportion of the scientific literature now available electronically, it would be possible (as well as desirable) to organize a systematic survey of journals, augmented by enquiries to the germplasm recipients. Such surveys should lead to an increase in the available information, thereby increasing the value of accessions.

Information on rice germplasm conserved in the AfricaRice gene bank is stored in a database called AfricaRice Genebank Information System (ARGIS) and made available to users online (www.africarice.org/argis). Likewise, information newly gained through internal research activities as well as that obtained from the recipients will be stored and managed in this system. Here, the accuracy of information and data for each accession is the most important. These data are now being integrated into IRIS (International Rice Information System, http://iris.irri.org). A key priority for the future development of IRIS will be to develop improved mechanisms for seamlessly sharing data, to facilitate collaboration among the many different partners involved in GRiSP.

Conclusions

The rational exploitation of the genetic diversity existing among the 20,000 accessions held in the AfricaRice gene bank represents one of the basic components of efforts to improve rice production in Africa. To enhance this exploitation, service to stakeholders needs to be improved in three areas:

- The information associated with each accession conserved in the gene bank: there is a need to complete the identification of key descriptors for each accession in collaboration with experts for target traits. This will enable the formation of sub-sets for targeted traits and users. There is also a need to genotype/sequence the accessions in the selected sub-sets and associate the genotypic and phenotypic data, to enhance their use.
- Conservation methods and equipment: regeneration methods that avoid genetic drift are needed to ensure the maintenance of the genetic integrity of accessions.
- Management and use of information from the gene bank: it is important to update the information on a regular basis. Within the framework of GRiSP, AfricaRice, IRRI and the International Center for Tropical Agriculture (CIAT) will collaborate to ensure the accessibility, through IRIS, of all available information on rice gene bank accessions worldwide.

References

Abifarin, A.O. (1988) West Africa Rice Development Association (WARDA) activities in rice germplasm collection, conservation and utilization. In: *Proceedings of an International Conference on Crop Genetic Resources of Africa*, Ibadan, Nigeria, 17–20 October 1988, pp. 35–40.

Albar, L., Angratz-Reyser, M., Hébrard, E., Ndjiondjop, M.-N., Jones, M. and Ghesquière, A. (2006) Mutations in the eIF(iso)4G translation initiation factor confer high resistance of rice to *Rice yellow mottle virus*. *The Plant Journal* 47, 417–426.

Barry, M.B., Pham, J.L., Noyer, A.J.L., Billot, A.C., Courtois, A.B. and Ahmad, A.N. (2007) Genetic diversity of the two cultivated rice species (*O. sativa* and *O. glaberrima*) in maritime Guinea. Evidence for inter-specifc recombination. *Euphytica* 154, 127–137.

Bimpong, I.K., Serraj, R., Chin, J.H., Mendoza, E.M.T., Hernandez, J. and Mendioro, M.S. (2011) Determination of genetic variability for physiological traits related to drought tolerance in African rice (*Oryza glaberrima*). *Journal of Plant Breeding and Crop Science* 3(4), 60–67.

Bioversity International, IRRI and WARDA (2007) *Descriptors for Wild and Cultivated Rice (*Oryza *spp.).* Bioversity International, Rome, Italy; International Rice Research Institute, Los Baños, Philippines; Africa Rice Center, Cotonou, Benin.

Bocco, R., Lorieux, M., Seck, P.A., Futakuchi, K., Manneh, B. and Baimey, H. (2012) Agro-morphological characterization of a population of introgression lines derived from crosses between IR 64 (*Oryza sativa indica*) and TOG 5681 (*Oryza glaberrima*) for drought tolerance. *Plant Science* 183, 65–76.

Brar, D.S. and Khush, G.S. (2003) Utilization of wild species of genus *Oryza* in rice improvement. In: Nanda, J.S. and Sharma, S.D. (eds) *Monograph on Genus* Oryza. Science Publishers, Enfield, New Hampshire, pp. 283–309.

Djedatin, G., Ndjiondjop, M., Mathieu, T., Vera Cruz, C.M., Sanni, A., Ghesquière, A. and Verdier, V. (2011) Evaluation of African cultivated rice *Oryza glaberrima* for resistance to bacterial blight. *Plant Disease* 95(4), 441–447.

Dramé, K.N., Sanchez, I., Gregorio, G. and Ndjiondjop, M. (2011) Suitability of a selected set of simple sequence repeats (SSR) markers for multiplexing and rapid molecular characterization of African rice (*Oryza glaberrima* Steud.). *African Journal of Biotechnology* 10(35), 6675–6685.

Ghesquière, A., Séquier, J., Second, G. and Lorieux, M. (1997) First steps towards a rational use of African rice, *Oryza glaberrima*, in rice breeding through a 'contig line' concept. *Euphytica* 96, 31–39.

Guei, R.G., Sanni, K.A., Abamu, F.J. and Fawole, I. (2005) Genetic diversity of rice (*Oryza sativa* [L.]). *Agronomie Africaine* 5, 17–28.

Guimaraes, E.P. (2002) Genetic diversity of rice production in Brazil. In: Nguyen, V.N. (ed.) *Genetic Diversity in Rice Production, Case Studies from Brazil, India and Nigeria.* Food and Agriculture Organization of the United Nations (FAO), Rome, Italy, pp. 11–35.

Harlan, J.R. and Stemler, A. (1976) The races of sorghum in Africa. In: Harlan, J.R., De Wet, J.M. and Stemler, A.B. (eds) *Origin of African Plant Domestication.* Mouton, The Hague, Netherlands.

Hay, F.R. and Sackville Hamilton, R. (2010) Global strategy for the *ex situ* conservation of rice genetic resources. International Rice Research Institute, Manila, Philippines. www.croptrust.org/documents/cropstrategies/Rice%20Strategy.pdf.

Hijmans, R.J., Guarino, L., Cruz, M. and Rojas, E. (2001) Computer tools for spatial analysis of plant genetic resources data: 1. DIVA-GIS. *Plant Genetic Resources Newsletter* 127, 15–19.

Jones, M.P., Nwanza [sic], K.F., Miezan, K.M., Singh, B.N. and Guei, R.G. (1999) Rice germplasm evaluation and enhancement at WARDA. In: Rutger, J.N., Robinson, J.F. and Dilday, R.H. (eds) *Proceedings of the International Symposium on Rice Germplasm Evaluation and Enhancement*, Stuttgart, Arkansas, USA, 30 August to 2 September 1998. *Special Report* 195. Arkansas Agricultural Experiment Station, Fayetteville, USA, pp. 29–37.

Kiambi, D.K., Newbury, H.J., Maxted, N. and Ford-Lloyd, BV. (2008) Molecular genetic variation in the African wild rice *Oryza longistaminata* A. Chev. et Roehr. and its association with environmental variables. *African Journal of Biotechnology* 7(10), 1446–1460.

Linnaeus, C. (1753) *Species plantarum: exhibentes plantas rite cognitas, ad genera relatas, cum differentiis specificis, nominibus trivialibus, synonymis selectis, locis natalibus, secundum systema sexuale digestas.* Laurentius Salvius, Stockholm, Sweden.

Mackay, M.C. and Street, K.A. (2004) Focused identification of germplasm strategy – FIGS. In: Black, C.K., Panozzo, J.F. and Rebetzke, G.J. (eds) *Cereals 2004: Proceedings of the 54th Australian Cereal Chemistry Conference and the 11th Wheat Breeders' Assembly*, Canberra, ACT, Australia, 21–24 September 2004. Royal Australian Chemical Institute, Melbourne, Australia, pp. 138–141.

McCouch, S.R., McNally, K.L., Wang, W. and Sackville Hamilton, R. (2012) Genomics of gene banks: a case study in rice. *American Journal of Botany* 99(2), 1–17.

Ndjiondjop, M.N., Manneh, B., Cissoko, M., Drame, N.K., Kakai, R.G., Bocco, R., Baimey, H. and Wopereis, M. (2010) Drought resistance in an interspecific backcross population of rice (*Oryza* spp.) derived from the cross WAB56-104 (*O. sativa*) × CG14 (*O. glaberrima*). *Plant Science* 179, 364–373.

Ng, N.Q., Chang, T.T., Vaughan, D.A. and Zunbo-Alto Versos, C. (1991) African rice diversity: conservation and prospects for crop improvement. In: Ng, N.Q., Perrino, P., Attere, F. and Zedan, H. (eds) *Crop*

Genetic Resources of Africa. Proceedings International Conference on Crop Genetic Resources of Africa, Ibadan, Nigeria, 17–20 October 1988, Vol. 2. International Institute of Tropical Agriculture, Ibadan, Nigeria, pp. 213–227.

Nwilene, F.E., Williams, C.T., Ukwangwu, M.N., Dakouo, D., Nacro, S., Hamadoun, A., Kamara, S.I., Okhidievbe, O., Abamu, F.J. and Adam, A. (2002) Reactions of differential genotypes to African gall midge in West Africa. *International Journal of Pest Managment* 48, 195–201.

Portères, R. (1962) Berceaux agricoles primaires sur le continent africain. *Journal of African History* 3(2), 195–210.

Portères, R. (1970) Primary cradles of agriculture in the African continent. In: Fage, J.D. and Olivier, R.A. (eds) *Papers in African Prehistory.* Cambridge University Press, Cambridge, UK, pp. 43–58.

Portères, R. (1976) African cereals: eleusine, fonio, black fonio, teff, *Brachiaria, Paspalum, Pennisetum*, and African rice. In: Harlan, J.R., de Wet, J.M.J. and Stemler, A.B.L. (eds) *The Origin of African Plant Domestication.* Mouton, The Hague, Netherlands, pp. 409–452.

Ramírez-Villegas, J., Khoury, C., Jarvis, A., Debouck, D.G. and Guarino, L. (2010) A gap analysis methodology for collecting crop genepools: a case study with *Phaseolus* beans. *PLoS ONE* 5(10), e13497. doi:10.1371/journal.pone.0013497 (accessed 13 February 2013).

Rodenburg, J., Saito, K., Kakai, R.G., Touré, A., Mariko, M. and Kiepe, P. (2009) Weed competitiveness of the lowland rice varieties of NERICA in the southern Guinea savanna. *Field Crops Research* 114(3), 411–418.

Sackville Hamilton, N.R., Skøt, L., Chorlton, K.H., Thomas, I.D. and Mizen, S. (2002) Molecular genecology of temperature response in *Lolium perenne*: 1. Preliminary analysis to reduce false positives. *Molecular Ecology* 11(9), 1855–1863.

Sahrawat, K.L. and Sika, M. (2002) Comparative tolerance of *O. sativa* and *O. glaberrima* rice cultivars for iron toxicity in West Africa. *International Rice Research Notes* 27, 30–31.

Sanni, K.A., Fawole, I., Guei, R.G., Ojo, D.K., Somado Eklou, A., Sanchez, I., Ogunbayo, S.A. and Tia, D.D. (2008) Geographical patterns of phenotypic diversity in *Oryza sativa* landraces of Côte d'Ivoire. *Euphytica* 160, 389–400.

Semon, M., Nielsen, R., Jones, M. and McCouch, S.R. (2004) The population structure of African cultivated rice *Oryza glaberrima* (Steud.): evidence for elevated levels of LD caused by admixture with *O. sativa* and ecological adaptation. *Genetics* 169, 1639–1647.

Shands, H., Hawtin, G. and MacNeil, G. (2011) The cost to the CGIAR Centers of maintaining and distributing germplasm. In: Consortium Board of Trustees. Financial Support to the CGIAR Center Genebanks in 2011. Proposal to the Fund Council, Annex 4, pp. 15–91. www.cgiarfund.org/cgiarfund/sites/cgiarfund.org/files/Documents/PDF/fc4_funding_proposal_CGIAR_Genebanks.pdf (accessed 13 June 2012).

Thiémélé, D., Boisnard, A., Ndjiondjop, M.-N., Chéron, S., Séré, Y., Aké, S., Ghesquière, A. and Albar, L. (2010) Identification of a second major resistance gene to *Rice yellow mottle virus*, *RYMV2*, in the African cultivated rice species, *O. glaberrima. Theoretical and Applied Genetics* 121, 169–179.

Thottappilly, G. and Rossel, H.W. (1993) Evaluation of resistance to *Rice yellow mottle virus* in *Oryza* species. *Indian Journal of Virology* 9, 65–73.

USDA, ARS, National Genetic Resources Program (2012) Germplasm Resources Information Network – (GRIN) [Online Database]. National Germplasm Resources Laboratory, Beltsville, Maryland. www.ars-grin.gov/cgi-bin/npgs/html/index.pl (accessed 11 June 2012).

van Hintum, T.J.L., Brown, A.H.D., Spillane, C. and Hodgkin, T. (2000) Core collections of plant genetic resources. *IPGRI Technical Bulletin* No. 3. International Plant Genetic Resources Institute, Rome, Italy.

Vaughan, D.A., Morishima, H. and Kadowaki, K. (2003) Diversity in the *Oryza* genus. *Current Opinion in Plant Molecular Biology* 6, 139–146.

8 Gene Flow in African Rice Farmers' Fields

Edwin Nuijten* and Paul Richards
Wageningen University, Wageningen, Netherlands

Introduction

West Africa is an important region for rice genetic resources. Here are found two species of wild rice (*Oryza barthii*, *O. longistaminata*) and an indigenous domesticated rice (*O. glaberrima*). In addition, the two subspecies of Asian cultivated rice (*O. sativa*) – *indica* and *japonica* – are extensively cultivated. Collection of wild rice in the Lake Chad basin has been dated to 3500 BP based on archaeological findings of wild rice remains (Klee *et al.*, 2004). It has been proposed that the domestication of African rice (*O. glaberrima*) began about this time, most likely in the upper Niger delta (Portères, 1962; Li *et al.*, 2011). Asian rice has been present in West Africa since at least the 15th century, introduced through cross-continental trading links with the Middle East during the Middle Ages (Carpenter, 1978) or coastal links with the Mediterranean from the arrival of the Portuguese on the West African coast (late 15th century). Since the colonial period or earlier, higher-yielding Asian rice has tended to predominate, especially in areas such as southern Nigeria, where rice is a recent crop. However, in older centres of cultivation – the major valleys of the West African savannah, the upper Guinean coastal region from Senegal to Sierra Leone, and some mountain outliers, such as the Togo Hills – African rice continues to be present, and is locally important (Barry *et al.*, 2007; Nuijten *et al.*, 2009; Okry, 2011, Teeken *et al.* 2011; Temudo, 2011). In these older, established areas of rice cultivation, wild rice is sometimes still gathered locally (Richards, 1986; Nuijten, 2005).

It is often assumed that in traditional agriculture crops are enriched by gene exchange with wild and weedy relatives (De Wet and Harlan, 1975; Altieri and Merrick, 1987; Prain, 1993), although there seems little hard evidence for the movement of valuable traits from wild relatives into crops (Wood and Lenné, 1997). This, however, does not mean that no new potentially valuable traits arise from introgression (Jarvis and Hodgkin, 1999). Discovering actual gene flow between wild and cultivated species in the field is complicated by the fact that, because of pairing of homologous genes and other processes during meiosis and fertilization, fertile interspecific progenies resemble either the wild or the cultivated species. This is illustrated by research on gene exchange between *O. sativa* and *O. glaberrima* that indicates that as the interspecific hybrids are sterile, backcrosses are needed to restore fertility. Therefore, the fertile hybrid derivatives resembled the parental phenotypes (Sano, 1989). This explains why the

* Corresponding author: Edwin.Nuijten@wur.nl

© CAB International 2013. *Realizing Africa's Rice Promise*
(eds M.C.S. Wopereis *et al.*)

interspecific rice types found by Nuijten *et al.* (2009) can be divided into two main forms, one morphologically similar to *O. glaberrima* and the other similar to *O. sativa*. This may also explain why, although it is often suggested that farmer crop management plays an important role in gene flow between wild and cultivated rice, the number of studies documenting recognition and selection by farmers of genetic variation arising from natural introgression is limited (Jarvis and Hodgkin, 1999). Maintaining gene flow between wild and cultivated species in farmers' fields may also meet concerns about the reduction of genetic diversity (see Chang, 2003; cf. Sang and Ge, 2007).

The significance of gene flow between cultivated and wild species is placed in a new light by work on a protracted model of crop domestication (Allaby *et al.*, 2008). This model emphasizes the need for more information on how populations of farmers influence gene flow between cultivated and wild species over long periods of time. It is wrong to assume that selection must be based on conscious management strategies by individual farmers. Even 'modern' management systems often depend on what is termed 'distributed cognition' (Hutchins, 1995). No single individual holds an overview of the direction in which collective action develops. Hutchins' (1995) example is the navigation of a modern warship. A model of this kind has been proposed for farmer crop selection, based on neural network engineering concepts of unsupervised learning. In unsupervised learning there is no master plan. The stable states of a network are a function of nodal weightings in a massively interconnected network. Rice farmers in a region such as West Africa, it has been proposed, approximate these network conditions (Richards *et al.*, 2009). From an empirical point of view, information is needed on the weights to be applied to the nodes. (Depending on the scale, nodes can be different actors, among others, farmers, varieties and plants.) Observation of gene flow rates and how farmers' actions influence the direction and amount of these flows is thus a crucial empirical step.

This chapter focuses on how farmers influence gene flow. Conscious management practices could be part of this influence, but not its sum total. We are equally interested in the unintended consequences of farm management practices unconnected to variety choice, such as decisions over labour allocation that result in a particular field mosaic of planted varieties. This way we try to avoid the pitfall of imposing a breeder's rationale on farmer gene-flow management. Breeders seek to control gene flow according to genetic theory. Farmers weigh the nodes in a gene-flow network in a wide variety of ways, only some of which will be connected with their own (possibly imperfect) ideas about descent and inheritance in planting materials. Since gene flow is, at a biological level, the mechanism of 'extended domestication', detailed information on gene flow patterns in 'distributed' networks is important. The information also has practical consequences for food security in regions where adaptation (e.g. to climate change) depends to a significant extent on farmer seed choice. We need to know whether 'extended domestication' and 'unsupervised learning' remain robust, or whether they have begun to break down due to network fragmentation linked to the spread of market-based seed supply.

The relevance of our focus on documenting gene flow at field level is emphasized by work on the existence of local rice varieties in West Africa with an interspecific *O. sativa* × *O. glaberrima* background, resulting from gene flow between the two species in farmers' fields (Jusu, 1999; Semon *et al.*, 2005; Barry *et al.*, 2007; Nuijten and Van Treuren, 2007; Nuijten *et al.*, 2009). It has been shown that there is regional variation in the occurrence of these local interspecifics. Hybrid-derived farmer varieties do not occur everywhere the two species occur together. For instance, interspecific hybrid-derived rice varieties (so-called farmer hybrids) are much less common in the Togo Hills than in the upper West African coastal areas – especially Casamance, Guinea-Bissau and Sierra Leone (Nuijten *et al.*, 2009). Perhaps the relatively earlier arrival of Asian rice on the upper Guinea coast (Angladette, 1966) has played a part in the development of interspecific varieties as it has increased the window of opportunity for interspecific hybridization. Topography has also played a part in keeping *O. sativa* and *O. glaberrima* physically apart on the Togo side of the Togo Hills (*O. sativa* is confined to lower elevations with more fertile soils and *O. glaberrima* is cultivated in poor upland soils at higher elevations). On the Ghanaian side of the Togo Hills, local cultural practices have also contributed to the physical separation of the two species. In general, local communities on the Ghanaian side of the Togo Hills favour Asian rice for commerce, and reserve African rice for ceremonies and ritual usage (Teeken *et al.*, 2011). This suggests that farmers' practices influencing gene flow need to be explained with reference to a range of cultural and environmental factors (Nuijten 2005, 2010; Teeken *et al.*, 2010, 2011; Nuijten and Richards, 2011). But only a handful of studies exist on how farmers 'manage' gene flow in these diverse circumstances (Dyer *et al.*, 2011). In particular, rice has been little studied in this regard (Jarvis and Hodgkin, 1999).

Variation in farmer strategies is known. The reasons why are not fully established, and ought to be given growing interest in endogenous processes of crop development (Semon *et al.*, 2005; Barry *et al.*, 2007; Nuijten *et al.*, 2009). For instance, in some cultures it is considered good practice to cultivate the two rice species mixed (among other reasons, to spread risk and ensure a good harvest), whereas in other cultures seed purity is emphasized. In cultures where seed purity is emphasized (and it may be for aesthetic as well as practical reasons), selection of seeds from the middle part of the field is often thought desirable (Longley and Richards, 1993), though there are few if any records on how often this ideal is met in practice. In other settings maintaining low levels of mixture is considered virtuous, since this is connected with the probability of finding new plants in the field and in-field diversity makes farming impractical (Richards, 1986). A local idea that mixture serves as a kind of gene bank has been reported (Nuijten, 2005). Temudo (2011) reports that mixture can help identify theft. The role of men and women in seed selection and seed exchange may also be very different across different ethnic groups. For example, in The Gambia both male and female farmers think that women have the skills and patience to develop new rice varieties by selecting off-types in the field and then testing them for several years (Nuijten, 2010). In Sierra Leone, by contrast, men and women are thought to be equally capable of developing new rice varieties (Richards, 1986).

Thus, we argue that knowing more about gene flow in farmers' fields is important for two reasons: (i) it supports management approaches that help to preserve processes underlying endogenous crop development, including *in situ* gene flow between wild and cultivated rice species, and the associated emergence of inter-specific progenies; and (ii) to understand how to link these 'unsupervised' processes with formal variety development. The need for a decentralized varietal development approach in which farmers have an essential role has become clearer, not only in Africa (Gridley *et al.*, 2002; Dorward *et al.*, 2007; Efisue *et al.*, 2008; Nuijten *et al.*, 2009; Mokuwa *et al.*, 2013), but worldwide (Almekinders and Elings, 2001; Ceccarelli *et al.*, 2001, 2010; Sperling *et al.*, 2001; Witcombe, 2002; Gyawali *et al.*, 2007; Desclaux, 2008; Østergård *et al.*, 2009; Joshi *et al.*, 2012). Conserving the 'unsupervised' system may offer a cost-efficient way of supporting a continued output of robust, adapted varieties for farmers not yet within reach of the formal system.

Methodology

The information provided in this chapter is based on a review of the literature and extensive fieldwork in The Gambia (2000–2003, 2007 and 2008), Senegal and Guinea-Bissau (2007 and 2008), and Sierra Leone (various times from 1981 to 2008). The research focused on upland rice farming systems and used interdisciplinary approaches integrating social science and biological science research methods (Nuijten, 2011). This approach is based on critical realism (Sayer, 2000), of which the Context Mechanism Outcome configuration is a central element (see Pawson and Tilley, 1997). Depending on the research question, candidate mechanisms are postulated, after which various research methods are used to test the candidate mechanisms. Critical realism does not exclude positivistic or post-modern research approaches and can be considered a framework for integrating methods from the social and biological sciences. For this research it implies describing the socio-economic and biological aspects of the farming context, and testing socio-economic, biological and complex socio-biological candidate mechanisms for explaining the dynamics of gene flow in rice-farming systems. The research methods used consisted of various qualitative and quantitative surveys with farmers (Richards, 1986; Nuijten, 2005), field mapping (for methods see Nuijten and Richards, 2011), field observations, field trials and molecular analysis for comparisons of seed materials collected from farmers at the phenotypic (Nuijten and Van Treuren, 2007) and molecular levels (Nuijten and Van Treuren, 2007; Nuijten *et al.*, 2009).

Gene Flow as a Mechanism

Although much research has been done on the genetics of the isolation barriers between various rice species found in West Africa (Sano, 1989; Heuer and Miézan, 2003; Garavito *et al.*, 2010), the literature is not conclusive on the possibility and level of introgression between species of the *O. sativa* complex. According to Second (1991), *O. barthii* and *O. longistaminata* are the only rice species of the *O. sativa* genome group between which no introgression is possible, although Chu and Oka (1969) suggest introgression is possible between these two species. No sterility barriers exist between *O. barthii* and *O. glaberrima*, which are closely related (Chu and Oka, 1969; Second, 1982; Wang *et al.*, 1992). There has also been some lack of clarity about the levels of introgression among *O. sativa*, *O. barthii* and *O. glaberrima*. Some have reported that *O. sativa* is completely isolated from *O. glaberrima* and *O. barthii* (Chu and Oka, 1969; Spillane and Gepts, 2001), but artificial hybridization has been successful in the laboratory (Sano, 1989; Pham and Bougerol, 1993; Jones *et al.*, 1997; Heuer and Miézan, 2003) and some level of introgression does seem to occur in the field (Semon *et al.*, 2005; Barry *et al.*, 2007; Nuijten *et al.*, 2009). Information about the direction of gene flow between the different species is contradictory, most likely because the levels of interspecific gene flow are low and, hence, difficult to measure. We assume that some gene flow is possible between all species in all directions, although the actual levels of gene flow are not known.

The only allogamous rice species known in West Africa is *O. longistaminata*. All other rice species in West Africa are strongly autogamous. The level of cross-pollination of *O. sativa* is well studied, in particular in relation to genetically modified (GMO) rice (for a historical overview see Gealy *et al.*, 2003), but reports on the level of cross-pollination in *O. barthii* and *O. glaberrima* are difficult to find (Grist, 1986; Oka, 1988). One interesting question is whether there may be more gene flow from the other rice species to *O. longistaminata* than vice versa. The low levels of cross-pollination together with the strong F_1 sterility barriers make introgression between the various rice species rather rare, and consequently difficult to quantify.

Chances of Co-flowering

Between cultivated and wild species

Oryza barthii × O. glaberrima *and* O. longistaminata × O. glaberrima

The extent of sympatry[1] of *O. glaberrima* with *O. longistaminata* and *O. barthii* varies widely over time and space across West Africa and is shaped by a complex interplay of agroecological, socio-economic and cultural factors. As a result the overlap in co-flowering varies accordingly. Both *O. barthii* and *O. longistaminata* are found throughout West Africa. *Oryza longistaminata* grows only in lowlands that are seasonally submerged. *Oryza barthii* is found over a wide range of lowlands, including the hydromorphic zone, while *O. glaberrima* is grown in uplands and lowlands. In Sierra Leone and Guinea, some plants resembling *O. barthii* grow as weeds in upland fields, and these are on occasion collected for seed and planted (Richards, 1986). Aside from being deliberately planted from time to time, the (semi) wild material persists when panicles are harvested along with the main cultivar and the seed is only incompletely rogued (Richards, 1986).

Oryza barthii × O. sativa *and* O. longistaminata × O. sativa

Almost all lowlands are today cultivated with *O. sativa* varieties. This suggests that, compared to *O. glaberrima*, there are more opportunities for co-flowering of *O. sativa* with *O. barthii* and *O. longistaminata*. In areas where farmers transplant rice, they weed out the wild species through ploughing (Nuijten, 2005). In lowland areas where rice is broadcast, some farmers are able to differentiate wild from cultivated rice at the vegetative stage whereas others cannot, resulting in diverse levels of wild rice in adjacent fields. In The Gambia, the wild rices tend to flower earlier than the cultivated rice (Nuijten, 2005). The overlap in co-flowering depends on the duration and the level of photoperiod sensitivity of the cultivated rice varieties. In lowland areas adjacent to the cultivated rice, farmers generally leave *O. barthii* to grow as it can be harvested for food (Richards, 1986; Nuijten 2005). Thus, some farmer practices (ploughing, weeding) reduce the chances of co-flowering, while other practices (cultural value of wild rice as a food) allow co-flowering (Nuijten, 2005). The genetic background of some semi-wild varieties resembling *O. barthii* collected in Guinea and Sierra Leone (Nuijten *et al.*, 2009) suggests that gene flow occurs between *O. sativa* and *O. barthii*. The exact implications and potential benefits of these findings need to be studied further.

Among cultivated species

Oryza sativa × O. glaberrima

Oryza sativa and *O. glaberrima* are often found in the same field across the upper Guinea coastal region, whereas this is not common in the Togo Hills (Teeken *et al.*, 2011). The reasons why farmers purposively plant the species in mixtures are agronomic (e.g. increase in yield, prevention of lodging, more simultaneous flowering, or a second harvest on the same land due to different flowering periods) and related to local beliefs (to encourage a good harvest or to prevent witchcraft). Often *O. glaberrima* is considered as the first (i.e. ancestral) rice (in historical terms) and hence good to have in a field with *O. sativa*, to guide it, much as an indigene guides a settler to understand the local terrain (Richards, 1986; Nuijten, 2005). But the importance of having pure *O. glaberrima* seed (with its distinctive grain shape in addition to its red seed coat) for rituals is the main reason why farmers in the Hohoe area (Ghana) do not mix *O. sativa* and *O. glaberrima* (Nuijten *et al.*, 2009; Teeken *et al.*, 2011). In countries like The Gambia and Senegal farmers have largely abandoned the cultivation of *O. glaberrima*, but it is still a frequent weed in farmers' fields, particularly in the transitional zones (Nuijten, 2005). Despite the fact that Gambian and Senegalese farmers do not like to have *O. glaberrima* in their fields, they do not bother to rogue it carefully from their seed (Teeken *et al.*, 2011). Elsewhere, for example in southern Guinea-Bissau, Guinea and Sierra Leone, *O. glaberrima* is liked and reputed to have better nutritional or therapeutic properties than *O. sativa* (Richards, 1996; Teeken *et al.*, 2010). The implication is that various cultural factors, together with agroecological factors, influence the level of gene flow between the two species in different ways.

Oryza sativa *subsp.* indica × O. sativa *subsp.* japonica

In general, in West Africa, varieties of *O. sativa* subsp. *indica* are grown in the lowlands and varieties of *O. sativa* subsp. *japonica* are grown in the uplands (De Kochko, 1987a; Barry *et al.*, 2007). However, some *indica* varieties can be found in upland fields, as a plot or mixed in plots with *japonica* (Nuijten and Van Treuren, 2007; Nuijten and Richards, 2011). For example, improved varieties developed for irrigated lowland cultivation, known locally as 'Peking' and 'CCA', were adopted by farmers for upland cultivation in The Gambia during the 1970s and 1980s (Nuijten, 2005). In various regions of Guinea-Bissau, varieties of both subspecies were found growing adjacently or mixed in upland areas and transitional zones (Nuijten, unpublished). Mixed stands of both subspecies may also be found in other regions in West Africa, such as in Guinea and Sierra Leone, as both subspecies are cultivated in lowland and upland areas (Okry, 2011; Teeken *et al.*, 2011; Mokuwa *et al.*, 2013). Whether *japonica* varieties are grown in lowland areas has not been studied. An F_1-hybrid sterility barrier exists between the two subspecies of *O. sativa*, although not as strong as the one between *O. sativa* and *O. glaberrima* (Oka, 1988). The presence in the field of varieties with an inter-subspecific background in West Africa has been reported (De Kochko, 1987b).

Field Settings

Adjacent plots may experience gene flow. Often farmers within a village will plant their rice in a single area. In the case of upland rice this is commonly done for sharing labour to clear vegetation, to make the burning process more effective and to reduce pest damage (Richards, 1986; Nuijten and Richards, 2011). In lowland areas fields are also situated in single areas, often because of limitations in land. Whether two adjacent rice varieties belonging to different species or subspecies then co-flower along a field boundary depends on various factors (Nuijten and Richards, 2011). The planting dates on adjacent plots may differ, especially if farmers share labour for planting. The duration of the rice varieties may differ, with no gene flow between varieties with different flowering periods if they are planted at the same time. Conversely, types with longer and shorter flowering periods may coincide in flowering if planted at different times. For that, topography is an important factor, since farmers often plant slopes (in inland valleys and hilly areas) with varieties with different durations to avoid labour bottlenecks and food scarcity (Richards, 1986). Rice varieties with the same flowering intervals may not exchange pollen if they have been planted at different levels on a slope. Those with different flowering intervals might overlap in flowering at a field boundary if the longer-duration variety is planted first on a lower slope. Thus, precise information is needed for assessing cross-boundary gene flow – on the number of rice varieties planted, their flowering period, planting dates, and location in the field, and the topography (Nuijten and Richards, 2011).

An important finding from rice farms in The Gambia, however, is that the chance of gene flow is much greater among co-flowering plants within a single field than across field boundaries. It was estimated that the average rate of cross-pollination between different varieties within the same field was 0.125%, some 25 times higher than the estimated average rate of cross-pollination between different varieties in neighbouring fields (0.005%) (Nuijten and Richards, 2011). Data collected by Okry (unpublished) suggest similar rates for upland rice farms in south-west maritime Guinea. If these findings are replicated for other rice-farming systems in West Africa then it seems that mixing of seed – whether deliberate planting of *O. sativa* and *O. glaberrima* within one field (Longley and Richards, 1993; Jusu, 1999; Teeken *et al.*, 2011), or due to a relaxed attitude to roguing off-types – is the more likely origin of varieties with an interspecific background than planting of *O. sativa* and *O. glaberrima* in adjacent plots. The findings of Nuijten and Richards (2011) also imply that the chances of cross-pollination between wild and cultivated species are higher when the wild rice plants are found within rice fields (i.e. when wild varieties remain unweeded), rather than as populations occurring on a field margin.

From F_1 Hybrid to a New Variety, Taking *O. sativa* × *O. glaberrima* F_1 Hybrids as Starting Point

A crucial factor in the development of interspecific varieties – after the emergence of rice plants with an interspecific background – is the natural backcrossing process in the field. *Oryza sativa* × *O. glaberrima* F_1 hybrids with the cytoplasmic DNA from *O. sativa* do not produce pollen (Sano, 1989), although there may be exceptions (Pham and Bougerol, 1993). F_1 hybrids can backcross to either species, provided co-flowering occurs. Various scenarios for backcrossing are illustrated in Fig. 8.1. Two main scenarios for backcrossing can be identified and are described in detail below.

First scenario: backcrossing to the F1 hybrid

F_1 hybrids are known to farmers – as 'useless' plants that flower but set no seeds. In very rare cases, F_1 hybrids do produce a few seeds (one or two per plant), as the result of pollination from surrounding 'normal' plants (Nuijten, 2005). These seeds will not be harvested (as the farmer will consider the mother plant useless) and, if not eaten by birds or other animals, may co-flower the next season if rice is planted in the same field. In the case where fertility is restored

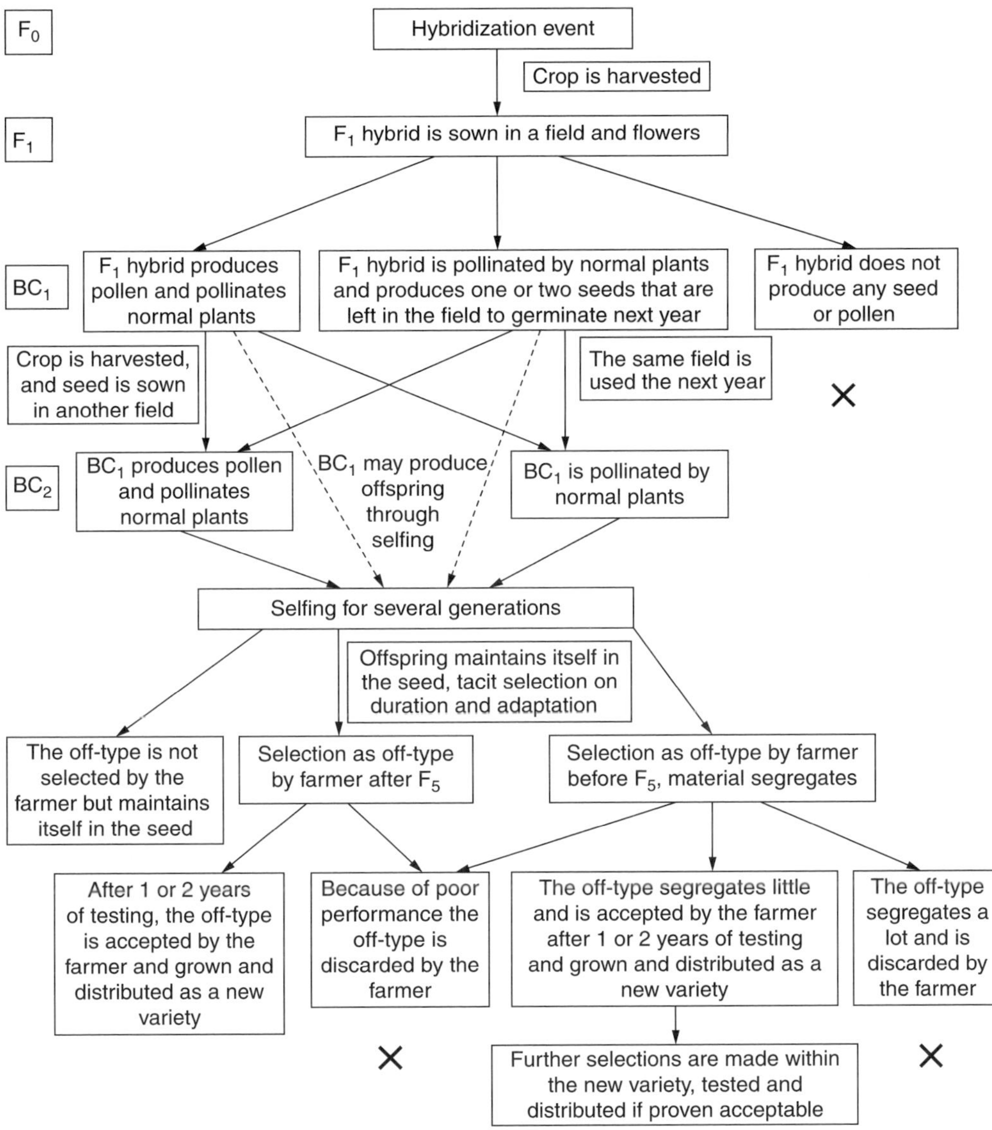

Fig. 8.1. Scenarios for the development of rice varieties with interspecific background.

with the first backcross, the progeny of the backcross may produce fully filled panicles which can then be included in the harvest by a farmer. Whether or not partly fertile panicles are included in the seed depends on the harvesting method used. If farmers use a sickle, a partly filled panicle is likely included in the harvest (Fig. 8.2), but this is less likely if farmers use a knife. Another possibility is that the fertility of the backcross is not fully restored, and that the pollen produced by the backcross pollinates a few flowers of surrounding normal plants. In both sub-scenarios the same field needs to be cultivated for at least two seasons. For upland rice a general rule is not to sow rice for a second season in the same field.

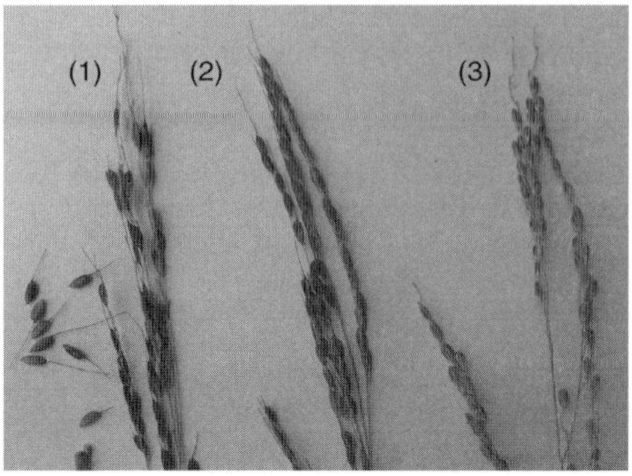

Fig. 8.2. Two off-types found in a field with the variety Pa Three Month (farmer hybrid with semi-droopy panicle, e.g. variety with interspecific background) in Kamba, Sierra Leone, 2009. Most of the flowers on panicle (1) are unfertilized and have awns similar in colour to those of *O. barthii* and interspecific F_1 hybrids. The seeds of panicle (1) dropped off very easily. The features of the (fertile) off-type (2) are intermediate between the off-type (1) and the variety Pa Three Month (3). Molecular analysis may clarify the genetic status of the two off-types.

Second scenario: backcrossing to the parent

The other scenario is that F_1 hybrids do produce some pollen that pollinates surrounding plants. So far, research on the fertility of crosses between *O. sativa* and *O. glaberrima* has not found an F_1 progeny that produces pollen, but it is suggested that wide compatibility varieties may be able to produce F_1 hybrids with fertile pollen (Pham and Bougerol, 1993). In this scenario, the backcrossed seed will not be recognized by a farmer (as it sits on a panicle of a normal plant), and will be included in the seed for the next season. Assuming two backcrosses are needed to restore full fertility, this in-field backcrossing needs to be repeated in the next two seasons, in which the backcross pollinates surrounding normal plants, of which some will be included in the seed for the following year. This scenario is not limited to rice-growing areas where the same fields are used for several years, or where rice ratoons, but is also possible in upland areas where farmers only grow rice for one year and then move to a new area to clear fields for the next season's rice farms.

More research is needed to decide which of the scenarios in Fig. 8.1 are more likely, and under what conditions. Among other observational requirements, it is necessary to observe – in detail over a season – a sample of different locations to work out the co-flowering intervals of the different rice types, including ratoons of the main cultivars.

After the backcrossing

When fertility is fully restored after two or three backcrosses, the offspring of the interspecific cross may maintain itself in the farmer's seed through self-pollination, if it is adapted to the local agroecological conditions, and if the farmer does not consider the offspring as clearly inferior.

When the level of mixture in the field is low, farmers apply negative selection, only roguing those plants with an inferior morphotype (Nuijten, 2005). The roguing is done either at harvest or when seed is being prepared for saving (Richards, 1986; Nuijten, 2005). The rigour of roguing also depends on the availability of time and labour. Only when the level of mixture is high do most farmers decide to remove all off-types. Women may object to pounding mixtures of varieties that have different threshabilities, cooking times, pericarp colours, etc. Conversely, when the level of mixture in the field is low, an off-type (being the progeny of a backcross) may readily remain in the seed. So long as it is not considered to be clearly inferior farmers will not bother to remove it (Nuijten, 2005).

Some farmers even prefer mixed material, since (with small plots) they may have time to sort and try out off-types (Richards, 1986), or be able to rediscover lost varieties (Nuijten and Richards, 2011). Depending on their socio-economic position, some farmers may have sufficient time – in their own words 'patience' – to look for off-types (Nuijten, 2010). Such farmers exercise positive selection for off-types, having learned the value of experimenting with 'new' types (often on difficult land). Eventually, off-types that seem valuable to the farmer (in terms of duration, panicle size, grain size, tillering, etc.), may be selected by farmers for testing to see whether the off-type may make a useful new variety. If after 1 or 2 years of testing the farmer finds the new variety useful, he or she gives some seed to friends and relatives for testing (Nuijten and Richards, 2011). If the new variety endures this cycle of testing successfully a new variety has been developed. A valuable off-type, once separated out and carefully tested, may bring prestige or gratitude to the farmer who selected it in the field, if it has some useful trait (e.g. taste, yield, duration, ability to grow on low-fertility soils, tillering ability). This prestige and gratitude is reflected in the newly developed variety being named after the farmer who selected it as an off-type (Nuijten and Almekinders, 2008).

Participatory research with farmers in The Gambia suggests that off-types that segregate a lot are abandoned very quickly. These farmers were given F_3 progenies, but they complained that the level of mixture was too high (Nuijten, 2005). In fact, various farmers who had developed new varieties from selected off-types said they never observed segregation in the progeny. Morphological observation of the farmer hybrids (rice varieties with an interspecific background as described by Nuijten *et al.*, 2009) suggests that some of these farmer hybrids look completely uniform, whereas others show some segregation for one or two traits (e.g. husk colour, awning, basal sheath colour).

The spread of new varieties depends on various factors. Important factors are climatic variability and climate change. Some farmer hybrids identified by Nuijten *et al.* (2009) are very short in duration and were quickly adopted in The Gambia and southern Senegal in the 1970s and 1980s when rainfall was reduced. War, such as the conflict in Sierra Leone, may force farmers to crop their lands more frequently (due to security concerns), and in these cases there is a greater need for varieties adapted to poor soils and low management. Cultural factors may determine patterns of out-migration (in times of peace and war) and hence influence the dispersion of new varieties over wider regions.

Macro, Regional Settings

It is not only the micro setting at field and village levels that shapes farmer activities in relation to crop development, but also the macro (regional) setting. The above illustrates that distances between the nodes of the networks farmers are working in vary from short to very long. How exactly the macro/regional setting influences farmer activities is not yet well understood. Portères (1962) thought that the domestication of *O. glaberrima* took place in the Inner Niger Delta in Mali because of the large diversity found there. Archaeological findings show that gathering of wild rice also took place in the Lake Chad basin (Klee *et al.*, 2004), and this therefore is a second possible site for initiation of the long-term domestication cycle as described by Alleby *et al.* (2008). Domestication of rice in West Africa began in a wetter phase, about 3500 BP (see McIntosh and McIntosh, 1981), so it is likely that early varieties were taken south by populations of gatherers following the gradual shift of the savannah and forest zones towards their present position (see Brooks, 1989). Earlier arid phases (enlargement of Dahomey Gap) served to split a number of West African animal and plant species into 'upper' and 'lower' West African populations. It is known that this happened with oil palm (Cochard *et al.*, 2009). But it is not known if there are similar differences among wild rice populations in the region. If so, the domestication of African rice based on a western and eastern focus (Inner Niger Delta and Lake Chad) might preserve, in a similar way to oil palm, some degree of ancient geographical and evolutionary differentiation in the genetic make-up of African rice.

Semon *et al.* (2005) suggest that *O. glaberrima* from northern Nigeria is genetically different from *O. glaberrima* from western West Africa. Our own data point to an upper Guinean coast gene pool being somewhat different from *O. glaberrima* collected in the Togo Hills (Nuijten *et al.*, 2009). There was once a debate among anthropologists and linguists about whether the people of the Togo Hills belonged to some kind of a remnant West African palaeolithic population (Westermann, 1954). Notwithstanding a strong argument based on anthropological and historical studies (Nugent, 1997), the idea that there may be some connections in language and material culture (e.g. African rice) across the chain of West African uplands north-eastwards from the Togo Hills, through the Atacora Mountains and Jos Plateau (central Nigeria), to the Adamawa Plateau, adjacent to the Lake Chad basin (Blench, 1989) is not entirely dead. To shed more light on this, further investigation, with (e.g.) molecular markers, with more extensive sampling, in particular from the western and eastern halves of West Africa, is needed (see Li *et al.*, 2011).

Linking the Macro Level with the Micro Level

In addition to biological data, better archaeological, historical and anthropological data will improve our understanding of the level of differentiation in genetic material on the upper and lower Guinea coast compared to Mali, central Nigeria and regions adjacent to the Lake Chad basin and, most importantly, the development of the wide range of rice-farming systems across West Africa, at micro and macro levels. The interspecific rice types in the upper Guinea coast region (from Senegal to Sierra Leone) identified by Nuijten *et al.* (2009) can be sub-divided into two types resulting from a combination of natural and cultural selection pressures. The earliest cultivation of the interspecific rice types appears to have been in Sierra Leone and Guinea-Bissau, which were connected through trade overseas (Mouser *et al.*, forthcoming). Both sub-types seem to have advantages under adverse conditions such as drought and war (Nuijten *et al.*, 2009). They are found cultivated adjacent to fields with *O. sativa* and *O. glaberrima* varieties, and within fields, mixed to various degrees, with varieties of *O. sativa* and *O. glaberrima* (Nuijten and Richards, 2011). Although it is not clear whether an F_1-hybrid sterility barrier exists between these farmer hybrids (varieties with an interspecific background) and *O. glaberrima* and *O. sativa*, it is likely that some hybridization, leading to the development of new varieties, takes place. As such, the farmer hybrids contribute in various ways to the portfolio of rice varieties managed by farmers working under sub-optimal farming conditions, indirectly increasing the coping strategies of farmers. They may also have potential relevance for exploitation by plant breeders (Nuijten *et al.*, 2009). Future advances in crop development could be achieved through better co-operation between scientists and rice farmers working under sub-optimal conditions in order to provide solutions in a world of increasing socio-political and climatic uncertainty.

Conclusion

Without an integrated approach for crop development it will not be possible to: (i) set up a management approach to maintain farmers' activities resulting in the development of new germplasm; or (ii) better understand how to link farmer crop development with scientific crop development (through participatory plant breeding). In this chapter we have described several scenarios for the development of interspecific rice varieties. Much information on how farmers influence gene flow is not yet known, both at micro and macro levels. From an empirical point of view, more information is needed on the weights applied by farmers to the nodes and the ways farmers weigh the nodes in a so-called gene-flow network. In order to do that, biological data on gene flow and the level of interspecific mixtures within farmers' fields need to be integrated with, among other things: (i) anthropological data on farmer practices in relation to field use and co-flowering of different rice species, and what type of farmers experiment with off-types and under what sorts of conditions; and (ii) with historical data on trade routes, cultural linkages between ethnic groups and time and place of introduction of various types of *O. sativa* in West Africa. Although such data sets seem very diverse, they can be analysed in an integrated way using an interdisciplinary approach, combining natural-science modelling approaches with social-science descriptive approaches (Nuijten, 2011; Nuijten *et al.*, Chapter 29, this volume).

Note

[1] 'Sympatry' is defined as 'species occupying the same geographical range without loss of identity from inter-breeding' (modified from www.merriam-webster.com/dictionary/sympatry, accessed 14 February 2013).

References

Allaby, R.G., Fuller, D.Q. and Brown, T.A. (2008) The genetic expectations of a protracted model for the origins of domesticated crops. *Proceedings of the National Academy of Sciences of the United States of America* 105, 13982–13986.

Almekinders, C.J.M. and Elings, A. (2001) Collaboration of farmers and breeders: participatory crop improvement in perspective. *Euphytica* 122, 425–438.

Altieri, M.A. and Merrick, L.C. (1987) *In situ* conservation of crop genetic resources through maintenance of traditional farming systems. *Economic Botany* 41(1), 86–96.

Angladette, A. (1966) *Le Riz.* Maissonneuve & Larose, Paris, France.

Barry, M.B., Pham, J.L., Noyer, J.L., Billot, C., Courtois, B. and Ahmadi, N. (2007) Genetic diversity of the two cultivated rice species (*O. sativa* & *O. glaberrima*) in maritime Guinea. Evidence for interspecific recombination. *Euphytica* 154, 127–137.

Blench, R. (1989) The evolution of the cultigen repertoire of the Nupe of west-central Nigeria, Azania. *Archaeological Research in Africa* 24(1), 51–63.

Brooks, G.E. (1989) Ecological perspectives on Mande population movements, commercial networks, and settlement patterns from the Atlantic Wet Phase (ca. 5500–2500 B.C.) to the Present. *History in Africa* 16, 23–40.

Carpenter, A.J. (1978) The history of rice in Africa. In: Buddenhagen, I.W. and Persley, G.J. (eds) *Rice in Africa.* International Institute of Tropical Agriculture, Ibadan, Nigeria.

Ceccarelli, S., Grando, S., Bailey, E., Amri, A., El-Felah, M., Nassif, F., Rezgui, S. and Yahyaoui, A. (2001) Farmer participation in barley breeding in Syria, Morocco and Tunisia. *Eyphytica* 122, 521–536.

Ceccarelli, S., Grando, S., Maatougui, M., Michael, M., Slash, M., Haghparast, R., Rahmanian, M., Taheri, A., Al-Yassin, A., Benbelkacem, A., Labdi, M., Mimoun, H. and Nachit, M. (2010) Plant breeding and climate changes. *Journal of Agricultural Science* 148, 627–637.

Chang, T.-T. (2003) Origin, domestication and diversification. In: Smith, C.W. and Dilday, R.H. (eds) *Rice: Origin, history, technology and production.* Wiley, Hoboken, New Jersey, pp. 3–25.

Chu, Y.-E. and Oka, H.I. (1969) Introgression across isolating barriers in wild and cultivated *Oryza* species. *Evolution* 24, 344–355.

Cochard, B., Adon, B., Rekima, S., Billotte, N., Desmier de Chenon, R., Koutou, A., Nouy, B., Omoré, A., Purba, A.R., Glazsmann, J.C. and Noyer, J.L. (2009) Geographic and genetic structure of African oil palm diversity suggests new approaches to breeding. *Tree Genetics & Genomes* 5, 493–504.

De Kochko, A. (1987a) Isozymic variability of traditional rice (*Oryza sativa* L.) in Africa. *Theoretical and Applied Genetics* 73, 675–682.

De Kochko, A. (1987b) A classification of traditional rice varieties (*Oryza sativa* L.) from Africa using isozymic variability. *Evolutionary Trends in Plants* 1(2), 105–110.

De Wet, J.M.J. and Harlan, J.R. (1975) Weeds and domesticates: evolution in the man-made habitat. *Economic Botany* 29, 99–107.

Desclaux, D., Nolot, J.M., Ciffoleau, Y., Gozé, E. and Leclerc, C.(2008) Changes in the concept of genotype × environment interactions to fit agriculture diversification and decentralized participatory plant breeding: pluridisciplinary point of view. *Euphytica* 163, 533–546.

Dorward, P., Craufurd, P., Marfo, K., Dogbe, W. and Bam, R. (2007) Improving participatory varietal selection processes: participatory varietal selection and the role of informal seed diffusion mechanisms for upland rice in Ghana. *Euphytica* 155, 315–327.

Dyer, G.A., González, C. and Lopera, D.C. (2011) Informal 'seed' systems and the management of gene flow in traditional agroecosystems: the case of cassava in Cauca, Colombia. *PLoS ONE* 6(12), e29067. doi:10.1371/journal.pone.0029067 (accessed 14 February 2013).

Efisue, A., Tongoona, P., Derera, J., Langyintuo, A., Laing, M. and Ubi, B. (2008) Farmers' perceptions on rice varieties in Sikasso region of Mali and their implications for rice breeding. *Journal of Agronomy and Crop Science* 194, 393–400.

Garavito, A., Guyot, R., Lozano, J., Gavory, F., Samain, S., Panaud, O., Thome, J., Ghesquière, A. and Lorieux, M. (2010) A genetic model for the female sterility barrier between Asian and African cultivated rice species. *Genetics* 185, 1425–1440.

Gealy, D.R., Mitten, D.H. and Rutger, J.N. (2003) Gene flow between red rice (*Oryza sativa*) and herbicide-resistant rice (*O. sativa*): implications for weed management. *Weed Technology* 17, 627–645.

Gridley, H.E., Jones, M.P. and Wopereis-Pura, M. (2002) Development of New Rice for Africa (NERICA) and participatory varietal selection. In: Witcombe, J.R., Parr, L.B. and Atlin, G.A. (eds) *Breeding*

Rainfed Rice for Drought-prone Environments: Integrating conventional and participatory plant breeding in South and Southeast Asia. Proceedings of a DFID Sciences Research Programme/IRRI conference, 12–15 March 2002, Los Baños, Laguna, Philippines. Centre for Arid Zone Studies, University of Wales, Gwynedd, UK, 23–28.

Grist, D.H. (1986) *Rice*, 6th edn. Longman, London.

Gyawali, S., Sunwar, S., Subedi, M., Tripathi, M., Joshi, K.D. and Witcombe, J.R. (2007) Collaborative breeding with farmers can be effective. *Field Crops Research* 101, 88–95.

Heuer, S. and Miézan, K.M. (2003) Assessing hybrid sterility in *Oryza glaberrima* × *O. sativa* hybrid progenies by PCR marker analysis and crossing with compatibility varieties. *Theoretical and Applied Genetics* 107, 902–909.

Hutchins, E. (1995) *Cognition in the Wild.* MIT Press, Cambridge, Massachusetts.

Jarvis, D.I. and Hodgkin, T. (1999) Wild relatives and crop cultivars: detecting natural introgression and farmer selection of new genetic combinations in agroecosystems. *Molecular Ecology* 8, 159–173.

Jones, M.P., Dingkuhn, M., Aluko, G.K. and Semon, M. (1997) Interspecific *Oryza sativa* L. × *O. glaberrima* Steud. progenies in upland rice improvement. *Euphytica* 94, 237–246.

Joshi, K.D., Devkota, K.P., Harris, D., Khanal, N.P., Paudyal, B., Sapkota, A. and Witcombe, J.R. (2012) Participatory research approaches rapidly improve household food security in Nepal and identify policy changes required for institutionalisation. *Field Crops Research* 131, 40–48.

Jusu, M.S. (1999) Management of genetic variability in rice (*Oryza sativa* L. and *O. glaberrima* Steud.) by breeders and farmers in Sierra Leone. PhD thesis, Wageningen University, Netherlands.

Klee, M., Zach, B. and Stika, H.P. (2004) Four thousand years of plant exploitation in the Lake Chad basin (Nigeria), part III: plant impressions in potsherds from the Final Stone Age Gajiganna Culture. *Vegetation History and Archaeobotany* 13, 131–142.

Li, Z.-M., Zheng, X.-M. and Ge, S. (2011) Genetic diversity and domestication history of African rice (*Oryza glaberrima*) as inferred from multiple gene sequences. *Theoretical and Applied Genetics* 123(1), 21–31.

Longley, C. and Richards, P. (1993) Selection strategies of rice farmers in Sierra Leone. In: De Boef, W., Amanor, K. and Wellard, K. (eds) *Cultivating Knowledge: Genetic diversity, farmer experimentation and crop research.* Intermediate Technology Publications, London, pp. 51–57.

McIntosh, S.K. and McIntosh, R.J. (1981) West African prehistory. *American Scientist* 69, 602–613.

Mokuwa, A., Nuijten, E., Okry, F., Teeken, B., Maat, H., Richards, P. and Struik, P.C. (2013) Robustness and strategies of adaptation within farmer varieties of African rice (*Oryza glaberrima*) and Asian rice (*Oryza sativa*) across West Africa. *PLoS One* 8(3), e34801. doi:10.1371/journal.pone.0034801 (accessed 4 March 2013).

Mouser, B., Nuijten, E., Okry, F. and Richards, P. (forthcoming) Red and white rice at Sierra Leone: linked histories of slavery, emancipation and seed selection. In: Bray, F., Schafer Dagma and Fields-Black, E. (eds) *Global Rice: Promiscuity, knowledge, and power.* Cambridge University Press, Cambridge (in press).

Nugent, P. (1997) Myths of origin and the origins of myth: local politics and the uses of history in Ghana's Volta Region. *Working Papers on African Societies* no. 22. Das Arabische Buch, Berlin, Germany.

Nuijten, E. (2005) Farmer management of gene flow: the impact of gender and breeding system on genetic diversity and crop improvement in The Gambia. PhD thesis, Wageningen University, Netherlands.

Nuijten, E. (2010) Gender and management of crop diversity in The Gambia. *Journal of Political Ecology* 17, 42–58.

Nuijten, E. (2011) Combining research styles of the natural and social sciences in agricultural research. *NJAS – Wageningen Journal of Life Sciences* 57, 197–205.

Nuijten, E. and Almekinders, C.J.M. (2008) Mechanisms explaining farmer variety naming and name consistency of rice and pearl millet varieties in The Gambia. *Economic Botany* 62(2), 148–160.

Nuijten, E. and Richards, P. (2011) Pollen flows within and between rice fields in relation to farmer variety development in The Gambia. *Plant Genetic Resources* 9, 361–374.

Nuijten, E. and Van Treuren, R. (2007) Spatial and temporal dynamics in genetic diversity in upland rice and late millet in The Gambia. *Genetic Resources and Crop Evolution* 54, 989–1009.

Nuijten, E., Van Treuren, R., Struik, P.C., Mokuwa, A., Okry, F., Teeken, B. and Richards, P. (2009) Evidence for the emergence of new rice types of interspecific hybrid origin in West African farmers' fields. *PLoS ONE* 4(10), e7335. doi:10.1371/journal.pone.0007335 (accessed 14 February 2013).

Oka, H.I. (1988) *Origin of Cultivated Rice.* Elsevier, Tokyo, Japan.

Okry, F. (2011) Strengthening rice seed systems and agro-biodiversity conservation in West Africa: a socio-technical focus on farmers' practices of rice seed development and diversity conservation in Susu cross border lands of Guinea and Sierra Leone. PhD thesis, Wageningen University, Netherlands.

Østergård, H., Finckh, M.R., Fontaine, L., Goldringer, I., Hoad, S.P., Kristensen, K., Lammerts van Bueren, E.T., Mascher, F., Munk, L. and Wolfe, M.S. (2009) Time for a shift in crop production: embracing complexity through diversity at all levels. *Journal of the Science of Food and Agriculture* 89, 1439–1445.

Pawson, R. and Tilley, N. (1997) *Realistic Evaluation.* Sage, London.

Pham, J.L. and Bougerol, B. (1993) Abnormal segregations in crosses between two cultivated rice species. *Heredity* 70, 466–471.

Portères, R. (1962) Berceaux agricoles primaires sur le continent africain. *Journal of African History* 3(2), 195–210.

Prain, G.D. (1993) Mobilizing local expertise in plant genetic resources research. In: De Boef, W., Amanor, K. and Wellard, K. (eds) *Cultivating Knowledge: Genetic diversity, farmer experimentation and crop research.* Intermediate Technology Publications, London, pp. 102–110.

Richards, P. (1986) *Coping with Hunger: Hazard and experiment in an African rice-farming system.* Allen and Unwin, London.

Richards, P. (1996) Culture and community values in the selection and maintenance of African rice. In: Brush, S. and Stabinsky, D. (eds) *Valuing Local Knowledge: Indigenous people and intellectual property rights.* Island Press, Washington, DC, pp. 209–229.

Richards, P., De Bruin-Hoekzema, M., Hughes, S.G., Kudadjie-Freeman, C., Offei, S.K., Struik, P.C. and Zannou, A. (2009) Seed systems for African food security: linking molecular genetic analysis and cultivator knowledge in West Africa. *International Journal of Technology Management* 45, 196–214.

Sang, T. and Ge, S. (2007) The puzzle of rice domestication. *Journal of Integrative Plant Biology* 49, 760–768.

Sano, Y. (1989) The direction of pollen flow between two co-occurring rice species, *Oryza sativa* and *O. glaberrima. Heredity* 63, 353–357.

Sayer, A. (2000) *Realism and Social Science.* Sage, London.

Second, G. (1982) Origin of the genetic diversity of cultivated rice (*Oryza* spp.): study of the polymorphism scored at 40 isozyme loci. *Japanese Journal of Genetics* 57, 25–57.

Second, G. (1991) Molecular markers in rice systematics and the evaluation of genetic resources. *Biotechnology in Agriculture and Forestry* 14, 468–494.

Semon, M., Nielsen, R., Jones, M.P. and McCouch, S.R. (2005) The population structure of African cultivated rice *Oryza glaberrima* (Steud.): evidence for elevated levels of linkage disequilibrium caused by admixture with *O. sativa* and ecological adaptation. *Genetics* 169, 1639–1647.

Sperling, L., Ashby, J.A., Smith, M.E., Weltzien, E. and McGuire, S. (2001) A framework for analyzing participatory plant breeding approaches and results. *Euphytica* 122, 439–450.

Spillane, C. and Gepts, P. (2001) Evolutionary and genetic perspectives on the dynamics of crop genepools. In: Cooper, D., Spillane, C. and Hodgkin, T. (eds) *Broadening the Genetic Base of Crop Production.* CAB International, Wallingford, UK, pp. 25–70.

Teeken, B., Nuijten, E., Okry, F., Mokuwa, A., Temudo, M., Struik, P.C. and Richards, R. (2010) Advocating the integration of socio-cultural factors in models for variety dissemination. Paper presented at the 3rd International Rice Congress, Hanoi, Vietnam, 8–12 November. http://ricecongress.com/previous/ext-Pdfs/OP03-4169-Teeken%20TP-edited.pdf (accessed 14 February 2013).

Teeken, B., Okry, F. and Nuijten, E. (2011) Improving rice based technology development and dissemination through a better understanding of local innovation systems. In: *Innovation and Partnerships to Realize Africa's Rice Potential.* Proceedings of the second Africa Rice Congress, Bamako, Mali, 22–26 March 2010. www.africarice.org/workshop/ARC/4.2%20Teeken%20fin.pdf (accessed 14 February 2013).

Temudo, M.P. (2011) Planting knowledge, harvesting agro-biodiversity: a case study on southern Guinea-Bissau rice farming. *Human Ecology* 39, 309–321.

Wang, Z.Y., Second, G. and Tanksley, S.D. (1992) Polymorphism and phylogenetic relationships among species in the genus *Oryza* as determined by analysis of nuclear RFLPs. *Theoretical and Applied Genetics* 83, 565–581.

Westermann, D. (1954) Die Togorestvölker und ihre Sprachen. *Tribus* 4–5, 63–68.

Witcombe, J.R. (2002) Participatory crop improvement strategies in rice in the DFID Plant Sciences Research Programme. In: Witcombe, J.R., Parr, L.B. and Atlin, G.N. (eds) *Breeding Rainfed Rice for Drought-prone Environments: Integrating conventional and participatory plant breeding in South and Southeast Asia.* Proceedings of a DFID Sciences Research Programme/IRRI conference, 12–15 March 2002, Los Baños, Laguna, Philippines. Centre for Arid Zone Studies, University of Wales, Gwynedd, UK, pp. 23–28.

Wood, D. and Lenné, J.M. (1997) The conservation of agrobiodiversity on-farm: questioning the emerging paradigm. *Biodiversity and Conservation* 6, 109–129.

9 Making Rice Genomics Work for Africa

Susan McCouch,[1]* Rod A. Wing,[2] Mandè Semon,[3] Ramaiah Venuprasad,[3] Gary Atlin,[4] Mark E. Sorrells[1] and Jean-Luc Jannink[5]

[1]Plant Breeding and Genetics Department, Cornell University, Ithaca, New York; [2]Arizona Genomics Institute, School of Plant Sciences, BIO5 Institute, University of Arizona, Tucson, Arizona; [3]Africa Rice Center (AfricaRice), Ibadan, Nigeria; [4]International Maize and Wheat Improvement Center (CIMMYT), Texcoco, Mexico; [5]United States Department of Agriculture Agricultural Research Service, Cornell University, Ithaca, New York

Introduction

In this chapter we explore opportunities and challenges associated with making rice genomics work for Africa. In the first section, we discuss how sequencing technology is being used to describe the extent, distribution and evolution of genetic variation in African rice. The sequence information is used to discover single nucleotide polymorphisms (SNPs), to inform germplasm conservation efforts, and to provide a framework for dissecting phenotype–genotype relationships. In the second section, we discuss strategies for identifying genotype–phenotype associations and using that information to enhance the efficiency of applied plant breeding programmes. The strategies include quantitative trait locus (QTL) mapping, genome-wide association mapping, marker-assisted selection (MAS) and genomic selection. We take note of some of the biological, logistical and institutional challenges that must be addressed, and envision the development of a functioning breeding pipeline that makes effective use of new tools.

Many of the tools and resources needed to develop genomics-based rice breeding programmes in Africa are already in place. These include diverse germplasm resources, reference genome sequences (RefSeq) for both wild and cultivated *Oryza* species, high- and low-density SNP arrays and sequence-based genotyping platforms, coordinated phenotyping networks, and models and strategies for using genomics information to predict phenotypic performance in a breeding programme. Given this foundation, the feasibility of implementing genomics-based breeding in African rice programmes ultimately depends on the ability to integrate large genotypic and phenotypic data sets in real time and the ability to practise rapid-cycle genotype-based selection. To meet this challenge, new types of data management, decision-support systems, and timely nursery management will be required, as well as new institutional partnerships, training programmes and coordinated research networks.

Finally, genomics-facilitated selection can only improve breeding efficiency if useful models are developed linking DNA polymorphisms with plant performance in appropriate field environments. The challenge facing the next generation of rice researchers and breeders is to implement

* Corresponding author: srm4@cornell.edu

© CAB International 2013. *Realizing Africa's Rice Promise* (eds M.C.S. Wopereis *et al.*)

cost-effective strategies that take advantage of high-throughput genotyping and increasingly efficient phenotyping strategies to make reliable predictions about plant performance in field environments that are relevant to African farmers.

What Do We Mean by Genomics?

Genomics is the study of the genomes of organisms, or the entirety of their hereditary information. It includes intensive efforts to determine the complete DNA sequence of individuals, to annotate that sequence by identifying structural features (such as chromosomes, genes, regulatory sequences, repetitive elements, and polymorphisms), and to interpret the functional significance or breeding value of these features. The development of increasingly high-throughput and low-cost sequencing methodologies makes it possible to identify and study genome-wide DNA variation in hundreds, thousands or millions of individuals in real time (Craig *et al.*, 2008; Schuster, 2008; Edwards and Batley, 2010).

These new sequencing technologies have dramatically changed the landscape for detecting and monitoring genetic variation in crop plants (Rafalski, 2002; Duran *et al.*, 2009; Edwards and Batley, 2010). SNPs are the most abundant form of genetic variation in plants and other eukaryotic genomes (Kwok *et al.*, 1996). As genetic markers, SNPs represent sites in the genome where the DNA sequence differs between individuals. SNPs are rapidly replacing simple sequence repeats (SSRs) as the DNA marker of choice for applications in plant breeding and genetics because they are more abundant, stable, amenable to automation, efficient and cost-effective to detect, and straightforward to manage in a database (Schuster, 2008; McCouch *et al.*, 2010). SNPs occur in both coding and non-coding regions of nuclear and plastid DNA, however not all SNPs are equally informative. Informative SNPs must first be polymorphic in the germplasm of interest. If they are polymorphic, they may be informative because they are in or near genes that contribute to a phenotype of interest or they may be useful simply because they provide a fingerprint that uniquely identifies a line or lineage. SNPs that are directly responsible for a phenotype (because they cause a change in the protein product of a gene or alter the way the gene is expressed) are called causative or functional SNPs, while SNPs that do not directly affect the way a gene functions are called indicative or linked SNPs. All types of SNPs are of interest to plant researchers; functional SNPs directly affect plant biology or plant performance, while linked SNPs can be used to track alleles of interest or to trace the evolutionary history of a gene, a region of the genome or the genome as a whole, and collections of genome-wide SNPs, even without knowing if they are linked to a gene or QTL of interest, are useful in genomics-assisted breeding applications (Gupta *et al.*, 2001; Rafalski, 2002).

SNPs are generally discovered by sequencing two or more genomes, aligning the sequences and comparing them. In 2005, rice became the first crop species to have its genome completely sequenced, and this accomplishment rapidly catalysed international interest in rice as a model genome for crop genetic research (IRGSP, 2005). For many years, the *Oryza sativa* subsp. *japonica* (temperate) variety Nipponbare was the only rice genome that had been sequenced to high accuracy via the sequencing of physically aligned bacterial artificial clones (BACs). Additional rice varieties were 're-sequenced' using a shotgun approach, where short sequence reads were aligned to the Nipponbare reference genome. Re-sequencing is a strategy that generates millions of short-sequence reads (50–150 base pairs [bp]) at random throughout the genome. These short sequences are aligned to the higher-quality, fully assembled Nipponbare sequence in order to identify SNPs, insertion/deletions (indels) and other forms of DNA variation that distinguish the re-sequenced genome from Nipponbare. This approach has been used to sequence dozens of diverse rice genomes from different subpopulations and species (Ammiraju *et al.*, 2010a; Huang *et al.*, 2010; McCouch *et al.* 2010; Xu *et al.*, 2011). Once the short reads are aligned to Nipponbare, DNA polymorphisms (SNPs, indels, copy number variants or CNVs) are individually identified based on their nucleotide position in the Nipponbare reference genome. Alignment to a reference genome that has been annotated to identify genes makes it possible to predict whether a SNP falls within or near a gene of interest, and whether a genic SNP is expected to cause a functional change in the protein product (i.e. non-synonymous change) that might alter the expression of the gene (Ondov *et al.*, 2008). This information can be very useful in predicting whether a particular SNP or indel is responsible for a phenotype of interest.

Sequencing of *Oryza* Species

Sequencing of *O. glaberrima*

Oyrza glaberrima cv. CG14 was the first African rice genome to be completely sequenced to high accuracy using a pooled BAC approach (Wing *et al.*, 2012, unpublished results). CG14 now serves as an independent reference genome for domesticated African rice, alongside Nipponbare for Asian rice. The *O. glaberrima* RefSeq is being annotated to identify genes, transposons and other functionally relevant motifs. It is also being compared with the Nipponbare sequence to identify structural variation, novel genes and selective sweeps. This analysis will allow us to identify specific genes and QTLs that distinguish *O. glaberrima* from *O. sativa*, and will provide insights about the traits and genomic regions that were under selection by early agriculturalists in West Africa.

Reference sequences for wild relatives

In addition to the *O. glaberrima* RefSeq, an international effort is under way to generate high-quality de-novo reference sequences from representative accessions of all 21 wild *Oryza* species under the organizational umbrella of the International *Oryza* Map Alignment Project (IOMAP) (Goicoechea *et al.*, 2010) (Fig. 9.1). The wild relatives of rice, although agronomically inferior, contain a virtually untapped reservoir of agriculturally important genes that could be used to improve the cultivated rice species. Once developed, these RefSeqs will be available for SNP calling to facilitate diversity analysis of rice and its wild relatives (Ammiraju *et al.*, 2010a). The IOMAP consortium is emphasizing the generation of high-quality de-novo RefSeqs rather than re-sequences, because of the significant amount of structural variation detected in targeted gene regions. So far, this has been demonstrated both for specific regions, such as *Adh*1, *Moc*1 and *Hd*1 (Ammiraju *et al.*, 2008, 2010b; Lu *et al.*, 2009; Sanyal *et al.*, 2010) and genome wide (Hurwitz *et al.*, 2010). Additionally, efforts are under way to generate de-novo RefSeqs for other AA genome wild relatives, as well as for each of the major subpopulations of *O. sativa*, because it is acknowledged that a single RefSeq is insufficient to comprehensively understand or capture genome diversity within *Oryza*.

Characterization of genes on chromosome 3, short arm

In addition to these full genome *Oryza* reference sequences, the IOMAP project aims to functionally characterize the majority of genes on the short arm of chromosome 3 as part of an international effort to functionally characterize all rice genes by 2020 (Zhang *et al.*, 2008). Interest in this region of the genome was related to the fact that the Wing laboratory led the US effort to sequence the short arm of chromosome 3 as part of the International Rice Genome Sequencing Project (IRGSP, 2005). A nearly complete set of chromosome 3 short arm RefSeqs is available for 12 *Oryza* species (indicated with an asterisk in Fig. 9.1) and the *Oryza* outgroup species *Leersia perrieri*.

Once these sequences have been annotated to identify the genes and regulatory sequences they contain, it will be possible to determine how much useful variation is found in distantly related wild species and to begin to introgress useful variation (from the AA genome species). It should also be possible to take advantage of the new knowledge contributed by the IOMAP initiative to modify domesticated genomes via transgenics in cases where novel genes are found in genomes that are not sexually compatible with *O. sativa* and *O. glaberrima*.

Diversity of *Oryza* in Africa

Africa contains the largest diversity of *Oryza* species in the world, with three AA genome species (*O. glaberrima, O. barthii* and *O. longistaminata*), the BB genome species (*O. punctata*), the CC genome species (*O. eichingeri*), the FF genome species (*O. brachyantha*), and finally the BBCC tetraploid species (*O. punctata*). These species span 10–15 million years of evolutionary history

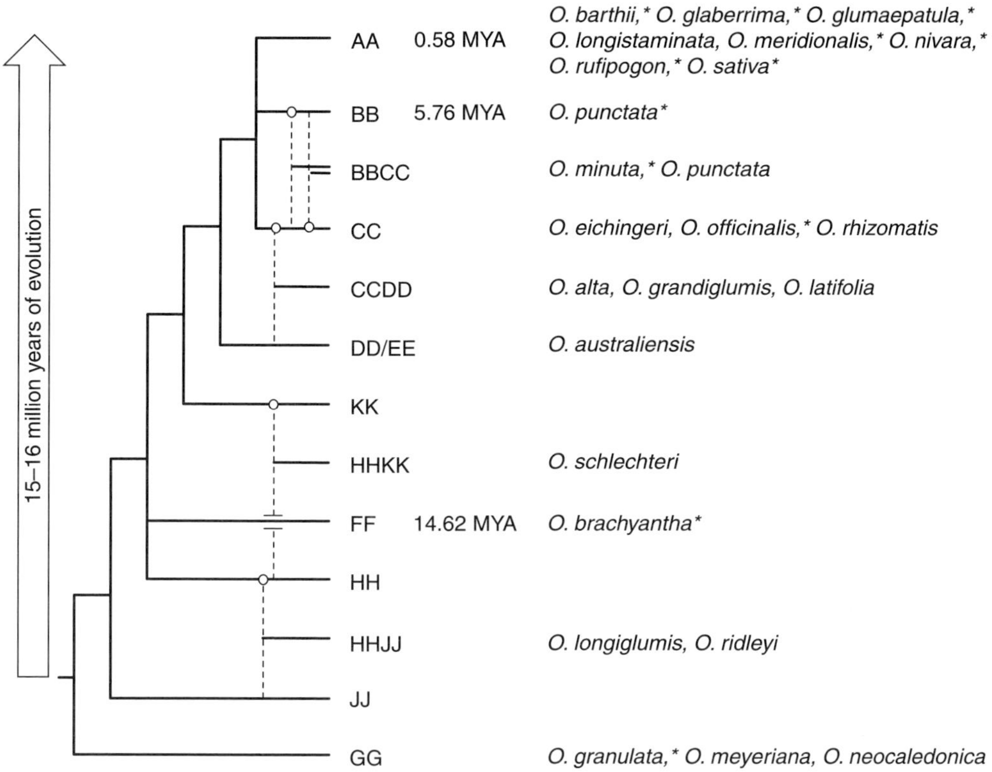

Fig. 9.1. *Oryza*: 24 species (two cultivated) – ten genome types. Asterisks indicate species for which a nearly complete set of chromosome 3 short arm reference sequences is available. (After Ge *et al.*, 1999; Ammiraju *et al.*, 2008; www.knowledgebank.irri.org/extension/index.php/wild-rice-taxonomy.) MYA, million years ago.

(Vaughan, 1994), making Africa a critical player in understanding and preserving the genomic diversity of the *Oryza* genus.

Brief descriptions of the African Oryza *species and their sequencing status*

Oryza glaberrima was domesticated approximately 3500 years ago from its wild ancestor, *O. barthi* in West Africa (Porteres, 1962; Klee *et al.*, 2000, 2004). There is some contention about whether the domestication process was truly independent of Asian rice (Nayar, 2010), but it is evident that *O. glaberrima* experienced a drastic domestication bottleneck and, compared to *O. barthii*, has a very narrow genetic base. This suggests that there is ample opportunity for African breeders to exploit the genetic diversity of its closely related wild ancestor.

The other closely related AA genome species, *O. longistaminata*, is a highly diverse species capable of reproducing both clonally, via rhizomes, and sexually, where it behaves primarily as an outcrossing species. To agriculturalists, *O. longistaminata* is a noxious weed because of the invasive nature of its rhizomatous habit which makes it very difficult to eradicate. The first disease-resistance gene cloned in rice was *Xa21*, which was derived from a backcross introgression line between *O. sativa* and *O. longistaminata* (Ikeda *et al.*, 1990; Song *et al.* 1995) (Table 9.1). RefSeqs for both *O. barthii* and *O. longistaminata* are in progress.

The diploid BB genome species, *O. punctata*, is a serious noxious weed in East Africa (Vaughan, 1994; Brink and Belay, 2006). The IOMAP consortium completed a RefSeq for this genome, which is slightly larger than the AA genome species at 425 megabases (Mb) (Ammiraju *et al.*, 2006). This genome will serve as an important evolutionary outgroup for all the AA genome species comparisons, as it is estimated to have

Table 9.1. African rice species as sources of resistance and tolerance to abiotic and biotic stresses.

Stress	*O. barthii* AA	*O. glaberrima* AA	*O. longistaminata* AA	*O. punctata* BB	*O. eichingeri* CC	*O. brachyantha* FF
Drought	Tolerance	Tolerance	Tolerance			
High mineral content in soil		Tolerance				Tolerance
Bacterial blight	Resistance	Resistance	Resistance			Resistance
Bacterial leaf streak	Resistance					
Stem rot				Resistance	Resistance	
Tungro virus						Resistance
RYMV		Resistance			Resistance	
Nematodes		Resistance	Resistance			
African rice gall midge		Resistance				
Brown planthopper			Resistance	Resistance	Resistance	
Green leafhopper					Resistance	
Stem borer						Resistance
Whorl maggot				Resistance		Resistance

diverged from the AA genome lineage about 2–3 million years ago. The classification of *O. punctata*'s tetraploid BBCC genome is still in question. Several important traits have been identified in *O. punctata*, including resistance to bacterial leaf blight and brown planthopper (Brink and Belay, 2006), which could potentially be used to improve cultivated rice.

Oryza eichingeri is one of three CC genome species and the only one endemic to Africa. Potentially useful traits that have been identified in specific accessions include resistances to rice stem rot (Figoni *et al.*, 1983), *Rice yellow mottle virus* (RYMV) (Brar and Khush, 1997), brown planthopper (Brar and Khush, 2003) and green leafhopper (Brar and Khush, 2003) (Table 9.1). The genome of *O. eichingeri* is about 650 Mb, twice that of *O. glaberrima* (Ammiraju *et al.*, 2006). Development of a full genome sequence for this species is still in the planning stage.

Oryza brachyantha (FF genome) is one of the most distantly related species of cultivated rice, having diverged from a common ancestor approximately 10 million years ago. It contains alleles for disease and insect resistance, as well as tolerance to laterite soils (Figoni *et al.*, 1983; Heinrichs *et al.*, 1985; Chaudhary and Khush, 1990; Khush and Brar, 2002; Brar and Khush, 2003) (Table 9.1). It has the smallest genome of any *Oryza* species at 362 Mb (Ammiraju *et al.*, 2006) and is being sequenced under the leadership of Mingsheng Chen (Chinese Academy of Sciences, Beijing).

The rapid accumulation of sequence information on diverse species opens the door to a new age of germplasm conservation, evaluation and utilization (McCouch *et al.*, 2012). It also challenges gene banks and the breeding community to join forces with geneticists, molecular biologists, computational biologists and evolutionists to better understand the relationship between DNA, RNA, phenotypic and environmental variation and to better utilize the wealth of natural variation for plant improvement.

Breeding Using Diverse Gene Pools of *Oryza*

Despite global awareness about the value of genetic variation and efforts to preserve genetic resources in the world's gene banks, only a small fraction of the naturally occurring wild and cultivated variation in rice has been explored to date. This is changing as genomics and sequencing-based activities begin to provide descriptions of germplasm resources at the molecular level (Sakai *et al.*, 2011). Since the mid-1990s, marker-based breeding strategies at Africa Rice Center (AfricaRice) have focused primarily on introgressing single genes/QTLs or simply inherited traits from diverse donors into elite recurrent parents (see Ndjiondjop *et al.*, Chapter 12, this volume). Marker-assisted selection (MAS) has been very productive in rice, where many genes and QTLs of large effect are known to confer agronomically useful phenotypes, and where SNPs, SSRs and other molecular markers are available for efficient backcross conversion (Fig. 9.2) (Collard and Mackill, 2008).

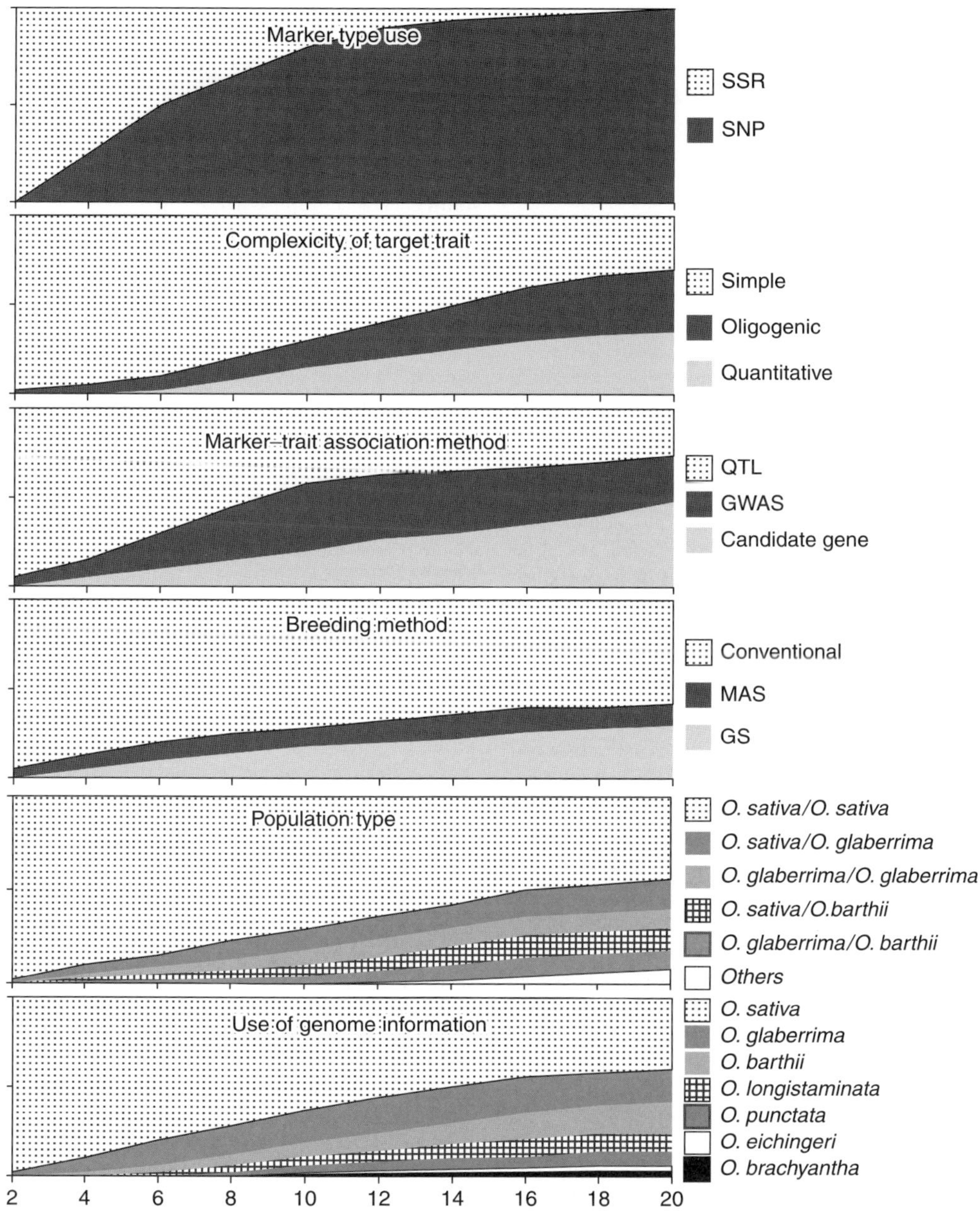

Fig. 9.2. Projected relative contribution of different factors to genomics-assisted rice breeding over a 20-year time period (2000–2020) in Africa. GS, genomic selection; GWAS, genome-wide association studies; MAS, marker-assisted selection; QTL, quantitative trait locus; SNP, single nucleotide polymorphism; SSR, simple sequence repeat.

Significant progress has been made using marker-assisted backcrossing to move genes and QTLs across species (e.g. *O. sativa* × *O. glaberrima*) or subspecies (*O. sativa* subsp. *indica* × *O. sativa* subsp. *japonica*) boundaries. In these cases, there are usually sterility barriers to contend with, and these must be overcome on a case-by-case basis. To address the sterility barrier, several forms of interspecific (*O. sativa* × *O. glaberrima*) bridging materials have been created (AfricaRice, 2010). These bridge materials make it easier to exploit the range of variation available in *O. glaberrima* and *O. barthii* and to recombine that variation with diverse forms of Asian rice (AfricaRice, 2010).

With the availability of new, high-resolution genomics platforms, there is growing interest in a molecular breeding strategy known as genomic selection. Genomic selection involves the use of genome-wide SNP scans to model phenotypic performance, and makes it possible for a breeder to select on multiple, complex (quantitatively inherited) traits simultaneously, without prior knowledge of the genes or QTLs involved. Genomic selection is most efficient when the objective is to make rapid genetic gain within a well-adapted, well-defined breeding pool, and where selection is being imposed on additive genetic variation (see 'Genomic selection' below).

To take advantage of the diverse gene pools available in rice, African rice breeders are likely to handle multiple streams of germplasm in their breeding programmes. One stream focuses on introgressing novel genetic variation from diverse and often unadapted donors, and the other focuses on recombining favourable alleles within adapted gene pools. The first is a form of pre-breeding and requires the development of inter-specific or inter-subspecific populations, where the objective is to enhance the performance of adapted, elite lines by introgressing selected novel alleles from divergent sources (Tanksley and McCouch, 1997; Xiao *et al.*, 1998; Moncada *et al.*, 2001; Li *et al.*, 2004; Sarla and Mallikarjuna Swamy, 2005; McCouch *et al.*, 2007; Venuprasad *et al.*, 2009, 2011a). The second breeding stream involves only elite, adapted materials that are inter-mated in an effort to improve their overall performance. This may involve the improvement of defensive traits (e.g. resistance to diseases, insects, weeds, abiotic stress), grain quality characteristics or the improvement of yield per se (Lamkey and Lee, 2006).

While the quantitative genetic models underlying these two streams are different (the first involves additive, epistatic and genotype-by-genotype interaction [G×G] effects and the second targets mostly additive genetic variation), in both cases genomics can help drive the efficiency of breeding programmes and enhance genetic gain per cycle of selection. In the case of genomic selection, information about the effects of single genes or specific introgressions that are known to be of value can be readily incorporated into the quantitative models used to predict performance. For complex traits like yield under drought, several potentially useful large-effect QTLs have been detected whose expression varies depending on the genetic background (e.g. Bernier *et al.*, 2007; Venuprasad *et al.*, 2009, 2011b,c). It is particularly important to address the question of genetic background effects in rice because of the number of deeply differentiated subpopulations and species that are used in most breeding programmes. Prediction of introgressed allele effects across a range of genetic backgrounds will be necessary to facilitate the efficient utilization of variation from wild and unadapted donor materials. Detailed sequence information can be used to model G×G interactions and to facilitate the identification of allelic series where different donor accessions may contribute subtle trait variation based on finely tuned differences among alleles at the same loci.

Understanding Genotype–Phenotype Relationships

Understanding the relationship between genotypic (SNP) variation and phenotypic variation is key to the application of genomics in plant improvement, and it is also an important area of basic study in the biological sciences. Using both 'forward genetics' (phenotype to genotype) and 'reverse genetics' (genotype to phenotype) approaches, researchers are investigating how variation at the DNA level contributes to an organism's phenotype in the context of normal growth and development, response to environment, and ability to withstand biotic or abiotic stress. Our understanding of genotype–phenotype relationships for rice in Africa will expand and deepen with the accumulation of data and information related to germplasm and growing environments of interest to African rice breeders. This will enable scientists to develop predictive models where genotypic information is used to predict phenotypic performance and to test these predictions in the field. As participants in this process, the new generation of plant breeders in Africa will collectively begin to transform plant breeding from a largely black-box activity into an increasingly predictive and hypothesis-driven science.

Genomics-based studies that examine genotype–phenotype relationships can be used to enhance the productivity and sustainability of rice production in Africa. Here we review four approaches: (i) quantitative trait locus (QTL) mapping (Tanksley, 1993; Mauricio, 2001); (ii) genome-wide association studies (GWAS) (Zhu *et al.*, 2008); (iii) marker-assisted selection (MAS) (Lande and Thompson, 1990; Collard and Mackill, 2008); and (iv) genomic selection (GS) (Meuwissen *et al.*, 2001; Jannink *et al.*, 2010). Some of these approaches aim to identify markers that uniquely tag genes or regions of the genome that condition traits or phenotypes of interest for MAS, while others utilize genome-wide polymorphisms to identify parents with complementary forms of variation for use in crossing, or to predict which offspring are most likely to outperform others in a breeding programme (as in GS).

These methodologies are of interest because the resulting associations or correlations between genotype and phenotype provide the basis for predicting the performance of new lines from DNA evidence. In turn, these predictions enable much more rapid and inexpensive selection of the most valuable lines, accelerating the breeding cycle and increasing selection intensity. All four of the genomics-based studies reviewed below require the same essential ingredients: a relevant population of plants, a set of genotypes for each member of the population, data on trait values or phenotypic performance for each member of the population, and appropriate analysis tools. The essential difference between studies focusing on genetic discovery and those focusing on plant improvement have to do with the finite nature of genetic studies, which are conducted as discrete projects, while genomic selection is integrated into a continuous product-development pipeline that has to be managed by the breeders. None the less, because both types of studies share the same ingredients and require similar attention to the quality of their execution, we begin by reviewing strategies that maximize their efficiency.

Genotyping and phenotyping strategies

To achieve the highest level of statistical power for identifying QTLs, experimental and environmental variation associated with both the genotypic and the phenotypic assays must be minimized. Genotyping with SNPs is currently the most reliable method for achieving this on the genotypic side, because SNP technology is highly automated and readily differentiates the four nucleotides (ACTG), eliminating a large degree of the experimental error and subjectivity involved in interpreting SSR polymorphisms and restriction fragment length polymorphisms (RFLPs), and greatly facilitating data integration across laboratories. In rice, numerous different types of SNP detection platform are available, including several low-density assays designed for specific types of populations (e.g. 384-SNPs; Thomson *et al.*, 2011), a medium-density fixed array (44,000 SNPs; Zhao *et al.*, 2011), a high-density fixed array (1M SNPs; McCouch *et al.*, 2010) and the genotyping by sequencing approach (Huang *et al.*, 2010; Elshire *et al.*, 2011; Wang *et al.*, 2011). In addition, re-sequencing data for hundreds or thousands of different rice genomes, including several African species (as outlined above), is being generated and will soon be publicly available as the basis for designing virtually any SNP assay of interest (Ammiraju *et al.*, 2010a; McCouch *et al.*, 2010; Xu *et al.*, 2011).

Fixed-array or uniplex SNP assays are well-suited to linkage and association mapping or single-gene introgression projects where the key cost concern is the price per marker data point. However, they remain too expensive for routine use in breeding where the key cost is the genotyping cost per selection unit (the line, family or individual under selection). Breeders are likely to begin to routinely acquire and use genome-wide marker data when the cost of genotyping is equivalent to the cost of phenotyping a new selection candidate in a single replicated field trial (about US$20–30 in many species and environments). Genotyping based on highly multiplexed next-generation sequencing of restriction-site associated DNA (RAD) (Baird *et al.*, 2008) – also referred to as genotyping by sequencing (GBS) (Elshire *et al.*, 2011) – is rapidly reducing genotyping costs to this level, and will shift the phenotyping:genotyping cost balance on a per-line basis strongly in favour of genotyping. At this cost, genotyping a line or population is a fraction of the cost of acquiring highly precise phenotypic information for low-heritability agronomic traits, which requires replicated field testing over several locations and years.

With the rapid evolution of high-throughput genotyping platforms and technologies, it is difficult for small laboratories and institutions to keep up with the pace of change. Thus, routine genotyping is almost always out-sourced to professional genotyping centres that offer competitive pricing based on economies of scale. Collectively, these centres can ensure ready access to the newest technologies, which are almost always cheaper, faster and technically more straightforward than older ones. Advanced laboratories in the EU, USA and Asia routinely out-source their genotyping to commercial centres for these reasons. In contrast, expertise on the phenotyping side, which is in critically short supply, is much more deeply rooted in local scientific communities that are well positioned to evaluate materials directly in environments of interest. This suggests that the competitive advantage of international agricultural research centres, such as AfricaRice, and national agricultural research systems (NARS) located in rice-growing environments, will be to invest in new phenotyping capabilities and to enhance the efficiency of phenotyping strategies, while out-sourcing most of their genotyping activities. Phenotyping capabilities are in great demand and a serious investment in sophisticated and efficient phenotyping and analytical capacity would drive new forms of collaborative research internationally, bringing renewed attention to the advantages of the international agricultural research centres.

In any phenotyping endeavour, variation due to genetics must be distinguished from variation due to environment. When phenotyping is done in the field, this has traditionally been addressed by replicating experiments over years and locations, as well as by controlling as many environmental variables at each site as possible. When experiments are conducted in a growth chamber or greenhouse, environmental variation may be further reduced and this may enhance a researcher's ability to identify meaningful genotype–phenotype associations. For example, traits such as disease or insect resistance are amenable to evaluation under controlled conditions because they often require that specific strains of a pathogen or pest be used for inoculation and that plants be protected from other stresses that may interfere with the evaluation of the disease response. Similarly, for some abiotic stresses, a specific amount, timing or type of stress may be critical to the evaluation, making it preferable to evaluate plants under controlled conditions. While controlled conditions can greatly accelerate the identification of genes and QTLs underlying complex phenotypes, these associations must ultimately be tested in the field to determine their relevance and reliability for breeding applications. For composite traits, such as yield, drought resistance and flowering time, where the genotype × environment interaction component is known to be high, and the complexity of environmental signals over the course of the growing season cannot be modelled in a greenhouse or growth chamber, phenotypic evaluation is best performed directly in the field. New tools and screening methodologies are being used to enhance the precision and efficiency of phenotyping, in concert with the use of genomics to enhance the power and efficiency of genetic characterization. For example, geographic information systems (GIS) are widely used to help identify and characterize target populations of environments (TPEs) selected to be representative of the larger spectrum of production environments, and remote-sensing technologies make it feasible to evaluate the performance of large populations of plants under diverse field conditions. Where possible, simultaneous phenotypic evaluation of a QTL-mapping population or GWAS diversity panel under both controlled and field conditions provides valuable data for determining whether the same or different SNPs are significantly associated with a phenotype in different environments. Ultimately, advances in phenotyping, genotyping and environmental characterization together make it possible for breeders to begin to tailor varieties to specific environments.

It is critical that when phenotyping a mapping population or association mapping panel, a high level of broad-sense heritability (H) or repeatability be achieved to ensure reliable estimation of effects (this H should be estimated on a marker or haplotype basis, as discussed below). This is especially important when alleles are at low frequency in the population, with the minor allele occurring in only a few lines. Broad-sense heritability is rarely reported for GWAS phenotypic information, but is the key parameter to monitor in any genetic analysis of quantitative traits that are affected by the environment. In association or QTL-mapping experiments, investments aimed at increasing phenotyping precision, in order to achieve a high level of H, provide high returns because the effects estimated in these experiments are the basis for subsequent marker-assisted breeding and introgression.

Existing phenotypic data can be leveraged at virtually any phase of a breeding programme to begin to identify markers or QTLs that are associated with a major effect on plant performance. Selective genotyping, including bulked segregant analysis (BSA) (Lander and Botstein, 1989; Darvasi and Soller, 1992; Navabi *et al.*, 2009; Sun *et al.*, 2010) may be performed on a population that has been evaluated for phenotypic performance in the field, greenhouse or growth chamber to identify marker alleles that exist at different frequencies in individuals with contrasting phenotypes. The breeder harvests leaf tissue (and ideally seed) from one or both phenotypic extremes of the population, extracts DNA and performs a genotyping assay using one of the many SNP assays available to the rice community (the density of SNPs required at this point depends on the population structure of the lines being evaluated) and compares SNP allele frequencies among the two phenotypic groups. This approach was successfully used by Bernier *et al.* (2007) and Venuprasad *et al.* (2009) to identify large-effect QTLs for grain yield under drought stress in rice, and by Nandi *et al.* (1997) to identify a submergence-tolerance QTL in lowland rice.

Finally, we note that to maximize the benefit of applying genomics to plant breeding, researchers need to evaluate the effects of **alleles** more so than those of lines. Traditional plant breeding research has developed effective tools for the identification and crossing of favourable lines. Genomics research, on the other hand, is providing tools to increase the efficiency of manipulating alleles, by identifying desirable alleles and pyramiding them or by using their estimated effects to predict the value of new lines. For the purpose of allele evaluation, the replication of **lines** is much less, or not at all, necessary: the alleles themselves are replicated across lines regardless of whether the lines themselves are replicated. Experiments with unreplicated lines can therefore have the greatest QTL detection power (Knapp and Bridges, 1990) and provide the best predictive ability (Zhong *et al.*, 2009). The same holds true for GWAS studies. Daetwyler *et al.* (2007) and Hayes *et al.* (2009) have described the relationship between trait heritability on an evaluation-unit basis (lines, families, etc.), the number of loci affecting the trait of interest, and the size of the GWAS population. Their analyses indicate that the precision of haplotype effect estimation increases with the size of the population, and that optimal designs consider the marker or haplotype as the unit of evaluation, rather than the line.

Earlier mapping experiments emphasized achieving a high repeatability of effect estimates for individual lines because genotyping was much more expensive than phenotyping. Now that the balance has shifted, optimal designs – i.e. those that maximize the repeatability of haplotype rather than line effects – are those that test as many lines as possible with little or no replication of individual lines. Designs maximizing the number of unreplicated lines used in the genetic analysis of populations, although theoretically superior, were impractical until genotyping costs dropped to the point where genotyping became inexpensive relative to phenotyping. We have now reached this point, which has far-reaching consequences for the implementation of field trials for phenotyping complex traits in the context of genomics-based crop improvement.

QTL mapping

A QTL is a region of the genome defined by molecular markers that is predicted to contain a gene or genes associated with a specific trait

(Mauricio, 2001; Bernardo, 2008). QTL mapping involves the analysis of a population (or populations) derived from a cross between two parents, where individual plants, lines or families within the population have been characterized for a set of well-distributed molecular-marker polymorphisms (RFLP, SSR, SNPs, etc.), as well as for one or more quantitative traits. A QTL is declared when there is a statistical association between the segregation of molecular polymorphism and a measurable phenotype, using the individual segregants within the population as replicates (Tanksley, 1993). The phenotype of interest may be a feature of the whole plant, an organ or a tissue, or it may be characterized as a feature associated with the DNA, RNA or protein. It may be evaluated under field conditions or experiments may be conducted under controlled conditions in a growth chamber or greenhouse. The objective of QTL analysis is to identify the position and relative importance of genetic factors that collectively determine a trait or phenotype of interest, and to identify sources of favourable and unfavourable alleles at each QTL (Bernardo, 2008).

QTL mapping is relevant to the agricultural community because it provides a way of genetically dissecting quantitative variation found in naturally occurring germplasm resources and offers insight into the linkage and epistatic relationships among genes and QTLs controlling diverse traits of interest. The QTL database in Gramene offers one of the largest repositories of QTL information for rice in the world (Ni *et al.*, 2009). Plant breeders are able to make direct use of QTL results for MAS as long as the genetic materials in their breeding programme are identical or closely related to the parents used to detect the QTLs. Molecular geneticists use QTLs as a first step in map-based cloning to identify genes underlying the QTL and to examine their molecular function and role in a biochemical or regulatory pathway. QTL information is valuable in both cases because it narrows the genomic search space for identifying genes underlying complex phenotypes, and it provides global information about the genetic architecture of those phenotypes, including the location and relative importance of each locus contributing to the quantitative trait.

It should be noted that QTL mapping works well when the genetic architecture of a trait is such that one or a few loci account for a large proportion of the genetic variance. All QTL effects are estimated with error, but when effects are very large, the 'signal-to-noise' ratio is large and effects are estimated with reasonable accuracy. When this is not the case, i.e. when the trait is actually controlled by many genes with small effects that are similar in size, QTL mapping experiments can provide misleading results due to the selection bias that results from the process of ranking QTLs by their effect sizes, and then selecting as significant only those that exceed a given threshold. If the true sizes of effects of genes contributing to a polygenic trait are small and similar, as is likely to be the case for many complex traits, then much of the difference in estimates will result from experimental 'noise' that will not recur in subsequent experiments. The situation is analogous to a cultivar trial in which differences are small, and error variance is large relative to genotypic differences; in such trials, when cultivars are ranked on the basis of their mean yields, and then the high tail of the distribution is selected, the effect sizes of the selected genotypes always shrink back towards the mean in subsequent trials, in rough proportion to their repeatability in the estimation experiment. This type of selection bias in QTL mapping leads to a more or less random subset of QTLs being declared significant in any particular experiment (Lande and Thompson, 1990; Beavis, 1994; Bernardo, 2008; Heffner *et al.*, 2009). MAS schemes based on such QTL effects are likely to give disappointing results. When the genetic control of traits is truly polygenic, approaches based on genome-wide association of phenotypes with haplotypes at many loci are likely to be superior (Heffner *et al.*, 2009).

Genome-wide association studies (GWAS)

GWAS, like QTL mapping, are based on the analysis of phenotypes and genotypes, but in GWAS, the genotyping and phenotyping are performed on a diverse collection of unrelated strains (referred to as a diversity panel) rather than on the progeny of a bi-parental cross (referred to as a QTL-mapping population) (Yu *et al.*, 2005; Zhu *et al.*, 2008). Genetic relationships among individuals that comprise association mapping panels vary widely. The use of very diverse panels may make it difficult to phenotype certain traits in the field, if some genotypes are so poorly adapted that trait expression is affected; in practice, association mapping panels should be composed of the most diverse set of adapted lines possible. GWAS analysis, like QTL mapping, relies on the strength of linkage disequilibrium (LD) between markers and phenotypes in the materials under study, and both aim to identify genes or regions of the genome that underlie complex phenotypes. While both approaches depend on identifying association between markers (i.e. SNPs) and traits of interest, GWAS does so in the context of evolutionary biology and population genetics, while QTL mapping does so in the context of inheritance genetics (Bernardo, 2008). GWAS generally provides greater resolution for the same population size, due to the fact that LD generally decays more quickly in a set of unrelated lines compared to bi-parental mapping populations (Flint-Garcia *et al.*, 2003; Famoso *et al.*, 2011). This is because there have been more generations of effective recombination separating lines in a diversity panel from their last common ancestor, than among segregants derived from a recent bi-parental cross. Despite these differences in resolution, the two approaches are complementary and are often pursued jointly (Yu *et al.*, 2008; Legarra and Fernando, 2009; McMullen *et al.*, 2009; Famoso *et al.*, 2011). They can be used to dissect complex traits in virtually any segregating population of interest as long as genotypes and phenotypes can be reliably assayed on the same individuals or families.

Marker-assisted selection (MAS)

MAS represents an early stage in the development of genomic-assisted breeding strategies (Lande and Thompson, 1990) and can be used as soon as marker-alleles are identified that can reliably predict a trait or a phenotype of interest in populations that are relevant to the breeder. The reliability of the association between a marker-allele and a phenotype of interest is key because MAS substitutes selection on phenotype with selection on specific marker-alleles. Most MAS programmes use small numbers of markers to select for genes or QTLs of relatively large effect and use two schemes: (i) backcrossing to introduce favourable alleles from a donor variety into one or more elite genetic backgrounds; and (ii) recurrent selection to cumulate alleles from several loci in a breeding population, using markers to enrich favourable alleles in each generation. This latter practice is called marker-assisted recurrent selection (MARS; Charmet *et al.*, 1999). A requirement of using markers identified in prior QTL studies is that these markers can be traced as identical by descent (IBD) to the parents carrying the favourable allele in the QTL analysis. Thus, there must be a clear line of descent from the QTL study to the breeding material, with loci being followed by marker genotyping throughout. In addition, when breeding populations are generated by crossing highly divergent materials, as in the case of *O. sativa* × *O. glaberrima* or *O. sativa* × *O. longistaminata* populations, breeders must ensure that the marker–trait relationship is not disrupted by dramatic changes in the genetic background as the favourable alleles are transferred to an elite agronomic background.

As reviewed by Collard and Mackill (2008), MAS has several advantages over conventional phenotypic selection in a breeding programme, as long as the marker genotypes can be inexpensively generated within the context of the normal breeding cycle and the information can be communicated to the breeder in a timely and cost-effective way. The practice saves the breeder's time, resources and effort because it allows selection to be carried out at the seedling stage (even for traits expressed late in the life of the plant), it permits the selection of heterozygous individuals without the need for progeny testing, it facilitates gene pyramiding for durable disease and insect resistance, and it greatly enhances the efficiency of backcross conversion by facilitating both foreground and background selection, thereby helping to break linkage between favourable and unfavourable alleles (known as linkage drag).

As the cost of genotyping continues to fall and the number of markers available for rice continues to rise, large populations of lines can now be genotyped at much lower cost than it would take to evaluate them in multi-location trials for key phenotypes associated with performance (Jannink, 2005; Heffner *et al.*, 2010). In rice, MAS has been used successfully to rapidly introgress genes and QTLs into elite breeding lines to enhance biotic and abiotic stress tolerance, grain quality and yield (Xiao *et al.*, 1998;

Ashikari and Matsuoka, 2006; McCouch *et al.*, 2007; Neeraja *et al.*, 2007; Septiningsih *et al.*, 2008; Shanti *et al.*, 2010; Lorieux *et al.*, Chapter 10, this volume; Ndjiondjop *et al.*, Chapter 12, this volume). MAS strategies require high confidence in marker–QTL allele association and therefore do not reliably handle quantitative traits with complex genetic architecture involving hundreds of genes or QTLs, where this confidence cannot be obtained. Even for traits with such architectures, however, a few large-effect loci may segregate that are amenable to these strategies.

Genomic selection (GS)

Genomic selection (GS) extends the use of markers to breeding highly polygenic traits when there can be little confidence in specific marker–QTL associations, and is complementary to conventional MAS. Genomic selection couples the relevant phenotyping of large plant breeding populations with high-density marker technologies to deal with quantitative traits such as yield, drought tolerance or flowering time (Meuwissen *et al.*, 2001; Buckler *et al.*, 2009; Jannink *et al.*, 2010). This coupling requires effective database solutions to handle the large amounts of phenotypic and genotypic (marker) data, as well as new statistical methods. The requisite genotyping capacity is already available in rice, as inexpensive genome-wide marker-detection platforms permit parallel scoring of a few hundred up to hundreds of thousands of polymorphisms across the rice genome (Huang *et al.*, 2010; McCouch *et al.*, 2010; Wang *et al.*, 2011). At the same time, algorithms have been developed that can simultaneously estimate the effects of all markers on phenotype, providing unbiased predictions of the performance of newly genotyped lines and capturing the influence of even the many small effects determining quantitative traits (Meuwissen *et al.*, 2001; Jannink *et al.*, 2010).

To generate accurate predictions, genomic selection proceeds as follows. A training population must be constituted and characterized both genotypically and phenotypically, generating a joint genotype–phenotype data set. This population should be closely related to the breeding lines on which genomic selection will be practised. For a breeding programme, a logical choice for the training population are experimental lines that have previously passed through the programme and are therefore already well characterized in the environments that the programme targets (Heffner *et al.*, 2009; Jannink *et al.*, 2010). In practice, this is likely to be the early-generation testing component of the breeding programme, with the model updated every time a test of new breeding materials is conducted. This joint genotype–phenotype data set is analysed using one of the statistical methods that have been proposed to train a prediction model (a comprehensive review is given by Lorenz *et al.*, 2011).

Models can be optimally designed to predict either the performance of the lines themselves or that of their progeny. In the latter case, the prediction is called the genomic estimated breeding value (GEBV) and it is the optimal predictor for selecting parents to initiate a new round of crosses and breeding. Prediction models can be developed for any trait for which there is sufficient phenotypic data, and these multivariate predictions can then be combined into a selection index, much like phenotypic measurements. A great value of GS methods is that, through evaluation of the effects of alleles that occur in many lines across a breeding programme, they leverage investment made in all phenotyping efforts for the purpose of improving the prediction for any given line. This leveraging is an important advancement over conventional phenotypic selection where a phenotypic measurement really provides information only on the line itself.

In conventional breeding systems, selection is based almost exclusively on the phenotype of a candidate. In GS schemes, selection (at least in the preliminary stages) is based on the candidate's genotype, which is quickly and inexpensively determined, rather than on its phenotype, which is very expensive to measure reliably, particularly for low-heritability traits. This 'decoupling' of selection from phenotyping can potentially increase breeding progress in three main ways:

1. It permits estimates of haplotype effects from previous testing to be used in predicting line performance, potentially increasing accuracy.
2. It allows selection intensity to be increased by permitting selection of lines that have been inexpensively genotyped, but not expensively phenotyped.

3. It permits recurrent selection to be conducted without intervening cycles of line extraction and phenotyping, allowing a radical shortening of the breeding cycle.

Developing the capacity for GS represents a significant investment of time, labour and organizational resources for a breeding programme (Eathington *et al.*, 2007), but it is an essential step forward in an era where genotyping hundreds or thousands of accessions can be accomplished more quickly and economically than evaluating those same materials for phenotypic performance in the field. We are witnessing a dramatic shift in how we value and invest in phenotypic evaluation. Previously, when molecular evaluation was slower and more expensive than phenotypic evaluation, genotyping of a line made it precious and, in order to leverage value from the genotype, candidate lines were extensively evaluated over replications, locations and years. The reverse is now true: the phenotyping of a line makes it precious, and inexpensive genotyping serves to extract value from that investment, whether the measurements have been highly replicated or not.

These changes require non-trivial improvements in bioinformatic and logistical support to jointly manage and analyse, on time, thousands of marker data points in hundreds or thousands of individuals every season. If appropriate coordination can be achieved between the genotyping and phenotyping activities, nursery management and decision support for selection, GS can serve as a valuable proxy for field-based evaluation, increasing genetic gains per unit time (Heffner *et al.*, 2010). Taking advantage of historical data available to a breeding programme, the accuracy of GS predictions can be validated relatively rapidly. Fully realizing the benefits of GS in breeding schemes across a programme is far easier said than done, however, and can be expected to take years. African rice programmes would be well advised to begin to integrate the use of genome-wide SNP panels to document the genotypic variation that is being utilized by breeding programmes throughout the continent. The information will provide the basis for all future efforts to assess genotype–phenotype associations and to assign GEBVs for materials of interest to breeders and farmers in Africa.

Genomics and adaptive breeding

Just as genomics tools are facilitating the mobilization of genetic resources from wild and cultivated relatives for use in breeding, they will also open the door to the utilization of elite germplasm resources coming from outside of Africa. Many elite lines have hitherto been inaccessible because of lack of adaptation (disease and insect resistance, tolerance to problem soils, etc.) or grain quality. In breeding terms, a mega-environment may be defined as a set of environments in which lines perform similarly. In rice, mega-environments are defined by crop management (e.g. transplanting versus direct seeding; irrigated versus rainfed), hydrology (upland versus lowland), soils and climate (e.g. tropical, subtropical or temperate), among other delineations. Particular combinations of hydrology, crop management and climate may recur across vast regions, and even on different continents; there are many rice-growing areas in Africa where conditions are similar to those in parts of South or South-East Asia, or South America.

In the past, movement of elite germplasm between mega-environments has been hampered by the occurrence of region-specific diseases or local grain-quality preferences. However, with over a decade of investment in identifying markers closely linked to or mapping within major genes controlling diseases important in Africa, such as blast, bacterial leaf blight and *Rice yellow mottle virus*, as well as grain quality traits such as grain shape and aroma, the use of MAS allows breeders to rapidly introgress specific traits of interest. This helps to 'adapt' elite but exotic lines coming from the same mega-environment(s) but different geographic regions, to the local biotic or socio-economic environment. GS approaches would also facilitate the movement of new, potentially elite germplasm among regions within the same mega-environment. If mega-environments are well characterized, breeders will be able to confidently identify multiple locations at which the agronomic performance of breeding lines is expected to be highly correlated with their own, with obvious implications for identifying new sources of useful elite materials. Investment in breeding-site characterization could connect distant but environmentally similar locations into breeding networks that could exchange promising new lines on the basis of GEBVs.

Summary of Genomic-assisted Breeding Potential

The key conceptual difference between conventional breeding and either MAS or GS approaches is that in the former, the line or family is the unit of evaluation, whereas in the latter, marker or haplotype alleles are evaluated (Meuwissen *et al.*, 2001). Because marker alleles recur across lines, additive effects may be estimated over many environments regardless of the identity of lines that contain them. Effects estimated previously in relevant training sets can be used to predict the performance of new lines containing those same alleles, despite the fact that the new lines may not have been phenotyped (Eathington *et al.*, 2007; Heffner *et al.*, 2009; Lorenzana and Bernardo, 2009). This concept is particularly important in the application of GS to breeding for stress-prone environments. It means that not every line needs to be evaluated in the target environment (e.g. drought-affected or N-deficient) to predict its performance in that environment. Rather, through allele effect estimates, lines evaluated in specific mega-environments 'share information' on a target environment with all members of a cohort.

This information sharing has exciting ramifications beyond breeding for improved adaptation to optimal and stress environments. It means that different breeding programmes may contribute to each other's success at picking broad-adaptation winners through data pooling – in effect, allowing a number of small but cooperating breeding programmes to leverage each other's efforts to increase their breeding efficiency.

Both scientific research and institutional innovations are needed to take advantage of these ideas and bring them to fruition. On the scientific side, a programme needs to be established to examine the relationships between allelic effects for environments targeted by different breeding programmes. When considering possible sources of difference between allelic effects important to national rice improvement programmes in Africa, Asia and America, two sources must be distinguished. First, the effect of the allele may be modulated by specifics of the environments targeted by the national programmes – for example, the allele may be beneficial under one rainfall pattern but not another. Second, less obviously, the effect of the allele may be modulated by genetic background: average allele frequencies across loci might differ sufficiently among national programmes to generate gene-interaction effects. Divergence in the average genetic background of African national programmes is an empirical issue of utmost importance that is, at present, poorly quantified. It is easily addressed by developing genome-wide SNP profiles for representative breeding lines and populations from each of the major rice-breeding programmes in Africa and assembling this information in a central diversity database. A central research body such as AfricaRice would be well positioned to address this question and provide a firm empirical foundation for future work seeking to enhance mutual benefit among African national programmes through cooperative genotype–phenotype analyses.

The mutual benefits that could be derived from cooperation through genomics will require central coordination and coordinated research. A far-reaching and transformative use of the potential to predict performance is referred to as 'open source' collaborative breeding. In this model, a central research body like AfricaRice collects and analyses genotype–phenotype information from a number of national programme partners or small seed and breeding companies in Africa, Asia and America. In exchange for the information delivered by the breeding programmes, AfricaRice would deliver analyses, providing predictions for new lines that combine both genome-wide estimates of value for highly polygenic traits and genotypes for large-effect QTL affecting oligogenic traits. These predictions contribute decision support for the selection of parents and specific crosses. AfricaRice's own breeding efforts would focus on validating those predictions, as well as developing pre-breeding lines and improving source populations to address deficiencies identified broadly across its 'client' breeding programmes. Deliverables would include genotyped lines tailored to specific environments and markets, as well as predictions of their performance. The sustainability and scaling up of this approach would be driven by continued decreases in genotyping cost per data point and per DNA sample, and by the needs for constant updating and validating of training sets, developing breeding plans and breeder-friendly software for implementation of GS, and trained personnel in applied breeding programmes to manage rapid-cycle genotype-based selection.

Institutional Implications for International Research Centres

The insight that alleles, rather than the lines themselves, need to be evaluated across environments opens opportunities for research programmes in Africa to benefit fully from the global drop in genotyping costs by increasing selection intensity. This can be achieved by investing early in the genotyping of large numbers of African rice accessions and breeding lines, using the genotyping information as the basis for selecting informative subsets of lines for phenotyping in target environments, and achieving a high level of precision in estimating GEBVs. It is important that the training population(s) be representative of the breeding population(s) (selection candidates) and that the training population(s) be adequately evaluated in the target environments.

International research centres, like AfricaRice, may position themselves in a key niche between upstream technology developers and downstream producers of finished varieties, serving an essential role as generators of pre-breeding materials, training populations and elite candidate lines with associated GEBVs, as well as serving a vital role in educating the next generation of African plant breeders. Generating reliable GEBVs will require ongoing collaboration with both public- and private-sector programmes on the ground in Africa to develop and evaluate appropriate training sets in target environments. AfricaRice's ability to support these activities (not to take exclusive responsibility for them) will be synergistic in the African context, enabling the system as a whole to deliver improved rice varieties to meet the diverse needs of African farmers.

It is important that researchers at AfricaRice stay abreast of the latest developments in genotyping technology, statistical analysis and information management through collaborations with universities and research institutions or companies worldwide. Collaboration is under way to re-sequence the contents of the AfricaRice gene bank, including both *O. glaberrima* and *O. sativa*, as well as related wild *Oryza* species, and to develop a public database of information about genetic variation in *Oryza* species in Africa. A similar investment is needed to genotype breeding populations that provide the foundation for selection by each of the national programmes as well as the international centre(s) on the African continent. This information can be used to systematically explore these materials and develop novel pre-breeding populations, develop SNP assays tailored to the needs of the African rice community, and to expand the rice gene pool available to breeders and researchers worldwide. This is likely to expand the rational use of *O. glaberrima*, *O. barthii* and other African rice germplasm, as well as to accelerate the adaptation of stress-tolerant *O. sativa* lines adapted to African conditions. Links with breeding programmes and seed companies will be essential to ensure that the generation of training sets and GEBV information is relevant to local needs; and iterative cycles of genotyping and phenotyping, along with consistent exchange of information and feedback, will be essential to fine-tune the process.

Finally, making rice genomics work for Africa will inevitably involve the evolution of a new institutional culture. Incorporating high-throughput genomics information into a breeding programme requires new management structures and reporting protocols. Vast amounts of genomic information are generated and processed with every cycle of selection, and this requires rapid and transparent communication and exchange of information among all members of a team. Successful integration of genomics can improve breeding efficiency by accelerating the selection of parents, reducing breeding cycle times, enhancing genetic gain per cycle of selection (due to improvements in the accuracy with which offspring with high breeding value are selected), and helping to match varieties/alleles to specific environments. However, it also imposes new requirements on the people involved, including formulation of explicit goals, objectives and time frames for each activity, transparency of work plans and transparent monitoring of both progress and impediments, open access to information at all levels of transaction, formalized reporting and communication pipelines, and streamlined teamwork among collaborating scientists. In many ways, these changes mimic the organization of many private-sector breeding organizations. For this reason, internships for young African plant breeders in the private sector may be an efficient way of helping to prepare them for a productive career in public service where they are expected to be able to integrate genomics into breeding programmes.

As discussed by Reece and Haribabu (2007), the sociological changes required to enable multi-disciplinary teams to work together with a shared vision are often the most difficult to achieve and require visionary and highly skilled research managers to guide the process. At every level, innovation will be required and adjustments will have to be made to match expectations with local realities, but the process of making genomics work for Africa has already begun, and a great deal will be learned in the years ahead as we explore the opportunities that unfold before us.

Acknowledgements

This work was supported in part by grant funding from the NSF Plant Genome Program (#0822284 and #1026200 to RAW and #0606461 and #1026555 to SMc), from the USDA-National Institute for Food and Agriculture (#2009-65300-05661 and #2011-68002-30029 to MS and JLJ; #2009-65300-05698 to SMc), the Bud Antle Endowed Chair (to RAW), the Ministry of Foreign Affairs, Japan (JAPN1 to AfricaRice) and the Rockefeller Foundation (RF #2000 FS 095, #ROCF6 to AfricaRice). We are grateful to Cheryl Utter for help with formatting the manuscript.

Note

[1] QTL analysis can detect many significant loci (QTLs), but attention is focused only on those QTLs with highly significant LOD scores (logarithm [base 10] of odds). This is to avoid type 1 error when using small (biased) sample/population sizes. However, it leads to an inflated estimate of the percentage variance explained (R^2 value) in the second step of the analysis where a model is built based only on the highly significant QTLs, ignoring the many QTLs of small effect. Thus, the model overestimates the effect (R^2 value) of the loci included in the model, and fails to acknowledge the 'minor' effect loci that collectively also contribute to the phenotype. For this reason, QTLs that are detected using small mapping populations (such as are normally used in QTL-mapping studies) appear to explain a greater portion of the phenotypic variance than they really do.

References

AfricaRice (2010) *New Breeding Directions at AfricaRice: Beyond NERICA*. Africa Rice Center, Cotonou, Benin.

Ammiraju, J.S., Luo, M., Goicoechea, J.L., Wang, W., Kudrna, D., Mueller, C., Talag, J., Kim, H.R., Sisneros, N.B., Blackmon, B., Fang, E., Tomkins, J.B., Brar, D., MacKill, D., McCouch, S., Kurate, N., Lambert, G., Galbraith, D.W., Arumuganathan, K., Rao, K., Walling, J.G., Gill, N., Yu, Y., SanMiguel, P., Soderlund, C., Jackson, S. and Wing, R.A. (2006) The *Oryza* bacterial artificial chromosome library resource: construction and analysis of 12 deep-coverage large-insert BAC libraries that represent the 10 genome types of the genus *Oryza*. *Genome Research* 16, 140–147.

Ammiraju, J.S., Lu, F., Sanyal, A., Yu, Y., Song, X., Jiang, N., Pontaroli, A.C., Rambo, T., Currie, J., Collura, K., Talag, J., Fan, C., Goicoechea, J.L., Zuccolo, A., Chen, J., Bennetzen, J.L., Chen, M., Jackson, S. and Wing, R.A. (2008) Dynamic evolution of *Oryza* genomes is revealed by comparative genomic analysis of a genus-wide vertical data set. *The Plant Cell* 20, 3191–3209.

Ammiraju, J.S., Song, X., Luo, M., Sisneros, N., Angelova, A., Kudrna, D., Kim, H.R., Yu, Y., Goicoechea, L.J., Lorieux, M., Kurata, N., Brar, D., Ware, D., Jackson, S. and Wing, R.A. (2010a) The OryzaBAC resource: a genus-wide and genome scale tool for exploring rice genome evolution and leveraging useful genetic diversity from wild relatives. *Breeding Science* 60, 536–545.

Ammiraju, J.S., Fan, C., Yu, Y., Song, X., Cranston, K., Pontaroli, A.C., Lu, F., Sanyal, A., Jiang, N., Rambo, T., Currie, J., Collura, K., Talag, J., Bennetzen, J.L., Chen, M., Jackson, S. and Wing, R. (2010b) Spatio-temporal patterns of genome evolution in allotetraploid species of the genus *Oryza*. *The Plant Journal* 63(3), 430–442.

Ashikari, M. and Matsuoka, M. (2006) Identification, isolation and pyramiding of quantitative trait loci for rice breeding. *Trends in Plant Science* 11, 344–350.

Baird, N.A., Etter, P.D., Atwood, T.S., Currey, M.C., Shiver, A.L., Lewis, Z.A., Selker, E.U., Cresko, W.A. and Johnson, E.A. (2008) Rapid SNP discovery and genetic mapping using sequenced RAD markers. *PLoS ONE* 3, e3376.

Beavis, W.D. (1994) The power and deceit of QTL experiments: lessons from comparative QTL studies. In: Wilkinson D.B. (ed.) *Proceedings of the 49th Annual Corn and Sorghum Research Conference*. American Seed Trade Association, Washington, DC, pp. 250–265.

Bernardo, R. (2008) Molecular markers and selection for complex traits in plants: learning from the last 20 years. *Crop Science* 48, 1649–1664.

Bernier, J., Kumar, A., Venuprasad, R., Spaner, D. and Atlin, G. (2007) A large-effect QTL for grain yield under reproductive-stage drought stress in upland rice. *Crop Science* 47, 507–516.

Brar, D.S. and Khush, G.S. (1997) Alien introgression in rice. *Plant Molecular Biology* 35, 35–47.

Brar, D.S. and Khush, G.S. (2003) Utilization of wild species of genus *Oryza* in rice improvement. In: Nanda, J.S. and Sharma, S.D. (eds) Monograph on genus *Oryza*. Science Publishers, Enfield, New Hampshire and Plymouth, UK, pp. 283–309.

Brink, M. and Belay, G. (Eds) (2006) *Plant Resources of Tropical Africa 1. Cereals and Pulses*. PROTA Foundation, Wageningen, Netherlands; Backhuys Publishers, Leiden, Netherlands; and Technical Centre for Agricultural and Rural Cooperation, Wageningen, Netherlands.

Buckler, E.S., Holland, J.B., Bradbury, P.J., Acharya, C.B., Brown, P.J., Browne, C., Ersoz, E., Flint-Garcia, S., Garcia, A., Glaubitz, J.C., Ma. Goodman, M., Harjes, C., Gulll, K., Kroon, D.E., Larsson, S., Lepak, N.K., Li, H., Mitchell, S.E., Pressoir, G., Peiffer, J.A., Oropeza Rosas, M., Rocheford, T.R., Cinta Romay, M., Romero, S., Slavo, S., Sanchez Villeda, H., Sofia da Silva, H., Sun, Q., Tian, F., Upadyayula, N., Ware, D., Yates, H., Yu, J., Zhang, Z., Kresovich, S. and McMullen, M.D. (2009) The genetics architecture of maize flowering time. *Science* 325(5941), 714–718.

Charmet, G., Robert, N., Perretant, M.R., Gay, G., Sourdille, P., Groos, C., Bernard, S. and Bernard, M. (1999) Marker-assisted recurrent selection for cumulating additive and interactive QTLs in recombinant inbred lines. *Theoretical and Applied Genetics* 99, 1143–1148.

Chaudhary, R.C. and Khush, G.S. (1990) Breeding rice varieties for resistance against *Chilo* spp. of stem borers in Asia and Africa. *Insect Science and Its Application* 11, 659–669.

Collard, B.C.Y. and Mackill, D.J. (2008) Marker-assisted selection: an approach for precision plant breeding in the twenty-first century. *Philosophical Transactions of the Royal Society B* 363, 557–572.

Craig, D.W., Pearson, J.V., Szelinger, S., Sekar, A., Redman, M., Corneveaux, J.J., Pawlowski, T.L., Laub, T., Nunn, G., Stephan, D.A., Homer, N. and Huentelman, M.J. (2008) Identification of genetic variants using bar-coded multiplexed sequencing. *Nature Methods* 5, 887–893.

Daetwyler, H.D., Villanueva, B., Bijma, P. and Woolliams, J.A. (2007) Inbreeding in genome-wide selection. *Journal of Animal Breeding and Genetics* 124, 369–376.

Darvasi, A. and Soller, M. (1992) Selective genotyping for determination of linkage between a marker locus and a quantitative trait locus. *Theoretical and Applied Genetics* 85, 353–359.

Duran, C., Appleby, N., Clark, T., Wood, D., Imelfort, M., Batley, J. and Edwards, D. (2009) AutoSNPdb: an annotated single nucleotide polymorphism database for crop plants. *Nucleic Acids Research* 37, D951–D953.

Eathington, S.R., Crosbie, T.M., Edwards, M.D., Reiter, R.S. and Bull, J.K. (2007) Molecular markers in a commercial breeding program. *Crop Science* 47, S154–S163.

Edwards, D. and Batley, J. (2010) Plant genome sequencing: applications for crop improvement. *Plant Biotechnology Journal* 8, 2–9.

Elshire, R.J., Glaubitz, J.C., Sun, Q., Poland, J.A., Kawamoto, K., Buckler, E.D. and Mitchell, S.E. (2011) A robust, simple genotyping-by-sequencing (GBS) approach for high diversity species. *PLoS ONE* 6(5), e19379.

Famoso, A.N., Zhao, K., Clark, R.T., Tung, C.-W., Wright, M.H., Bustamante, C., Kochian, L.V. and McCouch, S.R. (2011) Genetic architecture of aluminum tolerance in rice (*Oryza sativa*) determined through

genome-wide association analysis and QTL mapping. *PLoS Genetics* 7(8): e1002221. DOI: 10.1371/journal.pgen.1002221 (accessed 24 February 2012).
Figoni, R.A., Rutger, J.N. and Webster, R.K. (1983) Evaluation of wild *Oryza* species for stem rot (*Sclerotium oryzae*) resistance. *Plant Disease* 67, 998–1000.
Flint-Garcia, S.A., Thornsberry, J.M. and Buckler, E.S. IV (2003) Structure of linkage disequilibrium in plants. *Annual Review of Plant Biology* 54, 357–374.
Ge, S., Sang, T., Lu, B.R. and Hong, D.Y. (1999) Phylogeny of rice genomes with emphasis on origins of allotetraploid species. *Proceedings of the National Academy of Sciences USA* 96(25), 14400–14405.
Goicoechea, J.L., Ammiraju, J.S.S., Marri, P.R., Chen, M.S., Jackson, S., Yu, Y., Rounsley, S. and Wing, R.A. (2010) The future of rice genomics: sequencing the collective *Oryza* genome. *Rice* 3(2–3), 89–97.
Gupta, P.K., Roy, J.K. and Prasad, M. (2001) Single nucleotide polymorphisms: a new paradigm for molecular marker technology and DNA polymorphism detection with emphasis on their use in plants. *Current Science* 80, 524–535.
Hayes, B.J., Bowman, P.J., Chamberlain, A.J. and Goddard, M.E. (2009) Invited review: Genomic selection in dairy cattle: progress and challenges. *Journal of Dairy Science* 92, 433–443.
Heffner, E.L., Sorrells, M.E. and Jannink, J.L. (2009) Genomic selection for crop improvement. *Crop Science* 49, 1–12.
Heffner, E.L., Lorenz, A.J., Jannink, J.L. and Sorrells, J.E. (2010) Plant breeding with genomic selection: gain per unit time and cost. *Crop Science* 50, 1681.
Heinrichs, E.A., Viajante, V.D. and Romena, A.M. (1985) Resistance of wild rices, *Oryza* spp., to the whorl maggot *Hydrellia philippina ferino* (Diptera, Ephydridae). *Environmental Entomology* 14, 404–407.
Huang, X., Wei, X., Sang, T., Zhao, Q., Feng, Q., Zhao, Y., Li, C., Zhu, C., Lu, T. and Zhang, Z. (2010) Genome-wide association studies of 14 agronomic traits in rice landraces. *Nature Genetics* 42, 961–967.
Hurwitz, B.L., Kudrna, D., Yu, Y., Sebastian, A., Zuccolo, A., Jackson, S.A., Ware, D., Wing, R.A. and Stein, L. (2010) Rice structural variation: a comparative analysis of structural variation between rice and three of its closest relatives in the genus *Oryza*. *Plant Journal* 63, 990–1003.
Ikeda, R., Khush, G.S. and Tabien, R.E. (1990) A new resistance gene to bacterial blight derived from *O. longistaminata*. *Japanese Journal of Breeding* 40 (Suppl. 1), 280–281.
IRGSP (International Rice Genome Sequencing Project) (2005) The map-based sequence of the rice genome. *Nature* 436, 793–800.
Jannink, J.L. (2005) Selective phenotyping to accurately map quantitative trait loci. *Crop Science* 45, 901–908.
Jannink, J.L., Lorenz, A.J. and Iwata, H. (2010) Genomic selection in plant breeding: from theory to practice. *Briefings in Functional Genomics* 9, 166–177.
Khush, G.S. and Brar, D.S. (2002) Biotechnology for rice breeding: progress and potential impact. Paper presented at the 20th Session of the International Rice Commission, Bangkok, Thailand, 23–26 July.
Klee, M., Zach, B. and Neumann, K. (2000) Four thousand years of plant exploitation in the Chad Basin of northeast Nigeria I: the archaeobotany of Kursakata. *Vegetation History and Archaeobotany* 9(4), 223–237.
Klee, M., Zach, B. and Stika, H.-P. (2004) Four thousand years of plant exploitation in the Lake Chad Basin (Nigeria), part III: plant impressions in potsherds from the final stone age Gajiganna culture. *Vegetation History and Archaeobotany* 13(2), 131–142.
Knapp, S.J. and Bridges, W.C. (1990) Using molecular markers to estimate quantitative trait locus parameters: power and genetic variances for unreplicated and replicated progeny. *Genetics* 126, 769–777.
Kwok, P.Y., Deng, Q., Zakeri, H., Taylor, S.L. and Nickerson, D.A. (1996) Increasing the information content of STS-based genome maps: identifying polymorphisms in mapped STSs. *Genomics* 31, 123–126.
Lamkey, K.R. and Lee, M. (Eds) (2006) *Plant Breeding, The Arnel R. Hallauer International Symposium.* Blackwell, Ames, Iowa.
Lande, R. and Thompson, T. (1990) Efficiency of marker-assisted selection in the improvement of quantitative traits. *Genetics* 124, 743–756.
Lander, E.S. and Botstein, D. (1989) Mapping Mendelian factors underlying quantitative traits using RFLP linkage maps. *Genetics* 121, 185–199.
Legarra, A. and Fernando, R.L. (2009) Linear models for joint association and linkage QTL mapping. *Genetics Selection Evolution* 41, 43–60.

Li, J., Xiao, J., Grandillo, S., Jiang, L., Wan, Y., Deng, Q., Yuan, L. and McCouch, S. (2004) QTL detection for rice grain quality traits using an interspecific backcross population derived from cultivated Asian (*O. sativa* L.) and African (*O. glaberrima* S.) rice. *Genome* 47, 697–704.

Lorenz, A.J, Chao, S., Asoro, F.G., Heffner, E.L., Hayashi, T., Iwata, H., Smith, K.P., Sorrells, M.E. and Jannink, J.L. (2011) Genomic selection in plant breeding: knowledge and prospects. *Advances in Agronomy* 110, 77–123.

Lorenzana, R. and Bernardo, R. (2009) Accuracy of genotypic value predictions for marker-based selection in biparental plant populations. *Theoretical and Applied Genetics* 120, 151–161.

Lu, F., Ammiraju, J.S.S., Sanyal, A., Zhang, S., Song, R., Chen, J., Li, G., Sui, Y., Song, X., Cheng, Z., de Oliveira, A.C., Bennetzen, J.L., Jackson, S.A., Wing, R.A. and Chen, M. (2009) Comparative sequence analysis of *MONOCULM1*-orthologous regions in 14 *Oryza* genomes. *Proceedings of the National Academy of Sciences USA* 106(6), 2071–2076.

Mauricio, R. (2001) Mapping quantitative trait loci in plants: uses and caveats for evolutionary biology. *Nature Reviews Genetics* 2, 370–381.

McCouch, S.R., Sweeney, M., Li, J., Jiang, H., Thomson, M., Septinginsih, E., Edwards, J., Moncada, P., Xiao, J., Garris, A., Tai, T., Martinez, C., Tohme, J., Sugiono, M., McClung, A., Yuan, L.P. and Ahn, S.N. (2007) Through the genetic bottleneck: *O. rufipogon* as a source of trait-enhancing alleles for *O. sativa. Euphytica* 154, 317–339.

McCouch, S.R., Zhao, K., Wright, M., Tung, C.-W., Ebana, K., Thomson, M., Kimball, J., Reynolds, A., Tyagi, W., Wang, D., DeClerck, G., Ali, M.L., McClung, A., Eizenga, G. and Bustamante, C. (2010) Development of genome-wide SNP assays for rice. *Breeding Science* 60, 524–535.

McCouch, S.R., McNally, K.L., Wang, W. and Sackville-Hamilton, R. (2011) Genomics of gene banks: a case study in rice. *American Journal of Botany* 99(2), 407–423. DOI:_10.3732/ajb.1100385.

McMullen, M.D., Kresovich, S., Villeda, H.S., Bradbury, P., Li, H., Sun, Q., Flint-Garcia, S., Thornsberry, J., Acharya, C., Bottoms, C., Brown, P., Browne, C., Eller, M., Guill, K., Harjes, C., Kroon, D., Lepak, N., Mitchell, S.E., Peterson, B., Pressoir, G., Romero, S., Rosas, M.O., Salvo, S., Yates, H., Hanson, M., Jones, E., Smith, S., Glaubitz, J.C., Goodman, M., Ware, D., Holland, J.B. and Buckler, E.S. (2009) Genetic properties of the maize nested association mapping population. *Science* 325, 737–740.

Meuwissen, T.H.E., Hayes, B.J. and Goddard, M.E. (2001) Prediction of total genetic value using genome-wide dense marker maps. *Genetics* 157, 1819–1829.

Moncada, M., Martinez, C., Tohme, J., Chatel, M., Borrero, J., Gauch, H. and McCouch, S. (2001) Quantitative trait loci for yield and yield components in an *Oryza sativa* × *Oryza rufipogon* BC_2F_2 population evaluated in an upland environment. *Theoretical and Applied Genetics* 102, 41–52.

Nandi, S., Subudhi, P.K., Senadhira, D., Manigbas, N.L., Sen-Mandi, S. and Huang, N. (1997) Mapping QTLs for submergence tolerance in rice by ADLP analysis and selective genotyping. *Molecular and General Genetics* 255(1), 1–8.

Navabi, A., Mather, D.E, Bernier, J., Spaner, D.M. and Atlin, G.N. (2009) QTL detection with bidirectional and unidirectional selective genotyping: marker-based and trait-based analyses. *Theoretical and Applied Genetics* 118(2), 347–358.

Nayar, M.N. (2010) Origin of African rice from Asian rice. In: *Innovation and Partnerships to Realize Africa's Rice Potential.* Proceedings of the Second Africa Rice Congress. Africa Rice Center, Cotonou, Benin, pp. 1.18.1–1.18.13. Available at: www.africarice.org/workshop/ARC/1.18%20Nayar%20ed2.pdf (accessed 23 February 2012).

Neeraja, C., Maghirang-Rodriguez, R., Pamplona, A., Heuer, S., Collard, B., Septiningsih, E., Vergara, G., Sanchez, D., Xu, K. and Ismail, A.M. (2007) A marker-assisted backcross approach for developing submergence-tolerant rice cultivars. *Theoretical and Applied Genetics* 115, 767–776.

Ni, J., Pujar, A., Youens-Clark, K., Yap, I., Jaiswal, P., Tecle, I., Tung, C.-W., Ren, L., Spooner, W., Wei, X., Avraham, S., Ware, D., Stein, L. and McCouch. S.R. (2009) Gramene QTL database: development, content and applications. *Database* 2009, bap005. DOI:10.1093/database/bap005.

Ondov, B.D., Varadarajan, A., Passalacqua, K.D. and Bergman, N.H. (2008) Efficient mapping of Applied Biosystems SOLiD sequence data to a reference genome for functional genomic applications. *Bioinformatics* 24, 2776.

Porteres, R. (1962) Berceaux agricoles primaires sur le continent Africain. *Journal of African History* 3(2), 195–210.

Rafalski, A. (2002) Applications of single nucleotide polymorphisms in crop genetics. *Current Opinion in Plant Biology* 5, 94–100.

Reece, J.D. and Haribabu E. (2007) Genes to feed the world: the weakest link? *Food Policy* 32(4), 459–479.
Sakai, H., Ikawa, H., Tanaka, T., Numa, H., Minami, H., Fujisawa, M., Shibata, M., Kurita, K., Hamada, M., Kanamori, H., Namiki, N., Wu, J., Itoh, T., Matsumoto, T. and Sasaki, T. (2011) Distinct evolutionary patterns of *Oryza glaberrima* deciphered by genome sequencing and comparative analysis. *The Plant Journal* 66(5), 796–805.
Sanyal, A., Ammiraju, J.S., Lu, F., Yu, Y., Rambo, T., Currie, J., Kollura, K., Kim, H.R., Chen, J., Ma, J., San Miguel, P., Mingsheng, C., Wing, R.A. and Jackson, S.A. (2010) Orthologous comparisons of the *Hd1* region across genera reveal *Hd1* gene lability within diploid *Oryza* species and disruptions to microsynteny in sorghum. *Molecular Biology and Evolution* 27(11), 2487–2506.
Sarla, N. and Mallikarjuna Swamy, B.P. (2005) *Oryza glaberrima*: a source for the improvement of *Oryza sativa*. *Current Science* 89, 955.
Schuster, S.C. (2008) Next-generation sequencing transforms today's biology. *Nature Methods* 5, 16–18.
Septiningsih, E.M., Pamplona, A.M., Sanchez, D.L., Neeraja, C.N., Vergara, G.V., Heuer, S., Ismail, A.M. and Mackill, D.J. (2008) Development of submergence-tolerant rice cultivars: the *Sub1* locus and beyond. *Annals of Botany* 103, 151–160.
Shanti, M.L., Devi, G.L., Kumar, G.N. and Shashidhar, H.E. (2010) Molecular marker-assisted selection: a tool for insulating parental lines of hybrid rice against bacterial leaf blight. *International Journal of Plant Pathology* 1, 114–123.
Song, W.-Y., Wang, G.-L., Chen, L.-L., Pi, L.-Y., Holsten, T., Gardner, J., Wang, B., Zhai, W.-X., Zhu, L.-H., Fauquet, C. and Ronald, P. (1995) A receptor kinase-like protein encoded by the rice disease resistance gene, *Xa21*. *Science* 270(5243), 1804–1806.
Sun, Y., Wang, J., Crouch, J.H. and Xu, Y. (2010) Efficiency of selective genotyping for genetic analysis of complex traits and potential applications in crop improvement. *Molecular Breeding* 26, 493–511.
Tanksley, S.D. (1993) Mapping polygenes. *Annual Review of Genetics* 27, 205–233.
Tanksley, S.D. and McCouch, S.R. (1997) Seed banks and molecular maps: unlocking genetic potential from the wild. *Science* 277, 1063–1066.
Thomson, M.J., Zhao, K., Wright, M., McNally, K.L., Rey, J., Tung, C.-W., Reynolds, A., Scheffler, B., Eizenga, G., McClung, A., Ismail, A., de Ocampo, M., Mojica, C., Reveche, M.Y., Dilla, C.J., Mauleon, R., Leung, H., Bustamante, C. and McCouch, S.R. (2011) High-throughput single nucleotide polymorphism genotyping for breeding applications in rice using the BeadXpress platform. *Molecular Breeding*. DOI: 10.1007/s11032-011-9663-x (accessed 23 February 2012).
Vaughan, D.A. (1994) *The Wild Relatives of Rice: A genetic resources handbook*. International Rice Research Institute, Manila, Philippines.
Venuprasad, R., Dalid, C.O., Del Valle, M., Zhao, D., Espiritu, M., Sta Cruz, M.T., Amante, M., Kumar, A. and Atlin, G.N. (2009) Identification and characterization of large-effect quantitative trait loci for grain yield under lowland drought stress in rice using bulk-segregant analysis. *Theoretical and Applied Genetics* 120, 177–190.
Venuprasad, R., Impa, S., Veeresh Gowda, R.P., Atlin, G.N. and Serraj, R. (2011a) Rice near-isogenic-lines (NILs) contrasting for grain yield under lowland drought stress. *Field Crops Research* 123(1), 38–46.
Venuprasad, R., Bool, M.E., Quiatchon, L. and Atlin, G.N. (2011b) A QTL for rice grain yield in aerobic environments with large effects in three genetic backgrounds. *Theoretical and Applied Genetics*. DOI: 10.1007/s00122-011-1707-4 (accessed 23 February 2012).
Venuprasad, R., Bool, M.E., Quiatchon, L., Sta Cruz, M.T., Amante, M. and Atlin G.N. (2011c) A large-effect QTL for rice grain yield under upland drought stress on chromosome 1. *Molecular Breeding*. DOI: 10.1007/s11032-011-9642-2 (accessed 23 February 2012).
Wang, L., Wang, A., Huang, X., Zhao, Q., Dong, G., Qian, Q., Sang, T. and Han, B. (2011) Mapping 49 quantitative trait loci at high resolution through sequencing-based genotyping of rice recombinant inbred lines. *Theoretical and Applied Genetics* 122(2), 327–340.
Xiao, J., Li, J., Grandillo, S., Ahn, S., Yuan, L., Tanksley, S. and McCouch, S. (1998) Identification of trait-improving quantitative trait loci alleles from a wild rice relative, *Oryza rufipogon*. *Genetics* 150, 899–909.
Xu, X., Liu, X., Ge, S., Jensen, J.D., Hu, F., Li, X., Dong, Y., Gutenkunst, R.N., Fang, L., Huang, L., Li, J., He, W., Zhang, G., Zheng, X., Zhang, F., Li, Y., Yu, C., Kristiansen, K., Zhang, X., Wang, J., Wright, M., McCouch, S., Nielsen, R., Wang, J. and Wang, W. (2011) Resequencing 50 accessions of cultivated and wild rice yields markers for identifying agronomically important genes. *Nature Biotechnology* 30(1), 105–111.

Yu, J., Presspoir, G., Briggs, W.H., Bi, I.V., Yamasaki, M., Doebley, J.F., McMullen, M.D., Gaut, B.S., Nielsen, D.M., Holland, J.B., Kresovich, S. and Buckler, E.S. (2005) A unified mixed-model method for association mapping that accounts for multiple levels of relatedness. *Nature Genetics* 38(2), 203–208.

Yu, J., Holland, J.B., McMullen, M.S. and Buckler, E.S. (2008) Genetic design and statistical power of nested association mapping in maize. *Genetics* 178, 539–551.

Zhang, Q., Li, J., Xue, Y., Han, B. and Deng, X.W. (2008) Rice 2020: a call for an international coordinated effort in rice functional genomics. *Molecular Plant* 1, 715–719.

Zhao, K., Tung, C.-W., Eizenga, G.C., Wright, M.H., Ali, M.L., Price, A.H., Norton, G.J., Islam, M.R., Reynolds, A., Mezey, J., McClung, A.M., Bustamante, C.D. and McCouch, S.R. (2011) Genome-wide association mapping reveals rich genetic architecture of complex traits in *Oryza sativa. Nature Communications.* DOI: 10.1038/ncomms1467 (accessed 23 February 2012).

Zhong, S., Dekkers, J.C.M., Fernando, R.L. and Jannink, J.L. (2009) Factors affecting accuracy from genomic selection in populations derived from multiple inbred lines: a barley case study. *Genetics* 182, 355–364.

Zhu, C., Gore, M., Buckler, E.S. and Yu, J. (2008) Status and prospects of association mapping in plants. *Plant Genome* 1, 5–20.

10 Unlocking the *Oryza glaberrima* Treasure for Rice Breeding in Africa

Mathias Lorieux,[1,2]* Andrea Garavito,[1] Julie Bouniol,[1] Andres Gutiérrez,[2] Marie-Noëlle Ndjiondjop,[3] Romain Guyot,[1] César Pompilio Martinez,[2] Joe Tohme[2] and Alain Ghesquière[1]

[1]*Institut de recherche pour le développement (IRD), UMR Diversité, Adaptation et Développement des Plantes (DIADE), IRD-CIRAD-UMZ, Montpellier, France;* [2]*International Center for Tropical Agriculture (CIAT), Cali, Colombia;* [3]*Africa Rice Center (AfricaRice), Cotonou, Benin*

Domestication and Genetic Diversity of *Oryza glaberrima*

African cultivated rice (*Oryza glaberrima* Steud.) was domesticated independently of Asian cultivated rice (*Oryza sativa* L.) in Africa from the wild species *O. barthii* A.Chev. (syn. *O. breviligulata* A.Chev. & Roehr.) (Second, 1982), which diverged 0.6–0.7 million years ago from *O. sativa* (Zhu and Ge, 2005; Ammiraju *et al.*, 2008). While Asian rice has spread worldwide and become the most produced food crop in the world, African rice is grown only in tropical West Africa and there sporadically. *Oryza glaberrima* demonstrates a typical case of reduction of genetic diversity observed in crops compared to their wild progenitors as a result of the dual bottlenecks of domestication and breeding (Buckler *et al.*, 2001; Tenaillon *et al.*, 2004; Zeder *et al.*, 2006). Using isozyme, RFLP (restriction fragment length polymorphism), SSR (simple sequence repeat) and transposable element markers, previous studies have detected dramatic reduction of genetic diversity associated with the domestication of African rice and revealed substantially lower genetic diversity in *O. glaberrima* than in *O. sativa* (Second, 1982; Wang *et al.*, 1992; Ishii *et al.*, 2001). The domestication of *O. glaberrima* occurred much later than that of *O. sativa*, but prior to the introduction of the latter into Africa (Sweeney and McCouch, 2007). The most recent analysis of diversity, comparing nucleotide variation of 14 independent nuclear loci between *O. glaberrima* and *O. barthii*, showed that *O. glaberrima* has lost 76% of the nucleotide diversity of its wild progenitor (Li *et al.*, 2011a), while *O. barthii* itself harbours slightly less diversity than *O. sativa* subsp. *indica*. The low genetic diversity of *O. glaberrima* compared to *O. sativa* is likely to have resulted from a reduction of diversity during the migration of the wild progenitor of African *Oryza* species into Africa plus a severe genetic bottleneck during its domestication from small initial populations of *O. barthii*. The eco-geographical diversity seems to be so low that clustering analysis is unable to pinpoint the domestication places and dispersion of *O. glaberrima* (Li *et al.*, 2011a). Portères' (1970) hypothesis remains the most probable – that African rice was first domesticated in the inland delta of the upper Niger River and subsequently spread along Sahelian rivers and their tributaries to two

* Corresponding author: mathias.lorieux@ird.fr

© CAB International 2013. *Realizing Africa's Rice Promise* (eds M.C.S. Wopereis *et al.*)

secondary centres of diversity: one along the coast of The Gambia and Guinea-Bissau, and the other in the Guinea forest between Sierra Leone and western Côte d'Ivoire.

Although *O. glaberrima* was recognized as an interesting source of agronomically important traits for rice breeding by the French and international scientific community, its utilization was, for a long time, hampered by the strong reproductive isolation between the two cultivated rice species (Sano *et al.*, 1979) and by a lack of convenient tools and general strategy for rational introgression into *O. sativa*. Interspecific hybridization with formal breeding objectives was initiated in the 1990s by the Africa Rice Center (AfricaRice) and resulted in NERICA (New Rice for Africa) varieties adapted to upland and lowland growing conditions of Africa (Jones *et al.*, 1997; Saito *et al.*, 2010).

This chapter reviews recent advances of research on interspecific sterility in rice with emphasis on the reproductive barriers between the two cultivated rice species, and highlights how we can – with the advent of sequence information, genetic maps and markers – develop pre-breeding schemes suitable for gene and trait discovery, with the aim of easier transfer of *O. glaberrima* traits of agronomic interest into elite *O. sativa* cultivars.

Interspecific Sterility Between *O. sativa* and *O. glaberrima*

Two main types of reproductive barriers are observed in plants: pre-zygotic barriers that prevent the formation of hybrid zygotes and post-zygotic barriers (such as hybrid weakness, unviability and hybrid sterility) that hamper gene flow between species or subspecies, and which largely depend on divergence time between the species (Ouyang *et al.*, 2010). Hybrid sterility is the most common form of post-zygotic reproductive isolation. In rice, one of the best known examples is the hybrid sterility between the subspecies of *O. sativa* (*indica* and *japonica*), which show embryo-sac abortion and pollen sterility. A general gametic lethal model has been proposed in which two independent loci affect the gamete development and gametes carrying the recessive alleles at both loci are aborted during the development while gametes of other genotypes were normal (Oka, 1957, 1974). These negative interactions can be also observed within a single locus as the consequence of independent evolution of its alleles causing sterility when two incompatible alleles are brought together in hybrids. The (recently cloned) sterility gene *S5* (Ikehashi and Araki, 1986; Chen *et al.*, 2008) is an example where neutral alleles are found in wide compatibility varieties (WCVs) and suppress the negative effects of *indica* and *japonica* alleles in hybrids. Nevertheless, the molecular basis of hybrid incompatibility is usually complex and often involves accumulative effects and interactions of genes at multiple loci. It is critical that we obtain a better understanding of the genetic bases and biological mechanisms of hybrid sterility, since it hinders the transfer of useful genes between the two cultivated rice species and is a major obstacle for utilization of the strong heterosis exhibited in *O. sativa* F_1 hybrids.

Approximately 50 loci controlling hybrid fertility have been identified in the 'A genome' *Oryza* species, mainly from studies of *indica–japonica* crosses. Post-zygotic sterility is the most frequent and includes loci causing female-gamete abortion and others inducing pollen sterility. There is increasing documentation of interspecific sterility and the identification of sterility genes in *O. sativa–O. glaberrima* hybridization. Near complete pollen sterility is generally observed in F_1 hybrids of the two species. Pollen sterility is accompanied by a reduction in female fertility (Bouharmont *et al.*, 1985). Nevertheless, obtaining backcross progenies is feasible and allows us to derive advanced-backcross lines and near-isogenic lines suitable for identifying and mapping sterility genes (Garavito *et al.*, 2010). More than 12 pollen-sterility loci have been described, some of which are associated with embryo-sac sterility (see Table 10.1 for review). Transmission ratio distortion (TRD) of parental alleles at sterility loci is commonly observed in interspecific progenies and may be the result of many different post-zygotic effects. Considering, for example, a locus with two *a* and *A* alleles coming from *O. sativa* and *O. glaberrima*, respectively, a backcross between a hybrid plant with *Aa* genotype and an *O. sativa* plant with *aa* genotype is expected to give a 1:1 segregation between *aa* and *Aa* genotypes in the following generation. The TRD refers to the departure from this ratio and its origin is

Table 10.1. A non-exhaustive list of loci involved in reproductive barriers between the cultivated rice species.

Locus	Chromosome	Effects on sterility[a]	Reference(s)
S1	6	P, ES	Sano (1990); Koide *et al.* (2008c); Garavito *et al.* (2010)
S3	11	P, ES	Sano (1983, 1986)
S4	6	P, ES	Sano (1986)
qSS-6a	6	P	Li *et al.* (2011c)
S18	10	P	Doi *et al.* (1998a, 2003b)
S19/qSS-3	3	P, ES	Taguchi *et al.* (1999); Doi *et al.* (2003b); Li *et al.* (2011c)
S20/qSS-7a	7	P	Doi *et al.* (1999, 2003b); Li *et al.* (2008)
S21	7	P	Doi *et al.* (2003b)
S29(t)[b]*/qSS-2*	2	P	Hu *et al.* (2006); Li *et al.* (2008); Zhu *et al.* (2005)
S33(t)[b]	1	P, ES	Ren *et al.* (2005); Jing *et al.* (2007)
S34(t)[b]	3	P	Zhang *et al.* (2005); Jing *et al.* (2007)
S36(t)	2	P	Li *et al.* (2011b)
qSS-1	1	P	Li *et al.* (2008)
qSS-7b	7	P	Li *et al.* (2008, 2011c)

[a]ES: embryo sac sterility; P: pollen sterility.
[b]Locus name given in sterility studies of *O. sativa* subsp. *indica* × subsp. *japonica* crosses, but with different chromosome positions.

inferred according to the cross analysed. When an interspecific F_1 plant is used as the female parent, the TRD can be attributed to the female TRD (fTRD), since interspecific F_1 hybrids are almost completely male sterile. Reciprocally, when an advanced interspecific isoline is used as the male parent for crossing with an *O. sativa* parental line, we measure male TRD (mTRD).

A genetic map has been constructed from 125 individuals of an (*O. sativa* × *O. glaberrima*) × *O. sativa* (IR64/TOG5681//IR64) backcross population, using 140 SSR anchor markers derived from a Universal Core Genetic Map (UCGM; Orjuela *et al.*, 2010) to monitor loci involved in hybrid sterility through the detection of fTRD (Garavito *et al.*, 2010). A majority of sterility loci between the two species and previously identified by Li *et al.* (2008, 2011b,c) and Doi *et al.* (1998b) could be co-localized with fTRD regions mapped by Garavito *et al.* (2010) and confirmed the link between fTRD and pollen and spikelet sterility (Plate 2). In one case (S_{33}), sterility genes seemed specific for pollen sterility as no accompanying fTRD was found. Variations in direction and intensity of fTRD were also observed and may suggest the existence of different alleles involved in the sterility barriers. Occurrence of TRD in genomic regions where no sterility loci have been recorded could be explained by new sterility loci still to be discovered or other mechanisms not related to gamete elimination, such as pre-zygotic sterility favouring, for example, pollen germination on hybrid stigmas or post-zygotic effects leading to hybrid weakness or endosperm abortion which are also reported in interspecific hybridization (Matsubara *et al.*, 2003; Koide *et al.*, 2008a,b). Epistatic interactions are suspected to be acting between regions harbouring TRD and sterility locus, similar to barriers between *indica* and *japonica* (Li *et al.*, 2008). The strongest fTRD ($P < 10^{-5}$) was found at the S_1 locus and is characterized by an extreme distortion in favour of the *O. glaberrima* allele whatever the orientation of crosses or genetic background of the parents. Thus, the S_1 locus is presumed to be the major determinant of female sterility, while pollen sterility is the result of S_1 plus additive and epistatic effects of other sterility loci dispersed on the different chromosomes (Garavito *et al.*, 2010).

Focus on the S_1 Locus

The S_1 locus was first mapped on chromosome 6 thanks to its strong linkage with waxy (*Wx*) gene, which can be identified by pollen staining of F_1 hybrids (Sano, 1983, 1990). Sano's group has also suggested that the S_1 locus has at least two components, one controlling mTRD mapped in a 45 kb interval of the Nipponbare chromosome 6 sequence, and a second controlling the female sterility component (Koide *et al.*, 2008c). On the assumption that the degree of fTRD would be inversely proportional to the genetic distance between the sterility factor and the markers subjected to fTRD, several backcross populations between Nipponbare, IR64, Curinga (*O. sativa*) and CG14, MG12 (*O. glaberrima*, alias IRGC103544) were fine-mapped using SSR (Garavito *et al.*, 2010). Taken together, these maps enabled determination of a sharp peak of the fTRD value, corresponding to a maximum common fTRD, and leading to high-resolution mapping of S_1 in a region of a 27.8 kb on chromosome 6.

To compare the chromosomal regions between the two species, we used the large genomic library of the *O. glaberrima* cv. CG14 prepared in bacterial artificial chromosomes (BAC) for the *Oryza* Mapping Alignment project (OMAP; Kim *et al.*, 2008). The BAC clone OG-BBa0049I08 containing a 164,664-bp genomic sequence of *O. glaberrima* was selected as it carried the markers with the highest fTRD. Additional convenient markers were derived from the *O. glaberrima* sequence and were used to analyse the recombination events around the peak of fTRD in the rare BC_1 plants homozygous for the *O. sativa* allele of S_1 (S_1^s), since each of these plants originated from a viable S_1^s female gamete transmitted by the F_1 hybrids and the other S_1^s from the recurrent *O. sativa* parent. No recombination was observed in a chromosomal segment of at least 7.3 cM around S_1^s (886 kb in the Nipponbare genome) which was shared by all the S_1^s/S_1^s plants. Thus, results suggested the necessity of inheriting other *O. sativa* factors in the vicinity of S_1 to ensure the viability of the S_1^s gametes. The existence of an additional factor on each side of S_1 (hereafter denoted as S_1A and S_1B) was proposed to infer a sterility system based on three components (S_1, S_1A, S_1B) playing a role in female sterility of F_1 hybrids and the associated fTRD, through epistatic interactions. The frequency of S_1^s elimination and the length of the genomic block surrounding the S_1 locus that does not recombine differ between the subspecies – *indica* or *japonica* – used in interspecific combinations, but also according to the presence of another factor (S_1C) located approximately 50 cM from S_1 (Garavito *et al.*, 2010).

A General Model to Explain the Reproductive Barrier Between *O. glaberrima* and *O. sativa*

A general model to explain the reproductive barrier between the two species has been developed based on the assumption that the post-zygotic barriers in rice are due to the accumulation of incompatibilities in epistatic interactions between genes. It is based on the very general Bateson–Dobzhansky–Muller (BDM) model (Bateson, 1909). This model says that a specific epistasis among S_1A, S_1 and S_1B is the cause of the female hybrid sterility between the two species and explains the elimination of the S_1^s gamete. Considering that S_1A^s-S_1^s-S_1B^s and S_1A^g-S_1^g-S_1B^g are the original haplotypes observed in *O. sativa* and *O. glaberrima*, respectively, the epistatic interaction among the three *O. glaberrima* factors will cause the abortion of daughter cells carrying S_1^s during the two first meiotic divisions and will cause an fTRD. When recombination occurs in the mother cell and eliminates S_1B^g, S_1A^g or both, the epistatic incompatible interaction acting against S_1^s would cease, allowing the development of the corresponding S_1^s daughter cells and eliminating the fTRD. Finally, at the end of the megagametogenesis, only the S_1^s megaspores carrying the non-recombinant haplotype S_1A^s-S_1^s-S_1B^s could survive and would lead to a functional embryo sac (Fig. 10.1). Cytological observations support this genetic model and the resulting reduction of seed set, since different types of female-gamete abortion occur during female gametogenesis of the F_1 hybrids: some are characterized by a complete absence of embryo sacs and are probably due to an early abortion just after meiosis, and others are characterized by embryo sacs with fewer than seven cells (Bouharmont *et al.*, 1985; Koide *et al.*, 2008c).

The S_1 Locus as an Entry Point to Analyse Domestication and Evolution in Rice

The genomic sequence of the 50.38 kb interval presumed to contain the female component of the S_1 locus was carefully annotated to identify putative candidate gene for S_1. Seven genes and three pseudogenes were predicted and were

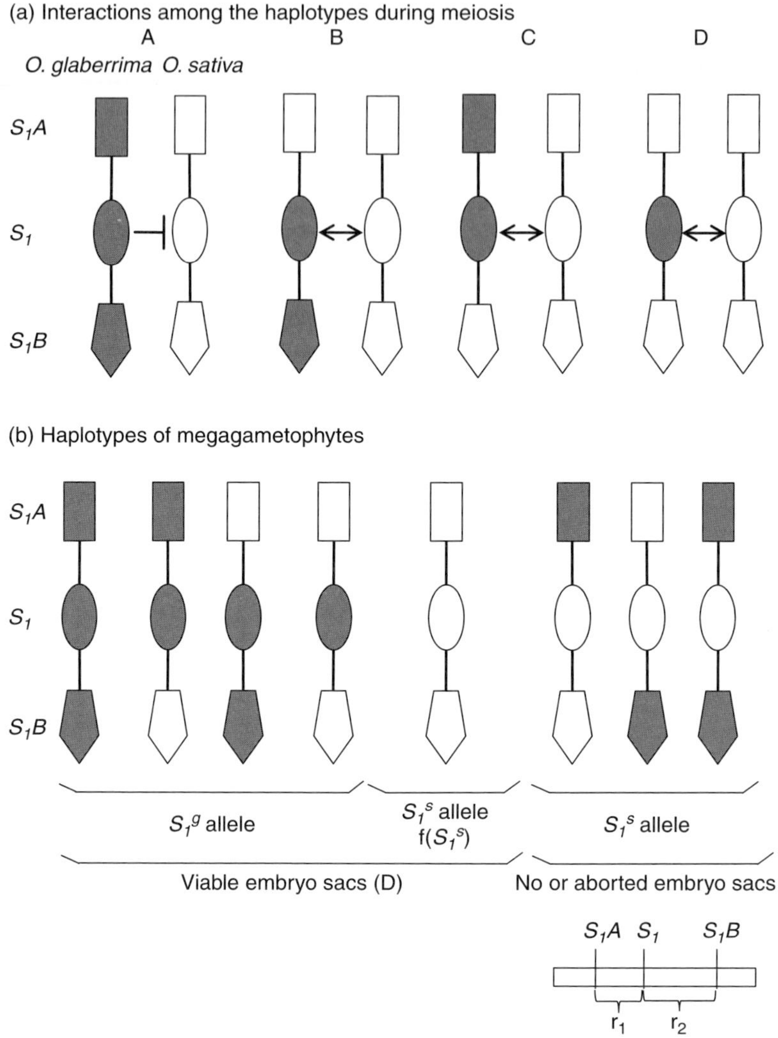

Fig. 10.1. Genetic model based on a Bateson–Dobzhansky–Muller (BDM) model of interspecific incompatibility (Bateson, 1909) and consequences on the elimination of S_1^s female gametes. (a) During the meiotic divisions of the mother cell, the epistatic interactions of the *O. glaberrima* S_1A^g-S_1^g-S_1B^g haplotype cause the abortion (⊣) of the daughter cells carrying S_1^s and will originate a female Transmission Distortion Ratio (fTRD). If a recombination eliminates S_1B^g (B), S_1A^g (C) or both (D), the epistatic incompatible interaction against S_1^s would cease (double arrow) allowing the development of S_1^s gametes and eliminating the fTRD. (b) Among the S_1^s gametes, only the small fraction corresponding to the non-recombinant *O. sativa* S_1A^s-S_1^s-S_1B^s haplotype (f(S_1^s)) can survive and develop in a normal embryo sac, since it is derived from a recombination event in the daughter cell during the meiotic process which suppresses the epistasis against S_1^s. All the other S_1^s allelic configurations will result in the abortion of megaspores or abnormal embryo sacs. Configurations giving viable embryo sacs (D) estimate the reduction of the seed set observed in F_1 hybrids. Therefore, the fTRD is directly dependent on the recombination fraction between S_1A–S_1 (r_1) and S_1–S_1B (r_2).

considered as candidates. Among the predicted genes, two deserved special attention as their putative functions, differential presence between *O. glaberrima* and *O. sativa*, and greater divergence between the two species, could justify relationships with reproductive effects (Garavito *et al.*, 2010). The first gene (*49108-10*) is carried by a Pack-*Mutator*-like transposable element (Pack-MULE) and is present only in the *O. glaberrima* S_1 sequence. This gene showed similarities with an APETALA (AP2) transcription factor and mutations in AP2 in *Arabidopsis* are known to stop mega-sporogenesis after the first meiotic division (Byzova *et al.*, 1999). Pack-MULEs are known to have an important role in rice genome evolution, as they can capture and relocate gene fragments to other genomic contexts (Jiang *et al.*, 2004) leading to modulation of both MULEs and paralogous gene expression (Hanada *et al.*, 2009). If the complete paralogous AP2 gene had a function similar to that of the *Arabidopsis* genes, this pack-MULE could affect its expression in the heterozygotes, in a dose-dependent fashion, causing abortion of female gametes, and would appear very close to hybrid dysgenesis mechanism (Michalak, 2009). The second candidate gene of interest (*49108-11*) belongs to the super-family of F-Box proteins. Members of this family constitute protein complexes known as SCF (Skp1–Cullin–F-Box) involved in the control of a wide range of processes (Xu *et al.*, 2009). Several F-box genes and their associated proteins have been related with the progression of the cell cycle, especially during sporogenesis and gametogenesis (Wang and Yang, 2006; Pesin and Orr-Weaver, 2008; Gusti *et al.*, 2009). A strong functional similarity can be seen with the complex composed by an F-box and a SUMO E3 ligase-like protein that controls the inter-subspecific hybrid sterility mediated by the *Sa* locus in *O. sativa* (Long *et al.*, 2008).

The structural analysis was extended to the 813 kb genomic sequence of *O. glaberrima* covering the S_1A and S_1 regions and most of the S_1B region, which has been compared to its orthologous regions in *O. sativa* (Guyot *et al.*, 2011). A strong structural conservation was observed all along the S_1 locus between the two cultivated species. This was in accordance with the relatively recent divergence (approximately 0.7 million years) of Asian and African rice lineages by geographical isolation (Ma and Bennetzen, 2004; Ammiraju *et al.*, 2008) and it reinforced the consistency of candidate genes provided by F-Box genes as they represented the most important structural genetic variation between the two cultivated rice species. Thus, effects of duplicated F-Box genes strengthened by a local inversion, which increases the epistatic cohesion of the factors, may contribute greatly to the reproductive isolation between the two species (Guyot *et al.*, 2011). Further experiments are being carried out at Institut de recherche pour le développement (IRD) to confirm candidate genes and to better understand the mechanism controlling the post-zygotic barrier at the S_1 locus.

Identification of the main locus involved in the reproductive barrier between the cultivated rice species reflects the divergence accumulated between their respective wild progenitors. The colocalization of sterility locus between *O. sativa* and *O. glaberrima* with the ones observed between *indica* and *japonica* accessions or between *O. rufipogon* and *O. sativa* can provide useful guidelines for in-depth analysis of the evolution of critical loci responsible for BDM reproductive barriers, which may have fixed different alleles of incompatibilities between the species (Plate 2). S_{10} and S_1, for example, or S_{21}–S_{23} on chromosome 2 and S_{22}–$S_{29(t)}$–S_{32}–*qSS-2* on chromosome 7 are the most striking colocalizations of sterility genes between different A genome species. Accumulation of effects of additional loci before and during the domestication of rices may explain why the reproductive barrier is so pronounced between the two cultivated species. Identification of sterility genes may also help in refining the analysis of the potential gene flow between the cultivated species, since they coexist in West Africa. Hence, specific transposable elements identified in the *O. glaberrima* and *O. sativa* S_1 sequences were used to determine Retrotransposon-based insertion polymorphism (R-BIP) markers, which proved to be very useful for the analysis of the integrity of the S_1 locus in a representative collection of wild and cultivated African rices. Results confirmed fairly well that no recombination occurred in the entire S_1 region even in very rare off-types or accessions supposed to have been introgressed elsewhere in the genome (our unpublished results).

Development of Chromosome Segment Substitution Lines (CSSLs) and Near-isogenic Lines (NILs) to Exploit the Genetic Diversity of *O. glaberrima*

Interspecific hybridization offers an attractive way of enlarging the genetic diversity for crop improvement, but it is hampered by reproductive barriers, increased sterility and reduced recombination between the cultivated and the related species. The development of markers and genetic mapping provides powerful tools to more systematically introgress a genome of a distant species into a cultivated one, and to characterize phenotypic variation through the development of specific populations. Among such populations, complete sets of chromosome segment substitution lines (CSSLs), which are based on the representation of small chromosome fragments coming from a donor species in another recipient species, are particularly suitable in the case of interspecific hybridization between *O. sativa* and *O. glaberrima* (Ghesquière *et al.*, 1997). The general scheme for developing a CSSL population is based on successive backcrosses and selection of individuals bearing only one or a small number of targeted chromosomal segments of the *O. glaberrima* donor in the *O. sativa* background, using well-distributed molecular markers (Plate 3). Usually a first set of 50–60 randomly chosen BC_1F_1 individuals is large enough to cover the complete genome of the donor parent. The BC_1F_1 plants are backcrossed again to generate the BC_2 families. Then 10–15 individuals from each BC_2F_1 family are completely genotyped and a total of about 150–200 BC_2F_1 plants showing a targeted segment are selected and used for the next backcrossing generation to produce BC_3F_1s. Again, 120–150 BC_3F_1 individuals are selected from 400–500 BC_3F_1 individuals after a second round of genotyping. These individuals are selfed to develop BC_3F_2s and then homozygous lines for the target segment generated through single-seed descent (SSD), or they can be backcrossed again to generate BC_4F_1 plants if additional isogenization is needed. To optimize the coverage of the entire genome of the donor in a minimum set of lines, graphical genotyping software such as GGT (van Berloo, 2008) and CSSL FINDER (http://mapdisto.free.fr/CSSLFinder) have been developed. A universal core genetic map is also available to select SSR markers anchored in the Nipponbare physical map and polymorphic across a large range of cultivars and species with the A genome (Orjuela *et al.*, 2010).

The CSSL approach has become popular for identifying and introgressing genes from one varietal group to another in rice, namely from *indica* to *japonica* and vice versa. Special effort has been made to use recipient varieties with available genomic sequences, such as Nipponbare (a temperate *japonica* variety), Zhenshen 97B (a hybrid rice maintainer line) and 93-11 (a restorer line sequenced by the Beijing Genomics Institute). The latter two lines are also used in an ambitious project to develop 14 CSSL libraries comprising seven A genome wild species accessions and included in the OMAP (Wing *et al.*, 2007). Using CSSL populations developed from 93-11 and Zhenshen 97B with Nipponbare as donor, quantitative trait loci (QTLs) have been identified for enhancing the cultivability of *indica* rice (Zhao *et al.*, 2009), number of panicles and grain yield under low nitrogen and phosphorus conditions (Wang *et al.*, 2009), and grain length and width (Zhu *et al.*, 2009). CSSLs from Kasalath (an *Aus* landrace from India) were developed in the *japonica* genetic background of Koshihikari to map QTLs affecting heading date (Ebitani *et al.*, 2005), cadmium concentration in brown rice (Ishikawa *et al.*, 2005), increased number of panicles per plant, increased number of grains per panicle and increased root mass (Maduoka *et al.*, 2008). For African rice species, several CSSL initiatives have been developed or are in progress (Table 10.2; Ali *et al.*, 2010). The most advanced libraries concern two CSSL populations using the *O. glaberrima* accession MG12 in the background of Caiapo (a tropical *japonica*) and the multipurpose *O. glaberrima* accession TOG5681 in IR64 (*indica*) (Gutiérrez *et al.*, 2010).

Identification, Mapping and Transfer of Valuable Genes from *O. glaberrima*

Systematic introgression is time- and labour-intensive, so it needs to be conceived as an investment for the future and focused on important recipient *O. sativa* varieties or lines and on multipurpose *O. glaberrima* donors. As different lines of a CSSL population show strong morphological similarities, they represent an ideal tool to identify QTLs from remote or unadapted donor germplasm with the opportunity to reveal small effects that are usually masked by major QTLs in direct F_2 or recombinant inbred line (RIL) primary populations. In addition, genotype × environment interactions can be precisely monitored through replicated trials over years, across different

Table 10.2. Summary of chromosome segment substitution lines (CSSLs), backcross inbred lines (BILs) and near-isogenic lines (NILs) developed or in development, using *O. glaberrima* or *O. barthii* as donors and *O. sativa* as recurrent parent.

Donor accession	Recurrent parent (*O. sativa*)	Potential traits targeted[a]	Reference(s)
O. glaberrima TOG 5675	*indica* IR64	BPH resistance (*Bph1*)	Ram *et al.* (2010)
O. glaberrima TOG 5681	*indica* IR64	RYMV resistance (*rymv1-2*; *rymv1-3*)	Albar *et al.* (2006)
O. glaberrima TOG 5681	*indica* IR64	Nematode resistance (*Meloidogyne* spp.)	S. Bellafiore (personal communication); Bimpong *et al.* (2010)
O. glaberrima IRGC 104589	*indica* IR64	RYMV resistance (*Rymv2*)	Thiémélé *et al.* (2010)
O. glaberrima IRGC 96726/TOG 5307	*indica* IR64	RYMV resistance (*Rymv3*)	L. Albar (personal communication)
O. glaberrima IRGC 103544/MG 12	Tropical *japonica* Caiapo	*Rice stripe necrosis virus* resistance	Guttiérrez *et al.* (2010)
O. glaberrima IRGC 103544/MG 12	*indica* Mylang 23	Yield and yield components	Kang *et al.* (2008)
O. glaberrima IRGC 104038	*japonica* Taichung 65	Pollen fertility and heading date	Doi *et al.* (1997) (2003a)
O. glaberrima IRGC 96717/CG 14	Tropical *japonica* WAB 56-104	Drought tolerance and weed competitiveness	Ndjiondjop *et al.* (2010)
O. glaberrima	Temperate *japonica* cv. Koshihikari	Glabrous gene	Angeles-Shim *et al.* (2009)
CSSLs under construction			
O. barthii IRGC 10193	Tropical *japonica* Curinga	To be evaluated for yield-related traits, grain traits, tolerance to biotic and abiotic stresses	M. Lorieux (cited in Ali *et al.*, 2010)
O. barthii IRGC 101937	*indica* Zhenshen 97B and 93-11	To be evaluated for yield-related traits, grain traits, tolerance to biotic and abiotic stresses	S.B. Yu (cited in Ali *et al.*, 2010)
O. glaberrima IRGC 96717/CG 14	*indica* Zhenshen 97B	Drought tolerance and grain quality	S.B. Yu (cited in Ali *et al.*, 2010)
O. glaberrima and *O. barthii*	Tropical *japonica* LaGrue and temperate *japonica* M-202	To be evaluated for yield-related traits, grain traits, tolerance to biotic and abiotic stresses	P. J. Sanchez and G. C. Eizenga (cited in Ali *et al.*, 2010)

[a]BPH, brown planthopper; RYMV, *Rice yellow mottle virus.*

environments and growing conditions. A good example of direct utilization of CSSLs was the identification of a major resistance gene to *Rice stripe necrosis virus* (RSNV) and yield-component QTLs in the (MG12 × Caiapo) CSSL population (Gutiérrez *et al.*, 2010; Fig. 10.2). CSSL lines usually contain more than their single targeted segment, but additional backcrossing can be performed to focus on a specific chromosome segment or on a confirmed line with special interest to derive true NILs ready for fine mapping and positional cloning. For instance, using the (IR64 × Tog5681) CSSL population, a NIL has been developed for the *Rice yellow mottle virus* (RYMV) resistance gene *rymv1-3*/IR64 (Albar *et al.*, 2006). Similarly, three confirmed IR64 NILs resistant to nematode *Meloidogyne incognita* have been derived from an (IR64 × Tog5681) CSSL population to map and to ultimately clone the resistance genes to *Meloidogyne*

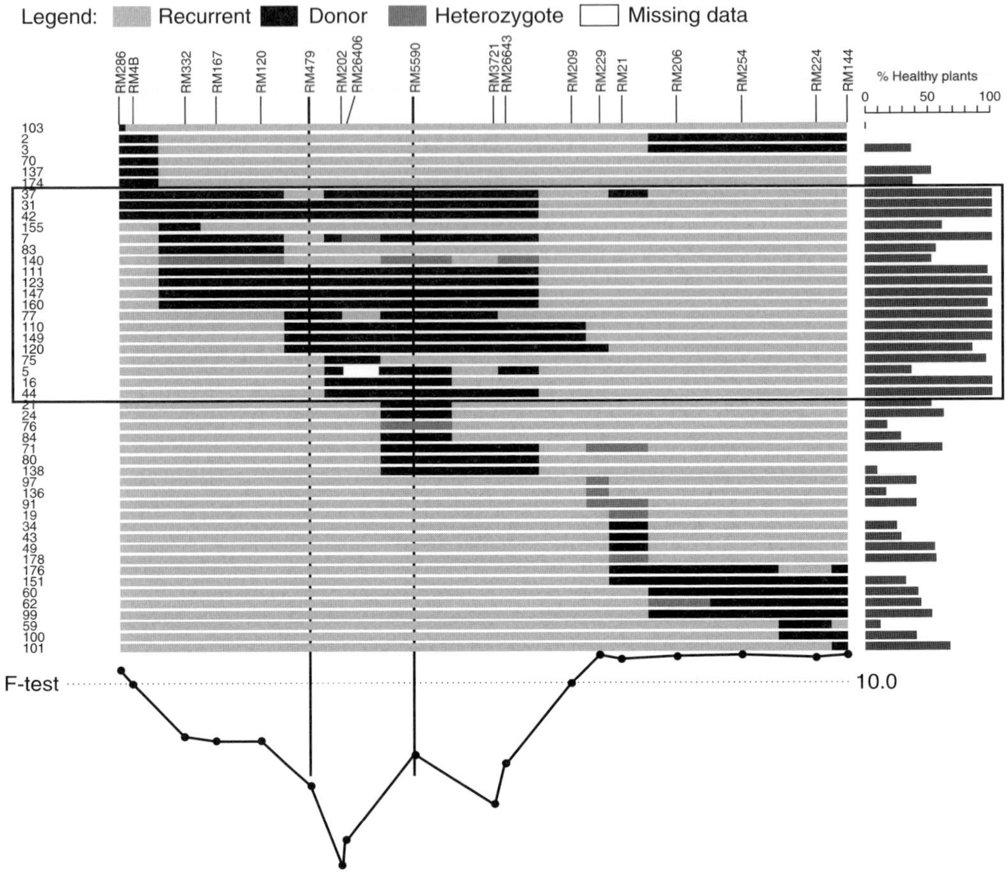

Fig. 10.2. Identification of a major QTL for resistance to *Rice stripe necrosis virus* (RSNV) in *O. glaberrima* MG12. A major resistance QTL is located on chromosome 11 between SSR markers RM479 and RM5590 (F = 70.63, $P \sim 0.0$, significance threshold = 10.0). On the right, solid bars indicate the percentage of healthy plants for each line. The resistant lines (more than 85% healthy plants) are located within the black frame. The most probable location of the resistance QTL is given by the intersection of the black frame and the positions of the markers RM479 and RM5590, which define a common introgressed region between the resistant lines.

spp. found in *O. glaberrima* (our unpublished results). CSSLs are also a starting point for dissecting complex traits, identifying lines with individual desirable QTLs, which can be reassembled efficiently in a common genetic background though a recurrent marker selection process. Pyramiding of QTLs for plant height and number of grains (Ashikari *et al.*, 2005) to efficiently increase those traits in the Koshihikari *japonica* cultivar demonstrated how introgressed lines could be efficiently connected to breeding objectives and varietal release (Ashikari and Matsuko, 2006).

Conclusion: Application for New Breeding Schemes Adapted for *O. glaberrima*

Since the early 2000s, two major breakthroughs have been made for better gene discovery in rice genetic resources: user-friendly mapped markers such as SSRs, and suitable populations for trait analysis. In addition, single-nucleotide polymorphism (SNP) markers derived from rice-sequencing projects offer a quasi-infinite number of markers. Specifically designed 384-SNP sets will increase genotyping efficiency of any segregating progeny for gene and QTL identification. Also, a 44,000 array is already available and a 700,000 SNP array is under development (Tung *et al.*, 2010; see also McCouch *et al.*, Chapter 9, this volume). Using the *O. glaberrima* CG14 genomic sequence, new sets of versatile SNPs well distributed on the rice genome and showing intra- and interspecific polymorphism are being developed by IRD for the study and use of *O. sativa* × *O. glaberrima* introgressions. Outsourcing of genotyping via dedicated platforms such as the Molecular Breeding Platform of the Generation Challenge Programme (GCP) to private companies will greatly alleviate activities in plant breeding, allowing breeders and geneticists to focus on phenotypic evaluation of the traits of interest. The first generation of upland NERICA varieties developed with *O. glaberrima* CG14 have been scored for their *O. glaberrima* and *O. sativa* contents (Semagn *et al.*, 2006). According to the S_1 locus genetic model described above, the presence of S_1^g allele in some of them is expected to restore the F_1 interspecific reproductive barrier if S_1^g NERICA varieties are used in crosses with *O. sativa* lines. The S_1 locus is not the only locus responsible for the reproductive barrier, but the S_1^s allele is an important prerequisite for favouring fertility restoration in backcross generations. The female transmission of the S_1^s allele in *O. sativa* backcross progenies may vary from 1.5% to 10% according to recipient parents and combinations (Garavito *et al.*, 2010; our unpublished results). We developed the concept of interspecific bridge lines (iBridges) based on marker-assisted selection of homozygous S_1^s/S_1^s individuals, which resulted in an increase of the proportion of fertile individuals in the first backcross generation (Table 10.3). Combined with CSSL or backcross inbred lines (BIL) strategies as described above, this process allows accelerated fertility restoration and more efficient development of fertile introgressed lines for phenotypic evaluation and gene discovery. Producing lines that are able to give fertile progeny when crossed with more diverse *O. sativa* elite lines is expected to greatly facilitate the incorporation of *O. glaberrima* germplasm in conventional rice breeding schemes. This concept is being applied through international collaborations (AfricaRice, International Center for Tropical Agriculture, IRD and national agricultural research systems) within the framework of the Global Rice Science Partnership (GRiSP) to enlarge the use of other important *O. glaberrima* donors and to diversify the target recurrent *O. sativa* varieties.

Table 10.3. First steps of the development of iBridges through marker-assisted selection for S_1^S allele and subsequent fertility restoration in the *O. sativa/O. glaberrima//O. sativa* backcross progenies. Off-types refer to some natural accessions showing evidence of introgression events based on SSR and SNP marker data.

Parent	No. comb[a]	No. BC_1 seed set	No. BC_1 screened by MAS[b]	No. S_1^S (RM190)	S_1^S allele transmission in F_1	Range	Fertile individuals (S_1^S/S_1^S genotype, %)	Fertile individuals (S_1^S/S_1^g genotype, %)
O. sativa line								
IR64	12	1536	831	45	0.055	0–19	54	2
Curinga	11	1633	797	33	0.041	0–16	45	5
WAB165	11	1989	669	33	0.049	0–10	16	3
Total	34	5158	2297	112	0.049	0–11	40	3
Off-types	10	1503	878	233	0.265	16–46	72	23

[a]comb, combinations; [b]MAS, marker-assisted selection.

References

Albar, L., Bangratz-Reyser, M., Hébrard, E., Ndjondjop, M.-N., Jones, M. and Ghesquière, A. (2006) Mutations in the eIF(iso)4G translation initiation factor confer high resistance of rice to *Rice yellow mottle virus*. *Plant Journal* 47, 417–426.

Ali, M.L., Sanchez, P.L., Yu, S.B., Lorieux, M. and Eizenga, G. (2010) Chromosome segment substitution lines: a powerful tool for the introgression of valuable genes from *Oryza* wild species into cultivated rice (*O. sativa*). *Rice* 3(4), 218–234.

Aluko, G., Martinez, C., Tohme, J., Castano, C., Bergman, C. and Oard, J. H. (2004) QTL mapping of grain quality traits from the interspecific cross *Oryza sativa* × *O. glaberrima*. *Theoretical and Applied Genetics* 109(3), 630–639.

Ammiraju, J.S., Lu, F., Sanyal, A., Yu, Y., Song, X., Jiang, N., Pontaroli, A.C., Rambo, T., Currie, J., Collura, K., Talaga, J. Fana, C., Goicoecheaa, J.L., Zuccoloa, A., Chenb, J., Bennetzene, J.L., Chenb, M., Scott Jackson, S. and Wing, R.A. (2008) Dynamic evolution of *Oryza* genomes is revealed by comparative genomic analysis of a genus-wide vertical data set. *The Plant Cell* 20(12), 3191–3209.

Angeles-Shim, R.B., Asano, K., Takashi, T., Kitano, H. and Ashikari, M. (2009) Mapping of the glabrous gene in rice using CSSLs derived from the cross *Oryza sativa* subsp. *japonica* cv. Koshihikari and *O. glaberrima*. Paper presented at the 6th International Rice Genetics Symposium, Manila, Philippines, 16–19 November.

Ashikari, M. and Matsuoka, M. (2006) Identification, isolation and pyramiding of quantitative trait loci for rice breeding. *Trends in Plant Science* 11, 344–350.

Ashikari, M., Sakakibara, H., Lin, S., Yamamoto, T., Takashi, T., Nishimura, A., Angeles, E.R., Qian, Q., Kitano, H. and Matsuoka, M. (2005) Cytokinin oxidase regulates rice grain production. *Science* 309, 741–745.

Bateson, W. (1909) Heredity and variation in modern lights. In: Seward, A.C. (ed.) *Darwin and Modern Science*. Cambridge University Press, Cambridge, pp. 81–101.

Bimpong, K.I., Carpena, A.L., Mendioro, M.S., Fernandez, L., Ramos, J., Reversat, G. and Brar, D.S. (2010) Evaluation of *Oryza sativa* × *O. glaberrima* derived progenies for resistance to rootknot nematode and identification of introgressed alien chromosome segments using SSR markers. *African Journal of Biotechnology* 9, 3988–3997.

Bouharmont, J., Olivier, M. and Chassart, M.D. (1985) Cytological observations in some hybrids between the rice species *Oryza sativa* L. and *O. glaberrima* Steud. *Euphytica* 34(1), 75–81.

Buckler, E.S, Thornsberry, J.M. and Kresovich, S. (2001) Molecular diversity, structure and domestication of grasses. *Genetic Research* 77, 213–218.

Byzova, M.V., Franken, J., Aarts, M.G.M., de Almeida-Engler, J., Engler, G., Mariani, C., Van Lookeren Campagne, M.M. and Angenent, G.C. (1999) *Arabidopsis* STERILE APETALA, a multifunctional gene regulating inflorescence, flower, and ovule development. *Genes & Development* 13(8), 1002–1014.

Chen, J., Ding, J., Ouyang, Y., Du, H., Yang, J., Cheng, K., Zhao, J., Qiu, S., Zhang, X., Yao, J., Liu, K., Wang, L., Xu, C., Li, X., Xue, Y., Xia, M., Jiǁ, Q., Luǁ, J., Xu, M. and Zhang, Q. (2008) A triallelic system of *S5* is a major regulator of the reproductive barrier and compatibility of *indica–japonica* hybrids in rice. *Proceedings of the National Academy of Sciences of the United States of America* 105, 11436–11441.

Doi, K., Iwata, N. and Yoshimura, A. (1997) The construction of chromosome substitution lines of African rice (*Oryza glaberrima* Steud.) in the background of *japonica* rice (*O. sativa* L.). *Rice Genetics Newsletter* 14, 39–41.

Doi, K., Taguchi, K. and Yoshimura, A. (1998a) A new locus affecting high F_1 pollen sterility found in backcross progenies of *japonica* rice and African rice. *Rice Genetics Newsletter* 15, 146–148.

Doi, K., Yoshimura, A. and Iwata, N. (1998b) RFLP mapping and QTL analysis of heading date and pollen sterility using backcross populations between *Oryza sativa* L. and *Oryza glaberrima* Steud. *Breeding Science* 48(4), 395–399.

Doi, K., Taguchi, K. and Yoshimura, A. (1999) RFLP mapping of *S20* and *S21* for F_1 pollen semi-sterility found in backcross progeny of *Oryza sativa* and *O. glaberrima*. *Rice Genetics Newsletter* 16, 65–68.

Doi, K., Sobrizal, K., Sanchez, T., Kurakazu, T., Nagai, Y. and Yoshimura, A. (2003a) Developing and evaluating rice chromosome segment substitution lines. In: Mew, T.W., Joshi, R.C. and Brar, D.S. (eds) *Rice Science: Innovations and impact of livelihood*. International Rice Research Institute, Los Baños, Philippines, pp. 289–296.

Doi, K., Taguchi, K. and Yoshimura, A. (2003b) Genetic analysis of pollen sterility loci found in breeding progeny between *O. sativa* and *O. glaberrima*. In: Khush, G.S., Brar, D.S. and Hardy, B. (eds) *Advances in Rice Genetics*. International Rice Research Institute, Los Baños, Philippines, pp. 149–151.

Ebitani, T., Takeuchi, Y., Nonoue, Y., Yamamoto, T., Takeuchi, K. and Yano, M. (2005) Construction and evaluation of chromosome segment substitution lines carrying overlapping chromosome segments of *indica* rice cultivar ‘Kasalath’ in a genetic background of *japonica* elite cultivar ‘Koshihikari’. *Breeding Science* 55, 65–73.

Garavito, A., Guyot, R., Lozano, J., Gavory, F., Samain, S., Panaud, O., Tohme, J., Ghesquière, A. and Lorieux, M. (2010) A genetic model for the female sterility barrier between Asian and African cultivated rice species. *Genetics* 185(4), 1425–1440.

Ghesquière, A., Séquier, J., Second, G. and Lorieux, M. (1997) First steps towards a rational use of African rice, *Oryza glaberrima*, in rice breeding through a 'contig line' concept. *Euphytica* 96(1), 31–39.

Gusti, A., Baumberger, N., Nowack, M., Pusch, S., Eisler, H., Potuschak, T., De Veylder, L., Schnittger, A. and Genschik, P. (2009) The *Arabidopsis thaliana* F-Box protein FBL17 is essential for progression through the second mitosis during pollen development. *PLoS ONE* 4(3), e4780.

Gutiérrez, A., Carabali, S., Giraldo, O., Martinez, C., Correa, F., Prado, G., Tohme, J. and Lorieux, M. (2010) Identification of a *Rice stripe necrosis virus* resistance locus and yield component QTLs using *Oryza sativa* × *O. glaberrima* introgression lines. *BMC Plant Biology* 10, 6.

Guyot, R., Garavito, A., Gavory, F., Samain, S., Tohme, J., Ghesquière, A. and Lorieux, M. (2011) Patterns of sequence divergence and evolution of the *S1* orthologous regions between Asian and African cultivated rice species *PLoS ONE* 6(3), e17726.

Hanada, K., Vallejo, V., Nobuta, K., Slotkin, R.K., Lisch, D., Meyers, B.C., Shiu, S-H. and Jiang, N. (2009) The functional role of Pack-MULEs in rice inferred from purifying selection and expression profile. *The Plant Cell* 21(1), 25–38.

Hu, F.Y., Xu, P., Deng, X.N., Zhou, J.W., Li, J. and Tao, D.Y. (2006) Molecular mapping of a pollen killer gene *S29(t)* in *Oryza glaberrima* and colinear analysis with *S22* in *O. glumaepatula*. *Euphytica* 151, 273–278.

Ikehashi, H. and Araki, H. (1986) Genetics of F_1 sterility in remote crosses of rice. In: *Rice Genetics.* Proceedings of the International Rice Genetics Symposium, 27–31 May 1985. International Rice Research Institute, Manila, Philippines, pp. 119–130.

Ishii, T., Xu, Y. and McCouch, S.R. (2001) Nuclear and chloroplast-microsatellite variation in A-genome species of rice. *Genome* 44, 658–666.

Ishikawa, S., Ae, N. and Yano, M. (2005) Chromosomal regions with quantitative trait loci controlling cadmium concentration in brown rice (*Oryza sativa*). *New Phytologist* 168, 345–350.

Jiang, N., Bao, Z., Zhang, X., Eddy, S.R. and Wessler, S.R. (2004) Pack-MULE transposable elements mediate gene evolution in plants. *Nature* 431, 569–573.

Jing, W., Zhang, W., Jiang, L., Chen, L., Zhai, H.Q. and Wan, J.M. (2007) Two novel loci for pollen sterility in hybrids between the weedy strain Ludao and the *japonica* variety Akihari of rice (*Oryza sativa* L.). *Theoretical and Applied Genetics* 114, 915–925.

Jones, M.P., Dingkuhn, M., Aluko, G.K. and Semon, M. (1997) Interspecific *O. sativa* L. × *O. glaberrima* Steud. progenies in upland rice improvement. *Euphytica* 92, 237–246.

Kang, J.W., Suh, J.P., Kim, D.M., Oh, C.S., Oh, J.M. and Ahn, S.N. (2008) QTL mapping of agronomic traits in an advanced backcross population from a cross between *Oryza sativa* L. cv. Milyang 23 and *O. glaberrima*. *Korean Journal of Breeding Science* 40, 243–249.

Kim, H., Hurwitz, B., Yu, Y., Collura, K., Gill, N., SanMiguel, P., Mullikin, J., Maher, C., Nelson, W., Wissotski, M., Braidotti, M., Kudrna, D., Goicoechea, J.L., Stein, L., Ware, D., Jackson, S.A. Soderlund, C. and Wing, R.A. (2008) Construction, alignment and analysis of twelve framework physical maps that represent the ten genome types of the genus *Oryza*. *Genome Biology* 9(2), R45.

Koide, Y., Ikenaga, M., Sawamura, N., Nishimoto, D., Matsubara, K., Onishi, K., Kanazawa, A. and Sano Y. (2008a) The evolution of sex-independent transmission ratio distortion involving multiple allelic interactions at a single locus in rice. *Genetics* 180(1), 409–420.

Koide, Y., Ikenaga, M., Shinya, Y., Matsubara, K. and Sano, Y. (2008b) Two loosely linked genes controlling the female specificity for cross-incompatibility in rice. *Euphytica* 164(3), 753–760.

Koide, Y., Onishi, K., Nishimoto, D., Baruah, A.R., Kanazawa, A. and Sano, Y. (2008c) Sex-independent transmission ratio distortion system responsible for reproductive barriers between Asian and African rice species. *New Phytologist* 179, 888–900.

Li, J., Xu, P., Deng, X., Zhou, J., Hu, F., Wan J. and Tao, D. (2008) Identification of four genes for stable hybrid sterility and an epistatic QTL from a cross between *Oryza sativa* and *Oryza glaberrima*. *Euphytica* 164, 699–708.

Li, Z.M., Zheng, X-M. and Ge, S. (2011a) Genetic diversity and domestication history of African rice (*Oryza glaberrima*) as inferred from multiple gene sequences. *Theoretical and Applied Genetics* 123(1), 21–31.

Li, F., Liu, F.H., Morinaga, D. and Zhao, Z. (2011b) A new gene for hybrid sterility from a cross between *Oryza sativa* and *O. glaberrima*. *Plant Breeding* 130, 165–171.

Li, J., Xu, P., Zhou, J., Hu, F., Deng, X., Chen, Z. and Tao, D. (2011c) Molecular mapping of sterility QTLs *qSS-3, qSS-6a* and *qSS-7* as single Mendelian factors via NIL strategy. *Rice Science* 18, 110–115.

Long, Y., Zhao, L., Niu, B., Su, J., Wu, H., Chen, Y., Zhang, Q., Guo, J., Zhuang, C., Mei, M., Xia, J., Wang, L., Wu, H. and Liu, Y.-G. (2008) Hybrid male sterility in rice controlled by interaction between divergent alleles of two adjacent genes. *Proceedings of the National Academy of Sciences of the United States of America* 105(48), 18871–18876.

Ma, J. and Bennetzen, J.L. (2004) Rapid recent growth and divergence of rice nuclear genomes. *Proceedings of the National Academy of Sciences of the United States of America* 101(34), 12404–12410.

Maduoka, Y., Kashiwagi, T., Hirotsu, N. and Ishimaru, K. (2008) Indian rice 'Kasalath' contains genes that improve traits of Japanese premium rice 'Koshihikari'. *Theoretical and Applied Genetics* 116, 603–612.

Matsubara, K., Khin, T. and Sano, Y. (2003) A gene block causing cross-incompatibility hidden in wild and cultivated rice. *Genetics* 165(1), 343–352.

Michalak, P. (2009) Epigenetic, transposon and small RNA determinants of hybrid dysfunctions. *Heredity* 102, 45–50.

Ndjiondjop, M.-N., Manneh, B., Cissoko, M., Drame, N.K., Kakai, G.K., Bocco, R., Baimey, H. and Wopereis, M. (2010) Drought resistance in an interspecific backcross population of rice (*Oryza* spp.) derived from the cross WAB56-104 (*O. sativa*) × CG14 (*O. glaberrima*). *Plant Science* 179, 364–373.

Oka, H.I. (1957) Genetic analysis for the sterility of hybrids between distantly related varieties of cultivated rice. *Journal of Genetics* 55, 397–409.

Oka, H.I. (1974) Analysis of genes controlling F_1 sterility in rice by the use of isogenic lines. *Genetics* 77, 521–534.

Orjuela, J., Garavito, A., Bouniol, M., Arbelaez, J.D., Moreno, L., Kimball, J., Wilson, G., Rami, J.F., Tohme, J., McCouch, S.R. and Lorieux, M. (2010) A universal core genetic map for rice. *Theoretical and Applied Genetics* 120, 563–572.

Ouyang, Y., Liu, Y.-G. and Zhang, Q. (2010) Hybrid sterility in plant; stories from rice. *Current Opinion in Plant Biology* 13, 186–192.

Pesin, J.A. and Orr-Weaver, T.L. (2008) Regulation of APC/C activators in mitosis and meiosis. *Annual Review of Cell and Developmental Biology* 24(1), 475–499.

Portères, R. (ed.) (1970) *Primary Cradles of Agriculture in the African Continent.* Cambridge University Press, Cambridge.

Ram, T., Deen, R., Gautam, S.K., Ramesh, K., Rao, Y.K. and Brar, D.S. (2010) Identification of new genes for brown planthopper resistance in rice introgressed from *O. glaberrima* and *O. minuta*. *Rice Genetics Newsletter* 25, 67.

Ren G., Xu, P., Deng, X., Zhou, J., Hu F., Li, J., Li, F., Zhang, Z. and Tao, D. (2005) A new gamete eliminator from *Oryza glaberrima*. *Rice Genetics Newsletter* 11, 106–108.

Saito, K., Azoma, K. and Sié, M. (2010) Grain yield performance of selected lowland NERICA and modern Asian rice genotypes in West Africa. *Crop Science* 50, 281–291.

Sano, Y. (1983) Genetic studies of speciation in cultivated rice. 3. A new gene controlling sterility in F_1 hybrids of 2 cultivated rice species: its association with photoperiod sensitivity. *Journal of Heredity* 74, 435–439.

Sano, Y. (1986) Sterility barriers between *Oryza sativa* and *O. glaberrima*. In: *Rice Genetics*. Proceedings of the International Rice Genetics Symposium, 27–31 May 1985. International Rice Research Institute, Manila, Philippines, pp. 109–118.

Sano, Y. (1990) The genic nature of gamete eliminator in rice. *Genetics* 125, 183–191.

Sano, Y., Chu, Y.-E. and Oka, H.-I. (1979) Genetic studies of speciation in cultivated rice. 1. Genetic analysis for the F_1 sterility between *O. sativa* L. and *O. glaberrima* Steud. *Japanese Journal of Genetics* 54(2), 121–132.

Second, G. (1982) Origin of the genic diversity of cultivated rice (*Oryza* spp.): study of the polymorphism scored at 40 isozyme loci. *Japanese Journal of Genetics* 57, 25–57.

Semagn, K., Ndjiondjop, M.-N. and Cissoko, M. (2006) Microsatellites and agronomic traits for assessing genetic relationships among 18 New Rice for Africa (NERICA) varieties. *African Journal of Biotechnology* 5, 800–810.

Sweeney, M. and McCouch, S. (2007) The complex history of the domestication of rice. *Annals of Botany* 100, 951–957.

Taguchi, K., Doi, K. and Yoshimura, A. (1999) RFLP mapping of *S19*, a gene for F_1 pollen semi-sterility found in backcross progeny of *Oryza sativa* and *O. glaberrima*. *Rice Genetics Newsletter* 16, 70–71.

Tenaillon, M.I., U'Ren, J., Tenaillon, O. and Gaut, B.S. (2004) Selection versus demography: a multilocus investigation of the domestication process in maize. *Molecular Biology and Evolution* 21, 1214–1225.

Tung, C.-W., Zhao, K., Wright, M., Ali, M.L., Jung, J., Kimball, J., Tyagi, W., Thomson, M., McNally, K., Leung, H., Kim, H., Ahn, S.-N., Reynolds, A., Scheffler, B., Eizenga, G., McClung, A., Bustamante, C.

and McCouch, S. (2010) Development of a research platform for dissecting phenotype–genotype associations in rice (*Oryza* spp.). *Rice* 3(4), 205–217.

van Berloo, R. (2008) GGT 2.0: versatile software for visualization and analysis of genetic data. *Journal of Heredity* 99, 232–236.

Wang, Y. and Yang, M. (2006) The ARABIDOPSIS SKP1-LIKE1 (ASK1) protein acts predominantly from leptotene to pachytène and represses homologous recombination in male meiosis. *Planta* 223, 613–617.

Wang, Z.Y., Second, G. and Tanksley, S.D. (1992) Polymorphism and phylogenetic relationships among species in the genus *Oryza* as determined by analysis of nuclear RFLPs. *Theoretical and Applied Genetics* 83, 565–581.

Wang, Y., Sun, Y.J., Chen, D.Y. and Yu, S.B. (2009) Analysis of quantitative trait loci in response to nitrogen and phosphorus deficiency in rice using chromosomal segment substitution lines. *Acta Agronomica Sinica* 35, 580–587.

Wing, R.W., Ammiraju, J.S.S., Luo, M., Kim, H.R., Yu, Y., Kudrna, D., Goicoechea, J.L., Wang, W., Nelson, W., Rao, K., Brar, D., Mackill, D.J., Han, B., Soderlund, C., Stein, L., SanMiguel, P. and Jackson, S. (2005) The *Oryza* Map Alignment Project: the golden path to unlocking the genetic potential of wild rice species. *Plant Molecular Biology* 59, 53–62.

Xu, G., Ma, H., Nei, M. and Kong, H. (2009) Evolution of F-box genes in plants: different modes of sequence divergence and their relationships with functional diversification. *Proceedings of the National Academy of Sciences of the United States of America* 106(3), 835–840.

Zeder, M.A., Emshwiller, E., Smith, B.D. and Bradley, D.G. (2006) Documenting domestication: the intersection of genetics and archaeology. *Trends in Genetics* 22, 139–155.

Zhang, Z., Xu, P., Hu, F., Zhou, J., Li, J., Deng, X., Ren, G., Li, F. and Tao, D. (2005) A new sterile gene from *Oryza glaberrima* on chromosome 3. *Rice Genetics Newsletter* 22, 26–29.

Zhao, L., Zhou, H., Lu, L., Liu, L., Li, X., Lin, Y. and Yu, S. (2009) Identification of quantitative trait loci controlling rice mature seed culturability using chromosomal segment substitution lines. *Plant Cell Reports* 28, 247–56.

Zhu, Q. and Ge, S. (2005) Phylogenetic relationships among A-genome species of the genus *Oryza* revealed by intron sequences of four nuclear genes. *New Phytolologist* 167, 249–265.

Zhu, S., Wang, C., Zheng, T., Zhao, Z., Ikehashi, H. and Wan, J. (2005) A new gene located on chromosome 2 causing hybrid sterility in remote cross of rice. *Plant Breeding* 124, 440–445.

Zhu, W., Lin, J., Yang, D., Zhao, L., Zhang, Y., Zhu, Z., Chen, T. and Wang, C. (2009) Development of chromosome segment substitution lines derived from backcross between two sequenced rice cultivars, *indica* recipient 93-11 and *japonica* donor Nipponbare. *Plant Molecular Biology Reporter* 27, 126–131.

11 Rice Genetic Improvement for Abiotic Stress Tolerance in Africa

Khady N. Dramé,[1]* Baboucarr Manneh[2] and Abdelbagi M. Ismail[3]
[1]*Africa Rice Center, Dar es Salaam, Tanzania;* [2]*Africa Rice Center, Saint-Louis, Senegal;* [3]*International Rice Research Institute, Los Baños, Philippines*

Introduction

In sub-Saharan Africa, biophysical constraints and high reliance on low-input and rainfed agriculture impose various abiotic stresses on rice crops (Defoer *et al.*, 2002). Most of these stresses are associated with water availability (drought and excess water), soil problems (salinity, nutrient deficiencies and toxicities) and extreme temperatures (heat and cold). These stresses fluctuate seasonally and vary spatially. They adversely affect rice growth and productivity, resulting in reduced yields. Abiotic stresses are limiting factors in almost all rice production environments – in both rainfed and irrigated areas (Defoer *et al.*, 2002; Saito *et al.*, Chapter 15, this volume; Table 11.1). They contribute to significantly lower on-farm yields compared to attainable yield. Results of farm-household surveys conducted by the Africa Rice Center (AfricaRice) in 12 sub-Saharan African countries showed that, for example, when drought and flooding occur, they induce average yield losses of 33% (AfricaRice, 2011a). Yield losses due to soil-related problems are higher in the lowlands than in the uplands. In Senegal and Uganda, yield losses due to the soil-related constraints salinity and iron toxicity are as high as 40% and 25%, respectively (AfricaRice, 2011b).

Crop and water management can be improved through efficient irrigation schemes, better infrastructure and hydrological control, and optimal use of fertilizers, which can alleviate some of the effects of abiotic stresses. These measures are effective in reducing yield losses, and most improved varieties are responsive to such measures, consequently stabilizing productivity and narrowing yield gaps (Ismail *et al.*, 2008). However, the costs of adopting these measures are prohibitive for most resource-poor African farmers and local governments. Alternatively, developing rice germplasm tolerant to the prevalent stresses involves no additional costs to farmers and can enhance and stabilize productivity considerably (Ismail and Tuong, 2009). Hence, varietal improvement should be given priority as an entry point for increasing and stabilizing rice production in sub-Saharan Africa. As much as possible, this should be combined with improved management practices to realize the full yield potential of the new tolerant varieties developed. This is all the more important given that the potential impact of research targeted to reduce yield loss due to stresses was estimated at a global cumulative benefit of US$ 32.9 million for seven countries over 3 years (AfricaRice, 2011b). This means

* Corresponding author: k.drame@cgiar.org

© CAB International 2013. *Realizing Africa's Rice Promise* (eds M.C.S. Wopereis *et al.*)

Table 11.1. Major rice production ecosystems in Africa: actual and potential yield and limiting constraints. (Adapted from Defoer *et al.*, 2002; Diagne *et al.*, Chapter 3, this volume).

Rice agro-ecosystem	Share of rice area (%)	Yield: Actual / Potential[a] (t/ha)	Abiotic production constraints	Input use
Rainfed upland	32	1.2/2–4	P and N deficiency, acidity, Al toxicity, drought, erosion declining soil fertility, cold in highlands	Very low
Rainfed lowland	38	1.9/3–6	Water control, N and P deficiency, Fe toxicity	Low
Irrigated Sahel/ savannah	26	3.7/6–11	N deficiency, salinity and alkalinity, extreme temperatures	High
Irrigated humid/ sub-humid zone		1.9/5–8	N deficiency, Fe toxicity	Medium
Mangrove		<1/2–4	Acid sulfate, salinity, Fe toxicity, excess water	Very low
Deepwater/ Floating	4	<1/2–4	Salinity, excess water, cold in highlands	Very low

[a]Low end of the range refers to potential yield at current input levels; high end refers to potential yields at increased input levels.

additional income for farmers and increased rice production in Africa.

In the rice gene pool – both cultivated and wild species – tremendous genetic variation exists for tolerance to abiotic stresses. This could be exploited to develop improved varieties that are more tolerant to the major abiotic stresses that constrain rice production in sub-Saharan Africa than existing varieties. Progress in developing rice varieties tolerant to abiotic stresses has been slow because of the complexity of the tolerance mechanisms, poor understanding of the inheritance of tolerance, low heritability and lack of efficient screening techniques (Lafitte *et al.*, 2006). However, advances in understanding stress physiology and identification of tolerance genes/QTLs (quantitative trait loci) (Ismail *et al.*, 2007; Jena and Mackill, 2008; Thomson *et al.*, 2010b) offer promising opportunities for rice improvement with regards to abiotic-stress tolerance. Both conventional and biotechnological approaches are being used by AfricaRice to exploit the rich reservoir of genetic resources present in the African germplasm pool, particularly *Oryza sativa* landraces and *O. glaberrima* for the improvement of local rice varieties. In this chapter, we present the major abiotic stresses that constrain rice production in sub-Saharan Africa and the breeding activities to develop rice varieties tolerant of these stresses and adapted to local conditions in Africa. We also highlight the challenges that should be addressed in order to sustainably increase rice productivity in sub-Saharan Africa.

Major Abiotic Stresses Limiting Rice Production in Sub-Saharan Africa

Sub-Saharan Africa is affected by numerous types of environmental stresses related to water availability, soil problems and climatic issues. In many cases, several stresses are experienced simultaneously (e.g. in mangrove-swamp rice fields where submergence, salinity and iron toxicity occur together). Rice production in such areas is severely constrained by abiotic stresses. Specific data on the extent of abiotic stresses in rice-growing areas of sub-Saharan Africa are scarce. Most of the data available are either not specific to rice-growing areas or limited to a few countries. This issue is being addressed by AfricaRice. Insight on farmer perceptions of abiotic constraints is provided by Diagne *et al.* (Chapter 4, this volume).

Drought

Drought is generally avoided in areas where irrigation water is available throughout the season, but it is a consistent feature across much of the 63.5 million hectares of rainfed rice sown annually, most of which is in tropical Asia, Africa and Latin America (Lafitte *et al.*, 2006). It can occur at any stage during the rice cropping season, but it is more devastating when it occurs just prior to flowering than it is during the vegetative stage, with substantial effects on grain yield (Boojung and Fukai, 1996). About 70% of the rice area in sub-Saharan Africa is rainfed (Diagne *et al.*, Chapter 3, this volume; Table 11.1). The spatial and temporal variability of rainfall in this region expose rice plants to frequent drought spells. Regardless of the total rainfall and distribution, the poor physical properties of highly weathered, coarse-textured soils in some parts of sub-Saharan Africa induce low water-holding capacity and establish water deficit as a major constraint to rainfed crop yields in sub-Saharan Africa (Hsiao *et al.*, 1980). This is particularly true for upland rice, which makes up 32% of rice-growing areas in sub-Saharan Africa (Table 11.1). Analysis of farmer perceptions in 18 countries in sub-Saharan Africa across rice environments provided evidence that in 2008 an estimated 10% of rice farmers experienced drought affecting 37% of their rice area, causing 29% of rice yield loss (Diagne *et al.*, Chapter 4, this volume). The diversity of affected production systems, variability of drought in terms of timing and severity, and the multiple traits involved in drought tolerance require strategic research to prioritize and define environment-specific approaches for developing drought-tolerant rice cultivars (Manneh *et al.*, 2007).

Salinity

Irrigation has the potential to ensure high rice yields and is a good strategy to offset recurrent droughts. Unfortunately, soils of most irrigated areas in sub-Saharan Africa have continued to be degraded as a result of poor irrigation practices and the absence of efficient drainage. These have led to a rapid rise in the water table and an increase in soil sodium/alkalinity and salinity (Bertrand *et al.*, 1993). In the Office du Niger (Mali), 50% of the water table is now saline and occasionally very saline despite low mineral content of the irrigation water (Bertrand *et al.*, 1993). In the Senegal River delta, marine-derived soil salinity is an inherent problem and sodicity is increasing in irrigated flood plains in inland areas due to high evaporation, rising groundwater tables and poor drainage (WARDA, 1993; Matlon *et al.*, 1998). Miézan and Dingkuhn (2001) observed that waters of the Niger and Senegal rivers carry substantial alkalinity, and the salt content of water sometimes increases markedly between the main rivers and the actual irrigation site. However, van Asten *et al.* (2003) show that salt accumulation in the soils of Sahelian countries has more to do with the underlying bedrock than with the irrigation system. Examining soils in the irrigated areas of Foum Gleita (Mauritania) they found that the geographical distribution of salt was not related to the presence of irrigation or drainage canals. Also the alkaline salts present in the upper soil layers in Foum Gleita did not come from irrigation water, but from the underlying bedrock. Additional to the salt stress itself, the high pH resulting from the sodification/alkalinization reduces the availability of plant nutrients such as zinc and increases nitrogen losses through volatilization (Bertrand *et al.*, 1993; Miézan and Dingkuhn, 2001). According to van Asten *et al.* (2004), using data from Sourou Valley (Burkina Faso), this shortage of nutrients in the soil is a bigger problem than soil degradation attributable to irrigation.

High rice productivity is also constrained by soil salinity in many mangrove-swamp areas. Mangrove rice is grown on about 200,000 ha of cleared mangrove swamps along the rivers and coastal estuaries of The Gambia, Guinea, Guinea-Bissau, Nigeria, Senegal and Sierra Leone (Matlon *et al.*, 1998). In these areas, rice cultivation depends on the length of the salt-free period, which is an interplay of *in situ* rainfall, the volume of freshwater flow, and salt intrusion from the sea. Swamps located along river banks farthest from the sea experience a salt-free period of 6–9 months during which rice can be grown. Closer to the sea, rice can be grown for only 3–4 months.

Rice is most sensitive to salinity stress at seedling and reproductive stages (Gregorio *et al.*, 2002). However, salinity tolerance at these two stages is only weakly associated (Moradi *et al.*, 2003). Discovering and combining suitable tolerance traits for both stages is essential for developing resilient salt-tolerant varieties. Moreover, the salt-tolerance level of cultivars depends on environmental conditions (Asch *et al.*, 1997). Generally, salinity effects on rice are more severe in arid climates than in humid ones. For example, salinity levels at an electric conductivity (EC) of 9.5 mS/cm were reported to cause a 50% yield loss in the humid tropics (Flowers and Yeo, 1981), whereas under the arid conditions of the Sahelian dry season an equivalent yield loss was observed at an EC of only 3.5 mS/cm (Dingkuhn *et al.*, 1993).

Excess water

Excess water can be of two types: transient flash floods or submergence that completely inundates the crop for a short duration (up to 2 weeks), and longer-term flooding where water stagnates for up to a few months at different water depths (e.g. 30–50 cm: partial/stagnant, semi-deep; >100 cm: deep-water; up to 3 or 4 m: very deep-water or floating rice) (Mackill *et al.*, 2012). In sub-Saharan Africa, the deep-water area accounts for less than 4% of the whole rice-growing area (Diagne *et al.*, Chapter 3, this volume), mainly in the flood plains of large rivers such as the Niger River. Excess water is also a common constraint throughout the rainfed and irrigated rice-production areas when high rainfall and/or impeded drainage occurs, particularly early in the season (Futakuchi, 2005). In the flood plains, sudden floods can temporarily submerge the rice crop in some areas, and waterlogging can continue through most of the season in others – sometimes both occur within the same season. In savannah and forest zones in West and Central Africa, inland valleys prevail and rice in valley bottoms often experiences waterlogging for several weeks after heavy rainfall (Manneh *et al.*, 2007). Analysis of farmer perceptions in 18 countries in sub-Saharan Africa provided evidence that in 2008 an estimated 5% of farmers experienced flooding, affecting 37% of their farmland, causing a production loss of 27% (Diagne *et al.*, Chapter 4, this volume).

While rice is adapted to waterlogged conditions, complete submergence for several days can be fatal. However, the extent of damage to rice is affected by several factors linked to floodwater conditions, including interference in normal gas exchange and light interception. These factors are largely affected by the duration of the flood, its depth, temperature, and the level of turbidity and turbulence of the floodwater, and vary considerably across locations and seasons (Das *et al.*, 2009). Rice yields in flood-prone areas are very low, mostly averaging below 1.5 t/ha, because farmers usually grow their low-yielding traditional varieties (Haefele *et al.*, 2010); and modern high-yielding varieties tolerant of these stresses are mostly non-existent. In some locations, such as Mopti (Mali) and the Sokoto Rima River flood plains (Nigeria), *O. glaberrima* (African rice) is still cultivated despite its low yield potential relative to the more widely cultivated *O. sativa* (Asian rice) (Jones *et al.*, 1997; Diarra *et al.*, 2004; Sakagami, 2012).

Phosphorus deficiency

Phosphorus (P) deficiency is one of the major limiting factors for crop production in highly weathered soils in the humid tropics. Total soil P is generally low, with only 2–4% of the total P available to plants. This is because of the high P-fixing capacity of fine-textured soils found in humid and sub-humid zones (Abekoe and Sahrawat, 2001). About 5.7 billion hectares worldwide lack sufficient plant-available P (Batjes, 1997) and 50% of potential arable land worldwide has acid soils, which are widely distributed in sub-Saharan Africa. The bioavailability of inorganic phosphate is reduced in acid soils.

Phosphorus deficiency can be corrected through fertilizer application, but the lack of locally available P sources and the high cost of importing and transporting fertilizers prevent many resource-poor rice farmers from applying P. Furthermore, some rice soils can quickly fix up to 90% of the added P fertilizer into less-soluble forms (Dobermann *et al.*, 1998) that cannot be used by plants. This tight binding of P in the soil is frequently the primary cause of P deficiency, rather than a low total P content.

Insufficient plant-available soil P is a major constraint for rice production. This is particularly apparent in sub-Saharan Africa under upland conditions, which are commonly characterized by infertile, highly acidic, P-fixing soils, usually in areas where little or no fertilizer is applied. Under lowland conditions, where P is more available than in drier areas, P deficiency is still a major factor limiting performance of modern rice varieties (Defoer *et al.*, 2002). Phosphorus deficiency is likely to become an increasingly important constraint, as P is removed from soils under intensive rice production (De Datta *et al.*, 1990) using high-yielding modern varieties.

Iron toxicity

Iron (Fe) toxicity is widely distributed in tropical lowlands and is frequently reported in many inland valleys (mangrove swamps, rainfed and irrigated lowlands) of sub-Saharan Africa (Masajo *et al.*, 1986). It is usually associated with poor drainage and the presence of iron in the parent rock or in the soils of adjacent slopes through which groundwater flows into the lowland (WARDA, 1988). A wide range of soil types can be Fe-toxic, including acid-sulfate soils, acid-clay soils, peat soils and valley-bottom soils receiving interflow water from adjacent slopes (Becker and Asch, 2005). A survey conducted during 2000–2001 in three West African countries (Côte d'Ivoire, Ghana and Guinea) showed that more than 50% of the lowlands studied and about 60% of the cultivated rice plots were affected by Fe toxicity, and 10% of lowland crop fields were abandoned due to high Fe-toxicity stress (Chérif *et al.*, 2009).

In anaerobic conditions and acidic conditions (pH <5), as found in waterlogged soils, the ferrous form is stabilized and readily taken up by plants. Rice is particularly sensitive to Fe toxicity at two growth stages – soon after transplanting up to tillering, and during heading/flowering (Prade *et al.*, 1990). Depending on the variety and the intensity of the toxicity, yield losses average about 30% and can accrue up to levels that can cause complete crop failure of susceptible varieties (Masajo *et al.*, 1986; Abifarin, 1989; WARDA, 1997). Excessive uptake of Fe is often accompanied with other nutrient deficiencies, especially potassium, phosphorus, calcium and magnesium (Benckiser *et al.*, 1984; Prade *et al.*, 1990). However, Sahrawat *et al.* (1996) report cases of Fe toxicity without apparent deficiency in other nutrients.

Extreme temperatures

Low-temperature stress is common for rice grown in temperate regions and at high elevations in the tropics. In Africa, cold temperatures occur in the highlands of East and Southern Africa and in some areas of the Sahel region of West Africa during the cold, dry harmattan season, which extends from November to February. For example, in Madagascar, mean temperatures at 1500 m vary from 17°C in October, the rice-sowing period, to 20°C during the reproductive stage (Zenna *et al.*, 2010). Minimum temperatures can fall below 10°C during early vegetative stage and below 14°C during reproductive stage (Zenna *et al.*, 2010). In the Sahel, there are significant temperature fluctuations during the year with low temperatures during panicle initiation in the wet season and high temperatures around flowering during the dry season. In both cases, spikelet sterility may occur, leading to substantial yield loss (Dingkuhn, 1995). In this region, day minimum temperatures fall below 20°C during November to March, while maximum temperatures regularly rise above 40°C from April to June (Fig. 11.1). Planting rice between mid-September and mid-November in the Sahel is generally associated with near total spikelet sterility, since the reproductive phase of the crop coincides with periods of low night temperatures (Dingkhun, 1995; Manneh *et al.*, 2007). Moreover, crop duration increases with low temperatures, considerably limiting the possibility of double-cropping in areas where water control is possible (Coly, 1980; Matlon *et al.*, 1998). However, the introduction of short-duration varieties can make double-cropping feasible. Dingkuhn and Miezan (1995) used a simulation model based on photothermal constants to show that short-duration genotypes with high plasticity of duration perform best

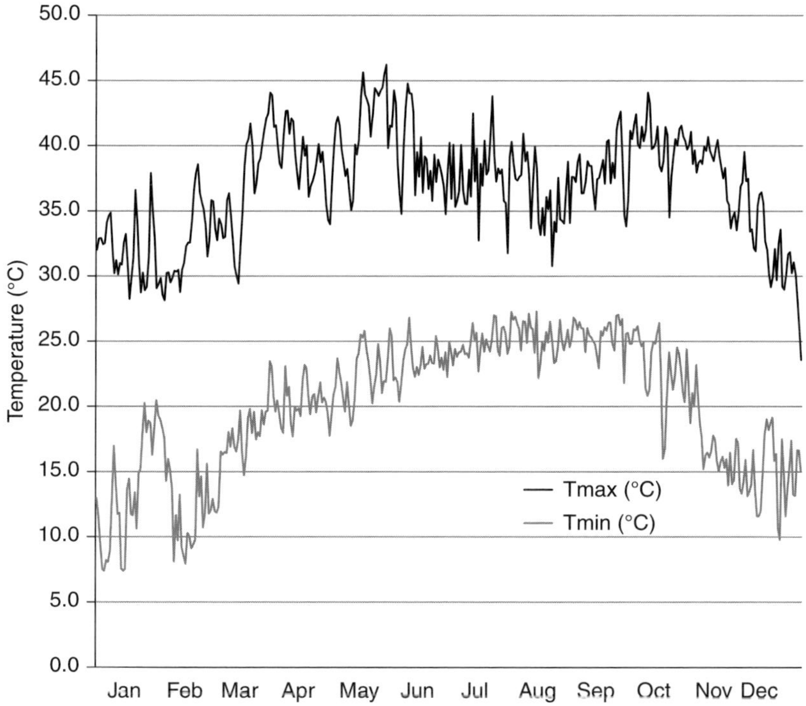

Fig. 11.1. Example of seasonal and diurnal temperature variations during two major cropping seasons in the Sahel (case of Fanaye, Senegal River valley, northern Senegal, 2011). Tmax, maximum temperature; Tmin, minimum temperature.

in the dry season, while medium-duration types with high plasticity perform well in the wet season, and short-duration types with low plasticity of duration perform well in both seasons. Rising temperatures, as a consequence of climate change, could have positive effects on the flexibility of cropping calendars, but the vulnerability of rice to high temperature is also expected to increase. Global mean temperatures have already risen by around 0.6°C over the last century and are projected to increase by 1.4°C to 5.8°C over the next century (IPCC, 2001).

Rice is sensitive to temperature fluctuations, but its sensitivity depends on the developmental stage. Flowering stage and 9–11 days before heading are the most sensitive stages to extremes of temperatures, with very cold or very hot weather leading to high spikelet sterility (Yoshida *et al.*, 1981; Andaya and Mackill, 2003a; Manneh *et al.*, 2007; Zenna *et al.*, 2010). Rice is particularly sensitive to temperatures below 15°C. The extent of damage depends on the ambient air or water temperature, cropping pattern, growth stage and variety (Zenna *et al.*, 2010). Crop failure can be observed when low temperature is manifested at different growth stages, such as germination, seedling, vegetative, reproductive and grain maturity (Andaya and Mackill, 2003a,b).

When rice is exposed to air temperatures higher than 35°C, heat injuries occur. By studying weather data at the International Rice Research Institute (IRRI) farm from 1979 to 2003, Peng *et al.* (2004) show that grain yield declined by 10% for each 1°C increase in minimum temperature during the dry season, whereas the effect of maximum temperature on crop yield was insignificant. In a different study, Ziska and Manalo (1996) found that, at a constant day temperature of 29°C, increasing night temperature did not significantly alter growth or yield; however, increasing night temperature at a day temperature of 33°C resulted in significant declines in grain-filling and grain yield.

Strategies to Improve Rice Tolerance to Abiotic Stresses for Africa

Various conventional and biotechnological approaches are being used to develop rice varieties tolerant of abiotic stresses. It is now possible to develop new tolerant rice varieties through conventional breeding in combination with marker-assisted selection or through direct transfer of tolerance genes into rice varieties via genetic engineering. At AfricaRice, two main strategies are being followed for genetic improvement of rice tolerance to abiotic stresses (Fig. 11.2): (i) the use of genetic diversity existing in rice, with particular focus on African rice (*O. glaberrima*) and related wild species, *O. barthii*; and (ii) the use of molecular markers to transfer tolerance QTLs to elite African germplasm, in addition to conventional breeding. Efforts are also devoted to identify new QTLs for tolerance.

To ensure farmers' access to these new technologies and their rapid adoption, AfricaRice is progressively introducing participatory approaches into the various breeding programmes. In different countries, breeding lines and varieties selected for tolerance to salt stress, drought, Fe toxicity and cold were evaluated by farmers through field visits and on-farm tests against their local varieties. By 2011, some 21 promising stress-tolerant lines had been identified for potential submission to national varietal release committees. In 2010, an Africa-wide Rice

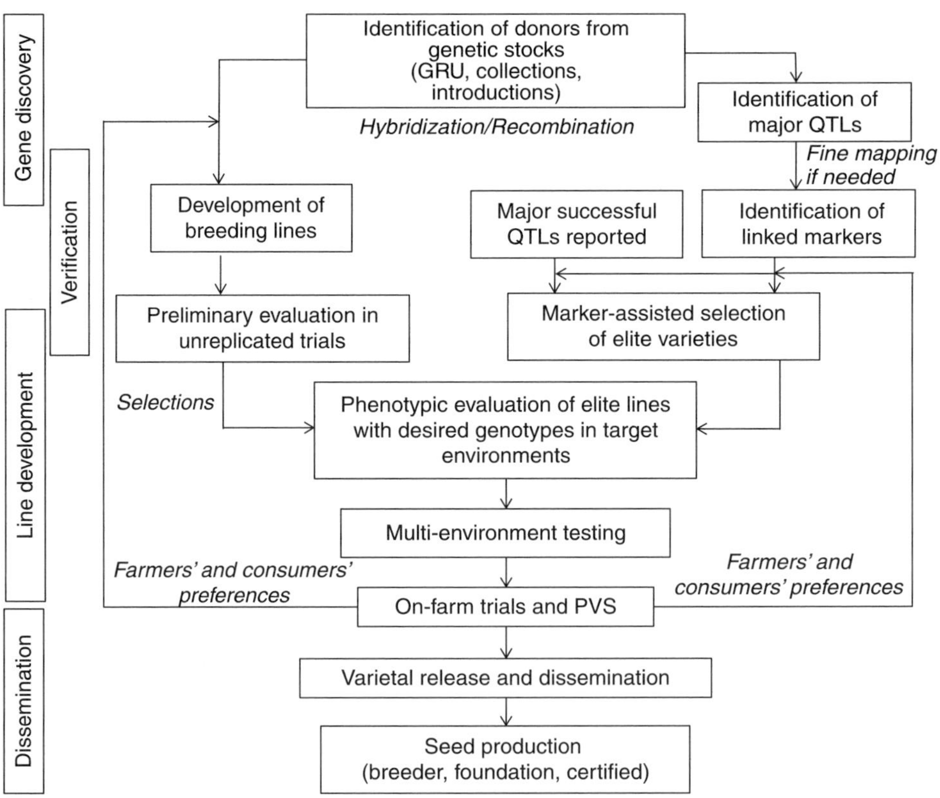

Fig. 11.2. Framework of a breeding process for genetic improvement of stress tolerance in rice. GRU, Genetic Resources Unit; PVS, participatory varietal selection; QTLs, quantitative trait loci.

Breeding Task Force was revived and this network is now responsible for multi-environment testing of new breeding lines developed for lowland, irrigated, upland, high-elevation and mangrove environments (see Kumashiro *et al.*, Chapter 5, this volume).

Tapping the gene pool of African rice

With the success of NERICA varieties, *O. glaberrima* gained renewed interest. This species is known for its good adaptation to sub-Saharan Africa environmental and soil conditions. It has tolerance to many abiotic stresses, including drought, Fe toxicity, acidity and low-input conditions (Sano *et al.*, 1984; Jones *et al.*, 1997) and has the ability to grow in a wide range of unfavourable ecosystems such as rainfed hilly areas, deepwater floating conditions and in coastal mangrove areas (Sarla and Swamy, 2005). While AfricaRice holds a collection of about 2500 *O. glaberrima* accessions, only four accessions were used to create the NERICA varieties for both upland and lowland environments. There is still a tremendous amount of unexploited genetic diversity in the primary gene pool of *O. glaberrima*. AfricaRice has embarked on the characterization of *O. glaberrima* accessions and their evaluation for different traits of interest, including tolerance to abiotic stresses.

Several *O. glaberrima* were collected in Mali (RAM series) and several donors of drought tolerance have been identified, including RAM3, RAM118 and RAM152 (Ndjiondjop *et al.*, 2007; Bimpong *et al.*, 2011a). The development of backcross interspecific lines (*O. sativa* × *O. glaberrima*) showed that drought tolerance from *O. glaberrima* is transferable to the progenies. Several introgression lines from different cross combinations were evaluated and lines tolerant of drought have been identified (Ndjiondjop *et al.*, 2007, 2010; Bimpong *et al.*, 2011b; Bocco *et al.*, 2011). Similarly, good sources of salinity tolerance have been identified within *O. glaberrima* (e.g. RAM121, TOG6224 and TOG7230), as well as some of the interspecific lowland NERICA-L varieties. These tolerant varieties do not have the tolerance allele of Pokkali at the *Saltol* locus. Further studies are assessing whether these alleles can effectively be combined with *Saltol* to breed highly tolerant varieties. Variability also exists in the interspecific lowland NERICA and *O. glaberrima* varieties with regards to Fe-toxicity tolerance. NERICA-L 19 and NERICA-L 49 combine both tolerance to iron toxicity and good agronomic characters (Nwilene *et al.*, unpublished). The tolerance was inherited from the *O. glaberrima* parent, TOG5681. It has also been reported that CG14, the *O. glaberrima* parent of upland NERICA varieties, has remarkable tolerance to Fe-toxicity stress (Sahrawat and Sika, 2002), while TOG7235, TOG6216, TOG7442 and TOG7291 show moderate tolerance (Mendoza *et al.*, 2000). About 500 accessions from the AfricaRice *O. glaberrima* collection are being evaluated at Vallée du Kou (Burkina Faso) to identify new sources of tolerance to Fe toxicity in this species. New research directions are also being considered in view of the current and anticipated rise in temperature resulting from climate change. New projects are in the pipeline to incorporate heat tolerance into local African varieties. For this trait, too, *O. glaberrima* could be exploited to confer early morning flowering in interspecific crosses to avoid spikelet sterility due to heat stress (Yoshida *et al.*, 1981; Manneh *et al.*, 2007).

One of the major limitations in the use of *O. glaberrima* for the improvement of *O. sativa* was the interspecific hybrid sterility (Ghesquiere *et al.*, 1997). Even though AfricaRice succeeded in the development of interspecific varieties – NERICA – the introgression of useful traits from *O. glaberrima* into *O. sativa* is still tedious and time-consuming. AfricaRice and Institut de recherche pour le développement (IRD) have developed interspecific bridges between the two cultivated species. These interspecific bridges comprise *O. sativa* lines carrying large introgressions of the *O. glaberrima* genome that are compatible with *O. sativa* in subsequent crosses. Marker-assisted selection was carried out on backcross progenies to select the homozygous lines bearing the *O. glaberrima* allele at the *S1* locus. The *S1* gene encodes a gamete eliminator which induces both male and female gamete abortion through allelic interaction in interspecific *O. sativa* × *O. glaberrima* crosses (Sano, 1986; Garavito *et al.*, 2010). The fertility restoration is monitored for three generations to derive fertile backcross inbred lines (BILs – BC_1F_3) with improved cross-ability with *O. sativa*. For more details see Lorieux *et al.* (Chapter 10, this volume).

In addition to *O. glaberrima*, genetic variability for tolerance to abiotic stresses is also present in *O. sativa*, especially in the traditional varieties, and this is being exploited. Through germplasm exchange, tolerant varieties identified elsewhere are included in breeding programmes to enhance tolerance to abiotic stresses in the local varieties. Meanwhile, agronomic and quality traits preferred by consumers are incorporated in the new tolerant lines to meet the market demand.

Speeding up the development of tolerant varieties through molecular breeding

With a better understanding of the mechanisms and genetics of stress tolerance, breeders are now using more precise breeding approaches – notably marker-assisted backcrossing (MABC) – to develop varieties with higher levels of tolerance and acceptable grain quality. For many abiotic stresses, major QTLs/genes have been identified in rice (Jena and Mackill, 2008; Ismail and Thomson, 2011). Some of these QTLs have been fine-mapped and a MABC system developed to incorporate them into high-yielding varieties. Incorporation of these QTLs into popular varieties demonstrates substantial impacts on rice productivity in farmers' fields (Singh *et al.*, 2009; Thomson *et al.*, 2010a,b; Ismail and Thomson, 2011; Mackill *et al.*, 2012). Likewise at AfricaRice, MABC has been initiated for all abiotic stresses for which major QTLs have been identified and successfully used. A certain set of QTLs is targeted for each rice production environment and research conducted at different AfricaRice stations.

Improving drought tolerance of rice has been hindered by the low level of genetic variability and the complex inheritance of the trait, as well as the difficulty of accurately measuring the level of tolerance. Molecular approaches to improve drought tolerance have been widely applied to rice, and numerous QTLs identified for secondary traits associated with drought response, including rooting traits (depth, volume, thinness, penetration ability), leaf rolling and death, membrane stability, and osmotic adjustment (Laffite *et al.*, 2006). However, very few studies have mapped QTLs related to better yield under drought. A few major QTLs for yield under drought have been identified that can result in substantial yield improvement under drought (Bernier *et al.*, 2007; Venuprasad *et al.*, 2009; Vikram *et al.*, 2011). These QTLs offer considerable opportunities for enhancing drought tolerance of rice varieties for Africa for both uplands and lowlands. Drought tolerance QTLs *qtl12.1* and *DTY3.1*, respectively associated with yield under upland stress and lowland stress (Bernier *et al.*, 2007; Venuprasad *et al.*, 2009), are targeted for marker-assisted selection (MAS) at AfricaRice. Near-isogenic lines (NILs) contrasting for grain yield under drought, developed at IRRI, are being tested under upland and lowland conditions in Benin and Nigeria. The *qtl12.1* co-localizes with the *Pup1* (P uptake-1) QTL that confers tolerance to P deficiency (Chin *et al.*, 2010). AfricaRice is aiming to transfer the *Pup1* QTL into the background of popular upland varieties. A preliminary survey using *Pup1* gene-specific markers showed that almost all upland varieties tested (including upland NERICA varieties) have Kasalath (*Pup1* donor) allele at all or part of the loci tested, but very few have the gene *OsPupK46-2*, which seems to be the major determinant of *Pup1* effect (Gamuyao *et al.*, 2012). This gives room for improvement of both drought and P deficiency.

For irrigated areas affected by salinity, the tolerance QTL *Saltol*, derived from the salt-tolerant cultivar Pokkali (Bonilla *et al.*, 2002; Thomson *et al.*, 2010a), is being incorporated into elite irrigated rice varieties, such as Sahel 108, Sahel 201 and ITA 212, to improve their tolerance to salinity. Advanced backcross breeding lines (BC_2–BC_3) have been obtained with *Saltol* introgression using FL478 and Pokkali as donor parents. In the Sahel, where rice is also affected by cold stress at the vegetative stage, AfricaRice scientists are targeting the cold tolerance QTLs (*Ctb-1*, *Ctb-2*) identified in Silewah (Saito *et al.*, 2004) and *qCTS12a* identified in M202 (Andaya and Mackill, 2003b). For submergence-prone areas, African *Sub1* lines were developed using WITA 4, a popular variety in Nigeria, as recipient (Gregorio *et al.*, unpublished) and new versions using NERICA-L 19, TOX4004, Kogoni, FARO44 and BW348-1 as recipients are in the pipeline. The major QTL *Sub1*, identified from FR13A (Xu and Mackill, 1996), provides tolerance to complete submergence for up to 2 weeks. Asian *Sub1* varieties have been extensively evaluated in South and South-east Asia and in Africa. In all cases, similar results were obtained (Table 11.2), with consistent yield advantages of 1 t/ha to over 3 t/ha compared with the original varieties following submergence for durations of about 4 to over 18 days (Mackill *et al.*, 2012). However, this QTL is not effective in areas that are likely to experience prolonged waterlogging or partial stagnant flooding of over 20 cm (Singh *et al.*, 2011). Therefore for these areas and for deep-water conditions, new materials should be developed targeting elongation ability.

Another powerful marker-assisted approach, marker-assisted recurrent selection (MARS), is being implemented for the improvement of drought tolerance in the lowlands. Allele diversity within bi-parental populations is exploited by MABC to increase the frequency of beneficial alleles for quantitative traits. Cyclical recombination of lines bearing interesting chromosomal segments will then be conducted and favourable alleles accumulated. Six populations have been developed and more than 200 lines are under evaluation for drought tolerance in Nigeria, Mali and Burkina Faso (M.-N. Ndjiondjop, Cotonou, Benin, 2012, personal communication).

With the use of molecular markers, AfricaRice expects to increase the efficiency and effectiveness of its breeding programmes compared to conventional breeding methods.

Challenges and Outlook

Anticipating the effects of climate change on rice production

Climate change is emerging as one of the most important challenges of the 21st century. Africa is particularly vulnerable to climate change, because of its high proportion of low-input, rainfed agriculture compared with other regions of the developing world (IPCC, 2001). The changes in key climatic variables (i.e. rainfall and temperature) will likely modify the distribution, frequency and severity of abiotic stresses. Prediction models in some cases provide contradictory conclusions, but what is clear is that increased flooding in low-lying areas, greater frequency and severity of droughts in arid and semiarid areas, rising of sea water level (which will increase salinity intrusion in

Table 11.2. Performances of rice lines bearing the resistance allele at *Sub1* locus after 21 days of submergence at AfricaRice station, Ibadan (Nigeria), in the 2009 dry season. Survival is expressed as the number of surviving hills under submergence as a percentage of survival without submergence. (From Gregorio, unpublished.)

Variety	Survival (%)	Yield under submerged conditions (t/ha)[a]	Yield under normal conditions (t/ha)	Yield reduction (%)
TDK1-Sub1(BC_3F_3)	95.45	3.62a	4.46	18.99
Swarna-Sub1(BC_3F_3)	90.95	3.56a	4.45	19.92
BR11-Sub1	95.34	2.60b	4.11	36.49
Samba Mahsuri-Sub1(BC_3F_3)	94.42	2.59b	3.94	34.16
Samba Mahsuri-Sub1(BC_2F_3)	96.45	2.37b	3.62	34.47
CR1009-Sub1	96.95	2.21b	3.61	29.78
IR64-Sub1(BC_3F_3)	96.92	2.21b	3.02	26.89
IR64	21.43	1.32c	3.19	58.46
Samba Mahsuri	17.44	0.81cd	3.83	78.73
Swarna	19.49	0.64de	4.29	85.16
FARO 35 (ITA 212)	20.41	0.41de	3.61	88.61
WITA 4	4.08	0.08e	3.77	97.99

[a]Numbers followed by the same letters are not significantly different from each other.

coastal areas) and excessive heat stress, will more likely be observed. An increase in CO_2 concentration in the absence of increases in temperature is expected to increase productivity of rice. However, if the temperature rises by about 1°C or more, the situation reverses, particularly in tropical climates (Ahmed *et al.*, 2010). Globally, climate-change effects will adversely affect rice production. Breeding programmes should, therefore, be adjusted in anticipation of these changes, to develop rice varieties adapted to future environments. Breeding for abiotic-stress tolerance is one important option, but other alternatives should also be considered. Research is being coordinated by IRRI to introduce the C_4 metabolic pathway into rice. This is expected to enhance yield potential of the current rice varieties by 30–50%. Besides, the new varieties are expected to require less nitrogen and water and also possess better adaptation to unfavourable climates, particularly to areas with higher temperatures and insufficient water resources (Sheehy *et al.*, 2007).

Changes in the climatic patterns will result in several other modifications in agricultural practices. As described by Satya and Maiti (2011), changes in the management practices for increasing crop productivity (such as water-saving and water-harvesting technologies), shift in cultivation patterns (rice to be replaced by wheat in drier and cooler areas and *vice versa* for wetter and hotter areas), changes in sowing and harvesting times to avoid stressful windows of time, reduction of crop growth duration, etc., are expected to be important determinants for shaping future agriculture to cope with worsening climate adversities. Additionally, a clear understanding of target environments and their evolution in terms of climate change will be an important objective to better set priorities and match new rice varieties to future climates in particular regions. The use of prediction models, such as RIDEV (Rice Development) developed by AfricaRice in the 1990s, will help in this exercise. This risk-assessment and planning model was developed to estimate the cycle duration and yield losses due to temperature-induced sterility for any combination of rice genotype, site, planting date and planting method in the Sahel (Dingkuhn, 1995). Additionally, it gives crop management recommendations based on the crop phenology. There are studies ongoing at AfricaRice, in collaboration with CIRAD scientists to improve RIDEV, thus allowing for predictions of crop responses to climate change.

Precise and high-throughput screening for tolerance of abiotic stresses

While high-throughput genome-wide genotyping is becoming increasingly accessible and is cost reducing, phenotyping of large germplasm collections and populations for stress tolerance in field trials is still laborious, imprecise and costly. The immediate difficulty of working with abiotic stresses still lies in reliably measuring the stress and quantifying the tolerance of the stress in large populations.

Field trials targeting climatic factors such as drought and temperature extremes are often unpredictable, while uniform stress conditions are difficult to achieve in trials for problem soils. This means that selection pressure is generally inconsistent and sometimes even contradictory across years and seasons (Lafitte *et al.*, 2006). Moreover, different stresses are often found together. For example, salinity problems are rarely due only to the common salt, NaCl; severity of Fe toxicity is dependent on other mineral deficiencies such as P and K; and during some periods both drought and heat stress can be concomitant. Most soil problems, such as Al and Fe toxicities and Zn deficiency, considerably hinder root growth and regeneration, increasing the crop's sensitivity to drought and nutrient deficiency. In general, abiotic stress conditions in the field are rarely due to a single factor and these are the typical farm conditions that are very difficult to reproduce in the laboratory or greenhouse. The complexity of the tolerance traits involved and the strong effect of the environment, and in most cases, the insufficient knowledge of the mechanisms involved in tolerance, add to these difficulties. There is therefore a need for high-throughput screening protocols to allow the evaluation of a large number of lines under the same conditions. For these reasons, phenotyping remains the current bottleneck for crop improvement and proper molecular mapping (Xu and Crouch, 2008; Serraj *et al.*, 2009).

Well-characterized environments and well-established selection criteria are prerequisites for developing a reliable and precise phenotyping system (Xu and Crouch, 2008). Developments in thermal, spectral reflectance, fluorescence and multi-sensor imaging may provide rapid, economic and non-invasive selection criteria (Chaerle and Van Der Straeten, 2000; Jones and Schofield, 2008). Once such reliable phenotyping methodologies are developed they can be transferred to national (NARS) partners to conduct multi-location/multi-environment trials using the same protocols to assess the responses of new breeding lines under a wider range of, or site-specific, stresses. This should be accompanied by effective data-acquisition systems to acquire and analyse the data in the shortest possible time. One major constraint for progress in Africa is the limited resources devoted to research on abiotic stresses, including lack of trained scientists and efficient phenotyping facilities. Greater investments in these areas would help build stronger breeding programmes to tackle existing challenges and ensure the capacity to develop adaptive and mitigation measures to cope with future adversities.

Breeding rice varieties with multiple stress tolerance

The desired features of new climate-ready rice varieties should include tolerance to various abiotic stresses that prevail in areas where these varieties are targeted for commercial use. This will entail combining tolerance of two or more stresses like drought and heat, salinity and excess water, cold and excess water, P deficiency and drought or Fe toxicity and excess water, into one genetic background. In most cases, breeders select for multiple stress tolerance by performing most of their selection process at actual field sites, as in participatory varietal selection (PVS) approaches. For example, studies of the distribution of the *Pup1* QTL in diverse germplasm have shown that most of the drought-tolerant lines and varieties developed for upland conditions also have the *Pup1* QTL (Chin *et al.*, 2010). With the progress made in the development of molecular tools, identification of QTLs and the use of DNA markers in breeding, it is now possible to make the precise desired combinations. For example, varieties combining submergence tolerance and salt tolerance are being developed by pyramiding *Sub1* and *Saltol* QTLs through MABC (Mackill *et al.*, 2012). These lines are needed in the coastal and mangrove areas of sub-Saharan Africa which can experience both salinity and submergence. The availability of high-throughput single-nucleotide polymorphism (SNP) genotyping platforms will also enable more efficient MABC by reducing the cost per marker and speeding up the process through 'multiplexing' (Thomson *et al.*, 2010b). AfricaRice, IRRI and Cornell University are developing SNP chips specific to African rice, *O. glaberrima* (M. Semon and M.-N. Ndjiondjop, Cotonou, Benin, 2012, personal communication).

Studies of the interactive effects of different stress combinations should be emphasized (Ismail *et al.*, 2007), since they will give greater insight into the positive and negative effects of combining genes for higher tolerance of the same or different abiotic stresses. The remaining challenge will be to incorporate these complex adaptive traits into high-yielding varieties through QTL pyramiding, while simultaneously retaining the adaptive features, high yield potential and grain quality of the recurrent varieties (Ismail *et al.*, 2008). Meanwhile, tolerance to pests and diseases should not be neglected. Climate-ready varieties should also possess sufficient resistance to major diseases that occur in the targeted rice-production zones of Africa.

Conclusions

Rice production in sub-Saharan Africa is constrained by various environmental stresses. Climate change adds more uncertainties, particularly for smallholder subsistence farmers. Stabilizing crop yields and limiting yield losses will undoubtedly reduce resource-poor African farmers' vulnerability to poverty and food insecurity. Several options exist with the most cost-effective one being crop improvement using different strategies and technologies as discussed in this chapter. The genetic variation observed in rice for tolerance to various abiotic stresses could be effectively exploited for crop improvement by using tolerant varieties and landraces as donors in breeding programmes using modern tools. Major QTLs have been identified for almost all abiotic stresses, including drought, providing considerable opportunity for enhancing rice tolerance of these stresses. These major QTLs are being used to improve varieties that are widely popular among farmers in the target regions.

More effort is, however, needed to collect and preserve local African germplasm, including wild species, in order to identify better donors for use in breeding and for tagging novel QTLs/ genes. Precise high-throughput phenotyping protocols also need to be in place for effective screening of breeding lines, as is a good knowledge of the target environments. The new tolerant varieties should perform well under both stress and non-stress conditions and also retain acceptable grain quality to ensure their adoption and acceptance by both farmers and consumers. Furthermore, the availability of rice cultivars tolerant to abiotic stresses will provide an incentive for farmers to invest in the costly processes of high input use and efficient postharvest technologies, and in land reclamation and preservation, besides other efficient agricultural practices. Overall, this will result in greater and more sustainable productivity in vulnerable regions of sub-Saharan Africa even with the anticipated climate change and its adverse effects on rice production.

References

Abekoe, M.K. and Sahrawat, K.L. (2001) Phosphate retention and extractability in soils of the humid zone in West Africa. *Geoderma* 102, 175–187.

Abifarin, A.O. (1989) Progress in breeding rice for tolerance to iron toxicity. In: *WARDA Annual Report.* West Africa Rice Development Association, Bouaké, Côte d'Ivoire, pp. 34–39.

AfricaRice (2011a) *Boosting Africa's Rice Sector: A research for development strategy 2011–2020.* Africa Rice Center, Cotonou, Benin.

AfricaRice (2011b) Overcoming abiotic stresses to increase production. In: *Africa Rice Center Annual Report 2010: Building African capacity on policy analysis and impact assessment.* Africa Rice Center, Cotonou, Benin, pp. 10–14.

Ahmed, A., Diwan, H. and Abrol, Y.P. (2010) Global climate change, stress and plant productivity. In: Pareek, A., Sopory, S.K., Bohnert, H.J. and Govind, J. (eds) *Abiotic Stress Adaptations in Plants: Physiological, molecular and genomic foundations.* Springer, Dordrecht, Netherlands, pp. 503–522.

Andaya, V.C. and Mackill, D.J. (2003a) QTLs conferring cold tolerance at the booting stage of rice using recombinant inbred lines from a *japonica* × *indica* cross. *Theoretical and Applied Genetics* 106, 1084–1090.

Andaya, V.C. and Mackill, D.J. (2003b) Mapping of QTLs associated with cold tolerance during the vegetative stage in rice. *Journal of Experimental Botany* 54, 2579–2585.

Asch, F., Dingkuhn, M. and Dorffling, K. (1997) Physiological stresses of irrigated rice caused by soil salinity in the Sahel. In: Miezan, K.M., Wopereis, M.C.S., Dingkuhn, M., Deckers, J. and Randolph, T.F. (eds) *Irrigated Rice in the Sahel: Prospects for sustainable development.* West Africa Rice Development Association, Bouaké, Côte d'Ivoire, pp. 247–273.

Batjes, N.H. (1997) A world data set of derived soil properties by FAO-UNESCO soil unit for global modeling. *Soil Use and Management* 13, 9–16.

Becker, M. and Asch, F. (2005) Iron toxicity in rice-conditions and management concepts. *Journal of Plant Nutrition and Soil Science* 168, 558–573.

Benckiser, G., Santiago, S., Neue, H.U., Watanabe, I. and Ottow, J.C.G. (1984) Effect of fertilization on exudation, dehydrogenase activity, iron-reducing populations and Fe^{++} formation in the rhzosphere of rice (*Oryza sativa* L.) in relation to iron toxicity. *Plant and Soil* 79, 305–316.

Bernier, J., Kumar, A., Ramaiah, V., Spaner, D. and Atlin, G. (2007) A large-effect QTL for grain yield under reproductive-stage drought stress in upland rice. *Crop Science* 47, 507–518.

Bertrand, R., Keita, B. and N'Diaye, M.K. (1993) Soil degradation in the irrigated areas of the major valleys of the southern Sahara (Case of the Niger Office in Mali). *Cahier Agricultures* 2, 318–329.

Bimpong, I.K., Serraj, R., Chin, J.-H., Mendoza, E.M.T, Hernandez, J. and Mendioro, M.S. (2011a) Determination of genetic variability for physiological traits related to drought tolerance in African rice (*Oryza glaberrima*). *Journal of Plant Breeding and Crop Science* 3(4), 60–67.

Bimpong, I.K., Serraj, R., Chin, J.-H., Ramos, J., Mendoza, E.M.T., Hernandez, J.E., Mendioro, M.S. and Brar, D.S. (2011b) Identification of QTLs for drought-related traits in alien introgression lines derived from crosses of rice (*Oryza sativa* cv. IR64) × *O. glaberrima* under lowland moisture stress. *Journal of Plant Biology* 54, 237–250.

Bocco, R., Lorieux, M., Seck, P.A., Futakuchi, K., Manneh, B., Baimey, H. and Ndjiondjop, M.N. (2011) Agro-morphological characterization of a population of introgression lines derived from crosses between IR 64 (*Oryza sativa indica*) and TOG 5681 (*Oryza glaberrima*) for drought tolerance. *Plant Science* 183, 65–76.

Bonilla, P., Dvorak, J., Mackill, D., Deal, K. and Gregorio, G. (2002) RFLP and SSLP mapping of salinity tolerance genes in chromosome 1 of rice (*Oryza sativa* L.) using recombinant inbred lines. *Philippine Journal of Agricultural Science* 85, 68–76.

Boojung, H. and Fukai, S. (1996) Effects of soil water deficit at different growth stages on rice growth and yield under upland conditions. 2. Phenology, biomass production and yield. *Field Crops Research* 48, 47–55.

Chaerle, L. and Van Der Straeten, D. (2000) Imaging techniques and the early detection of plant stress. *Trends in Plant Science* 5(11), 495–501.

Chérif, M., Audebert, A., Fofana, M. and Zouzou, M. (2009) Evaluation of iron toxicity on lowland irrigated rice in West Africa. *Tropicultura* 27(2), 88–92.

Chin, J.H., Lu, X., Haefele, S.M., Gamuyao, R., Ismail, A., Wissuwa, M. and Heuer, S. (2010) Development and application of gene-based markers for the major rice QTL Phosphorus uptake 1. *Theoretical and Applied Genetics* 120, 1073–1086.

Coly, A. (1980) Testing rice varieties of *indica* type for cold tolerance at seedling stage. *Japanese Journal of Crop Science* 49(4), 543–547.

Das, K.K., Panda, D., Sarkar, R.K., Reddy, J.N. and Ismail, A.M. (2009) Submergence tolerance in relation to variable floodwater conditions in rice. *Environmental and Experimental Botany* 66(3), 425–434.

De Datta, S., Biswas, T. and Charoenchamratcheep, C. (1990) Phosphorus requirements and management for lowland rice. In: *Phosphorus Requirements for Sustainable Agriculture in Asia and Oceania*, International Rice Research Institute, Manila, Philippines, pp. 307–323.

Defoer, T., Wopereis, M.C.S., Jones, M.P., Lançon, F. and Erenstein, O. (2002) Challenges, innovation and change: towards rice-based food security in sub-Saharan Africa. Paper presented at the 20th session of the International Rice Commission, Bangkok, 22–25 July.

Diarra, S., Kuper, M. and Mahé, G. (2004) *Mali: Flood Management – Niger River inland delta. Integrated Flood Management case study.* World Meteorological Organization and Global Water Partnership Associated Programme on Flood Management, Geneva, Switzerland.

Dingkuhn, M. (1995) Climatic determinants of irrigated rice performance in the Sahel – III. Characterizing environments by simulating crop phenology. *Agricultural Systems* 48(4), 435–456.

Dingkuhn, M. and Miezan, K.M. (1995) Climatic determinants of irrigated rice performance in the Sahel – II. Validation of photothermal constants and characterization of genotypes. *Agricultural Systems* 48(4), 411–433.

Dingkuhn, M., Asch, F. and Miezan, K. (1993) Salt tolerance of rice varieties under irrigated conditions in the Sahel. In: *WARDA Annual Report 1992.* West Africa Rice Development Association, Bouaké, Côte d'Ivoire.

Dobermann, A., Cassman, K.G., Mamaril, C.P. and Sheehy, J.E. (1998) Management of phosphorus, potassium, and sulfur in intensive, irrigated lowland rice. *Field Crops Research* 56, 113–138.

Flowers, T. and Yeo, A. (1981) Variability in the resistance of sodium chloride salinity within rice (*Oryza sativa* L.) varieties. *New Phytologist* 88, 363–373.

Futakuchi, K. (2005) Submergence damage in rice and challenges in expanding the crop's adaptability to submerged conditions in West and Central Africa. In: *Rice is Life: Scientific perspectives for the 21st century.* International Rice Research Institute and Japan International Research Center for Agricultural Sciences, Los Baños, Philippines, pp. 445–447.

Gamuyao, R., Chin, J.H., Pariasca-Tanaka, J., Pesaresi, P., Catausan, S., Dalid, C., Slamet-Loedin, I., Tecson-Mendoza, E.M., Wissuwa, M. and Heuer, S. (2012) The protein kinase Pstol1 from traditional rice confers tolerance of phosphorus deficiency. *Nature* 488, 535–541.

Garavito, A., Guyot, R., Lozano, J., Gavory, F., Samain, S., Panaud, O., Tohme, J., Ghesquière, A. and Lorieux, M. (2010) A genetic model for the female sterility barrier between Asian and African cultivated rice species. *Genetics* 185, 1425–1440.

Gregorio, G.B., Senadhira, D., Mendoza, R.D., Manigbas, N.L., Roxas, J.P. and Guerta, C.Q. (2002) Progress in breeding for salinity tolerance and associated abiotic stresses in rice. *Field Crops Research* 76, 91–101.

Ghesquière, A., Séquier, J., Second, G. and Lorieux, M. (1997) First steps towards a rational use of African rice, *Oryza glaberrima*, in rice breeding through a 'contig line' concept. *Euphytica* 96, 31–39.

Haefele, S.M., Ismail, A.M., Johnson, D.E., Vera Cruz, C. and Samson, B. (2010) Crop and natural resource management for climate-ready rice in unfavorable environments: coping with adverse conditions and creating opportunities. Presentation at the CURE workshop on Climate Change, Siem Reap, Cambodia, 4 May.

Hsiao, T.C., O'Toole, J.C. and Tomar, V.S. (1980) Water stress as a constraint to crop production in the tropics. In: *Proceedings Symposium on Priorities for Alleviating Soil-related Constraints to Food Production in the Tropics.* International Rice Research Institute and Cornell University, Los Baños, Philippines, pp. 339–369.

IPCC (Intergovernmental Panel on Climate Change) (2001) *Climate Change 2001: The scientific basis.* Cambridge University Press, Cambridge.

Ismail, A.M. and Thomson, M.J. (2011) Molecular breeding of rice for problem soils. In: de Oliveira, A.C. and Varshney, R.K. (eds) *Root Genomics.* Springer, Dordrecht, Netherlands, pp. 289–312.

Ismail, A.M. and Tuong, T.P. (2009) Brackish water coastal zones of the monsoon tropics: challenges and opportunities. In: Haefele, S.M. and Ismail, A.M. (eds) *Natural Resource Management for Poverty Reduction and Environmental Sustainability in Rice-based Systems. Proceedings* No. 15. International Rice Research Institute, Los Baños, Philippines, pp. 113–121.

Ismail, A.M., Heuer, S., Thomson, M.J. and Wissuwa, M. (2007) Genetic and genomic approaches to develop rice germplasm for problem soils. *Plant Molecular Biology* 65, 547–570.

Ismail, A.M., Thomson, M.J., Singh, R.K., Gregorio, G.B. and Mackill, D.J. (2008) Designing rice varieties adapted to coastal areas of South and Southeast Asia. *Journal of the Indian Society of Coastal Agricultural Research* 26, 69–73.

Jena, K.K. and Mackill, D.J. (2008) Molecular markers and their use in marker-assisted selection in rice. *Crop Science* 48, 1266–1276.

Jones, H.G. and Schofield, P. (2008) Thermal and other remote sensing of plant stress. *General and Applied Plant Physiology* 34(1–2), 19–32.

Jones, M.P., Dingkuhn, M., Aluko, G.K. and Semon, M. (1997) Interspecific *Oryza sativa* L. × *O. glaberrima* Steud progenies in upland rice improvement. *Euphytica* 92, 237–246.

Lafitte, R.H., Ismail, A.M. and Bennett, J. (2006) Abiotic stress tolerance in tropical rice: progress and the future. *Oryza* 43, 171–186.

Mackill, D.J., Ismail, A.M., Singh, U.S., Labios, R.V. and Paris, T.R. (2012) Development and rapid adoption of submergence-tolerant (*Sub1*) rice varieties. *Advances in Agronomy* 115, 303–356.

Manneh, B., Kiepe, P., Sie, M., Ndjiondjop, M., Drame, K.N., Traore, K., Rodenburg, J., Somado, E.A., Narteh, L., Youm, O., Diagne, A. and Futakuchi, K. (2007) Exploiting partnerships in research and development to help African rice farmers cope with climate variability. *SAT eJournal* 4(1), 24 pages. http://www.icrisat.org/journal/SpecialProject/sp16.pdf.

Masajo, T.M., Alluri, K., Abifarin, A.O. and Janakiram, D. (1986) Breeding for high and stable yields in Africa. In: Juo, A.S.R. and Lowe, J.A. (eds) *The Wetlands and Rice in Sub-Saharan Africa.* International Institute for Tropical Agriculture, Ibadan, Nigeria, pp. 107–114.

Matlon, P., Randolph, T. and Guei, R. (1998) Impact of rice research in West Africa. In: *Impact of Rice Research.* Proceedings of the International Conference on the Impact of Rice Research, Bangkok, Thailand, 3–5 Jun 1996. Thailand Development Research Institute, Bangkok, Thailand and International Rice Research Institute, Manila, Philippines, pp. 383–404.

Mendoza, R.D., Moliñawe, J.A., Gregorio, G.B., Guerta, C.Q. and Brar, D.S. (2000) Genetic variability of tolerance for iron toxicity in different species of *Oryza* and their derivatives. In: Khush, G.S., Brar, D.S. and Hardy, B. (eds) *Advances in Rice Genetics.* World Scientific and Imperial College Press, Hackensack, New Jersey, pp. 154–157.

Miézan, K. and Dingkuhn, M. (1994) Rice and salinity in the Sahel. *Ag-Sieve* 6(6), 4. http://fadr.msu.ru/rodale/agsieve/txt/vol6/6/art4.html.

Moradi, F., Ismail, A., Gregorio, G. and Egdane, J. (2003) Salinity tolerance of rice during reproductive development and association with tolerance at the seedling stage. *Indian Journal of Plant Physiology* 8, 105–116.

Ndjiondjop, M.N., Manneh, B., Semon, M., Efisue, A., Cisse, F., Futakuchi, K., Gridley, H., Bocco, R. and Cissoko, M. (2007) Gene mining of rice germplasm (*Oryza glaberrima* and *Oryza sativa*) to improve drought resistance in rainfed production systems for resource-poor farmers of Africa. In: *Abstracts of the Rockefeller Foundation Meeting on Biotechnology, Breeding and Seed Systems for African Crops*, Research and product development that reaches farmers, Maputo, Mozambique, 26–29 March, p. 170.

Ndjiondjop, M.N., Manneh, B., Cissoko, M., Dramé, K.N., Glele, R.K., Bocco, R., Baimey, H. and Wopereis, M. (2010) Drought resistance in an interspecific backcross population of rice (*Oryza* spp.) derived from the cross WAB56-104 (*O. sativa*) × CG14 (*O. glaberrima*). *Plant Science* 179, 364–373.

Peng, S., Huang, J., Sheehy, J.E., Laza, R.C., Visperas, R.M., Zhong, X., Centeno, G.S., Khush, G.S. and Cassman, K.G. (2004) Rice yields decline with higher night temperature from global warming. *Proceedings of the National Academy of Sciences USA* 101(27), 9971–9975.

Prade, K., Ottow, J.C.G., Jacq, V.A., Malouf, G. and Loyeri J.Y. (1990) Relationships between the properties of flooded rice soils and iron toxicity in lower Casamance (Senegal). *Cahiers ORSTOM, Série Pédologie* 25(4), 453–474.

Sahrawat, K.L. and Sika, M. (2002) Comparative tolerance of *Oryza sativa* and *O. glaberrima* rice cultivars for iron toxicity in West Africa. *International Rice Research Notes* 27(2), 30–31.

Sahrawat, K.L., Mulbah, C.K., Diatta, S., Delaune, R.D., Patrick, W.H. Jr, Singh, B.N. and Jones, M.P. (1996) The role of tolerant genotypes and plant nutrients in the management of iron toxicity in lowland rice. *Journal of Agricultural Sciences* 126, 143–149.

Saito, K., Hayano-Saito, Y., Maruyama-Funatsuki, W., Sato, Y. and Kato, A. (2004) Physical mapping and putative candidate gene identification of a quantitative trait locus *Ctb1* for cold tolerance at booting stage of rice. *Theoretical and Applied Genetics* 109, 515–522.

Sakagami, J.-I. (2012) Submergence tolerance of rice species, *Oryza glaberrima*. In: Najafpour, M.M. (ed.) *Applied Photosynthesis*. InTech, Rijeka, Croatia.

Sano, Y. (1986) Sterility barriers between *Oryza sativa* and *O. glaberrima*. In: Kush, G.S. (ed.) *Rice Genetics*. Proceedings of the International Rice Genetics Symposium. International Rice Research Institute, Manila, Philippines, pp. 109–118.

Sano, Y., Sano, R. and Morishima, H. (1984) Neighbour effects between two occurring rice species *O. sativa* and *O. glaberrima*. *Journal of Applied Ecology* 21, 245–254.

Sarla, N. and Swamy, B.P.M. (2005) *Oryza glaberrima*: a source for the improvement of *Oryza sativa*. *Current Science* 89(6), 955–963.

Satya, P. and Maiti, R.K. (2011) Molecular breeding for mitigation of abiotic stress in projected climate change scenario: potentials and limitations for developing and underdeveloped countries. *International Journal of Bio-resource and Stress Management* 2(1), I–V.

Serraj, R., Kumar, A., McNally, K.L., Slamet-Loedin, I., Bruskiewich, R., Mauleon, R., Cairns, J. and Hijmans, R.J. (2009) Improvement of drought resistance in rice. *Advances in Agronomy* 103, 41–99.

Sheehy, J.E., Mitchell, P.L. and Hardy, B. (2007) *Charting New Pathway to C_4 Rice*. International Rice Research Institute, Los Baños, Philippines.

Singh, S., Mackill, D.J. and Ismail, A.M. (2009) Responses of *Sub1* rice introgression lines to submergence in the field: yield and grain quality. *Field Crops Research* 113, 12–23.

Singh, S., Mackill, D.J. and Ismail, A.M. (2011) Tolerance of longer-term partial stagnant flooding is independent of the *Sub1* locus in rice. *Field Crops Research* 121, 311–323.

Thomson, M.J., de Ocampo, M., Egdane, J., Rahman, M.R., Sajise, A.G., Adorada, D.L., Tumimbang-Raiz, E., Blumwald, E., Seraj, Z.I., Singh, R.K., Gregorio, G.B. and Ismail, A.M. (2010a) Characterizing the *Saltol* quantitative trait locus for salinity tolerance in rice. *Rice* 3(2/3), 148–160.

Thomson, M.J., Ismail, A.M., McCouch, S.R. and Mackill, M.J. (2010b) Marker assisted breeding. In: Pareek, A., Sopory, S.K., Bohnert, H.J. and Govindjee (eds) *Abiotic Stress Adaptation in Plants: Physiological, molecular and genomic foundation*. Springer, Dordrecht, Netherlands, pp. 451–469.

Van Asten, P.J.A., Barbiéro, L., Wopereis, M.C.S, Maeght, J.L. and van der Zee, S.E.A.T.M. (2003) Actual and potential salt-related soil degradation in an irrigated rice scheme in the Sahelian zone of Mauritania. *Agricultural Water Management* 60(1), 13–32.

Van Asten, P.J.A., Barro, S.E., Wopereis, M.C.S. and Defoer, T. (2004) Using farmer knowledge to combat low productive spots in rice fields of a Sahelian irrigation scheme. *Land Degradation & Development* 15(4), 383–396.

Venuprasad, R., Dalid, C.O., del Valle, M., Zhao, D., Espiritu, M., Sta Cruz, M.T., Amante, M., Kumar, A. and Atlin, G. (2009) Identification and characterization of large-effect quantitative trait loci for grain yield under lowland drought stress in rice using bulk-segregant analysis. *Theoretical and Applied Genetics* 120, 177–190.

Vikram, P., Mallikarjuna-Swamy, B.P., Dixit, S., Ahmed, H.U., Sta Cruz, M.T., Singh, A.K. and Kumar, A. (2011) *qDTY1.1*, a major QTL for rice grain yield under reproductive-stage drought stress with a consistent effect in multiple elite genetic backgrounds. *BMC Genetics* 12, 89.

WARDA (1988) *Annual Report.* West Africa Rice Development Association, Monrovia, Liberia, pp. 18–19.

WARDA (1993) *Annual Report.* West Africa Rice Development Association, Bouaké, Côte d'Ivoire, pp. 13–28.

WARDA (1997) Tougher plants for the lowlands. In: *Annual Report.* West Africa Rice Development Association, Mbé, Côte d'Ivoire, pp. 19–23.

Xu, K. and Mackill, D. (1996) A major locus for submergence tolerance mapped on rice chromosome 9. *Molecular Breeding* 2, 219–224.

Xu, Y. and Crouch, J.H. (2008) Marker-assisted selection in plant breeding: from publications to practice. *Crop Science* 48, 391–407.

Yoshida, S., Satake, T. and Mackill, D.S. (1981) High-temperature stress in rice. *IRRI Research Paper Series* no. 67.

Zenna, N., Luzi-Kihupi, A., Manneh, B., Rabeson, R., Gasore, E.R. and Traore, K. (2010) Weathering the cold. *Rice Today* (January–March), pp. 26–27.

Ziska, L.H. and Manalo, P.A. (1996) Increasing night temperature can reduce seed set and potential yield of tropical rice. *Australian Journal of Plant Physiology* 23, 791–794.

12 Integration of Molecular Markers in Rice Improvement: A Case Study on Resistance to *Rice yellow mottle virus*

Marie-Noëlle Ndjiondjop,[1]* Laurence Albar,[2] Mounirou Sow,[1] Nasser Yao,[1] Gustave Djedatin,[1] Deless Thiémélé[3] and Alain Ghesquière[2]

[1]Africa Rice Center (AfricaRice), Cotonou, Benin; [2]Institut de Recherche pour le Développement (IRD), UMR Diversité, Adaptation et Développement des Plantes (DIADE), IRD-CIRAD-UM2, Montpellier, France; [3]Centre National de la Recherche Agronomique (CNRA), Station de Recherche Bimbresso, Côte d'Ivoire

Introduction

The development of molecular markers in the 1980s was considered as highly promising for plant breeding (Young and Tanksley, 1989), as they offer to partly replace phenotypic selection (which is often expensive and fastidious) with genotypic selection. A wide range of molecular breeding methods have been described, from marker-assisted selection (MAS) of one major gene to large-scale genotypic selection for multiple traits or genetic background. The main advantages of MAS have been reviewed extensively (Collard and Mackill, 2008; Xu and Crouch, 2008); they generally concern traits for which phenotyping is associated with particular stages of development or environmental conditions, traits with low heritability, recessive characters, or when pyramiding would mask the effect of individual genes. As soon as markers associated with traits of interest are available, selection for major genes is theoretically accessible to any laboratory with basic molecular-biology equipment (Van Damme *et al.*, 2011). However, the impact of molecular markers in breeding appears to be limited in comparison with the number of publications that report genetic linkage between markers and genes of interest. This may be explained by the delay observed in the theoretical development of a methodology and its practical outputs or absence of information on the methodology used (Collard and Mackill, 2008). Van Damme *et al.* (2011) hypothesize that it may result from a lack of access to the information in some developing countries and these authors therefore developed a database that synthesizes all the markers available for MAS in 19 species.

Rice, the staple food of half of the world's population, has two cultivated species, *Oryza sativa* that originated from Asia and is cultivated worldwide, and *Oryza glaberrima*, which is endemic to Africa. *Oryza sativa* is high yielding, but lacks sources of resistance to some strains of rice diseases and insect pests specific to African rice-cropping agro-ecosystems. Conversely, *O. glaberrima* is low yielding, shatters spontaneously and has few panicle branches, but often constitutes a source of resistance to African pests and diseases. Some of its 'rustic' characteristics were successfully transferred into

* Corresponding author: m.ndjiondjop@cgiar.org

© CAB International 2013. *Realizing Africa's Rice Promise* (eds M.C.S. Wopereis *et al.*)

the *O. sativa* background by Africa Rice Center (AfricaRice) scientists during the 1990s to develop the set of varieties referred to as NERICA (Jones *et al.*, 1997; Futakuchi and Sié, 2009). Given its small genome, rice was considered as a model species among monocotyledons during the 1990s. Important molecular and genomic tools were developed much earlier for rice than for other crops, and facilitated the identification of molecular markers associated with traits of interest. Jena and Mackill (2008) counted 1488 genes with known chromosomal position in rice databases (www.gramene.org/), including about 100 genes for resistance to biotic stresses. For the most important of these genes, closely linked or allele-specific markers are available. In addition, simple sequence repeat (SSR) markers evenly distributed over the rice genome are available and represent highly efficient tools for MAS (Orjuela *et al.*, 2010; Narshimulu *et al.*, 2011).

So far, MAS has been used in rice mainly to introgress the favourable alleles of genes involved in resistance to diseases and pests. In India and the Philippines, bacterial blight (BB) resistance alleles of genes *Xa4*, *xa5*, *xa13* and *Xa21* have been successfully introgressed into hybrid rice KMR3, PRR78, IR58025B and Pusa 6B (Shanti *et al.*, 2010). Different combinations of favourable alleles of either two or three resistance genes have also been incorporated into elite varieties using MAS, providing high level of resistance to ten highly virulent bacterial isolates (Singh *et al.*, 2001; Leung *et al.*, 2004; Agarcio *et al.*, 2007; Borines *et al.*, 2008; Collard and Mackill, 2008; Perez *et al.*, 2008; Sundaram *et al.*, 2008, 2009; Vera Cruz *et al.*, 2009). Likewise, Chinese elite restorer lines Minghui 63, Shanyou 63 and 6078 were improved for BB resistance using *Xa21* gene (Chen *et al.*, 2000, 2001), while two Indonesian cultivars 'Angke' and 'Conde', containing *Xa4*, were improved using the resistance gene *xa5*. The two BB-resistance genes, *xa13* and *Xa21*, present in IRBB55 were combined with the Basmati quality traits of Pusa Basmati-1 (PB-1), the most popular high-yielding Basmati rice variety (Joseph *et al.*, 2004). Unlike BB, blast and rice gall midge have received less attention with respect to markers and MAS. However, two blast-resistance genes, *Pi-1* and *Pi-2*, were successfully pyramided in the susceptible genotype CO39 and popular rice varieties IR36, IR64, IR72 and Jaya. Meanwhile, *Pi-1*, *Piz-5* and *Pita* were combined into a single genotype (Hittalmani *et al.*, 2000). Chinese varieties Digu, BL-1 and Pi-4, which carry resistance genes *Pi-d(t)*, *Pib* and *Pita2*, respectively, were crossed with G46B to pyramid all these genes in the same background (Chen *et al.*, 2004). Katiyar *et al.* (2001) successfully pyramided two Asian gall-midge-resistance genes *Gm2* and *Gm6t*, using two complementary donors. Kumaravadivel *et al.* (2006) report similar work targeting genes *Gm1* and *Gm4*.

In this chapter, we report on the application of molecular tools for the improvement of resistance to *Rice yellow mottle virus* (RYMV), which is endemic to Africa and can cause up to 80% yield loss in some rice-cropping systems. We specifically focus on the first major gene identified, *RYMV1*.

Rice yellow mottle virus

RYMV was first reported in Kenya (Bakker, 1970), and since the early 1990s, it has been reported in all rice-growing regions of Africa, including Madagascar (Kouassi *et al.*, 2005). The main symptoms are yellowing and mottling, with susceptible varieties showing stunting and sterility (Plate 4). The disease was first observed in lowland rice agro-ecosystems, but it has subsequently been reported in upland agro-ecosystems (Awoderu *et al.*, 1987). Yield losses in affected rice fields ranged from 64% to 100% in Mali (Sy *et al.*, 1994), up to 82% in Sierra Leone (Taylor *et al.*, 1990) and up to 71% in Niger (Issaka *et al.*, 2012). In Central Africa, the virus has been reported in Cameroon and Chad (Traoré *et al.*, 2001); in East Africa, its centre of origin, the disease has been detected in Kenya, Malawi, Rwanda and Tanzania (Thottappilly, 1992; Banwo, 2003; Ndikumana *et al.*, 2011). Madagascar has also been seriously affected, to the point where some farmers have abandoned their fields (Reckhaus and Randrianangaly, 1990).

RYMV is a *Sobemovirus* and its virions are icosahedral particles. The genome is composed of a single-stranded, positive-sense RNA, of about 4451±1 nucleotides (depending on the strain), and contains four open reading frames (ORF1, ORF2a&b, ORF3 and ORF4) (Kouassi *et al.*, 2005). The P1 protein coded by ORF1 is involved in the plant-infection process, virus spread in the plant

(Bonneau *et al.*, 1998) and has been described as a suppressor of virus-induced gene-silencing (VIGS; Voinnet *et al.*, 1999). The polyprotein encoded by ORF2a and ORF2b contains a protease, the genome-linked viral protein (VPg) and an RNA-dependent RNA polymerase, which are involved in virus replication. The coat-protein encoded by ORF3 is implicated in virus encapsidation, cell-to-cell and long-distance movement, and systemic infection (Brugidou *et al.*, 1995).

Both cultivated rice species, *O. glaberrima* and *O. sativa*, can be infected by RYMV. Additionally, the African wild rice species, *O. longistaminata* and *O. barthii*, and several wild grasses (e.g. *Echinochloa colona, Panicum repens, Erasgrotis tenuifilia* and *Dinebra retroflexa*) are hosts of the virus (Bakker, 1971; Konate *et al.*, 1997). Mechanical inoculation is very efficient for transmitting the virus in the laboratory and must also play a role in the field; for example, virus dispersal is facilitated by the wind through contact between plant leaves (Sarra *et al.*, 2004) and by farmers and farming tools during transplanting operations (Abo and Sy, 1998; Abo *et al.*, 2000). Beetles belonging to the family Chrysomelidea can transmit the virus 1–8 days after feeding on an infected plant (Bakker, 1971; Hibino, 1996; Abo and Sy, 1998). Cattle and rodents are also reported as occasional secondary vectors of the virus (Sarra and Peters, 2003). Finally, although RYMV has been detected at high rates (65–100%) in all parts of the seed in several rice genotypes, there is no evidence of seed transmission (Konate *et al.*, 2001). However, rice seedbeds have been identified as primary sources of RYMV infections (N'Guessan *et al.*, 2000; Traoré *et al.*, 2006). Therefore, good cultural practices – including weed elimination, early control of insect populations by chemicals if necessary, as well as adoption of direct seeding or appropriate management of rice seedbeds – are seen as efficient complementary means of reducing the impact of RYMV in rice fields (see also Séré *et al.*, Chapter 17, this volume).

Rice Resistance to RYMV

Studies on resistance to RYMV generally distinguish two types of resistance: partial and high resistance. Partial resistance is expressed as a delay in virus accumulation and symptom expression (Ioannidou *et al.*, 2000). Over time, virus content eventually reaches the same concentration as in susceptible varieties, but plant growth and development are generally much less affected in partially resistant accessions (Albar *et al.*, 1998; Zouzou *et al.*, 2008). High resistance is characterized by the absence of symptoms, irrespective of the rice development stage at inoculation time (Ndjiondjop *et al.*, 1999). The virus is only detected erratically using enzyme-linked immunosorbent assay (ELISA) serological detection method. Viral RNA concentration, estimated by real-time PCR, is approximately 10^6 times lower in resistant plants compared to susceptible ones (Poulicard *et al.*, 2010). High resistance is associated with the failure of virus cell-to-cell movement (Ndjiondjop *et al.*, 2001).

Partial resistance to RYMV and its use in breeding

The delay of symptom development depends greatly on the virus isolate and on the rice variety. Under controlled conditions, using 2-week-old seedlings infected mechanically, the development of highly susceptible lines stops 2–3 weeks after inoculation and plants may die. In field evaluation, using later inoculation (50 days), symptom progression is less severe, but affects more or less severely plant height, time to flowering, panicle exsertion and spikelet sterility (Awoderu, 1991).

Partial resistance is widely distributed in varieties belonging to the tropical *japonica* group of *O. sativa*, such as Azucena or Moroberekan, which are grown in upland conditions. It has not been reported in *indica* varieties cultivated in lowland conditions. Quantitative trait loci (QTLs) analysis using virus accumulation and symptom-expression data, pinpointed seven chromosomic regions involved in the control of partial resistance. Three of them, located on chromosomes 1, 2 and 12, were consistent across environments and responsible for a major part of the phenotypic variations (Albar *et al.*, 1998; Boisnard *et al.*, 2007). From the late 1980s, the International Institute of Tropical Agriculture (IITA, Nigeria) and AfricaRice crossed partially resistant upland varieties with high-yielding susceptible lowland varieties (ITA212, ITA22, ITA304, ITA230 and ITA306) to improve RYMV resistance (IITA, 1986), but limited positive results were obtained. This is probably related to the strong genetic relationship between plant morphology and resistance (Albar *et al.*, 1998), impairing (to some extent) the full use of partial resistance in breeding for lowland conditions. Partial resistance to RYMV is also present in some *O. glaberrima* accessions with no direct relationship with plant morphology (Thiemélé *et al.*, 2010). This provides a better model to elucidate the genetic basis of partial resistance to RYMV.

High resistance to RYMV, its diversity and genetic basis

High resistance was first described in *O. sativa* accession Gigante originating from East Africa (Ndjiondjop *et al.*, 1999). The genetic basis of resistance in Gigante was identified in the late 1990s: it is recessive and under the control of a single gene, named *RYMV1*. *RYMV1* was mapped on the long arm of chromosome 4 thanks to closely linked SSR markers providing tools for further MAS (Ndjiondjop, 1999; Albar *et al.*, 2003). Positional cloning of *RYMV1* revealed that the gene encodes the translation-initiation factor eIF(iso)4G (Albar *et al.*, 2006) and the resistance is thus associated with a translation-initiation complex, like many other recessive resistances against plant viruses (Robaglia and Caranta, 2006). The resistance allele, *rymv1-2*, is characterized by a single nucleotide polymorphism (SNP), leading to an amino-acid substitution in the conserved central domain of the protein (Table 12.1).

Only one other *O. sativa* accession has been reported that harbours the *rymv1-2* allele: Bekarosaka from Madagascar (Rakotomalala *et al.*, 2008). Conversely, a much larger number of highly resistant *O. glaberrima* accessions have been reported (Thottappilly and Rossel, 1993; Albar *et al.*, 2006; Thiémélé *et al.*, 2010). The survey of a collection of 337 accessions representative of the diversity of *O. glaberrima* identified 29 highly resistant accessions. Analysis of *RYMV1* diversity among these accessions revealed that they did not bear the *rymv1-2* resistance allele. The resistance of 14 of these accessions was due to three new alleles at the *RYMV1* locus, while the remaining 15 accessions did not show any difference from the susceptibility allele, *Rymv1-1*.

The resistance allele *rymv1-3*, found in Tog5681, was characterized by a deletion of three amino acids; the *rymv1-4* allele, found in Tog5672, resulted from one SNP leading to an amino-acid substitution different from that of *rymv1-2*; the *rymv1-5* allele, in Tog5674, presented an amino-acid substitution and a three-amino-acid deletion. The *rymv1-3* and *rymv1-4* alleles were the most frequent with 9 and 4 accessions, respectively, while *rymv1-5* was found in only one accession. The four mutations conferring resistance occurred in the same domain of the eIF(iso)4G factor. Functional analyses involving *rymv1-2* strongly suggested that the mutations impair the interaction between this eIF(iso)4G factor and the VPg of RYMV, a viral protein linked to the 5'-end of the viral RNA, and thus impair the infection cycle of

Table 12.1. Variability of the *RYMV1* gene: alignment of the 299–339 region of the *RYMV1* product (part of the conserved MIF4G domain) from susceptible and resistant *O. sativa* and *O. glaberrima* accessions. (From Thiemélé *et al.*, 2010, with kind permission from Springer Science and Business Media.)

Allele	Phenotype[a]	Species	Amino-acid sequence[b]
Rymv1-1	S	*O. sativa*	AFEGAESLRAEIAKLTGPDQEMERRDKERIVKLRTLGNIRL
rymv1-2	R	*O. sativa*	----------K------------------------------
Rymv1-1	R/S	*O. glaberrima*	------D----------------------------------
rymv1-3	R	*O. glaberrima*	------D--------------------***-----------
rymv1-4	R	*O. glaberrima*	------D---------------------K------------
rymv1-5	R	*O. glaberrima*	------D---------N***---------------------

[a]S and R refer to susceptible and resistant phenotypes, respectively; R/S indicates the presence of the variant in both susceptible and resistant accessions.
[b]Only amino acids that are different from the first sequence are reported. Deletions are represented by '*'.

RYMV (Hébrard *et al.*, 2010). A set of allele-specific molecular markers was developed to help MAS (Thiémélé *et al.*, 2010).

The detection of 15 highly resistant accessions that did not show any difference from the susceptibility allele *Rymv1-1* in the central conserved domain of *RYMV1* suggested the existence of a different genetic control of their resistance, especially as six of those accessions did not differ from the susceptibility allele in the whole *RYMV1* sequence. Among these accessions, Tog7291 revealed a second recessive resistance gene, *RYMV2*, recently mapped on the long arm of chromosome 1. Characterization of candidate genes is under way and should allow the identification of the gene (unpublished results). Preliminary results also suggest that a third resistance may control the high resistance of some other *O. glaberrima* accessions.

Resistance and Resistance Breakdown

One of the main problems encountered by breeders in the use of major resistance genes is the ability of pathogens to rapidly evolve and overcome such resistance (Lecoq *et al.*, 2004). Resistance-breakdown frequency is dependent on characteristics of the pathogen (including its mutation rate), the nature of resistance genes, and the number and nature of mutations necessary for virulence acquisition, which may vary depending on the virus isolates.

The breakdown of *RYMV1*-mediated resistance and its molecular mechanisms have been studied by comparing the viral sequences from plants showing generalized symptoms and/or high virus content with the sequences of the same isolate propagated on susceptible lines. Both *rymv1-2* and *rymv1-3* have been overcome experimentally by RYMV isolates (Hébrard *et al.*, 2006; Traoré *et al.*, 2006). Resistance breakdown generally involves mutations in the viral protein genome-linked VPg. Such mutations restore the interaction between the eIF(iso)4G factor and VPg (Hébrard *et al.*, 2010). Most *rymv1-2* resistance-breakdown mutations have occurred on amino-acid 48 of the VPg (Pinel-Galzi *et al.*, 2007), while the *rymv1-3* allele was mostly overcome by mutations occurring at positions 41 and/or 52 of the VPg (Traoré *et al.*, 2010). RYMV strains have contrasting abilities to overcome *rymv1-2* and *rymv1-3* resistance alleles according to the presence of a glutamic acid (E) / threonine (T) at codon 49 of the VPg. Isolates with an E at position 49 (E pathotype) of the VPg are only able to break the *rymv1-2* allele-mediated resistance, while isolates with a T at position 49 (T pathotype) overcome with high frequency the *rymv1-3* allele (Poulicard *et al.*, 2012) and at low frequency the *rymv1-2* allele using specific mutational pathway (Traoré *et al.*, 2010). No direct resistance-breaking was observed in using natural isolates collected under virulent conditions (i.e. on susceptible varieties). This may be explained by the fact that resistance genes are not yet deployed on a large enough scale to give sufficient selective value to resistance-breaking viral strains. It could be also a consequence of possible fitness losses of virulent mutants compared to avirulent ones under natural conditions. Competition experiments using co-inoculations of E and T pathotypes under virulent conditions showed a much better adaptation of T pathotype on *O. glaberrima* accessions compared to *O. sativa* (Poulicard *et al.*, 2010).

Introgression of *rymv1-2* Resistance Allele into Elite Cultivars of West Africa

Four varieties (IR64, FKR28, Sahelika and IR47686), widely grown in West Africa but susceptible to RYMV, were selected as recipient parents of the resistance allele *rymv1-2*, the donor parent being Gigante. IR64 is a popular and improved *indica* variety from the Philippines, FKR28 and Sahelika are popular and improved *indica* varieties from Burkina Faso and Mali. IR47686 is a *japonica* variety showing partial resistance to RYMV, and therefore a good candidate for the introgression of the major resistance gene with the final goal of durable resistance to RYMV. These varieties were recommended by the national agricultural research systems (NARS) of Mali, Burkina Faso, Guinea and The Gambia, respectively. The marker RM252 (mapping 1.8 cM downstream from *RYMV1* and polymorphic among the donor and recipient parents) was selected for marker-assisted introgression (Fig. 12.1).

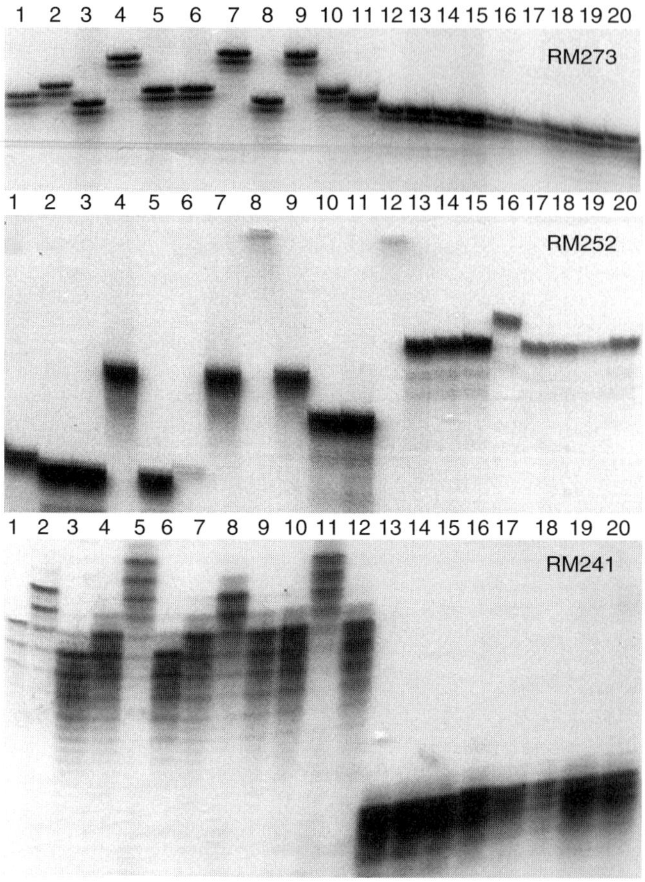

Fig. 12.1. Genetic diversity observed at three SSR loci spanning the *RYMV1* resistance gene in different varieties of rice (*O. sativa* and *O. glaberrima*). *Oryza sativa* subsp. *japonica* (1–6): 1, Haw Mom; 2, Cana Roxa; 3, Carolina Gold; 4, Pate Blanc; 5, Azucena; 6, Nipponbare. *Oryza sativa* subsp. *indica* (7–12): 7, IR64; 8, Gigante; 9, IR5; 10, MTU9; 11, ASD1; 12, Carreon. *Oryza glaberrima* (13–20): 13, Tog5681; 14, Tog5672; 15, Tog5673; 16, IG10; 17, SG329; 18, CG14; 19, CG17; 20, CG20. Resistance alleles of *RYMV1* are present in accessions Gigante, Tog5681 and Tog5672. (From Albar *et al.*, 2003, with kind permission from Springer Science and Business Media.)

DNA extraction and polymerase chain reaction (PCR) were carried out at AfricaRice, Cotonou (Benin), using the protocol described by Ikeda *et al.* (2001) for rice SSR. Amplified products were separated by electrophoresis in 5% polyacrylamide gels using a SequiGEN 38 × 50 cm gel apparatus (BioRad Laboratories) and the banding patterns were visualized using silver staining.

F_1 hybrids obtained by crossing Gigante (RYMV-resistant donor parent) with the four RYMV-susceptible elite lines (recurrent parent) were used to develop BC_1F_1 populations. The BC_1F_1 plants were screened for the presence/absence of the *RYMV1* resistance allele using RM252. Thirty BC_1F_1 plants heterozygous for RM252 were obtained from the four crosses: 3 for IR64/Gigante//IR64, 6 for Sahelika/Gigante//Sahelika, 8 for IR47/Gigante//IR47 and 13 for FKR28/Gigante//FKR28. Twenty-five of those BC_1F_1 plants were used to continue the backcrossing procedure (Table 12.2). BC_2F_1 plants were screened for heterozygous status on RM252 and 35 BC_2F_1 were selected and backcrossed to produce the BC_3F_1. The BC_3F_1 plants

Table 12.2. Workflow summary of the plant material developed for creating RYMV-resistant isolines by marker-assisted selection (no. plants derived at each generation).

		Foreground (RM252)	Foreground (RM252)	Foreground (RM252–RM241–RM73) and background	Foreground + phenotypic	Phenotypic
Recurrent parent	Selection generation	BC_1	BC_2	BC_3	BC_3F_2	BC_3F_5 (NILs)
FKR28		9	4	45	3	5
IR47		8	12	4	5	9
IR64		2	10	80	7	12
Sahelika		6	9	10	3	4
Total		25	35	139	18	30

were screened for RM252. Next, foreground screening was completed with two additional SSR markers – RM241 and RM273 located on opposite sides of RM252 and RYMV1 (Fig. 12.1) for the combinations involving IR64, Sahelika and FKR28; and RM255 and RM451 for the combination of Gigante and IR47. This step confirmed the efficiency of the selection on RM252 and avoided the selection of plants that may have recombined between RM252 and *RYMV1*. One hundred and thirty-nine BC_3F_1 plants heterozygous for the three markers surrounding *RYMV1* were selected for background selection. The return to recurrent parent in genome regions other than the *RYMV1* locus, was surveyed using 69–80 SSR markers well distributed among the chromosomes. The average number of markers per non-carrier chromosome ranged from 5 to 7, depending on the cross combination. Four SSR markers were used for the carrier chromosome. Finally, 24 BC_3F_1 individuals were selected with more than 94% return to the recurrent parent. From these individuals, BC_3F_2 plants were developed and 18 of these, bearing the homozygous *rymv1-2* allele (checked using RM252), were selected. At this stage, selection was also completed by visual observation to keep the individuals as close as possible to the recipient parent. These lines were advanced to F_3 (Plate 5). After a final round of phenotypic selection, 30 BC_3F_3 lines homozygous for *rymv1-2* were selected and used to develop BC_3F_5 lines.

The 30 BC_3F_5 lines were checked for the effectiveness of high resistance to RYMV under controlled conditions at AfricaRice, Cotonou, using artificial infection with virus isolate B27 (S1 strain). Resistance to RYMV was visually scored as described by Konate *et al.* (1997) and virus content was measured by ELISA as described by Sere *et al.* (2007). All BC_3F_5 lines displayed resistant phenotype, similar to that of the donor Gigante (Table 12.3), whereas IR64 and Sahelika were susceptible and IR47 and FKR28 showed intermediate phenotype. The BC_3F_5 lines were also evaluated in field trials, in Guinea and Mali, for general adaptability traits (plant height, phenology, grain weight) and resistance to RYMV. In Guinea, the evaluation was conducted under natural infection in a field often infected by RYMV. In Mali, only 12 IR64 BC_3F_5 lines were evaluated: 2-week-old seedlings of lines and the two parents were first inoculated artificially with a local virus strain, and then transplanted into an often-infected field. For each trial (laid in a randomized complete block design with four replications), a local variety was added as control. In both trials, the BC_3F_5 lines did not show any visual symptoms of RYMV disease. Five BC_3F_5 lines were discarded because of poor agronomic performance. The selected BC_3F_5 lines also showed resistance to RYMV under controlled conditions in Burkina Faso, Côte d'Ivoire and under initial field testing in other countries, including Côte d'Ivoire, Ghana, Nigeria and Sierra Leone.

Conclusion and Perspectives

The fine-mapping and positional cloning of the gene *RYMV1* in the *O. sativa* variety Gigante

Table 12.3. Evaluation of BC_3F_5 progenies for resistance to RYMV.

Near-isogenic line (NIL)/Parental line	Recurrent parent	Greenhouse evaluation (Benin)	
		OD ELISA[a]	Visual symptom
NIL2	FRK28	0.01937 a	R
NIL4	FRK28	0.02661 a	R
NIL5	FRK28	0.02956 a	R
NIL16	Sahelika	0.04689 a	R
NIL36	IR47686	0.01628 a	R
NIL46	IR47686	0.034 a	R
NIL48	IR47686	0.03433 a	R
NIL49	IR47686	0.02772 a	R
NIL54	IR47686	0.03667 a	R
NIL56	IR47686	0.03314 a	R
NIL58	IR47686	0.03589 a	R
NIL59	IR47686	0.022 a	R
NIL127	IR64	0.02022 a	R
NIL129	IR64	0.01633 a	R
NIL130	IR64	0.02872 a	R
NIL132	IR64	0.02367 a	R
NIL133	IR64	0.03467 a	R
NIL135	IR64	0.03544 a	R
NIL145	IR64	0.03433 a	R
NIL147	IR64	0.03489 a	R
NIL157	IR64	0.03422 a	R
Sahelika	Recurrent parent	0.10133b	S
FKR28	Recurrent parent	0.03689 a	PR
IR47	Recurrent parent	0.02678 a	PR
IR64	Recurrent parent	0.09867 b	S
Gigante	Donor	0.008 a	R

[a]OD ELISA, Optical density after enzyme-linked immunosorbent assay; OD values followed by same letter are not significantly different at 5% (Tukey's test); PR, partially resistant; R, resistant; S, susceptible.

have facilitated the introgression of the resistance allele, *rymv1-2*, into four elite cultivars from AfricaRice's NARS partners. Thirty rice lines bearing the *rymv1-2* allele were successfully obtained through MAS and distributed to NARS breeders. The transfer of the high-resistance allele and its functionality (i.e. absence of symptoms or virus in the leaves) in the new genetic backgrounds was confirmed through phenotypic evaluation after inoculation. These lines will be put through Africa-wide Rice Breeding Task Force multi-location evaluation, testing the durability of resistance in various geographical areas (see Kumashiro *et al.*, Chapter 5, this volume).

As *O. glaberrima* shows a greater diversity of *RYMV1* alleles, a similar process of resistance-transferring can be undertaken to diversify the resistance alleles according to the distribution of E and T pathotypes of the virus. For instance, deployment of resistant lines carrying *rymv1-3* is expected to be durable in East Africa, since the E pathotype (the only pathotype present in this region) is unable to break *rymv1-3* resistance. A second resistance gene (*RYMV2*) to RYMV was found in *O. glaberrima* and the existence of a third gene is suspected in the same species. Confirmation of the existence of these genes will pave the way for pyramiding different resistance genes and combining different resistance alleles, to improve the durability of resistance.

AfricaRice's strategy is to build upon the new RYMV-resistant lines by pyramiding with well-characterized and validated resistance genes for other constraints (biotic and abiotic) prevalent in African farmers' fields. The development, by Institut de recherche pour le développement (IRD) and AfricaRice, of fertile interspecific bridge lines bearing favourable genes from *O. glaberrima* associated with useful traits will enhance the sustainability of this approach and its extension to other rice cultivars. AfricaRice also deems that capacity-building for MAS and more generally for modern plant breeding is of great importance. The Center is planning to establish a genomic platform to allow systematic use of molecular markers in breeding for a wide range of traits. New horizons are also emerging with the availability of the sequence of the *O. glaberrima* genome.

References

Abo, M.E. and Sy, A.A. (1998) Rice virus diseases: epidemiology and management strategies. *Journal of Sustainable Agriculture* 11, 113–134.

Abo, M.E., Alegbejo, M.D., Sy, A.A. and Misari, S.M. (2000) An overview of the mode of transmission, host plants and methods of detection of *Rice yellow mottle virus. Journal of Sustainable Agriculture* 17, 19–36.

Agarcio, J.S., Borines, L.M., Tabanao, D.A., Ordonez, S.A., Baliuag, N.M., Dela Cruz, A., Rico, E.P., Natural, M.P., Porter, B.W., White, F.F., Leung, H., Vera Cruz, C.M. and Redona, E.D. (2007) Improving resistance of hybrid rice parental lines to bacterial blight. In: *Hybrid Rice and Agroecosystem.* Proceedings of the JSPS International Seminar 2007, 22–25 November 2007, Hanoi University of Agriculture, Vietnam. Japan Society for the Promotion of Science, pp. 67–74.

Albar, L., Lorieux, M., Ahmadi, N., Rimbault, I., Pinel, A., Sy, A.A., Fargette, D. and Ghesquière, A. (1998) Genetic basis and mapping of the resistance to *Rice yellow mottle virus.* I. QTLs identification and relationship between resistance and plant morphology. *Theoretical and Applied Genetics* 97, 1145–1154.

Albar, L., Ndjiondjop M.-N., Esshak, Z., Berger, A., Pinel, A., Jones, M. and Fargette, D. (2003) Fine genetic mapping of a gene required for *Rice yellow mottle virus* cell-to-cell movement. *Theoretical and Applied Genetics* 107, 371–378.

Albar, L., Bangratz-Reyser, M., Hebrard, E., Ndjiondjop, M.-N., Jones, M. and Ghesquière, A. (2006) Mutations in the eIF(iso)4G translation initiation factor confer high resistance of rice to *Rice yellow mottle virus. The Plant Journal* 47, 417–426.

Awoderu, V.A. (1991) *Rice yellow mottle virus* in West Africa. *Tropical Pest Management* 37, 356–362.

Awoderu, V.A., Alam, M.S., Thottappilly, G. and Alluri, K. (1987) Outbreaks and new records. Ivory Coast. *Rice yellow mottle virus* in upland rice. *FAO Plant Protection Bulletin* 35, 32–33.

Bakker, W. (1970) Rice yellow mottle, a mechanically transmissible virus disease of rice in Kenya. *Netherlands Journal of Plant Pathology* 76, 53–63.

Bakker, W. (1971) Three new beetle vectors of *Rice yellow mottle virus* in Kenya. *European Journal of Plant Pathology* 77, 201–206.

Banwo, O.O. (2003) *Rice yellow mottle virus* (RYMV) disease: a national problem in Tanzania. *Acta Phytopathologica et Entomologica Hungarica* 38, 99–107.

Boisnard, A., Albar, L., Thiémélé, D., Rondeau, M. and Ghesquière, A. (2007) Evaluation of genes from eIF4E and eIF4G multigenic families as potential candidates for partial resistance QTLs to *Rice yellow mottle virus* in rice. *Theoretical and Applied Genetics* 116, 53–62.

Bonneau, C., Brugidou, C., Chen, L., Beachy, R.N. and Fauquet, C. (1998) Expression of the *Rice yellow mottle virus* P1 protein *in-vitro* and *in-vivo* and its involvement in virus spread. *Virology* 244, 79–86.

Borines, L.M., Espejo, E.O., Piamonte, R.T., Vera Cruz, C.M. and Redona, E. (2008) Development of bacterial blight resistance Mestizo hybrid maintainer and restorer lines through marker-aided backcrossing. *Annals of Tropical Research* 30, 1–21.

Brugidou, C., Holt, C., N'Gon a Yassi, M., Zhang, S., Beachy, R.N. and Fauquet, C. (1995) Synthesis of an infectious full-length cDNA clone of *Rice yellow mottle virus* and mutagenesis of the coat protein. *Virology* 206, 108–115.

Chen, S., Xu, C.G., Lin, X.H. and Zhang, Q. (2000) Improvement of bacterial blight resistance of 'Minghui 63', an elite restorer line of hybrid rice, by molecular marker-assisted selection. *Crop Science* 40, 239–244.

Chen, S., Xu, C.G., Lin, X.H. and Zhang, Q. (2001) Improving bacterial blight resistance of '6078', an elite restorer line of hybrid rice, by molecular marker-assisted selection. *Plant Breeding* 120, 133–137.

Chen, X.W, Li, S.G., Ma, Y.Q., Li, H.Y., Zhou, K.D. and Zhu, L.H. (2004) Marker-assisted selection and pyramiding for three blast resistance genes, *Pi-d(t)1*, *Pi-b*, *Pi-ta2*, in rice. *Sheng wu Gong Cheng xue bao (Chinese journal of biotechnology)* 5, 708–714.

Collard, B.C.Y. and Mackill, D.J. (2008) Marker-assisted selection: an approach for precision plant breeding in the twenty-first century. *Philosophical Transactions of the Royal Society B: Biological Sciences* 36, 557–572.

Futakuchi, K. and Sié, M. (2009) Better exploitation of African rice (*Oryza glaberrima* Steud.) in varietal development for resource-poor farmers in West and Central Africa. *Agicultural Journal* 4, 96–102.

Hébrard, E., Pinel-Galzi, A., Bersoult, A., Sire, C. and Fargette, D. (2006) Emergence of a resistance-breaking isolate of *Rice yellow mottle virus* during serial inoculations is due to a single substitution in the genome-linked viral protein VPg. *The Journal of General Virology* 87, 1369–1373.

Hébrard, E., Poulicard, N., Traoré, O., Wu, H.-C., Albar, L. and Fargette, D. (2010) Direct interaction between the *Rice yellow mottle virus* (RYMV) VPg and the central domain of the rice eIF(iso)4G1 factor correlates with rice susceptibility and RYMV virulence. *Molecular Plant–Microbe Interactions* 23, 1506–1513.

Hibino, H. (1996) Biology and epidemiology of rice viruses. *Annual Review of Phytopathology* 34, 249–274.

Hittalmani, S., Parco, A., Mew, T.W., Zeigler, R.S. and Huang, N. (2000) Fine mapping and DNA marker-assisted pyramiding of the three major genes for blast resistance in rice. *Theoretical and Applied Genetics* 100, 1121–1128.

IITA (1986) *Annual Report and Research Highlights.* International Institute for Tropical Agriculture, Ibadan, Nigeria.

Ikeda, N., Bautista, N.S., Yamada, T., Kamijima, O. and Ishii, T. (2001) Ultra-simple DNA extraction method for marker-assisted selection using microsatellite markers in rice. *Plant Molecular Biology Reporter* 19, 27–32.

Ioannidou, D., Lett, J.M., Pinel, A., Assigbetse, K., Brugidou, C., Ghesquière, A. and Nicole, M. (2000) Responses of *Oryza sativa japonica* sub-species to infection with *Rice yellow mottle virus*. *Physiological and Molecular Plant Pathology* 57, 177–188.

Issaka, S., Basso, A., Sorho, F., Onasanya, A., Haougui, A., Sido, A.Y., Ake, S., Fargette, D. and Séré, Y. (2012) Diagnosis and importance of rice yellow mottle disease epidemics in Niger Republic. *Journal of Applied Biosciences* 50, 3501–3511.

Jena, K.K. and Mackill, D.J. (2008) Molecular markers and their use in marker-assisted selection in rice. *Crop Science* 48, 1266–1276.

Jones, M.P., Audebert, A., Mande, S. and Aluko, K. (1997) Characterization and utilization of *Oryza glaberrima* Steud. in upland rice breeding. In: Jones, M.P., Dingkuhn, M., Johnson, D.E. and Fagade, S.O. (eds) *Interspecific Hybridization: Progress and prospects.* Proceedings of the workshop: Africa/Asia Joint Research on Interspecific Hybridization between the Africa and Asian Rice Species (*O. glaberrima* and *O. sativa*). West Africa Rice Development Association, Bouaké, Côte d'Ivoire, pp. 43–59.

Joseph, M., Gopalakrishnan, S., Sharma, R.K., Singh, V.P., Singh, A.K., Singh, N.K and Mohapatra, T. (2004) Combining bacterial blight resistance and Basmati quality characteristics by phenotypic and molecular marker-assisted selection in rice. *Molecular Breeding* 13, 377–387.

Katiyar, S., Verulkar, S., Chandel, G., Zhang, Y., Huang, B. and Bennett, J. (2001) Genetic analysis and pyramiding of two gall midge resistance genes (*Gm-2* and *Gm-6t*) in rice (*Oryza sativa* L.). *Euphytica* 122, 327–334.

Konate, G., Traore, O. and Coulibaly, M.M. (1997) Characterization of *Rice yellow mottle virus* isolates in Sudano-Sahelian areas. *Archives of Virology* 142, 1117–1124.

Konate, G., Sarra, S. and Traore, O. (2001) *Rice yellow mottle virus* is seed-borne but not seed transmitted in rice seeds. *European Journal of Plant Pathology* 107, 361–364.

Kouassi, N.K., N'Guessan, P., Albar, L., Fauquet, C.M. and Brugidou, C. (2005) Distribution and characterization of *Rice yellow mottle virus*: a threat to African farmers. *Plant Disease* 89, 124–133.

Kumaravadivel, N., Uma, M.D., Saravanan, P.A. and Suresh, H. (2006) Molecular marker-assisted selection and pyramiding genes for gall midge resistance in rice suitable for Tamil Nadu Region. In: *Abstracts: 2nd International Rice Congress 2006*, New Delhi, 9–13 October 2006. p. 257.

Lecoq, H., Moury, B., Desbiez, C., Palloix, A. and Pitrat, M. (2004) Durable virus resistance in plants through conventional approaches: a challenge. *Virus Research* 100, 31–39.

Leung, H., Wu, J., Liu, B., Bustaman, M., Sridhar, R., Singh, K., Redona, E.D., Quang, V.D., Zheng, K., Bernardo, M., Wang, G., Leach, J., Choi, I.R. and Vera Cruz, C.M. (2004) Sustainable disease resistance in rice: current and future strategies. In: Fischer, T., Turner, N., Angus, J., McIntyre, L., Robertson, M., Borrell, A. and Lloyd, D. (eds) *New Directions for a Diverse Planet.* Proceedings of the 4th International Crop Science Congress, Brisbane, Australia, 26 September to 1 October (CD-ROM and online). The Regional Institute, Gosford, Australia.

Narshimulu, G., Jamaloddin, M., Vemireddy, L.R., Anuradha, G. and Siddiq, E. (2011) Potentiality of evenly distributed hypervariable microsatellite markers in marker-assisted breeding of rice. *Plant Breeding* 130, 314–320.

Ndikumana, I., Gasoré, R., Issaka, S., Pinel-Galzi, A., Onasanya, A., Hassani-Mehraban, A., Fargette, D., Peters, D. and Séré, Y. (2011) *Rice yellow mottle virus* in rice in Rwanda: first report and evidence of strain circulation. *New Disease Reports* 23, 18.

Ndjiondjop, M.N. (1999) Heredity, mechanism and mapping of a marker linked to resistance gene of *Rice yellow mottle virus* (RYMV) in rice. PhD thesis, Université des Sciences et Techniques du Languedoc, Montpellier, France.

Ndjiondjop, M.N., Albar, L., Fargette, D., Fauquet, C. and Ghesquière, A. (1999) The genetic basis of high resistance to *Rice yellow mottle virus* (RYMV) in cultivars of two cultivated rice species. *Plant Disease* 83, 931–935.

Ndjiondjop, M.N., Brugidou, C., Zang, S., Fargette, D., Ghesquière, A. and Fauquet, C. (2001) High resistance to *Rice yellow mottle virus* (RYMV) in two cultivated rice cultivars is correlated to the failure of cell-to-cell movement. *Physiological and Molecular Plant Pathology* 59, 309–316.

N'Guessan, P., Pinel, A., Caruana, M.L., Frutos, R., Sy, A., Ghesquière, A. and Fargette, D. (2000) Evidence of the presence of two serotypes of *Rice yellow mottle Sobemovirus* in Côte d'Ivoire. *European Journal of Plant Pathology* 106, 167–178.

Orjuela, J., Garavito, A., Bouniol, M., Arbelaez, J.D., Moreno, L., Kimball, J. and Wilson, G. (2010) A universal core genetic map for rice. *Theoretical and Applied Genetics* 120, 563–572.

Perez, L.M., Redona, E.D., Mendioro, M.S., Vera Cruz, C.M. and Leung, H. (2008) Introgression of *Xa4*, *Xa7* and *Xa21* for resistance to bacterial blight in thermosensitive genetic male sterile rice (*Oryza sativa* L.) for the development of two-line hybrids. *Euphytica* 164, 627–636.

Pinel-Galzi, A., Rakotomalala, M., Sango, E., Sorho, F., Kanyeka, Z., Traoré, O., Séréme, D., Poulicard, N., Rabenantoandro, Y., Séré, Y., Konaté, G., Ghesquière, A., Hébrard, E. and Fargette, D. (2007) Theme and variations in the evolutionary pathways to virulence of an RNA plant virus species. *PLoS Pathogens* 3, 1761–1770.

Poulicard, N., Pinel-Galzi, A., Hébrard, E. and Fargette, D. (2010) Why *Rice yellow mottle virus*, a rapidly evolving RNA plant virus, is not efficient at breaking *rymv1-2* resistance. *Molecular Plant Pathology* 11, 145–154.

Poulicard, N., Pinel-Galzi, A., Traore, O., Vignols, F., Ghesquiere, A., Konate, G., Hebrard, E and Fargette, D. (2012) Historical contingencies modulate the adaptability of *Rice yellow mottle virus*. *PLoS Pathogens* 8, 2482–2490.

Rakotomalala, M., Pinel-Galzi, A., Albar, L., Ghesquière, A., Rabenantoandro, Y., Ramavovololona, P. and Fargette, D. (2008) Resistance to *Rice yellow mottle virus* in rice germplasm in Madagascar. *European Journal of Plant Pathology* 122, 277–286.

Reckhaus, P.M. and Randrianangaly, S. (1990) *Rice yellow mottle virus* (RYMV) on rice in Madagascar. *International Rice Research Newsletter* 15, 30.

Robaglia, C. and Caranta, C. (2006) Translation initiation factors: a weak link in plant RNA virus infection. *Trends of Plant Science* 11, 40–45.

Sarra, S. and Peters, D. (2003) *Rice yellow mottle virus* is transmitted by cows, donkeys, and grass rats in irrigated rice crops. *Plant Disease* 87, 804–808.

Sarra, S., Oevering, P., Guindo, S. and Peters, D. (2004) Wind-mediated spread of *Rice yellow mottle virus* (RYMV) in irrigated rice crops. *Plant Pathology* 53, 148–153.

Sere, Y., Onasanya, A., Akator, K., Afolabi, A. and Abo, M.E. (2007) Serological Differentiation Indices (SDI) and phylogenetic analysis of *Rice yellow mottle virus* isolates in Côte d'Ivoire. *Journal of Biological Science* 7, 1147–1154.

Shanti, M.L., Devi, G.L., Kumar, G.N. and Shashidhar, H.E. (2010) Molecular marker-assisted selection: a tool for insulating parental lines of hybrid rice against bacterial leaf blight. *International Journal of Plant Pathology* 1, 114–123.

Singh, S., Sidhu, J.S., Huang, N., Vikal, Y., Li, Z., Brar, D.S., Dhaliwal, H.S. and Khush, G.S. (2001) Pyramiding three bacterial blight resistance genes (*xa5*, *xa13* and *Xa21*) using marker-assisted selection into *indica* rice cultivar PR 106. *Theoretical and Applied Genetics* 102, 1011–1015.

Sundaram, R.M., Vishnupriya, M.R., Biradar, S.K., Laha, G.S., Shobharani, N., Sarma, M.P. and Sonti, R.V. (2008) Marker assisted introgression of bacterial blight resistance in Sambha Mahsuri, an elite *indica* rice variety. *Euphytica* 160, 411–422.

Sundaram, R.M., Vishnupriya, M.R., Laha, G.S., Rani, N.S., Rao, P.S., Balachandran, S.M., Reddy, G.A., Sarma, N.P. and Sonti, R.V. (2009) Introduction of bacterial blight resistance into Triguna, a high yielding, mid-early duration rice variety. *Biotechnology Journal* 4, 400–407.

Sy, A.A., Alluri, K. and Akator, K. (1994) Selection of RYMV resistant cultivars and estimation of RYMV-induced yield loss under natural and artificial pressure. In: *Annual Report 1993.* West Africa Rice Development Association (WARDA), Bouaké, Côte d'Ivoire. pp. 44–45.

Taylor, D.R., Fofi, A.S. and Suma, M. (1990) Natural infection of *Rice yellow mottle virus* disease (RYMV) on rice in Sierra Leone. *International Rice Research Newsletter* 15, 5–19.

Thiémélé, D., Boisnard, A., Ndjiondjop, M.-N., Chéron, S., Séré, Y., Aké, S., Ghesquière, A. and Albar, L. (2010) Identification of a second major resistance gene to *Rice yellow mottle virus, RYMV2,* in the African cultivated rice species, *O. glaberrima. Theoretical and Applied Genetics* 121, 169–179.

Thottappilly, G. (1992) Plant virus diseases of importance to African agriculture. *Journal of Phytopathology* 134, 265–288.

Thottappilly, G. and Rossel, H.W. (1993) Evaluation of resistance to *Rice yellow mottle virus* in *Oryza* species. *Indian Journal of Virology* 9, 65–73.

Traoré, O., Pinel, A., Fargette, D. and Konaté G. (2001) First report and characterization of *Rice yellow mottle virus* in Central Africa. *Plant Disease* 85, 920.

Traoré, O., Pinel, A., Hébrard, E., Gumedzoé, M.Y.D., Fargette, D., Traoré, A.S. and Konaté, G. (2006) Occurrence of resistance-breaking isolates of *Rice yellow mottle virus* in West and Central Africa. *Plant Disease* 90, 259–263.

Traoré, O., Pinel-Galzi, A., Issaka, S., Poulicard, N., Aribi, J., Aké, S., Ghesquière, A., Séré, Y., Konaté, G., Hébrard, E. and Fargette, D. (2010) The adaptation of *Rice yellow mottle virus* to the eIF(iso)4G-mediated rice resistance. *Virology* 408, 103–108.

Van Damme, V., Gómez-Paniagua, H. and de Vicente, M.C. (2011) The GCP molecular marker toolkit, an instrument for use in breeding food security crops. *Molecular Breeding* 28, 597–610.

Vera Cruz, C.M., Jena, K.K., Ryong, I.C., Brar, D.S. and Leung, H. (2009) Marker-assisted selection for biotic stress tolerance in rice: current status and emerging needs. In: *[Proceedings of the] International Symposium on Rice Research in the Era of Global Warming,* Taichung, Taiwan, October 6–7, 2009. Taiwan Agricultural Research Institute, Council of Agriculture, Taichung, Taiwan.

Voinnet, O., Pinto, Y.M. and Baulcombe, D. (1999) Suppression of gene silencing: a general strategy used by diverse DNA and RNA viruses of plants. *Proceedings of the National Academy of Sciences USA* 23, 14147–14152.

Xu, Y. and Crouch, J.H. (2008) Marker-assisted selection in plant breeding: from publications to practice. *Crop Science* 48, 391–407.

Young, N.D. and Tanksley, S.D. (1989) RFLP analysis of the size of chromosomal segments retained around the *Tm-2* locus of tomato during backcross breeding. *Theoretical and Applied Genetics* 77, 353–359.

Zouzou, M., Hilaire, K.T., Mongomaké, K. and Souley, I. (2008) Screening rice (*Oryza sativa* L.) varieties for resistance to *Rice yellow mottle virus. Scientific Research and Essay* 3, 416–424.

13 Hybrid Rice in Africa: Challenges and Prospects

Raafat A. El-Namaky* and Matty Demont
Africa Rice Center (AfricaRice), Saint-Louis, Senegal

Introduction

The 2008 food crisis and the increasing gap between rice consumption and production in Africa underline the need to accelerate the introduction and adoption of higher-yielding rice varieties. Hybrid rice technology has contributed significantly to food security, environmental protection, and employment opportunities in China since the early 1980s (Yan *et al.*, 2010). Hybrid technology may contribute to food security in Africa through two avenues: (i) exploitation of heterosis in order to increase crop productivity, and (ii) attracting and encouraging private-sector involvement in seed production research and development (R&D) (Naseem *et al.*, 2010). Heterosis, or hybrid vigour, is the occurrence of a superior offspring from mixing the genetic contributions of its parents. In hybrid rice, the extent of heterosis depends on the relationship between the parental lines, and heterosis for grain yield ranges between less than 10% and more than 40% (El-Namaky, 2008). Availability of suitable pollination control systems and the extent of out-crossing between female and male parents are the key factors determining the success of commercial exploitation of heterosis (Mao *et al.*, 1998).

Since the early 1990s, several other countries (Bangladesh, Egypt, India and Vietnam) have developed and introduced hybrid rice technology to their farmers. The availability of adequately trained human resources is an essential prerequisite for developing and using hybrid rice technology. Hybrid rice breeding uses several concepts, skills and procedures that are strikingly different from those used for 'traditional' inbred rice breeding (Virmani *et al.*, 1997).

During the first decade of the 2000s, several African countries (Côte d'Ivoire, Liberia, Madagascar, Mozambique, Nigeria, Tanzania and Uganda) started to evaluate and cultivate rice hybrids from China. Only Egypt has succeeded in developing a hybrid rice breeding programme and produces Egyptian hybrids on a commercial scale. Grain yields obtained with hybrids have been in the order of 12–14 tonnes per hectare (Bastawisi *et al.*, 2003). The programme developed a seed production system for the multiplication of cytoplasmic male sterile (CMS) lines (A/B) and hybrid rice (A/R). Cropping practices and recommendations for CMS multiplication, hybrid seed production and hybrid rice cultivation were developed by the national programme. In addition, the programme features a special strategy to develop

* Corresponding author: r.elnamaky@cgiar.org

© CAB International 2013. *Realizing Africa's Rice Promise* (eds M.C.S. Wopereis *et al.*)

aromatic hybrid rice and hybrids tolerant to salinity and drought.

In this chapter, we present the Africa Rice Center (AfricaRice) strategy for developing hybrid rice for sub-Saharan Africa and its preliminary results.

Hybrid Rice Breeding and Distribution Strategy

Breeding strategy

Building on hybrids developed by partners outside of Africa: the Green Super Rice project

The Green Super Rice (GSR) project (2009–2011) was jointly coordinated by the Chinese Academy of Agricultural Sciences (CAAS), AfricaRice and the International Rice Research Institute (IRRI), and implemented in seven countries in Asia, six provinces in China and eight African countries (Liberia, Mali, Mozambique, Nigeria, Rwanda, Senegal, Tanzania and Uganda). The development and cultivation of GSR hybrid and inbred cultivars with high nutrient use efficiency and stress resistances are expected to provide a sustainable way of reducing food insecurity and poverty in sub-Saharan Africa and Asia. Building on the tremendous capacity in breeding and genomic technology of China, the short-term (3-year) goals of the project were to enter at least 15 rice varieties in national trials in project countries and develop capacity in hybrid seed production.

The **general strategy** was to identify Chinese hybrid lines that have good adaptability and high yield potential for the rainfed lowland and irrigated areas of sub-Saharan Africa, and further improve their tolerance to major stresses – including drought, nitrogen (N) and phosphorus (P) deficiencies, pests (e.g. African rice gall midge) and diseases (e.g. blast, bacterial leaf blight, *Rice yellow mottle virus*) – using conventional and molecular breeding approaches.

Developing an in-house hybrid rice breeding programme

Started in 2010 dry season, the programme is based at AfricaRice Sahel station in Saint-Louis, Senegal. The aim is to: (i) evaluate the performance of rice hybrids in multi-locational yield trials under African conditions; (ii) develop new parental lines from local varieties; (iii) determine adaptability of some CMS lines in Africa; and (iv) establish a hybrid rice seed-production system in, among others, Senegal and Mali.

Hybrid seed production and marketing strategy

The technology of hybrid rice seed production has been developed and practised successfully not only in China, but also in many other countries with temperate and tropical conditions. Technologically, there should be no serious barrier to hybrid rice seed production in Africa. The 1–2 t/ha seed yield obtained in many countries under different conditions is economically viable. National partners involved in the seed sector in Senegal and Mali are being trained in hybrid seed production. Furthermore, farmers, private sector and NGOs involved in the rice sector will be sensitized about hybrid rice technology and trained in hybrid rice technology and seed production. Various models will be established for hybrid rice seed production and industry in different countries and regions.

Preliminary Results

Multi-locational evaluation of hybrids

Experimental methods

In 2010, some 122 hybrids and eight checks were evaluated in observational yield trials in six African countries (Mali, Mozambique, Nigeria, Senegal, Tanzania and Uganda) using an augmented design and local fertilizer recommendations. Each block consisted of 17 plots with replicated checks included in 10 blocks. At the same time, the varieties were also evaluated under greenhouse conditions for resistance to the main rice diseases and insect pests in Africa – namely bacterial leaf blight (BLB), blast and *Rice yellow mottle virus* (RYMV) in Benin, and African rice gall midge (AfRGM) in Ibadan, Nigeria.

Yield, days to maturity and yield-component data were analysed using SAS mixed model under SAS/STAT 9.2 (Littell *et al.*, 1996). Multiple comparison adjustment for the *P*-values was then performed to test differences between hybrids and the best check for each country at a significance level of $P < 0.05$.

Results

Very highly significant genotype (G) by environment (E) or country interactions were observed for yield, yield components and crop duration, indicating that the hybrids and check varieties performed differently across environments (Table 13.1). Given this G×E interaction, hybrid performance was analysed on country-by-country basis.

The mean grain yield of the 122 hybrids was lower than the check variety in all countries except Mozambique (Table 13.2), while mean grain yield of the top ten hybrids in each country was higher than the yield of the best check variety in the same country. Non-significant differences for days to 50% flowering were observed between the mean of the 122 hybrids, the mean of the top ten hybrids and the local check, except in Mozambique where the local check was of longer duration (101 days) and in Nigeria where mean duration of the ten top hybrids was 106 days. QY1, HanF1-35, 3LYR24, HLY9348, GLYR24 and HXY836 were the best hybrids in Senegal, Mali, Nigeria, Tanzania, Uganda and Mozambique, respectively, with grain yields of 10.55, 12.50, 5.12, 8.20, 8.56 and 3.24 t/ha, respectively. Milling rate for the best hybrids ranged between 61.7% and 68.0%. All of the promising hybrids were long grained (grains 6.3–7.2 mm long). Regarding disease and insect resistance, all promising hybrids were susceptible to AfRGM under natural conditions, but only QY1 was highly susceptible to blast under natural infestation. All promising hybrids were susceptible to RYMV under artificial inoculation in screen house.

These preliminary results confirmed the fact that while some hybrids developed in Asia can perform well in Africa for yield potential, given their susceptibility to African diseases it would not be prudent to rely on them for commercial rice production. Developing local hybrids with high yield potential and resistance to biotic stresses will be more efficient. Many countries, such as Egypt, India, Pakistan, Philippines and Vietnam, have succeeded in establishing hybrid rice programmes and releasing local hybrids.

Table 13.1. Analysis of variance results for yield of 122 F_1 hybrid rice varieties across six countries (Mali, Mozambique, Nigeria, Senegal, Tanzania and Uganda). (Data from GSR project.)

Effect	Num DF[a]	Den DF[b]	F-Value	Prob F
Genotype (hybrids)	136	199	3.4529	<0.0001
Environment (country)	5	30	118.2648	<0.0001
G×E (hybrids × country)	596	199	2.2987	<0.0001

[a]Numerator and [b]denominator degrees of freedom associated with the *F* statistic (SAS mixed model uses a likelihood-based estimation scheme, it does not directly calculate or display sums of squares).

Developing new hybrids

In order to develop new hybrids with high yield potential and tolerance to biotic and abiotic stresses at AfricaRice, 300 test crosses were performed between three CMS lines (IR68886A, IR69625A and IR58025A) and various African varieties with the aim of identifying new restorer and maintainer varieties from local materials resistant to insects and diseases. The preliminary results indicated that there were 15 promising restorer lines with good phenotypic acceptability and with grain yield of 25.8–70.7 g/plant for the F_1 fertile combinations. At the same time, hybrids with about ten varieties showed complete sterility with the CMS line, which indicates that these varieties could be used as potential maintainer lines. Backcrosses were performed with three NERICA-L varieties to transfer cytoplasmic male sterility into these varieties. However, sterility of the NERICA-L varieties was not stable, so they were replaced by three breeding lines (ARSH 46-6-3-B, ARSH-23-1-2-2 and ARSH -23 3-1-2). The same three CMS lines were used with some local restorers like Sahel 108, Sahel 134, Giza 178 and Giza 182 to establish small plots for hybrid seed production (Fig.13.1). To increase cross-pollination, flag leaves were cut, a rope was pulled across the rice plants (Fig. 13.2) and gibberillic acid was applied. These seed-production experiments yielded 1–2.5 tonnes of hybrid seeds per hectare.

Table 13.2. Comparison of mean grain yield, days to 50% flowering, grain quality, insect and disease resistance of 122 hybrids, top ten hybrids, best hybrid and best check varieties in six sub-Saharan African countries.

	Check variety			122 hybrids				Best ten hybrids[a]				Best hybrid							
				Yield (t/ha)		DTF		Yield (t/ha)		DTF							Resistances to diseases and pests		
Country	Name	Yield (t/ha)	DTF	Mean	SD	Mean	SD	Mean	SD	Mean	SD	Name	Yield (t/ha)	DTF	Milling rate (%)	Grain length (mm)	AfRGM	Blast	RYMV
Senegal	Sahel 108	7.35	87.5	7.09	1.68	83.6	3.9	9.78	6.5	84	4.6	QYI	10.55	80.0	68.0	6.30	S	HS	S
Mali	WITA9	10.23	96.5	6.33	2.85	95.8	7.8	11.20	0.75	93.9	7.4	HanF1-35	12.50	88.1	66.4	6.72	S	MR	S
Nigeria	WITA4	3.23	81.5	2.55	1.01	92.8	5.4	4.61	0.63	106	7.7	3LYR24	5.12	112.0	61.7	7.19	S	R	S
Tanzania	WITA9	5.13	88	4.08	1.49	81	8.3	6.98	0.76	84	9.8	HYL9348	8.20	85.0	65.3	6.50	S	R	S
Uganda	NERICA-L 19	5.54	81.7	5.00	1.55	80.1	1.4	7.73	0.47	80.0	0.0	GLYR24	8.56	80.5	63.2	7.17	S	R	S
Mozambique	WITA9	1.83	100.7	1.84	0.45	86.91	4.2	2.74	0.25	85.2	5.8	HXY836	3.24	87.0	64.7	6.85	S	R	S

DTF, days to 50% flowering; SD, standard deviation; MR, moderately resistant; HS, highly susceptible; R, resistant; S, susceptible.
[a]Average of best ten hybrids.

Fig. 13.1. Small plots of hybrid seed production, Ndiaye, Senegal, 2010. (Photo: R. El-Namaky.)

Fig. 13.2. Using a rope for supplementary pollination in a small plot for hybrid seed production, Fanaye, Senegal, 2010. (Photo: R. El-Namaky.)

Challenges and Prospects

Hybrids are one of the key forms of biological intellectual property rights (IPR) in agriculture. Because the yield gains conferred by heterosis tend to decline dramatically after the first generation (F_1) of seed, farmers must purchase new F_1 seed each season to continually capture such yield gains. The major challenges of hybrid rice are high seed cost, necessity of changing seed every crop season, and continuous dependence on the external source for supply of seed. Establishing efficient infrastructure for seed production, processing, certification and distribution are very important. This cannot be done successfully without an efficient seed industry, very-high-performing hybrids (compared to the best inbred varieties available), and a robust training programme for farmers, and technicians in the public and private sectors.

Testing of Chinese hybrid lines in Africa showed promise in terms of high yields but susceptibility to African insect pests and diseases. Progress made in the hybrid rice research programme at AfricaRice, Senegal will allow testing of AfricaRice developed hybrids in 2014.

We conclude that hybrid technology not only offers the potential of increasing rice productivity, but can also serve as a mechanism for leveraging private-sector investment in rice R&D in Africa.

References

Bastawisi, A.O., El-Mowafi, H.F., Abo Yousef, M.I., Draz, A.E., Aidy, I.R., Maximos, M.A. and Badawi, A.T. (2003) Hybrid rice research and development in Egypt. In: Virmani, S.S., Mao, C.X. and Hardy, B. (eds) *Hybrid Rice for Food Security, Poverty Alleviation and Environmental Protection*, Proceedings of the 4th International Symposium on Hybrid Rice, Hanoi, Vietnam, 14–17 May 2002. International Rice Research Institute, Los Baños, Philippines, pp. 257–263.

El-Namaky, R.A. (2008) Genetic studies on using wide compatibility gene in hybrid rice breeding. PhD thesis, Faculty of Agriculture, Mansoura University, Egypt.

Littell, R.C., Milliken, G.A., Stroup, W.W. and Wolfinger, R.D. (1996). *SAS System for Mixed Models*. SAS Institute, Cary, North Carolina.

Mao, C.X., Virmani, S.S. and Kumar, I. (1998) Technological innovations to lower the costs of hybrid rice seed production. In: Virmani, S.S., Siddiq, E.A. and Muralidharan, K. (eds) *Advances in Hybrid Rice Technology*, Proceedings of the 3rd International Symposium on Hybrid Rice, Hyderabad, India, 14–16 November 1996. International Rice Research Institute, Los Baños, Philippines, pp. 111–128.

Naseem, A., Spielman, D.J. and Omamo, S.W. (2010) Private-sector investment in R&D: a review of policy options to promote its growth in developing-country agriculture. *Agribusiness* 26, 143–173.

Virmani, S.S., Viraktamath, B.C., Casal, C.L., Toledo, R.S., Lopez, M.T. and Manalo, J.O. (1997) *Hybrid Rice Breeding Manual*. International Rice Research Institute, Los Baños, Philippines.

Yan, W., Chen Zhixin and Hong Haoru (2010) A short review of the hybrid rice seed industry in China. In: Xie, F. and Hardy, B. (eds) *Accelerating Hybrid Rice Development*. International Rice Research Institute, Los Baños, Philippines, pp. 517–522.

14 Development of an Integrated Rice Seed Sector in Sub-Saharan Africa: Meeting the Needs of Farmers

Amadou M. Bèye,[1]* Thomas Remington,[2] Marco C.S. Wopereis[3] and Aliou Diagne[3]

[1]*Africa Rice Center (AfricaRice), Abidjan, Côte d'Ivoire;* [2]*Catholic Relief Services, Bujumbura, Burundi;* [3]*Africa Rice Center (AfricaRice), Cotonou, Benin*

Introduction

Seeds are the backbone of agricultural production. Despite this importance, however, rice farmers in Africa lack assured access to sufficient, good-quality seed of preferred varieties in time for sowing. In the 1970s and 1980s, public-sector seed programmes in sub-Saharan Africa generally promoted the dissemination of improved rice varieties. With the structural reforms of the 1990s, the seed sector was liberalized, though the private sector has only partly replaced the public sector in providing seed to farmers. With the growing awareness that promoting rice production in Africa is crucial for economic growth, food security and social stability, 'seed' is firmly back on the agenda of many governments and technical and financial rice development partners. This became particularly evident after the 2008 food crisis, which was manifested as a 'rice crisis' in many African countries (Viatte *et al.*, 2009).

Farmers acquire rice seed through 'formal' and 'informal' channels. The formal system includes both the public, or government, and the private, or commercial, seed sectors (Louwaars, 1994; Bay, 1998; Louwaars and De Boef, 2012). The 'informal' or 'farmer seed' system includes farmers managing their own seed, but also informal seed trade among farmers and purchase from the paddy grain market.

Establishing commercially viable seed systems for rice is particularly challenging in sub-Saharan Africa, because of the predominance of farm-saved seed for rice crops, the easy production of rice seed as it is self-pollinating, the complexity of African rice cropping systems and the great diversity of rice varieties.

In most of West and Central Africa, formal rice seed systems have been more development- than market-oriented. Such seed initiatives have the development goal of assisting farmers in accessing seed of new varieties, rather than a commercial goal of creating profitable seed enterprises. They are often characterized by heavy and inefficient bureaucratic structures within classic seed regulatory frameworks. These constraints are exacerbated by slow processes of variety development and evaluation (addressed in Kumashiro *et al.*, Chapter 5, this volume), and slow variety release and registration (addressed in Sanni *et al.*, Chapter 6, this volume).

In East and Southern Africa, examples exist of well-functioning commercially viable seed

* Corresponding author: a.beye@cgiar.org

© CAB International 2013. *Realizing Africa's Rice Promise* (eds M.C.S. Wopereis *et al.*)

companies producing quality rice seed. Such seed companies often already produce hybrid maize seed and other cash crops and have added rice to their market portfolio, responding to a rapidly growing commercial rice sector in the region.

As part of structural adjustment reforms, formal seed systems have gone through several changes across sub-Saharan Africa. Public-sector seed services were reduced or eliminated and the private sector encouraged to assume a greater role, particularly in production and marketing of certified seed. This drive to 'privatize' the seed system has not succeeded and has rather left a void that has been partly filled by NGOs, development projects and farmers' associations, which have been providing smallholder farmers with free or subsidized seed. In the aftermath of the 2008 food crisis, there are renewed efforts to establish well-functioning seed systems. Though often not explicitly stated, this shift from a public to a private seed sector is also a shift from a 'development driver' to get new varieties to farmers to a 'commercial driver' to create profitable and sustainable seed enterprises (Louwaars, 2007).

This chapter discusses formal and informal seed systems and those with attributes of both, and describes the concept of integrated seed-sector development for rice based on Louwaars and De Boef (2012). This framework is discussed in the context of rice-sector development in Africa, highlighting roles and responsibilities of different actors and stakeholders.

African Seed Systems

African rice seed systems are characterized by the coexistence of a formal sector where seed is produced and commercialized by government agencies and private seed companies, and an informal sector where seed is produced by and exchanged among farmers.

Formal seed system

The formal seed system (also called the conventional seed system) is designed to provide certified seed. It relies on a well-organized seed system in which the original breeder-supplied seed is multiplied through a series of stages (from breeder seed to foundation seed to certified seed) to obtain sufficient commercial seed. The use of release and seed certification procedures and the intervention of the processing industry[1] are the backbones of this system.

The functioning of the seed system is influenced by a large number of regulations in research (research protocols and variety maintenance), variety release procedures (DUS and VCU[2] evaluation procedures), seed production rules (production norms – isolation, presence of off-types, weeds, plant health protection, harvesting, etc.), trade (traceability), economics (marketability and profitability) and regulatory organs (seed boards, national release committees, national seed services).

Quality control aims to protect the interests of farmer customers. It is monitored by periodically inspecting seed fields and the seed dealer points (markets and other sales points), and has a constant vigil on the seed marketed by collecting seed samples to be analysed. Seed certification is a legally sanctioned system for the quality control of seed for sale. It is carried out by national seed services (NSS) through field inspections and laboratory analyses. Norms are adapted from international rules set by the International Seed Testing Association (ISTA). The NSS control the functioning of the whole system (delivery of agreements to seed growers and private seed companies, supervision of seed production and conditioning, certification, monitoring of commercialization, update of catalogues of varieties, etc.).

The system is typically managed by the ministries of agriculture. It faces many constraints, including: (i) limited supply of breeder seed; (ii) poor seed quality control; (iii) poor demand estimation; (iv) inadequate marketing and distribution systems; and (v) reluctance of small-scale rice farmers to pay premium prices for certified seed. This has led to a serious 'rice seed production gap' in many countries, especially in West and Central Africa.

Informal seed systems

Informal seed systems are traditional systems operating at the local and village level through farmer seed production and seed exchange mechanisms based on local considerations without public-sector regulation or support. Within the household it is often women who manage the rice seed. Harvesting by hand, with panicle-selection, facilitates positive selection and seed can then be stored as panicle bunches, which prevents accidental physical mixture of varieties. Often, farmers manage seed of multiple varieties of different durations and stature suitable to different landscape positions and hydrological levels.

Individual farmers, and sometimes farmers' groups, obtain seed from their own harvest, their family, friends and relatives, or purchase seed on the local village market or from local paddy traders. These are individual transactions where the farmer usually knows the seed seller and is able to verify the origin of the seed being considered for purchase. Confidence in the seller is a proxy for seed certification and can perhaps be referred to as 'social certification'. The rate of adoption of improved varieties is generally low and ultimately dependent on access to seed of these varieties from the formal seed system. However, farmer seed systems offer a range of traditional and improved varieties that are accessible, of acceptable quality, and affordable and obtainable without cash transactions.

Informal seed systems do not respond to public norms and consequently do not receive the needed attention from policy makers, scientists and the general public. It is important to develop policies that recognize and support informal seed systems, their contribution to *in-situ* conservation and on-farm management of agro-biodiversity, the efficient diffusion of varieties, and the appropriate protection of farmers' and communities' rights.

Where are African farmers getting their rice seed?

Surveys conducted in 2009 by the Africa Rice Center (AfricaRice) and national (NARS) partners in 16 countries in sub-Saharan Africa, involving more than 30,000 farming households, provide a good source of information on seed access by rice farmers (Bonou *et al.*, 2012). About 90% of the farmers use traditional varieties, and 75% of those using improved varieties use 'informal' seed systems (Fig. 14.1).

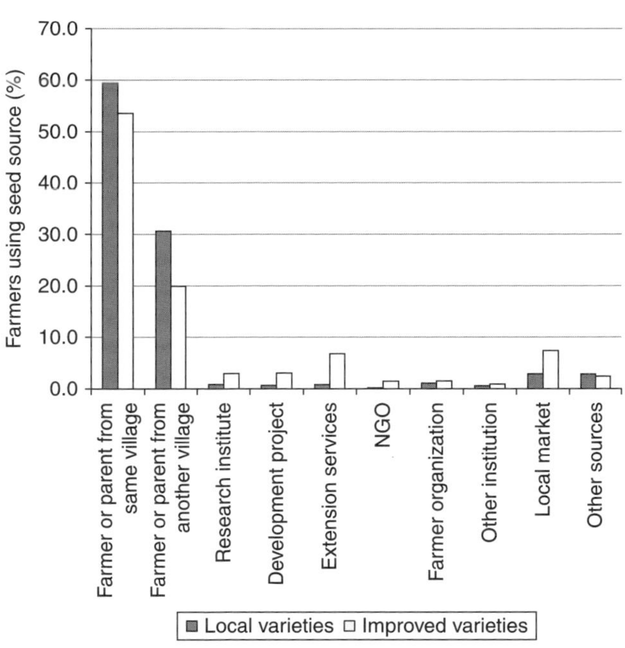

Fig. 14.1. Sources of access to seed by rice farmers in Africa. (Adapted from Bonou *et al.*, 2012.)

These farmers obtained seed from their previous harvest or they bought, exchanged or received seed from other farmers within their own village or from neighbouring villages. Other possible seed sources included development projects, NGOs and government extension agencies. Less than 10% of the farmers indicated that they had obtained seed of improved rice varieties from a local market. Percentages were even lower for traditional varieties (Fig. 14.1). These figures show the informal sector to be the dominant source of seed for African rice farmers.

Bonou *et al.* (2012) also analysed farmer involvement in different types of seed transactions, i.e. the extent to which farmers use their own saved seed (farmer-saved seed) and are engaged in different market and non-market seed transaction activities. Use of farmer-saved seed was reported by more than 52% of farmers cultivating traditional varieties and 44% of farmers cultivating improved varieties; purchasing of seed was reported by more than 23% of farmers cultivating traditional varieties and more than 26% of farmers cultivating improved varieties; selling seed was reported by more than 24% of farmers cultivating traditional varieties and more than 23% of farmers cultivating improved varieties; giving seed free to other farmers was reported by about 35% of farmers cultivating traditional varieties and more than 34% of farmers cultivating improved varieties; receipt of free seed was reported by more than 20% of farmers cultivating traditional varieties and 25% of farmers cultivating improved varieties; and exchange of seed was reported by 20% of farmers cultivating traditional varieties and more than 19% of farmers cultivating improved varieties. Thus, consistent with the figures on the sources of seed, the majority of farmer seed transactions are informal – for both traditional and improved varieties. Nevertheless, the fact that about 24% of the farmers indicated that they had bought at least part of their seed needs indicates there is a market for seed. A deeper analysis of data on purchased seed sources shows that, apart from the local market, farmers purchase seeds from other commercial routes. More than 70% of purchased seed is from their colleagues within their own village or from neighbouring villages.

Towards Integrated Rice Seed Sector Development

Louwaars and De Boef (2012) highlight the importance of facilitating interactions between the formal and informal seed systems to encourage development of the seed sector; and accepting the pluralistic nature of seed-sector development with roles for the public and commercial sector, as well as for farmers, community seed producers and NGOs. Rice seed sector development in Africa needs to address issues of availability, accessibility, seed quality, varietal quality and purity, and resilience to effectively contribute to increasing productivity and sustainability of rice seed systems in Africa (Table 14.1).

Figure 14.2 is a proposed framework for rice seed sector development in sub-Saharan Africa. The informal seed system is central, with the rice farmer as both a 'customer and manager' of seed (depicted in the central hexagon). Farmers use their own saved seed to produce paddy, or seed obtained from neighbours or other sources without either support or interference from the public or private sectors. However, farmers may be able

Table 14.1. Key issues for rice seed sector development in Africa. (Adapted from Remington *et al.*, 2002.)

Issue	Description
Availability	The supply of seed from all sources is adequate to meet the demand and needs of rice farmers
Access	Farmers as seed customers are able to acquire the seed they want through cash transactions or barter
Seed quality	Seed is of good quality in terms of cleanliness (analytical purity) and viability (germination and emergence)
Varietal quality and purity	The seed is of the varieties preferred by farmer customers and of adequate purity to meet production and marketing requirements
Resilience	There are diverse seed sources available to farmers to meet their needs after shocks such as drought, flood or conflict

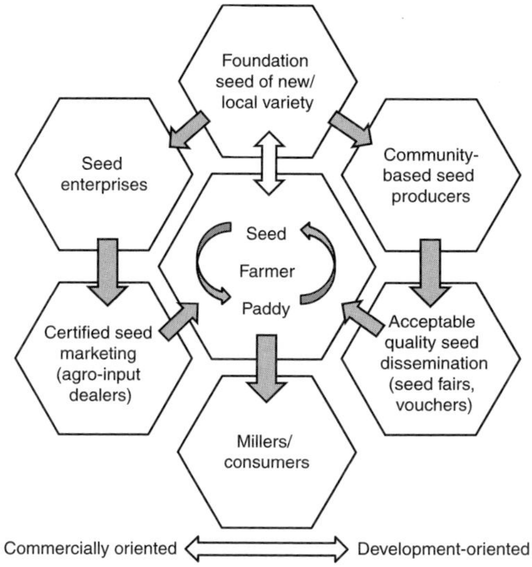

Fig. 14.2. Integrated rice seed sector development framework.

to access seed of a new or local variety from 'outside' the farming community, either through a commercially oriented seed value chain (left side of figure) or a development-oriented seed value chain (right side of figure).

In the absence of a reliable seed market, farmers are likely to rely on their own saved seed and informal seed systems. Farmers may profit from development-oriented interventions that stimulate community-based seed production and the production of 'acceptable quality' seed through seed fairs and donations of seed in emergency situations or purchase of seed through an exchange of vouchers. Development-oriented interventions can also reinforce farmers' capacities with respect to selecting and storing good-quality seed from their own production.

On the other hand, in cases where, for example, input subsidies for mineral fertilizer are linked with purchase of certified seed, many farmers may rely on specialized seed production systems, purchasing certified seed from seed enterprises, seed unions and agro-input dealers. Such commercially oriented seed value chains are more likely to emerge around irrigated systems, whereas rainfed upland and lowland systems often need development-oriented seed value chains to get access to new varieties. The public, development-oriented chain focuses on seed production to meet national targets, whereas the private, commercially oriented value chain focuses on profits and predictions (Louwaars and De Boef, 2012).

Breeder or foundation seed of new or widely grown local varieties provides the starting point of the proposed framework (Fig. 14.2), as this is the source for commercial- and development-oriented value chains. In some countries, foundation-seed production has already been devolved to the private sector. The delivery of breeder seed is assured by the national research systems. In the past, AfricaRice has often assisted with the production of breeder and foundation seed, especially following disasters, both natural and conflict (e.g. Liberia and Sierra Leone in 1999 and 2000). In general, to ensure seed security, there is a need to have sufficient stocks of foundation seed of major rice varieties at hand, either at country or regional level.

National and local government authorities in collaboration with private sector and development partners need to plan seed needs well ahead of time to ensure that sufficient amounts of seed of the right varieties are available and accessible to farmers.

Strengthening Farmer Seed Systems

The informal or farmer rice seed systems are the main source of seed for the majority of farmers in the rainfed lowland and upland rice systems in Africa. The farmer seed system is an efficient way of delivering seed to farmers because of its low transaction costs. 'Once farmers have the variety, the economic gains from using certified seed of the self-pollinated crops usually does not justify the investment. Moreover, certified seed is not always of better quality than the seed reproduced by a farmer' (Almekinders and Louwaars, 1999).

It is widely expected that farmer-saved seed and farmer-to-farmer seed exchange will continue to be the primary source of seed supply for the majority of farmers for a long time. Farm-saved seed is a common feature of agricultural systems worldwide, particularly in self-pollinated crops and in systems where hybrids are not used (Bay, 1998). Bay (1998) estimates that, in the mid-1980s, some 75% of seeds of self-pollinated crops in the USA (e.g. wheat, barley and oats) were farm-saved; similar figures might be found in Europe.

Important aspects in upgrading the quality and the diversity of seed produced on-farm include (Bèye *et al.*, 2011): (i) sensitization for local quality control; (ii) training on maintenance of varietal purity; (iii) information sharing on existing market opportunities; and (iv) promotion of hermetic rice seed storage. A study in Uganda showed that farmers with training were able to maintain seed purity comparable to that of certified seed (J. Lamo, Saint-Louis, Senegal, 2012, personal communication).

Strengthening Commercially Oriented Seed Value Chains

Investments in commercially oriented seed systems in Africa have focused mainly on hybrid maize and vegetable production, where farmer value addition is high. The focus on maize hybrids has greater prospects for seed enterprise profitability (Scoones and Thompson, 2011). The 'hybrid maize model', however, is not directly applicable to rice, which is self-pollinated and farmers are able to retain seeds from harvest for subsequent crops without a significant loss in performance. Nevertheless, farmers may purchase certified seed to gain access to a new variety or to overcome loss of varietal purity, problems with seed storage, and insufficient farmer-saved seed due to poor harvest.

National seed legislation regulates varietal release, seed production, certification and seed commercialization, including seed marketing, packaging and labelling. The Food and Agriculture Organization of the United Nations (FAO) in collaboration with the African Seed Trade Association (AFSTA) and CGIAR centres, sub-regional organizations (SADC, ASARECA, CORAF/WECARD[3]), European, US (USDA[3]) and Japanese governments (JICA[3]) and organizations such as West Africa Seed Alliance (WASA) and Alliance for a Green Revolution in Africa (AGRA) have helped most African countries in West, Central, East and Southern Africa to develop national legislation, train inspectors and laboratory technicians, and establish commercial seed companies.

In addition, harmonized regulatory frameworks and regional varietal catalogues have been adopted at regional levels. Nevertheless, the harmonized procedures remain difficult to implement for two main reasons. First, statutory seed standards, derived from international legislation, appear to be too strict. Consequently, the standards are difficult for most small-scale seed producers to meet. Second, most countries do not have operational national seed boards and seed-control infrastructures as they lack funding and trained personnel. As a result, current seed legislation and regulations do not facilitate field inspections and seed certification, and often act as barriers to the entry of small commercial seed companies and to informal seed production by farmers and farmer groups rather than as a support and encouragement.

Major challenges also exist for commercial seed enterprises at both institutional and infrastructure levels. Entrepreneurs face cumbersome seed legislation and out-dated infrastructure, and they lack training and access to inventory credit and business support services.

In their book *African Seed Enterprises*, Van Mele *et al.* (2011) discuss conditions under which seed enterprises can perform best:

- Equipment and infrastructure: enterprises should have seed-conditioning facilities, storage and market access.
- Policy environment: support should be given to assist new seed enterprises to start small and grow.
- Inventory and operating capital: seed producers require capital to invest in equipment, operate a seed business and also inventory credit to be able to store seed for later sale.
- Diversity of product: seed enterprises are advised to produce and sell diverse crop seed in addition to rice seed. This increases possibilities for income generation while diminishing risks. Often it will be a more profitable product like hybrid maize seed that will serve as the foundation for the enterprise to build on by adding new products over time. The potential of hybrid rice seed playing that role is discussed by El-Namaky and Demont (Chapter 13, this volume).
- Management: train and retain staff along with investment in new product research and market development.
- Quality control: enterprises need to build trust and reputation for quality products with their farmer customers. For this, it is important to build up strong internal support for in-house quality-control systems.

- Seed certification: certification is useful if it improves quality control, but costs need to be controlled. 'Quality Declared' and 'Truthfully Labelled' seed may be a more feasible way forward with more emphasis on branding.
- Marketing strategies: seed enterprises need to use diverse and innovative marketing strategies, including radio campaigns and TV adverts, demonstrations, seed fairs, billboards and field days, appropriate pricing, packaging of small quantities and attractive labelling.
- Enterprise cooperation: smaller seed enterprises can strengthen themselves by joining together in associations, federations or unions to achieve economies of scale, protect the market from counterfeiters, and share equipment with each other (including inter-enterprise hiring).

In East and Southern Africa, where the maize-seed industry is well-developed, there is a trend towards a deregulation of formal seed system regulations. These can be considered as 'emerging formal' systems (Louwaars and De Boef, 2012). In Tanzania and Zambia, for example, Quality Declared Seed (QDS) is largely used instead of certified seed. This has enabled farmers to access seeds of relatively good quality without the constraints and expense of formal seed certification. The QDS was developed by FAO, which looks for 'softer' seed legislation for countries that are not able to meet the standards of ISTA. Seed controls are conducted in 10% of seed-producing areas. Other countries in East and Southern Africa have similar systems referred to as Guaranteed Seed (Mozambique), Standard Seed (Botswana), Commercial Seed (Kenya and Uganda) and Approved Seed (Malawi). In addition there is 'Truthfully Labelled Seed', where certification is made voluntary and seed producers are allowed to attach their own 'truthful' label on the supplied seed bags or packets.

A promising development in terms of linking seed producers with potential buyers through the internet is being tested in Côte d'Ivoire under the responsibility of the NSS through a central website ('cyber seed', www.ci-semence.org). This involves accredited technicians and quality-control farmers trained in quality-control techniques, who monitor seed-production activities within a specific zone. They organize seed-production activities, including planning, control of seed-production fields, management of seed stocks, and support to seed commercialization. Quality-control farmers are farmers selected by their colleagues to monitor seed activities. They are trained in quality-control techniques and are monitored by the accredited technicians. They control seed-production activities in their respective cooperatives or associations, and monitor seed moisture content, germination rate and physical purity. Information on seed-production activities (seed producers, quantities, prices, etc.) are communicated to the central website for wide diffusion by mobile (cell) phone and other mass-media tools. Through this system, traceability is ensured and the required information about seed quality is available for each seed lot.

The cyber-seed concept was tested successfully in Daloa, Côte d'Ivoire, in 2006 by Coopérative de Commercialisation des Produits Vivriers de Daloa (COPROCOVIDA) and the Ministry of Agriculture. In 2007, four additional centres were created in Issia (Centre), Korhogo (North), Zouan-Hounien (West) and N'Zécrézessou (East). These centres helped farmers sell increasing amounts of seed: from about 300 tonnes of rice seed in 2007 to 863 tonnes in 2010.

Strengthening Development-oriented Seed Value Chains

Development-oriented seed value chains are more important in rainfed systems than they are in irrigated systems, as the former usually have difficult market access. In these systems, there is a need to build capacity among farmers on how best to manage and save their own seed to maintain varietal purity and seed quality because they may only sporadically get access to new varieties or new 'clean' sources of a particular variety through development-oriented seed value chains. Gradually, with farmers gaining access to markets, such systems may evolve towards commercially oriented seed systems, where seed quality is becoming an increasingly important issue.

The community-based seed system (CBSS) approach, used extensively by AfricaRice in many of its seed projects, is an example of an integrated seed system that aims to integrate the strengths and opportunities in both the formal (for the production of breeder and foundation seeds) and the informal seed systems (for diffusion of improved as well as traditional varieties) (Bèye *et al.*, 2011). CBSS is designed to enable smallholders to meet their seed requirements by improving their know-how in basic seed production and quality constraints. This decentralized system relies on individual entrepreneurial farmers and farmers' groups, who are trained to produce seed of acceptable quality that is disseminated through development projects, seed fairs and vouchers.

Successful examples of functional CBSS for rice can be found in many sub-Saharan African countries, including: Cameroon, Chad, Côte d'Ivoire, Ethiopia, The Gambia, Guinea, Kenya, Madagascar, Mali, Nigeria and Tanzania. In Côte d'Ivoire, more emphasis is being placed on professionalization of farmers' groups/associations through the development of online commercialization of rice seed with traceable information about available seed stocks and their characteristics (germination rate, moisture content, physical purity and the presence of weed seeds), the proposed prices, production trends by variety, etc. This new orientation opens new areas of collaboration among farmers' groups, agro-dealers and private seed companies. In addition, it helps to drive the production of quality seed with the objective of generating more revenue, and moves CBSS from a development orientation towards a commercial orientation.

The involvement of NGOs in seed systems is often under-appreciated as they tend to operate independently, especially in situations of seed insecurity due to natural disaster or conflict (all too common in Africa). Depending on the NGO and the context, they may support community seed production or carry out direct purchase and distribution of seed. NGOs are increasingly using vouchers to facilitate farmer access to seed rather than intervening directly in the seed supply chain. This could be via 'cash' vouchers that recipients can redeem at special fairs for a wide range of seeds, or 'commodity' vouchers or coupons that can only be redeemed for certified or quality-declared seed.

Conclusions

A viable and sustainable integrated rice seed sector requires integration of the formal and informal systems, with knowledgeable and empowered farmer seed customers and managers. Tripp (2003) describes this need eloquently:

> Finally, it is worth repeating that the development of a commercial seed sector is not in competition with, or an alternative to, the strengthening of farm-level seed management capacities. Indeed, the emergence of a commercial seed sector will occur only where farmer seed systems are strong, where farmers know a great deal about what varieties are available, are engaged in widespread seed and information exchange and are confident and knowledgeable consumers. Any aspirations for commercial seed sector development need to begin with attention to farmers.

There is no blueprint solution for seed system development in Africa, and the best possible approaches are likely to be specific to rice agroecosystems and value chain, and will evolve over time – for example, with emerging formal systems becoming 'more formal', catering to specific market niches. Governments need to strengthen both commercially and development-oriented rice value chains and support farmer seed systems. This will require supporting different kinds of entrepreneurship, such as small- to medium-scale seed companies producing certified inbred and hybrid rice seed for commercial purposes, possibly with some level of time-bound exclusivity; local seed businesses of groups of farmers that produce seed of 'acceptable quality' as part of development projects enabling farmers to get access to new varieties (eventually moving to a more commercial type of operation); local or community-based initiatives targeting the promotion of biodiversity conservation and utilization; and local seed businesses that evolve around rice mills and agro-industries, requiring strict control of seed to deliver specific products.

At the regional level, seed legislation needs to be harmonized or complemented by workable implementation guidelines to facilitate seed flows across borders and to stimulate the emergence of large-scale seed companies that can operate internationally. Adequate rice seed security stocks must be maintained to respond to emergency situations. At both national and regional levels, there is a need to adopt an integrated rice seed sector development approach, aiming to promote diversified seed systems, meeting the seed needs of all of Africa's rice farmers.

Notes

[1] The processing industry enables seeds to be calibrated, so that they are of uniform size and weight, and the delivery of disease-free seeds.
[2] DUS, Distinctness Uniformity and Stability; VCU, Value for Cultivation and Use.
[3] SADC, Southern African Development Community; ASARECA, Association for Strengthening Agricultural Research in Eastern and Central Africa; CORAF/WECARD, West and Central African Council for Agricultural Research and Development; USDA, United States Department of Agriculture; JICA, Japan International Cooperation Agency.

References

Almekinders, C. and Louwaars, N. (1999) *Farmers' Seed Production: New approaches and practices.* Intermediate Technology Publications, London, UK.
Bay, A.P.M. (1998) The seed sector in sub-Saharan Africa: alternative strategies. Paper presented at the Meeting on Seed Policy and Programmes in sub-Saharan Africa, Abidjan, Côte d'Ivoire, 27–28 November.
Bèye, A.M., Jones, M. and Simpson, B. (2011) *The Community-based Seed System: The case of traditional rice farming system. The Technician's Manual.* AfricaRice Training Course Collection. Africa Rice Center, Cotonou, Benin.
Bonou, A., Seck, P.A., Diagne, A., Kinkingninhoun, F.M. and Amovin-Assagba, E. (2012) La problématique semencière dans la riziculture africaine: accès et demande de semence de qualité par les paysans et perspectives pour améliorer le système. *Cahiers Agricultures* (in press).
Louwaars, N.P. (1994) Integrated seed supply: a flexible approach. Paper presented at the 4th Research Planning Workshop, Seed Unit, International Livestock Centre for Africa, Addis Ababa, 12–15 June.
Louwaars, N.P. (2007) *Seeds of Confusion: The impact of policies on seed systems.* Published PhD thesis, Wageningen University, Netherlands.
Louwaars, N. and De Boef, W. (2012) Integrated seed sector development in Africa: a conceptual framework for creating coherence between practice, programs and policies. *Journal of Crop Improvement* 26, 39–59.
Remington, T., Maroko, J., Walsh, S., Omanga, P. and Charles, E. (2002) Getting off the seeds and tools treadmill with CRS seed vouchers and fairs. *The Journal of Disaster Studies, Policy and Management* 26(4), 316–328.
Scoones, I. and Thompson, J. (2011) The politics of seed in Africa's green revolution: alternative narratives and competing pathways. *IDS Bulletin* 42(4), 1–23.
Tripp, R. (2003) How to cultivate a commercial seed sector. Paper presented at the Symposium on Sustainable Agriculture in the Sahel, Bamako, Mali, 1–5 December. www.syngentafoundation.org/db/1/447.pdf.
USAID Economic Growth Project (2009) *La chaîne de valeur: Options stratégiques pour la relance de la filière semencière au Sénégal.* US Agency for International Development, [Washington, DC].
Van Mele, P., Bentley, J.W. and Guei, R.G. (Eds) (2011) *African Seed Enterprises: Sowing the seeds of food security.* CAB International, Wallingford, UK.
Viatte, G., De Graaf, J., Demeke, M., Takahatake, T. and Rey de Arce, M. (2009) *Responding to the Food Crisis: Synthesis of medium-term measures proposed in inter-agency assessments.* Food and Agriculture Organization of the United Nations, Rome, Italy.

15 Towards a Better Understanding of Biophysical Determinants of Yield Gaps and the Potential for Expansion of the Rice Area in Africa

Kazuki Saito,[1]* Andrew Nelson,[2] Sander J. Zwart,[1] Abibou Niang,[1] Abdoulaye Sow,[3] Hiroe Yoshida[4] and Marco C.S. Wopereis[1]

[1]*Africa Rice Center (AfricaRice), Cotonou, Benin;* [2]*International Rice Research Institute (IRRI), Los Baños, Philippines;* [3]*Africa Rice Center (AfricaRice), Saint-Louis, Senegal;* [4]*National Agriculture and Food Research Organization (NARO), Tsukuba, Ibaraki, Japan*

Introduction

Rice is grown in diverse environments in Africa, and this is reflected in farmers' yields. These range from less than 1 t/ha in low-input, rainfed systems to more than 9 t/ha in high-input, irrigated systems. As highlighted by Seck *et al.* (Chapter 2, this volume), Africa's rice production needs to be augmented substantially to reduce the current heavy reliance on imports.

Increasing rice production is possible through increasing rice yield per unit of land and through expansion of rice harvested area. To raise rice productivity per unit of land, there is a need to better understand which biophysical factors limit productivity in farmers' fields, and to what extent productivity could be increased via improved crop management. 'Potential yield' is defined as the **maximum** yield that can be obtained from a crop in a given environment as determined by simulation models with plausible physiological and agronomic assumptions (Evans and Fischer, 1999). Under irrigated conditions, potential yield is determined by climate (solar radiation and temperature), varietal characteristics and crop establishment methods including sowing date and density. Under rainfed conditions, potential yield is also affected by water availability.

Validated crop-simulation models are rarely available in Africa and, if they are, input values to run them, such as long-term weather data, are usually lacking. A proxy for potential yield can be the maximum yield obtained with good agricultural practices in an experimental field or in a high-yielding farmer's field. The 'yield gap' is defined as the difference between potential yield and average on-farm yield obtained by farmers (Becker *et al.*, 2003). Several 'yield gaps' can be defined (Fig. 15.1). Because of diminishing returns on investment, yields in farmers' fields do not in general exceed 80% of the potential yield estimated by validated simulation models.

A good understanding of potential yield and yield gaps enables us to identify opportunities for yield improvement in farmers' fields.

* Corresponding author: k.saito@cgiar.org

© CAB International 2013. *Realizing Africa's Rice Promise* (eds M.C.S. Wopereis *et al.*)

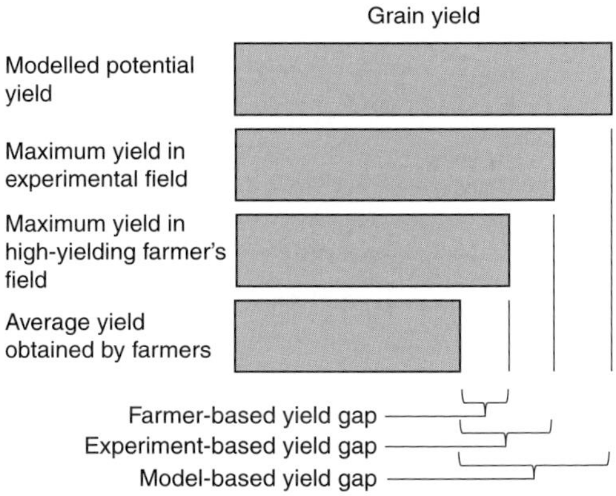

Fig. 15.1. Framework for estimation of yield gap based on three different potential yields (model-based, experimental field maximum and on-farm maximum). (Adapted from Lobell *et al.*, 2009.)

Closing these yield gaps may require adoption of alternative crop management practices or capital investment in, for example, bunding, land levelling or irrigation. To determine which of these alternative crop management or investments is needed, it is necessary to understand the biophysical determinants of rice productivity. What are the major yield-limiting factors (e.g. drought or excess water, nutrient deficiencies, and extreme temperature) and yield-reducing factors (e.g. insects, diseases, weeds, birds)? This type of information can be obtained via detailed studies in farmers' fields (often referred to as 'yield gap surveys'), as demonstrated by (e.g.) Wopereis *et al.* (1999) and Becker *et al.* (2003) for irrigated rice systems in West Africa. A complementary 'diagnostic survey' can provide information on, for example, production orientation (subsistence, market), household wealth, access to input and output markets, and access to rice knowledge and technologies. Combining information from the two surveys allows us to develop pathways to raise rice productivity in a sustainable manner.

In this chapter, we provide an overview of rice production systems in Africa and their distribution as a function of spatial variability of climate (agroecological zones) and soils on the continent. We also discuss the current status of knowledge with respect to spatial variability of disease epidemics. Climate factors such as temperature, humidity and rainfall strongly affect rice production through indirect effects on the incidence of pests and diseases. Next we discuss the current status of knowledge about rice yield gaps in Africa. We analyse the potential to enhance rice production in Africa by increasing rice harvested area (by bringing more land under cultivation or by increasing cropping intensity) and approaches to identify 'best-bet areas' for expansion. We conclude with a discussion of the challenges that hinder a better overview of determinants of yield gaps and areas with the best potential for sustainable expansion of rice harvested area in Africa.

Rice production systems

In Africa, five major rice production systems are distinguished: rainfed upland, rainfed lowland, irrigated lowland, deep water and mangrove swamp; the last two are of relatively minor importance in terms of surface area (Maclean *et al.*, 2002; Balasubramanian *et al.*, 2007; Seck *et al.*, 2012). New estimates of surface area under each rice production system are provided by Diagne *et al.* (Chapter 3, this volume).

Rice fields are usually flooded during part or all of the growing season, except in the case of rainfed upland rice. Surface-water regimes and water sources (e.g. irrigation, rainfall, water table) distinguish the rice-production systems.

- Irrigated lowland rice is generally grown in bunded fields with assured irrigation for one or two crops per year. Dam-based irrigation, water diversion from rivers, and pump irrigation from wells are major sources of irrigation water.
- Rainfed lowland rice is grown on level to slightly sloping, unbunded or bunded fields in lower parts of the toposequence and in inland valleys, which are defined as flat-floored, relatively shallow valleys and are widespread in the undulating landscape (see Rodenburg, Chapter 22, this volume). Fields are flooded by rains and groundwater for part of the rice-growing season, although in some seasons fields may not be flooded due to lack of rainfall. Rainfed lowland rice is also grown in flash-flood areas, where water level is suddenly increased during the

rice-growing season, causing short-term submergence. A fuzzy transition exists between rainfed and irrigated lowland rice production systems, where a water-management continuum exists ranging from strictly rainfed (no water control) to fully irrigated lowlands, which may evolve with investments in water-control measures.

- Deep-water rice production systems are found in the flood plains along the major rivers such as the Niger River, inland valleys, and coastal wetlands (Kawano and Sakagami, 2008). In the 'floating-rice' area, water depth remains high (up to 3 m) for an extended period (up to 5 months). In the 'deep-water' rice area, water remains in the fields for several months, but water is not as deep as in the floating-rice area. Rice varieties that are 140–180 cm tall are required for this system.
- Rice fields in the mangrove-swamp production system are located on tidal estuaries close to the sea. Rice can be grown during the period when freshwater floods wash the land and displace tidal flows. Tall rice varieties or varieties adapted to flash flooding are grown in this system.
- Rainfed upland rice is generally grown on level or sloping, unbunded fields. Flooding is rare in this system. In some cases (e.g. Uganda) supplementary irrigation may be used.

Distribution of Rice Production Systems in Relation to Agroecological Zones in Africa

You *et al.* (2009a,b) provide estimates of crop production, area and yield for 10 km ×10 km grid cells for 20 crops, including rice, based on downscaling from sub-national production, statistical or survey data by taking other spatial data such as land use, population density and crop suitability into account. Plate 6 shows the distribution of rice-production systems in Africa classified as: (i) rainfed, high-input/commercial; (ii) rainfed, low-input/subsistence; and (iii) irrigated, following the definition of global agroecological zones developed by the Food and Agriculture Organization of the United Nations and the International Institute for Applied Systems Analysis (FAO and IIASA, 2000). The distribution was calculated from sub-national production data in 2000; updated data are not yet available. Data on rice cultivation at sub-national scale are not available in Balasubramanian *et al.* (2007), Seck *et al.* (2012), FAOSTAT (http://faostat.fao.org/) or Diagne *et al.* (Chapter 3, this volume), and so could not be used in our assessment. Rice areas described in this chapter refer to areas where rice was cultivated in the particular system in 2000. If rice was grown and harvested more than once in the year, the physical area is not increased.

Rainfed, high-input/commercial systems use high-yielding varieties and some animal traction and mechanization (FAO and IIASA, 2000). Farmers apply some fertilizer and use pesticides. This system is not widespread in Africa, but can be found in countries such as Côte d'Ivoire, Madagascar and Uganda. The rainfed, low-input/subsistence system uses traditional varieties and mainly manual labour without (or with little) application of fertilizer or pesticides. This classification does not distinguish between rainfed lowland and upland rice systems.

The irrigated system refers to rice areas provided with either full or partial irrigation infrastructure. In general, modern varieties and relatively high fertilizer inputs are used, together with advanced management options such as soil- and water-conservation measures.

Rice is grown in 15 of the 16 agroecological zones distinguished in Africa (HarvestChoice, 2009) (Plate 7; Table 15.1). These agroecological zones are grouped according to temperature (tropical or sub-tropical), elevation (warm or cool) and moisture (arid, semiarid, sub-humid and humid) as explained in the footnote to Table 15.1. Thus, these groups include climate factors that strongly affect rice growth through direct effects on physiological processes and through indirect effects on the incidence of pests and diseases.

In West Africa, the terms 'Sahel', 'Sudan savannah', 'Guinea savannah' and 'Equatorial forest' are frequently used for agroecological zoning (e.g. Windmeijer and Andriesse, 1993; Defoer *et al.*, 2004). In this chapter, 'Sahel' is 'tropical – warm / arid'; 'Sudan savannah' is 'tropical – warm / semiarid'; 'Guinea savannah' is 'tropical – warm / sub-humid'; and 'Equatorial forest' is 'tropical – warm / humid'.

Table 15.1. Estimated share (%) of rice area by agroecological zone and production system in Africa in 2000. (Spatial analysis of data from HarvestChoice, 2009; You *et al.*, 2009a,b.)

Agroecological zone[a]	Rainfed, low-input/ subsistence rice system	Rainfed, high input/ commercial rice system	Irrigated rice system	Total
Tropical – warm / humid	11	1	4	16
Tropical – warm / sub-humid	36	2	11	49
Tropical – warm / semiarid	10	0.3	7	17
Tropical – warm / arid	0.3	0.1	1	2
Tropical – cool / humid	1	0.2	1	2
Tropical – cool / sub-humid	4	0.1	1	5
Tropical – cool / semiarid	0.2	0	<0.1	0.2
Tropical – cool / arid	<0.1	0	<0.1	<0.1
Sub-tropical – warm / humid	<0.1	0	0	<0.1
Sub-tropical – warm / sub-humid	<0.1	0	0.1	0.1
Sub-tropical – warm / semiarid	<0.1	0	<0.1	<0.1
Sub-tropical – warm / arid	<0.1	0	8	8
Sub-tropical – cool / humid	0	0	0	0
Sub-tropical – cool / sub-humid	<0.1	0	<0.1	<0.1
Sub-tropical – cool / semiarid	<0.1	0	0	<0.1
Sub-tropical – cool / arid	<0.1	0	<0.1	<0.1
Total	62	4	34	100

[a]'Tropical': monthly temperature adjusted to sea level greater than 18°C for all months; 'Sub-tropical': monthly temperature adjusted to sea level less than 18°C for one or more months.
'Cool': elevation greater than 1200 m in the tropical zone and greater than 800 m in the sub-tropical zone; otherwise classified as 'warm'.
'Arid': less than 70 days of growing period, which is defined as the period during the year when average temperatures are greater than or equal to 5°C and rainfall plus moisture stored in the soil exceed half the potential evapotranspiration; 'semiarid': 70–180 day growing period; 'sub-humid': 180–270 day growing period; 'humid': growing period more than 270 days.

Spatial analyses of agroecological zones and rice-distribution data (Plates 6 and 7) show that the largest proportion of rice area (49% of total) is within the tropical – warm / sub-humid zone, while 17% is within the tropical – warm / semiarid zone and 16% is in the tropical – warm / humid zone (Table 15.1). In the humid, sub-humid and semiarid zones, rainfed rice systems occupy a larger area than irrigated rice-production systems. Irrigated rice-production systems are dominant in the arid zone, with 8% (of the total rice area) in the sub-tropical – warm / arid zone and 1% in the tropical – warm / arid zone. Tropical – cool / humid and sub-humid zones (elevation greater than 1200 m) have 7% of the total rice area.

Rainfed low-input/subsistence rice farming is practised on 62% of the total rice surface area in Africa, while 34% is irrigated rice areas and 4% is rainfed, high-input/commercial systems. The latter includes not only tropical – warm / sub-humid in West Africa, but also 'Other zones including all the production systems' in the other regions (Table 15.2). The total area for irrigated rice is higher than the 26% reported by Diagne *et al.* (Chapter 3, this volume), probably as a result of the fact that the data were collected in different years and using different collection methods.

In West and East Africa, the tropical – warm / sub-humid zone has the largest share of rice area in both irrigated and rainfed, low-input systems, followed by tropical – warm / semiarid zone in West Africa and tropical – warm / humid zone (irrigated rice) and tropical – cool / sub-humid zone (rainfed, low-input) in East Africa (Table 15.2). The estimated share of irrigated rice in the tropical – warm / arid zone in West Africa is small (1%). In North Africa, the sub-tropical – warm / arid zone is restricted to Egypt. The rainfed, low-input system in tropical – warm / humid zone is predominant in Central Africa, including the Democratic Republic of Congo (DRC).

Table 15.2. Estimated distribution of rice area by region, agroecological zone and production system in Africa in 2000. (Spatial analysis of data from HarvestChoice, 2009; You *et al.*, 2009a,b.)

Agroecological zone	Production system	Rice area (×1000 ha) in 2000	Estimated share (%) of rice area in Africa
West Africa[a]			
Tropical – warm / sub-humid	Rainfed, low-input	1973	26
Tropical – warm / semiarid	Rainfed, low-input	619	8
Tropical – warm / sub-humid	Irrigated	423	6
Tropical – warm / humid	Rainfed, low-input	380	5
Tropical – warm / semiarid	Irrigated rice	366	5
Tropical – warm / sub-humid	Rainfed, high-input	101	1
Tropical – warm / arid	Irrigated	94	1
Other zones including all the production systems		138	2
Subtotal		4093	55
North Africa[b]			
Sub-tropical – warm / arid	Irrigated	610	8
Other zones including all the production systems		31	0.4
Subtotal		642	9
East Africa[c]			
Tropical – warm / sub-humid	Rainfed, low-input	584	8
Tropical – warm / sub-humid	Irrigated	423	6
Tropical – warm / humid	Irrigated	262	3
Tropical – cool / sub-humid	Rainfed, low-input	196	3
Tropical – warm / humid	Rainfed, low-input	157	2
Tropical – warm / semiarid	Irrigated	123	2
Tropical – cool / humid	Irrigated	94	1
Tropical – cool / sub-humid	Irrigated	90	1
Other zones including all the production systems		245	3
Subtotal		2174	29
Central Africa[d]			
Tropical – warm / humid	Rainfed, low-input	279	4
Tropical – warm / sub-humid	Rainfed, low-input	141	2
Other zones including all the production systems		170	2
Subtotal		590	8
Southern Africa[e]			
Subtotal		1	<0.1
Total		7500	100

[a]West Africa: Benin, Burkina Faso, Côte d'Ivoire, The Gambia, Ghana, Guinea, Guinea-Bissau, Liberia, Mali, Mauritania, Niger, Nigeria, Senegal, Sierra Leone and Togo.
[b]North Africa: Algeria, Egypt, Morocco, Sudan (now Sudan and South Sudan) and Tunisia.
[c]East Africa: Burundi, Comoros, Eritrea, Ethiopia, Kenya, Madagascar, Malawi, Mozambique, Rwanda, Somalia, Tanzania, Uganda and Zambia.
[d]Central Africa: Angola, Cameroon, Central African Republic, Chad, Congo, DRC and Gabon.
[e]Southern Africa: Lesotho, Namibia, South Africa, Swaziland and Zimbabwe.

Distribution of Rice Production Systems in Relation to Soil Constraints

African soils generally have inherently poor fertility as they are old, often strongly weathered and leached (Bationo *et al.*, 2006). Furthermore, inadequate land-use and crop-management practices have led to increased soil erosion and depletion of nutrients and, consequently, a decline in rice productivity (Becker and Johnson, 2001a). In all rice-production systems, soil-related abiotic stresses may occur in rice (Defoer *et al.*, 2004; Balasubramanian *et al.*, 2007). In this section, we present a quantitative characterization of soil fertility within rice-growing environments in Africa. The approach follows that of Haefele and Hijmans (2007), who studied rainfed lowland rice in Asia, by combining spatial databases of soils and rice area. Soil information was derived from the digital version of the Soil Map of the World (FAO, 1995), and we used four soil groups with different levels of soil fertility and severity of soil constraints based on the interpretation of the modifiers of the soil fertility capability classification (FCC) system (Sanchez and Buol, 1985). The following are descriptions of the four soil-fertility groups (Haefele and Hijmans, 2007).

- **Good, fertile soils with no major soil constraints:** topsoils not designated with any of the FCC modifiers h, k, e, a, i, s, c, O, n or b (see below). Absence of modifiers a, h and b indicates soil pH values in the optimum range between 6.0 and 7.3. Soils included here might be designated with the FCC modifiers x (volcanic materials) or v (vertic soil properties). Soils in this group have a range of indigenous soil fertility, but are generally less weathered than the next two soil groupings.
- **Poor soils with no major soil constraints:** topsoils are designated with no FCC modifier other than h (10–60% Al saturation of the effective CEC or pH between 5 and 6). Crop growth is not limited by any major soil constraint, although severe P deficiency may occur. However, the acid soil reaction and the high Al saturation indicate highly weathered soils with limited indigenous nutrient supply and low nutrient-retention capacity.
- **Very poor soils with considerable soil constraints:** topsoils are designated with one or several of the FCC modifiers k (<10% weatherable minerals in silt and sand fraction or exchangeable K <0.20 meq per 100 g soil), e (effective CEC <4 meq per 100 g soil), a (>60% Al saturation) or i (percentage of free Fe_2O_3 divided by percentage clay >0.15, and more than 35% clay or hues of 7.5 YR [yellow–red] or redder, and granular structure). Crop growth on these soils is potentially limited by combinations of low nutrient reserves (k), low CEC (e), Al toxicity (a) and high P fixation (i). Generally, these are highly weathered soils with very limited indigenous nutrient supply, low nutrient-retention capacity, frequent and often severe P deficiency, acidic to very acidic soil reaction (pH <5), and Fe and Al toxicities.
- **Problem soils:** topsoils are designated with the FCC modifiers s (saline soils), c (acid-sulfate soils), O (organic soils), n (sodic soils) or b (alkaline soils). Crop growth on these soils is likely to be limited by salinity (s), very low pH, P deficiency, and Fe, S or Al toxicity (c), nutrient deficiencies of N, Zn, K, P, Cu and Mo (O), or high pH and P, Fe, Zn deficiency (n, b).

Comparative distribution of rice areas by soil-fertility group for temperature zones (subtropical / tropical), elevation zone (warm / cool), moisture zone (humid / sub-humid / semiarid / arid), and production system (irrigated rice / rainfed, high-input / rainfed, low-input) is presented in Table 15.3.

The sub-tropical zone is characterized by either good or problem soils, with virtually nothing in between (Table 15.3). In contrast, problem soils are uncommon in the tropical zone, where poor and very poor soils predominate. The arid zone is also characterized by high percentages of good and problem soils. Good and poor soils are common in the semiarid zone, whereas the sub-humid zone has large areas of poor and very poor soils. Very poor soils are widespread in the humid zone.

Irrigated systems have higher percentages of good and problem soils than rainfed systems, and lower proportions of poor and very poor soils. About 67% and 25% of problem soils are designated with the FCC modifiers s (saline soils) and n (sodic soils), respectively, which are therefore more common than those designated with other FCC modifiers (c, O and b: 0.4–12%). When the distribution of rice areas by soil-fertility group is compared among Africa, South-east Asia and South Asia, the rainfed, low-input system in Africa tends to have poorer soils than the rainfed lowland systems in South Asia, but comparable soil fertility to rainfed lowland systems in South-east Asia.

Distribution of rice areas by soil-fertility group for the different regions, agroecological

Table 15.3. Comparative distribution of rice areas by soil-fertility group for temperature zone (sub-tropical / tropical), elevation zone (warm / cool), moisture zone (humid / sub-humid / semiarid / arid) and production system (irrigated rice / rainfed, high-input / rainfed, low-input). (Spatial analysis of data from FAO, 1995; HarvestChoice, 2009; You *et al.*, 2009a,b.)

	Estimated share (%)			
Zone / system	Good soils	Poor soils	Very poor soils	Problem soils
Temperature				
Sub-tropical	52	1	1	46
Tropical	21	32	43	5
Elevation				
Warm	23	30	38	9
Cool	22	20	55	3
Moisture				
Humid	7	19	73	1
Sub-humid	20	35	41	4
Semi-arid	35	38	16	10
Arid	53	1	4	42
Production system				
Irrigated rice	30	25	27	18
Rainfed, high-input	9	29	60	2
Rainfed, low-input	21	31	45	3
Rainfed lowland in South-east Asia[a]	25	18	47	10
Rainfed lowland in South Asia[a]	45	33	14	8

[a]Source: Haefele and Hijman (2007). Data on intermediate and shallow rainfed rice areas were combined.

zones and production systems is given in Table 15.4. Very poor soils are dominant in the humid and sub-humid zones. In the arid and semiarid zones, very poor soils account for less than 20%. High percentages of problem soils (>20%) are found in the irrigated rice system in the arid zone in West and North Africa, and more than 70% of them are designated with the FCC modifier s (saline soils). These findings are consistent with previous reports, which showed that rice productivity is limited by salinity in these areas (FAO, 2002; Defoer *et al.*, 2004). There is a high percentage of problem soils (26%) in the irrigated rice system in the tropical – warm / semiarid zone in East Africa, and 77% of these are designated with the FCC modifier n (sodic soils).

As soil tests commonly used for nitrogen are insufficiently reliable to be used as FCC parameters (Sanchez *et al.*, 2003; Haefele and Hijmans, 2007), nitrogen deficiencies could not be included as soil constraints in our spatial analysis. However, soils without major nutrient limitations will likely develop nitrogen deficiency in continuous cropping systems without nutrient inputs, and nitrogen deficiency may cause other nutrient deficiencies (Sanchez *et al.*, 2003). Many previous studies in both lowland and upland conditions across all agroecological zones in West Africa have shown substantial increases in rice yield in response to nutrient inputs via inorganic nitrogen fertilizer or legumes grown before rice cultivation, although response to nitrogen inputs is highly variable, depending on (e.g.) rice growing condition, variety used, production systems (e.g. Becker and Johnson, 1998, 2001a,b; Wopereis *et al.*, 1999; Akanvou *et al.*, 2000; Becker *et al.*, 2003). Indigenous soil nitrogen supply limited rice yield in most cases. See also Haefele *et al.* (Chapter 20, this volume).

At the landscape level (e.g. watershed), natural resources (particularly water and soil resources) are strongly correlated with their position in the toposequence (Andriesse and

Table 15.4. Estimated share of rice physical area by region, agroecological zone, production system and soil-fertility group. Estimated share with >1% of total rice physical area is presented.

		Estimated share (%)			
Agroecological zone	Production system	Good soils	Poor soils	Very poor soils	Problem soils
West Africa					
Tropical – warm / sub-humid	Rainfed, low-input	19	40	40	1
Tropical – warm / semiarid	Rainfed, low-input	34	43	15	9
Tropical – warm / sub-humid	Irrigated	13	54	27	6
Tropical – warm / semiarid	Irrigated	40	37	11	13
Tropical – warm / humid	Rainfed, low-input	12	19	68	1
Tropical – warm / sub-humid	Rainfed, high-input	12	33	52	3
Tropical – warm / arid	Irrigated	59	0.3	19	22
North Africa					
Subtropical – warm / arid	Irrigated	52	1	1	46
East Africa					
Tropical – warm / sub-humid	Rainfed, low-input	25	30	38	7
Tropical – warm / sub-humid	Irrigated	20	24	43	13
Tropical – warm / humid	Rainfed, low-input	10	20	70	0.5
Tropical – warm / humid	Irrigated	4	32	63	0.5
Tropical – cool / sub-humid	Rainfed, low-input	35	18	40	7
Tropical – warm / semiarid	Irrigated	30	18	26	26
Tropical – cool / humid	Irrigated	5	25	69	0.3
Tropical – cool / sub-humid	Irrigated	15	30	54	1
Central Africa					
Tropical – warm / humid	Rainfed, low-input	2	4	93	1
Tropical – warm / sub-humid	Rainfed, low-input	13	7	77	2

Fresco, 1991; Windmeijer and Andriesse, 1993). In inland valley systems, soil fertility tends to improve as one moves down the slope to the valley bottom, calling for soil-fertility recommendations at the plot level that differ with toposequence position, such as proposed for farming systems in South-east Asia (e.g. Homma *et al.*, 2003). Such variations strongly affect rice-cropping systems, farmers' crop and varietal choices, and rice productivity (Andriesse and Fresco, 1991; Audebert *et al.*, 1999; Touré *et al.*, 2009). These variations could not be captured in the work reported in this chapter.

Mapping Potential Epidemics of Rice Diseases

Climate factors such as temperature, humidity and rainfall strongly affect rice production through indirect effects on the incidence of pests and diseases. Savary *et al.* (2012) developed global risk maps of potential epidemics involving five rice diseases, using a simple generic model 'EPIRICE'. Diseases considered were leaf blast, brown spot, bacterial blight, sheath blight and tungro virus (which occurs only in Asia). The following is a summary of the results for Africa.

- High potential risk areas for leaf blast epidemics in Africa include West Africa (e.g. Guinea, Liberia, Nigeria, Sierra Leone), Central Africa (e.g. Cameroon, Gabon), and East Africa (e.g. western Tanzania, Madagascar, the Ethiopian highlands).
- Potential epidemics of brown spot and bacterial blight in Africa have similar patterns, and high-risk areas for both are in the humid and sub-humid zones of West Africa.
- Simulation predicted potential sheath blight epidemics in the humid and sub-humid zones of West Africa, Central Africa (e.g. Cameroon, Gabon), and in eastern Madagascar.
- Tungro virus has not been reported in Africa as there is no vector. However, tungro

potential epidemics were simulated in humid and sub-humid zones in West Africa and Madagascar, indicating that, should the vector become established in these areas, there is a potential risk of tungro occurring.

Although Savary *et al.* (2012) indicate that more research is needed to improve the EPIRICE model, this fairly crude information suggests where major epidemics may be expected and thus where research efforts need to be focused. The humid and sub-humid zones of West Africa are areas at risk of epidemics of all the diseases included in the analysis. Madagascar is also prone to rice disease epidemics. Such areas will need more attention to develop integrated crop-health management strategies for avoiding disease epidemics. At the landscape level (e.g. watershed), the occurrence and intensity of biotic stresses also differs within the toposequence (Defoer *et al.*, 2004), so the results of the model will need to be validated through field observations on disease incidence and impact on yield.

Analysing Rice Yield Gaps in Rice-growing Environments in Africa

While agroecological zoning and soil mapping in relation to rice distribution provide useful information, these approaches are not directly linked to rice productivity. In this section, we describe crop models for estimating potential yield, and review previous 'yield gap surveys' in major rice-production systems in Africa.

Estimating potential yield using crop-simulation models

Combining crop-simulation models, which simulate crop phenology, growth and yield, with geographic information systems (GIS) is a potentially powerful approach to estimate potential yield as well as characterize rice-growing environments in Africa. However, past simulation efforts for rice in Africa have focused mainly on the potential yield of irrigated lowland rice in arid and semiarid environments in West Africa (tropical – warm / arid and tropical – warm / semiarid zones), where diurnal and seasonal variation in temperature is a major determinant of rice production (Dingkuhn and Sow, 1995). Dingkuhn and Sow (1995) identified areas where there is potential risk of spikelet sterility occurring due to extreme (high and low) temperatures. In addition to this model (ORYZA_S), several other models have been used for simulating potential yield of irrigated and rainfed lowland rice in Africa (Sheehy *et al.*, 2004; Hijmans and Serraj, 2008).

Hijmans and Serraj (2008) used the ORYZA2000 model (Bouman *et al.*, 2003) to determine relative yield reduction of rainfed lowland rice attributable to drought stress, using the weather database of the US National Aeronautics and Space Administration (NASA) Langley Research Center Atmospheric Sciences Data Center POWER (Prediction Of Worldwide Energy Resource) Project, and estimated planting date for each degree resolution, and some soil parameters including water table depth and percolation rate, with global coverage. The results showed that although variation in relative yield reduction is large across areas with similar total rainfall during the rice-growing season, relative yield reduction is negatively related to total rainfall during the rice-growing season. Rainfed lowland rice production is considered possible only with total rainfall above 450 mm. When total rainfall during rice growing season is 750–850 mm, relative yield reduction (median value) is about 50%. East African countries (except for Madagascar) tend to have a larger yield reduction than West African countries (except for areas where rice cannot be grown without irrigation, such as in the tropical – warm / arid zone).

If suitable crop-simulation models are developed or adapted for Africa, high-quality long-term climate data at coarse resolution for use as input for such models are commonly lacking. This will constrain GIS-based assessment of growth environments using crop models. Satellite-based climate data, such as Hijmans and Serraj (2008) used, can be useful as an alternative for the assessment. However, while ground-based and NASA satellite-based data were correlated, there were differences in climatic parameters between them (Yang *et al.*, 2007; White *et al.*, 2008; Bai *et al.*, 2010).

These differences are due to the fact that both data sources have inherent errors and uncertainties. Errors related to ground-based parameters include poor maintenance of climate-observation facilities, resulting in inadequate data. Uncertainties related to satellite-based parameters include pixel size, sensor resolution, navigation time, algorithm accuracy and geographical coincidence of instantaneous information recorded by a satellite with measurements on the ground (Bai *et al.*, 2010). If there are no long-term ground-based data or a lot of missing data, the use of relationships between ground-based and NASA satellite-based data may be a means for estimating missing ground-based data to be used in crop-simulation models.

Yield gaps and determinants in major rice production systems

Becker *et al.* (2003) reported that average on-farm yields of irrigated lowland rice in different agroecological zones in West Africa range from 3.4 t/ha to 5.4 t/ha, and potential yields range from 6.9 t/ha to 9.8 t/ha. The potential yield is highest in the Sahel zone (Senegal) and lowest in the humid forest zone (Côte d'Ivoire). The yield gaps range from 3.2 t/ha to 5.9 t/ha, showing considerable scope for increasing yields.

Becker and Johnson (1999) conducted surveys in irrigated systems of the forest zone of Côte d'Ivoire. Yields varied between 0.2 t/ha and 7.3 t/ha, with average yields of 3.2 t/ha under partial irrigation and 4.2 t/ha in fully irrigated systems. Age of seedlings at transplanting, timeliness of operations and application of P fertilizer explained 60% of observed variability.

Wopereis *et al.* (1999) highlighted the low recovery rates of fertilizer N applied to the crop in farmers' fields in irrigated systems in Burkina Faso, Mali and Senegal. Farmers can, therefore, improve efficiency and profitability by improving the recovery rate of applied nutrients, especially N, through better crop management in general, without major increases in investment in fertilizers. The most important constraints that resulted in low N recovery rates were (Wopereis *et al.*, 1999): timing of N fertilizer application not coincident with critical growth stages of the rice plant; use of relatively old (>40 days) seedlings at transplanting; unreliable irrigation water supply; weed problems; and late harvesting (Senegal River delta). Haefele *et al.* (2000, 2001) showed that rice yields in farmers' fields in Mauritania and Senegal could be raised by 2 t/ha through improved weed and soil-fertility management.

Potential yields of irrigated lowland rice in Madagascar are estimated at about 11.4–14.9 t/ha (Sheehy *et al.*, 2004), while on-farm yields range from 2.6 t/ha to 9.9 t/ha (Tsujimoto *et al.*, 2009), which suggests yield gaps range from 1.5 t/ha to 12.3 t/ha. While yield-gap studies have not been carried out for irrigated lowland rice in other East and North African countries, trials managed by researchers achieved more than 11 t/ha in Egypt and Kenya, and more than 9 t/ha in Mozambique (Matsushima *et al.*, 1994; Namba, 2003, 2005; Menete *et al.*, 2008). Thus, potential rice yields in Egypt, Kenya and Madagascar seem to be higher than those in West Africa.

Studies in West Africa show average farm yields for rainfed lowland rice range from 1.0 t/ha to 2.2 t/ha (Becker and Johnson, 2001b). Given that potential yields of rainfed lowland rice are assumed to be similar to those of irrigated lowland rice, the yield gaps are 4.8–7.6 t/ha (Becker and Johnson, 2001b; Becker *et al.*, 2003). Becker and Johnson (2001b) studied the effects of improved water control and crop management on lowland rice productivity in West Africa. Retaining flood water with field bunds increased rice yield by about 40% and improved weed control (about 25% less weed biomass in bunded than in open plots). Application of mineral fertilizer N increased rice yields by almost 20% in bunded fields, but resulted in no increase in open fields. Land levelling together with bunding facilitated improved water management which decreased weed growth and increased nutrient use efficiencies.

Rice yield measurements for rainfed upland rice, including intensive and extensive systems, showed a range in farmers' fields of 0.8–1.6 t/ha (Becker and Johnson, 2001a). While potential yields have not been estimated for upland rice in Africa, trials managed by researchers have given rice yields of 4.0–5.6 t/ha with nutrient input and also with supplementary irrigation in two of five studies (Dingkuhn *et al.*, 1998; Oikeh *et al.*, 2008; Ekeleme *et al.*, 2009; Saito and Futakuchi, 2009; Kamara *et al.*, 2010). Thus, yield gaps also appear to be high under rainfed upland conditions, but not as large as those under irrigated and rainfed lowland conditions. Becker and Johnson (2001a) showed that increased cropping intensity and reduced fallow duration were associated with yield reduction: intensification-induced yield loss was about 25% (a drop from an average of 1.5 t/ha to 1.1 t/ha) and was mainly related to increased weed infestation and declining soil quality.

Few of these yield-gap surveys quantified losses due to biotic stresses with the exception of weeds. More information on potential losses due to biotic stresses in rice fields in Africa can be found in Chapters 16 (weeds), 17 (diseases), 18 (insect pests) and 19 (birds).

There is a dearth of up-to-date information on magnitude and determinants of yield gaps across major rice-production regions in Africa. The yield-gap surveys mentioned above were mainly conducted in the 1990s and crop models available at the time were only validated for irrigated growing conditions in West Africa. Diagne *et al.* (Chapter 4, this volume) provide information on losses due to biotic and abiotic stresses based on farmer perceptions of the occurrence and relative importance of the stresses from a survey conducted in 2009 in several African countries.

In 2011, the Africa Rice Center (AfricaRice) launched the Africa-wide Rice Agronomy Task Force, a collaborative effort of (initially) 15 African countries, to be expanded gradually to include at least all of AfricaRice's member countries. The Task Force has launched a collective effort to analyse major determinants of rice productivity in different rice systems through yield-gap and diagnostic surveys.

Potential for Sustainable Expansion in Rice Harvested Area

In terms of the potential for expansion of cultivated area and increased production, lowlands or wetlands show the greatest promise, with an estimated total of 239 million ha across sub-Saharan Africa (Andriesse, 1986). Less than 5% of the lowlands are currently planted with rice (Balasubramanian *et al.*, 2007). Wetlands can be defined as areas where the soil is saturated with water either permanently or seasonally. The wetlands of sub-Saharan Africa include coastal plains, including deltas, estuaries and tidal flats; inland basins, comprising extensive drainage depressions; river flood plains, consisting of recent alluvial deposits bordering rivers; and inland valleys. Inland valleys are known as *dambos* in East and Central Africa, *fadamas* in northern Nigeria and Chad, *bas-fonds* or *marigots* in francophone countries, and 'inland valley swamps' in Sierra Leone (Andriesse, 1986).

There is a large untapped potential for irrigation in Africa, extending to about 24 Mha or 1.8 times greater than the existing irrigation area (You *et al.*, 2011; Fig. 15.2). Nigeria has the largest potential for large- and small-scale irrigation investments, with about 5.7 Mha, followed by Benin, Guinea, Mozambique, Uganda and Tanzania, which each have more than 1 Mha potential. There is an estimated further 2 Mha of irrigated land that could be rehabilitated. Algeria, Egypt, Somalia and Sudan (now Sudan and South Sudan) share more than 70% (about 1.5 Mha) of disused irrigated area (i.e. that could be rehabilitated) (You *et al.*, 2009c); with the exception of Egypt, rice production is not popular in these countries. Water scarcity is a constraint to rice production in Egypt. Double cropping of rice is often possible with the introduction of irrigation, which can augment the rice harvest area. However, introducing intensive rice cultivation using irrigation may increase water scarcity in the future. Proper land-development and crop-management practices are essential for sustaining productivity of irrigated lowland rice systems.

The above assessment of irrigation potential could be used to guide the distribution of investment funds across countries (You *et al.*, 2009c, 2011). But, as the next step, in-depth local-level assessments are essential. AfricaRice has started using remotely sensed imagery and advanced algorithms to map inland valleys at national level. This technology allows more precise estimation of the total area of inland valleys using a standardized method and also mapping of their exact location. The methodology that has been developed and is being evaluated uses a digital elevation model (DEM). The DEM is generated using stereo-pair images collected by the Advanced Spaceborne Thermal Emission and

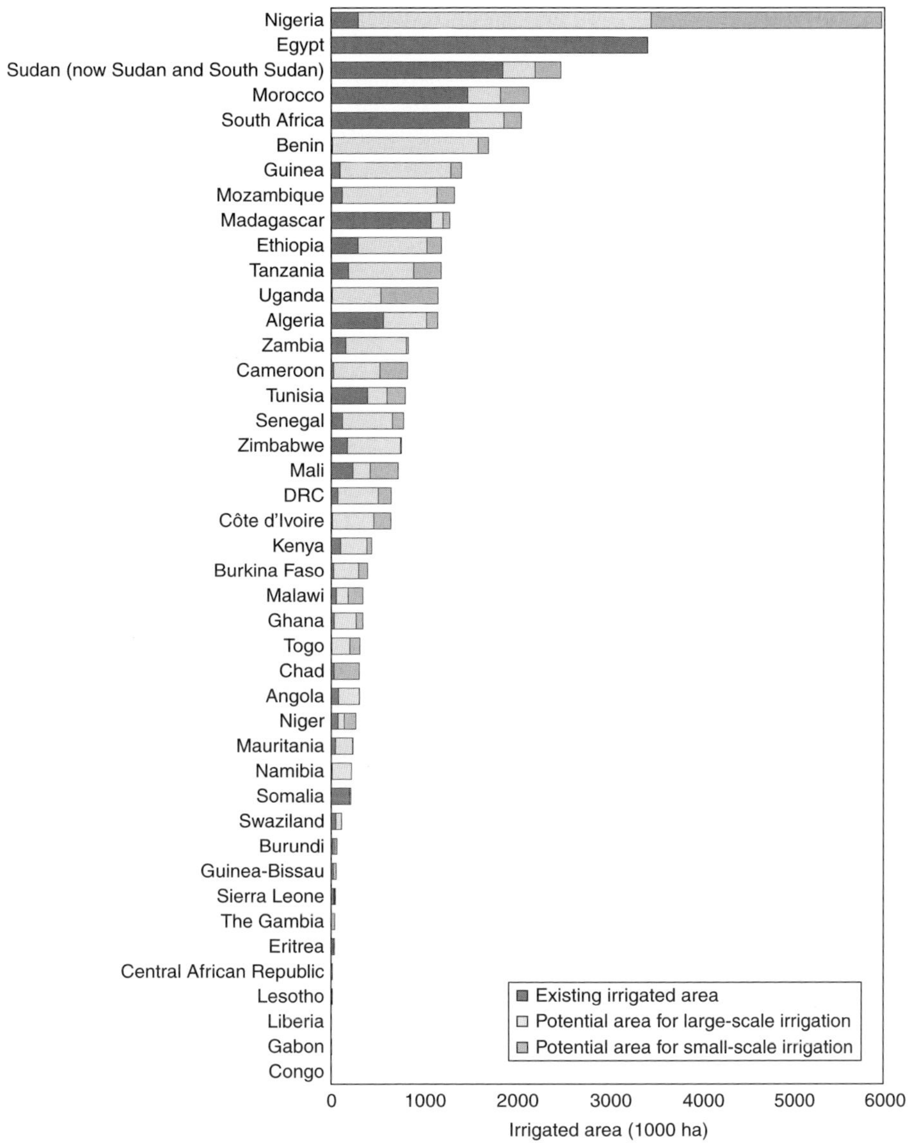

Fig. 15.2. Existing irrigated area and irrigation potential in Africa. (From You *et al.*, 2011, with permission from Elsevier.) The countries in footnotes in Table 15.2 are included except for Comoros and Rwanda. Definitions of large-scale and small-scale irrigation refer to You *et al.* (2011).

Reflection Radiometer (ASTER) instrument – a joint project of NASA and the Ministry of Economy, Trade and Industry (METI) of Japan – on-board the Terra satellite. These images, which have a spatial resolution of 30 m and 1 m accuracy, are freely available for download. The algorithm follows two steps: first, the streams are determined from the DEM using a standardized procedure in ArcGIS; second, a calculation procedure creates transects along each section of the stream. The elevation of the stream is assessed (in metres above sea level) and the areas along the transect that have the same elevation (±2 m) are identified as the inland valley. The procedure has been applied for Benin and Togo, and validated with a digital map of inland valleys from the IMPETUS (An Integrated Approach to the Efficient Management of Scarce Water Resources in West Africa) project (Giertz *et al.*, 2008). The first findings are shown in Plate 8 (A–D). Initial results show the majority of inland valleys to be well mapped, but modifications are required in the algorithm to improve mapping of first-order inland valleys.

Expansion of upland rice areas for increasing rice production may be possible in Africa, through replacing (or rotating with) other upland crops or growing rice in more favourable uplands, where there is sufficient rainfall (low risk of drought for rice cultivation) or supplementary irrigation available. Even in these cases, crop-rotation systems with other crops such as legumes, fallowing or crop–livestock systems is essential for sustainable production, as continuous rice cropping tends to reduce rice yield and sustainability. (See also Haefele *et al.*, Chapter 20, this volume.)

Conclusions

The rice-growing environments in Africa are highly diverse. Rainfed, low-input systems account for more than 60% of the total rice area. In West and East Africa, which have larger rice areas than the other regions, the tropical – warm / sub-humid zone is predominant for irrigated rice and rainfed, low-input rice production systems, whereas in North Africa irrigated rice is grown in the sub-tropical – warm / arid zone. Extreme temperature is an important abiotic constraint to rice production in tropical – warm / semiarid and arid zones in West Africa (high and low temperature) and in the highlands of East Africa (low temperature). In rainfed systems, drought risk is likely to be high in East Africa; the spatial and temporal effects of drought, flooding or a combination of the two needs to be analysed by taking into account local geo-morphological and hydrological information, as well as farmers' rice-cropping practices. Potential yields tend to be higher in East and North Africa than in West Africa. A high percentage of problem (mainly saline and sodic) soils is found in irrigated rice systems in arid and semiarid zones; however, soil constraints in general are more common in humid and sub-humid zones than in arid and semiarid zones across West, Central and East Africa. Epidemics of major diseases are expected to occur in humid and sub-humid zones in West Africa and Madagascar. This information is a first step towards determining research priorities and targeting development and diffusion of rice technologies for each region or country.

The mapping of potential pest epidemics did not cover biotic stresses such as *Rice yellow mottle virus*, or potential outbreaks of insect pests. In the upland production system, stem borers, rice bugs, nematodes and termites are important biotic stresses, while in lowland systems, African rice gall midge and *Rice yellow mottle virus* are also important. Rodents and birds are other major biotic stresses occurring across rice-production systems (Balasubramanian *et al.*, 2007; de Mey and Demont, Chapter 19, this volume; Diagne *et al.*, Chapter 4, this volume). Moreover, problems with weeds are extremely common across major rice-production systems (Rodenburg and Johnson, 2009; Rodenburg and Johnson, Chapter 16, this volume; Diagne *et al.*, Chapter 4, this volume). However, knowledge of the spatial and temporal extent and severity of these biotic stresses is still limited. Thus, further research is needed to determine where, when and how rice production might be affected by biotic constraints in Africa in the future.

In this chapter, we describe the GIS-based characterization of rice-growing environments and potential irrigation areas at the continental or regional level, and promising results for the identification of inland valleys through satellite imagery. However, these will need to be complemented with data and analyses of socio-economic factors such as distance to markets, road conditions and land-tenure issues (Erenstein *et al.*, 2006; You *et al.*, 2011).

More work is clearly needed in developing crop-simulation models that work for African growth conditions. Combined with locally collected data on crop management practices, such models are expected to facilitate estimation of exploitable yield- and water-productivity gaps, identification of risks of outbreaks of pests and diseases, and identification of regions with the greatest potential for enhanced rice production and expansion of rice-cropped area – all issues that are of great importance to Africa's food security.

Acknowledgements

Figure 15.2 is reprinted from *Food Policy*, volume 36, L. You, C. Ringler, U. Wood-Sichra, R. Robertson, S. Wood, T. Zhu, G. Nelson, G. Zhe and S. Yan, 'What is the irrigation potential for Africa? A combined biophysical and socioeconomic approach', pages 770–782, Copyright (2011), with permission from Elsevier.

References

Akanvou, R., Becker, M., Chano, M., Johnson, D.E., Gbaka-Tcheche, H. and Touré, A. (2000) Fallow residue management effects on upland rice in three agroecological zones of West Africa. *Biology and Fertility of Soils* 31, 501–507.

Andriesse, W. (1986) Area and distribution. In: Juo, A.R.S. and Lowe J.A. (eds) *The Wetlands and Rice in Sub-Saharan Africa.* Proceedings of an International Conference, Ibadan, Nigeria, November 1985. International Institute of Tropical Agriculture, Ibadan, Nigeria, pp. 15–30.

Andriesse, W. and Fresco, L.O. (1991) A characterization of rice-growing environments in West Africa. *Agriculture, Ecosystems and Environment* 33, 377–395.

Audebert, A.Y., Becker, M. and Johnson, D.E. (1999) Differential response of rice hydrological conditions and agronomic management. *African Crop Science Conference Proceedings* 4, 107–111.

Bai, J., Chen, X., Dobermann, A., Yang, H., Cassman, K.G. and Zhang, F. (2010) Evaluation of NASA satellite- and model-derived weather data for simulation of maize yield potential in China. *Agronomy Journal* 102, 9–16.

Balasubramanian, V., Sié, M., Hijmans, R.J. and Otsuka, K. (2007) Increasing rice production in sub-Saharan Africa: challenges and opportunities. *Advances in Agronomy* 94, 55–133.

Bationo, A., Hartemink, A., Lungu, O., Naimi, M., Okoth, P., Smaling, E. and Thiombiano, L. (2006) African soils: their productivity and profitability of fertilizer use. Background paper prepared for the African Fertilizer Summit, Abuja, Nigeria, 9–13 June 2006.

Becker, M. and Johnson, D.E. (1998) Legumes as dry season fallow in upland rice-based systems of West Africa. *Biology and Fertility of Soils* 27, 358–367.

Becker, M. and Johnson, D.E. (1999) Rice yield and productivity gaps in irrigated systems of the forest zone of Côte d'Ivoire. *Field Crops Research* 60, 201–208.

Becker, M. and Johnson, D.E. (2001a) Cropping intensity effects on upland rice yield and sustainability in West Africa. *Nutrient Cycling in Agroecosystems* 59, 107–117.

Becker, M. and Johnson, D.E. (2001b) Improved water control and crop management effects on lowland rice productivity in West Africa. *Nutrient Cycling in Agroecosystems* 59, 119–127.

Becker, M., Johnson, D.E., Wopereis, M.C.S. and Sow, A. (2003) Rice yield gaps in irrigated systems along an agro-ecological gradient in West Africa. *Journal of Plant Nutrition and Soil Science* 166, 61–67.

Bouman, B.A.M., Kropff, M.J., Tuong, T.P., Wopereis, M.C.S., ten Berge, H.F.M. and van Laar, H.H. (2003) *ORYZA2000 Growth Model.* International Rice Research Institute, Los Baños, Philippines. Available at www.knowledgebank.irri.org/oryza2000.

de Mey, Y., Demont, M. and Diagne, M. (2012) Estimating bird damage to rice in Africa: evidence from the Senegal River valley. *Journal of Agricultural Economics* 63, 175–200.

Defoer, T., Wopereis, M.C.S., Jones, M.P., Lançon, F., Erenstein, O. and Guei, R.G. (2004) Rice-based production systems for food security and poverty alleviation in sub-Saharan Africa. *International Rice Commission Newsletter* 53, 85–96.

Dingkuhn, M. and Sow, A. (1995) Potential yields of irrigated rice in the Sahel. In: Miezan, K.M., Wopereis, M.C.S., Dingkuhn, M., Deckers, J. and Randolph, T.F. (eds) *Irrigated Rice in the Sahel: Prospects for sustainable development.* West Africa Rice Development Association, Bouaké, Côte d'Ivoire, pp. 361–379.

Dingkuhn, M., Jones, M.P., Johnson, D.E. and Sow, A. (1998) Growth and yield potential of *Oryza sativa* and *O. glaberrima* upland rice cultivars and their interspecific progenies. *Field Crops Research* 57, 57–69.

Ekeleme, F., Kamara, A.Y., Oikeh, S.O., Omoigui, L.O., Amaza, P., Abdoulaye, T. and Chikoye, D. (2009) Response of upland rice cultivars to weed competition in the savannas of West Africa. *Crop Protection* 28, 90–96.

Erenstein, O., Oswald, A. and Moussa, M. (2006) Determinants of lowland use close to urban markets along an agro-ecological gradient in West Africa. *Agriculture, Ecosystems and Environment* 117, 205–217.

Evans, L.T. and Fischer, R.A. (1999) Yield potential: its definition, measurement, and significance. *Crop Science* 39, 1544–1551.
FAO (1995) *The Digital Soil Map of the World.* Version 3.5, completed 3 March 1995, 1:5,000,000 scale. Food and Agriculture Organization of the United Nations, Rome, Italy.
FAO (2002) *FAO Rice Information*, Vol. 3. Food and Agriculture Organization of the United Nations, Rome, Italy.
FAO and IIASA (2000) Global Agro-Ecological Zones 2000. www.iiasa.ac.at/Research/LUC/GAEZ/index.htm.
Giertz, S., Steup, G., Sintondji, L., Gbaguidi, F. and Schönbrodt, S. (2008) Survey of inland valleys in the Upper Ouémé catchment. In: Judex, M. and Thamm, H.P. (eds) *IMPETUS Atlas Benin, Research Results 2000–2007*, 3rd edn. University of Bonn, Bonn, Germany, pp. 99–100.
Haefele, S.M. and Hijmans R.J. (2007) Soil quality in rice-based rainfed lowlands of Asia: characterization and distribution. In: Aggarwal, P.K., Ladha, J.K., Singh, R.K., Devakumar, C. and Hardy, B. (eds) *Science, Technology, and Trade for Peace and Prosperity.* Proceedings of the 26th International Rice Research Conference, New Delhi, India, 9–12 October 2006. International Rice Research Institute, Los Baños, Philippines; Indian Council of Agricultural Research, and National Academy of Agricultural Sciences, New Delhi, India, pp. 297–308.
Haefele, S., Johnson, D.E., Diallo, S., Wopereis, M.C.S. and Janin, I. (2000) Improved soil fertility and weed management is profitable for irrigated rice farmers in Sahelian Africa. *Field Crops Research* 66, 101–113.
Haefele, S., Wopereis, M.C.S., Donovan, C. and Maubuisson, J. (2001) Improving productivity and profitability of irrigated rice production in Mauritania. *European Journal of Agronomy* 14, 181–196.
HarvestChoice (2009) Agroecological zones. http://harvestchoice.org/maps/aez-16-class (accessed 6 September 2012).
Hijmans, R.J. and Serraj, R. (2008) Modeling spatial and temporal variation of drought in rice production. In: Serraj, R., Bennett, J. and Hardy, B. (eds) *Drought Frontiers in Rice: Crop improvement for increased rainfed production.* World Scientific Publishing, Singapore and International Rice Research Institute, Los Baños, Philippines, pp. 19–31.
Homma, K., Horie, T., Shiraiwa, T., Supapoj, N., Matsumoto, N. and Kabaki, N. (2003) Toposequential variation in soil fertility and rice productivity of rainfed lowland paddy fields in mini-watershed (Nong) in northeast Thailand. *Plant Production Science* 6, 147–153.
Kamara, A.Y., Ekeleme, F., Omoigui, L.O., Oikeh, S.O., Chikoye, D. and Tegbaru, A. (2010) Response of upland rice cultivars to nitrogen fertilizer in the savanna of Nigeria. *Agronomy Journal* 102, 333–339.
Kawano, N. and Sakagami, J. (2008) Submergence stress to rice plants in West Africa. In: Sakagami, J. and Ito, O. (eds) Forefront of Rice Cultivation in Africa. *JIRCAS Working Report* No. 57. Japan International Research Center for Agricultural Sciences, Ibaraki, Japan, pp. 25–36.
Lobell, D., Cassman, K. and Field, B.C. (2009) Crop yield gaps: their importance, magnitudes, and causes. *Annual Review of Environment and Resources* 34, 179–204.
Maclean, J.L., Dawe, D.C., Hardy, B. and Hettel, G.P. (eds) (2002) *Rice Almanac.* International Rice Research Institute, Los Baños, Philippines; West Africa Rice Development Association, Bouaké, Côte d'Ivoire; International Center for Tropical Agriculture, Cali, Colombia; Food and Agriculture Organization of the United Nations, Rome, Italy.
Matsushima, S., Koyama, M. and Ishizaki, Y. (1994) Practical studies on low-cost rice cultivation by direct sowing under flooded conditions in the tropics. 15. Establishment of a new method of direct sowing rice cultivation under deeply flooded conditions and a challenge for maximizing yield with it. *Japan Journal of Crop Science* 63 (extra 1), 6–7.
Menete, M.Z.L., van Es, H.M., Brito, R.M.L., DeGloria, S.D. and Famba, S. (2008) Evaluation of System of Rice Intensification (SRI) component practices and their synergies on salt-affected soils. *Field Crops Research* 109, 34–44.
Namba, T. (2003) [Optimum planting density and nitrogen application rate for maximizing rice yield in the Nile delta]. *Japan Journal of Crop Science* 72, 133–141. [Japanese with English abstract.]
Namba, T. (2005) [A method of nitrogen application for maximizing rice yield in the Nile delta]. *Japan Journal of Crop Science* 74, 253–259. [Japanese with English abstract.]
Oikeh, S.O., Nwilene, F., Diatta, S., Osiname, A., Toure, A. and Okeleye, K.A. (2008) Responses of upland NERICA rice to nitrogen and phosphorus in forest agroecosystems. *Agronomy Journal* 100, 735–741.
Rodenburg, J. and Johnson, D.E. (2009) Weed management in rice-based cropping systems in Africa. *Advances in Agronomy* 103, 149–218.

Saito, K. and Futakuchi, K. (2009) Performance of diverse upland rice cultivars in low and high soil fertility conditions in West Africa. *Field Crops Research* 111, 243–250.
Sanchez, P.A. and Buol, S.W. (1985) Agronomic taxonomy for wetland soils. In: *Wetland Soils: Characterization, classification, and utilization.* Proceedings of a workshop held 26 March to 5 April 1984. International Rice Research Institute, Los Baños, Philippines, pp. 207–229.
Sanchez, P., Palm, C.A. and Buol, S.W. (2003) Fertility capability soil classification: a tool to help assess soil quality in the tropics. *Geoderma* 114, 157–185.
Savary, S., Nelson, A., Willocquet, L., Pangga, I. and Aunario, J. (2012) Modeling and mapping potential epidemics of rice diseases globally. *Crop Protection* 34, 6–17.
Seck, P.A., Diagne, A., Mohanty, S. and Wopereis, M.C.S. (2012) Crops that feed the world 7: rice. *Food Security* 4, 7–24.
Sheehy, J.E., Peng, S., Dobermann, A., Mitchell, P.L., Ferrer, A., Yang, J., Zou, Y., Zhong, X. and Huang, J. (2004) Fantastic yields in the System of Rice Intensification: fact or fallacy? *Field Crops Research* 88, 1–8.
Touré, A., Becker, M., Johnson D.E., Koné, B., Kossou D.K. and Kiepe, P. (2009) Response of lowland rice to agronomic management under different hydrological regimes in an inland valley of Ivory Coast. *Field Crops Research* 114, 304–310.
Tsujimoto, Y., Horie, T., Randriamihary, H., Shiraiwa, T. and Homma, K. (2009) Soil management: the key factors for higher productivity in the fields utilizing the System of Rice Intensification (SRI) in the central highland of Madagascar. *Agricultural Systems* 100, 61–71.
White, J.W., Hoogenboom, G., Stackhouse, P.W. and Hoell, J.M. (2008) Evaluation of NASA satellite- and assimilation model-derived long-term daily temperature data over the continental US. *Agricultural and Forest Meteorology* 148, 1574–1584.
Windmeijer, P.N. and Andriesse, W. (eds) (1993) *Inland Valleys in West Africa: An agro-ecological characterization of rice growing environments.* (ILRI Publication 52). International Institute for Land Reclamation and Improvement, Wageningen, Netherlands.
Wopereis, M.C.S., Donovan, C., Nebie, B., Guindo, D. and N'Diaye, M.K. (1999) Soil fertility management in irrigated rice systems in the Sahel and savanna regions of West Africa. *Field Crops Research* 61, 125–145.
Yang, H., Cassman, K., Stackhouse, P.W. and Hoell, J.M. (2007) Usability of NASA satellite imagery-based daily solar radiation for crop yield simulation and management decisions. Poster presented at the Fall American Geophysical Union Meeting, San Francisco, California, 10–14 December.
You, L., Wood, S. and Wood-Sichra, U. (2009a) Generating plausible crop distribution maps for sub-Saharan Africa using a spatially disaggregated data fusion and optimization approach. *Agricultural Systems* 99, 126–140.
You, L., Guo, Z., Koo, J., Ojo, W., Sebastian, K., Tenorio, M.T., Wood, S. and Wood-Sichra, U. (2009b) *Spatial Produciton Allocation Model (SPAM) 2000.* Version 3, Release 1. http://MapSPAM.info/.
You, L., Ringler, C., Nelson, G., Wood-Sichra, U., Robertson, R., Wood, S., Zhe, G., Zhu, T. and Sun, Y. (2009c) Torrents and trickles: irrigation spending needs in Africa. Background Paper 9, African Infrastructure Country Diagnostic. World Bank Group, Washington, DC.
You, L., Ringler, C., Wood-Sichra, U., Robertson, R., Wood, S., Zhu, T., Nelson, G., Zhe, G. and Yan, S. (2011) What is the irrigation potential for Africa? A combined biophysical and socioeconomic approach. *Food Policy* 36, 770–782.

16 Managing Weeds of Rice in Africa

Jonne Rodenburg[1]* and David E. Johnson[2]
[1]*Africa Rice Center (AfricaRice), Dar es Salaam, Tanzania;*
[2]*International Rice Research Institute (IRRI), Los Baños, Philippines*

Introduction

Weeds are the most frequent and widespread biotic constraint to productivity throughout the rice environments of Africa (Balasubramanian *et al.*, 2007; Diagne *et al.*, Chapter 4, this volume). Rice is a weak competitor against weeds and the majority of African farmers have few options and resources available for effective weed control (Rodenburg and Johnson, 2009). Weed competition may increase with inadequate land preparation (soil tillage and levelling), cropping intensification, the introduction of weed seeds (through manure, machinery or as a contaminant of rice seed), use of poor-quality rice seeds, use of old rice seedlings for transplanting, inadequate water management, improper fertilizer management, labour shortages for hand weeding and delayed or inappropriate herbicide applications (e.g. Becker and Johnson, 1999, 2001b). Further, as no single intervention is likely to resolve the challenges of sustainably managing weeds, several measures usually need to be combined to achieve adequate control. This chapter provides an overview of crop and weed management practices currently available, and also considers some issues likely to become relevant to African rice production systems.

Weeds in Rice in Africa

Economic importance

In sub-Saharan Africa (SSA), weeds are estimated to account for rice yield losses of at least 2.2 million tonnes (Mt) per year (Rodenburg and Johnson, 2009). Combined with costs of weed control, the financial losses easily surpass half the cost of current regional rice imports. If not controlled, weeds cause yield losses in the range of 28–74% in transplanted lowland rice, 28–89% in direct-seeded lowland rice and 48–100% in upland rice (Rodenburg and Johnson, 2009). In West Africa, it has been shown that farmers can increase their rice yields by 15–23% by applying relatively basic measures to improve weed control, such as bunding of fields to retain flood water, and timely interventions such as herbicide applications and hand weeding (e.g. Haefele *et al.*, 2000; Becker and Johnson, 2001b; Becker *et al.*, 2003).

African farmers do not, however, perceive weeds as solely undesirable. Many species often considered as weeds also feature in traditional pharmacopoeias (e.g. Stepp and Moerman, 2001) or are collected for domestic use, crop and postharvest pest control functions or as an additional source of food (Rodenburg *et al.*, 2012).

* Corresponding author: j.rodenburg@cgiar.org

© CAB International 2013. *Realizing Africa's Rice Promise* (eds M.C.S. Wopereis *et al.*)

Problem weed species

Certain weed species are particularly difficult to control or have an acute effect on the crop through strong competition for resources (nutrients, water and light) or parasitic nature. Weeds can also have secondary negative effects on a crop, crop-management or postharvest operations, and incur costs to control, complicate access to the field (with or without equipment or machinery), act as vectors of rice diseases, attract other important rice pests (like insects, rodents and birds), impede water flow in irrigation or drainage canals, or lower product quality if their seeds mix with rice grains.

Problem weeds in rice-based production systems in Africa can be characterized in a number of ways depending on their biology or on the problems they pose. Perennial rhizotomous or tuber-bearing species include *Imperata cylindrica*, *Oryza longistaminata*, *Leersia hexandra*, *Bolboschoenus maritimus*, *Sacciolepis africana*, *Cyperus halpan*, *C. esculentus* and *C. rotundus*. Examples of competitive, fast-growing or prolific-seed-producing annual weeds of rice include *Euphorbia heterophylla*, *Cyperus difformis*, *Ageratum conyzoides*, *Eleusine indica*, *Sphenoclea zeylanica* and *Ammannia prieureana*. Some grass species, like *O. barthii*, *O. longistaminata*, *Echinochloa colona* and *E. crus-pavonis*, are difficult to target in the field because of their resemblance to rice, while other species – such as *Ischaemum rugosum* and *Rottboellia cochinchinensis* – cause postharvest problems as they produce seeds that are similar in size and shape to rice grains. Weed species can be alternative hosts for diseases and pests such as *Rice yellow mottle virus* and African rice gall midge (e.g. *O. barthii*, *O. longistaminata*, *E. crus-pavonis*) or attract birds (e.g. *E. colona*), while deep-water or aquatic plants may infest and block drainage and irrigation canals (e.g. *Typha domingensis*, *Acroceras zizanioides*, *Ipomoea aquatica*, *Eichhornia crassipes*, *Salvinia nymphellula*, *Pistia stratiotes*). Another group is the parasitic weeds (e.g. *Striga asiatica*, *S. hermonthica*, *Rhamphicarpa fistulosa*), which parasitize the roots of cereal crops like rice to survive (*Striga* spp.) or to enhance their reproductive success (e.g. *R. fistulosa*), and are increasing in importance according to several reports (e.g. Rodenburg *et al.*, 2010, 2011b). Some important weeds in different rice environments are listed in Table 16.1.

Table 16.1. Important weed species in upland, hydromorphic and lowland rice production systems in Africa. (Adapted from Rodenburg and Johnson, 2009.)

Upland		Hydromorphic		Lowland	
Rottboellia cochinchinensis	A, g	*Ageratum conyzoides*	A, b	*Sphenoclea zeylanica*	A, b
Digitaria horizontalis	A, g	*Panicum laxum*	A, g	*Cyperus difformis*	A, s
Ageratum conyzoides	A, b	*Leersia hexandra*	P, g	*Fimbristylis littoralis*	A, s
Euphorbia heterophylla	A, b	*Cyperus rotundus*	P, s	*Oryza longistaminata*	P, g
Imperata cylindrica	P, g	*Digitaria horizontalis*	A, g	*Echinochloa colona*	A, g
Paspalum scrobiculatum	P, g	*Eclipta prostrata*	A, b	*Echinochloa crus-pavonis*	A, g
Mariscus cylindristachyus	P, s	*Spilanthes uliginosa*	A, b	*Leersia hexandra*	P, g
Trianthema portulacastrum	A, b	*Commelina benghalensis*	A, b	*Oryza barthii*	A, g
Striga hermonthica	A, p[a]	*Fimbristylis littoralis*	A, s	*Cyperus iria*	A, s
Striga asiatica	A, p[a]	*Echinochloa colona*	A, g	*Bolboschoenus maritimus*	P, s
Cynodon dactylon	P, g	*Cyperus esculentus*	P, s	*Ischaemum rugosum*	A, g
Commelina benghalensis	A, b	*Cynodon dactylon*	P, g	*Panicum laxum*	A, g
Brachiaria lata	A, g	*Rhamphicarpa fistulosa*	A, p[b]	*Ludwigia abyssinica*	A, b
Cyperus rotundus	P, s			*Ammania prieureana*	A, b
Chromolaena odorata	P, b			*Cyperus esculentus*	P, s
Panicum laxum	A, g			*Cyperus halpan*	P, s
				Eclipta prostrata	A, b
				Rhynchospora corymbosa	P, s

A = annual, P = perennial; g = grass, b = broadleaved, s = sedge; p = parasitic (a = obligate hemi-parasitic, b = facultative hemi-parasitic).

Weed Management in African Rice-based Cropping Systems

Across ecosystems

Vigorous early crop growth with rapid canopy closure is imperative for rice to compete well with weeds, particularly for light. Many factors contribute to rapid early rice growth and good crop establishment, including, for example, good land preparation comprising soil tillage, bunding and levelling to enable uniform flooding depth (in continuous or temporarily flooded systems), the use of good-quality rice seed or healthy rice seedlings for transplanting and, depending on the system, timely water and nutrient management. Optimizing such components through integrated crop management (ICM) practices has been shown to reduce weed problems and increase productivity by up to 25% on farmers' fields (Becker and Johnson, 1999; Haefele *et al.*, 2000).

Hand weeding is the most widely applied intervention against weeds across rice systems. While it is effective in reducing direct competition from weeds and in preventing weeds from producing and shedding seeds, it is extremely labour demanding, requiring 250 to 780 work hours per hectare (Rodenburg and Johnson, 2009). For farm households at subsistence level, the burden of hand weeding is commonly borne by women and school-age children. Hand hoes or push weeders are often used in row-planted crops. Such tools, however, are difficult to use to control weeds in the crop row and they may also cause crop damage (Navasero and Khan, 1970). The use of power tillers or tractors for mechanical weeding is not common in SSA, but in appropriate locations (with favourable soil and hydrological conditions) such machines could alleviate some of the labour burdens associated with hand weeding. Fires, either pre- or post-season, are widely used by farmers to clear weeds from fields, and can save a considerable amount of labour where mechanization is not available. The frequent use of fire can, however, result in a shift in the weed composition towards more tolerant species such as *I. cylindrica*.

Herbicides provide an economically attractive alternative to hand weeding by reducing overall weeding time and enabling farmers to use time- and labour-saving crop establishment methods such as direct (broadcast) seeding rather than transplanting (e.g. Akobundu and Fagade, 1978). Herbicides are commonly used in combination with other control options. In the irrigated schemes in the north of Senegal, for instance, in direct-seeded rice most farmers rely on chemical weed control followed by hand weeding (e.g. Haefele *et al.*, 2002). However, effective and safe herbicide use requires farmers to use the appropriate product, application equipment, rates (Zimdahl, 2007) and timing (e.g. Haefele *et al.*, 2000). Subsistence farmers in SSA frequently lack sources of information or are unable to read the use and safety instructions; they also have limited market access while, in turn, markets often have a limited product range and intermittent supplies. In addition, farmers often lack sufficient financial means for the purchase of the product as well as the required application and protection equipment (Balasubramanian *et al.*, 2007). Incorrect use of herbicides may result in poor weed control (Haefele *et al.*, 2000), increased costs and phytotoxicity damage to the crop (e.g. Johnson *et al.*, 2004) or accelerate the evolution of herbicide resistance in weeds.

Choice of rice cultivar by farmers is also often, at least partly, influenced by the cultivar's ability to suppress or compete with weeds. Examples of where farmers favour competitive cultivars include the choice of Jaya in the irrigated schemes of the Senegal River valley (e.g. Poussin *et al.*, 2005) and the frequently observed use of the vigorous landraces of the African rice species *O. glaberrima* (e.g. Sarla and Swamy, 2005). Some of the newly developed lowland NERICA cultivars have useful weed-competitive traits too (Rodenburg *et al.*, 2009). Apart from these cases, however, cultivars with confirmed weed-competitive traits and good adaptation to African rice ecosystems are still relatively rare (e.g. Rodenburg and Johnson, 2009). Poor competitiveness of 'improved' rice cultivars against weeds may be a contributory factor to their limited adoption by farmers in the upland and rainfed rice areas where weeds are often a serious constraint. On the other hand, superior weed competitiveness alone is unlikely to be sufficient reason for adoption. Grain quality characteristics including colour, cooking and eating qualities, and characteristics such as cycle length and height are probably more important for cultivars to be accepted by farmers (Dalton, 2004) and there are local stresses (both biotic and abiotic) against which farmers will judge a cultivar's suitability. The challenge remains therefore to combine weed competitiveness with other desirable qualities.

Given the wide range of potential weed species in rice cropping systems, and the diverse conditions under which they germinate and grow, integrated approaches to weed management are likely to be the most effective and sustainable. Multiple strategies are required to provide control across the various periods when different weeds will be able to establish, such as when the fields dry out or are drained. Integrated weed management may also be more compatible with farmers' resources than single-component technologies that may require a high level of external inputs. The choice of different practices will depend on the production system (irrigated, rainfed upland or rainfed lowland). An overview of strategies, their application environments, and advantages and disadvantages, is provided in Table 16.2.

Table 16.2. Overview of weed control strategies in rice in Africa.

Weed control method	Target systems	Advantage	Main disadvantages	R&D priority
Hand weeding	Mainly in rainfed systems	Highly effective, prevents weed seed production; important in providing 'spot control' of problem weeds	Labour intensive, negative effects on women and children	Low
Controlled flooding	Irrigated systems	Controls most weed species	Requires large amounts of water, good infrastructure and equipment	Medium
Pre- or post-season fires	Across systems	Can reduce seed production and soil seed bank	Ineffective for species like *C. odorata* and *I. cylindrica*; can cause soil degradation	Low
Mechanical weeding	Across systems	Effective, prevents weed seed production, relatively quick	Requires availability of equipment, less effective in controlling weeds in the crop row	High
Chemical weed control, including resistance management	Mainly in larger-scale irrigated systems	Effective when applied well, labour-saving	High market dependence, requires equipment and know-how, risk of development of herbicide-resistant weeds	High
Improved rice cultivars (weed competitive or parasite resistant)	Upland systems, direct-seeded lowland systems, but not broadly applied	Effective, cheap and labour-saving	Requires combination with other genetic traits (e.g. grain quality, stress resilience)	Medium
Crop rotations, intercropping, improved fallow	Mainly in rainfed upland systems	Provides basis of resilient systems	Requires land area; risk of competition with rice crop	Medium
Integrated weed management	General	Effective and putatively sustainable	Labour and knowledge intensive	High

Irrigated ecosystems

Maintaining standing water in the field is a critical weed-management strategy in irrigated rice systems. Flooding the soil to 5–10 cm water depth or more reduces the emergence and establishment rate of most weed species (e.g. Akobundu, 1987). For this to be effective, however, fields need to be well levelled, to ensure an even depth of flooding, which in turn requires skills and equipment not commonly available to resource-poor farmers in this region. Herbicides are also important means of weed control in the direct-seeded lowland systems (Johnson, 1997). Chemical control methods in irrigated lowland rice systems involve the pre-emergence applications of oxadiazon or pendimethalin, pre- or (early) post-emergence applications of butachlor, glyphosate or paraquat, or post-emergence spraying of 2,4-D, bentazon, propanil or bensulfuron (Table 16.3).

In some irrigated systems, a dry-season upland crop (e.g. maize, legumes or cotton) is grown. Such rotations provide an opportunity for alternative control measures such as a selective herbicide (Zimdahl, 2007) and the management and conditions associated with a non-rice crop may not favour troublesome weeds such as *Echinochloa* spp. A range of methods such as land preparation, hand weeding, herbicides and flooding are commonly used in an integrated manner to address problem weeds such as the perennial wild rice species *O. longistaminata* (for examples, see Rodenburg and Johnson, 2009).

Rainfed upland ecosystems

In the humid forest zone in West Africa, rice farmers traditionally managed soil fertility, weeds and other biotic stresses through shifting

Table 16.3. Common herbicides and their application range – combinations are often used to control a wider range of weed species. (Adapted from Johnson, 1997.)

	Active ingredient	Lowland	Upland	Example products	Target weeds	Known exceptions[a]
Post-emergence	2,4-D	+	+	Herbazol	B,S	*C. benghalensis, E. heterophylla*
	bentazon	+		Basagran	B,S	
	MCPA	+		Herbit	B,S	
	molinate	+		Ordram	G,S,(B)	*I. rugosum*
	propanil	+	+	Stam	G,(B,S)	*O. barthii, R. cochinchinensis, C. benghalensis, E. prostrata, T. portulacastrum*
Pre-/post-emergence	triclopyr		+	Garlon	B,S	
	bensulfuron	+		Londax	B,S	
	butachlor	+		Machete	B,G,S	*L. hexandra, O. barthii, R. cochinchinensis, C. benghalensis, E. prostrata, T. portulacastrum*
	glyphosate	+	+	Round-up	B,S,G	
	paraquat	+	+	Gramoxone	B,G,S	
	piperophos	+		Rilof	G,S	*F. littoralis, E. indica*
	quinclorac	+		Facet	G	
	thiobencarb	+		Saturn	G,S,B	*L. hexandra, O. barthii, R. cochinchinensis, A. conyzoides, C. benghalensis, E. prostrata*
Pre-emergence	fluorodifen	+	+	Preforan	B	
	oxadiazon	+	+	Ronstar	B,S,G	*O. barthii, C. benghalensis, C. odorata, E. prostrata*
	pendimethalin	+	+	Stomp	G,B,S	*L. hexandra, O. barthii, C. benghalensis, E. heterophylla*
		+		Rifit	S,G,B	

B = broadleaved; G = grass; S = sedge; [a]weed species with known resistance to the specific herbicide.

cultivation – long fallows (>10 years) alternated with short (1–3 seasons) cropping periods (e.g. de Rouw, 1995). In such systems, the fallow vegetation and a portion of the seed bank are killed by 'slash-and-burn' fires and the populations of pioneer species including crop weeds are limited by the short length of the cropping period. Such systems are still common in some areas, though they have become less widespread (Ampong-Nyarko, 1996; Johnson, 1997). Greater demand for rice and increasing population density in SSA, however, have resulted in intensification of rice production (Balasubramanian *et al.*, 2007) with reduced fallow periods leading to increased weed problems (Becker and Johnson, 2001a).

As upland cropping systems have intensified, herbicides have sometimes become an important control method, such as in rice–cotton rotation systems in the savannah zone (Johnson, 1997). Herbicides used in rainfed upland rice systems include fluorodifen, oxadiazon and pendimethalin applied pre-emergence, butachlor, piperophos and thiobencarb applied pre- or early post-emergence, and 2,4-D, bentazon, MCPA, propanil and triclopyr sprayed post-emergence (Table 16.3).

Rotations of rice with non-cereal crops like cowpea, soybean and groundnut are common in subsistence systems, and changing of cropping practices may aid the management of problem rice weeds (Rodenburg and Johnson, 2009). While such practices impact on weed species composition in rice (Kent *et al.*, 2001), studies on rotations and intercropping in rice-based systems in Africa are scarce. Such methods, for example, have been advocated in the control of *Striga* spp., for which rotations or the use of 'trap crops' have been proposed and tested in cereal crops other than rice (as reviewed by Rodenburg *et al.*, 2010). Such approaches could be validated and promoted for rice-based cropping systems in the future.

Improving the 'quality of fallow vegetation' in rotation systems has been proposed to reduce weed growth and improve soil fertility (Becker and Johnson, 1998). Such improved fallows use weed-suppressing legumes that continue to grow after rice harvest, thereby reducing weed growth and build-up of the weed seed bank during the off-season. Such 'short fallow' rotation systems in the forest and savannah zones of West Africa have been shown to increase rice yields by 20–30% and lower weed growth in the crop (e.g. Becker and Johnson, 1998). The choice of fallow species, plant population density, planting date and crop management, however, needs to be carefully considered to avoid competition for resources between the legume and the rice. Despite possible advantages, intercropping, improved fallow systems or relay cropping with legumes show low farmer adoption rates in Africa. This lack of adoption has been suggested as being due to the additional labour and energy required for clearing and incorporation of the legume into the soil, unfavourable land-tenure agreements, poor crop establishment, additional costs of inputs, and a lack of direct economic benefit of the legume (e.g. Tarawali *et al.*, 1999).

Rainfed lowland ecosystems

Bunding of fields in rainfed lowland systems improves water management, and increases the periods for which the soil remains flooded, and has been shown to decrease weed biomass by 25% (Becker and Johnson, 2001b). Extending the period of flooding reduces opportunities for weeds to germinate and establish. In addition, most of the weed control strategies described above can also be applied in rainfed lowland systems, although the use of rotations with non-rice crops is usually restricted to the dry season due to the limited range of crops tolerant of the flooded or waterlogged conditions of the wet season.

Future Weed Management Issues

Crop intensification, labour and water shortages, changing environmental conditions, and the evolution of herbicide-resistant weed ecotypes will limit the options and 'set the agenda' for weed management issues in rice in Africa in the future (e.g. Rodenburg and Johnson, 2009). Decreasing labour–crop area ratios expected in rice-production areas in SSA may cause farmers to gradually shift from transplanting to direct seeding. In Asia, the transition to direct seeding, coupled with an increased reliance on herbicides, resulted in a shift in the weed population structure to one dominated by grass species such as *Echinochloa* spp., *Leptochloa chinensis*, *I. rugosum*,

L. hexandra and 'weedy' rices (Rao *et al.*, 2007). If rice farmers in Africa follow a similar model, such weeds, which are already present in African rice systems, are expected to rapidly become dominant. While there are as yet no confirmed cases of herbicide resistance among important rice weeds in Africa (Rodenburg and Johnson, 2009), populations of herbicide-resistant weed ecotypes are likely to develop. The problem may, however, already exist. Propanil, for example, has been observed to be less effective in controlling *E. colona* in Senegal than in the past (Haefele *et al.*, 2000), and the continuous use of a single product is quite common in SSA due to the limited range of products available on the local market. Changes in environmental conditions such as increased levels of atmospheric CO_2, increased temperatures and rainfall irregularities may also render weeds more resistant (e.g. Patterson *et al.*, 1999). This may have serious consequences for systems where herbicides are the main means of weed control.

In lowland rice, where increases in temperature or rainfall variability will likely have a smaller direct impact, CO_2 increases may cause a crop that uses the C_3 photosynthesis pathway (like rice) to become more competitive against weeds that use the C_4 pathway. At the same time, C_3 weed species, such as the perennial rhizotomous *O. longistaminata*, *L. hexandra*, *B. maritimus*, *S. africana* and *C. halpan* may benefit and become even more difficult to control (Rodenburg *et al.*, 2011a) as increased CO_2 has a stimulating effect on belowground growth (Oechel and Strain, 1985). If, however, in response to water scarcity, water-saving production methods are implemented in irrigated systems, overall weed competition will increase (e.g. Krupnik *et al.*, 2012) and a shift can be expected to species favouring hydromorphic conditions such as *A. amplectans*, *E. colona*, *E. indica*, *Panicum repens*, *C. esculentus*, *Eleocharis* spp., *Scirpus maritimus*, *Ageratum conyzoides* and *Eclipta prostrata*.

Climate change is expected to have a greater impact on rainfed production systems as these are most vulnerable to rainfall irregularities and are populated by most of the C_4 weeds and all of the parasitic weed species. In rainfed systems, the area infested with parasitic weeds could increase, particularly in places where soil degradation and erratic rainfall become prevalent (Rodenburg *et al.*, 2010). Furthermore, because of their likely greater drought and heat tolerance, C_4 species like the perennial grasses *I. cylindrica*, *Paspalum scrobiculatum* and *Cynodon dactylon*, the annual grasses *R. cochinchinensis*, *Digitaria horizontalis*, *E. indica*, *Dactyloctenium aegyptium*, *Pennisetum purpureum* and *E. colona*, and the sedges *Fimbristylis littoralis*, *C. rotundus* and *C. esculentus* are likely to become more competitive with rice (Rodenburg *et al.*, 2011a).

A relatively recent approach to weed control is the combined use of non-selective herbicides with herbicide-resistant rice cultivars. These approaches may have the labour-saving benefits of conventional chemical control without the concomitant phytotoxicity risks, and would provide technically sound solutions for the control of important yield-reducing weeds, such as wild or weedy rices in irrigated and rainfed lowlands, and parasitic weeds in uplands (Rodenburg and Demont, 2009). These technologies, however, will require careful stewardship to reduce the incidence of gene-flow to wild or weedy rice relatives causing them to develop herbicide resistance. Also, the reliance on a limited range of herbicide molecules would inevitably result in the evolution of herbicide resistance among a wider range of weed species. In addition, effective seed and microcredit systems would be a prerequisite to successful application of these technologies (Demont *et al.*, 2009). Meeting such prerequisites, however, is likely to be a challenge for most rice-growing areas in SSA.

Likely changes in the economic environment for rice production in Africa will mean that weed management systems must evolve to improve labour productivity. Farmers are likely to make increased use of herbicides and integrated measures. They are also required to improve the timing of operations and targeting of certain growth stages for given species to give the best chance of control. Such changes will inevitably mean that weed control will become more 'knowledge intensive', and for this farmers will require better and more timely access to information resources than they have at present.

Concluding Remarks

Additional options to manage weeds are required to augment the limited range currently available to rice farmers in Africa. Approaches are needed to target local problem weeds and these must be tailored to local practice and resource availability. Strategies also need to anticipate likely future changes in soil fertility, water availability and population pressure, while addressing environmental concerns by avoiding excessive dependency on herbicides. Emerging weed problems include species such as hemi-parasitic, annual grass and perennial rhizotomous weeds. These weeds are likely to get worse, and effective strategies are required to control them. Moreover, future climate change adaptation strategies for rice-based production systems, such as alternative cropping system designs or improved stress-tolerant cultivars, should simultaneously address possible implications for weed competition.

There are many gaps in our knowledge of the crop and weed interactions in relation to the environment and management practice. A better understanding of these interactions would support predictions of likely changes in weed communities and strengthen the development of appropriate and effective knowledge-based technologies as part of an integrated weed management approach. Development efforts should include training farmers in integrated crop and weed management, optimal and safe herbicide use, and the establishment of practical and low-risk finance systems (e.g. microcredit). Meanwhile, national and regional policies to stimulate favourable and functional input markets, infrastructure and biosafety regulations would help African rice farmers to effectively manage weeds.

References

Akobundu, I.O. (1987) *Weed Science in the Tropics – Principles and practices.* Wiley, Chichester, UK.

Akobundu, I.O. and Fagade, S.O. (1978) Weed problems of African rice lands. In: Buddenhagen, I.W. and Persley, G.J. (eds) *Rice in Africa.* Academic Press, London, pp. 181–192.

Ampong-Nyarko, K. (1996) Weed management in rice in Africa. In: Auld, B.A. and Kim, K.U. (eds) *Weed Management in Rice.* Food and Agriculture Organization of the United Nations, Rome, Italy, pp. 183–191.

Balasubramanian, V., Sie, M., Hijmans, R.J. and Otsuka, K. (2007) Increasing rice production in sub-Saharan Africa: challenges and opportunities. *Advances in Agronomy* 94, 55–133.

Becker, M. and Johnson, D.E. (1998) Legumes as dry season fallow in upland rice-based systems of West Africa. *Biology and Fertility of Soils* 27, 358–367.

Becker, M. and Johnson, D.E. (1999) Rice yield and productivity gaps in irrigated systems of the forest zone of Côte d'Ivoire. *Field Crops Research* 60, 201–208.

Becker, M. and Johnson, D.E. (2001a) Cropping intensity effects on upland rice yield and sustainability in West Africa. *Nutrient Cycling in Agroecosystems* 59, 107–117.

Becker, M. and Johnson, D.E. (2001b) Improved water control and crop management effects on lowland rice productivity in West Africa. *Nutrient Cycling in Agroecosystems* 59, 119–127.

Becker, M., Johnson, D.E., Wopereis, M.C.S. and Sow, A. (2003) Rice yield gaps in irrigated systems along an agro-ecological gradient in West Africa. *Journal of Plant Nutrition and Soil Science* 166, 61–67.

Dalton, T.J. (2004) A household hedonic model of rice traits: economic values from farmers in West Africa. *Agricultural Economics* 31, 149–159.

de Rouw, A. (1995) The fallow period as a weed-break in shifting cultivation (tropical wet forests). *Agriculture, Ecosystems and Environment* 54, 31–43.

Demont, M., Rodenburg, J., Diagne, M. and Diallo, S. (2009) *Ex ante* impact assessment of herbicide resistant rice in the Sahel. *Crop Protection* 28, 728–736.

Haefele, S.M., Johnson, D.E., Diallo, S., Wopereis, M.C.S. and Janin, I. (2000) Improved soil fertility and weed management is profitable for irrigated rice farmers in Sahelian West Africa. *Field Crops Research* 66, 101–113.

Haefele, S., Wopereis, M.C.S. and Donovan, C. (2002) Farmers' perceptions, practices and performance in a Sahelian irrigated rice scheme. *Experimental Agriculture* 38, 197–210.

Johnson, D.E. (1997) *Weeds of Rice in West Africa.* West Africa Rice Development Association, Bouaké, Côte d'Ivoire.

Johnson, D.E., Wopereis, M.C.S., Mbodj, D., Diallo, S., Powers, S. and Haefele, S.M. (2004) Timing of weed management and yield losses due to weeds in irrigated rice in the Sahel. *Field Crops Research* 85, 31–42.

Kent, R.J., Johnson, D.E. and Becker, M. (2001) The influences of cropping system on weed communities of rice in Côte d'Ivoire, West Africa. *Agriculture, Ecosystems and Environment* 87, 299–307.

Krupnik, T.J., Rodenburg, J., Haden, V.R., Mbaye, D. and Shennan, C. (2012) Genotypic trade-offs between water productivity and weed competition under the System of Rice Intensification in the Sahel. *Agricultural Water Management* 115, 156–166.

Navasero, N.C. and Khan, A.U. (1970) Use of mechanical power for rotary weeding. *Pans* 16, 87–92.

Oechel, W.C. and Strain, B.R. (1985) Native species responses to increased atmospheric carbon dioxide concentration. In: Strain, B.R. and Cure, J.D. (eds) *Direct Effects of Increasing Carbon Dioxide on Vegetation.* US Department of Energy, Office of Basic Energy Sciences, Carbon Dioxide Research Division, Washington, DC. [Available from the National Technical Information Service, US Department of Commerce, Springfield, Virginia.]

Patterson, D.T., Westbrook, J.K., Joyce, R.J.V., Lingren, P.D. and Rogasik, J. (1999) Weeds, insects, and diseases. *Climatic Change* 43, 711–727.

Poussin, J.C., Diallo, Y., Legoupil, J.C. and Sow, A. (2005) Increase in rice productivity in the Senegal River valley due to improved collective management of irrigation schemes. *Agronomy for Sustainable Development* 25, 225–236.

Rao, A.N., Johnson, D.E., Sivaprasad, B., Ladha, J.K. and Mortimer, A.M. (2007) Weed management in direct-seeded rice. *Advances in Agronomy* 93, 153–255.

Rodenburg, J. and Demont, M. (2009) Potential of herbicide resistant rice technologies for sub-Saharan Africa. *AgBioForum* 12, 313–325.

Rodenburg, J. and Johnson, D.E. (2009) Weed management in rice-based cropping systems in Africa. *Advances in Agronomy* 103, 149–218.

Rodenburg, J., Saito, K., Kakai, R.G., Toure, A., Mariko, M. and Kiepe, P. (2009) Weed competitiveness of the lowland rice varieties of NERICA in the southern Guinea savanna. *Field Crops Research* 114, 411–418.

Rodenburg, J., Riches, C.R. and Kayeke, J.M. (2010) Addressing current and future problems of parasitic weeds in rice. *Crop Protection* 29, 210–221.

Rodenburg, J., Meinke, H. and Johnson, D.E. (2011a) Challenges for weed management in African rice systems in a changing climate. *Journal of Agricultural Science* 149, 427–435.

Rodenburg, J., Zossou, N., Gbehounou, G., Ahanchede, A., Touré, A., Kyalo, G. and Kiepe, P. (2011b) *Rhamphicarpa fistulosa*, a parasitic weed threatening rain-fed lowland rice production in sub-Saharan Africa – a case study from Benin. *Crop Protection* 30, 1306–1314.

Rodenburg, J., Both, J., Heitkönig, I.M.A., van Koppen, C.S.A., Sinsin, B., Van Mele, P. and Kiepe, P. (2012) Land-use and biodiversity in unprotected landscapes: the case of non-cultivated plant use and management by rural communities in Benin and Togo. *Society and Natural Resources* 25(12), 1221–1240.

Sarla, N. and Swamy, B.P.M. (2005) *Oryza glaberrima*: a source for the improvement of *Oryza sativa. Current Science* 89, 955–963.

Stepp, J.R. and Moerman, D.E. (2001) The importance of weeds in ethnopharmacology. *Journal of Ethnopharmacology* 75, 19–23.

Tarawali, G., Manyong, V.M., Carsky, R.J., Vissoh, P.V., Osei-Bonsu, P. and Galiba, M. (1999) Adoption of improved fallows in West Africa: lessons from mucuna and stylo case studies. *Agroforestry Systems* 47, 93–122.

Zimdahl, R.L. (2007) *Fundamentals of Weed Science*, 3rd edn. Academic Press, London.

17 Managing the Major Diseases of Rice in Africa

Yacouba Séré,[1]* Denis Fargette,[2] Myimaorga Emanuel Abo,[3] Kerstin Wydra,[4] Mohamed Bimerew,[4] Amos Onasanya[1] and Salomon Kofi Akator[5]

[1]Africa Rice Center (AfricaRice), Dar es Salaam, Tanzania; [2]Institut de recherche pour le développement (IRD), Montpellier, France; [3]National Cereals Research Institute (NCRI), Badegi, Nigeria; [4]Centre for Tropical and Subtropical Agriculture and Forestry (CeTSAF), Georg-August Universität, Goettingen, Germany; [5]Africa Rice Center (AfricaRice), Cotonou, Benin

Introduction

Rice is the principal food grain consumed by almost half of the world's population (Khush, 2005), making it the most important food crop currently produced (Cottyn *et al.*, 2001). Rice is increasingly becoming a regular staple for the populations of sub-Saharan Africa (SSA). Rice availability and prices impact directly on the welfare of the poorest consumers in the region, many of whom are resource-poor farmers depending on rice as both a staple food and a source of income. It is therefore not surprising that rice is a major component of the food-security and poverty-alleviation strategies of many SSA countries. Against this background, any improvement in rice productivity will contribute significantly to achieving a higher level of regional and household food security, while responding to the needs of the poorest by enhancing their diet both quantitatively and qualitatively and by providing additional income opportunities (Seck *et al.*, 2012).

The 2007–2008 crisis in rice availability and price prompted African countries to develop initiatives to increase their domestic production by increasing the area under rice cultivation or increasing the productivity per unit area by using high-yielding varieties and fertilizers. However, both development of new areas for rice cultivation and intensification face prevailing and unpredictable challenges, among which diseases are likely to feature prominently.

This chapter gives an overview of the rice diseases identified in Africa and their importance, and then focuses on the three major ones. We describe what is known about pathogen variability and show how that knowledge can be used to create varietal resistance as part of an integrated approach to disease management.

Major Rice Diseases and Their Importance in Africa

Diseases of rice

Various studies have inventoried rice diseases in SSA (Roger, 1958, Notteghem and Baudin, 1981; Akinsola *et al.*, 1982, Mbodj *et al.*, 1984; Séré, 1988a,b,c; Sy and Séré, 1996).

* Corresponding author: y.sere@cgiar.org

© CAB International 2013. *Realizing Africa's Rice Promise* (eds M.C.S. Wopereis *et al.*)

At a very early stage, research on rice diseases was conducted in the framework of an integrated pest management (IPM) approach. As early as 1979, a seminar on the integrated management of rice pests was organized by the Africa Rice Center (AfricaRice) in Bobo-Dioulasso (Burkina Faso). A further seminar in 1981 at Fendall, Liberia (ADRAO, 1982), and a series of training workshops gave impetus to the IPM approach.

Three categories of rice pathogens are identified (Sy and Séré, 1996): (i) **major pathogens** (Plates 9a, 9b, 10 and 11) – blast fungus (*Magnaporthe oryzae*), *Rice yellow mottle virus* (RYMV) and the bacterium responsible for leaf blight (*Xanthomonas oryzae* pv. *oryzae*), the prevalence of which (by rice ecosystem) was defined by the Integrated Pest Management (IPM) Task Force (Table 17.1); (ii) **secondary pathogens** – responsible for brown spot (*Bipolaris oryzae*), leaf scald (*Gerlachia oryzae*) and sheath blight (*Rhizoctonia solani*); and (iii) other **pathogens classified as minor** – responsible for false smut (*Ustilaginoides virens*), narrow brown spot (*Cercospora jansenea*), sheat rot (*Sarocladium oryzae*), bakanae disease (*Fusarium moniliforme*), bacterial leaf streak (*Xanthomonas oryzae* pv. *oryzicola*) and grain discoloration (caused by a complex of fungi). Two further pathogens which inflict minor yield losses – *Sclerophtora macrospora* and *Corallocytostroma oryzae* – have also been identified in Africa (Séré, 1988a,c). Moreover a pathogen responsible for red stripe (El-Namaky, 2011) was recently described (Plate 12).

Despite many efforts to develop integrated management of rice diseases based on varietal resistance, unexpected epidemic explosions still appear – such as blast in Kenya in 2008 (Séré *et al.*, 2011) and RYMV in Rwanda in 2009 (Ndikumana *et al.*, 2011).

Table 17.1. Importance of major rice diseases across rice ecosystems in West Africa. (Adapted and corrected from Fakorede and Yoboué, 2001).

	Rice ecosystem			
			Irrigated	
Disease	Upland	Lowland	Forest and savannah	Sahel
Blast	++	+	+	–
RYMV	–	+	+	++
Bacterial blight	–	–	+	+

Key: from – no disease to ++ high pressure.

Rice blast

Rice blast is caused by an ascomycete fungus *Magnaporthe oryzae* (anamorph: *Pyricularia oryzae*) (Couch and Kohn, 2002). The symptoms (elliptical grey-white lesions) appear on the aboveground organs of the rice plant: the most frequently described are leaf, node and neck blast. Neck blast (Plate 9b) is considered more destructive than leaf blast (Plate 9a) (Zhu *et al.*, 2005).

Rice blast has been widely and intensively studied both globally and in Africa, because the interaction between rice and the blast pathogen has both practical and theoretical interest. The practical interest is related to the importance of rice in human nutrition and the importance of unpredictable yields losses caused by blast worldwide (Jia *et al.*, 2009). The theoretical interest is linked to the fact that the blast pathogen is considered as a model for investigation by plant pathologists (Valent, 1990).

According to Bidaux (1978) rice blast was first reported in Africa in 1922. It is the most widespread disease in SSA. In Burkina Faso, surveys in farmers' fields indicated that intensifying rice cultivation (use of fertilizer and modern, but susceptible, varieties) may lead to increased yield losses due to blast, reducing an important part of the benefit created by intensifying rice cultivation (Séré *et al.*, 2011). Yield losses of 1–22% were recorded in rainfed lowland, and 4–45% in irrigated systems in the south and west of the country. Yield losses of up to 44% (equivalent to 2 t/ha) were recorded in the irrigated perimeter of Vallée du Kou (Séré *et al.*, 2011).

In many countries, blast inflicts significant damage: heavy yield losses (up to 100%) were reported by farmers in Ghana (Nutsugah *et al.*, 2004) and in some locations in The Gambia (Jobe *et al.*, 2002); in Sierra Leone, losses in excess of 80% were reported in susceptible cultivars and accessions in experimental plots (Fomba and Taylor, 1994). In Nigeria, blast outbreaks have been reported to cause rice yield losses of about

35–50% and, in a serious outbreak of the disease, up to 100% of yield may be lost (WARDA, 1999a,b). Yield losses of 20–30% have been recorded in Benin (Vodouhe *et al.*, 1981), 36–63% in Burkina Faso (Séré, 1981), 64% in Togo (Akator *et al.*, 1981) and up to 80% in Côte d'Ivoire (Delassus, 1973).

Rice yellow mottle disease

First recorded in 1966 at Otonglo near Lake Victoria, Kenya (Bakker, 1970), *Rice yellow mottle virus* (RYMV) (genus *Sobemovirus*) is now a major biotic constraint of rice, present in most of the rice-growing countries in Africa (Abo *et al.*, 1998; Kouassi *et al.*, 2005; Séré *et al.*, 2008b; Traoré *et al.*, 2009; Ndikumana *et al.*, 2011).

RYMV disease is characterized by mottling and yellowing symptoms of various intensities depending on genotype and time of infection. Infected plants show pale yellow mottling on their leaves, stunted growth, fewer tillers, asynchronous flower formation, poor panicle exsertion, spikelet discoloration and sterility (Plate 10). In severe cases, affected plants may die. Yield loss ranges from 10% to 100%, depending on the timing of the infection and the type of variety (Abo *et al.*, 1998; Kouassi *et al.*, 2005).

RYMV disease is transmitted when the sap of infected leaves comes into contact with the cells of healthy leaves – for example, leaf contact in closely spaced plants, contaminated hands of field workers, rice stubble incorporated into the soil, and intertwining of rice roots. RYMV is not transmitted through rice seeds (Konate *et al.*, 2001; Abo *et al.*, 2004). Many insects are vectors of RYMV, including *Dicladispa gestroi*, *Trichispa sericea*, *Chaetocnema* spp., *Sessilia pussila*, *Chnootriba similis*, other beetles belonging to the family Chrysomelidae; the long-horned grasshoppers *Conocephalus merumontanous* and *Con. longipennis*; the short-horned grasshoppers, *Oxya hyla*, *Paratettix* sp., *Zonocerus variegatus*, *Euscyrtus* sp., *Cofana spectra*, *Cof. nimacuata*, *Locris rubra* and *L. maculate* (Abo et al., 1998; Nwilene, 1999).

The increasing incidence and importance of RYMV in Africa is attributed to the cultivation of new highly susceptible exotic rice varieties mostly from Asia (Thresh *et al.*, 2001) and the availability of water through irrigation which allows for sequential planting and maintenance of higher crop intensity without dry-season gaps, which favours increase of both insect vectors and alternative hosts (Traoré *et al.*, 2009). The RYMV epidemic at the beginning of the 1990s in West Africa was the result of intensification of rice cultivation (Traoré *et al.*, 2009).

Bacterial blight

Bacterial blight of rice, caused by *Xanthomonas oryzae* pv. *oryzae*, is another major biotic constraint to rice production and productivity. The disease was first observed in Africa (Mali) by Buddenhagen *et al.* (1979). In the following years, it was reported from Senegal (Trinh, 1980), Cameroon (Notteghem and Baudin, 1981), Niger (Reckhaus, 1983), Madagascar and Nigeria (Buddenhagen, 1985), Burkina Faso (Séré and Nacro, 1992) and Tanzania (Ashura *et al.*, 1999), and later from Benin, Guinea, The Gambia, Mozambique, Rwanda and Uganda (Onasanya *et al.*, 2009; El-Namaky, 2011).

The pathogen enters the host plant through natural openings at the leaf tip or margins (Ou, 1985) or through wounds. The pathogen reaches the xylem, where it multiplies and spreads throughout the plant, resulting in systemic infection (Huang and Cleene, 1989; Gnanamanickam *et al.*, 1999).

Typical bacterial blight symptoms (Plate 11) include leaf blight, pale yellow leaves and wilting (named 'kresek' symptom). Leaf blight is most common between maximum tillering and maturity stages. However, kresek – the most devastating manifestation of the pathogen – is most commonly observed at seedling stage, with seedlings being most susceptible in the 21 days after transplanting.

The pathogen mainly attacks rice, but also infects other *Oryza* species and wild hosts (Li *et al.*, 1985). In Niger, an extensive study at 23 sites revealed bacterial blight infection on *Brachiaria* sp., *Cyperus esculentus*, *C. rotondus*, *Dactyloctenium aegyptium*, *Echinocloa* sp., *Eulesine indica*, *Kyllinga squamulata*, *Leersia hexandra*, *Oryza barthii*, *O. longistaminata*, *Panicum lactum*, *P. repens* and *Pennisetum pedicelatum* (AfricaRice, 2010).

A survey carried out in several West African countries revealed yield loses of 2.7–41.0%

(Awoderu *et al.*, 1991) and a disease incidence of 70–85% in farmers' fields (Séré *et al.*, 2005). The introduction of a high-yielding but susceptible variety from Taiwan to the bacterial blight pathogen in the mid-1990s drew scientists' attention to the importance of this disease (Ouedraogo *et al.*, 2007). Complete crop failures have occurred in Burkina Faso (Ouedraogo *et al.*, 2007).

Variability of Rice Disease Pathogens in Africa

Rationale

In addition to the identification and prioritization of rice pathogens in Africa, research activities were pursued to develop IPM, mainly through the collaborative network implemented by AfricaRice and its national (NARS) partners called the IPM Task Force (WARDA, 2002). Priority was placed on varietal resistance as the main component of an IPM strategy (WARDA, 1999a,b). As information on pathogen diversity is essential for adequate utilization of resistant varieties and for developing strategies to increase the durability of resistance (Xia *et al.*, 2000), research activities were undertaken to better understand the structure of pathogen populations.

Rice blast

Blast pathogen diversity at pathological level is usually analysed by infecting 'differential' varieties with different isolates of the pathogen. However, in Africa, blast-trapping nurseries were developed and implemented, not only to identify efficient resistance genes, but also to characterize rice-growing areas (especially screening sites) in terms of the structure of their blast pathogen populations (Séré *et al.*, 2007, 2011). Such nurseries appeared to be an effective tool for characterizing the virulence spectrum of blast populations using limited equipment and labour (Séré *et al.*, 2007). The best sites for screening for durable resistance were identified (Séré *et al.*, 2007) and efficient resistance genes were found. For instance, the virulence genes that overcome the blast-resistance genes of rice *Pi9*, *Pit* and *Piz-5* are not present or are extremely rare in five countries – Benin, Burkina Faso, Guinea, Mali and Nigeria (Séré *et al.*, 2011). The results from field screening in Africa of varieties with resistance genes to Asian blast pathogens suggested a difference between blast pathotypes on the two continents (Séré *et al.*, 2007).

The development of molecular tools offered new opportunities for analysing the blast fungus diversity to help effective deployment of resistance and to identify shifts in races or population structures (Javan-Nikkhah *et al.*, 2004; Chen *et al.*, 2006). A tremendous amount of knowledge on blast pathogen population diversity has been accumulated throughout the world, especially with the identification of disperse-repetitive DNA sequences called MGR (Hamer and Givan, 1990). Studies in West Africa focused on describing the extent of blast pathogen diversity (lineages and pathotypes) in and around key sites in Burkina Faso, Côte d'Ivoire, Ghana and Nigeria: the persistent dominant lineages and major pathotype groups were identified (Chipili *et al.*, 1999). However, such studies were conducted in few countries in Africa (Chipili *et al.*, 1999; Nutsugah *et al.*, 2008) in comparison to what was done elsewhere in the world. Consequently, the research needs to be continued on a larger scale throughout Africa.

RYMV

Several hundred RYMV isolates from cultivated rice and wild *Poaceae* were collected in more than ten countries and stored in isolate banks mainly at the Institut de recherche pour le développement (IRD, France) and AfricaRice. All the isolates were serologically typed with both polyclonal and monoclonal antibodies. The coat protein gene of some isolates representative of the geographic distribution and of the serological variability was sequenced. These studies indicated that RYMV is a variable virus, and that there are several strains with different geographical distributions and pathogenic properties (N'Guessan *et al.*, 2000; Pinel *et al.*, 2000). Five major serotypes were described and named Ser1, Ser2, Ser3, Ser4 and Ser5. Comparing the molecular and immunological typing of RYMV isolates, Fargette *et al.* (2002b) found that molecular typing is consistent with immunological typing. However, Ser5 includes two strains (S5 and S6), leading to the identification of six strains: S1, S2 and S3 are West African isolates, while S4, S5 and S6 are from East Africa.

Phylogenetic analyses were performed, not only to analyse the genetic relationships between the isolates, but also to assess the links between geographic and genetic distances. The most basal strains were in East Africa. Phylogenetic inferences showed that the centre of origin of RYMV was in East Africa, possibly within the Eastern-Arc mountains biodiversity hot spot, and that successive strain radiations had occurred from the east to the west of the continent. Altogether, the data suggested that RYMV originated from wild *Poaceae* and infected cultivated rice only recently (Fargette *et al.*, 2004).

The evolution rate of RYMV was calculated from sequences of the coat protein gene of isolates collected from rice over a 40-year period in different parts of Africa. The results show that an RNA plant virus such as RYMV evolves as rapidly as most RNA animal viruses (Fargette *et al.*, 2008).

Bacterial blight

Xanthomonas oryzae pv. *oryzae* is a Gamma proteobacterium. The pathogen is a Gram-negative rod with round ends of 0.5–0.8 × 1.0–2.0 μm. The pathogen fails to grow on L-alanine as exclusive carbon source and 0.2% vitamin-free casamino acids, but it is insensitive to 0.001% cupric nitrate, which differentiates it from *X. oryzae* pv. *oryzicola* (the cause of rice bacterial leaf streak) (Vera Cruz *et al.*, 1984).

Although studies have not been as extensive in Africa as in Asia, the pathogen does show high pathogenic variability in Africa. Séré *et al.* (2005) found that four Malian isolates from different origins reacted differently on four varieties (Bouaké 189, BG90-2, NERICA 1 and NERICA 4). Isolate 1 (from Molodo) is virulent on three varieties and isolate 3 (from Nango) was not virulent on any of the four varieties. Two isolates (2 and 4, from Ndebougou and Niono, respectively) were virulent on two varieties. None of the four isolates developed a compatible reaction on NERICA 1.

Another pathotyping analysis carried out with 50 strains of *X. oryzae* pv. *oryzae* isolated from seven West African countries on 18 near-isogenic rice lines (NILs) resulted in the description of two pathotypes of the pathogen, *Pta* and *Ptb*, having three and two pathotype sub-groups, respectively (Onasanya *et al.*, 2009). A similar study carried out on 47 isolates from 25 locations in Niger revealed three major pathogroups (AfricaRice, 2010).

Gonzalez *et al.* (2007) evaluated 16 African strains of *X. oryzae* pv. *oryzae*, using differential varieties. They identified three new races (A1, A2 and A3) of *X. oryzae* pv. *oryzae* among African isolates, none of which has been described in Asian populations of the pathogen.

A pathotyping analysis carried out with 23 strains collected from West Africa and with reference strains from Asia against seven rice differential lines with monogenic resistance genes identified two pathogroups, PI and PII. The latter comprised the more virulent strains of the pathogen, which comprised 61% of the strains tested (Bimerew, 2010).

Trapping nurseries were used in a bacterial blight-prone environment in West Africa to analyse the interaction between resistance genes and a natural pathogen population (AfricaRice, 2010). None of the Malian races of *X. oryzae* pv. *oryzae* induced susceptible reactions in the rice lines bearing resistance genes *xa5*, *Xa7*, *Xa14*, *Xa18* or *Xa21*, or one of the associations *Xa4+xa5*, *xa5+Xa21* or *xa13+Xa21*. The *X. oryzae* pv. *oryzae* races that induce susceptible reaction to *Xa1*, *Xa2*, *xa7*, *xa8*, *Xa11*, *Xa14*, IR24 (*Xa18* and other), *Xa21*, the associations *Xa4+Xa21* and *Xa4+xa5+xa13+Xa21* were rare in Niger. In Burkina Faso, lines with resistance gene *Xa7* or the four pyramided genes (*Xa4+xa5+xa13+Xa21*) were resistant or moderately resistant to the natural population of *X. oryzae* pv. *oryzae* over 2 years, as was Gigante, while lines with *xa3*, *Xa10*, *Xa11*, *Xa4+xa5*, *Xa4+xa13* or *xa5+xa13* were moderately resistant.

Molecular techniques were used to characterize the genetic diversity among African strains of *X. oryzae* pv. *oryzae*. A study carried out using multilocus sequence analysis of *X. oryzae* pv. *oryzae* strains isolated from Africa based on *atpD*, *dnaK*, *gyrB* and *efp* housekeeping genes demonstrated that African strains are distinct from Asian ones (Bimerew, 2010), indicating that African strains are genetically distant from Asian ones. White *et al.* (1995) and Gonzalez *et al.* (2007) previously reported a difference between African and Asian strains of *X. oryzae* pv. *oryzae*.

Integrated Management of Rice Diseases in Africa

Background

Genetic control receives the most attention among the control measures to be used against rice pathogens, because it is the easiest method for farmers to adopt and the principal component of the IPM strategy (WARDA, 1999a,b). Based on the known interactions between pathogens and plant hosts, screening methods were developed and used to characterize rice varieties developed by breeders. In order to reinforce the resistance, other control measures were also investigated (WARDA, 1999a,b).

Rice blast

Genetic control

HOST–PATHOGEN RELATIONSHIP. Two kinds of host–pathogen relationship are described for the blast pathogen and rice.

In the vertical system, the relationship between virulence genes of the rice pathogens and resistance genes of rice varieties is explained by the gene-for-gene theory (Kiyosawa, 1980; Silué *et al.*, 1992).

The vertical resistance is controlled by few genes of major effect (Wang *et al.*, 1994; Liu *et al.*, 2002; Sallaud *et al.*, 2003; Chen *et al.*, 2004). They are responsible for a qualitative, complete and non-durable resistance. More than 70 *Pi* genes have been identified (Dai *et al.*, 2007; Lin *et al.*, 2007; Ballini *et al.*, 2008), some of which have been molecularly characterized (Suh *et al.*, 2009). Markers for many blast resistance genes (e.g. *Pil*, *Pi2* [or *Piz-5*], *Pita*, *Pit*, *Pi7*, *Pi9*, *Pi1* and *Pib*) are now available (Khush and Brar, 2004).

Although no comprehensive study of the durability of the major genes has been carried out in Africa, an analysis of the virulence of the blast population in Benin by planting lines and varieties bearing known major resistance genes each month from September 2007 to October 2008, gives an indication of the efficiency of some resistance genes over a year. For instance, the variety 75-1-127 (with *Pi9*) and Moroberekan (*Pi7* + *Pi5* and other major genes and quantitative trait loci (QTLs) according to Chen *et al.*, 1997) remained efficient in each monthly trial, while IRBLz5-CA/CO (*Piz5*) failed in the 11th month (August 2008) and other NILs were efficient only during the first 2 to 4 months (Table 17.2). Resistance of *Pi9* and *Piz5* was confirmed in trials conducted at Ouèdèmè (Benin) in 2009 and 2011 (Table 17.3). However, in the hot spot of Longorola (Sikasso) in Mali, *Pi9* was susceptible while *Piz5* was resistant (Ayeko, 2012).

When the vertical resistance fails against a blast strain, the severity of the disease will depend on the ability of the variety to slow the epidemic either by reducing the size of the lesions or by reducing the production of new spores. This aptitude is conferred by what is known as 'horizontal resistance'. Scientists agree that this system is stable and durable (Wang *et al.*, 1994). However, this type of resistance has low heritability because of strong environmental influence on the expression of the resistance genes (Wang *et al.*, 1989). It is generally polygenic, but durable resistance has also been shown to be also conferred by major genes like *Pi40* (Suh *et al.*, 2009) and the recessive gene *pi21* (Fukuoka and Okuno, 2001).

MANAGEMENT OF THE RESISTANCE. The two kinds of resistance can exist together in the same cultivar. For example, Moroberekan, a traditional West African cultivar, is known to have a durable resistance to blast. It possesses two vertical resistance genes (*Pi5* on chromosome 4 and *Pi7* on chromosome 11), but also QTLs for partial resistance on eight chromosomes (Wang *et al.*, 1994). Moreover, at least six major blast resistance loci have been identified in Moroberekan (Chen *et al.*, 1997). It is therefore important, when screening varieties for blast resistance, to know whether the resistance is complete and non-durable or partial and long-lasting.

Table 17.2. Efficiency of blast resistance genes in 13 month planting dates of varieties and NILs with different resistance genes

Variety or NIL	Resistance gene	Planting date						
		Oct 07	Nov 07	Dec 07	Jan 08	Feb–Aug 08	Sep 08	Oct 08
IRBL5-M/CO	*Pi5(t)*	S	S	S	S	S	S	S
Co39	*Pia*	S	S	S	S	S	S	S
IRBLkp-K60/CO	*Pik-p*	S	S	S	S	S	S	S
IRBLks-CO/CO	*Piks*	S	S	S	S	S	S	S
C102 TTP	*Pita*	S	S	S	S	S	S	S
C104 PKT	*Pi3*	**R**	S	S	S	S	S	S
IRBLzt-IR56/CO	*Piz-t*	**R**	S	S	S	S	S	S
IRBL7-M/CO	*Pi7*	**R**	S	S	S	S	S	S
Nato	*Pii*	**R**	S	S	S	S	S	S
IRBL1-CL/CO	*Pi1*	**R**	**R**	S	S	S	S	S
C101 LAC	*Pi1 + Pi1b + Pi33*	**R**	**R**	S	S	S	S	S
C101 A51	*Pi2 (= Piz5)*	**R**	**R**	S	S	S	S	S
St 1	*Pif*	**R**	**R**	S	S	S	S	S
IRBLkh-K3/CO	*Pik-h*	**R**	**R**	S	S	S	S	S
IRBLb-IT13/CO	*Pib*	**R**	**R**	S	S	S	S	S
IRBLta2-IR64/CO	*Pita-2*	**R**	**R**	S	S	S	S	S
IRBLsh-Fu/CO	*Piz*	**R**	**R**	S	S	S	S	S
IRBLkm-Ts/CO	*Pik-m*	**R**	**R**	S	S	S	S	S
IR1529	*Pi33*	**R**	**R**	**R**	**R**	S	S	S
K 59	*Pit*	**R**	**R**	**R**	**R**	S	S	S
Shwetasoke	*?*	**R**	**R**	**R**	**R**	S	S	S
IRBLz5-CA/CO	*Piz-5*	**R**	**R**	**R**	**R**	**R**	S	S
Moroberekan	*Pi5 + Pi7 + Pi10(t) + Pi157 +* QTLs	**R**	**R**	**R**	**R**	**R**	**R**	**R**
75-1-127	*Pi9*	**R**	**R**	**R**	**R**	**R**	**R**	**R**

R, resistant; S, susceptible.

Table 17.3. Reaction of blast resistance genes against natural population at Ouèdèmè in Benin. (Adapted from Ayeko, 2012.)

Variety or monogenic line	Resistance gene(s)	Reaction under natural conditions	
		2009	2011
Moroberekan	*Pi5 + Pi7 + Pi10(t) + Pi157 +* QTLs	R	R
IRBL 9-W	*Pi9*	R	R
IRBLZ 5-CA (R)	*Piz-5*	R	R
IRBLZT-T	*Pizt*	R	R

A procedure to characterize the nature of the resistance of rice varieties was designed at AfricaRice (Séré *et al.*, 2004, 2011). It consists of, first, evaluating a large number of varieties in order to identify entries with low disease score and, second, characterizing the nature of the resistance of those entries in order to ascertain whether their low disease score is due to vertical or horizontal resistance.

In order to make the resistance to blast durable, it is possible to pyramid several vertical resistance genes (Hittalmani *et al.*, 2000). Moreover the utilization of multi-lines composed of a mixture of varieties with different resistance genes (Wolfe, 1985; Zhu *et al.*, 2000)

enables the reduction of disease pressure and ensures stability of blast control. The combination of vertical resistance genes and partial ones is what makes the durability of resistance possible, as in the case of Moroberekan (Wang *et al.*, 1994).

Other control measures

In Africa, several efficient fungicides have been identified: benomyl and edifenphos (Delassus, 1973), tricyclazol (Mbodj, 1989), kitazin and thiophanate-methyl (Séré, 1981). In Burkina Faso (Séré *et al.*, 2011), seed treatment with systemic products can ensure efficient protection at low cost, mainly when associated with foliar treatment. However, such control measures should be used only when it is necessary to save a crop – for instance, in experiments where it is essential for any reason to secure the yield of a susceptible variety or line in a blast-prone environment.

Environmental factors influence the expression of blast, including temperature, humidity, leaf wetness, nitrogen fertilization and drought. Crop management practices that minimize the negative impact of such factors can be used in blast management. For instance, in farmers' fields, blast incidence increases with increasing nitrogen supply (Séré *et al.*, 2011). To ensure improved management of nitrogen, split application is better than single application (Kürschner *et al.*, 1992).

Moreover, longer duration of leaf wetness (which is related to duration of high relative humidity of the atmosphere) appeared to increase neck blast damage in farmers' fields in Burkina Faso (Séré *et al.*, 2011). Therefore, planting rice so that the reproductive stage occurs in late October, when the relative humidity becomes low, was recommended and used to reduce neck blast. Comparing a susceptible variety (FKR 16) with a resistant one (FKR 48) at different planting dates with and without chemical protection (Séré *et al.*, 2011) indicated that using chemical products on the susceptible variety reduced the incidence of blast from 22.9% to 5.2%. Without any chemical, it is possible to reduce disease incidence by planting the same susceptible cultivar earlier (3.6% incidence) or later (6.4% incidence).

RYMV

Genetic control

Two kinds of resistance to RYMV have been described. A partial resistance associated with delayed virus accumulation in the host was found in *Oryza sativa* subsp. *japonica* cultivars (Albar *et al.*, 1998). Partial resistance in these cultivars is associated with tolerance, which is characterized by low symptom severity despite the high virus content at the late stage of infection (Ioannidou *et al.*, 2003). Partial resistance is polygenic and linked to major QTLs on chromosomes 7 and 12 (Pressoir *et al.*, 1998).

The second type is a high resistance characterized by the absence of symptoms, a low amount of virus and a limited impact on yield (around 5%). This high resistance is controlled by a recessive gene (Ndjiondjop *et al.*, 1999). Four alleles of the locus *RYMV1* have been identified: *rymv1-2* in *O. sativa* cv. Gigante (Ndjiondjop *et al.*, 1999; Albar *et al.*, 2003) and cv. Bekarosaka (Rakotomalala *et al.*, 2008), *rymv1-3* in *O. glaberrima* accession TOG 5681 (Albar *et al.*, 2006), *rymv1-4* (Albar *et al.*, 2006) in *O. glaberrima* accession TOG 5672, and *rymv1-5* (Thiémélé *et al.*, 2010) in *O. glaberrima* accession TOG 5674. TOG 5672, known to possess *rymv1-4*, also carries a second resistance gene on the locus *RYMV2* (Thiémélé *et al.*, 2010).

Recurrent backcrossing has been used by AfricaRice breeders to introgress the resistance genes into the background of elite varieties. However, the emergence of resistance-breaking RYMV isolates is a matter of concern. Such resistance-breaking isolates are found in natural populations of RYMV (Konate *et al.*, 1997; Sorho *et al.*, 2005; Traoré *et al.*, 2006). Moreover, resistance-breaking strains can emerge after serial inoculation of virus-resistant plants, illustrating the ability of RYMV for host adaptation (Fargette *et al.*, 2002a).

The genetic basis of the resistance-breaking phenomenon has been analysed (Fargette *et al.*, 2002a; Hébrard *et al.*, 2006, 2008; Pinel-Galzi *et al.*, 2007; Poulicard *et al.*, 2010; Traoré *et al.*, 2006). The Virus Protein genome link (VPg) encoded by the ORF2a of RYMV was identified as the virulence factor (Hébrard *et al.*, 2006). A single mutation in the VPg of a strain is sufficient for it to break the resistance. For instance, the ability to overcome the *rymv1-2* or *rymv1-3* alleles appeared to be associated with polymorphism in the VPg sequence at position 49, a site that is under very strong positive selection (Pinel-Galzi *et al.*, 2007). A threonine residue confers a strong ability to break *rymv1-3* resistance, whereas strains possessing glutamic acid at this position are more adapted to *rymv1-2* resistance breaking (Traoré *et al.*, 2010).

At AfricaRice, screening for resistance is conducted in a screen house through mechanical inoculation of the last expanded leaves of 21-day-old seedlings. The disease symptoms are evaluated on all the rice plant's leaves, to determine the ability of the varieties to control virus movement within the plant. Susceptibility of rice cultivars can vary with plant age, with plants developing a kind of resistance at adult stage manifested by recovery ability (e.g. variety PNA 647F4-56) (Soko *et al.*, 2010). Even in the field, the recovery ability of PNA 647F4-56 was observed by farmers in Mali who indicated that the growth cycle became longer (Séré, 2005, unpublished observation).

The potential of using three insect vectors (*Oxya hyla*, *Locris rubra* and *Cnootriba similes*) to screen rice varieties was investigated. The results revealed that although there are some differences between the mechanical screening and the insect vector methods, both methods screen the varieties in the same way and, therefore, insects can be used to screen for RYMV resistance (Séré *et al.*, 2008a).

Integrated management

The amount of inoculum is important in determining the impact of a disease (Sorho *et al.*, 2005; Traoré *et al.*, 2009). Therefore, phytosanitation involving isolation of nurseries and removal of infected weeds and rice ratoons can reduce disease incidence and decrease the risk of emergence of virulent isolates (mainly resistance-breaking isolates) (Sorho *et al.*, 2005). Resistant cultivars should be associated with prophylactic measures within an integrated disease management approach (Traoré *et al.*, 2009). The insect vectors play an important role in transferring RYMV from surrounding contaminated rice or weeds to new rice fields (Nwilene, 1999) – their control should be part of the integrated approach to RYMV control.

Bacterial blight

Genetic control

RICE–PATHOGEN INTERACTION. Resistance gene products of the host recognize the presence of avirulence (*avr*) gene products of the pathogen, resulting in a rapid defence response and an incompatible interaction, restricting the pathogen to the site of inoculation.

Host plants have strategies to recognize pathogen attacks and activate defence mechanisms, resulting in the expression of various degrees of resistance to infection, manifested as vertical resistance controlled by major genes specific to a particular pathogen race, or quantitative (horizontal) resistance controlled by many genes and effective against a number of races.

HOST RESISTANCE. More than 30 resistance genes have been identified worldwide from cultivated and wild rice (Sun *et al.*, 2004; Niño-Liu *et al.*, 2006). These resistance genes encode different classes of proteins, such as nucleotide-binding site leucine-rich repeat (NBS-LRR) proteins and receptor kinase (RK). Additionally, the differential expression of resistance genes after infection indicates that rice has developed different strategies to overcome bacterial blight infection (Song *et al.*, 1995; Yoshimura *et al.*, 1998; Gu *et al.*, 2005).

Africa is endowed with some indigenous species of *Oryza* (including *O. barthii*, *O. longistaminata* and *O. glaberrima*), which can be used as potential sources of resistance to bacterial leaf blight. Vikal *et al.* (2007) report that out of 84 accessions of *O. glaberrima* evaluated against seven pathotypes of *X. oryzae* pv. *oryzae* in Punjab (India) over a period of 3–4 years, 13 showed a resistant to moderate resistant reaction to all pathotypes. Similarly, five and four accessions of *O. barthii* and *O. longistaminata*, respectively, also showed resistance to moderate resistance to these pathotypes. The broad-spectrum resistance gene, *Xa21*, is also found in *O. longistaminata* (Yoshimura *et al.*, 1998). Besides looking for other sources of resistance genes, it is also possible to combine two or more resistance genes by gene pyramiding, thereby improving the resistance spectrum and durability of a cultivar. Huang *et al.* (1997) report that rice varieties with two, three and four resistance genes have a wider spectrum and a higher level of resistance than varieties with only a single resistance gene.

Curiously, IR24 – the recurrent parent of the NILs developed by the International Rice Research Institute (IRRI) to analyse the pathological diversity of the bacterial blight pathogen and which was susceptible in Asia – is resistant to some African populations (AfricaRice, 2010). This means that the *Xa18* and other resistance genes that IR24 harbours (Liu *et al.*, 2007; Wu *et al.*, 2007) are efficient against some African bacterial blight pathogen strains.

The susceptibility of Gigante to bacterial blight is to be taken into consideration in breeding for resistance to RYMV, because in breeding programmes Gigante is used as parent to transfer RYMV resistance into many varieties.

Integrated management

The management of bacterial blight is also based on using resistant varieties. However, as the major genes alone are not able to ensure durable resistance, it is important to pyramid several genes (Gnanamanickam *et al.*, 1999) and/or add genes for partial resistance (Huang *et al.*, 1997).

Moreover, cultural practices can reduce bacterial blight severity. For instance, farmers in Niger used to burn crop residues after harvesting heavily infected fields, destroying the surrounding weeds that serve as a reservoir of the pathogen and thereby reducing inoculum (AfricaRice, 2010). Management of fertilizers, particularly nitrogen, is another cultural practice to be used, because increasing the level of N increases bacterial blight severity. A survey in farmers' fields in northern Benin indicated that the higher the quantity of urea applied, the greater the disease severity (AfricaRice, 2010). However, the impact of N depends on the time it is applied and the time infection occurs (Basso, 2010). When applied after infection, there is no increase in disease severity (Basso, 2010).

Conclusion: Challenges and Opportunities for Integrated Management of Rice Diseases in Africa

In order to develop integrated management strategies against rice diseases based on the utilization of resistant varieties, research on rice diseases in Africa focused on the following questions:

- What pathogens infect rice in Africa?
- Which are the most important of these that need particular attention?
- What control measures can be used in an IPM approach to reduce their impact on yield?

Next, research zoomed into the three major pathogens that cause greatest yield losses in the absence of any control measure. Although pertinent information was generated, some gaps remain to be filled.

For blast, it will be important to describe the blast population structure throughout Africa to better advise on the development and deployment of resistant cultivars. Moreover, the possible existence of new resistance genes in local cultivars has to be examined. Partial resistance needs to be considered, particularly in association with vertical resistance, as the stability of the resistance of varieties like Moreberekan and Tetep seems to be linked with such associations. It will be important to investigate the possibility of pyramiding major resistance genes (Hittalmani *et al.*, 2000). Moreover, the utilization of multilines to stabilize the resistance to blast disease (Wolfe, 1985; Zhu *et al.*, 2000) needs to be evaluated under African conditions. The impact of management practices on the disease needs to be evaluated in order to develop an IPM system.

There have been tremendous achievements in studies related to RYMV through collaborative research involving NARS, AfricaRice and advanced institutions (mainly IRD). The diversity in countries not yet sampled has to be investigated. It will be important to look for new resistance genes in local cultivars, mainly *O. sativa* subsp. *indica* and *O. glaberrima*, and in related species. The genetic background of RYMV pathogenicity needs deeper investigation, especially the resistance-breaking mechanism. It will also be important to assess the genetic background of the recovery ability, in order to ascertain whether it is a plant resistance mechanism or an environment-dependent condition and to see whether it can be used as a component of integrated management of RYMV.

Relatively less progress has been made in research on bacterial blight compared to advancement in the two other main rice diseases in Africa. However, a capacity in bacterial

include termites, caseworms, vectors of *Rice yellow mottle virus* (RYMV; see Séré *et al.*, Chapter 17, this volume), and grain-sucking bugs.

The vast majority of insects that are pests of stored grain belong to just two orders: Coleoptera and Lepidoptera (Reed, 2010). According to their pest status, they can be divided into two groups: primary and secondary storage insect pests. Primary storage pests are able to destroy stored rice (paddy or milled grains) independently. They are responsible for severe losses of stored rice and other cereals in Africa. Secondary storage insect pests can only attack grains that have been damaged by primary insect pests (Table 18.2). Most of the storage pests have been dispersed across the world by international trade (Youm *et al.*, 2011) and lead to quantitative and qualitative losses, as well as price reduction in most African markets. A quantitative loss of 18% was reported on farmers' stored rice in Benin over 4 months of storage (Togola *et al.*, 2010). The quantitative loss was estimated on damaged and undamaged grains collected from farmers' stored paddy grains. The damage is higher when the storage exceeds 4 months, due to the high reproduction rate of these insects and their short life cycle of just 45 days (Togola, unpublished data). Moreover, these pests produce heat and moisture and contaminate rice grain with their waste products and secretions (Walker and Farrell, 2003). The damaged grains are inappropriate for trade, or for use as food or seeds.

Managing Insect Pests

Varietal resistance

Resistant and tolerant rice cultivars play an important role in the reduction of yield losses due to insect pests and assessment of different rice varieties for insect resistance is an integral component of pest management. Because of its unique advantages (e.g. generally compatible with other control measures), host-plant resistance is a key component in the integrated control of rice insect pests in Africa. Success in identifying resistant material depends to a large extent on the ability to adequately evaluate germplasm and improved genotypes. Screening germplasm under artificial and natural pest infestations is essential for identifying better sources of resistance to major insect pests of rice. Knowledge of the mechanisms and factors contributing to host-plant resistance to insects is useful in selecting suitable criteria and breeding methodology for the genetic improvement of rice plants for insect resistance. Some of the factors associated with resistance, such as silica content and longer internode elongation in *Oryza sativa* varieties, can be used as 'marker traits' to screen and select for resistance to pests. Considerable progress has been made by the Africa Rice Center (AfricaRice) in the development of NERICA varieties that combine the high yield potential of Asian rice (*Oryza sativa*) with many useful traits from the African cultivated

Table 18.2. Key postharvest insect pests of stored paddy and milled rice in Africa.

Main species	Order (Family)	Status
Rice weevil, *Sitophilus oryzae* and Maize weevil, *Sitophilus zeamais*	Coleoptera (Curculionidae)	Primary pests of rice (paddy and milled grains)
Lesser grain borer, *Rhizopertha dominica*	Coleoptera (Bostrichidae)	Primary pest of rice (paddy and milled grains)
Angoumois grain moth, *Sitotroga cerealella*	Lepidoptera (Gelechiidae)	Primary pest of rice (paddy grain)
Red flour beetles, *Tribolium castaneum* and *T. confusum*	Coleoptera (Tenebrionidae)	Primary pest of rice (paddy and milled grains)
Rice moth, *Corcyra cephalonica*	Lepidoptera (Pyralidae)	Secondary insect pest of rice (paddy and milled grains)
Merchant grain beetle, *Oryzaephilus mercator*	Coleoptera (Sylvanidae)	Secondary insect pest of rice (paddy and milled grains)
Rust-red grain beetle, *Cryptolestes ferrugineus*	Coleoptera (Cucujidae)	Secondary insect pest of rice (paddy and milled grains)

species (*Oryza glaberrima*). Today, there are both upland and lowland NERICA varieties and traditional *O. sativa* varieties that are resistant or tolerant to some key pests of rice in Africa. Systematic evaluation of rice germplasm for AfRGM resulted in identification of over 50 primary sources of resistance among *O. glaberrima* and traditional *O. sativa* varieties (Nwilene *et al.*, 2002). Several gall-midge-tolerant rice varieties (e.g. NERICA-L 19, NERICA-L 49, Cisadane, BW 348-1, Leizhung) have been developed and released for commercial cultivation in Africa. Nwilene *et al.* (2009b) identified anti-xenotic and antibiotic traits associated with resistance to AfRGM in some of these rice varieties, but the traits have yet to be utilized in breeding. The quantitative trait loci (QTLs) or genes conferring resistance to AfRGM have also been identified from (*O. sativa* × *O. glaberrima*) crosses, ITA306 × TOS14519 and ITA306 × TOG7106. Some progress has also been made in identifying suitable upland NERICA varieties (NERICA 1, 2, 5, 7 and 14) that are resistant or tolerant to rice stem borers (Rodenburg *et al.*, 2006; Nwilene *et al.*, 2008b). Some NERICA varieties (NERICA 5, 14, 18) are suitable for cultivation in termite-prone areas of north-central Nigeria (Agunbiade *et al.*, 2009). Crop improvement programmes need to place emphasis on developing germplasm with multiple resistance to key insect pests using biotechnological tools (e.g. marker-assisted selection), because there are often two or more stresses in most rice production environments in Africa.

Bt-rice

Rice transformation with a *Bt*-gene (delta-endotoxins from *Bacillus thuringiensis*) was first targeted against Asian yellow stem borer (YSB, *Scirpophaga incertulas*), Asian striped stem borer (*Chilo suppressalis*) and leaffolder (*Cnaphalocrocis medinalis*) in Asia. Genetically modified rice plants (*Bt*-rice) resistant to striped stem borer, leaffolders and other insects have been developed (Fujimoto *et al.*, 1993; Chen *et al.*, 2005a), and two *Bt*-rice varieties (Huahui 1 and Xianyou 63) were authorized for marketing especially in China, in 2009 (Chen *et al.*, 2011). Review of the literature on the entomological situation of *Bt*-rice in Asia and the current rice insect pest status in Africa led to the conclusion that, in the current state of knowledge, it seems inappropriate to introduce *Bt*-rice to control the diversity of rice insect pests in Africa (Silvie *et al.*, 2012). Moreover, the high diversity of insects found on rice in Africa (including Madagascar) militates against the use of *Bt*-rice due to the problem of resistance and that the toxins only affect a small number of stem borer species (van Rensburg, 2007). Response to the toxins may be quite different from one species to another. The introduction of *Bt*-rice in Africa is currently inappropriate because most African countries have no containment facilities to test transgenic rice for insect resistance. Regulatory frameworks are needed that promote quality control and informed use of transgenic crops and plant protection products.

Field-level integrated pest management

A majority of smallholder farmers in Africa rely on their traditional knowledge to manage pest problems, mainly by the use of botanical pesticides (Mugisha-Kamatenesi *et al.*, 2008). Apart from the traditional use of local herbs as biopesticides, indigenous knowledge systems also promote the use of sustainable soil and crop management practices which indirectly favour biologically-based pest regulation by natural enemies (Kuponiyi and Bamigboye, 2009). However, in a period of unprecedented change, farmers find that their indigenous knowledge provides only limited guidance. For farmers to more effectively manage pests on an ecological and sustainable basis, they need a combination of indigenous and scientific knowledge. The success of pest management programmes will depend largely on how well farmers understand and combine knowledge of biological and ecological processes with their own farming experiences. There is a growing realization that future agricultural growth hinges on smallholder farmers, who must be knowledgeable and exposed to a learning process that involves continuous field observation, agroecological analysis and management practices under local conditions, and that enhances decision-making capacity. A widespread constraint to the development of improved control methods is farmers' lack of information on the behaviour and reproductive cycles of target pests. When farmers misidentify the causes of damage observed in their fields, they may spray an insecticide to combat a pest when the damage is not caused by a pest. Also, by the time a pest is observed, it may be too late to do much about it except maybe spray with an insecticide. This is often the case in Africa. The smallholder farmers need to be taught to target the life-cycle stage that is susceptible to control measures. This is an example of how knowledge can benefit smallholder farmers and help protect rice from insect attacks. The farmer field school for IPM is an innovative model for community-based farmer education that uses non-formal or 'discovery learning' methods (William and Garba, 2011). The participatory learning and action-research (PLAR) for integrated rice management developed by AfricaRice may also serve as a vehicle to reach out to farmers, because it is a bottom-up social learning process to promote technological change through improving farmers' capacity to exchange knowledge, experiences and practices, to better observe, analyse and take appropriate decisions for action, and to get organized for action. Through the PLAR process, groups of farmers are able to find adaptive responses to site-specific problems and make the best use of available resources and local knowledge as well as research-based understanding of underlying processes (Defoer *et al.*, 2009; Defoer and Wopereis, chapter 31, this volume).

Landscape-level integrated pest management

There is considerable evidence that as agricultural production systems are intensified by increased use of external inputs to increase yield, and structural changes occur at landscape level, they tend to lose biodiversity and become destabilized, with increased frequency and extent of pest outbreaks (Swift *et al.*, 1996; Knops *et al.*, 1999). However, we know relatively little about the ecological mechanisms that result in this destabilization, or how important natural enemy diversity is in maintaining pest-control functions. In West Africa, for example, there is good circumstantial evidence that heavy infestations of AfRGM have become more common since the 1970s, with outbreaks associated with rice–rice double cropping and increases in nitrogen fertilizer application (Williams *et al.*, 1997). Studies in West Africa have also shown that the abundance of generalist herbivores and predators is positively correlated with weed biomass, whereas specialist herbivore abundance is positively correlated with rice biomass (Afun *et al.*, 1999). Understanding the complexity of these interactions and determining the positive and negative effects of intensification on the functioning of the different components of crop-associated biodiversity is essential for future development of sustainable production technologies.

The overall strategy in insect pest management for rice is to use a holistic 'plant health' approach, which should ideally start from considering soil-health aspects. A healthy plant might tolerate a higher population of noxious insects, while a carefully managed soil should provide improved ecosystem services such as beneficial organisms able to attack herbivores. Because of poor rotations, overuse of chemical pesticides, and inappropriate agronomic practices, soils might lose their ability to provide these important services, and hence need to be supplemented with appropriate amendments. For instance, the use of mycorrhyzal or endophytic fungi increases plants' ability to withstand attacks by pathogens, nematodes and insect pests (Secilia and Bagyaraj, 1992; Tian *et al.*, 2004). Also, complex interactions between soil fertility, insect attack and parasitoids have been documented in cereals such as maize (Chabi-Olaye *et al.*, 2005).

At the landscape level, the importance of alternative host plants both as an off-season resource for the different rice pests, but also as refugia for natural enemies cannot be overemphasized. Some studies in West Africa have identified possible innovations in this regard. For instance, AfRGM is attacked by parasitoids such as *Platygaster diplosisae* and *Aprostocetus procerae* in rice-production systems (Nwilene *et al.*, 2008c). However, the level of parasitization is low because the parasitoid populations build up too late to prevent heavy AfRGM infestation. Meanwhile, a related gall midge (paspalum gall midge, PGM, *Orseolia bonzii*) that forms galls on *Paspalum* grass is an important alternative host for the two parasitoids of AfRGM. Hence, habitat manipulation to increase the carry-over of parasitoids from PGM (which does not attack rice) to AfRGM, such as dry-season cultivation to encourage *Paspalum scrobiculatum* abundance early in the wet season, could improve the natural biological control of AfRGM. Similarly, ecological engineering using push–pull approaches, such as the one developed for maize in East Africa (Khan *et al.*, 1997), could be modified against cereal stem borers in West Africa, and particularly those attacking rice. Here the challenge is not just technical: any innovations requiring additional farmer inputs, such as seeds of companion crops and labour, need to be tested for their economic profitability and sociocultural acceptability in West Africa. As shown in the next section, more complex management options require a robust linkage to and integration with available indigenous knowledge. The next challenge facing the development and deployment of ecosystem services is to use modern tools in population genetics to assess the presence of locally distinct populations of both rice pests and their natural enemies. Molecular markers can be used to characterize pest populations and to gather detailed information about their specific location and migration, as well as those of locally available and introduced beneficial organisms. These data, combined with ecological and biological characteristics, are critical for more appropriate targeting of biological control interventions (Agunbiade *et al.*, 2012).

Postharvest integrated insect pest management

Postharvest IPM begins with good postharvest handling and management practices right from the field to the storage environment. The prerequisites and options for good storage pest management include: (i) harvesting on time; (ii) maintenance and protection of the site and storage environment from birds, rodents and the weather (controlling grain and air moisture), and basic hygiene using thermal disinfestation of the site by solar heat or treatment with traditional additives; and (iii) commodity management (cleaning and drying of appropriate packaging facilities) using hermetic storage (pits or metal drums) or treatment with synthetic insecticides/pesticides. Because of the danger of pesticides affecting the quality of rice stored for human consumption, research has made it possible to develop alternative measures to minimize pesticide risks to human and agroecosystem health. Plant products such as botanical extracts, essential oils and vegetable oils are being explored as potential pest management tools because they are not toxic to plants, are systemic, biodegrade easily and stimulate the host's metabolism (Dubey *et al.*, 2008). These measures need to be explored more effectively in order to preserve the quantity and quality of stored rice in Africa. Moreover, technologies such as rendering males infertile by using ultraviolet rays, pheromone traps and baits can be explored as future options to manage postharvest insects. Finally, quarantine measures should be rigorously applied in order to minimize the dispersion of invasive pests across national borders.

Climate Change and Pest Movements

Dwindling and erratic rainfall patterns, rising air temperature and extreme heat are having an impact on the spatial and temporal distribution and proliferation of insect populations. This may alter host plant–insect interactions and will thus require new IPM strategies. Climate change can increase the risk of pest outbreaks leading to greater yield losses with inherent negative consequences for food security in Africa. The current distribution and shifts in the relative importance of major pests of rice and their natural enemies in the upland–lowland rice agroecosystem continuum needs to be established using the phenology models and risk maps in geographic information systems (GIS). Parasitoids – the key natural enemies of the most important pests of rice – depend on a series of adaptations to the ecosystem and physiology of their hosts and host plants for survival and are thus likely to be highly sensitive to changes in environmental conditions. Climate change affecting pest and natural-enemy profiles might disrupt *Bt*-crops, increase the number of pest generations, prompt immigration of new invasive species, and enable existing species to expand their ranges (Chen *et al.*, 2005b). It is likely that insect

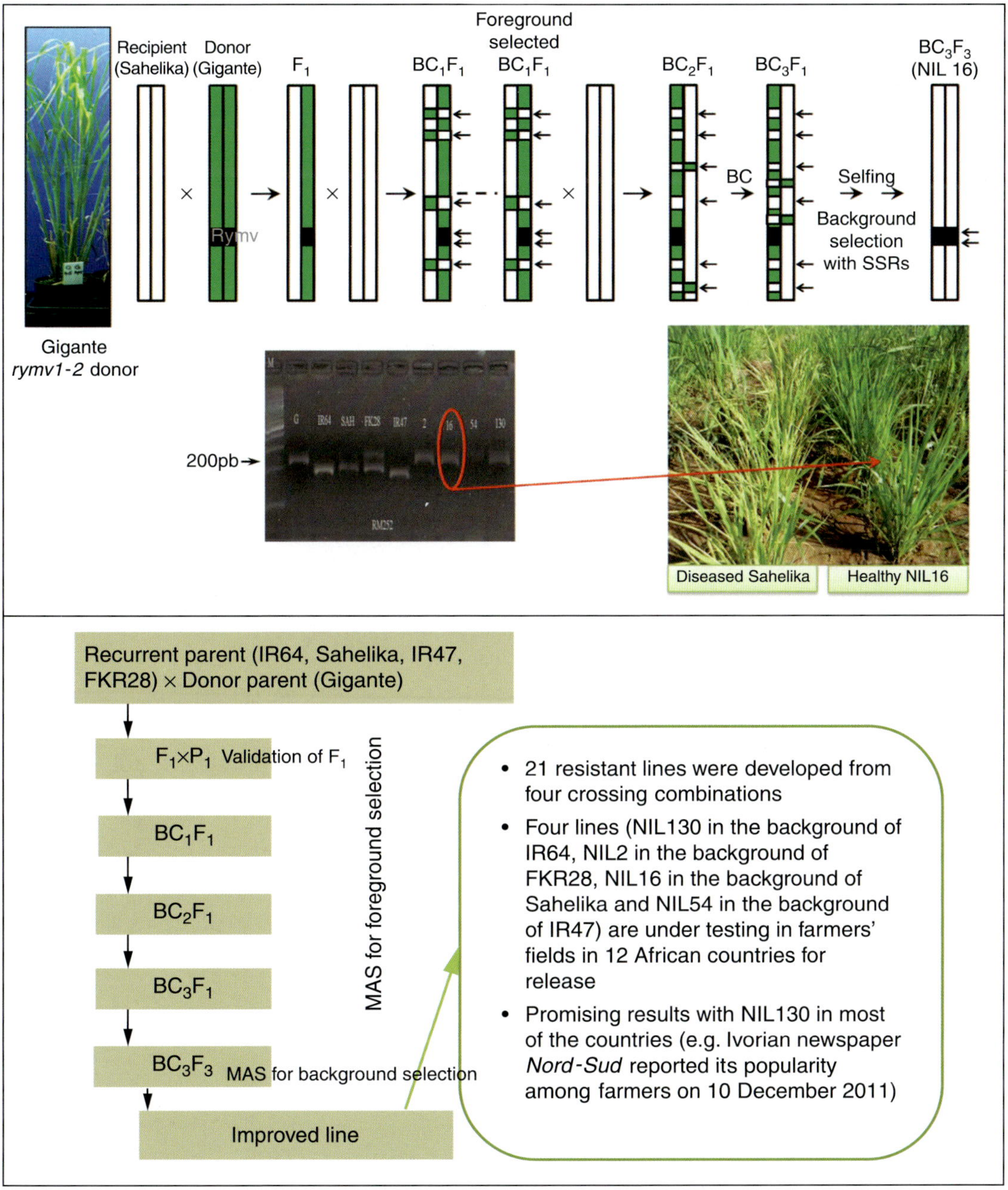

Plate 5. Marker-assisted selection schematic diagram used to develop the near-isogenic lines (NILs); case of NIL16 derived from variety Sahelika.

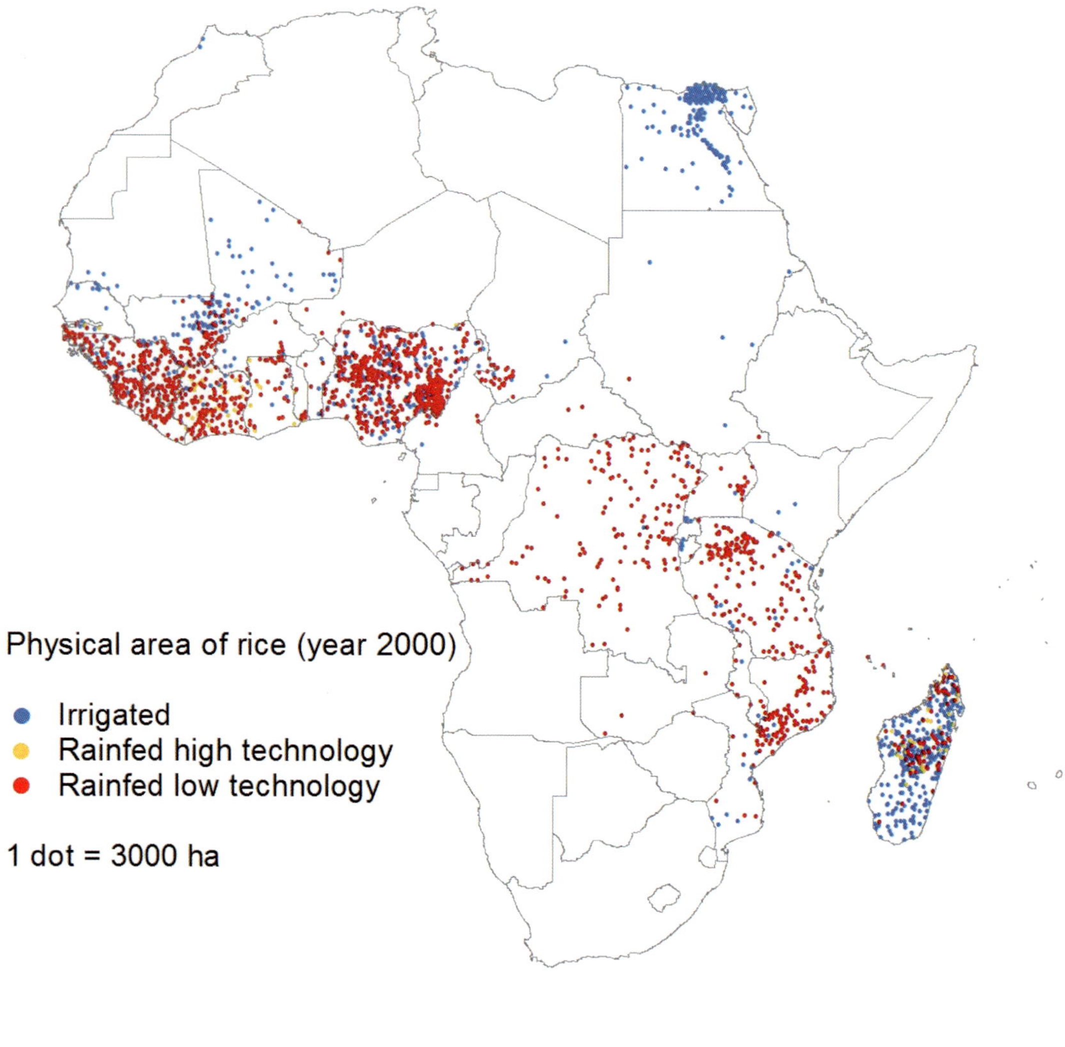

Plate 6. Distribution of rice physical area in Africa. (Data from You *et al.*, 2009a,b.) Note: Rice is not grown in northern parts of Tombouctou and Kidal regions (based on local experts' knowledge of rice in Mali), so the points in northern Mali must be due to analytical error.

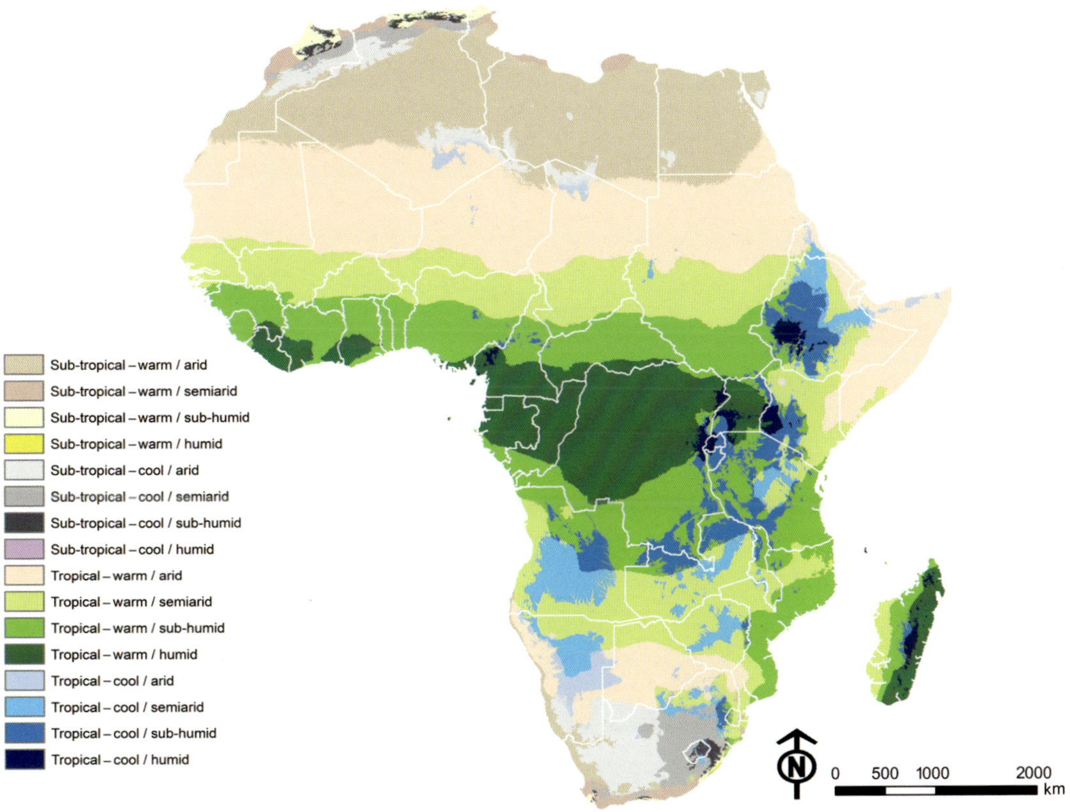

Plate 7. Agroecological zones in Africa. (Adapted from HarvestChoice, 2009.)

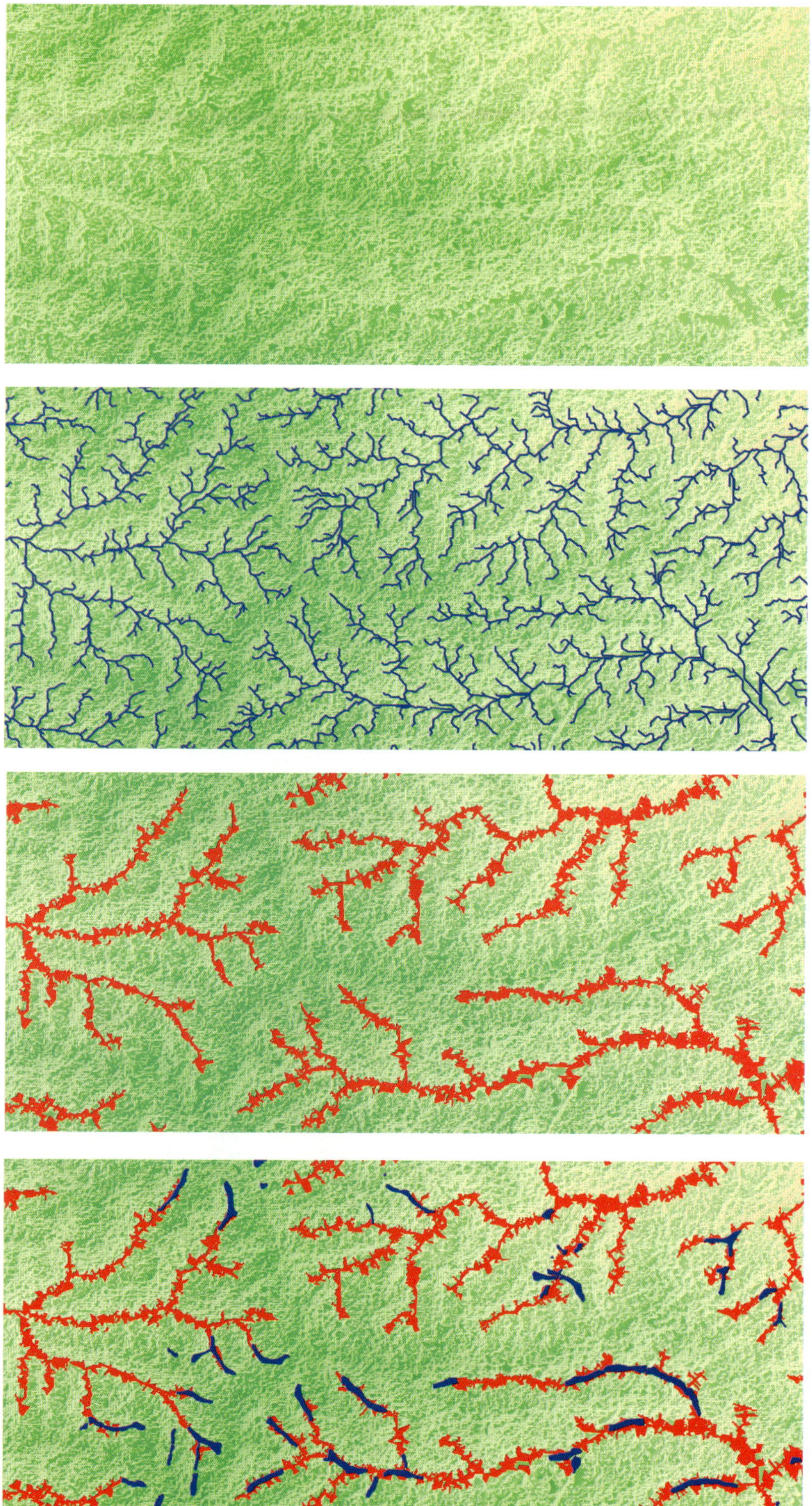

(A) Digital elevation model

(B) Streams

(C) Inland valleys

(D) Validation with IMPETUS data set (blue)

Plate 8. Methodological steps followed to obtain and validate inland valleys from a Digital Elevation Model. The example shown is around Djougou, central Benin: (A) digitial elevation model; (B) streams; (C) inland valleys; and (D) validation with IMPETUS data set (blue).

A B

Plate 9. Blast symptoms. (**A**) Leaf blast at Bagre (Burkina Faso) in 2008. (**B**) Neck blast at Banicoara (Benin) in 2009. (From Sere *et al.*, 2011, with permission from AfricaRice.)

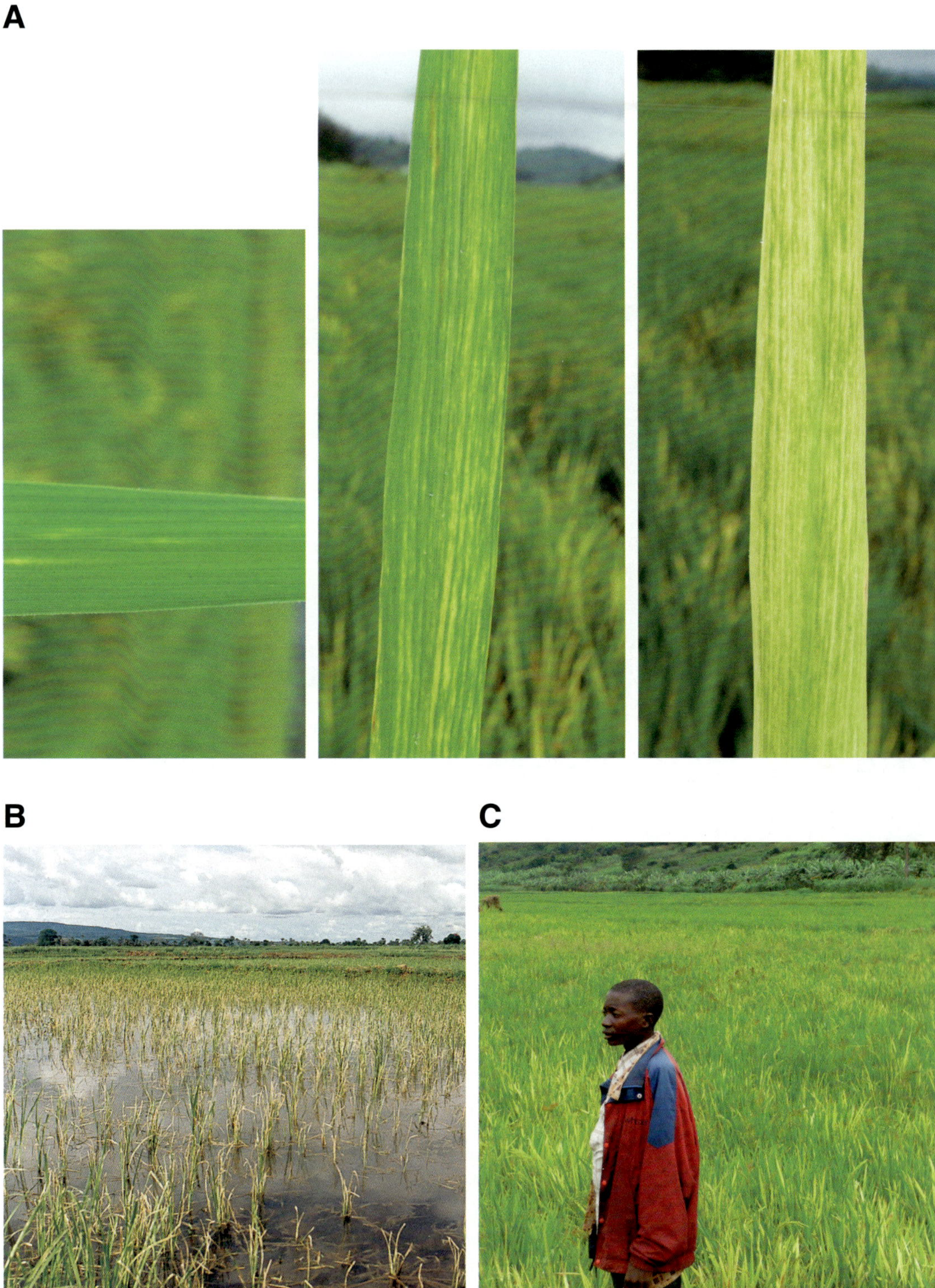

Plate 10. RYMV symptoms. (**A**) Leaves with RYMV symptoms (increasing in severity from left to right), Cyabaya (Rwanda), 2009. (**B**) Heavy infestation of RYMV in farmers' field, Karfiguela (Burkina Faso), 1989. (**C**) At Cyabayaga (Rwanda) 20 years later. (Photos: Y. Séré.)

11A

11B

12

Plate 11. Bacterial leaf blight. (**A**) Infected rice of susceptible variety (right), Office du Niger (Mali). (**B**) Close up of infected leaves. (Photos: Y. Séré (**A**) and S. Gibert and I. Wonni (**B**).)
Plate 12. Red stripe, Kayo, Office du Niger (Mali). (Photo: Y. Séré.)

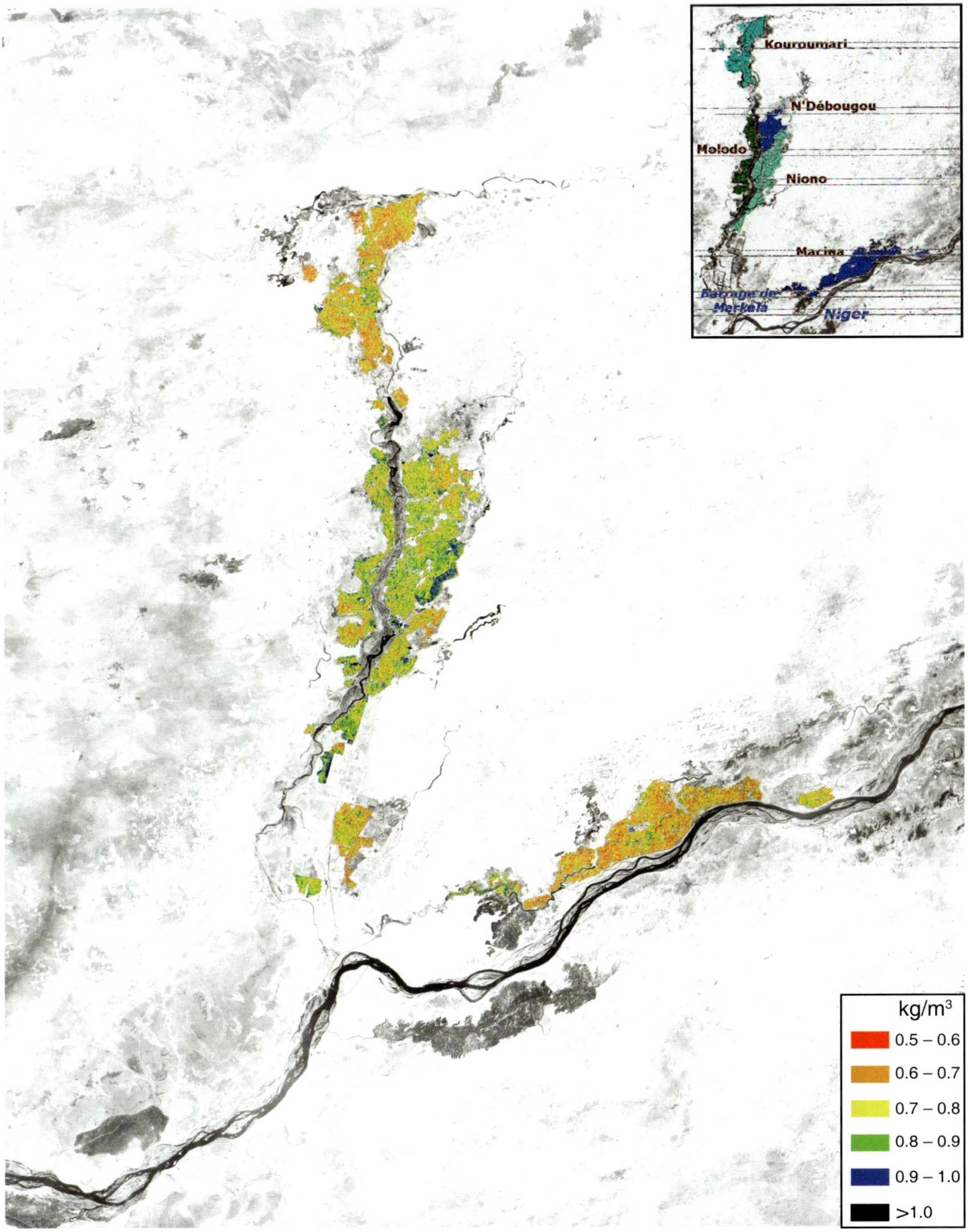

Plate 13. Water productivity of rice in the Office du Niger during the rainy season 2006 as assessed through remote sensing. Inset: Location of the five management zones. (From Zwart and Leclert, 2010, with kind permission from Springer Science and Business Media.)

management strategies developed in the past will also be affected by changes in climate. Thus, better knowledge and understanding of pest behaviour and actual farmer practices under different projected scenarios are needed to adapt existing or develop new IPM technologies to respond to possible threats resulting from climate change. There is, therefore, a need to test and disseminate a set of monitoring and forecasting tools, including simulation models to national-programme scientists, and decision makers in IPM research and extension. It will also be useful to document how farmers adapt their insect pest management strategies to climate change.

In the warm and humid tropics of Africa, various species of insects remain active year-round and populations fluctuate according to the availability of food plants, presence of natural enemies and environmental conditions. For example, in the humid forest zone of Nigeria, AfRGM persisted through the short dry season on ratoons of cultivated rice at a rainfed site and on dry-season crops at an irrigated one. In contrast, at rainfed sites in the moist savannah zone the pest survived the longer dry season on the perennial wild rice *Oryza longistaminata*, while ratoons and volunteers of *Oryza sativa* provided 'bridges' between the wild host and wet-season rice crops (Williams *et al.*, 1999). Some pests of rice known to be restricted to the lower part of the slope are gradually moving to the upper portion. The reason for this sudden movement pattern is unknown. In order to aid pest-management decision making and understand spatial dynamics of pests in agroecosystems, including cultivated and non-cultivated areas, we need to build new tools for tracking insect movement in Africa. Such tools can be used to forecast infestation risks, target geographical variations in biotype and movement of insect vectors from alternative host plants to rice and vice versa, and target better management of pests at hot-spot locations and agroecosystems in cultivated lands. Some of the available new tools that can be used are: (i) stable isotopes (of hydrogen, carbon, etc.); (ii) phytochemical markers (gossypol, tomatine, etc.); and (iii) bacterial molecular markers.

Stable isotopes represent a powerful tool to determine the trophic (with carbon or nitrogen isotopes) or geographic origin (with hydrogen or oxygen isotopes) of any animal (Hood-Novotny and Knols, 2007; IAEA, 2009). Although they require costly equipment for analysis (gas chromatography/mass spectrometry), their use is growing, and the cost of the analyses is decreasing. In entomology, stable hydrogen and carbon isotopes have been used to track migrations of monarch butterfly (*Danaus plexippus*) in the USA (Hobson *et al.*, 1999). In France, Ponsard *et al.* (2004) showed that stable carbon isotopes are indicators of the photosynthetic type (C_3 or C_4) of the host plants of *Ostrinia nubilalis*, regardless of the feeding habits and metabolism intensity of adults. Analysis of the isotopic composition of stable isotopes of hydrogen can reveal the movements of animals such as fish and birds (Hobson and Wassenaar, 2008). This tool, which is being widely used in ecology, has been used on the Lepidoptera *D. plexippus* (Hobson *et al.*, 1999) and *Helicoverpa armigera* (Menozzi *et al.*, 2007) to determine the geographical origin of migratory adults.

Detection of plant-specific compounds in adult tissues is an excellent indicator of a larval feeding source. Orth *et al.* (2007) designed a test to detect gossypol, a cotton-specific alkaloid, as well as cotinine, a metabolite of nicotine, in the adipose tissues of *Heliothis virescens* moths. Coupled with the analysis of the carbon isotopic composition, the analysis of ingested gossypol gave better understanding of the trophic origin of *Helicoverpa zea* adults collected in pheromone traps (Head *et al.*, 2010).

Another technique for decoding the migration and geographical origin of an insect is **bacterial molecular markers**. This technique has been used to determine the geographic origin of fish (Le Nguyen, 2008a) and collected fruits (Le Nguyen, 2008b). For insects, analysis of bacterial communities is still in its infancy. The advantage is that the technique is applicable to every insect and does not require any background knowledge on the bacterial flora hosted by the target insect. The analysis of the bacterial flora on or within an insect using molecular markers to locate the geographical origin of the insect is an innovative approach. Insects harbour endosymbiotic bacteria such as *Buchnera* and *Wolbachia*, or Enterobacteriaceae such as *Citrobacter freundii* and *Klebsiella proteus* (Carletto *et al.*, 2008). There may be other bacteria hosted by the insect that might be specific to a given location. The technique most commonly used for insects is polymorphism analysis of 16S rDNA V3 region of bacterial communities by PCR (polymerase chain reaction) using the technique DGGE (denaturing gradient gel electrophoresis) by means of primers common to many bacteria. This technique has been used to identify the habitats of *Aedes albopictus* and *Ae. aegypti* (Zouache *et al.*, 2010), *Anopheles stephensi* (Rani *et al.*, 2009) and the carabid *Poecilus chalcites* (Lehman *et al.*, 2009).

Some rice pests, such as *Chilo zacconius* and AfRGM, can have alternative host plants (grasses, sorghum, wild rice). These tools can be useful to know their trophic and geographic origin with the aim of better pest control.

New Approaches to Characterization of Rice Pest Constraints and Quantification of Yield Losses

An individual crop stand is usually not exposed to a single organism, but rather to multiple harmful ones (pathogens, insect pests, weeds, etc.). It is then more relevant to consider injury profiles rather than individual injuries. In order to characterize patterns of rice cropping practices and injuries, key concepts and approaches have been developed in tropical Asia based on 'injury profiles' and 'production situations' (Savary *et al.*, 2000a). 'Injury profiles' are defined as a given combination of injury levels caused by a range of pests during a crop cycle, while a 'production situation' is considered as the physical, biological, social and economic context in which agricultural production takes place. Characterization of injury profiles in relation to production situations has the potential for developing pest management strategies that can be adapted throughout the region rather than being site-specific. Some injuries or their combinations have a stronger or weaker yield-reducing effect depending on the level of attainable yield, i.e. the yield performance of a crop that has not been exposed to yield-reducing factors, especially pests (Savary *et al.*, 2000b). While the information gathered pertains to the individual field, analysis of the data should aim at conclusions that have relevance to the region. The fact that a particular injury prevails in nearly all sites or seasons is not necessarily an indication of its importance. By contrast, some injuries may occur sporadically and cause considerable yield reductions. A combination of information on occurrences of injuries (through surveys) and on experimental measurement of yield losses they may cause is necessary to assess the importance of a given injury (Savary *et al.*, 2000a). Such a combination provided quantitative background to set priorities for rice pest management in Asia (Savary *et al.*, 2000b). This is a powerful tool to analyse yield losses at large scales. Survey data provide information on the occurrence of pests and diseases, while experiments quantitatively highlight their impact in terms of yield loss (Savary *et al.*, 2006). For example, RICEPEST, a simulation model of rice yield losses, enables the simultaneous handling of production situations and injury profiles, as well as the modelling of management strategies (Willocquet *et al.*, 2002). These concepts, approaches and models could be adapted to rice cropping under African conditions.

Conclusions

Insect pests, diseases and weeds inflict enormous losses to rice production in Africa. If rice production is to keep pace with increasing demand, effective and sustainable management strategies are urgently needed to tackle these important biotic constraints. Sustainable and efficient pest management practices require scientific expertise to develop, through research, the various IPM strategies (insect resistant or tolerant rice varieties, biological control, crop management practices and substitution of inorganic pesticides with biopesticides) and to effectively disseminate these technologies to farmers for adoption. Marker-assisted selection and other biotechnological techniques are providing new ways of manipulating plant resistance to insect pests. Efforts by AfricaRice to develop the interspecific hybrid progenies combining the hardiness of African *Oryza glaberrima* with the high productivity of Asian *Oryza sativa*, using embryo-rescue and double-haploidy, have paved the way for improving the resistance of rice varieties to stresses such as rice stem borers and other insect pests. Biological control shows great promise for regulating damaging populations of rice-feeding insects in Africa. For example, there is an abundance and diversity of natural enemies (parasitoids, predators and pathogens) in West African rice ecosystems and natural biological control is playing a major role in managing pests. Thus, every effort should be made to conserve and enhance the activities of natural enemies. High levels of natural biological control seen in Africa may be due to the low use of natural-enemy-destroying pesticides, where the incidence of pest resurgence, which is currently in the crisis stage in Asian rice, has not been experienced.

Some important available natural enemies (e.g. *Platygaster diplosisae* and *Aprostocetus procerae* reported to be effective against AfRGM, *Cotesia sesamiae* and *Xanthopimpla stemmator* against rice stem borers, *Metarhizium anisopliae* against termites) and plant products (such as azadirachtin [neem] and nicotine against stem borers and termites) have not been packaged commercially for use by smallholder farmers. There are no commercial biocontrol laboratories in Africa, whereas over 400 biocontrol laboratories exist in India to cater for the location-specific needs of farmers. There is a need to develop private–public partnerships with biological control/biopesticide producers for the development, scaling up and commercialization of technologies.

Rice IPM development in Africa is confronted by many challenges, such as national government policies that result in a lack of trained rice scientists to develop IPM components and the absence of effective information- and technology-dissemination programmes. Consequently, the IPM packages that have been developed are reaching only a few farmers and are having little impact at the farm level. The adoption of IPM technologies is low in Africa as a result of socio-economic, institutional and policy constraints. The lack of appropriate institutional technology-transfer mechanisms is a critical impediment to increased application of IPM. It has become increasingly evident that future agricultural growth hinges on African smallholder farmers, who must be knowledgeable in all aspects of rice production, including the management of pests. It is thus important that agricultural technology dissemination activities be upgraded and receive high priority in the national budgets of African countries, and not be dependent on external resources, if we are to reduce rural poverty and achieve food security and sustainable rural livelihoods in Africa.

References

Afun, J.V.K., Johnson, D.E. and Russell-Smith, A. (1999) Weeds and natural enemy regulation of insect pests in upland rice: a case study from West Africa. *Bulletin of Entomological Research* 89, 391–402.

Agunbiade, T.A., Nwilene, F.E., Onasanya, A., Semon, M., Togola, A., Tamò M. and Falola, O.O. (2009) Resistance status of upland NERICA rice varieties to termite damage in northcentral Nigeria. *Journal of Applied Sciences* 9(21), 3864–3869.

Agunbiade, T., Steele, L., Coates, B.S., Gassmann, A., Margam, V.M., Ba, M., Dabire, C., Baoua, I., Bello-Bravo, J., Seufferheld, F., Sun, W., Tamò, M. and Pittendrigh, B.R. (2012) IPM-omics: from genomics to extension for integrated pest management of cowpea. In: Boukar, O., Coulibaly, O., Fatokun, C., Lopez, K. and Tamò M. (eds) *Innovative Research Along the Cowpea Value Chain.* Proceedings of the 5th World Cowpea Research Conference, 26 September–1 October 2010, Saly, Senegal, pp. 231–248.

Brenière, J. (1983) *The Principal Insect Pests of Rice in West Africa and Their Control*, 2nd edn. West Africa Rice Development Association, Monrovia, Liberia.

Carletto, J., Gueguen, G., Fleury, F. and Vanlerberghe-Masutti, F. (2008) Screening the bacterial endosymbiotic community of sap-feeding insects by terminal-restriction fragment length polymorphism analysis. *Entomologia Experimentalis et Applicata* 129, 228–234.

Chabi-Olaye, A., Nolte, C., Schulthess, F. and Borgemeister, C. (2005) Abundance, dispersion and parasitism of the noctuid stemborer *Busseola fusca* (Fuller) in mono- and intercropped maize in the humid forest zone of the southern Cameroon. *Bulletin of Entomological Research* 95, 169–177.

Chen, H., Tang, W., Xu, C.G., Li, X.H., Lin, Y.J. and Zhang, Q.F. (2005a) Transgenic *indica* rice plants harboring a synthetic cry2A* gene of *Bacillus thuringiensis* exhibit enhanced resistance against lepidopteran rice pests. *Theoretical Applied Genetics* 111, 1330–1337.

Chen, D., Ye, G., Yang, C., Chen, Y. and Wu, Y. (2005b) The effect of high temperature on the insecticidal properties of *Bt* cotton. *Environmental and Experimental Botany* 53, 333–342.

Chen, M., Shelton, A. and Ye, G.Y. (2011) Insect-resistant genetically modified rice in China: from research to commercialization. *Annual Review of Entomology* 56, 81–101.

Defoer, T., Wopereis, M.C.S., Idinoba, P., Kadisha, K.L., Diack, S. and Gaye, M. (2009) *Curriculum for Participatory Learning and Action Research (PLAR) for Integrated Rice Management (IRM) in Inland Valleys of Sub-Saharan Africa: Facilitator's manual.* Africa Rice Center, Cotonou, Benin.

Dhaliwal, G.S., Vikas Jindal and Dhawan, A.K. (2010) Insect pest problems and crop losses: changing trends. Indian *Journal of Ecology* 37(1), 1–7.

Dubey, N.K., Srivastava B. and Kumar A. (2008) Current status of plant products as botanical pesticides in storage pest management. *Journal of Biopesticides* 1(2),182–186.

FAOSTAT (2006) Statistical database on agricultural production. Food and Agriculture Organization of the United Nations, Rome, Italy.

Fujimoto, H., Itoh, K., Yamamoto, M., Kyozuka, J. and Shimamoto, K. (1993). Insect resistant rice generated by introduction of a modified δ-endotoxin gene of *Bacillus thuringiensis. Nature Biotechnology* 11, 1151–1155.

Head, G., Jackson, R.E., Adamczyk, J., Bradley, J.R., Van Duyn, J., Gore, J., Hardee, D.D., Leonard, B.R., Luttrell, R., Ruberson, J., Mullins, J.W., Orth, R.G., Sivasupramaniam, S. and Voth, R. (2010) Spatial and temporal variability in host use by *Helicoverpa zea* as measured by analyses of stable carbon isotope ratios and gossypol residues. *Journal of Applied Ecology* 47, 583–592.

Hobson, K.A. and Wassenaar, L.I. (Eds) (2008) *Tracking Animal Migration with Stable Isotopes.* Academic Press, London.

Hobson, K.A., Wassenaar, L.I. and Taylor, O.R. (1999) Stable isotopes (δD and δ^{13}C) are geographic indicators of natal origins of monarch butterflies in eastern North America. *Oecologia* 120, 397–404.

Hood-Nowotny, R. and Knols, B.G.J. (2007) Stable isotope methods in biological and ecological studies of arthropods. *Entomologia Experimentalis et Applicata* 124, 3–16.

IAEA (2009) *Manual for the Use of Stable Isotopes in Entomology.* International Atomic Energy Agency, Vienna, Austria.

IRRI, AfricaRice and CIAT (2010) *Global Rice Science Partnership (GRiSP).* International Rice Research Institute, Los Baños, Philippines; Africa Rice Center, Cotonou, Benin; and International Center for Tropical Agriculture, Cali, Colombia.

Khan, Z.R., Chiliswa, K., Ampong-Nyarko, K., Smart, L.E., Polaszek, A., Wandera, J. and Mulaa, M.A. (1997) Utilization of wild gramineous plants for the management of cereal stem borers in Africa. *Insect Science and its Application* 17(1), 143–150.

Knops, J.M.H., Tilman, D., Haddad, N.M., Naeem, S., Mitchell, C.E., Haarstad, J., Ritchie, M.E., Howe, K.M., Reich, P.B., Siemann, E. and Groth, J. (1999) Effects of plant species richness on invasion dynamics, disease outbreaks, insect abundances and diversity. *Ecology Letters* 2, 286–293.

Kuponiyi, F.A. and Bamigboye, E.O. (2009). The use of indigenous knowledge systems (IKS) in rice production by farmers in Ekiti State, Nigeria. *International Journal of Agricultural Economics and Rural Development* 2(2), 67–74.

Lehman, R., Lundgren, J. and Petzke, L. (2009) Bacterial communities associated with the digestive tract of the predatory ground beetle, *Poecilus chalcites*, and their modification by laboratory rearing and antibiotic treatment. *Microbial Ecology* 57, 349–358.

Le Nguyen, D.D., Ha, N.H., Dijoux, D., Loiseau, G. and Montet, D. (2008a) Determination of fish origin by using 16S rDNA fingerprinting of bacterial communities by PCR–DGGE: an application on *Pangasius* fish from Vietnam. *Food Control* 19(5), 454–460.

Le Nguyen, D.D., Gemrot, E., Loiseau, G. and Montet, D. (2008b) Determination of citrus fruit origin by using 16S rDNA fingerprinting of bacterial communities by PCR- DGGE: an application to clementine from Morocco and Spain. *Fruits* 63, 75–84.

Menozzi, P., Ouin, A., Achaleke, J., Coulon, M., Sarthou, J.P. and Ponsard, S. (2007) Using stable isotopes to trace origin and host plants of an African polyphagous pest and a European beneficial insect. *Communications in Agricultural and Biological Sciences* 72(3), 375–380.

Mugisha-Kamatenesi, M., Deng, A.L., Ogendo, J.O., Omolo, E.O., Mihale, M.J., Otim, M., Buyungo, J.P. and Bett, P.K. (2008) Indigenous knowledge of field insect pests and their management around lake Victoria basin in Uganda. *African Journal of Environmental Science and Technology* 2(10), 342–348.

Nacro, S., Heinrichs, E.A. and Dakouo, D. (1996) Estimation of rice yield losses due to the African rice gall midge, *Orseolia oryzivora* Harris and Gagné. *International Journal of Pest Management* 42, 331–334.

Nwilene, F.E., Williams, C.T., Ukwungwu, M.N., Dakouo, D., Nacro, S., Hamadoun, A., Kamara, S.I., Okhidievbie, O., Abamu, F.J. and Adam, A. (2002) Reactions of differential rice genotypes to African rice gall midge in West Africa. *International Journal of Pest Management* 48(3), 195–201.

Nwilene, F.E., Nwanze, K.F. and Okhidievbie, O. (2006) *African Rice Gall Midge: Biology, ecology and control – Field Guide and Technical Manual.* Africa Rice Center, Cotonou, Benin.

Nwilene F.E., Nwanze, K.F. and Youdeowei, A. (2008a) Impact of IPM on food and horticultural crops in Africa – a review. *Entomologia Experimentalis et Applicata* 128(3), 355–363.

Nwilene, F.E., Jones, M.P., Brar, D.S., Youm, O., Togola, A., Kehinde, A., Ukwungwu, M.N., Kamara, S.I. and Hamadoun, A. (2008b) Integrated pest management (IPM) strategies for NERICA varieties. Module 8. In: Africa Rice Center (WARDA), FAO and SAA (eds) *NERICA®: The New Rice for Africa – A compendium* (Somado, E.A., Guei, R.G. and Keya, S.O. eds). Africa Rice Center, Cotonou, Benin; Food and Agriculture Organization of the United Nations, Rome, Italy; and Sasakawa Africa Association, Tokyo, Japan, pp. 83–94.

Nwilene, F.E., Togola, A., Agunbiade, T.A., Ogah, E.O., Ukwungwu, M.N., Hamadoun, A., Kamara, S.I. and Dakouo, D. (2008c) Parasitoid biodiversity conservation for sustainable management of the African rice gall midge, *Orseolia oryzivora* (Diptera: Cecidomyiidae) in lowland rice. *Biocontrol Science and Technology* 18(10): 1075–1081.

Nwilene, F.E., Sanyang, S., Traore, A.K., Togola, A., Goergen, G., Agunbiade, T.A., Tamo, M., Akintayo, I., Cissé, B. and Ogah, E. (2009a) *Rice Stem Borers: Biology, ecology and control – Field Guide and Technical Manual.* Africa Rice Center, Cotonou, Benin.

Nwilene, F.E., Okhidievbie, O., Agunbiade, T.A., Traore, A.K., Gaston, L.N., Togola, M.A. and Youm, O. (2009b) Antixenosis component of rice resistance to African rice gall midge, *Orseolia oryzivora. International Rice Research Notes* 34: 1–5.

Oerke, E.C. (2006) Centenary review: crop losses to pests. *Journal of Agricultural Science* 144, 31–43.

Orth, R.G., Head, G. and Mierkowski, M. (2007) Determining larval host plant use by a polyphagous lepidopteran through analysis of adult moths for plant secondary metabolites. *Journal of Chemical Ecology* 33, 1131–1148.

Pimentel, D. (2007) Area-wide pest management: environmental, economic and food issues. In: Vreysen, M.J.B., Robinson, A.S. and Hendrichs, J. (eds) *Area-wide Control of Insect Pests: From research to field implementation.* Springer, Dordrecht, Netherlands, pp. 35–47.

Ponsard, S., Bethenod, M.T., Bontemps, A., Pélozuelo, L., Souqual, M.C. and Bourguet, D. (2004) Carbon stable isotopes: a tool for studying the mating, oviposition, and spatial distribution of races of European corn borer, *Ostrinia nubilalis*, among host plants in the field. *Canadian Journal of Zoology* 82, 1177–1185.

Rani, A., Sharma, A., Rajagopal, R., Adak, T. and Bhatnagar, R.K. (2009) Bacterial diversity analysis of larvae and adult midgut microflora using culture-dependent and culture-independent methods in lab-reared and field-collected *Anopheles stephensi* – an Asian malarial vector. *BMC Microbiology* 9, 96–118.

Reed, C.R. (2010) *Managing Stored Grain to Preserve Quality and Value.* AACC International, Saint Paul, Minnesota.

Rodenburg, J., Diagne, A., Oikeh, S., Futakuchi, K., Kormawa, P.M., Semon, M., Akintayo, I., Cissé, B., Sié, M., Narteh, L., Nwilene, F., Diatta, S., Sere, Y., Ndjiondjop, M.N., Youm, O. and Keya, S.O. (2006) Achievements and impact of NERICA on sustainable rice production in sub-Saharan Africa. *International Rice Commission Newsletter* 55, 45–58.

Savary, S., Willocquet, L., Elazegui, F.A., Teng, P.S., Du, P.V., Zhu, D., Tang, Q., Lin, X., Singh, H.M. and Srivastava, R.K. (2000a) Rice pest constraints in tropical Asia: characterization of injury profiles in relation to production situations. *Plant Disease* 84, 341–356.

Savary, S., Willocquet, L., Elazegui, F.A., Castilla, N.P. and Teng, P.S. (2000b) Rice pest constraints in tropical Asia: quantification of yield losses due to rice pests in a range of production situations. *Plant Disease* 84, 357–369.

Savary, S., Teng, P.S., Willocquet, L. and Nutter, F.W. Jr. (2006) Quantification and modeling of crop losses: a review of purposes. *Annual Review of Phytopathology* 44, 89–112.

Secilia, J. and Bagyaraj, D.J. (1992) Selection of efficient vesicular-arbuscular mycorrhizal fungi for wetland rice. *Biology and Fertility of Soils* 13, 108–111.

Silvie, P., Togola, A., Adda, C., Nwilene, F., Ravaomanarivo, L., Randriamanantsoa, R. and Menozzi, P. (2012) Le riz *Bt* dans le contexte de la riziculture en Afrique et à Madagascar. *Cahier Agricultures* (numéro spécial riz).

Swift, M.J., Vandermeer, J., Ramakrishnan, P.S., Anderson, J.M., Ong, C.K. and Hawkins, B.A. (1996) Biodiversity and agroecosystem function. In: Mooney, H.A., Cushman, J.H., Medina, E., Sala, O.E. and Schulze, E.D. (eds) *Functional Roles of Biodiversity: A global perspective.* Wiley, New York, pp. 261–297.

Tian, X.L., Cao, L.X., Tan, H.M., Zeng, Q.G., Jia, Y.Y., Han, W.Q. and Zhou, S.N. (2004) Study on the communities of endophytic fungi and endophytic actinomycetes from rice and their antipathogenic activities *in vitro*. *World Journal of Microbiology and Biotechnology* 20, 303–309.

Togola, A., Nwilene, F.E., Chougourou, D.C. and Agunbiade, T. (2010) Presence, populations and damage of the angoumois grain moth, *Sitotroga cerealella* (Olivier) (Lepidoptera, Gelechiidae), on rice stocks in Benin. *Cahiers Agricultures* 19(3), 205–209.

Van Rensburg, J.B.J. (2007) First report of field resistance by the stem borer, *Busseola fusca* (Fuller) to *Bt*-transgenic maize. *South African Journal of Plant and Soil* 24, 147–151.

von Braun, J., Swaminathan, M.S. and Rosegrant, M.W. (2004) Agriculture, food security, nutrition and the Millenium Development Goals. In: *2003–2004 IFPRI Annual Report.* International Food Policy Research Institute, Washington, DC, pp. 5–19. http://www.ifpri.org/publication/agriculture-food-security-nutrition-and-millenium-development-goals.

Walker, D.J. and Farrell, G. (2003) *Food Storage Manual.* Natural Resources Institute, Chatham, UK, and World Food Programme, Rome, Italy.

William, S. and Garba, M.H. (2011) Sustainable crop production intensification in the Senegal and the Niger River basins of francophone West Africa. *International Journal of Agricultural Sustainability* 9(1), 171–185.

Williams, C.T., Harris, K.M. and Okhidievbie, O. (1997) Management of the African rice gall midge (*Orseolia oryzivora*) in West Africa. Completion Report for ODA Holdback Project R5619(H). CAB International, Wallingford, UK.

Williams, C.T., Okhidievbie, O., Harris, K.M. and Ukwungwu, M.N. (1999) The host range, annual cycle and parasitoids of the African rice gall midge *Orseolia oryzivora* (Diptera: Cecidomyiidae) in central and southeast Nigeria. *Bulletin of Entomological Research* 89, 589–597.

Willocquet, L., Savary, S., Fernandez, L., Elazegui, F.A., Castilla, N., Zhu, D., Tang, Q., Huang, S., Lin, X., Singh, H.M. and Srivastava, R.K. (2002) Structure and validation of RICEPEST, a production situation-driven, crop growth model simulating rice yield response to multiple pest injuries for tropical Asia. *Ecological Modelling* 153, 247–268.

Youdeowei, A. (2004) Fighting hunger and poverty: IPM contributions in Africa. Plenary Lecture, 15th International Plant Protection Congress, Beijing, China, May.

Youm, O., Vayssieres, J.F., Togola, A., Robertson, S.P. and Nwilene, F.E. (2011) International trade and exotic pests: the risks for biodiversity and African economies. *Outlook on Agriculture* 40(1), 59–70.

Zouache, K., Raharimalala, F.N., Raquin, V., Tran-Van, V., Raveloson, L.H.R., Ravelonandro, P. and Mavingui, P. (2011) Bacterial diversity of field-caught mosquitoes, *Aedes albopictus* and *Aedes aegypti*, from different geographic regions of Madagascar. *FEMS Microbiology Ecology* 75, 377–389.

19 Bird Damage to Rice in Africa: Evidence and Control

Yann de Mey[1] and Matty Demont[2*]
[1]*Hasselt University, Centre for Environmental Sciences, Diepenbeek, Belgium;* [2]*Africa Rice Center (AfricaRice), Saint-Louis, Senegal*

Introduction

Rice, one of the most important cereal crops worldwide, has the potential to play a significant role in achieving global food security. However, several biotic and abiotic stresses seriously jeopardize this potential. According to Oerke (2005), some 15% of global rice production is lost to animal pests (arthropods, nematodes, rodents, birds, slugs and snails). The Global Rice Science Partnership (GRiSP) identifies birds as the second most important biotic constraint in African rice production after weeds, based on farmer surveys in 20 African countries (IRRI *et al.*, 2010). Despite current control practices, birds cause substantial losses to the African rice sector. Diagne *et al.* (Chapter 4, this volume) provide an *ex-ante* assessment of the benefits of rice research. They indicate discounted cumulative benefits of US$292 million for research aimed at reducing yield loss to birds and rodents, and $2.679 billion for total benefit derived from the research agenda – thus, the contribution of research on reducing losses to birds and rodents comprises 10.9% of the total benefit of the whole research agenda.

Rice is mainly affected by birds in the humid zone and, to various degrees, in the Sahel and Sudanian savannah zones (Manikowski, 1984; FAO, 1991). The most serious pest birds are gregarious and migratory, such as Red-billed Quelea, but locally other types such as water birds (e.g. ducks and geese) can also be of importance. Rice crops are vulnerable to bird attacks at early crop establishment stages (mostly by water birds) and highly susceptible from the milky stage up to maturation (exclusively by land birds). Most damage occurs during the dry season; in the rainy season there is usually an abundance of seeds from wild grasses available as alternative food sources (Ruelle and Bruggers, 1982).

Aside from physical yield losses – which are the main focus of this chapter – other problems caused by pest birds include extensive labour requirements for bird scaring, the associated use of child labour, possible health or environmental hazards resulting from the use of chemical poisons, and the discouragement of farmers from dry-season rice cultivation.

Although the adverse impact of birds on rice has received much international attention in the past and is still generally recognized, little research on bird damage or control is currently conducted. Given the increasing importance of bird damage in some regions such as the Sahel (de Mey, 2009), this chapter reviews the available evidence of bird damage to rice in Africa, the

* Corresponding author: m.demont@irri.org

© CAB International 2013. *Realizing Africa's Rice Promise* (eds M.C.S. Wopereis *et al.*)

bird species involved, and the measures commonly used to control them.

Pest Bird Species Injurious to Cereal Crops in Africa

On a global scale, only a handful of birds are serious pests of cereal crops. Birds can become serious pests when large flocks migrate seasonally and concentrate in large populations. In a review of literature, Manikowski (1984) lists 36 bird species as 'known to cause damage' among the approximately 1390 bird species in West Africa (van Perlo, 2002). The seven most important species are Spur-winged Goose (*Plectropterus gambensis*), Knob-billed Goose (*Sarkidiornis melanotos*), Village Weaver (*Ploceus cucullatus*), Black-headed Weaver (*Ploceus melanocephalus*), Red-billed Quelea (*Quelea quelea*), Red-headed Quelea (*Q. erythrops*) and Golden Sparrow (*Passer luteus*). More background information on granivorous pest birds in sub-Saharan Africa (their identification, biology and feeding habits) is provided by Allen (1997).

Red-billed Quelea is one of the most notorious pest bird species in the world, injurious to various cereal crops such as rice, millet, sorghum and wheat (FAO, 1991). It occurs throughout sub-Saharan Africa. It gathers in flocks of several million birds and breeds in colonies that can cover more than 100 hectares (with about 30,000 nests per hectare). It is considered the most numerous bird worldwide with population numbers totalling about 1.5 billion at the end of the breeding season (Elliott, 1989). Red-billed Quelea has been studied extensively and there are many publications describing its pest status and control strategies in African agriculture (see Oschadleus, 2001).

Evidence of Bird Damage to Rice in Africa

Factors influencing bird damage

Birds differ in nature to many other pests of rice in that they can migrate over long distances and have a flexible diet in which agricultural crops may play only an incidental role (birds prefer wild seeds, see below). Great variability exists in the occurrence and extent of the crop damage inflicted by birds that farmers experience (in both space and time) due to many biological, environmental and management factors (described below).

Fields close to breeding or roosting sites are most susceptible to damage from birds (FAO, 1991). The presence of trees, bushes or reeds in the vicinity of the field increases vulnerability because these provide birds with perches and nesting sites. Fields close to water sources (e.g. rivers or large irrigation canals) are more frequently damaged by birds because – as a supply of drinking water – they attract birds (Manikowski, 1984) and provide habitats for water birds such as geese that are potentially harmful to rice. These factors in the immediate surroundings of rice fields vary spatially, which makes bird damage a region-dependent problem with some regions having serious bird problems, while others experience almost no bird attacks at all. Furthermore, as birds are migratory the problem needs to be managed cooperatively between neighbouring countries.

Field size influences bird damage. Large fields have longer borders (in absolute terms) which are the zones preferentially attacked by birds. Moreover, a large field requires a large labour force for bird scaring, which is particularly challenging to assemble during peak periods (Manikowski and Da Camara-Smeets, 1979).

The timing of farming operations during the season is a crucial factor that can influence the incidence of bird damage. Sowing too late during the wet season predisposes the crop to damage by migrating birds from Europe arriving at the time when the crop matures (Tréca, 1985). Therefore, synchronized cropping among farmers may dilute bird-inflicted losses to some extent. Damage is expected to be higher during the dry season than during the wet season due to the lack of available wild seeds (Ruelle and Bruggers, 1982). Damage also differs with respect to the growth stage of the rice crop (Tréca, 1977). Rice is susceptible to bird damage at the early crop establishment stages (see discussion below), but subsequently invulnerable during the vegetative and booting stages. Throughout maturation (milk to hard-dough stages) the rice crop is highly susceptible to bird damage (Ruelle and Bruggers, 1982). Postharvest losses can also occur during drying, storage or trading because at these stages the rice grains are frequently left exposed in the open. This is particularly true during drying, especially if it is carried out by simply laying the rice grains on a sheet in the sun.

The effectiveness of weed control in a field and on the banks of irrigation canals plays an indirect role in bird damage as weedy fields attract birds. The bulk of the diet of Red-billed Quelea does not consist of cultivated cereals; they prefer smaller seeds of various wild grasses (Ward, 1965). This implies that if these wild grasses are widely available in or near a farmer's fields, bird flocks will be attracted to that area and occasionally feed on the cereal crops. This way, the birds become familiar with the region and will return for feeding purposes. A study on wheat showed that weedy patches in fields are clearly associated with greater bird damage (Luder, 1985) and several studies on rice cite the same interaction between weeds and birds (e.g. Bruggers, 1979; Tréca, 1985; FAO, 1991). A survey in the Senegal River valley revealed that almost 80% of rice farmers were aware of the interaction between weeds and birds, and 57% plan their weeding efforts in relation to bird pressure (de Mey, 2009). This is confirmed by a field trial conducted by Africa Rice Center (AfricaRice), which concludes that weed control reduces bird damage in irrigated rice (Rodenburg *et al.*, 2010). Weeds infesting rice fields attract granivorous birds, which leads to increased crop losses due to the effects of weed competition being compounded by the losses to birds.

Bird damage is also affected by the crop establishment method (Tréca, 1977). In the case of direct seeding, seeds are highly vulnerable to bird attack when they are sown straight on to the soil. A common practice, however, is to cover seeds with soil to protect them. Young, emerging seedlings are still vulnerable to birds; however, damage at this stage is rare during the rainy season when most granivorous migratory birds have flown to Europe and Asia (Tréca, 1977). On the other hand, if farmers transplant rice, the vulnerability of freshly sown rice seeds in the nursery bed to bird attacks can be greatly reduced by protecting the nurseries with nets. Planting density in the field is also an important factor. Uneven crop plant densities can result from improper sowing or planting, a badly levelled field or other factors leading to heterogeneous field conditions. Water birds are attracted to zones with plant densities much lower than in the immediate vicinity and can cause substantial damage to the surrounding rice plants (Tréca, 1977). The choice of rice variety also affects damage as birds have a preference for particular varieties and will completely ignore others (e.g. awned varieties) if sufficient alternative food sources are available. Varieties with weak stems are also more susceptible to mechanical damage if they cannot support the birds' weight (FAO, 1991).

As mentioned above, birds prefer seeds of wild grass species. As soon as these natural grain stocks start to decline at the end of the rainy season, Red-billed Queleas gather in inundation zones where wild seed production is greatest. Once these natural reserves of grains are exhausted and at the onset of the next rains (which cause the remaining seeds to start geminating – hence making them unavailable as a food source) the pest birds are threatened with starvation. This forces the birds into a nomadic migration to areas where the rains came earlier and where fresh crop and weed seeds are available in abundance. These huge flocks of birds are feared by farmers because they consist of large numbers of starving birds which see rice fields as oases of food abundance in an otherwise food-scarce landscape (Ward, 1971; Jones, 1989). Given the unpredictable nature of rainfall patterns, Red-billed Quelea movement patterns are difficult to predict. General trends show, however, that damage tends to be less severe in years of high rainfall, which probably relates to the availability of seeds of wild species as explained above (Bruggers, 1980). Life cycle is another biological factor, as, for example, young Red-billed Queleas will inflict damage in the immediate vicinity of the roosts, whereas mature individuals will fly out and cause damage further away. Feeding behaviour of different bird species varies with time of the day and the maturation stage of the rice crop (Bruggers, 1979). The presence of some (grain-eating) birds in a field can also attract others (FAO, 2001).

Review of the available evidence

Birds cause visual damage patterns to the rice crop such as: (i) direct damage when seeds or grains are eaten by birds; and (ii) indirect (mechanical) damage that occurs when a flock of birds is active in a rice field, resulting in grains that fall to the ground (Tréca, 1987). Both are important and should be taken into account when assessing bird damage.

There are many **direct** and **indirect** techniques to assess crop losses due to birds. **Direct** measures include field sampling and visual estimation techniques. The former is based on the counting or weighing of a representative sample of rice panicles (e.g. Bruggers and Ruelle, 1981), while the latter comprises experienced observers estimating losses by visual inspection of a field (e.g. Tracey and Saunders, 2010). Frequently, sampling techniques are also used to calibrate visual assessments. Due to the heterogeneity of the bird problem, however, large sample sizes are needed across several production seasons, which renders these methods labour intensive and expensive. **Indirect** methods include questionnaire surveys, monitoring of bird numbers, energetics models, and damage-abatement models. While questionnaire surveys are obviously subjective, they do serve to demonstrate the general levels of damage and the relative seriousness of the problem according to farmers' perceptions (e.g. Sidibé *et al.*, 2003). Monitoring bird numbers enables the species involved and its behaviour to be determined, which, in turn, enhances the understanding of the bird problem (e.g. Bruggers, 1980). Estimating bird damage is based on using the correlation between bird density and damage – a relationship that is (unfortunately) rarely known and difficult to obtain. Energetics models are more advanced models that take pest-bird population estimates and make use of the determinants of energy requirement of these populations (e.g. age, temperature and body weight) to predict the amount of damage they will inflict (e.g. Weatherhead *et al.*, 1982). de Mey *et al.* (2012) propose an indirect method relying on a production function with a damage-abatement component that enables bird damage to be econometrically isolated from overall rice crop productivity.

The latest extensive overview of world bird damage problems is provided by De Grazio (1978). Aggregating results of 32 studies (reporting damage in 626 plots), Manikowski (1984) suggests an average loss of 6.9% of cereal harvest in West Africa. The highest damage figures were found for millet in the Sahel zone (25%) and rice in the humid zone (19%). No more recent overviews are available. Table 19.1 provides a non-exhaustive literature overview and shows that bird damage is an important loss factor for cereal crop production in Africa. Table 19.1 suggests that bird damage on cereal crops is on average about 15–20% of production, but considerable spatial and temporal variability exists. Much research on bird damage on cereal crops was carried out by international research organizations prior to the 1980s, e.g. the FAO/UNDP Quelea Project (Jackson, 1973), but subsequently few studies have been conducted. Furthermore, much material is only available in internal documents or unpublished research reports; recent, peer-reviewed damage estimates are scarce.

Control of Bird Damage to Rice in Africa

Many techniques are available to protect crops from bird damage. A literature review of the effectiveness of existing bird control techniques is provided by Bishop *et al.* (2003). The main conclusions of this study are that the effectiveness of each technique varies with the bird species involved, and that optimal bird control methods combine several techniques or use them in a random fashion. Human-operated scaring techniques were the most effective; lethal methods are of only short-term benefit. In Africa, traditional, low-cost methods are mainly used. Two general classes of approaches can be distinguished. **Preventive** methods aim at not attracting birds to the field, while **protective** methods focus on protecting the rice crop when birds do visit the field. **Preventive** methods can be subdivided into lethal and non-lethal techniques. Lethal techniques are aimed at suppressing pest bird populations and are primarily implemented by national or regional (governmental) crop protection units. They include manual nest destruction, treatment with avicides (chemical substances lethal to certain bird species) and the use of explosives or flamethrowers. Non-lethal techniques include agronomic practices such as vegetation management, good weed management (as weeds attract birds), specific planning of the production season and choosing a variety

Table 19.1. Overview of studies reporting bird damage estimates on cereal crops in Africa. (From de Mey *et al.*, 2012, with permission from John Wiley and Sons, © 2012 The Agricultural Economics Society, with small modifications.)

Region	Year (period)	Crop	Estimation of losses	Estimation method (sample size[a])	Major species	Source	Reference
Africa, East	n.a.	Cereals	'A total of at least $15 million annually in eastern Africa'	Literature review (5 studies)	Red-billed Quelea and other ploceid species	Conference proceedings	Bruggers and Jaeger (1981)
Africa, Sahel	n.a.	Cereals	'Damage to cereal crops by quelea birds in specific locations could be as much as 100 percent but was estimated to be about 5 percent nationally in Sahel countries and at about 1 percent for all of savannah Africa'	n.a.	Red-billed Quelea	Book chapter	FAO (2001)
Africa, West	n.a.	Cereals	'Out of a total of 626 plots of cereals, birds damaged 6.9% of the harvest ... The average loss in the plots which were damaged by birds (71% of examined fields) was 13.7%[b]	Literature review (32 studies – 626 plots)	Red-billed Quelea, Village Weaver, Red-headed Quelea, Golden Weaver, Black-headed Weaver	Journal article (*Tropical Pest Management*)	Manikowski (1984)
Chad, Cameroon	1975–1977	Rice	'Damage varied from 13% to 26% when harvest overlapped with birds' arrival'	Sampling (10 plots – 2000 panicles)	Red-billed Quelea	Journal article (*Journal of Applied Ecology*)	Elliott (1979)
Ethiopia	1975–1979	Sorghum	Damage averaged between 9% and 51% in 1975–1977 before control measures. In 1978–1979, when control was applied, damage averaged between 0.6% and 22.1%	Visual estimation (5 sites – 1000 panicles)	Red-billed Quelea	Conference proceedings	Jaegar and Erickson (1980)
Kenya	1952	Wheat, rice, millet, sorghum	$0.7 million loss; 100,000 bags	n.a.	Red-billed Quelea, 'Doves', 'Weavers', 'Starlings'	Report	De Grazio (1973); De Grazio and Dehaven (1974); Pearson (1967) cited by De Grazio (1978)

Mali	1983–1986	Rice	'Damages are very variable among years (from 0.76 to 14% of the harvest, by mean), but they are not uniformly distributed, some fields being very heavily destroyed, when other are untouched'	Visual estimation (80 plots – 10,000 panicles)	'Ducks', 'Ruffs'	Journal article (*Journal d'agriculture traditionnelle et de botanique appliquée*)	Tréca (1987)
	2001–2002	Rice	22% of production	Survey (n = 280)	'Grain eating'	Report	Sidibé *et al.* (2003)
Morocco	n.a.	Wheat, rice	Wheat loss most severe; $4.0 million loss	n.a.	House Sparrow, Spanish Sparrow	Report	De Grazio *et al.* (1971) cited by De Grazio (1978)
Nigeria	n.a.	Rice, wheat, millet, sorghum, maize	$2.8 million/year loss in one province	n.a.	Red-billed Quelea, Golden Sparrow, 'Starlings', 'Doves', 'Weavers'	Report	De Grazio *et al.* (1971); Pearson (1967); Ward and Jones (1977) cited by De Grazio (1978)
Senegal, country	1976–1977	Cereals	5% of total cereal production (the equivalent of $4–5 million)	Sampling (375 plots)	'Weavers', Red-billed Quelea, Golden Sparrow	Journal article (*Protection Ecology*)	Bruggers and Ruelle (1981)
Senegal, Senegal River valley	2003–2007	Rice	Annual crop losses during the wet seasons of 2003–2007 were econometrically estimated at 3.4%, 10.2%, 11.5%, 30.3% and 13.4%, respectively, and 0.6%, 6.0%, 10.7%, 51.4% and 12.1% according to surveyed farmers' perceptions	Survey (n = 111) and damage abatement production function	Red-billed Quelea, Golden Sparrow	Journal article (*Journal of Agricultural Economics*)	de Mey *et al.* (2012)
Somalia	1975–1979	Rice, sorghum, maize	Damage averaged between 1% and 78% depending on crop and field location	Discussions and sampling	Village Weaver, Golden Palm Weaver, Chestnut Weaver, Red Bishop	Conference proceedings	Bruggers (1980)
South Africa	1953	Cereals	$1.4 million loss of sorghum	n.a.	Red-billed Quelea	Conference proceedings	Hey (1964); Pearson (1967) cited by De Grazio (1978)
Sudan	n.a.	Sorghum	$0.9 million loss annually	n.a.	Red-billed Quelea	Report	Schmutterer (1969) cited by De Grazio (1978)

Notes: Monetary values are in nominal terms. n.a. = not available.

[a]No. panicles always refers to the total sample size, not the number of panicles per plot. [b]This aggregate study uses results of some of the other studies presented in this table.

with bird-resistant characteristics. Religious techniques such as shamanism and fetishes are also still widely adopted in Africa. **Protective** methods include the use of repellents (chemical substances aimed at deterring birds), protecting fields or nurseries with nets or wires, covering individual heads of ripening crops with grass or cloth, and manual bird-scaring efforts. The latter may consist of a combination of auditory (e.g. noise-making devices, whips, shouting), visual (scarecrows, flags, reflective tape) and physical measures (e.g. throwing rocks or mud).

In general, traditional protective methods such as manual bird scaring, flags and scarecrows can provide satisfactory protection on small-scale, privately owned farms when bird numbers are low. However, when pest bird pressure is elevated, these methods become ineffective (Ruelle and Bruggers, 1982; de Mey *et al.*, 2012). Also, on large-scale, governmental production schemes, these methods are impractical, costly and ineffective (Bruggers, 1980). This suggests that the development of bird populations needs to be monitored and farmers need to be protected against the consequences of massive bird invasions through insurance systems (PINORD, 2009). In particular, there is a need to link bird invasions to climatic factors that influence the development of birds in Africa. Such information can be used to create observatories that monitor the diffusion of birds in order to prevent or anticipate massive bird invasions.

Large-scale, chemical control techniques to reduce population levels of pest birds to non-pest levels are still widely adopted in many African countries (Mullié *et al.*, 1999). Literature suggests that these are inefficient, however, because they do not significantly lower populations due to the pest birds' high reproductive potential and high mobility (Ward, 1979). The frequent inaccessibility of areas in which birds reside also creates a barrier to efficient application of these chemicals. In contrast, lethal control to locally reduce pest bird numbers in the vicinity of important cereal production areas to give temporary relief has been applied successfully. However, the success of this approach varies regionally and by the control method deployed (Bruggers and Jaeger, 1981). It is worth noting that these approaches entail severe environmental hazards when avicides are applied (Mullié *et al.*, 1999).

Conclusion

Birds, and especially Red-billed Quelea, inflict substantial losses on rice in Africa. Many factors influence the incidence of crop losses, some of which can be controlled by farmers. Accurate bird damage figures are important in order to put the bird problem into perspective and provide useful information for future research, allocation of research funding at the governmental level, and farm management decision making. The available evidence on physical crop losses is somewhat out-dated and suggests an average loss of 15–20% of total cereal production due to bird damage with substantial spatial and temporal heterogeneity.

Estimates of physical losses, however, underestimate true economic losses as they do not account for costs entailed by the supply-response adjustments that farmers make when faced with pests (Chambers *et al.*, 2010). This is supported by farmer surveys in Senegal that suggest that the mere presence of the avian risk is a block to intensification in irrigated rice cultivation (de Mey *et al.*, 2012). Moreover, aside from direct economic impacts, bird damage also has substantial social consequences. On the one hand, farmers who scare birds in the field are socially separated from their family for a long time. On the other hand, traditional bird scaring is frequently undertaken by children who sometimes miss school in doing so, which jeopardizes meeting key education objectives such as universal primary enrolment (de Mey *et al.*, 2012).

Finally, a review of the adopted techniques for controlling bird damage on rice in Africa suggests that mainly traditional, low-cost methods are used. These methods may be adequate under low pest pressure but under high pest pressure they become ineffective. This suggests that research is needed on (i) developing alternative bird-damage control measures that are low cost, environmentally friendly and can be easily adopted by farmers; (ii) predicting massive bird invasions through observatories; and (iii) developing index-based insurance systems that can protect farmers against avian risk. The public good nature of bird control – if one farmer scares birds from his or her field, these birds are merely displaced to adjacent fields – justifies government intervention.

References

Allen, R. (1997) *The Grain-eating Birds of Sub-Saharan Africa: Identification, biology and management*, 1st edn. Natural Resources Institute, Chatham, UK.

Bishop, J., McKay, H., Parrott, D. and Allan, J. (2003) Review of international research literature regarding the effectiveness of auditory bird scaring techniques and potential alternatives. Department for Environment, Food and Rural Affairs, London.

Bruggers, R.L. (1979) Summary of methiocarb trials against pest birds in Senegal. In: Jackson, W.B. (ed.) *Wildlife Damage Management*, Internet Center for Bird Control Seminars Proceedings. Bowling Green State University, Ohio, pp. 172–184.

Bruggers, R.L. (1980) The situation of grain-eating birds in Somalia. In: Clark, J.P. (ed.) *Proceedings of the 9th Vertebrate Pest Conference*, Vertebrate Pest Conference Proceedings Collection. University of California, California, pp. 4–16.

Bruggers, R.L. and Jaeger, M.M. (1981) Bird pests and crop protection strategies for cereals of the semi-arid African tropics. Paper presented at International Symposium on Sorghum, Pantancheru, India, 2–7 November.

Bruggers, R.L. and Ruelle, P. (1981) Economic impact of pest birds on ripening cereals in Senegal. *Protection Ecology* 3, 7–16.

Chambers, R.G., Karagiannis, G. and Tzouvelekas, V. (2010) Another look at pesticide productivity and pest damage. *American Journal of Agricultural Economics* 92, 1401–1419.

De Grazio, J.W. (1978) World bird damage problems. In: Howard, W.E. (ed.) *Proceedings of the 8th Vertebrate Pest Conference*, Vertebrate Pest Conference Proceedings Collection. University of California, California, pp. 8–24.

de Mey, Y. (2009) Economics of bird control in irrigated rice in the Senegal River valley. MSc thesis, Katholieke Universiteit Leuven, Belgium.

de Mey, Y., Demont, M. and Diagne, M. (2012) Estimating bird damage to rice in Africa: evidence from the Senegal River valley. *Journal of Agricultural Economics* 63, 175–200.

Elliott, C.C.H. (1979) The harvest time method as a means of avoiding quelea damage to irrigated rice in Chad/Cameroun. *Journal of Applied Ecology* 16, 23–35.

Elliott, C.C.H. (1989) The pest status of the quelea. In: Bruggers, R.L. and Elliott, C.C.H. (eds) *Quelea quelea: Africa's bird pest.* Oxford University Press, Oxford, pp. 17–34.

FAO (1991) *Manuel de protection des cultures contre les dégâts d'oiseaux.* Food and Agriculture Organization of the United Nations, Dakar, Senegal.

FAO (2001) Economic impacts of transboundary plant pests and animal diseases. In: FAO (ed.) *The State of Food and Agriculture 2001.* Food and Agriculture Organization of the United Nations, Rome, pp. 198–280.

IRRI, AfricaRice and CIAT (2010) *Global Rice Science Partnership (GRiSP).* International Rice Research Institute, Manila, Philippines; Africa Rice Center, Cotonou, Benin; and International Center for Tropical Agriculture, Cali, Colombia.

Jackson, J.J. (1973) FAO quelea research in Africa. In: Cones, H.N. and Jackson, W.B. (eds) *Wildlife Damage Management*, Internet Center for Bird Control Seminars Proceedings. Bowling Green State University, Ohio, pp. 220–222.

Jaeger, M.M. and Erickson, W.A. (1980) Levels of bird damage to sorghum in the Awash basin of Ethiopia and the effects of the control of quelea nesting colonies. In: Clark, J.P. (ed.) *Proceedings of the 9th Vertebrate Pest Conference*, Vertebrate Pest Conference Proceedings Collection. University of California, California, pp. 20–28.

Jones, P.J. (1989) General aspects of quelea movement. In: Bruggers, R. and Elliott, C.C.H. (eds) *Quelea quelea: Africa's bird pest.* Oxford University Press, Oxford, pp. 102–112.

Luder, R. (1985) Weeds influence Red-Billed Quelea damage to ripening wheat in Tanzania. *Journal of Wildlife Management* 49, 646–647.

Manikowski, S. (1984) Birds injurious to crops in West Africa. *Tropical Pest Management* 30, 379–387.

Manikowski, S. and Da Camara-Smeets, M. (1979) Estimating bird damage to sorghum and millet in Chad. *Journal of Wildlife Management* 43, 540–544.

Mullié, W.C., Diallo, A.O., Gadji, B. and Ndiaye, M.D. (1999) Environmental hazards of mobile ground spraying with cyanophos and fenthion for quelea control in Senegal. *Ecotoxicology and Environmental Safety* 43, 1–10.

Oerke, E.-C. (2005) Crop losses to pests. *Journal of Agricultural Science* 143, 31–43.

Oschadleus, H.D. (2001) *Bibliography of the African Quelea Species*, 1st edn. Avian Demography Unit, University of Cape Town, Cape Town, South Africa.

PINORD (2009) *Study of the Agricultural Risks in the Senegal River Valley.* Platforme d'Appui aux Initiatives du Nord, Ross-Bethio, Senegal.

Rodenburg, J., Demont, M. and Sow, A. (2010) Weed control reduces bird damage in irrigated rice. Poster presented at the Second Africa Rice Congress, Bamako, Mali, 22–26 March.

Ruelle, P. and Bruggers, R.L. (1982) Traditional approaches for protecting cereal crops from birds in Africa. In: Marsh, R.E. (ed.) *Proceedings of the Tenth Vertebrate Pest Conference,* Vertebrate Pest Conference Proceedings Collection. University of California, California, pp. 80–86.

Sidibé, F., Diarra, L.S., Diallo, A.K. and Sylla, C.H. (2003) Caractéristiques et impacts de la lutte antiaviaire dans les zones rizicoles de l'Office du Niger. Rapports techniques. Food and Agriculture Organization of the United Nations and Comité permanent Inter-Etats de Lutte contre la Sécheresse dans le Sahel, Bamako, Mali.

Tracey, J.P. and Saunders, G.R. (2010) A technique to estimate bird damage in wine grapes. *Crop Protection* 29, 435–439.

Tréca, B. (1977) Le problème des oiseaux d'eau pour la culture du riz au Sénégal. *Bulletin de l'Institut Fondamental d'Afrique Noire (IFAN) série A* 39, 682–692.

Tréca, B. (1985) Les possibilités de lutte contre les oiseaux d'eau pour protéger les rizières en Afrique de l'Ouest. *Journal d'agriculture traditionnelle et de botanique appliquée* 32, 191–213.

Tréca, B. (1987) Les dégâts d'oiseaux d'eau sur les rizières aménagées du delta central du Niger au Mali. *Journal d'agriculture traditionnelle et de botanique appliquée* 34, 153–170.

van Perlo, B. (2002) *Collins Illustrated Checklist: Birds of Western & Central Africa.* HarperCollins, London.

Ward, P. (1965) Feeding ecology of the black-faced dioch *Quelea quelea* in Nigeria. *Ibis* 107, 326–349.

Ward, P. (1971) Migration patterns of *Quelea-quelea* in Africa. *Ibis* 113, 275–297.

Ward, P. (1979) Rational strategies for the control of queleas and other migrant bird pests in Africa. *Philosophical Transactions of the Royal Society B: Biological Sciences* 287, 289–300.

Weatherhead, P.J., Tinker, S. and Greenwood, H. (1982) Indirect assessment of avian damage to agriculture. *Journal of Applied Ecology* 19, 773–782.

20 Increasing Rice Productivity through Improved Nutrient Use in Africa

Stephan M. Haefele,[1]* Kazuki Saito,[2] Kabirou M. N'Diaye,[3] Frank Mussgnug,[2] Andrew Nelson[1] and Marco C.S. Wopereis[2]
[1]*International Rice Research Institute (IRRI), Los Baños, Philippines;*
[2]*Africa Rice Center (AfricaRice), Cotonou, Benin;*
[3]*Africa Rice Center (AfricaRice), Saint-Louis, Senegal*

Introduction

Smallholder farmers in sub-Saharan Africa generally obtain production levels that are far below what would be possible under favourable conditions. The vast majority of farmers hardly use external inputs and they are, therefore, strongly dependent on native soil fertility. Soil fertility, defined as a mixture of soil chemical, physical and biological factors that affect land potential, is inherently low in sub-Saharan Africa, where nutrient-impoverished granites, basement sediments and sands cover about 90% of the land surface (Smaling, 2005). Low soil fertility and the often unfavourable climate create intense pressure on land, even at relatively low population densities.

Since the early 1990s there has been growing concern about the fertility of soils and, consequently, the sustainability of land use in Africa. Many studies suggest that soils are rapidly degrading. For example, Sanchez *et al.* (1997) stated that 'soil fertility depletion in smallholder farms is the fundamental biophysical root cause for declining per capita food production in sub-Saharan Africa'. Soil degradation seems to be more important in the Sudano-Sahelian regions of West Africa and in some East African countries, like Ethiopia, Kenya, Somalia and Sudan. Stoorvogel and Smaling (1990) analysed the nutrient balances of different cropping systems in Africa. The nutrient balances include, on one hand, major nutrient inflows from rainfall, organic manure, mineral fertilizers, symbiotic N-fixation and sedimentation; and, on the other hand, nutrient outflows through harvested produce and losses due to erosion, leaching, etc. They conclude that soil nutrient depletion is quite severe in Africa. Estimates of net losses were of the order of 10 kg N/ha, 4 kg P_2O_5/ha and 19 kg K_2O/ha per year.

However, such nutrient balances cannot be used to indicate sustainability or fertilizer requirements without consideration of the stocks of nutrients in the soil. Haefele *et al.* (2004) developed a nutrient-application strategy for irrigated systems in the Senegal River valley, placing emphasis on N and P and allowing for mining of the soil's K reserves, as they estimated that these K reserves would be sufficient for decades. A second important consideration is that even in resource-limited smallholder agriculture not all fields are continuously mined. Some fields may have very positive nutrient balances, especially those near the homestead (infields), while other more distant ones (outfields) may have negative

* Corresponding author: stephan.haefele@acpfg.com.au

© CAB International 2013. *Realizing Africa's Rice Promise* (eds M.C.S. Wopereis *et al.*)

balances. These typical soil-fertility gradients are due to preferential application of scarce nutrients from animal manure and other nutrient resources to infields. This ensures at least good yields in these limited areas and saves labour. Soil-fertility management strategies need to consider such gradients.

This chapter starts with a few key definitions related to nutrient management and a general overview of the quality of soil resources and fertilizer use in Africa. Next, innovations to increase rice productivity through improved nutrient use are discussed for production systems at different stages of intensification in sub-Saharan Africa (see also Saito *et al.*, Chapter 15, this volume), i.e. for rainfed, low-input/subsistence rice systems and for high-input irrigated lowland rice systems. Finally, we discuss challenges and opportunities for managing soil fertility and adapting crop management in general in rice-based systems in Africa.

Definitions

Nutrient resources are allocated by farmers through decisions made at the field and farm scales, and their use is constrained by the other principal resources for agricultural production available at the farm level – land, water, labour and capital. Within a given production system, farmers may manipulate soil fertility in several ways. They may: (i) add nutrients to replenish stocks and flows in the soil; (ii) block nutrient outflows from the field or farm; (iii) recycle nutrients that are not optimally used within the farm; or (iv) increase the efficiency with which nutrients are used by the various production systems (Hilhorst and Muchena, 2000).

The yield a rice farmer will obtain from a particular field will depend on the quantities of nutrients that are taken up by the plant during the growth cycle, either from the soil's indigenous nutrients (natural reserves) or from external inputs, such as mineral fertilizer, and whether this nutrient uptake is balanced. For example, for irrigated rice in Africa it has been shown that a balanced nutrient uptake would mean 14.1 kg N/ha, 6.9 kg P_2O_5/ha and 21.7 kg K_2O/ha for a 1 t/ha increase in yield (Haefele *et al.*, 2003), if yield levels are less than 80% of the maximum yield that is possible given climatic conditions (i.e. of the potential yield). Aiming for a higher yield level is not economical because crop response to nutrients is not linear beyond this point, i.e. greater and greater fertilizer quantities are required to obtain the same absolute increase in yield from higher starting points. A whole range of other factors intervenes as well, most notably variability in weather and choice of crop management practices, such as sowing date, variety, and weed management strategy. Looking beyond a growing season, it is important to know whether application of soil amendments will have a 'lasting' effect, i.e. will contribute to soil organic-matter build up, and eventually to increased soil nutrient-supplying capacity or improved recovery of fertilizer nutrients, or whether it mainly acts as a mineral fertilizer, i.e. giving a one-time boost to crop growth.

Installation of water-harvesting structures or irrigation and drainage infrastructure aim at increasing the yield potential in a given environment and reducing risk, lifting the production system to a higher level and opening up opportunities for intensification and diversification. This may also be possible through the introduction of new production systems, such as minimum tillage and fallow crops.

Rice Soil Resources and Mineral Fertilizer Use in Africa

Rice in Africa is grown in a wide range of agro-ecological zones from the humid forest to desert areas. Within these regional agro-ecological zones, five main systems of rice cultivation are distinguished with respect to water supply, soil hydrology and topography (adapted from Windmeijer *et al.*, 1994):

- **Irrigated** rice systems with relatively good water control, which have anaerobic soils for most of the season – mostly in deltas and flood plains.
- **Lowland rainfed** rice systems with varying degrees of water control, which have aerobic soils for a considerable part of the season – mostly in valley bottoms and flood plains.
- **Rainfed upland** rice systems, which have aerobic soils for most of the season – situated on plateaux and slopes.
- **Deep-water** and **mangrove** systems, rarely having water control, which have anaerobic

soils for most of the season – along river beds and in tidal areas in lagoons, deltas and in coastal areas.

A distribution of soil quality for the total rice area of most African countries is given in Table 20.1. These estimates were achieved by overlaying the rice area map of Africa (IRRI, 2012 unpublished) with the digital version of the *Soil Map of the World* (FAO, 1995) (see Haefele and Hijmans, 2007, for details). Soil-fertility constraints were classified and mapped according to the Fertility Capability Soil Classification (FCC) system (Sanchez and Buol, 1985; Sanchez *et al.*, 2003). Four groups of soils were distinguished. The first two groups ('good' and 'poor' soils) do not have major soil chemical constraints, but differ in their degree of weathering and, therefore, their indigenous soil fertility. The third group ('very poor' soils) represents highly weathered soils with very low nutrient availability and a high probability of soil chemical constraints to crop growth (acid, low nutrient reserves, low CEC, Al toxicity, high P fixation). The last group combines the most frequently cited 'problem soils', i.e. acid-sulfate

Table 20.1. Distribution of the total rice area in Africa on four major soil-quality groups (see the definition of the groups in the text). (Data from FAO, 1995; IRRI, 2012.)

Country	Good soils (%)	Poor soils (%)	Very poor soils (%)	Problem soils (%)
Angola	29	11	59	1
Benin	15	79	5	1
Burkina Faso	44	50	3	3
Burundi	40	5	48	7
Cameroon	30	24	35	11
Central African Republic	5	15	80	0
Chad	35	55	10	0
Côte d'Ivoire	8	26	66	0
Democratic Republic of Congo	7	3	90	0
Egypt	41	0	0	59
Equatorial Guinea	0	32	68	0
Ethiopia	66	8	26	0
The Gambia	51	38	0	11
Ghana	21	51	24	4
Guinea	27	17	51	5
Guinea-Bissau	16	43	21	20
Kenya	30	38	25	7
Liberia	8	19	69	4
Madagascar	7	26	63	4
Malawi	27	27	45	1
Mali	51	19	27	3
Mauritania	57	0	14	29
Mozambique	33	35	23	9
Niger	28	12	52	8
Nigeria	25	46	26	3
Rwanda	58	4	38	0
Senegal	10	44	18	28
Sierra Leone	16	18	63	3
Tanzania	26	18	53	3
Togo	7	92	1	0
Uganda	12	19	63	6
Zambia	30	21	42	7
Others (9 countries)	48	15	25	12
Average	27	28	37	8

soils, peat soils, saline and alkaline soils (Sanchez and Buol, 1985), which are characterized by specific and severe soil chemical constraints. (See Saito *et al.*, Chapter 15, this volume, for more details.)

The results (Table 20.1) show an abundance of very poor rice soils (37%), followed by equal fractions of poor (28%) and good soils (27%). Overall, problem soils are not common and make up 'only' 8% of all rice soils in Africa. However, the distribution of rice soil quality and constraints within countries is far from even. Some countries have a substantial proportion of good rice soils – for example, Burkina Faso, Egypt, Ethiopia, The Gambia, Mali and Mauritania. However, in many more countries more than 50% of the rice is grown on very poor soils – for example, Côte d'Ivoire, Democratic Republic of Congo (DRC), Equatorial Guinea, Guinea, Liberia, Madagascar, Sierra Leone, Tanzania and Uganda. Rice on problem soils is abundant in some countries, including Egypt, Guinea-Bissau, Mauritania and Senegal.

Africa is the continent with the lowest fertilizer use, estimated at 23 kg/ha agricultural land per year for the whole continent and at only 9 kg/ha agricultural land per year in West Africa (CEDEAO, 2006). It can be generally assumed that some of the increase in fertilizer use during the last 30–40 years occurred in rice cultivation, but little data is available on how much inorganic fertilizer is used for rice in Africa. Fertilizer availability is a problem in many places: in many African countries, fertilizer is only available where the supply chain for cash crops like cotton, cocoa or banana is well established. Only in some regions with large irrigation schemes are such supply chains established mainly for rice production. The only countries with fertilizer consumption above 400,000 t/year are Egypt, South Africa, Nigeria and Morocco. The only two countries in Africa with a significant rice area and considerable total fertilizer consumption are Egypt (2.0 million tonnes (Mt) in 2008) and Nigeria (0.5 Mt in 2008), and there are several countries with considerable rice areas and very low total fertilizer consumption (e.g. DRC, Guinea, Madagascar and Tanzania).

Regarding specific cases, relatively high fertilizer use of 75–143 kg N/ha has been reported for intensive irrigated rice cropping in the Sahel (Wopereis *et al.*, 1998; Haefele *et al.*, 2001). Becker *et al.* (2003) found much lower average rates in the humid forest zone (23 kg N/ha) and in the Guinea savannah zone (17 kg N/ha). High fertilizer rates are also common in Egypt and some intensive rice production schemes in East Africa (FAO, 2005). Mineral fertilizer use is not very common in rainfed lowlands (Meertens *et al.*, 1999; Sakurai, 2010; Kamara *et al.*, 2011), and rare in traditional upland rice cultivation (Balasubramanian *et al.*, 2007; Oikeh *et al.*, 2008; Mghase *et al.*, 2010). These differences are mainly determined by fertilizer availability (high in irrigated lowlands, low in rainfed uplands) and production risk (low in irrigated lowlands, high in rainfed uplands).

Innovations for Improving Nutrient Use

In this section, we provide (i) an overview of farmers' practices and constraints related to soil fertility and nutrient use; and (ii) technological options for improving nutrient use in the three major rice-based systems (irrigated lowland, rainfed upland and rainfed lowland), with a focus on West Africa.

Irrigated lowland rice-based systems

Farmers' practices, challenges and opportunities

Wopereis *et al.* (1999) and Donovan *et al.* (1999) conducted surveys in Senegal (Senegal River delta and middle valley), Mali (Office du Niger) and Burkina Faso (Kou Valley) to identify farmers' practices and determinants of rice productivity with a focus on nutrient management. Average farmers' yields ranged between 3.8 t/ha and 7.2 t/ha, resulting in an overall average of 4.5 t/ha. Yields of individual farmers were highly variable, ranging from almost complete crop failure (0.3 t/ha, due to weeds) to very high yields (8.7 t/ha). High average yields and low yield variability were found in relatively old irrigation schemes, e.g. in the Office du Niger. Maximum yields reached by farmers were only 40–60% of ten-year averages of simulated potential yield (limited by climate only). The difference between average farmers' yields and highest farmers' yield was between 0.7 t/ha and 4.1 t/ha, with an average of 2.6 t/ha, indicating considerable scope for improving yields.

Fertilizer use did not follow recommended rates and was quite variable, but quantities were generally high, with average fertilizer N input ranging from 73 kg/ha (Burkina Faso) to 143 kg/ha (Mali), average fertilizer P input from 15 kg/ha (Senegal) to 22 kg/ha (Burkina Faso) and average fertilizer K input from none (Mali, Senegal) to 22 kg/ha (Burkina Faso). Grain yields were highly variable, both within and across sites.

Only a few farmers in Mali and Burkina Faso applied organic inputs (compost, manure). Without exception, farmers applied N fertilizer, and most farmers also applied P fertilizer. Potassium fertilizer use was mainly restricted to sites where use of NPK compound fertilizer was recommended. Nitrogen was always the most limiting nutrient for rice growth. Average N fertilizer recovery was relatively low and in the range 18–50%, i.e. fertilizer N losses ranged from 50% to 82%.

Sustainability issues

The sustainability of intensive irrigated rice cropping in Africa has been studied through long-term fertility experiments (Haefele *et al.*, 2002; Bado *et al.*, 2010). Haefele *et al.* (2002) conclude that intensive irrigated rice cultivation is sustainable if N and P are applied as inorganic fertilizer. Bado *et al.* (2010) found that double cropping of irrigated rice does not decrease soil organic matter and the application of the recommended doses of NPK fertilizer (120 kg N/ha, 26 kg P/ha, 50 kg K/ha) maintained rice yields for 18 years. In arid or semi-arid environments like the Sahel, salinization or alkalinization, sometimes followed by sodification, may affect soil quality, especially if drainage is limited, because the water table rises close to the surface or the irrigation water is rich in dissolved inorganic ions. Reports of micronutrient deficiencies are rare and often concern problem soils (e.g. saline and sodic soils) or organic soils. However, where high yields are achieved regularly and perhaps even twice a year, micronutrient (e.g. Zn, S) deficiencies can appear quickly (van Asten *et al.*, 2004).

Improving crop management

It is clear that farmers could improve efficiency and profit by improving the recovery rate of applied nutrients, especially N, through better crop management in general, without major increases in investment in fertilizers. The most important constraints that resulted in low N recovery rates were: timing of N fertilizer application that did not coincide with critical growth stages of the rice plant; use of relatively old (>40 days) seedlings at transplanting; unreliable irrigation water supply; weed problems; and late harvesting (Senegal River delta) (Wopereis *et al.*, 1999). Similar results were obtained by Haefele *et al.* (2000, 2001) for the Senegal River delta and Segda *et al.* (2004) for the Bagré irrigation scheme in Burkina Faso.

Discussions with farmers in Burkina Faso and Senegal revealed that they lacked knowledge on (i) optimal timing, dosage and mode of fertilizer application; (ii) optimal sowing dates to avoid yield loss due to cold- or heat-induced sterility; and (iii) the importance of N as the main yield-limiting factor. Other factors included problems with collective and individual planning of the cropping calendar for double cropping of rice (two rice crops in the same field in one year) and the need to attend to rainfed crops outside the scheme.

Working with Senegalese and Mauritanian farmers, improved nutrient management (application of 20 kg P/ha and 150 kg N/ha in three splits at early tillering, panicle initiation and booting) increased yields by about 1 t/ha. Improving weed management (application of 6 l propanil/ha and 2 l 2,4-D-amine/ha at 2–3-leaf stage of weeds) also raised yields by about 1 t/ha over farmers' practice. The combined effect of improved nutrient and weed management was additive: improving both nutrient and weed management raised yields by almost 2 t/ha over average farmers' yields of 3.9 t/ha, i.e. an increase of about 50%. Value–cost ratios were between 2.1 and 4.6 for improved soil fertility and weed management, respectively, resulting in an increase in net revenues of 40–85% compared to farmers' practice. The results of the learning plots were amazingly consistent and were obtained for both smallholder farmers in Senegal (Haefele *et al.*, 2000) and large-scale farmers in Mauritania (Haefele *et al.*, 2001). The yield increases obtained are considerably larger than those obtained for similar work in intensive rice-cropping systems in Asia (Dobermann *et al.*, 2002).

Site- and season-specific mineral fertilizer recommendations

Recommended N rates for lowland rice usually range from 60 kg/ha to 120 kg/ha, applied in 2–3 splits at planting, early tillering and panicle initiation. An additional split at booting can be beneficial in very high-yielding systems (Wopereis-Pura *et al.*, 2002). Upper rates in the wet season under irrigated conditions are 90–120 kg N/ha. Very high N rates up to 150 kg/ha can be recommended in irrigated rice during the dry season, if high solar radiation enables potential grain yields of up to 12 t/ha (Haefele and Wopereis, 2004). Application of urea super granules (USGs) is promoted in some irrigated systems (Fofana *et al.*, 2010). Conditions necessary for USG technology are full water control (aerobic soil phases should be avoided) and (clayey) soils with limited percolation. USGs are pressed into the soft surface soil between four hills, usually in one or two applications. USG technology has been shown to reduce N losses considerably, but its disadvantage is the labour needed for application.

Phosphorus deficiency is common, especially if higher yields are targeted. The incorporation of P fertilizers during land preparation or surface application up to 20 days after transplanting is good practice for flooded rice crops.

Potassium fertilizers should be applied along with N and P on poor soils, if higher yields are targeted, and especially if two (high-yielding) crops are grown per year regularly. The amount of K that needs to be applied also depends on K inputs from the irrigation water and from dust depositions. In West Africa, the dust deposition (dry deposition) is highest at the northern fringes of the Sahel (e.g. in the Senegal River valley and in the Office du Niger, Mali) and decreases towards the south.

Simulation tools can be used to develop site- and season-specific mineral fertilizer recommendations, as demonstrated for irrigated rice in the Office du Niger (Haefele *et al.*, 2003). The dynamic ecophysiological ORYZA_S model provided potential rice yields under irrigation, based on weather conditions, cultivar choice and sowing date. This yield potential was then used in the static FERRIZ model, together with on-farm data on recovery efficiency of applied N, P and K, indigenous soil N, P and K supply, and maximum N, P and K accumulation and dilution in rice dry matter. Resulting outputs were fertilizer rates necessary to obtain different target yields depending on yield potential and soil nutrient supply. Adding current fertilizer and paddy prices into the analyses then allowed an agro-economic evaluation of different fertilizer options. In a last step, the dynamic decision tool RIDEV was used to simulate optimal timing of different management actions, such as fertilizer application, weeding and harvest. This approach showed that (i) current uniform recommendations for the wet season performed well except on low-K soils where the application of K was profitable; and (ii) adjusting fertilizer doses to the lower yield potential in the dry season reduced costs and risk without reducing profit. Based on the analysis, the existing recommendation could be adjusted for the wet and dry seasons, keeping fertilizer costs and risk low, and having net benefits close to optimal.

Based on this experience, Segda *et al.* (2005) used a combination of two simulation models and selected field data to develop alternative fertilizer recommendations (AFR) for irrigated rice in the irrigation scheme of Bagré (Burkina Faso). Existing fertilizer recommendations (EFR) in Bagré were 82 kg N/ha (wet season) or 105 kg N/ha (dry season), plus 31 kg/ha P and 30 kg K/ha, using urea (100–150 kg/ha) and a compound NPK fertilizer (300 kg/ha, containing 12% N, 24% P_2O_5 and 12% K_2O). RIDEV was used to improve timing of sowing to avoid cold-induced sterility and timing of N fertilizer applications (i.e. improved crop management). FERRIZ was used to determine AFR, based on estimations of indigenous nutrient supply for N, P and K, yield potential (Y_{pot}), internal N, P and K efficiency of rice, fertilizer N, P and K recovery fractions, and fertilizer and rice prices. Simulations suggested decreasing P and K doses to 21 kg/ha and 20 kg/ha, respectively, but increasing the N dose to 116 kg/ha in the wet season (Y_{pot} = 8 t/ha) and 139 kg/ha in the dry season (Y_{pot} = 9 t/ha). This translates to substituting two bags (100 kg) of compound NPK fertilizer for two bags of urea per hectare. AFR keeps the P-balance neutral, but a negative K balance was tolerated on the basis of the high soil K supply. Compared to existing recommendations, yield gains of up to 0.5 t/ha were simulated at equal costs. These yield gains were exceeded in farmers' fields during four consecutive growing seasons. AFR increased gross returns above fertilizer costs by an average of about US$160 per season as compared to both farmers' practice and existing recommendations. Although it is unlikely that farmers will follow these recommendations exactly, the basic message from this study – to apply more urea and less compound fertilizer – was adopted quickly by farmers (Segda *et al.*, 2010).

Organic inputs

In lowland rice-based systems, the aquatic fern *Azolla* (a genus of seven species in the family *Salviniaceae*) has been used as a green manure for rice in northern Vietnam, and central to southern China for centuries (Watanabe and Van Hove, 1996). Given its high N content, *Azolla* boosts growth directly, i.e. it acts as a substitute for mineral fertilizer. Depending on the ecological conditions, a 4–10-week old 'crop' of *Azolla* can accumulate an average of 70 kg N/ha (range 20–146 kg N/ha), of which about 80% is derived from atmospheric nitrogen (Roger and Ladha, 1992). Accordingly, Watanabe *et al.* (1989) estimated that the use of *Azolla* can replace 30–60 kg of mineral N fertilizer depending on frequency and time of incorporation. Experiences in Senegal (Diara *et al.*, 1987; Van Hove and Diara, 1987; Van Hove, 1989) and Egypt (Yanni *et al.*, 1994) show that its adoption is blocked by two main factors: (i) high labour cost needed for incorporation of *Azolla*, and (ii) contamination of irrigation and drainage canals with *Azolla*.

Rainfed upland rice-based systems

Farmers' practices, challenges and opportunities

Upland rice production in Africa is dominated by subsistence-oriented farm households, which do not use external inputs mainly due to high production risk and poverty. However, in Uganda upland rice is grown by farmers for income generation (Oonyu, 2011). But even there, most farmers do not apply inorganic fertilizer due to the high price of fertilizer and considerable production risks (Kijima *et al.*, 2011; Miyamoto *et al.*, 2012). Upland rice yields average about 1 t/ha, but there are large differences between farms, mostly because of differences in the quality of the land, rainfall patterns, and in crop management practices. Up to 2 t/ha can be obtained under favourable conditions – high soil fertility, good hydrological conditions, or following long fallows or appropriate previous crops in rotation systems (Sokei *et al.*, 2010; Kijima *et al.*, 2011; Miyamoto *et al.*, 2012). Upland rice is grown as a sole crop or mixed with other crops such as maize and beans both in slash-and-burn systems and in intensified systems, where upland rice is rotated with other crops on permanently cultivated lands. In slash-and-burn systems, increased cropping intensity (reduced fallow period and increased cropping period) have aggravated weed pressure and led to a general decline in land quality through soil nutrient depletion. Expansion of slash-and-burn upland rice-based systems should generally not be encouraged because of the fragility of these agroecosystems.

Many upland soils have low N availability and are highly P-fixing (Becker and Johnson, 2001; Sahrawat *et al.*, 2003). Becker and Johnson (2001) analysed cropping intensity effects on upland rice yield and sustainability in four agroecological zones (Guinea savannah, derived savannah, bimodal forest, monomodal forest) in Côte d'Ivoire. Increased cropping intensity was associated with yield reduction. Cropping intensification-induced yield loss was about 25% (a drop from an average 1.5 t/ha to 1.1 t/ha) and was mainly related to increased weed infestation and declining soil quality. Miyamoto *et al.* (2012) report that in intensified systems in Uganda, continuous rice cropping reduced rice yield, compared with rotation systems.

Use of organic fertilizer such as farmyard manure and cow dung is still limited except in a few countries, such as Madagascar and Uganda. In Namulonge (central Uganda), Miyamoto *et al.* (2012) report that 19% of farmers use chicken manure.

Organic inputs

As most farmers cannot afford mineral fertilizer, organic inputs have been considered as potential options for sustainable production in intensified upland rice-based systems. Promising alternative cropping systems include the use of weed-suppressing and multi-purpose legumes as short-term fallow crops. Some 54 legumes were evaluated as dry-season fallow crops in the mid-1990s in four agroecological zones (Guinea savannah, derived savannah, bimodal forest, monomodal forest) in Côte d'Ivoire (Becker and Johnson, 1998, 1999). Legumes were introduced into the upland rice crop one month before harvesting, and allowed to grow until the end of the dry season, when their N accumulation and weed suppression were evaluated. Fallow vegetation was cleared according to the practice of local farmers, and upland rice was seeded to evaluate the effect of the previous legume crop on weed growth and rice yield. Biomass accumulation at the end of the dry season was generally greater in legume fallows than in natural fallows (control), and several legume species suppressed weed growth. Nitrogen accumulation by legumes ranged from 1 kg/ha to 270 kg/ha, with 30–90% of the accumulated N derived from the atmosphere. Across sites, *Mucuna* spp., *Canavalia* spp. and *Stylosanthes guianensis* showed consistently high N accumulation. Rice yields following legume fallows were on average 0.2 t/ha greater than those following a natural weedy fallow. On average across four sites, *Tephrosia villosa* and *Stylosanthes guianensis* fallows increased yields by 0.5 t/ha and 0.4 t/ha, respectively, in comparison with a natural fallow. Furthermore, to increase benefits from the systems, the effects of timing of legume establishment in relation to rice and fallow-residue management (removing, burning, mulching or incorporating) prior to the rice crop on rice and weed growth were determined (Becker and Johnson, 1999; Akanvou *et al.*, 2000; Saito *et al.*, 2010). Timing of legume establishment in upland rice depended on legume, rice variety and their crop densities, and environmental conditions. Incorporating or mulching of fallow residues provided no significant yield advantage compared to burning. Africa Rice Center (AfricaRice) also conducted on-farm participatory legume evaluation and selected legumes were grown by farmers. Through this work, researchers learned that solutions such as improved fallows must consider the biophysical and socio-economic specifics of prevailing systems in order to be successful. However, adoption rates by farmers have been very low because of:

- Lack of seed availability and poor seed quality
- Insufficient adaptability of most of the legumes to diverse site conditions
- High variability in the performance of legumes (especially in marginal environments)
- High cost of land and labour.

Some of these constraints to green-manure adaptation, especially the high costs for land and labour, can be overcome by integrating grain legumes into rice-based systems, as these can provide (at least in part) the benefits of green manure (e.g. biological N fixation and weed suppression), while also giving harvestable product with market value or for home consumption (for this reason, they are known as 'dual-purpose legumes'). This can be achieved using different approaches:

- Intercropping upland rice with grain legumes like cowpea or soybean
- Developing upland rice–grain legume rotations
- Integrating early maturing legumes as pre- or post-rice crops into the existing upland or lowland monocrop rice systems.

Oikeh *et al.* (2008) compared dual-purpose soybean and upland rice rotation with continuous upland rice cropping in the Guinea savannah of Benin. Rice yield following soybean was 0.7 t/ha higher than in continuous rice cropping. However, in local farmers' practice, the aboveground biomass of legumes (except for litter that falls before harvest) is often entirely removed from the field for use as animal fodder. This practice will further contribute to negative N balances and might even accelerate soil fertility degradation.

Variable performance of legumes in diverse conditions may also be related to P deficiency in the uplands, because most legumes have considerable P requirements. Somado *et al.* (2003, 2011) suggest integrated approaches including application of rock phosphate (which is a relatively cheap source of P) to legumes and rice cultivation following legumes for enhancing both N and P use.

In uplands in Madagascar, conservation agriculture has been tested by collaboration works among Centre National de la Recherche Appliquée au Développement Rural (FOFIFA),

Université d'Antananarivo and Centre de coopération internationale en recherche agronomique pour le développement (CIRAD). Improved fallow systems using *Stylosanthes guianensis*, combined with mulching as residue management in a zero-tillage system has been introduced to farmers.

Potential use of mineral fertilizer in the uplands

Across Africa, transition from traditional slash-and-burn to intensified upland rice-based systems is occurring, or will occur, depending on the population density and land pressure. Such a transition requires mineral fertilizer application to upland rice for improved and sustainable productivity, ideally with organic inputs as described above. Knowledge of rainfall distribution and general soil fertility will help determine where small quantities of mineral fertilizer can make a difference. Data from Sokei *et al.* (2010) on trials in four countries in West Africa showed a positive relationship between rainfall during the period from around flowering to before maturity and yield increase due to fertilizer application. Thus, fertilizer application is more effective in high-rainfall conditions, whereas fertilizer application is less beneficial in drought-prone systems, or alternative fertilizer application methods are needed in such conditions. This indicates that in areas with large year-to-year variation in rainfall pattern, such as West Africa, site- and season-specific fertilizer recommendations would benefit tremendously from a weather-forecasting system. The positioning of the rice field on the toposequence can also affect soil fertility and hydrological conditions, and thus fertilizer strategies. Touré *et al.* (2009) demonstrated that in the top-most positions, water availability limited rice growth and its response to fertilizer application, whereas in relatively favourable uplands and hydromorphic areas in the middle positions, the response to fertilizer application was high. Farmers in central Benin distinguish soil types in their fields on the basis of water availability – such knowledge could be used for nutrient management recommendations (Takemura *et al.*, unpublished). Distance to markets, distance from village to fields, and farmers' socio-economic status should be taken into account for developing site-specific nutrient recommendations. Adesina (1996), for example, reports that fertilizer use is extremely low to zero for fields far from the village.

Rainfed lowland rice-based systems

Farmers' practices, challenges and opportunities

In general, soils in rainfed lowlands in sub-Saharan Africa are relatively less fragile and better able to support continuous cultivation than those in the uplands. Rainfed lowlands have considerable potential for intensification. Rainfed rice is mainly grown once a year in West Africa and, following the rice, vegetables or legumes are grown in relatively favourable conditions, where residual water or supplementary irrigation is available for these crops (Adigbo *et al.*, 2007; Balasubramanian *et al.*, 2007). In high-rainfall areas in Uganda, double rice cropping is common practice. In West Africa, rainfed lowland rice is often grown without any bunds or proper land levelling. In such (less-favourable) environments, farmers often do not use external inputs mainly because of the high production risk and poverty (Adesina, 1996; Becker and Johnson, 2001; Kijima *et al.*, 2010). Erenstein (2006) reports that fertilizer use was associated with market access and agroecological zone in West Africa. Rice yields in lowlands are generally low, averaging only 2–3 t/ha, although yields obtained by individual farmers may reach 6 t/ha (Becker and Johnson, 2001; Kijima *et al.*, 2010). The difference between the highest farmers' yield and average yield indicates that there is considerable potential for improvement. Major constraints related to soils in this system are N and P deficiency and Fe toxicity. Direct and indirect effects of Fe toxicity in inland valleys can lead to 40–45% rice yield reductions in lowlands, depending on the extent of the problem, water, soil and crop management (e.g. cultivar choice), as well as on the availability of other soil nutrients (Becker and Asch, 2005; Audebert and Fofana, 2009).

Improving crop management

Improved water control is a first step towards improving the productivity of rainfed lowlands.

Bunding and levelling facilitate water management and generally increase nutrient-use efficiencies, particularly in well-drained fields. Water management and regular drainage will also avoid problems with Fe toxicity in inland valley lowlands. Becker and Johnson (2001) report that mineral fertilizer N application significantly increases rice yields (18% on average across sites) in bunded fields only.

Site- and field-specific fertilizer management recommendations

As described for upland rice systems, the combination of rainfall distribution and general soil fertility, distance to markets, distance from village to fields, the positioning of the rice field in the toposequence, farmers' knowledge of their field characteristics and farmers' socio-economic conditions should be taken into account for developing site-specific nutrient recommendations. Given the diversity and dynamics of farmer reality and growing conditions in rainfed lowland systems, on-farm testing in the target environments is crucial for identifying agronomic practices that should result in reduced economic risk and increased fertilizer adoption by farmers (Posner and Crawford, 1992). In the upper positions of the toposequence drought might limit rice growth, and in the bottom the risk from submergence is high. Therefore, fertilizer-use efficiency in both situations can be limited. Nitrogen fertilizer application is more effective in middle parts on the toposequence than in drought-prone upper positions (Touré *et al.*, 2009). Organic inputs are most useful in upper parts because they help to increase the soil water retention, while soils at the bottom usually have a higher organic-matter content naturally. Furthermore, mobile nutrients like N and K are transported down the toposequence, thereby minimizing the need to apply N and K fertilizer in the bottom parts (Bognonkpe and Becker, 2009). The toposequence and related transport processes are also the reason for the widespread occurrence of Fe toxicity at the edges of inland valleys, where the Fe-rich groundwater surfaces. From numerous studies, Becker and Asch (2005) identified three environmental and soil conditions that are typically associated with the occurrence of Fe toxicity, and proposed different management options for each condition.

In rainfed lowland rice systems with favourable water availability (e.g. inland valleys with suitable topography), introduction of post-rice grain legumes or vegetables to the areas where their cultivation is not common practice will allow diversification and enhance organic inputs. That will raise productivity and profitability, and may also have beneficial effects on the rice crop grown in the wet season. Carsky and Ajayi (1992) report on efforts to introduce legumes into the inland valley rice-based cropping systems in West Africa.

Getting It Together

The review of principles and technologies to improve nutrient use in rice presented in this chapter shows that there is a wealth of knowledge available across rice environments. There is an urgent need to capture the basic principles of nutrient management (e.g. the importance of the soil nutrient-supplying capacity, the role of mineral fertilizers and their importance to plant growth, the role of organic inputs, the importance of N in lowland systems and of P and N in upland systems, and the fact that K usually only becomes important at higher yield levels and if most straw is exported from the field) in formats that can be widely distributed to farmers and other rice-development stakeholders, for example via videos. AfricaRice has produced a video on nutrient management in lowland systems that has been translated into 30 African local languages and has been widely distributed (see Van Mele *et al.*, Chapter 30, this volume).

Farmers' fields are generally highly heterogeneous in terms of growth conditions for the rice crop, and that complexity increases from the irrigated systems to the rainfed upland and lowland systems. Moreover, rice may be just one component of a farm, especially in rainfed systems. Going beyond conveying 'best-bet principles for nutrient management' often requires in-depth knowledge of these complex systems. Stratification of soils or fields in types that are likely to be responsive or non-responsive to mineral fertilizer is one important element of site-specific nutrient management in rainfed systems. Knowledge of seasonal differences in rainfall in rainfed systems, and of temperature and solar radiation in irrigated systems allows for the development of season-specific nutrient management recommendations. Moreover, recommendations should always be placed in the general crop and production system context, because so many other yield-limiting or yield-reducing factors may mask nutrient uptake in Africa.

A second important feature to consider is the heterogeneity in farmers' access to resources. Options to improve nutrient use and agricultural productivity need to take into account socio-economic factors such as household wealth, family structure, production orientation (subsistence, market), main source of income, and main types of constraints faced by farmers.

Conclusions, Looking Ahead

Mineral fertilizers will continue to play a key role in boosting rice productivity given the current very low level of fertilizer use in Africa. Where possible their use should be combined with organic inputs. Mineral fertilizers are the most direct supply for plant nutrients, while organic inputs help to maintain or rebuild the soil organic-matter reserve. Another, more practical, reason for advocating the combined use of organic inputs and mineral fertilizers is that 'very often neither of them is widely available or affordable in sufficient quantities' (Vanlauwe and Giller, 2006).

The most pressing challenge for rice-based systems is to ensure that knowledge of basic principles of integrated nutrient management for rice (rather than 'blanket fertilizer recommendations') is communicated to those that are most in need – Africa's rice farmers. This may be done through video or other means, such as rural radio station broadcasts, distribution of simple cropping calendar posters in village meetings and new information technology tools like mobile phones.

The capacity on the African continent to work on soil nutrient management in general is in desperate need of rebuilding, in both research and extension communities, after decades of under-investment in agriculture. There is a particular need to train a new generation of hands-on rice experts, through season-long training in rice management. AfricaRice has built a new training facility near its regional station in Senegal for this purpose. Japan International Cooperation Agency (JICA) has also established training facilities for rice in several countries in sub-Saharan Africa to rebuild rice extension capacity.

To date, blanket fertilizer recommendations are all that are available to most rice farmers in Asia and Africa, despite the fact that optimal amounts and sources of nutrients to meet the needs of the crop and optimize crop production can vary considerably, even across short distances within and among fields (Haefele and Wopereis, 2005). Major factors affecting the actual fertilizer needs include indigenous fertility of the site, season, variety used, residue management, type of fertilizer available, cropping system, nutrient inputs from irrigation water, atmospheric depositions, and yield target. But rice farmers who want to adjust the general recommendation to their specific conditions or needs have to depend on their own experience for the most part. To address this situation, AfricaRice and the International Rice Research Institute (IRRI) are in the process of developing a nutrient management tool that will allow individual rice farmers to manage their fields with greater precision. This approach relies on the scientific principles determined during 15 years of site-specific nutrient management (SSNM) research across Asia and Africa (Dobermann *et al.*, 2002; Haefele *et al.*, 2003b; Buresh *et al.*, 2010). Based on 15–20 simple questions about a specific field, the nutrient manager will provide farmers with a balanced fertilizer recommendation that aims to increase their profits in a sustainable manner. The nutrient manager is being tested for irrigated and favourable rainfed lowland rice systems in Senegal, Mali, Ghana and Nigeria. A mobile-phone version of the nutrient manager is already available in the Philippines.

Finally, there is also a need to address the challenges that lie ahead in terms of nutrient-management research: (i) for the new varieties that AfricaRice and partners are currently developing for different growth environments (see Kumashiro *et al.*, Chapter 5, this volume); (ii) for new production systems that are likely to become prominent in the near future, such as highly mechanized, direct-seeded rice systems and new crop rotations, such as rice–cotton or irrigated rice–wheat systems; and (iii) to anticipate the consequences of increasing water scarcity for irrigated rice-based systems due to climate change and increased demand for water for agricultural, industrial or urban use.

References

Adesina, A.A. (1996) Factors affecting the adoption of fertilizers by rice farmers in Côte d'Ivoire. *Nutrient Cycling in Agroecosystems* 46, 29–39.

Adigbo, S.O., Okeleye, K.A. and Adigbo, V.B. (2007) Performance of upland rice fitted into lowland rice–vegtable/cowpea sequence in rainfed inland valley. *Agronomy Journal* 99, 377–383.

Akanvou, R., Becker, M., Chano, M., Johnson, D.E., Gbaka-Tcheche, H. and Toure, A. (2000) Fallow residue management effects on upland rice in three agroecological zones of West Africa. *Biology and Fertility of Soils* 31, 501–507.

Audebert, A. and Fofana, M. (2009) Rice yield gap due to iron toxicity in West Africa. *Journal of Agronomy and Crop Science* 195, 66–76.

Bado, V.B., Aw, A. and Ndiaye, M. (2010) Long-term effect of continuous cropping of irrigated rice on soil and yield trends in the Sahel of West Africa. *Nutrient Cycling in Agroecosystems* 88, 133–141.

Balasubramanian, V., Sie, M., Hijmans, R.J. and Otsuka, K. (2007) Increasing rice production in sub-Saharan Africa: challenges and opportunities. *Advances in Agronomy* 94, 55–133.

Becker, M. and Asch, F. (2005) Iron toxicity in rice – conditions and management concepts. *Journal of Plant Nutrition and Soil Science* 168, 558–573.

Becker, M. and Johnson, D.E. (1998) Legumes as dry season fallow in upland rice-based systems of West Africa. *Biology and Fertility of Soils* 27, 358–367.

Becker, M. and Johnson, D.E. (1999) The role of legume fallows in intensified upland rice-based systems of West Africa. *Nutrient Cycling in Agroecosystems* 53, 71–81.

Becker, M. and Johnson, D.E. (2001) Cropping intensity effects on upland rice yield and sustainability in West Africa. *Nutrient Cycling in Agroecosystems* 59, 107–117.

Becker, M., Johnson, D.E., Wopereis, M.C.S. and Sow, A. (2003) Rice yield gaps in irrigated systems along an agro-ecological gradient in West Africa. *Journal of Plant Nutrition and Soil Science* 166, 61–67.

Bognonkpe, J.P. and Becker, M. (2009) Land use and dynamics of water and native soil N in inland valleys of Côte d'Ivoire. *European Journal of Scientific Research* 3, 342–356.

Buresh, R.J., Pampolino, M.F. and Witt, C. (2010) Field-specific potassium and phosphorus balances and fertilizer requirements for irrigated rice-based cropping systems. *Plant and Soil* 335, 35–64.

Carsky, R.J. and Ajayi, E.O. (1992) Fitting soil-improving legumes into inland valley rice-based cropping systems in West Africa. In: Mulongoy, M., Gueye, M. and Spencer, D.S.C. (eds) *Biological Nitrogen Fixation and Sustainability of Tropical Agriculture*. International Institute of Tropical Agriculture, Ibadan, Nigeria, pp. 395–404.

CEDEAO (2006) Stratégie Régionale de Promotion des Engrais en Afrique de l'Ouest. Pepared at the Sommet Africain sur les Engrais, Abuja, Nigeria, 9–13 June. Economic Community of West African States, Abuja, Nigeria.

Diara, H.F., Van Brandt, H., Diop, A.M. and Van Hove, C. (1987) *Azolla* and its use in rice culture in West Africa. In: *Azolla Utilization – Proceedings of Workshop on Azolla Use*. Fujian, China, 1985. International Rice Research Institute, Los Baños, Philippines, pp. 147–152.

Dobermann, A., Witt, C., Dawe, D., Gines, G.C., Nagarajan, R., Satawathananont, S., Son, T.T., Tan, P.S., Wang, G.H., Chien, N.V., Thoa, V.T.K., Phung, C.V., Stalin, P., Muthukrishnan, P., Ravi, V., Babu, M., Chatuporn, S., Kongchum, M., Sun, Q., Fu, R., Simbahan, G.C. and Adviento, M.A.A. (2002) Site-specific nutrient management for intensive rice cropping systems in Asia. *Field Crops Research* 74, 37–66.

Donovan, C., Wopereis, M.C.S., Guindo, D. and Nébié, B. (1999) Soil fertility management in irrigated rice systems in the Sahel and savanna regions of West Africa. Part II. Profitability and risk analysis. *Field Crops Research* 61, 147–162.

Erenstein, O. (2006) Intensification or extensification? Factors affecting technology use in peri-urban lowlands along an agro-ecological gradient in West Africa. *Agricultural Systems* 90, 132–158.

FAO (1995) *The Digital Soil Map of the World*, Version 3.5, 1:5,000,000 scale. Food and Agriculture Organization of the United Nations, Rome, Italy.

FAO (2005) *Fertilizer Use by Crop in Egypt*. Food and Agriculture Organization of the United Nations, Rome, Italy.

Fofana, B., Bandaogo, A., Sanou, S., Youl, S. and Mando, A. (2010) Improving rice productivity in Africa through deep placement of urea supergranules. Presentation at the 3rd International Rice Congress, Hanoi, Vietnam, 8–12 November.

Haefele, S.M. and Hijmans, R.J. (2007) Soil quality in rice-based rainfed lowlands of Asia: characterization and distribution. In: Aggarwal, P.K., Ladha, J.K., Singh, R.K., Devakumar, C. and Hardy, B. (eds) *Science, Technology, and Trade for Peace and Prosperity.* Proceedings of the 26th International Rice Research Conference, New Delhi, India, 9–12 October 2006. International Rice Research Institute, Los Baños, Philippines, and Indian Council of Agricultural Research and National Academy of Agricultural Sciences, New Delhi, India, pp. 297–308.

Haefele, S.M. and Wopereis, M.C.S. (2004) Combining field and simulation studies to improve fertilizer recommendations for irrigated rice in the Senegal River valley. In: Doberman, A., Witt, C. and Dawe, D. (eds) *Increasing Productivity of Intensive Rice Systems Through Site-specific Nutrient Management.* Science Publishers, Enfield, New Hampshire and International Rice Research Institute, Los Baños, Philippines, pp. 265–286.

Haefele, S.M. and Wopereis, M.C.S. (2005) Spatial variability of indigenous supplies for N, P and K and its impact on fertilizer strategies for irrigated rice in West Africa. *Plant and Soil* 274, 57–72.

Haefele, S.M., Johnson, D.E., Diallo, S., Wopereis, M.C.S. and Janin, I. (2000) Improved soil fertility and weed management is profitable for irrigated rice farmers in Sahelian West Africa. *Field Crops Research* 66, 101–113.

Haefele, S.M., Wopereis, M.C.S., Donovan, C. and Maubuisson, J. (2001) Improving the productivity and profitability of irrigated rice production in Mauritania. *European Journal of Agronomy* 14, 181–196.

Haefele, S.M., Wopereis, M.C.S. and Wiechmann, H. (2002) Long-term fertility experiments for irrigated rice in the West African Sahel: agronomic results. *Field Crops Research* 78, 119–131.

Haefele, S.M., Wopereis, M.C.S., Ndiaye, M.K., Barro, S.E. and Ould Isselmou, M. (2003) Internal nutrient efficiencies, fertilizer recovery rates and indigenous nutrient supply of irrigated lowland rice in Sahelian West Africa. *Field Crops Research* 80, 19–32.

Haefele, S.M., Wopereis, M.C.S., Schloebohm, A. and Wiechmann, H. (2004) Long-term fertility experiments for irrigated rice in the West African Sahel: effect on soil characteristics. *Field Crops Research* 85, 61–77.

Hilhorst, T. and Muchena, F. (2000) *Nutrients on the Move: Soil fertility dynamics in African farming systems.* International Institute for Environment and Development, London.

IRRI (2012) Rice Area Database of the World. Unpublished. International Rice Research Institute, Los Baños, Philippines.

Kamara, A.Y., Ekeleme, F., Omoigui, L.O. and Chikoye, D. (2011) Influence of nitrogen fertilization on yield and yield components of rain-fed lowland NERICA® rice in the northern Guinea savanna of Nigeria. *African Journal of Agricultural Research* 6, 3092–3097.

Kijima, Y., Ito, Y. and Otsuka, K. (2010) An Empirical Analysis of Expanding Rice Production in Sub-Saharan Africa. On the Possibility of a Lowland Rice Green Revolution in Sub-Saharan Africa: Evidence from the Sustainable Irrigated Agricultural Development (SIAD) Project in Eastern Uganda. *JICA Research Institute Working Paper* No. 25. Japan International Cooperation Agency, Tokyo, Japan.

Kijima, Y., Otsuka, K. and Sserunkuuma, D. (2011) An inquiry into constraints on a green revolution in sub-Saharan Africa: the case of NERICA rice in Uganda. *World Development* 39, 77–86.

Meertens, H.C.C., Ndege, L.J. and Lupeja, P.M. (1999) The cultivation of rainfed lowland rice in Sukumaland, Tanzania. *Agriculture, Ecosystems & Environment* 76, 31–45.

Mghase, J.J., Shiwachi, H., Nakasone, K. and Takahashi, H. (2010) Agronomic and socio-economic constraints to high yield of upland rice in Tanzania. *African Journal of Agricultural Research* 5, 150–158.

Miyamoto, K., Maruyama, A., Haneishi, Y., Matsumoto, S., Tsuboi, T., Asea, G., Okello, S., Takagaki, M. and Kikuchi, M. (2012) NERICA cultivation and its yield determinants: the case of upland rice farmers in Namulonge, central Uganda. *Journal of Agricultural Science* 4, 120–135.

Oikeh, S.O., Nwilene, F., Diatta, S., Osiname, O., Touré, A. and Okeleye, K.A. (2008) Responses of upland NERICA rice to nitrogen and phosphorus in forest agroecosystems. *Agronomy Journal* 100, 735–741.

Oonyu, J. (2011) Upland rice growing: a potential solution to declining crop yields and the degradation of the Doho wetlands, Butaleja district – Uganda. *African Journal of Agricultural Research* 6, 2774–2783.

Posner, J.L. and Crawford, E.W. (1992) Improving fertilizer recommendations for subsistance farmers in West Africa: the use of agro-economic analysis of on-farm trials. *Nutrient Cycling in Agroecosystems* 32, 333–342.

Roger, P.A. and Ladha, J.K. (1992) Biological N_2 fixation in wetland rice fields: estimation and contribution to nitrogen balance. *Plant and Soil* 141, 41–55.

Sahrawat, K.L., Jones, M.P., Diatta, S. and Sika, M. (2003) Long-term phosphorus fertilizer effects on phosphorus uptake, efficiency, and recovery by dryland rice on an Ultisol. *Communications in Soil Science and Plant Analysis* 14, 999–1011.

Saito, K., Azoma, K. and Oikeh, S.O. (2010) Combined effects of *Stylosanthes guianensis* fallow and tillage management on upland rice yield, weeds and soils in southern Benin. *Soil & Tillage Research* 107, 57–63.

Sakurai, T. (2010) Expansion, intensification, and sustainability of rice production in West Africa: the case of rainfed lowland rice in Côte d'Ivoire and Ghana. *International Cooperation in Agriculture* 8, 202–214.

Sanchez, P.A. and Buol, S.W. (1985) Agronomic taxonomy for wetland soils. In: *Wetland Soils: Characterization, classification, and utilization.* Proceedings of a workshop held 26 March to 5 April 1984. International Rice Research Institute, Los Baños, Philippines, pp. 207–229.

Sanchez, P.A., Shepherd, K.D., Soule, M.J., Place, F.M., Buresh, R.J., Izac, A.N., Mokwunye, A.U., Kwesiga, F.R., Ndiritu, C.G. and Woomer, P.L. (1997) Soil fertility replenishment in Africa: an investment in natural resource capital. In: Buresh, R., Sánchez, P.A. and Calhoun, F.C. (eds) *Replenishing Soil Fertility in Africa.* Soil Science Society of America, Madison, Wisconsin.

Sanchez, P.A., Palm, C.A. and Buol, S.W. (2003) Fertility capability soil classification: a tool to help assess soil quality in the tropics. *Geoderma* 114, 157–185.

Segda, Z., Haefele, S.M., Wopereis, M.C.S., Sedogo, M.P. and Guinko, S. (2004) Agro-economic characterization of rice production in a typical irrigation scheme in Burkina Faso. *Agronomy Journal* 96, 1314–1322.

Segda, Z., Haefele, S.M., Wopereis, M.C.S., Sedogo, M.P. and Guinko, S. (2005) Combining field and simulation studies to improve fertilizer recommendations for irrigated rice in Burkina Faso. *Agronomy Journal* 97, 1429–1437.

Segda, Z., Mando, A., Haefele, S.M., Sedogo, M.P., Guinko, S. and Wopereis, M.C.S. (2010) Closing yield gaps through partnerships and good agronomy in Africa. Presentation at the 3rd International Rice Congress, Hanoi, Vietnam, 8–12 November.

Smaling, E.M.A. (2005) Harvest for the world. Presentation at ITC, Enschede, Netherlands, 2 November.

Sokei, Y., Akintayo, I., Doumbia, Y., Gibba, A., Keita, S. and Assigbe, P. (2010) Growth and yield performance of upland NERICA varieties in West Africa. *Japanese Journal of Crop Science* 79(Extra issue 2), 2–3.

Somado, E.A., Becker, M., Kuehne, R.F., Sahrawat, K.L. and Vlek, P.LG. (2003) Combined effects of legume with rock phosphate on rice in West Africa. *Agronomy Journal* 95, 1170–1178.

Somado, E.A., Kiepe, P. and Niang, A. (2011) Alleviating phosphorus deficiency in rice-based systems in humid Africa. In: Yanagihara, S. (ed.) Next Challenges in Rice Development for Africa: Workshop for New Collaboration between JIRCAS and AfricaRice. *JIRCAS Working Report* 70. Japan International Research Center for Agricultural Sciences, Tsukuba, Japan, pp. 29–35.

Stoorvogel, J. and Smaling, E.M.A. (1990) *Assessment of Soil Nutrient Depletion in Sub-Saharan Africa.* Winand Staring Centre, Wageningen, Netherlands.

Touré, A., Becker, M., Johnson, D.E., Koné, B., Kossou, D.K. and Kiepe, P. (2009) Response of lowland rice to agronomic management under different hydrological regimes in an inland valley of Ivory Coast. *Field Crops Research* 114, 304–310.

van Asten, P.J.A., Barro, S.E., Wopereis, M.C.S. and Defoer, T. (2004) Using farmer knowledge to combat low productive spots in rice fields of a Sahelian irrigation scheme. *Land Degradation & Development* 15, 383–396.

Van Hove, C. (1989) *Azolla. Ses emplois multiples. Son intérêt en Afrique.* Food and Agriculture Organization of the United Nations, Rome, Italy.

Van Hove, C. and Diara, H.F. (1987) *Azolla* introduction in African agriculture: progress and problems. *International Rice Commission Newsletter* 36, 1–4.

Vanlauwe, B. and Giller, K.E. (2006) Popular myths around soil fertility management in sub-Saharan Africa. *Agriculture, Ecosystems and Environment* 116, 34–46.

Watanabe, I. and Van Hove, C. (1996) Phylogenetic, molecular and breeding aspects of *Azolla–Anabaena* symbiosis. In: Camus, J.M., Gibby, M. and Johns, R.J. (eds) *Pteridology in Perspective.* Royal Botanic Gardens, Kew, London, pp. 611–619.

Watanabe, I., Ventura, W., Mascarina, G. and Eskew, D.L. (1989) Fate of *Azolla* and urea nitrogen applied to wetland rice. *Biology and Fertility of Soils* 8, 102–110.

Windmeijer, P.N., Duivenbooden, N.V. and Andriesse, W. (eds) (1994) *Characterization of Rice-growing Agro-ecosystems in West Africa: Semi-detailed characterization of inland valleys in Côte d'Ivoire.* Winand Staring Centre (SC-DLO) and Wageningen Agricultural University, Wageningen, Netherlands.

Wopereis, M.C.S., Donovan, C., Nébié, B., Guindo, D., Ndiaye, M.K. and Häfele, S.M. (1998) Nitrogen management, soil nitrogen supply and farmers' yields in Sahelian rice based irrigation systems. *Advances in GeoEcology* 31, 1261–1266.

Wopereis, M.C.S., Donovan, C., Nebié, B., Guindo, D. and N'Diaye, M.K. (1999) Soil fertility management in irrigated rice systems in the Sahel and savanna regions of West Africa. Part I. Agronomic analysis. *Field Crops Research* 61, 125–145.

Wopereis-Pura, M.M., Watanabe, H., Moreira, J. and Wopereis, M.C.S. (2002) Effect of late nitrogen application on rice yield, grain quality and profitability in the Senegal River valley. *European Journal of Agronomy* 17, 191–198.

Yanni, Y.G., Shalaan, S.N. and El-Haddad, M. (1994) Potential role of *Azolla* as green manure for rice in Nile delta under different levels of inorganic fertilization. In: Hegazi, N.A., Fayez, M. and Monib, M. (eds) *Nitrogen Fixation with Non-legumes*. The American University in Cairo Press, Cairo, Egypt, pp. 127–132.

21 Assessing and Improving Water Productivity of Irrigated Rice Systems in Africa

Sander J. Zwart*
Africa Rice Center (AfricaRice), Cotonou, Benin

Introduction

Fresh water resources are becoming increasingly scarce in many parts of the world (Molden, 2007). Africa, however, is endowed with large available fresh water resources. It is estimated that total water available from rivers, lakes and wetlands on the continent amount to 31,766 km^3, which is more than Asia and North America, and total groundwater resources are estimated at 5.5 million km^3 (WRI, 2005). However, the sources are unevenly distributed across the continent and its agroecosystems. In large areas of North Africa overexploitation of ground and surface water is common. Many river basins in Egypt, Libya, Tunisia, Algeria and Morocco are classified as heavily exploited or overexploited (UNEP, 2008; FAO, 2011) and consequently fresh water is scarce (less than 1000 m^3 per person per year). Across sub-Saharan Africa (SSA), available fresh water resources per capita average 6322 m^3, but vary from 508 m^3 in Burundi to 218,000 m^3 in Congo (UNEP, 2008).

The agricultural sector is the largest consumer of fresh water resources. In SSA, approximately 88% of total annual water withdrawals in 2000 were destined for agriculture, 4% for industry and 9% for domestic use (WRI, 2005). In general, SSA river basins are still able to supply sufficient water for drinking water and irrigation. Exceptions are the Volta River basin in West Africa and the Orange and Limpopo River basins in Southern Africa, where population density and large-scale irrigation systems put great strains on water resources availability (Ravenga *et al.*, 2000).

There is large untapped irrigation potential in SSA. IFPRI (2010) estimates that the largest potential for small- and large-scale irrigation systems is in Nigeria (5.7 million ha). For SSA as a whole, this potential amounts to 21 million ha, of which the Gulf of Guinea sub-region has almost 50% (10 million ha). However, efforts to manage water and to make it available (e.g. for agriculture) are hindered by the undeveloped state of institutions in terms of low levels of expertise, knowledge and capacity to develop and manage irrigation, and by the prevalence of subsistence farming. Other challenges for development are related to the absence of an adequate policy and strategic framework, the often disappointing results of previous irrigation development, the need for continued support to cover recurrent costs from the public sector, and the relatively high costs of conventional irrigation development (IFPRI, 2010).

The agricultural sector faces increased competition from other users, such as the industrial

* Corresponding author: s.zwart@cgiar.org

© CAB International 2013. *Realizing Africa's Rice Promise*
(eds M.C.S. Wopereis *et al.*)

and domestic sectors. Due to growing demand for water resources from all sectors, it is projected that by 2025, some 13 countries in SSA (including Nigeria, Ghana and Madagascar) will face water stress (less than 1700 m^3 per capita per year) and another 10 countries (including Egypt, Malawi and Rwanda) will suffer from water scarcity (less than 1000 m^3 per capita per year) (UNEP, 2008). Countries that are currently on the safe side may face water stress in the future.

Water stress will be exacerbated by the impacts of climate change, which will have a large impact on river-basin hydrology. Sub-Saharan Africa is particularly vulnerable to the impacts of climate change (IPCC, 2007). Water availability from rainfall and rivers will decrease in many river basins as rainfall and runoff decline, while the evaporative demand will increase due to predicted elevated temperatures (e.g. Jackson *et al.* 2001; Smakhtin *et al.*, 2004; De Wit and Stankiewicz, 2006). Large areas in West Africa (Senegal River basin), North Africa (Morocco, Algeria, Tunisia) and Southern Africa (e.g. South Africa, Botswana) are projected to suffer an increase of 20% or more in water shortage by 2050 (measured in cubic metres of water per capita). On the other hand, the Nile River basin seems likely to benefit from changes in climate and increased water availability is foreseen (Arnell, 2004).

The increasing population, together with the anticipated impacts of climate change and expansion of irrigated areas, are putting pressure on water availability for African agriculture. Decreasing availability of water in SSA threatens the sustained realization of current levels of rice production in irrigated systems and will affect the food security and livelihoods of many. It is within this context that African agriculture has to become more efficient in the use of water to sustain and where possible enhance food production. It is therefore inevitable that agriculture, being the largest consumer of water, must make the largest gains in productive use of water resources (Molden, 2007).

Rice production is often claimed to be the largest consumer of water in an agricultural context. However, this depends on the definition of water use and the scale at which it is considered. Certainly, rice is the largest consumer of fresh water resources when we consider irrigation water supplies diverted to a single field. In irrigated rice systems, significant volumes of water are used for land preparation and farmers keep fields continuously inundated during the growing season (Bouman *et al.*, 2007). The water productivity, defined as the crop yield divided by the total water applied by irrigation (WP_I), at field level is generally low because of the water needed for land preparation and losses to seepage and lateral flow. The variation in magnitude in applied irrigation water is high and varies between fields and systems due to differences in land preparation (direct seeding or transplanting, puddling of soil), water management and variations in environmental conditions such as soil type and groundwater levels.

However, if we place the rice field in the context of an irrigation system or a river basin, the situation changes. The excess water that has not evaporated, transpired or percolated to depths where it can no longer be retrieved, will return to the system where it is available for re-use (Seckler, 1996). WP_I can be calculated for individual rice fields, but also at the level of irrigation systems or administrative units within the systems. At higher levels, the WP_I is expected to improve due to the re-use of 'lost' water from groundwater and drainage effluents. The best example is Egypt's Nile River delta: only a small share of total water discharge is effectively released into the Mediterranean Sea, because water is re-used several times to irrigate rice and other crops.

In the context of the water-scarcity debate that has evolved, the importance of evapotranspiration (the combination of transpiration from plants and evaporation from soil and water) has been acknowledged. Evapotranspiration – along with pollution and deep drainage – represents the water that is really lost from a basin and is not available for re-use in productive processes. Saving water by reducing evaporation during the growing season is therefore considered a real water saving, since more water will remain available for other uses within a river basin. Reducing the amount of irrigation water applied to a rice field will only save water from a river basin perspective if this water would otherwise be lost for re-use. For example, in the case of the Office du Niger irrigation scheme (Mali), drainage water is not re-used and cannot return to the Niger River; drainage water evaporates or flows to deeper groundwater layers and is, therefore, lost for re-use.

A review of 13 experiments found that evapotranspiration in irrigated rice varies from 400 mm to 800 mm seasonally (Zwart and Bastiaanssen, 2004). This is high compared to crops such as wheat (200–500 mm), but not compared to cotton (400–900 mm) or maize (200–1000 mm). If water productivity, defined as yield divided by the amount of water consumed by evapotranspiration (WP_{ET}), of rice is compared to other major crops, only maize has significantly higher values (1.1–2.7 kg/m^3) as it is a C_4 crop. The range of water productivity of rice is more or less equal to that of wheat: 0.6–1.6 kg/m^3 (Bouman and Tuong, 2001; Zwart and Bastiaanssen, 2004).

This chapter provides an overview of water productivity of irrigated rice at field and scheme levels in Africa based on a literature survey; reviews studies on assessment of water productivity at field and scheme levels for irrigated rice in Senegal and Mali; and ends with recommendations for improving the productive use of scarce water resources for rice production, and for further research.

Water Productivity in Rice in Africa: An Overview

A literature review was conducted to assess the levels of water productivity of rice by synthesizing the results of experimental trials and regional studies that have been conducted throughout Africa (Table 21.1). Two definitions of water productivity were considered: (i) rice yield divided by seasonal evapotranspiration (WP_{ET}), and (ii) yields divided by total water input from irrigation and precipitation (WP_I). Results from field experiments and (remote-sensing) modelling were included. All the studies are related to irrigated rice systems whether in SSA (Senegal, Nigeria, Ghana, Mali, Burkina Faso) or North Africa (Morocco, Egypt). We did not find any study on water productivity related to rainfed rice or any crop-modelling studies on rice water productivity on the African continent. In the reported studies, WP_{ET} values are higher in Egypt (1.25–1.65 kg/m^3) than in SSA countries, where WP_{ET} remains well below 1.0 kg/m^3. Yields in Egypt are among the highest in the world due to extremely favourable growing conditions. However, the values for WP_{ET} measured in Egypt are comparable with the global average of 1.1 kg/m^3 that was based on a review of 13 publications. The levels of water productivity under SSA conditions appear well below the world average levels (Zwart and Bastiaanssen, 2004).

High values of WP_{ET} in one country do not necessarily mean that others are performing below their potential. There are many reasons, including climate and soils, which cause variation in water productivity from country to country or from region to region, and these are not manageable by human intervention (Dawe, 2005). Comparing water productivity between different experiments must therefore be conducted with extreme caution and non-manageable factors must be excluded. For example, the high values obtained in Egypt must not be set as a benchmark value for a rainfed rice system in Togo, rather local optimal values of water productivity must be used.

Assessment and Improvement of Water Productivity of Rice at Field Level

The most direct way to raise water productivity in irrigated rice cropping in Africa is to improve crop management in general. Yields per unit of land or per unit of water consumed by evapotranspiration (expressed as WP_{ET}) or irrigation water applied (WP_I) are still far below what would be possible with improved management. It is also possible to increase water productivity, expressed as yield per unit of water applied (WP_I), by reducing the application of irrigation water during the growing season for land preparation and growth of the rice crop. However, it then becomes important to maintain rice yields. One option to reduce water intake at field level (and thus improve WP_I) is the practice of intermittent flooding or alternate wetting and drying (AWD). Rice is grown without a permanent layer of standing water on the field, and irrigation water is applied to obtain flooded conditions after a certain time has passed after the disappearance of ponded water (Bouman *et al.*, 2007). Bouman and Tuong (2001) report that such practices can reduce irrigation water input while maintaining rice production.

Table 21.1. Levels of water productivity in African countries.

Definition[a]	Method	Condition	WP range (kg/m^3)	Location and country	Years	Source
Y/ET	Field experiments	Irrigated	0.53–0.64	Ndiaye and Pont-Gendarme, Senegal	1990	Raes *et al.* (1992)
Y/ET	Field experiments	Irrigated	0.50–0.79	Kadawa, Nigeria	1991–1992	Nwadukwe and Chude (1998)
Y/ET	Regional estimate	Irrigated	0.56	Tono, Ghana	2005–2006	Mdemu *et al.* (2009)
Y/ET	Field experiments	Irrigated	0.93–1.01	Nile delta, Egypt	Not given	Kheira (2009)
Y/ET	Remote sensing	Irrigated	1.25–1.65	Nile delta, Egypt	2006	Zwart and Bastiaanssen (2007)
Y/ET	Remote sensing	Irrigated	0.53–1.03	Office du Niger, Mali	2005	Zwart and Leclert (2010)
Y/ET	Literature review	Irrigated	0.6–1.6	Globally	1979–2004	Zwart and Bastiaanssen (2004)
Y/(I+P)	Field experiments	Irrigated	0.23–1.28	Ndiaye and Fanaye, Senegal	2005–2006	De Vries *et al.* (2010)
Y/(I+P)	Field experiments	Irrigated	0.27–1.07	Fanaye, Senegal	2007	Krupnik *et al.* (2012a)
Y/(I+P)	Field experiments	Irrigated	0.22–1.43	Ndiaye, Senegal	2008	Schlegel (2010)
Y/(I+P)	Field experiments	Irrigated	0.34–0.88	Gorgo, Mogtedo & Itenga, Burkina Faso	1993–1994	Dembélé *et al.* (2001)
Y/I[b]	Field experiments	Irrigated	0.60–1.17	Kafr El-Sheikh, North Delta, Egypt	2006–2007	El-Bably *et al.* (2008)
Y/I[b]	Field experiments	Irrigated	0.52–0.99	Giza, Egypt	Not given	Nour *et al.* (1997)
Y/I[b]	Field experiments	Irrigated	0.50–0.80	Gharb, Morocco	1995–1997	Lage *et al.* (2004)

[a]Y = rice yield, ET = evapotranspiration, I = irrigation water deliveries, P = precipitation.
[b]In these studies the contribution of water from precipitation was nil or close to nil.

In Senegal, Africa Rice Center (AfricaRice) conducted water-saving experiments in irrigated rice schemes in the Senegal River valley (De Vries *et al.*, 2010; Krupnik *et al.*, 2012a). The specific objective of the study by De Vries *et al.* (2010) was to test the possibility of saving water in rice production in a Sahelian environment by quantifying the effects of different water regimes on rice yield and irrigation water productivity under weed-free conditions with ample nitrogen. Five field trials were conducted in 2005 (dry season, DS2005, and wet season, WS2005) and 2006 (dry season, DS2006) at two research stations – Ndiaye (delta, 16°11′N, 16°15′W) and Fanaye (middle valley, 16°32′N, 15°11′W). The experiment in DS2005 was only conducted at Ndiaye. Irrigation was measured at inlets at the entrance of the field; rice yield, irrigation water delivery and irrigation water productivity (WP_I) were determined for four irrigation regimes: alternate wetting and drying (AWD) throughout the season; AWD until panicle initiation and the rest of the season flooded (AWD-flooded); flooded until panicle initiation followed by AWD (flooded-AWD);

and continuously flooded throughout the season (flooded). Weeds were controlled at 21 days after sowing, and N was applied at the recommended rate of 150 kg/ha (see the publications for complete details of the experiment).

The amount of irrigation water used across all trials ranged from 480 mm in the AWD treatment during WS2005 in Ndiaye to 1490 mm in the flooded treatment during DS2005 in Ndiaye.

Rice yields ranged from 2.3 t/ha to 11.8 t/ha in the water-saving treatments. They ranged from 3.7 t/ha to 11.7 t/ha in the flooded treatments. In the wet season, the treatments in which AWD was applied during part of the season resulted in the highest yields at both sites. In the dry season, the continuously flooded treatment out-yielded other treatments, with the exception of AWD in Fanaye. At the particularly weed-infested Ndiaye site, the control of weeds increased yields from an average of 2.0 t/ha to 7.4 t/ha in the dry season and from 1.4 t/ha to 4.9 t/ha in the wet season. No weed control in combination with AWD during the vegetative stage reduced yields to below 1.0 t/ha. However, when weeds were controlled, crop yields obtained with a combination of AWD and flooding were comparable with those obtained in fully flooded plots receiving the same weed management treatment.

These results demonstrate that it is possible to attain major savings of irrigation water with little loss of yield in a Sahelian environment: during the wet season, irrigation water savings of 22–39% are possible for rice with no or little yield loss, while maintaining high water productivity. An important pre-condition, however, is good weed control and the application of sufficient mineral fertilizer.

Krupnik *et al.* (2012a) compared recommended farming practices (RFP) against an adapted form of the System of Rice Intensification (SRI) in two regions of the Senegal River valley (delta and middle valley). SRI is claimed as a means for saving water in rice systems, and in the Sahel it has been tested in The Gambia (Ceesay, 2010) and Mali (Styger *et al.*, 2010). Experiments were laid out according to a split-split-plot design that allowed evaluation of rice yields, weed competition, water savings and water productivity in both SRI and RFP. In RFP, a continuous floodwater layer was maintained in the fields until 2 weeks before harvest, whereas in SRI the system of AWD was applied in which a shallow water layer (2–3 cm depth) was reapplied only when the soil surface had begun to dry and hairline cracks became visible (no more than twice a week). Two weed management treatments, weedy and weed-free, were implemented. A full description of the experimental layout and the measurements is provided by Krupnik *et al.* (2012a).

Two major conclusions stem from this work. First (and this confirms previous claims), substantial field-level water savings and significant increases in water productivity can be obtained with SRI compared to RFP, through AWD practices. Table 21.2 shows that in both locations and for both seasons SRI resulted in higher values of WP_I than RFP under weed-free conditions. For example, in the delta area average WP_I for SRI was 0.74 kg/m^3 and 0.81 kg/m^3 in the dry and wet seasons, respectively, while WP_I for RFP was 0.60 kg/m^3 and 0.64 kg/m^3. The second major conclusion is that the positive impact of SRI on water productivity only holds when weeds are adequately controlled. When subject to weed

Table 21.2. Main effects on water productivity (WP_I in kg/m^3) for each crop management system and across sites and seasons. (Adapted from Krupnik *et al.*, 2012a.)

	Senegal River delta				Senegal River middle valley			
	Dry season 2008		Wet season 2008		Dry season 2008		Wet season 2009	
Water management	Weed-free	Weedy	Weed-free	Weedy	Weed-free	Weedy	Weed-free	Weedy
Recommended Farmer Practice	0.60b	0.34a	0.64b	0.42b	0.43b	0.23a	0.73b	0.70b
System of Rice Intensification	0.74a	0.22b	0.81a	0.35a	0.53a	0.25a	1.31a	0.95a

Values in a column sharing the same letter are not significantly different according to the least-significant means T-test (α=0.05).

competition, weed growth was consistently greater under SRI than RFP, in both the dry and wet seasons. Yield losses in the SRI trials without proper weed management were greater than in the RFP in the majority of cases. If farmers have sufficient resources to control weeds and fields are well-levelled, 'SRI-type water management' can, therefore, support water saving in rice-based systems.

These examples show the importance of keeping a close eye on crop management in general for enhancing water productivity in rice fields in Africa if attempts are made to reduce irrigation inputs at field level.

When trying to improve water productivity, the scale as well as its definition needs to be considered in order to ensure that the desired impact is achieved. Improving water productivity at field scale by reducing the water applications may not necessarily contribute to overall water productivity enhancement of an irrigation system. Often water is recovered and re-used by placing check dams or by pumping groundwater, and the impact of reducing water application may be much smaller than anticipated (Hafeez *et al.*, 2007). On the other hand, if a farmer aims at improving the benefit from his or her fields by reducing the cost of water, then reducing irrigation water applications may improve water productivity. Thus, farmers who use pumps to get water to their fields from either groundwater or rivers, and farmers who pay for water by volume may have an incentive to improve and optimize the water productivity in their fields. A practical way to implement AWD for increasing the water productivity is to monitor the depth of the water table in the field using a piezometer, or a simple perforated water tube placed inside the rice field. After an irrigation application the water depth will gradually decrease. When the water level reaches more than 15 cm below the soil surface, irrigation must be applied again to flood the soil with a depth of around 5 cm (IRRI, 2012).

Assessment of Water Productivity of Rice in a Large-scale Irrigation System Using Remote-sensing Techniques

The Office du Niger, situated in the Ségou region of Mali (13.7–14.9°N, 5.3–6.3°W), is one of the oldest and largest irrigation schemes in West Africa. It was originally intended as a large-scale cotton system, but today rice is cultivated on 99% of the land. The management of the irrigation system is sub-divided into five zones, of which Macina is the only one located along the Niger River (Plate 13 inset). The water for irrigation is diverted from the Niger River at the Markala dam and then flows into two former river branches from where the water is supplied to the fields through a hierarchical network of canals (Ertsen, 2006). Water is abundantly available to irrigators during the main cropping season and fields are likely to be overirrigated causing issues with waterlogging and drainage. Measurements of water supply to fields show large differences ranging from 8 m^3/ha to 30 m^3/ha per season (Vandersypen *et al.*, 2006). Future trends suggest that less water will be available for irrigation due to an expansion of the system, as well as reduced discharges in the Niger River and higher demands induced by climate change. There is a need to improve the irrigation performance in the Office du Niger in order to expand the system and sustainably provide water to the water users.

Irrigation performance assessment is considered an important management tool to implement, monitor and evaluate activities for water delivery services (Molden *et al.*, 2001). These assessments can focus on physical performance indicators as well as economic and institutional performance indicators. In this section, focus is on physical performance indicators.

Inputs that are required to assess the physical irrigation performance include measurements of different terms of the water balance such as discharge, evapotranspiration, effective precipitation, as well as measurements of crop yields, and estimates of irrigated area and cropping intensities. The application of indicators at the lower scale in an irrigation system, such as tertiary units, requires expensive and labour-intensive field campaigns (Vandersypen *et al.*, 2006). Moreover, it is virtually impossible to obtain a data set that systematically covers the whole system. This limits the possibilities for analysing, for example, the equity of water distribution among users in different parts of the system. The possibilities of using spatial remote-sensing data have been investigated (see Bastiaanssen and Bos, 1999, for a review).

Major advantages of remote-sensing-derived data over field-measured data are that they are objective, collected systematically and system-wide, and the information can be analysed at different scales (Bastiaanssen *et al.*, 2000).

A remote-sensing-based study was performed to analyse the irrigation performance of the Office du Niger irrigation system (Zwart and Leclert, 2010). The indicators assessed were water productivity (WP_{ET}), defined as the crop production divided by the seasonal water consumption from evaporation (kg/m^3), and the head–tail performance indicator (%). The latter is used to assess the uniformity of water distribution across the irrigation system or a sub-unit by assessing the spatial pattern of water consumption, rice yields and water productivity among irrigators in head and tail reaches of an irrigation unit or the entire system. Often, farmers close to the inlet of the system (head end) have better access to water, while irrigation water may fail to reach the irrigators near the tail end of a canal due to poor condition of control structures, illegal water use by irrigators upstream, or poor irrigation management.

The Surface Energy Balance Algorithm for Land (SEBAL) model (Bastiaanssen *et al.*, 1998) was developed to estimate the components of the surface energy balance, and thus the actual evapotranspiration, spatially from remote-sensing images and standard meteorological measurements (air temperature, wind speed, relative humidity and solar radiation). Remote-sensing images are used to calculate the Normalized Difference Vegetation Index (NDVI), the surface reflection and the surface temperature, which are inputs for the model. The model is equipped with a module to estimate biomass production spatially from the same images (Bastiaanssen and Ali, 2003). For this study, high-resolution Landsat images were used to produce detailed maps of estimated rice yield, seasonal water consumption and water productivity, from which the indicators were evaluated. The head–tail differences were assessed for the entire system, excluding the Macina zone. This area located in the former river bed was divided into 10 areas of equal size starting from the first irrigation water intake (head end) until the end of the system (tail end). The average yield, evapotranspiration and water productivity were calculated for each of the 10 areas (see Zwart and Leclert, 2010, for a full description of the methodology and the inputs used).

The average WP_{ET} in the 82,666 ha of rice cultivated in the Office du Niger during 2006 season was 0.78 kg/m^3, with a standard deviation of 0.12 kg/m^3. Certain areas in the system showed significantly higher water productivity, with values of up to 1.1 kg/m^3 (the blue and black areas in Plate 13). The zones of Macina and Kouroumari had lower average values (0.72 kg/m^3 and 0.75 kg/m^3, respectively) than the zones of Niono (0.83 kg/m^3), Molodo (0.81 kg/m^3) and N'Débougou (0.78 kg/m^3). However, the system average was low compared to the global range for water productivity of rice of 0.6–1.6 kg/m^3 (based on the experimental results of 13 sources worldwide; Zwart and Bastiaanssen, 2004). Slightly lower values (0.53–0.64 kg/m^3) were measured in a rice-based system in Senegal (Raes *et al.*, 1992); in Nigeria, water productivities for rice were in the range 0.50–0.79 kg/m^3 (Nwadukwe and Chude, 1998).

Significant differences were found in water consumption and estimated yields between the head and tail ends of the system (excluding Macina) (Fig. 21.1). Average water consumption amounted to approximately 780 mm per season at the head end of the system, but was 5% lower at the tail end of the system (743 mm per season). However, estimated rice yields at the tail end were 18% lower (5.4 t/ha compared to 6.5 t/ha at the head end). It was suggested by the irrigation managers that lower water quality and higher groundwater levels towards the tail end of the system could have been responsible for the lower yields. A potential limitation in the application of the model is the use of a fixed harvest index to estimate the crop yields from season biomass, as biomass estimation may be biased if fields are affected by weeds.

The low average water productivity throughout the Office du Niger (0.78 kg/m^3) and decreasing rice yields towards the tail end of the system (−18%) show that there is scope for improvement in the productive use of water resources. Remote-sensing data provided new insights into patterns of rice yields and water productivity. Remote sensing can further support the management of the system in strategic planning of water sources and in monitoring and evaluating where water resources are beneficially used and where water resources are wasted.

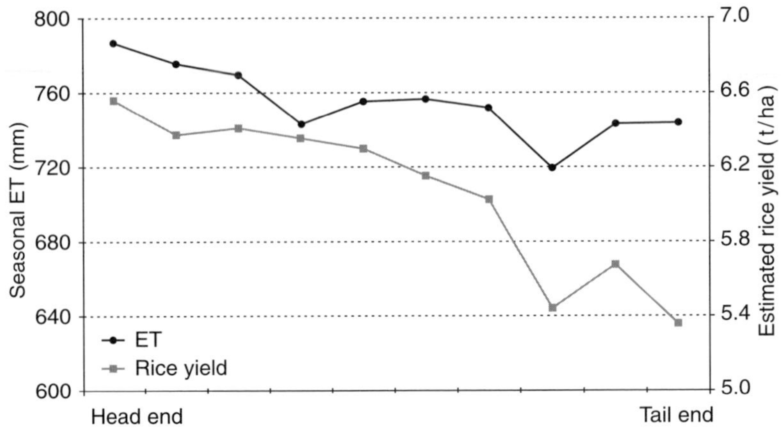

Fig. 21.1. The average seasonal water consumption from evapotranspiration and estimated rice yields in the wet season for the head and tail ends and eight areas in between in the Office du Niger, Mali. The assessment was made for the entire system, excluding the Macina zone (which receives water from a different canal). The distance between the head and tail of the system is approximately 100 km. (Adapted from Zwart and Leclert, 2010.)

By combining these data into existing data sets, the causes of low productivity can be researched and appropriate interventions can be developed. Other indicators can also be spatially analysed, so that water consumption and crop yields can be related to irrigation water supply: examples are cropping intensity, irrigation efficiency, and relative water supply (Bastiaanssen and Bos, 1999).

Conclusions

Africa is claimed to have abundant water and land resources, which would potentially allow the continent to expand its agricultural areas, increase food production and feed its own population. While on a continental scale this statement may hold true, on a regional scale large differences exist in availability of water resources for agriculture as the result of socio-economic and natural conditions.

Physical water shortages already exist in many river basins in Africa (e.g. the basins of the Volta, Orange and Limpopo rivers, and many smaller basins in North Africa) and the situation is likely to get worse (Arnell, 2004; Smakhtin *et al.*, 2004; De Wit and Stankiewicz, 2006). Climate change predictions show that in several regions the physical water availability will decrease significantly due to changes in rainfall, runoff and evapotranspiration. All countries in North Africa are already water scarce, while in SSA a total of 23 countries that are currently on the safe side will become either water stressed or water scarce in the near future (by 2025) as a result of population pressure and climate change effects. Projected water scarce or stressed countries by 2025 include the major rice-producing countries in Africa: Nigeria, Egypt and Madagascar (UNEP, 2008).

Less water will be available in future rice-farming environments in many regions. This will affect all rice-growing environments, including uplands, rainfed lowlands and irrigated rice systems. Spells of drought will occur more frequently and may become longer. Changing precipitation patterns will affect runoff and alter the hydrological regime of rivers and streams reducing river flows, increasing peak flows, shortening the period of water availability, and making the arrival of the peak discharges earlier or later. Rice farmers need to adjust their farming systems to the greater uncertainties and reduced supply of water from irrigation and rainfall in order not to become vulnerable to crop failure. Water-saving regimes, rainwater harvesting and drought-resistant varieties with shorter growing cycles are options that farmers can choose from.

Tools are being developed to enable farmers and irrigation managers to be more productive with scarce water resources and resilient to water scarcity. The field-scale studies in the Senegal River valley by De Vries *et al.* (2010) and Krupnik *et al.* (2012a) show that irrigation water supply can be reduced with limited yield penalties. Thus, the water productivity of rice (WP_I) can be improved by adopting new ways of managing irrigation water through SRI or AWD. Farmers can make significant water savings, but only when field management is optimal. Without proper weed management, such water-saving techniques may even have an adverse effect on water productivity and crop yields. The adoptability of SRI, which also requires land levelling and high nutrient inputs, is therefore believed to be low in SSA, since farmers have poor access to credit and markets (Krupnik *et al.*, 2012a). However, promising results were found where in a participatory on-farm experiment farmer-optimized practices (based on a selection by farmers of components of SRI and RFP) were evaluated (Krupnik *et al.*, 2012b). The on-farm evaluation revealed similar yields to RFP and SRI, but higher benefits due to different weed management. This confirms that simple blueprints that work in one place cannot simply be applied in another region without adjustments to the new situation.

The regional case study in Mali showed the large variation of water productivity (WP_{ET}) and yields that exists throughout an irrigation system. The variation is high between fields, but also between irrigation management units such as zones and tertiary units. Units close to the water inlet of the system outperform units at the tail end of the system. The remote-sensing analysis showed that improvements are possible and provides a valuable tool for irrigation managers to locate underperforming areas. However, it does not provide information on the underlying causes of high or low performance. Additional analysis on, for example, water quality, groundwater levels and water supply to irrigation units, must be conducted to pinpoint the causes of underperformance and to propose measures to improve yields and make the use of the water resources more productive. Spatial analysis in a GIS by combining remote-sensing-obtained data on crop yield and evapotranspiration with field measurements of (e.g.) water deliveries and groundwater levels in piezometers is an essential tool for such a purpose. Once such analysis is performed, measures can be targeted to specific locations. If, for example, the saline groundwater quality is an issue, salt-tolerant rice varieties can be introduced. Other management measures could include constructing and rehabilitating drainage systems, reducing water supply to match the demand, and taking land out of production.

In the case of the Office du Niger, farmers themselves have limited incentives to reduce water supply to their fields; yields are relatively high and they pay a fixed amount per season to irrigate their fields. Moreover, salinity is a serious issue and high water inputs ensure that salt is leached out of the root zone. However, waterlogging and limited drainage possibilities at the end of the season cause significant postharvest losses. The introduction of water-saving measures may be feasible, but it requires an in-depth analysis of possible effects on soil quality.

Potential tools for improving the productive use of scarce water resources have been identified, but need further research. Salinity and soil degradation are important issues in irrigation systems in the Sahel, and the effects of water-saving techniques must be investigated. In the case of the Senegal River valley, water is directly pumped from the river. Given high fuel prices, farmers and irrigation managers have concrete incentives to reduce water application to fields and increase profitability. However, in the case of the Office du Niger, a gravity-based system, farmers pay a fixed amount ('*redevance*') that permits them to irrigate during one season. Unlike systems with volumetric water pricing or pumps, there is no incentive for farmers to reduce water use; instead farmers will utilize all water available.

New research is proposed for irrigated rice systems focusing on testing promising water-saving techniques, such as intermittent irrigation, AWD and SRI, under a large range of conditions that prevail in Africa and including long-term effects on soils. The major challenge for research related to water-saving technologies will be creating impact and ensuring that tools meet farmers' needs. The right incentives must be created for farmers and irrigation managers to improve water management and sustain or improve productive use of water resources. On-farm participatory experiments are essential to optimize and improve a technology and to investigate the profitability and socio-economic constraints of proposed products and services. The impact pathways require high involvement of the stakeholders through an inter-disciplinary and participatory research process, which involves the farmers, as well as irrigation managers, development organizations and NGOs. Research must not only be implemented in experimental plots, but also in farmers' fields and irrigation blocks and schemes to assess the opportunities and limitations of water-saving options at different scales.

Acknowledgements

Plate 13 is reproduced from *Irrigation Science*, 28, 2010, 371–385, 'A remote sensing based irrigation performance assessment: a case study of the Office du Niger in Mali', S.J. Zwart and L.M.C. Leclert, Figs 1 and 6, © S.J. Zwart and L.M.C. Leclert, with kind permission from Springer Science and Business Media.

References

Arnell, N.W. (2004) Climate change and global water resources: SRES emissions and socio-economic scenarios. *Global Environmental Change* 14, 31–52.
Bastiaanssen, W.G.M. and Ali, S. (2003) A new crop forecasting model based on satellite measurements applied across the Indus basin, Pakistan. *Agriculture Ecosystems and Environment* 94, 321–340.
Bastiaanssen, W.G.M. and Bos, M.G. (1999) Irrigation performance indicators based on remotely sensed data: a review of literature. *Irrigation and Drainage Systems* 13, 291–311.
Bastiaanssen, W.G.M., Menenti, M., Feddes, R. and Holtslag, A.A.M. (1998) A remote sensing surface energy balance algorithm for land (SEBAL). 1. Formulation. *Journal of Hydrology* 212/213, 198–212.
Bastiaanssen, W.G.M., Molden, D.J. and Makin, I.W. (2000) Remote sensing for irrigated agriculture: examples from research and possible applications. *Agricultural Water Management* 46, 137–155.
Bouman, B.A.M. and Tuong, T.P. (2001) Field water management to save water and increase its productivity in irrigated lowland rice. *Agricultural Water Management* 49, 11–30.
Bouman, B.A.M., Humphreys, E., Tuong, T.P. and Barker, R. (2007) Rice and water. *Advances in Agronomy* 92, 187–237.
Ceesay, M. (2010) An opportunity for increasing factor productivity for rice cultivation in The Gambia through SRI. *Paddy and Water Environment* 9, 129–135.
Dawe, D. (2005) Increasing water productivity in rice-based systems in Asia – past trends, current problems, and future prospects. *Plant Production Science* 8, 221–230.
De Vries, M.E., Rodenburg, J., Bado, B.V., Sow, A., Leffelaar, P.A. and Giller, K.E. (2010) Rice production with less irrigation water is possible in a Sahelian environment. *Field Crops Research* 116, 154–164.
De Wit, M. and Stankiewicz, J. (2006) Changes in surface water supply across Africa with predicted climate change. *Science* 311, 1917–1921.
Dembélé, Y., Ouattara, S. and Keïta, A. (2001) Application des indicateurs 'approvisionnement relatif en eau' et 'productivité de l'eau' à l'analyse des performances des petits périmètres irrigués au Burkina Faso. *Irrigation and Drainage* 50, 309–321.
El-Bably, A.Z., Meleha, M.E., Abd Allah, A.A. and El-Khoby, W.M. (2008) Increasing rice productivity, water use efficiency, water saving and rice productivity in North Delta, Egypt. Paper presented at the 3rd International Conference on Water Resources and Arid Environments and the 1st Arab Water Forum, Riyadh, Saudi Arabia, 16–19 November.
Ertsen, M.W. (2006) Colonial irrigation: myths of emptiness. *Landscape Research* 31, 147–167.
FAO (2011) AquaStat. http://www.fao.org/nr/water/aquastat/main/index.stm (accessed October 2011).
Hafeez, M.M., Bouman, B.A.M., Van de Giesen, N. and Vlek, P. (2007) Scale effects on water use and water productivity in a rice-based irrigation system (UPRIIS) in the Philippines. *Agricultural Water Management* 92, 81–89.
IFPRI (2010) What is the irrigation potential for Africa? A combined biophysical and socioeconomic approach. *IFPRI Discussion Paper* 00993. International Food Policy Research Institute, Washington, DC.
IPCC (2007) *Climate Change 2007: Impacts, adaptation and vulnerability.* Contribution of Working Group II to the Fourth Assessment Report of the Intergovernmental Panel on Climate Change (Parry, M.L., Canziani, O.F., Palutikof, J.P., van der Linden, P.J. and Hanson, C.E. eds). Cambridge University Press, Cambridge.
IRRI (2012) Rice knowledge bank: water management: alternate wetting and drying (AWD). International Rice Research Institute, Los Baños, Philippines. http://www.knowledgebank.irri.org/watermanagement/coping-with-water-scarcity/alternate-wetting-and-drying-awd.html (accessed 14 May 2012).

Jackson, R.B., Carpenter, S.R., Dahm, C.N., McKnight, D.M., Naiman, R.J., Postel, S.L. and Running, S.W. (2001) Water in a changing world. *Ecological Applications* 11, 1027–1045.

Kheira, A.A.A. (2009) Crop water productivity as influenced by irrigation improvement in the Nile Delta. In: *American Society of Agricultural and Biological Engineers Annual International Meeting 2009*, Reno, Nevada, USA, 21–24 June 2009, Volume 3. Curran Associates, Red Hook, New York, pp. 1465–1482.

Krupnik, T.J., Rodenburg, J., Haden, V.R., Mbaye, D., Shennan, C. (2012a) Genotypic trade-offs between water productivity and weed competition under the System of Rice Intensification in the Sahel. *Agricultural Water Management* 115, 156–166.

Krupnik, T.J., Shennan, C., Settle, W.H., Demont, M., Ndiaye, A.B. and Rodenburg, J. (2012b) Improving irrigated rice production in the Senegal River valley through experiential learning and innovation. *Agricultural Systems* 109, 101–112.

Lage, M., Bamouh, A., Badawi, T. and El Mourid, M. (2004) Productivité de l'eau pour une culture du riz irrigué (*Oryza sativa* L.) conduit sous différents modes d'irrigation dans la région du Gharb (Maroc). Projet INCO-WADEMED Actes du Séminaire, Modernisation de l'Agriculture Irriguée, Rabat, Morocco, 19–23 April 2004.

Mdemu, M.V., Rodgers, C., Vlek, P.L.G. and Borgadi, J.J. (2009) Water productivity (WP) in reservoir irrigated schemes in the upper east region (UER) of Ghana. *Physics and Chemistry of the Earth* 34, 324–328.

Molden, D. (ed.) (2007) *Water for Food, Water for Life: A comprehensive assessment of water management in agriculture*. Earthscan, London and International Water Management Institute, Colombo, Sri Lanka.

Molden, D.J., Burton, M. and Bos, M.G. (2001) Performance assessment, irrigation service delivery and poverty reduction: benefits of improved system management. *Irrigation and Drainage* 56, 307–320.

Nour, M.A., Abd El-Wahab, A.E., El-Kady, A.A. and Ebaid, R.A. (1997) Productivity of some rice varieties under different irrigation intervals and potassium level. *Egyptian Journal of Applied Science* 12, 137–154.

Nwadukwe, P.O. and Chude, V.O. (1998) Manipulation of the irrigation schedule of rice (*Oryza sativa* L.) as a means of maximizing water use efficiency and irrigation efficiency in the semi-arid tropics. *Journal of Arid Environments* 40, 331–339.

Raes, D., Sy, B. and Van Passel, L. (1992) The water balance of rice irrigation schemes in the Senegal river delta. In: Feyen, J., Mwendera, E. and Badji, M. (eds) *Advances in Planning Design and Management of Irrigation Systems as Related to Sustainable Land Use.* Proceedings of an international conference, 14–17 September 1992, Leuven, Belgium, Vol. 2. Center for Irrigation Engineering, Leuven, Belgium, pp. 835–844.

Ravenga, C., Brunner, J., Henninger, N., Kassem, K. and Payne, R. (2000) *Pilot Analysis of Global Ecosystems: Wetland ecosystems.* World Resources Institute, Washington, DC.

Schlegel, I. (2010) Physiological and agronomic responses of selected rice genotypes to irrigation management in the Sahel. MSc thesis, Hohenheim University, Hohenheim, Germany.

Seckler, D. (1996) The new era of water resources management: from 'dry' to 'wet' water savings. *Research Report* 1. International Irrigation Management Institute, Colombo, Sri Lanka.

Smakhtin, V., Revenga, C. and Döll, P. (2004) Taking into account environmental water requirements in global-scale water resources assessments. *Comprehensive Assessment Research Report* 2. International Water Management Institute, Colombo, Sri Lanka.

Styger, E., Ag Atter, M., Guindo, H., Ibriham, H., Diaty, M., Abba, I. and Traore, M. (2010) Application of System of Rice Intensification practices in the arid environment of the Timbuktu region in Mali. *Paddy and Water Environment* 9, 137–144.

UNEP (2008) *Vital Water Graphics – An overview of the state of the world's fresh and marine waters*, 2nd edn. United Nations Environment Programme, Nairobi, Kenya.

Vandersypen, K., Bengaly, K., Keita, A.C.T., Sidibe, S., Raes, D. and Jamin, J.-Y. (2006) Irrigation performance at tertiary level in the rice schemes of the Office du Niger (Mali): adequate water delivery through over-supply. *Agricultural Water Management* 83, 144–152.

WRI (2005) *Ecosystems and Human Well Being: Biodiversity synthesis*. World Resources Institute, Washington, DC.

Zwart, S.J. and Bastiaanssen, W.G.M. (2004) Review of measured crop water productivity values for irrigated wheat, rice, cotton and maize. *Agricultural Water Management* 69, 115–133.

Zwart, S.J. and Bastiaanssen, W.G.M. (2007) SEBAL for detecting spatial variation of water productivity and scope for improvement in eight irrigated wheat systems. *Agricultural Water Management* 89, 287–296.

Zwart, S.J. and Leclert, L.M.C. (2010) A remote sensing based irrigation performance assessment: a case study of the Office du Niger in Mali. *Irrigation Science* 28, 371–385.

22 Inland Valleys: Africa's Future Food Baskets

Jonne Rodenburg*
Africa Rice Center (AfricaRice), Dar es Salaam, Tanzania

Introduction

Inland-valley ecosystems are expected to play a crucial role in boosting Africa's rice production. They are defined as the upper parts of river drainage systems, comprising the whole upland–lowland continuum (Windmeijer and Andriesse, 1993), from the rainfed uplands (*pluvial*) to rainfed, flooded and intensified lowlands in the valley bottom (*fluxial*), with the hydromorphic fringes (*phreatic*) as the (sloping) transition zone between them (Fig. 22.1).

The morphology of inland valleys can vary as a result of climate, geology and geomorphology. There are many shapes of inland valleys, but the three most frequently observed morphology types are: (i) rectilinear, broad valleys with gentle (<3%) and straight slopes; (ii) concave, relatively narrow valleys with concave side slopes (3–8%); and (iii) convex, with moderately steep convex side slopes and flat narrow (20–400 m) valley bottoms (Fig. 22.2). Raunet (1985) and Windmeijer and Andriesse (1993) report that in West Africa, rectilinear inland valleys (Fig. 22.3a) are located in areas with mean annual precipitation of 800–1100 mm as observed in the Sudan savannah to Guinea savannah zones; concave inland valleys (Fig. 22.3b) are associated with the Guinea savannah zone with intermediate rainfall regimes ranging from 1100 mm to 1400 mm per year, while convex inland valleys (Fig. 22.3c) are formed under the high-rainfall regimes of 1400 mm and above, typical of the equatorial forest zone.

With an estimated land area of 190 Mha,[1] inland valleys are common landscapes serving a multitude of ecosystem functions in many parts of Africa. In general, wetland environments and particularly valley bottoms – commonly referred to as *bas-fonds*, *fadamas* and *inland swamps* in West Africa; *mbuga* in East Africa, and *vleis*, *dambos*, *mapani*, *matoro*, *inuta* or *amaxhaphozi* in Southern Africa (Acres *et al.*, 1985) – generally have high agricultural production potential (Andriesse *et al.*, 1994), although the only major crop that can be grown under the temporary flooded conditions in these ecosystems is rice (e.g. Andriesse and Fresco, 1991). The development of inland valleys for rice production can be accomplished with relatively small-scale technologies that require moderate investments (Roberts, 1988). Thus, inland valleys are strategically important for realizing Africa's rice promise (e.g. Sakurai, 2006; e.g. Balasubramanian *et al.*, 2007).

Wetlands, including inland valleys, are particularly important assets for the rural poor as they can fulfil many crucial services

* Corresponding author: j.rodenburg@cgiar.org

© CAB International 2013. *Realizing Africa's Rice Promise* (eds M.C.S. Wopereis *et al.*)

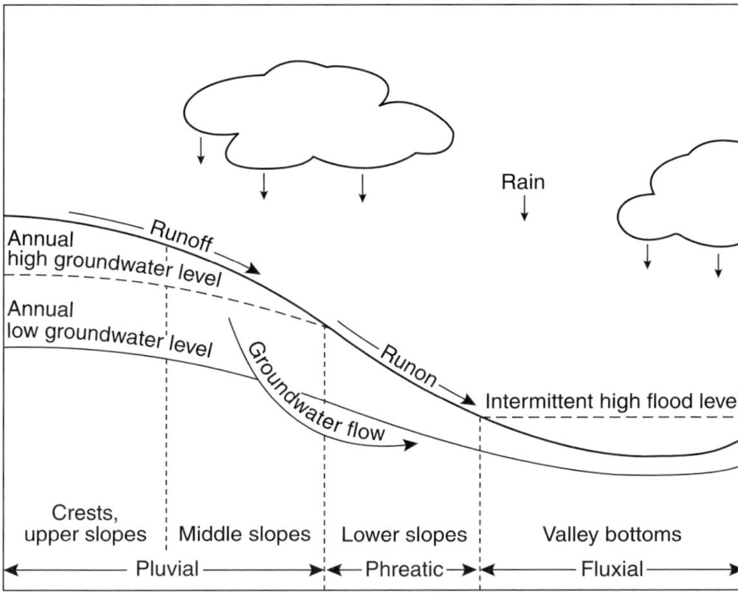

Fig. 22.1. Schematic landscape presentation of rice production environments along the upland–lowland continuum, and their hydrological regimes. (Adapted from Windmeijer and Andriesse, 1993.)

(Turner *et al.*, 2000). Apart from agricultural production, these ecosystems supply local communities with hunting, fishing, forest and forage resources (e.g. Roberts, 1988; Scoones, 1991; Adams, 1993) and they are local hot spots for biodiversity (Chapman *et al.*, 2001). As different inland-valley ecosystem functions may conflict with agricultural objectives, and because there are large area-specific differences in development suitability and risks, indiscriminate development should be avoided (McCartney and Houghton-Carr, 2009). Agricultural developments implemented without proper impact assessments can affect local livelihoods and environments negatively (e.g. Whitlow, 1983), as other functions, like biodiversity and water buffering, are inevitably lost, at least to a certain extent. Benefiting sustainably from the potential of inland valleys therefore requires aligning food production with biodiversity, soil and water conservation such that local rural livelihoods, and hence regional objectives of reducing poverty, are achieved while inherent ecosystem services of local and regional importance are safeguarded. This chapter focuses on how inland valleys can be efficiently and sustainably used to boost Africa's rice production.

Inland Valley Development Opportunities and Challenges

Ecological and economic importance

Inland valleys are not obvious ecosystems for agricultural production, and traditionally they have not often been used for agriculture in Africa (Adams, 1993; Verhoeven and Setter, 2010). This is partly because inland-valley bottoms are difficult to manage and they are also often associated with water-borne diseases such as bilharzia (schistosomiasis – *Schistosoma haematobium* and *S. mansoni*), river blindness (onchocerciasis – vector: *Onchocerca volvulus*, cause: *Wolbachia pipientis*), sleeping sickness (trypanosomiasis – *Trypanosoma brucei gambiense* or *T. brucei rhodesiense*) and malaria (e.g. *Plasmodium falciparum*, *P. malariae* and *P. ovale*) (Gbakima, 1994; Yapi *et al.*, 2005). Moreover, inland-valley exploitation is often complicated by unfavourable land-tenure arrangements (e.g. Fu *et al.*, 2010; Oladele *et al.*, 2010) or prohibitive customary beliefs. Despite such challenges, inland valleys have increasingly been put under production by several generations in areas where they exist. Global changes, such as population growth and

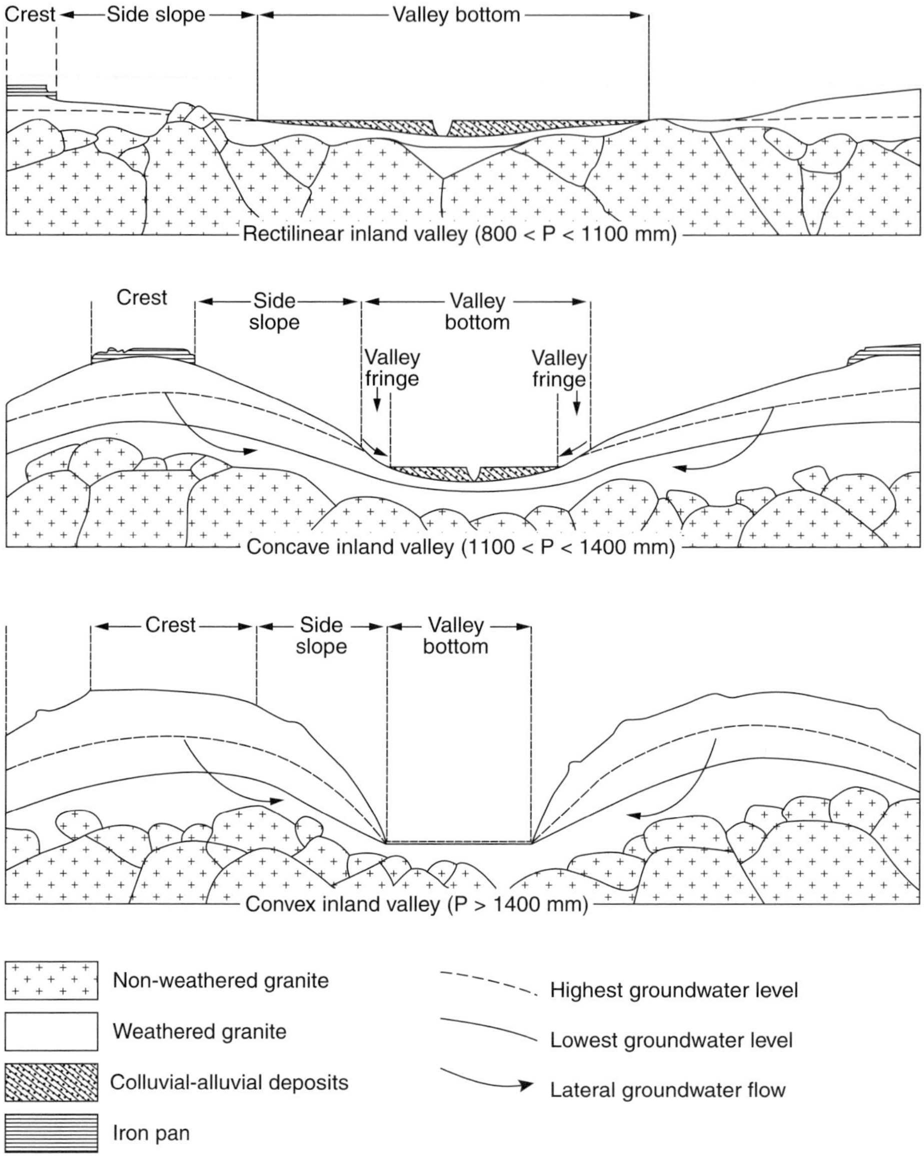

Fig. 22.2. The three most common inland-valley morphology types in West Africa, developed on granite–gneiss complexes under different rainfall regimes. (From Windmeijer and Andriesse, 1993, with permission from Alterra (Wageningen UR); adapted from Raunet, 1985.)

climate change, provide new incentives for inland-valley development. Valley bottoms and hydromorphic fringes generally have higher water availability and soil-fertility levels compared to the degraded upland soils (e.g. Andriesse *et al.*, 1994; van der Heyden and New, 2003), even though soil fertility can still be sub-optimal. A secure harvest from a wetland-produced crop becomes of invaluable importance in dry and unreliable agricultural environments (e.g. Scoones, 1991; Sakané *et al.*, 2011). Therefore, inland valleys are expected to become increasingly important in terms of food security in sub-Saharan Africa. However, climate change also poses a hydrological and therefore ecological threat to inland valleys. Ecological functioning

(a) (b) (c)

Fig. 22.3. Three inland-valley types: rectilinear inland valley Mwanza, north-east Tanzania (a), concave inland valley near Banfora, south-west Burkina Faso (b) and convex inland valley near Umuahia, south-east Nigeria (c). (Photos: J. Rodenburg.)

of inland valleys is sensitive to changes in water supply, and preventing degradation of these ecosystems requires adaptive management strategies (Erwin, 2009).

Inland-valley development also has an economic driver. About 10 million tonnes (Mt) of milled rice, approximately 40% of the annual regional consumption, is imported into Africa (mainly from Asian countries) each year, worth about US$5 billion (Seck *et al.*, 2010, 2012). Regional production has, however, increased steeply since the early 2000s due to a declining availability of global rice stocks for export, and consequently an increase in regional farm-gate prices from an estimated average $285 per tonne in 1999 to $564/t in 2009 (based on available data from 20 rice-producing countries in sub-Saharan Africa; FAO, 2010). These significant price changes have encouraged many small-scale farmers to take up rice production. Increased inland-valley use for rice (and vegetable) production can, for instance, be observed around large urban centres as a result of increasing population density and the attraction of the urban market (e.g. Erenstein, 2006; Erenstein *et al.*, 2006; Sakurai, 2006).

Inland valley development: water management

Developing inland valleys for effective rice-based production systems (i.e. cropping systems with rice as the dominant crop in association with other staple crops or vegetables) primarily means establishing water management structures – for example, to control flooding, optimize irrigation and conserve water for late-season use. As inland valleys in Africa are socio-economically and biophysically diverse and complex (Sakané *et al.*, 2011), the development of these landscapes for crop production requires a flexible and careful approach (e.g. Andriesse *et al.*, 1994). Every inland valley is unique in terms of its biophysical characteristics and there is no 'off-the-shelf' technology with a broad and indiscriminate application range. Therefore, technologies for inland valleys need to be locally adapted to be effective. Local morphology, hydrology and climate, for instance, will determine the depth, duration and frequency of flooding of the valley bottom, which in turn will determine the suitability of the valley bottom for rice-based production.

These physical conditions need to be considered in the design of the most effective water management system. There are five main systems: (i) the traditional random-basin system; (ii) the central-drain system; (iii) the interceptor-canal system; (iv) the head-bund system; and (v) the contour-bund system (Oosterbaan *et al.*, 1987). In the traditional random-basin system, the inland valley (the lower parts of the slope and the valley bottom) is partitioned into rectangular plots by small bunds. Farmers regulate the water level within these plots by opening the bunds. In the central-drain system, the valley-bottom drainage is improved by a central drainage canal. The remainder of the inland valley can still be divided by small bunds as in the traditional system. The interceptor-canal system has two interception canals along the valley fringes parallel to the central stream. Water from the central stream is carried to these interceptor canals by contour drains. Rice fields are flooded from the interceptor canals and these canals protect the rice fields from uncontrollable floods or runoff from the uplands and can ensure irrigation during short droughts. The head-bund system comprises head bunds that are built across the stream to create small reservoirs or ponds that can provide the rice fields with water through contour canals. In the contour-bund system, the valley bottom is divided by contour bunds across the stream. Within the space between two bunds (the rice field) the land can be levelled. Water from the stream is blocked by the contour bund and distributed over the field. Each contour bund has an outlet or spillway to enable water to flow from one field to another. To improve drainage of the lower fields, interceptor canals can be dug along the valley slopes (Windmeijer and Andriesse, 1993; Windmeijer *et al.*, 2002). The *sawah* systems in Madagascar represent a perfected example of a contour-bund system (Fig. 22.4 a and b).

Besides biophysical conditions, socio-economic and institutional factors should also be considered when inland valleys are to be developed. Like the biophysical environment, the socio-economic context in inland valleys is also diverse and often complex. Land tenure arrangements, for instance, vary between locations (e.g. Fu *et al.*, 2010), and within and between countries, ranging from ownership by families or individuals through single villages to whole states; state-owned tenure may discourage investment. Farmers working in the inland valleys do not often possess the required rights over the land and therefore do not always benefit from inland-valley development investments. Land-tenure arrangements also often have gender implications. Land is mostly owned by men but cultivated by women, particularly when the value of the land is low (e.g. low soil fertility, rainfed). After development, when the value of the land is raised, men may claim their rights again. Such social realities should be considered when inland-valley development projects are designed with the aim of benefiting the poor and empowering disadvantaged groups like women.

The Pegnasso inland valley, near Sikasso (south-east Mali), was developed in 1994 by Agence Française de Développement (AFD). The project deliberately opted for a partial development rather than a complete development to avoid land redistribution among farmers. The logic was that large investments would increase the land value and cause conflicts between men and women. Women who had used the land prior to the development would then risk being denied access to it, and would not benefit from the project. The partial development proposed and implemented consisted of the construction of a modest water-retention structure (Fig. 22.5a) and one central water inlet and irrigation canal to redirect water to the neighbouring farmers' plots. The plots were surrounded by simple bunds for within-plot water management. These modest improvements enabled farmers to make better use of the available water for a prolonged period of time and thereby increased rice productivity. Since rice production is mainly the responsibility of women, the project succeeded in its twin goals of benefiting the community while strengthening the position of women.

Fig. 22.4. Perfect examples of the contour-bund system: the *sawah* systems of Madagascar. (Photos: J. Rodenburg.)

Fig. 22.5. Water-retention structure of the Pegnasso inland valley, southern Mali (a) and laterite-rock bunds with inlet at the Blétou inland valley, southern Burkina Faso (b). (Photos: J. Rodenburg.)

In the Blétou valley, near Blédougou (southwest Burkina Faso), rice production was limited by the lack of water retention. Water was only available during and shortly after a rainfall event, and quickly drained to lower parts of the larger catchment area. Droughts were the main production constraint and yields were erratic and low (less than 1 t/ha). A project funded by the Common Fund for Commodities (CFC) was implemented between 2006 and 2009 with the aim of improving the water availability in the part of the inland valley with the highest production potential. A contour-bund system consisting of small water-retention bunds along the contour lines, covered with impermeable cloth and laterite rocks, was installed (Fig. 22.5b). A community-participatory development approach was used, whereby plots would be distributed among farmers according to their participation in the construction of these water-management structures. The active participation of farmers in the construction reduced the costs of the investment and, more importantly, provided a basis for ownership by the community. The farmers participating in the bund construction (e.g. collection of laterite rocks) were predominantly female (104 women out of a total of 121 farmers) and they were indeed rewarded when the plots were partitioned upon completion of the inland-valley development. The plots were assigned in a participatory manner with group consensus, respecting individual time investments and disregarding gender or age.

Many water-management infrastructures built in the 1970s have been abandoned. Such failures are thought to have resulted from the lack of local community participation during the selection, design and planning of the developments (e.g. Dries, 1991; Maconachie, 2008), or because traditional local land-tenure arrangements were overlooked (Brautigam, 1992). In the 1960s and 1970s, many irrigation-scheme developments in West Africa were funded by public investment corporations and development projects. The Benin–China Cooperation, for instance, developed a total of 1400 ha of inland valleys and flood plains into medium-sized (25–150 ha) irrigation schemes by equipping them with water-retention structures, irrigation and drainage canals, inlets and outlets, cofferdams and small bridges. Most of these irrigation schemes are under-utilized or abandoned. Examples of abandoned schemes can be found in Bamè and Zommon in Zagnanado (Ouémé valley, Benin), where broken and silted water management structures are common (Fig. 22.6 a, b and c). One exception is the nearby irrigation scheme of Koussin-Lélé, where farmers grow rice on 106 ha of developed land using gravity irrigation. A comparison between this scheme and the schemes of Bamè and Zonmon (33 ha and 84 ha, respectively) showed that careful selection of the valley and local stakeholder participation in planning, design, implementation and use of the developments are prerequisites for successful development efforts (Djagba *et al.*, 2013).

Production constraints

Estimated actual rice yields in inland valleys across Africa (1.4 t/ha according to Rodenburg and Demont, 2009) are much lower than the attainable yield in these production systems (>6 t/ha). The top ten production constraints in inland valleys, based on a survey among rainfed lowland rice experts in eight countries in West Africa, include biophysical (e.g. weed competition, lack of water control, poor soil

Fig. 22.6. Broken and silted water management infrastructures at Zommon, Benin. (Photos: J. Rodenburg (a and b); P. Kiepe (c).)

fertility) and socio-economic and institutional factors (e.g. land tenure and lack of inputs, labour and credit), and human health problems (Thiombiano *et al.*, 1996; Fig. 22.7). Overarching biophysical production constraints in the ecosystem (from hydromorphic fringes to intensified lowlands) are weeds, pests and diseases, nitrogen and phosphorus deficiencies, and iron toxicity (Table 22.1).

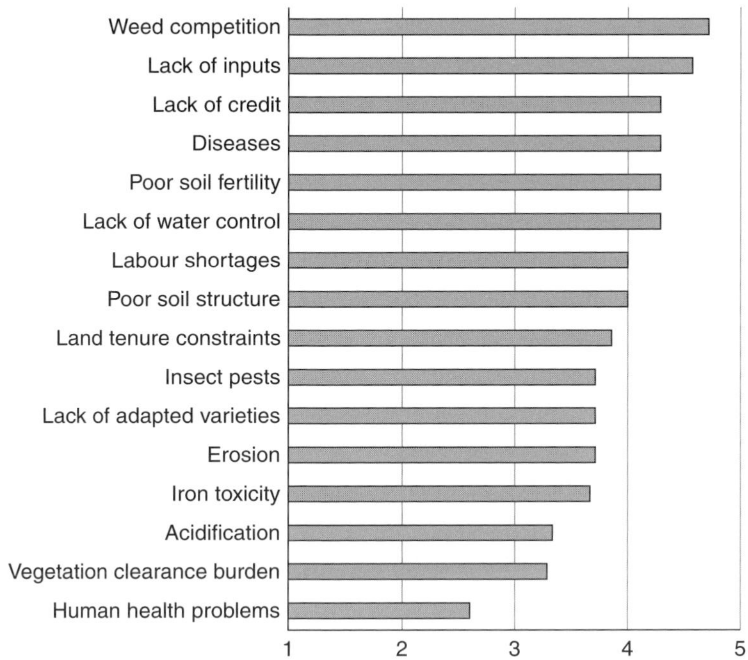

Fig. 22.7. Ranking of constraints[a] to the use of inland valleys for rice production as perceived by rainfed-lowland rice experts from eight countries in West Africa in 1993. (Adapted from Thiombiano *et al.*, 1996.)
[a]Rating of constraints on a scale of 1 (not important) to 5 (very important).

Table 22.1. Rice-growing ecosystem characterization (water supply, agroecological zone, and main biophysical production constraints). (Data from Andriesse *et al.*, 1994; Thiombiano *et al.*, 1996; Kiepe, 2006; Wopereis *et al.*, 2007.)

Ecosystem	Upland	Hydromorphic fringes	Lowland	Intensified lowland	Irrigated lowland
Main water supply	Rainfall	Rainfall + water table	Rainfall + water table + unregulated floods	Regulated floods	Full irrigation
Agroecological zones	Guinea savannah to humid forest	Guinea savannah to humid forest	Sudan savannah to humid forest	Sudan savannah to humid forest	Sahel to humid forest
Main biophysical production constraints	Drought, weeds, pests and diseases, P and N deficiencies, soil erosion, soil acidity	Drought, weeds, pests and diseases, P and N deficiencies, soil erosion, soil acidity, iron toxicity	Drought/flooding, weeds, pests and diseases, P and N deficiencies, iron toxicity	Drought/flooding, weeds, pests and diseases, P and N deficiencies, iron toxicity	Weeds, pests and diseases, salinity/ alkalinity, P and N deficiencies, iron toxicity

Weed competition is a major constraint. Dominant weeds in inland-valley rice are grasses like *Echinochloa colona*, *E. crus-pavonis*, *Oryza longistaminata*, *O. barthii*, *Ischaemum rugosum* and *Leersia hexandra* and sedges such as *Fimbristylis littoralis*, *Bolboschoenus maritimus*, *Kyllinga pumila*, *Cyperus difformis* and *C. iria*. Frequently encountered broad-leaved weeds in inland valleys are *Sphenoclea zeylanica*, *Ludwigia abyssinica*, *Heteranthera callifolia* and *Ipomoea aquatica* (Rodenburg and Johnson, 2009; Rodenburg and Johnson, Chapter 16, this volume). Another emerging problem in inland valleys across Africa, particularly the ones with poor water control, is the parasitic weed *Rhamphicarpa fistulosa* (Rodenburg *et al.*, 2010). This facultative parasitic weed, from the same family as the better known *Striga* spp., can grow independently like any of the other weeds mentioned, but in the vicinity of a suitable host plant like rice, it can develop into a parasite by establishing underground root-to-root connections, extracting host-derived carbohydrates and nutrients (Ouédraogo *et al.*, 1999). Parasitic infection causes crop yield losses in infested fields in excess of 60%, and fields are sometimes abandoned by farmers because of high infestations (Rodenburg *et al.*, 2011).

Other important biotic production constraints to rice in inland valleys are insect pests such as African rice gall midge, stem borers and rice bugs (see Nwilene *et al.*, Chapter 18, this volume), and diseases such as *Rice yellow mottle virus*, leaf blast, bacterial leaf blight and brownspot (see Séré *et al.*, Chapter 17, this volume); rats and birds can also cause significant yield losses (Balasubramanian *et al.*, 2007). African rice gall midge is common in West and East Africa. It damages rice tillers, and each 1% of damage is estimated to result in 2% of yield loss – final yield losses can reach 65% (Nacro *et al.*, 1996). *Rice yellow mottle virus*, endemic to Africa, is transmitted by beetles (order Coleoptera, family Chrysomelidae, e.g. *Chaetocnema pulla*, *Sesselia pusilla*, *Trichispa sericea* and *Dicladispa viridicyanea*) and can lead to total yield losses (ranging from 5% to 100% depending on agro-climatic zone) in rainfed lowland rice in Africa (Kouassi *et al.*, 2005).

Soil fertility in inland valleys is often far from optimal for sustainable and profitable crop production. While soil fertility varies across agroecological zones (Issaka *et al.*, 1997), soil sampling in inland valleys across West Africa revealed low to very low levels of nitrogen, available phosphorus, pH, CEC and total carbon (Issaka *et al.*, 1996), deficiencies in micronutrients like sulfur and zinc (Buri *et al.*, 2000; Abe *et al.*, 2010) and poor clay mineralogy (Abe *et al.*, 2006). A problem commonly associated with low soil fertility in inland valleys is iron toxicity (Becker and Asch, 2005; Audebert and Fofana, 2009), a complex nutrient disorder caused by excessive iron in the soil solution under specific but typical waterlogged conditions of inland valleys (Narteh and Sahrawat, 1999). A plant growing under such conditions takes up more soluble iron (Fe^{2+}) than it needs, resulting in iron accumulation in the leaves beyond the critical level, shown by reddish-brown or yellow coloration (leaf 'bronzing') and high leaf mortality, which in turn negatively impacts crop yield (e.g. Becker and Asch, 2005). Direct and indirect effects of iron toxicity can lead to 40–45% rice yield reductions in lowlands, depending on the extent of the problem, water, soil and crop management (e.g. cultivar choice), and the availability of other soil nutrients (Audebert and Fofana, 2009).

Ecosystem functions of inland valleys

Inland valleys deliver a range of associated ecosystem functions (Adams, 1993). Inland valleys are important for local flood and erosion control, water storage, nutrient retention, stabilization of the micro-climate, as well as for recreation and tourism and for retrieving water, clay and sand for crafts and construction. While the main crop is often rice, inland valleys and their fringes are used to grow a variety of other crops (e.g. maize, vegetables, fruit trees), and are also often used for cattle grazing (Fig. 22.8) – particularly during the dry season when the water table recedes below the soil surface of the valley bottoms, but with sufficient residual moisture to support crop growth. Furthermore, these environments provide important forest, wildlife and fisheries resources, and contribute to biological diversity as well as local cultural heritage (Dugan, 1990; Adams, 1993). The water resources available in inland valleys are often used by rural communities to fulfil a variety of daily household needs (Fig. 22.9 a and b). Besides the water resources, biological diversity

Fig. 22.8. Multiple uses of inland valleys: rice production is often combined with fruit trees (background) and small-scale vegetable and maize production and free-browsing cattle (foreground). Buganda, near Mwanza, north-west Tanzania. (Photo: J. Rodenburg.)

Fig. 22.9. Using local inland-valley water resources for daily household needs in Benin (a) and Burkina Faso (b). (Photos: J. Rodenburg.)

of inland valleys is probably among the most important functions for the local communities: inland valleys are important locations for the collection of non-agricultural plant resources, and local communities have considerable knowledge of the useful plant species, their use, abundance and collection places (Rodenburg *et al.*, 2012).

Because of their multifunctional character, inland valleys are attractive for exploitation and therefore vulnerable to degradation. The economic opportunities of inland valleys have been widely recognized and investments have been made to make these areas more accessible and profitable. Indiscriminate development of these vulnerable environments, however, will lead to degradation of the natural resources they harbour, and thereby jeopardize their unique and diverse ecosystem functions (e.g. Dixon and Wood, 2003). For a long time, the important functions of wetlands such as inland valleys for local communities have often been ignored in policy planning (Silvius *et al.*, 2000). Understanding the use and management of ecosystem functions by local communities is the first necessary step in generating recommendations for their sustainable use (Rodenburg *et al.*, 2012).

Different ecosystem services do not always conflict. For instance, while local community members in Togo and Benin perceive that agriculture is an important cause of a decline in plant biodiversity, agricultural fields were also considered as one of the most important locations for finding useful non-cultivated plants (Rodenburg *et al.*, 2012). Farmers recognize the useful weed species during weeding and leave them untouched or keep them apart after uprooting (see references in Rodenburg and Johnson, 2009). Useful species (predominantly trees) are also often maintained during field clearing (e.g. Leach, 1991; Madge, 1995; Kristensen and Lykke, 2003). In fact, this is a common strategy to cope with declining forests

(Shepherd, 1992). Other strategies, observed by Rodenburg *et al.* (2012) around inland valleys in Togo and Benin, include the establishment of a community garden with (about 300) useful species and the conservation of a small community forest. These observations show that local communities that depend on natural resources in and around inland valleys are able to exploit these landscapes synergistically, balancing agricultural production with biodiversity conservation, use and management. However, aligning the multiple interests in inland-valley resources requires participation of local communities in any development or conservation initiative, as they depend entirely on these natural resources. As primary stakeholders, their knowledge and needs should be taken into account when the objective is to achieve sustainable land use in inland valleys.

Turning Inland Valleys into Africa's Food Baskets

The Inland Valley Community of practice

The Consortium for the Sustainable Use of Inland Valley Agro-Ecosystems in Sub-Saharan Africa (Inland Valley Consortium, IVC), convened by the Africa Rice Center (AfricaRice) and composed of ten West African national agricultural research systems (NARS) and a number of international (IITA, ILRI, IWMI, FAO and CORAF/WECARD[2]) and advanced research institutes (CIRAD,[3] Wageningen University), was founded in 1993 with the objective to develop, in a concerted and coordinated manner, technologies and operational support systems for intensified but sustainable use of inland valleys in sub-Saharan Africa. IVC uses a multidisciplinary scientific approach aiming at: (i) determining the agroecosystem potential of inland valleys, based on an integrated characterization and classification; (ii) identifying means to achieve this potential by targeting research activities, by developing technological innovations and by transferring them to the plots, landscapes and watersheds; and (iii) capitalizing on available resources such as inland-valley ecosystems, inland-valley developments and local knowledge and innovations. IVC projects since the early 2000s have focused on rehabilitation of abandoned or sub-optimal functioning inland-valley systems; participatory valuation of inland-valley ecosystem goods and services; sustainable productivity improvement for rice, targeting water and weed management; exploring possibilities for the integration of rice–fish and rice–vegetable production; and enhancing the productivity and competitiveness of inland-valley lowlands through sustainable intensification and diversification and product value chain development, while conserving land and water resources. In 2011, IVC became the Inland Valley Community of practice, which better reflects its new *modus operandi*.

A strategy for sustainable inland valley selection, development and use

Based on almost 20 years of experience under the umbrella of IVC, complemented by insights gained from other initiatives, a strategy for inland-valley selection, development and use can be outlined. While regional food security is an important goal, and inland-valley development could be an effective way to achieve this, the selection, development and exploitation of these environments should be approached with care. Not all inland valleys are necessarily suitable for crop production (e.g. Kotze, 2011; Sakané *et al.*, 2011). If only 9.1% of all inland valleys in Africa were set aside for rice production (i.e. 17.30 Mha) and average rice productivity could be raised by 1 t/ha (from 1.4 to 2.4 t/ha, which is feasible according to Becker and Johnson, 2001), this would produce 41.5 Mt of paddy, about the same as the current sum of total rice production and imports in Africa.[4] Hence, only a fraction of the total inland-valley area should suffice to produce enough rice to feed the entire continent and the remainder should be safeguarded for other purposes such as pastoralism, biodiversity and wildlife sanctuaries, and natural (excess) water buffers. This strategy would, however, require systematic approaches and methodologies for: (i) selecting the 'best-bet' (most suitable and low-risk) inland valleys for agricultural development to avoid investment failures or unnecessary destruction of wetlands; (ii) land-use planning within the inland valley;

(iii) designing and implementing the 'best-fit' water management development infrastructure; and (iv) optimizing crop management practices for increased crop productivity.

Inland valley selection and land-use planning

Potential inland valleys (for agricultural development) can be identified with GIS (geographic information system) and remote-sensing tools (e.g. Thenkabail and Nolte, 1996; Gumma *et al.*, 2009; Chabi *et al.*, 2010). The necessary on-the-ground evaluation and confirmation ('ground truthing') can be done at the semi-detailed level using the Integrated Transect Method (ITM) proposed by van Duivenbooden *et al.* (1996). Based on geo-morphological, hydrological, soil and land-use characteristics, different inland-valley types can be distinguished, which will be helpful in identifying their (potential) 'best-bet' use (e.g. Andriesse *et al.*, 1994; Sakané *et al.*, 2011). Alongside these biophysical and agronomic assessments, socio-economic variables, such as market access and population densities, are important for feasibility studies to assess which valleys can be developed for agriculture (e.g. Narteh *et al.*, 2007). The 'best-bet' inland valleys for rice-based production systems should score high on (agricultural) production and marketing potential and low on environmental and social risks, and preferably also low on other ecosystem functions such as biodiversity. Hence, a thorough assessment of the valley's economic value for local communities, including direct, indirect and non-use benefits is required (Scoones, 1991). In addition, local stakeholders should play a role in the land-use planning within a selected inland valley. So-called multi-stakeholder platforms (MSP) could be created to reach consensus on economic, social and environmental gains and risks, and to identify hot spots for specific ecosystem services within the inland valley under consideration.

Inland valley development

After selection and land-use planning, the actual development can start, consisting mainly of clearing of vegetation and establishment of water management structures to increase water control. As mentioned above, various physical factors, such as the size of the catchment area, the valley morphology and soil texture (determining hydrological behaviour), need to be considered for the design of the most suitable water management system. This requires a thorough diagnostic study on the spatial and dynamic water movements within the valley (Wopereis *et al.*, 2007) and in-depth discussions with the farming community to ensure full buy-in of the community and that land tenure issues are identified and solved beforehand. Discussions on the land- and water-development options that could be put in place should also consider consequences in terms of water availability for downstream users.

To aid in deciding what water control infrastructure to develop, the diagnostic interactive tool called DIARPA (*diagnostic rapide de pré-aménagement*) may be used (Lidon *et al.*, 1998). This rapid hydraulic diagnosis method works as a decision tree and helps to assess the 'best-bet' type of intervention at a given location and a given level of investment, to optimize agricultural production with limited hydraulic risks. Explanatory indicators for water dynamics are categorized under topographic, pedological and hydrological indicators and have been selected, characterized and validated for use as an index for the selection of the best interventions (Table 22.2).

The proper functioning, management and maintenance of the water control requires the actual users (farmers) to understand the basic principles. They should be involved in the development of water control infrastructure as much as possible to enable them to acquire ownership. As in the case of the Blétou valley (south-west Burkina Faso) (see 'Inland valley development: water management' above), information on the extent of participation in the development work provided by an individual stakeholder can be used to guide plot partitioning once the inland-valley development is finalized, and the personal time investments at these stages will also ensure user commitment to future management and maintenance of the development structures, and thereby benefit the sustainability of the inland-valley production system. After completion of the water management structures, a performance assessment, similar to the one suggested by Dembele *et al.* (2011), should be carried out on a seasonal basis to enable farmers to make necessary adjustments and thereby further improve water productivity.

Table 22.2. DIARPA indicators and their threshold values for each type of intervention. (Adapted from Legoupil *et al.*, 1998.)

	Type of intervention					
Indicator	Contour bunding	Contour bunding with spill over	Water-retention dykes without seepage barrier	Water-retention dykes with seepage barrier	Diversion barrier to diffuse flow	Diversion barrier for the re-infiltration and restocking of the water table
Pedological						
Permeability (m/s)	$<10^{-4}$	$<10^{-4}$	$<10^{-4}$	$<10^{-4}$	$<10^{-4}$	$<10^{-4}$
Impermeable layer depth	NI	NI	NI	<2 m	NI	NI
Topographic						
Average longitudinal slope of the valley	<1%	<1%	<0. 5%	<0. 5%	<1%	<1%
Flow axis	No flow axis	NI	Noticeable flow axis	Noticeable flow axis	Narrow flow axis	Narrow flow axis
Hydrological						
Peak flow by metre of valley-bottom width	3×10^{-3} m^3/s	25×10^{-3} m^3/s	0. 25–0. 6 m^3/s	0. 25–0. 6 m^3/s	50×10^{-3} m^3/s	50×10^{-3} m^3/s
Depth of the valley-bottom groundwater flow at the start of the dry season	NI	NI	NI	NI	NI	<2 m

NI, not important.

The first steps towards improved water management in inland valleys in Africa will entail the construction of main and secondary drainage pathways and identification, bunding and levelling of individual fields with minimal soil disturbance. Slightly sloping valleys will be divided into relatively large bunded fields, whereas valleys with steep slopes will be divided into smaller, 'terraced', bunded fields. Worou *et al.* (2013) provide guidelines for the development of such 'partial water control' structures that can be constructed entirely by the farming community. Farmers involved in the Japan-funded SAWAH-IV project implemented by AfricaRice and partners in Benin and Togo obtain very good results in inland valley settings, introducing such relatively simple, low-cost water management structures (drainage canal development, bunding, levelling) that can be constructed and maintained entirely by farmers themselves. Use of power tillers is not essential at the first stage for land development and rice cultivation, but their introduction can substantially speed up land development once farmers are familiar with the technique. Similar observations were reported in Ghana by JICA staff (K. Saito, Cotonou, Benin, 2013, personal communication).

The introduction of such very simple water management structures will already lead to substantial yield gains (1–2 t/ha), especially if accompanied with good crop management practices (see below).

Locally adjusted production strategies

Local constraints need to be tackled in order to benefit from the inherent production potential of the inland valley.

Relevant modules of the *Curriculum for Participatory Learning and Action Research (PLAR) for Integrated Rice Management (IRM)* can be used to raise rice productivity. PLAR-IRM was developed by AfricaRice, based on the insight that a locally adapted and integrated approach is required to increase rice productivity in inland-valley production systems in Africa (Wopereis and Defoer, 2007). It is essentially a farmer-participatory, step-wise approach to put inland valleys under rice production using good agricultural practices (Defoer *et al.*, 2004; Wopereis *et al.*, 2007; see also Defoer and Wopereis, Chapter 31, this volume).

Following improved water management, key factors for raising productivity in inland valleys are weed and soil-fertility management (Wopereis and Defoer, 2007) and to a lesser extent pest and disease control (Table 22.1). However, the order of importance of production constraints needs to be assessed locally for each inland valley. Data collected during the detailed characterizations should be helpful in this respect (e.g. Sakané *et al.*, 2011). The PLAR-IRM curriculum also provides a useful method to identify key production constraints, as well as locally researchable issues. PLAR-IRM further stimulates farmer experimentation in order to test 'what works best' under the given local (biophysical and socio-economic) conditions, using an integrated management approach. The available modules of the PLAR curriculum provide guidelines for such approaches.

Through integrated water, soil-fertility and weed management in inland valleys, rice yields can be increased considerably. Bunding, puddling (if possible) and levelling, for instance, facilitate water management and decrease weed competition – as many weed species are not well adapted to permanently flooded conditions (e.g. Kent and Johnson, 2001) – and generally increase nutrient-use efficiencies, especially in well-drained fields. These relatively simple technologies have been shown to increase rice yields by 40%, and reduce weed infestation by 25% across agroecological zones (Becker and Johnson, 2001; Toure *et al.*, 2009). Yield improvements can then be achieved by using improved rice cultivars. For instance, some NERICA cultivars (New Rice for Africa) adapted to lowland conditions have an inherent high weed competitiveness (Rodenburg *et al.*, 2009) and high yield potential (Sié *et al.*, 2008).

Concluding Remarks

Inland valleys are the future food baskets of Africa and play a pivotal role in realizing the region's rice promise. Based on nearly 20 years of experience with work carried out by IVC, a locally adapted, step-wise and bottom-up approach for site selection, land-use planning, water-management design and implementation, and crop management is proposed for the sustainable exploitation of the inland-valley potential. This approach includes: (i) the selection of 'best-bet' inland valleys using multi-scale characterizations based on climate and geomorphological data and using GIS and remote-sensing tools followed by semi-detailed and detailed typologies using ITM and socio-economic assessments; (ii) a stakeholder-participatory land-use planning within the inland valley, based on the earlier characterizations and using MSPs; (iii) participatory inland-valley development (e.g. clearing, levelling and construction of water management structures) following relevant modules of the PLAR curriculum and the pre-development diagnostic tool DIARPA, followed by regular performance assessments of the water-control system; and (iv) improving rice productivity and resource-use efficiency through farmer-participatory identification of local production constraints and adapting management practices following the principles of IRM using PLAR.

It is essential to use systematic analysis approaches for the selection of 'best-bet' inland valleys for rice-based production systems, as only a fraction of the available inland valleys in Africa would need to be put under production in order to achieve regional self-sufficiency in rice. The remaining inland valleys could then be safeguarded to fulfil other ecosystem services. However, for this strategy to be effective, conservation regulations and monitoring and evaluation mechanisms need to be established to help protect those inland valleys that are either too vulnerable (to, e.g., soil and water degradation or social conflicts) or too valuable (because of other ecosystem functions such as biodiversity) to be subjected to agricultural development. Selection of 'best-bet' production valleys should be based on both biophysical and socio-economic criteria and be broadly supported by the local communities depending on them. The same approach is proposed for the identification of locations within the inland valley that should be used for crop production and those that should continue to fulfil other ecosystem functions. This again requires involvement of local stakeholders.

Following these steps, the next challenge is the actual development. The right choice of water-management system is of vital importance and depends largely on the valley morphology and the local soil and hydrological characteristics. Development and implementation of such water management systems and the agricultural production practices following such development should not negatively impact the water quality and availability downstream. For the actual crop production, high-yielding and stress-resistant lowland rice cultivars and locally adapted and integrated crop management practices are required. Postharvest facilities for drying, threshing, milling, storage and transport should also be included in inland-valley development plans.

Full local stakeholder participation in all stages, from decision making to development and implementation, should result in consensus on the selection and land-use plans of inland valleys, the implementation of broadly supported interventions, and flexible, locally adaptable and acceptable solutions to the numerous technical, socio-economic and institutional constraints encountered by resource-poor farmers working in the inland-valley systems of Africa. This should create a solid basis for the required sustainable use of inland valleys and for turning these valuable resources into Africa's food baskets.

Acknowledgement

The author gratefully acknowledges permission granted by Alterra (Wageningen UR) to reproduce copyrighted material from Windmeijer and Andriesse (1993) as Fig. 22.2 in this chapter.

Notes

[1] Based on FAO and national databases, in particular FAO TERRASTAT (2003).
[2] IITA, International Institute of Tropical Agriculture; ILRI, International Livestock Research Institute; IWMI, International Water Management Institute; FAO, Food and Agriculture Organization of the United Nations; CORAF/WECARD, West and Central African Council for Agricultural Research and Development.
[3] CIRAD, Centre de coopération internationale en recherche agronomique pour le développement.
[4] 10 Mt of milled rice was imported into Africa in 2010, this translates in to 16.7 Mt of paddy (using a conservative paddy–milled rice conversion rate of 60%); in the same year an estimated 24.7 Mt of paddy was produced in Africa (FAO, 2012), the sum is 41.4 Mt of paddy.

References

Abe, S.S., Masunaga, T., Yamamoto, S., Honna, T. and Wakatsuki, T. (2006) Comprehensive assessment of the clay mineralogical composition of lowland soils in West Africa. *Soil Science and Plant Nutrition* 52, 479–488.

Abe, S.S., Buri, M.M., Issaka, R.N., Kiepe, P. and Wakatsuki, T. (2010) Soil fertility potential for rice production in West African lowlands. *JARQ – Japan Agricultural Research Quarterly* 44, 343–355.

Acres, B.D., Rains, A.B., King, R.B., Lawton, R.M., Mitchell, A.J.B. and Rackham, L.J. (1985) African *dambos*: their distribution, characteristics and use. *Zeitschrift für Geomorphologie* Supplementbände 52, 63–86.

Adams, W.M. (1993) Indigenous use of wetlands and sustainable development in West-Africa. *The Geographical Journal* 159, 209–218.

Andriesse, W. and Fresco, L.O. (1991) A characterization of rice-growing environments in West Africa. *Agriculture, Ecosystems and Environment* 33, 377–395.

Andriesse, W., Fresco, L.O., Duivenbooden, N. and Windmeijer, P.N. (1994) Multi-scale characterization of inland valley agro-ecosystems in West Africa. *Netherlands Journal of Agricultural Science* 42, 159–179.

Audebert, A. and Fofana, M. (2009) Rice yield gap due to iron toxicity in West Africa. *Journal of Agronomy and Crop Science* 195, 66–76.

Balasubramanian, V., Sie, M., Hijmans, R.J. and Otsuka, K. (2007) Increasing rice production in sub-Saharan Africa: challenges and opportunities. *Advances in Agronomy* 94, 55–133.

Becker, M. and Asch, F. (2005) Iron toxicity in rice – conditions and management concepts. *Journal of Plant Nutrition and Soil Science* 168, 558–573.

Becker, M. and Johnson, D.E. (2001) Improved water control and crop management effects on lowland rice productivity in West Africa. *Nutrient Cycling in Agroecosystems* 59, 119–127.
Brautigam, D. (1992) Land rights and agricultural development in West Africa – a case study of 2 Chinese projects. *Journal of Developing Areas* 27, 21–32.
Buri, M.M., Masunaga, T. and Wakatsuki, T. (2000) Sulfur and zinc levels as limiting factors to rice production in West Africa lowlands. *Geoderma* 94, 23–42.
Chabi, A., Oloukoi, J., Mama, V.J. and Kiepe, P. (2010) Inventory by remote sensing of inland valleys agroecosystems in central Benin. *Cahiers Agricultures* 19, 446–453.
Chapman, L.J., Balirwa, J., Bugenyi, F.W.B., Chapman, C. and Crisman, T.L. (2001) Wetlands of East Africa: biodiversity, exploitation and policy perspectives. In: Gopal, B., Junk, W.J. and Davis, J.A. (eds) *Biodiversity in Wetlands: Assessment, function and conservation.* Blackhuys, Leiden, Netherlands, pp. 101–132.
Defoer, T., Wopereis, M.C.S., Idinoba, P.A., Kadisha, T.K.L., Diack, S. and Gaye, M. (2004) *Curriculum for Participatory Learning and Action Research (PLAR) for Integrated Rice Management (IRM) in Inland Valleys of Sub-Saharan Africa: Facilitators' manual.* Africa Rice Center (WARDA), Cotonou, Benin.
Dembele, Y., Yacouba, H., Keïta, A. and Sally, H. (2011) Assessment of irrigation system performance in south-western Burkina Faso. *Irrigation and Drainage* 61(3), 306–315.
Dixon, A.B. and Wood, A.P. (2003) Wetland cultivation and hydrological management in eastern Africa: matching community and hydrological needs through sustainable wetland use. *Natural Resources Forum* 27, 117–129.
Djagba, J.F., Rodenburg, J., Zwart, S.J., Houndagba, C.J. and Kiepe, P. (2013) Failure and success factors of irrigation system developments – a case study from the Ouémé and Zou valleys in Benin. *Irrigation and Drainage* (in press).
Dries, I. (1991) Development of wetlands in Sierra-Leone: farmers' rationality opposed to government policy. *Landscape and Urban Planning* 20, 223–229.
Dugan, P.J. (1990) *Wetland Conservation: A review of current issues and action.* IUCN, Gland, Switzerland.
Erenstein, O. (2006) Intensification or extensification? Factors affecting technology use in peri-urban lowlands along an agro-ecological gradient in West Africa. *Agricultural Systems* 90, 132–158.
Erenstein, O., Oswald, A. and Mahaman, M. (2006) Determinants of lowland use close to urban markets along an agro-ecological gradient in West Africa. *Agriculture, Ecosystems and Environment* 117, 205–217.
Erwin, K. (2009) Wetlands and global climate change: the role of wetland restoration in a changing world. *Wetlands Ecology and Management* 17, 71–84.
FAO (2010) FAOSTAT. http://faostat. fao. org/ (accessed 7 October 2010).
FAO (2012) *Rice Market Monitor*, vol. 15, issue 1. Food and Agriculture Organization of the United Nations, Rome, Italy, p. 33.
FAO TERRASTAT (2003) www.fao.org/ag/agl/agll/terrastat/ (accessed February 2003).
Fu, R.H.Y., Abe, S.S., Wakatsuki, T. and Maruyama, M. (2010) Traditional farmer-managed irrigation system in central Nigeria. *JARQ – Japan Agricultural Research Quarterly* 44, 53–60.
Gbakima, A.A. (1994) Inland valley swamp rice development – malaria, schistosomiasis, onchocerciasis in south central Sierra-Leone. *Public Health* 108, 149–157.
Gumma, M., Thenkabail, P.S., Fujii, H. and Namara, R. (2009) Spatial models for selecting the most suitable areas of rice cultivation in the inland valley wetlands of Ghana using remote sensing and geographic information systems. *Journal of Applied Remote Sensing* 3(1), 033537. doi:10.1117/1.3182847 (accessed 14 February 2013).
Issaka, R.N., Masunaga, T., Kosaki, T. and Wakatsuki, T. (1996) Soils of inland valleys of West Africa: general fertility parameters. *Soil Science and Plant Nutrition* 42, 71–80.
Issaka, R.N., Ishida, F., Kubota, D. and Wakatsuki, T. (1997) Geographical distribution of selected soil fertility parameters of inland valleys in West Africa. *Geoderma* 75, 99–116.
Kent, R.J. and Johnson, D.E. (2001) Influence of flood depth and duration on growth of lowland rice weeds, Côte d'Ivoire. *Crop Protection* 20, 691–694.
Kiepe, P. (2006) Characterization of three key environments for integrated irrigation–aquaculture and their local names. In: Halwart, M. and van Dam, A. A. (eds) *Integrated Irrigation and Aquaculture in West Africa: Concepts, practices and potential.* Food and Agriculture Organization of the United Nations, Rome, Italy, pp. 1–5.
Kotze, D.C. (2011) The application of a framework for assessing ecological condition and sustainability of use to three wetlands in Malawi. *Wetlands Ecology and Management* 19, 507–520.
Kouassi, N.K., N'Guessan, P., Albar, L., Fauquet, C.M. and Brugidou, C. (2005) Distribution and characterization of *Rice yellow mottle virus*: a threat to African farmers. *Plant Disease* 89, 124–133.

Kristensen, M.K. and Lykke, A.M. (2003) Informant-based valuation of use and conservation preferences of savanna trees in Burkina Faso. *Economic Botany* 57, 203–217.

Leach, M. (1991) Engendered environments: understanding natural resource management in the West African forest zone. *IDS Bulletin* 22, 17–24. Institute of Development Studies, Brighton, UK.

Legoupil, J.C., Ahmadi, N., Blanchet, F. and Lidon, B. (1998) *Etat de l'art : La mis en valeur agricole des zones de bas fonds en Afrique de l'Ouest – Caractérisation, fonctionnement et intensification agricole des zones de bas-fonds*. Centre de coopération internationale en recherche agronomique pour le développement, Paris, France.

Lidon, B., Legoupil, J.C., Blanchet, F., Simpara, M. and Sanogo, I. (1998) Le diagnostic rapide de pré-aménagement (DIARPA). *Agriculture et Développement* 20, 61–80.

Maconachie, R. (2008) New agricultural frontiers in post-conflict Sierra Leone? Exploring institutional challenges for wetland management in the Eastern Province. *Journal of Modern African Studies* 46, 235–266.

Madge, C. (1995) Ethnography and agroforestry research: a case study from The Gambia. *Agroforestry Systems* 32, 127–146.

McCartney, M.P. and Houghton-Carr, H.A. (2009) Working wetland potential: an index to guide the sustainable development of African wetlands. *Natural Resources Forum* 33, 99–110.

Nacro, S., Heinrichs, E.A. and Dakouo, D. (1996) Estimation of rice yield losses due to the African rice gall midge, *Orseolia oryzivora* Harris and Gagne. *International Journal of Pest Management* 42, 331–334.

Narteh, L.T. and Sahrawat, K.L. (1999) Influence of flooding on electrochemical and chemical properties of West African soils. *Geoderma* 87, 179–207.

Narteh, L.T., Moussa, M., Otoo, E., Andah, W.E.I. and Asubonteng, K.O. (2007) Evaluating inland valley agro-ecosystems in Ghana using a multi-scale characterization approach. *Ghana Journal of Agricultural Science* 40, 141–157.

Oladele, O.I., Bam, R.K., Buri, M.M. and Wakatsuki, T. (2010) Missing prerequisites for green revolution in Africa: lessons and challenges of *sawah* rice eco-technology development and dissemination in Nigeria and Ghana. *Journal of Food, Agriculture & Environment* 8, 1014–1018.

Oosterbaan, R.J., Gunneweg, H.A. and Huizing, A. (1987) Water control for rice cultivation in small valleys of West Africa. In: *ILRI Annual Report*. International Institute for Land Reclamation and Improvement, Wageningen, Netherlands, pp. 30–49.

Ouédraogo, O., Neumann, U., Raynal-Roques, A., Sallé, G., Tuquet, C. and Dembélé, B. (1999) New insights concerning the ecology and the biology of *Rhamphicarpa fistulosa* (*Scrophulariaceae*). *Weed Research* 39, 159–169.

Raunet, M. (1985) Lowland rice farming in Africa – approach by structural comparison [Bas-fonds et riziculture en Afrique : approche structurale comparative]. *Agronomie Tropicale* 40, 181–201.

Roberts, N. (1988) *Dambos* in development – management of a fragile ecological resource. *Journal of Biogeography* 15, 141–148.

Rodenburg, J. and Demont, M. (2009) Potential of herbicide resistant rice technologies for sub-Saharan Africa. *AgBioForum* 12, 313–325.

Rodenburg, J. and Johnson, D.E. (2009) Weed management in rice-based cropping systems in Africa. *Advances in Agronomy* 103, 149–218.

Rodenburg, J., Saito, K., Kakai, R.G., Toure, A., Mariko, M. and Kiepe, P. (2009) Weed competitiveness of the lowland rice varieties of NERICA in the southern Guinea savanna. *Field Crops Research* 114, 411–418.

Rodenburg, J., Riches, C.R. and Kayeke, J.M. (2010) Addressing current and future problems of parasitic weeds in rice. *Crop Protection* 29, 210–221.

Rodenburg, J., Zossou-Kouderin, N., Gbèhounou, G., Ahanchede, A., Touré, A., Kyalo, G. and Kiepe, P. (2011) *Rhamphicarpa fistulosa*, a parasitic weed threatening rain-fed lowland rice production in sub-Saharan Africa – a case study from Benin. *Crop Protection* 30, 1306–1314.

Rodenburg, J., Both, J., Heitkonig, I., van Koppen, K., Sinsin, B., Van Mele, P. and Kiepe, P. (2012) Land-use and biodiversity in unprotected landscapes: the case of non-cultivated plant use and management by rural communities in Benin and Togo. *Society & Natural Resources* 25(12), 1221–1240.

Sakané, N., Alvarez, M., Becker, M., Böhme, B., Handa, C., Kamiri, H.W., Langensiepen, M., Menz, G., Misana, S., Mogha, N.G., Möseler, B.M., Mwita, E.J., Oyieke, H.A. and van Wijk, M.T. (2011) Classification, characterisation, and use of small wetlands in East Africa. *Wetlands* 31, 1103–1116.

Sakurai, T. (2006) Intensification of rainfed lowland rice production in West Africa: present status and potential green revolution. *The Developing Economies* 44, 232–251.

Scoones, I. (1991) Wetlands in drylands: key resources for agricultural and pastoral production in Africa. *Ambio* 20, 366–371.

Seck, P.A., Tollens, E., Wopereis, M.C.S., Diagne, A. and Bamba, I. (2010) Rising trends and variability of rice prices: threats and opportunities for sub-Saharan Africa. *Food Policy* 35, 403–411.

Seck, P.A., Diagne, A., Mohanty, S. and Wopereis, M.C.S. (2012) Crops that feed the world 7: rice. *Food Security* 4, 7–24.

Shepherd, G. (1992) *Managing Africa's Tropical Dry Forests: A review of indigenous methods.* Overseas Development Institute, London.

Sié, M., Séré, Y., Sanyang, S., Narteh, L.T., Dogbe, S., Coulibaly, M.M., Sido, A., Cissé, F., Drammeh, E., Ogunbayo, S.A., Zadji, L., Ndri, B. and Toulou, B. (2008) Regional yield evaluation of the interspecific hybrids (*O. glaberrima* × *O. sativa*) and intraspecific (*O. sativa* × *O. sativa*) lowland rice. *Asian Journal of Plant Sciences* 7, 130–139.

Silvius, M.J., Oneka, M. and Verhagen, A. (2000) Wetlands: lifeline for people at the edge. *Physics and Chemistry of the Earth, Part B: Hydrology, Oceans and Atmosphere* 25, 645–652.

Thenkabail, P.S. and Nolte, C. (1996) Capabilities of Landsat-5 Thematic Mapper (TM) data in regional mapping and characterization of inland valley agroecosystems in West Africa. *International Journal of Remote Sensing* 17, 1505–1538.

Thiombiano, L., Jamin, J.-Y. and Windmeijer, P.N. (1996) *Inland Valley Development in West-Africa: Regional state of the art.* West Africa Rice Development Association, Bouaké, Côte d'Ivoire.

Touré, A., Becker, M., Johnson, D.E., Koné, B., Kossou, D.K. and Kiepe, P. (2009) Response of lowland rice to agronomic management under different hydrological regimes in an inland valley of Ivory Coast. *Field Crops Research* 114, 304–310.

Turner, R.K., van den Bergh, J.C.J.M., Soderqvist, T., Barendregt, A., van der Straaten, J., Maltby, E. and van Ierland, E.C. (2000) Ecological–economic analysis of wetlands: scientific integration for management and policy. *Ecological Economics* 35, 7–23.

van der Heyden, C.J. and New, M.G. (2003) The role of a *dambo* in the hydrology of a catchment and the river network downstream. *Hydrology and Earth Systems Science* 7, 339–357.

Van Duivenbooden, N., Windmeijer, P.N., Andriesse, W. and Fresco, L.O. (1996) The integrated transect method as a tool for land use characterisation, with special reference to inland valley agroecosystems in West Africa. *Landscape and Urban Planning* 34, 143–160.

Verhoeven, J.T.A. and Setter, T.L. (2010) Agricultural use of wetlands: opportunities and limitations. *Annals of Botany* 105, 155–163.

Whitlow, J.R. (1983) *Vlei* cultivation in Zimbabwe: reflections on the past. *Zimbabwe Agricultural Journal* 80, 123–135.

Windmeijer, P.N. and Andriesse, W. (1993) *Inland Valleys in West Africa: An agro-ecological characterization of rice growing environments.* International Institute for Land Reclamation and Improvement, Wageningen, Netherlands.

Windmeijer, P.N., Dugué, M.J., Jamin, J.-Y. and Van de Giesen, N. (Eds) (2002) *Describing Hydrological Characteristics for Inland Valley Development.* West Africa Rice Development Association, Bouaké, Côte d'Ivoire.

Wopereis, M.C.S., Defoer, T., Idinoba, M.E., Diack, S. and Dugué, M.J. (2007) *Participatory Learning and Action Research (PLAR) for Integrated Rice Management (IRM) in Inland Valleys of Sub-Saharan Africa: Technical manual.* Africa Rice Center (WARDA), Cotonou, Benin.

Worou, S.K. (2013) Approche simple et participative d'aménagement de bas-fonds. Africa Rice Center, Cotonou, Benin (in preparation).

Yapi, Y.G., Briet, O.J.T., Diabate, S., Vounatsou, P., Akodo, E., Tanner, M. and Teuscher, T. (2005) Rice irrigation and schistosomiasis in savannah and forest areas of Côte d'Ivoire. *Acta Tropica* 93, 201–211.

23 Consumer Preferences for Rice in Africa

Pieter Rutsaert,[1] Matty Demont[2]* and Wim Verbeke[1]
[1]*Department of Agricultural Economics, Ghent University, Ghent, Belgium;*
[2]*Africa Rice Center (AfricaRice), Saint-Louis, Senegal*

Introduction

Since the 1960s, African appetite for rice has increased at an average annual rate of +4%, i.e. twice as fast as in the world as a whole, to reach a total consumption level of 20 million tonnes (Mt) in 2009 (Table 23.1). Growth of rice consumption since the 1960s has been fastest in East Africa (e.g. Burundi, Ethiopia, Kenya, Rwanda and Zimbabwe), but since 2000, high growth rates are also recorded in North Africa (e.g. Algeria), West Africa (e.g. Benin) and Central Africa (e.g. Cameroon). This long-term structural trend is mainly linked to urbanization, which affects rice consumption in two ways. First, it provides women with increased opportunities to work outside the home, which increases the opportunity cost of their time for food shopping and cooking. This leads to a shift in consumer preferences in favour of a 'fast food' that can be easily cooked and needs less preparation time. Rice is a staple food that meets such needs (Reardon, 1993; Diagana *et al.*, 1999). Second, urbanization entails **urban bias**, i.e. the inefficient and systemic bias against agriculture and the rural economy in the allocation of developmental resources (Lipton, 1977; Bezemer and Headey, 2008). From a political perspective, urban dwellers are the most important voters in post-colonial Africa. Therefore, agricultural policies have mainly favoured cheap imports of food, rather than local food production. The result is that African policy makers now face the effects of the historical urban bias in rice markets. Specifically, urban consumers in Africa have developed a marked preference for imported rice and associated purchasing and eating habits (Demont *et al.*, 2013b).

In the early 1980s, agricultural economists warned that 'the problem of consumer tastes and responsiveness to the locally grown varieties which would replace rice imports' was largely ignored in the debate on the policy and economics of rice in West Africa (Pearson *et al.*, 1981; Ross, 1983). The same criticism is raised in rice breeding, which has 'not incorporated desirable consumption attributes and non-yield production traits into new varieties' (Dalton, 2004, p. 149). More recently, USAID reiterated this call for demand-focused research in its West Africa rice value-chain report, arguing that 'support for research into consumer demand and preferences for local rice' should be prioritized (USAID, 2009a, p. iii).

This chapter discusses the situation of rice in Africa from a consumer-behaviour perspective. Consumer behaviour is defined as 'those actions directly involved in obtaining, consuming,

* Corresponding author: m.demont@irri.org

© CAB International 2013. *Realizing Africa's Rice Promise* (eds M.C.S. Wopereis *et al.*)

Table 23.1. Evolution of annual rice consumption (thousands of tonnes) in the world and in Africa since the 1960s. (Data from FAO, 2012.)

Year[a]	1960s	1970s	1980s	1990s	2000	2001	2002	2003	2004	2005	2006	2007	2008	2009	Growth since 1960s[c]	Growth since 2000[d]
Algeria	6	11	22	38	54	25	96	60	65	51	66	70	66	73	+14%	+25%
Angola	15	28	57	63	57	68	73	80	84	85	83	85	95	98	+10%	+6%
Benin	6	12	42	98	66	102	130	154	193	246	290	272	269	296	+16%	+19%
Botswana		1	4	10	10	19	14	19	21	22	24	27	26	19	+14%	+11%
Burkina Faso	22	31	94	186	239	263	282	214	236	266	263	191	260	408	+9%	+9%
Burundi	2	4	11	25	30	36	35	33	39	39	49	49	45	52	+22%	+7%
Cameroon	15	38	84	131	158	283	289	319	355	395	447	422	461	581	+11%	+17%
Cape Verde	2	4	7	13	17	17	17	22	22	25	24	24	23	26	+8%	+5%
Central African Republic	3	7	9	10	16	19	20	21	14	23	26	28	30	30	+9%	+10%
Chad	19	22	30	64	74	68	81	66	68	76	66	63	103	78	+7%	+3%
Comoros	14	20	33	40	40	39	46	42	42	44	46	48	48	49	+3%	+3%
DR Congo[b]	5	4	8	28	37	41	45	33	30	31	32	30	31	32	+12%	–1%
Côte d'Ivoire	171	314	606	764	817	879	871	941	1,083	1,061	1,098	1,230	1,224	1,303	+6%	+5%
Djibouti	4	7	15	27	30	31	31	31	33	32	34	37	37	37	+6%	+2%
Egypt	734	1,057	1,309	2,356	2,684	2,680	3,070	3,005	2,905	2,641	2,859	2,960	3,227	3,077	+3%	+2%
Ethiopia	2	1	15	14	13	15	18	29	26	26	40	53	39	47	+52%	+19%
Gabon	2	5	12	37	46	47	48	50	51	51	50	49	55	51	+10%	+1%
The Gambia	27	37	63	73	53	64	57	44	48	57	66	85	92	102	+4%	+9%
Ghana	58	71	97	229	301	449	481	460	476	488	551	572	603	640	+8%	+10%
Guinea	154	229	407	680	759	798	809	780	779	810	857	909	955	1,032	+5%	+4%
Guinea-Bissau	35	48	82	119	123	126	130	135	120	113	107	116	131	147	+4%	+2%
Kenya	11	25	49	61	139	167	168	218	268	261	298	285	274	315	+17%	+10%
Liberia	111	182	257	134	163	179	180	160	161	191	256	255	314	369	+5%	+10%
Libya	10	30	49	74	81	83	84	83	84	83	84	83	84	84	+6%	+0%
Madagascar	723	981	1,157	1,271	1,508	1,603	1,684	1,713	1,784	1,850	1,913	1,983	2,043	2,123	+3%	+4%
Malawi	4	33	21	34	48	46	47	51	53	54	61	65	76	86	+12%	+7%
Mali	86	118	197	361	520	528	565	552	609	643	670	707	780	835	+7%	+5%
Mauritania	10	28	81	116	128	72	72	78	88	96	109	103	102	103	+7%	–1%
Mauritius	62	67	70	68	66	68	63	58	68	62	62	63	66	65	+0%	+0%
Morocco	9	9	18	26	26	26	19	12	20	28	22	31	32	38	+13%	+10%

Mozambique	50	84	127	126	165	162	179	228	264	294	376	428	407	433	+6%	+12%
Niger	12	25	70	96	125	150	236	244	257	301	221	198	154	124	+9%	+3%
Nigeria	148	449	1,347	2,251	2,699	3,120	2,862	2,976	2,929	2,840	3,019	3,275	2,987	3,222	+12%	+2%
Rwanda	1	3	10	15	9	34	25	29	30	51	54	55	62	99	+25%	+49%
Senegal	204	261	421	536	722	654	739	775	774	784	778	877	895	866	+4%	+2%
Sierra Leone	211	301	343	360	407	378	405	405	434	452	486	493	517	530	+3%	+3%
South Africa	66	97	205	441	550	560	659	745	743	754	788	959	628	737	+7%	+5%
Sudan	6	11	33	25	33	47	35	42	71	88	129	70	28	50	+9%	+16%
Togo	13	14	32	67	110	85	99	98	154	144	121	127	121	124	+8%	+3%
Tunisia	2	2	5	12	14	10	11	18	14	17	24	17	19	9	+4%	+2%
Uganda	9	14	19	63	114	88	113	125	124	142	125	144	138	147	+14%	+4%
Tanzania	81	189	344	442	616	638	647	773	801	764	811	831	890	873	+9%	+4%
Zambia	3	7	10	13	15	20	25	35	33	27	34	18	30	35	+14%	+15%
Zimbabwe	5	5	10	20	14	16	14	12	22	24	25	42	36	33	+29%	+14%
Others	5	8	13	26	35	35	38	43	46	51	48	47	45	44	+5%	+3%
Africa	3,157	4,926	7,973	11,720	14,015	14,933	15,720	16,122	16,614	16,675	17,707	18,606	18,668	19,646	+4%	+4%
World	142,939	195,283	256,010	297,974	322,991	323,654	326,065	328,069	333,956	339,399	343,128	345,370	353,613	354,603	+2%	+1%

[a]1960s, 1970s, 1980s and 1990s are 10-year averages.
[b]Formerly Zaire.
[c]Average annual growth rates since the 1960s.
[d]Average annual growth rates since 2000.

and disposing of products and services, including the decision processes that precede and follow these actions' (Engel *et al.*, 1995, p. 4). From a micro-economics point of view, much emphasis has traditionally been placed on consumer decision-making and choice behaviour, building on demand and utility maximization theories. Most of the presented schemes are so-called 'stage models', which assume that consumers move through a problem-solving process, ranging from the recognition of needs, over information search and the evaluation of alternatives, to reach the final stage of choice or purchase.

After realizing a need, consumers can start searching for information about potential solutions to satisfy the need that has been recognized. Both internal and external sources of information can be consulted. Internal sources typically pertain to previous experience and memory, whereas external sources include commercial or non-commercial stimuli in the consumers' environment. The following step is the evaluation of alternative solutions on criteria that are relevant for the individual consumer in the specific situation. Such criteria are referred to as **attributes**, about which consumers hold specific beliefs. Beliefs about attributes, combined with attribute-importance weights, result in product preference, which is further translated into purchasing intentions.

Rice Attributes from a Consumer Perspective in Africa

Consumer surveys in Côte d'Ivoire and Nigeria show that rice has become an ordinary good for urban consumers. Rice is regularly consumed by all socio-economic groups (Lançon *et al.*, 2003, 2004), but the view of rice as a uniform commodity is clearly out-dated. The West African rice market consists of individual country markets that differ greatly in terms of their size, importance of rice in food consumption and consumer preference patterns.

Attributes are product characteristics that are either **intrinsic**, like taste, texture or colour, or **extrinsic** to the product, like packaging, brand or label. Another attribute classification distinguishes between **search**, **experience** and **credence** attributes. Search attributes are available for product evaluation before purchase. Typical examples are price, appearance, brand and packaging. Experience attributes can only be evaluated upon product experience, thus after purchase or product use – examples are taste and texture. Credence attributes are attributes that consumers cannot evaluate or verify themselves. Instead they have to put trust in people or institutions, like government controls or industry claims. Attributes relating to production, processing and product contents are typical examples of the credence-type attributes (Nelson, 1970, 1974; Darby and Karni, 1973).

Search attributes

Relevant search attributes for rice in the African market include: rice type, price, financial service and cleanliness, each of which is discussed in the following paragraphs. On the West and Central African market there are five major rice types: (i) long-grain white rice with an intermediate level of starch; (ii) broken rice; (iii) parboiled rice; (iv) aromatic (mostly jasmine) rice; and (v) round-grain (*japonica*) rice (USAID, 2009b).

Table 23.2 provides an overview of consumer preferences for alternative rice types in West and Central Africa.

Long-grain white rice dominates the markets in most of West Africa, except for those markets that prefer parboiled or broken rice. Broken rice is a by-product of rice processing. In international markets, broken rice is considered an inferior product and is therefore much cheaper than whole rice. However, urban Senegalese, Gambian and Mauritanian consumers have developed a marked preference for broken rice (Brüntrup *et al.*, 2006). Most imported rice is either broken or milled rice, and is consumed more in the coastal regions, especially in the larger cities. In countries where the primary urban centre is also a port, with easier access to food imports than food grown in the country's hinterland, the food purchases of this non-agricultural population are typically biased towards imports (Saverimuttu and Rempel, 2004).

In some parts of Africa – such as south-west Mali, Guinea, Sierra Leone, northern Côte d'Ivoire, parts of Benin, Liberia and Nigeria –

Table 23.2. Consumer preferences for alternative rice types in West and Central Africa. (Adapted from USAID, 2009b, based on own observations.)

Country	Rice market characteristics
Benin	• Primarily an importer of high-quality white rice, but also buys aromatic rice and some '25% broken rice' (mix of 25% broken rice and 75% whole-grain rice). High-quality white and aromatic rice are preferred in urban areas. • Some consumers also prefer parboiled rice, especially in rural areas.
Burkina Faso	• Mainly 25% broken rice. Rice is not an essential staple food, but is becoming more popular. Rice is mostly consumed in urban areas, whereas rural populations consume local cereals (millet, sorghum and fonio) in addition to rice. • Although local rice is currently cheaper than imported rice, urban consumers prefer imported rice. Local rice is believed to be of good quality, but harder to access in urban areas than imported rice.
Cameroon	• Imports 25% broken rice, but also significant quantities of high-quality white rice.
Chad	• Market for 25% broken rice.
Côte d'Ivoire	• The overall market is dominated by 25% broken white rice, followed by high-quality white rice, aromatic broken and aromatic 100% whole-grain rice (an emerging product category). • Imported 25% broken rice is mostly sold on rural markets. In urban areas, the preference is for white long-grain 100% whole-grain rice, including aromatic rice. • Local rice is mostly consumed in rural areas.
The Gambia	• Price-conscious market; consumer preference is for 100% broken rice. • Some 25% broken white rice is imported as well. • Local rice is available only at the retail level but is sold at a price premium relative to imported rice.
Ghana	• Rice is not an essential staple food. Consumer preference is for high-quality white and aromatic rice. Aromatic rice is sold at a premium and Ghana is Africa's largest importer of aromatic rice. Some aromatic broken rice is imported as well. • In the north, consumers used to prefer parboiled rice, but fieldwork by USAID (2009b) revealed that preferences have shifted to white rice. • Local rice is sold at a lower price than the cheapest imported rice.
Guinea	• Imports are mostly low-quality 25% broken rice. Rural consumers prefer locally parboiled rice. • Some varieties of local rice are popular and sold at a premium to imported rice.
Guinea-Bissau	• Rice is the key staple food, but consumers readily switch between local and imported rice.
Liberia	• Consumer preference is for round-grain Chinese rice and low-quality parboiled rice.
Mali	• Consumer preference is mostly for 25% broken rice, but imports also include broken rice, both white and aromatic. • Rice is not a key staple food, but it is becoming more popular. • Some varieties of local rice (e.g. Gambiaka) are popular and sold at a premium to imported rice.
Mauritania	• Consumer preference is for 100% broken rice, both aromatic and white. • Some consumers prefer local rice.
Niger	• Consumer preference is for white 25% broken rice and for locally parboiled rice. The market is split equally between these two types of rice.
Nigeria	• In northern Nigeria the preference is for rice flour (97% share), as opposed to grain. In the southern part, the preference is for high-quality parboiled, mostly imported rice.
Senegal	• Consumer preference is for 100% broken rice, both white and aromatic, but there is a slow trend towards whole-grain rice. • Rice is a staple food. • In rice-production areas, local rice is preferred. In urban areas, consumers prefer imported rice; aromatic 100% broken rice is preferred in Dakar.
Sierra Leone	• Price-conscious market, imports 25% broken rice and white broken rice.
Togo	• Primarily imports high-quality rice, both white and aromatic rice, but also 25% broken rice.

parboiled rice is preferred. Parboiling is a transformation process that enhances the physical, chemical and organoleptic (involving substances that influence taste) qualities of rice. It falls into two categories: high quality (with a golden tinge) and low quality (with a dark colour and sometimes an off-flavour or off-odour). Nigeria is one of the largest importers of fully milled, high-quality parboiled rice and Liberia is one of West Africa's top importers of low-quality parboiled rice. Aromatic rice is an emerging and growing rice market segment that commands premium prices and is increasing in popularity throughout urban West Africa. Imported aromatic rice comes mainly from Thailand and Vietnam, and Ghana was one of the first countries to adopt aromatic rice. Round-grain rice is the preferred rice in Liberia and is used for making porridge because the grains stick together when cooked (Table 23.2; USAID, 2009b).

A second search attribute is price. It is important to analyse how consumers respond to price changes of rice. A large body of literature indirectly addresses this question by analysing the effectiveness of price policies in reversing the trend from traditional local grains to wheat and rice (Reardon, 1993; Kelly *et al.*, 1995; Reardon *et al.*, 1997; Akindes, 1999; Diagana and Reardon, 1999; Diagana *et al.*, 1999; Singare *et al.*, 1999). The devaluation of the CFA franc in 1994 provided a unique test case to observe the impact of a doubling of imported rice prices on consumption. However, instead of reducing imported rice intake, consumers responded to the devaluation by de-diversifying their consumption patterns due to the larger importance they attributed to non-price attributes such as availability and ease of cooking. Both Senegal and Côte d'Ivoire (the biggest rice importers) actually increased rice imports after devaluation; only Burkina Faso and Mali (both small rice importers) had small reductions in or stagnation of rice imports (Diagana *et al.*, 1999). The conclusion from this body of research was that price policies did not work due to the low responsiveness of West African rice consumers to price changes, especially in countries where rice is the main staple crop. More recent consumer-preference studies of Fall and Diagne (2008) and Lançon *et al.* (2004) confirm this conclusion.

A third search attribute is services associated with the purchase of rice, such as credit, or delay-payment mechanisms. A consumer survey in Côte d'Ivoire showed that this criterion is much more important than price (Lançon *et al.*, 2001). These services are often provided by retailers of imported rice and, less frequently, by dealers of local rice.

A fourth search attribute is cleanliness and market presentation of rice. The historical switch of urban consumption from local coarse grains to imported wheat and rice instead of local rice can be explained by consumers' perception that local rice is of inferior quality. Owing to a large percentage of foreign matter and low levels of postharvest grading and sorting, local rice fails to meet expectations concerning reduced workload and time spent on sorting and cooking rice, and hence falls short relative to imported rice in this convenience dimension. Several consumer-preference surveys in Benin, Burkina Faso, Côte d'Ivoire, Nigeria and Senegal confirm this (Lançon *et al.*, 2001, 2003; Konkobo *et al.*, 2002; Lançon and Benz, 2007; Fall and Diagne, 2008; Demont *et al.*, 2012, 2013a,b). This critically explains why imported rice is preferred in many countries over local rice, with Mali, The Gambia and Guinea as exceptions (USAID, 2009b).

Experience attributes

Experience attributes can be verified only after use of the product – for rice the most important ones are sensory characteristics, swelling capacity and cooking time. From a marketing perspective, lack of consistency between pre- and post-consumption evaluations can significantly affect satisfaction and repeat-purchase decisions. The evaluation of taste is subjective and taste preferences differ among countries, regions, households and even for types of meals. Consumer-preference studies show that taste is an important attribute that tends to favour local rice, but it is not the most decisive attribute in many cases (Lançon *et al.*, 2001, 2003; Konkobo *et al.*, 2002; Lançon and Benz, 2007; Fall and Diagne, 2008; Moseley *et al.*, 2010; Demont *et al.*, 2012).

In some countries, there are local rice varieties with unique taste characteristics that are particularly appreciated by consumers. For example,

'Ofada' rice in Nigeria is liked by consumers of all income classes for its distinct taste and aroma (PrOpCom, 2007). In Guinea and Mali, local rice is preferred for its taste and freshness (USAID, 2009b). In Senegal, sensory tests provide evidence of the existence of a market segment (14%) of older and less-educated consumers with a preference for traditional local rice (Demont *et al.*, 2013b). In Ghana, Tomlins *et al.* (2005, 2007) similarly observed that although the majority (86%) of consumers prefer imported raw and parboiled rice to that produced locally, due to the poor quality of local rice, there is a niche market segment (14% of consumers) that mostly prefers traditional local rice. Apart from the above-mentioned studies, sensory studies dealing with rice in the African market are scarce.

Swelling capacity is related to the physical and chemical (starch content) properties of the grain. Rice that swells more satisfies a larger number of consumers in terms of satiety for a given weight of rice because of its potential volume increase. This is an important experience attribute for large families with financial constraints (Lançon *et al.*, 2001, 2003; Dalton, 2004; Fall *et al.*, 2007; Fall and Diagne, 2008). The swelling capacity of rice is not only characteristic of the rice variety, but it is also influenced by the management of postharvest operations. While (according to rice breeders) there is a relation between paddy variety and swelling capacity, food technologists explain that the high swelling capacity of imported rice is due to its longer storage period (up to one year or more). In contrast to imported rice, the bulk of local rice is usually traded within months after the harvest (Lançon *et al.*, 2003). Storage of milled rice (ageing) affects rice quality: it leads to lower cohesiveness, drier grain surfaces, higher swelling capacity and firmer texture during cooking. Finally, cooking time also tends to become longer with increased storage time.

Credence attributes

Credence attributes generally gain importance as evaluative criteria in food-purchasing decision making. As indicated above, credence attributes are characteristics that are not directly revealed by experience and consumption, and where the consumer is reliant on third-party or external information to evaluate the attribute in the product. The market provision of quality is notoriously fraught with difficulties under asymmetric information: when producers cannot credibly signal the quality of their products, consumers' choices are predicated on the perceived average quality on the market, and this pooling equilibrium leads to market failures. Labelling and branding can be used to counteract the effects of quality uncertainty, as it not only indicates quality but also gives consumers a means of retaliation if the quality does not meet expectations (Akerlof, 1970). It has been established that origin and quality labelling entail market differentiation potential, which might be noticeable for consumers depending on what the labels appeal to (e.g. health, safety, taste differentiation) (Verbeke and Roosen, 2009). Country-of-origin labelling (COOL) can therefore serve as an extrinsic cue that supplements the use of intrinsic quality cues to form quality expectations (Verlegh and Steenkamp, 1999).

Batra *et al.* (2000) found that among consumers in developing countries – for reasons that go beyond brand quality assessments – brands perceived as having a non-local origin are attitudinally preferred to brands seen as local. This suggests that COOL not only serves as a quality cue, but also possesses an additional dimension of foreignness that contributes to attitude formation and liking for status-enhancing reasons. The greatest challenge for rice marketeers in West and Central Africa will be to reverse this trend, especially in urban markets. Recent evidence in this research area is promising. Tomlins *et al.* (2005) found evidence for COOL effects in Ghana in that consumers tended to like rice types that came from their own region. Demont *et al.* (2013b) show that under experimental conditions, urban Senegalese consumers are willing to pay an average price premium of 35% – relative to the price of traditional local rice – for enhanced-quality local rice and further added 6% if the latter was labelled and branded as Rival® (a trademarked COOL launched by the Oxfam-funded PINORD platform in 2007) (PINORD, 2009). This suggests that COOL has the potential to increase consumer acceptance of a food product by translating a typical credence characteristic (origin) into a searchable characteristic (an origin label).

Conclusion

Historically, strong urbanization and urban bias have marked urban markets in West Africa with strong consumption preferences for imported rice. Understanding consumer preferences for rice in Africa will become increasingly important for reversing this trend in order to increasingly meet the growing demand for rice with local rice and reduce the import bill. The available evidence shows that rice is far from a homogeneous commodity in West and Central Africa – different market segments can be identified, based on search, experience and credence attributes. Beliefs about attributes result in product preference, which is translated into purchasing intentions. Although consumer preferences are deeply embedded in societal norms and market behaviour, they are not unalterable. Credence attributes can increase acceptance of a food product, if they can be successfully transformed into search attributes.

As argued by Tollens *et al.* (Chapter 1, this volume), consumer research on rice in Africa should be part of an optimal mix between supply-shifting and demand-lifting research and development. The information generated by this research can be used by policy makers and stakeholders to tailor domestic rice value chains to market standards, in order to increase competitiveness of the domestic rice sector to imported rice, illustrated by Demont and Neven (Chapter 24, this volume). Experimental economics is a particularly useful tool for conducting consumer research as it allows testing of alternative marketing strategies in real or laboratory market conditions. In 2008, Africa Rice Center (AfricaRice) started a series of experimental auctions in Benin, Burkina Faso, Cameroon, The Gambia, Mauritania, Senegal and Uganda, and is planning similar experiments to assess consumers' willingness to pay for rice quality and marketing attributes in African markets (Demont *et al.*, 2012, 2013a,b).

References

Akerlof, G.A. (1970) The market for 'lemons': quality, uncertainty and the market mechanism. *Quarterly Journal of Economics* 84, 488–500.

Akindes, F. (1999) Food strategies of urban households in Côte d'Ivoire following the 1994 CFA franc devaluation. *Food Policy* 24, 479–493.

Batra, R., Ramaswamy, V., Alden, D.L., Steenkamp, J.B.E.M. and Ramachander, S. (2000) Effects of brand local and nonlocal origin on consumer attitudes in developing countries. *Journal of Consumer Psychology* 9, 83–95.

Bezemer, D. and Headey, D. (2008) Agriculture, development, and urban bias. *World Development* 36, 1342–1364.

Brüntrup, M., Nguyen, T. and Kaps, C. (2006) The rice market in Senegal. *Agriculture & Rural Development* 1, 22–25.

Dalton, T.J. (2004) A household hedonic model of rice traits: economic values from farmers in West Africa. *Agricultural Economics* 31, 149–159.

Darby, M.R. and Karni, E. (1973) Free competition and optimal amount of fraud. *Journal of Law & Economics* 16, 67–88.

Demont, M., Zossou, E., Rutsaert, P., Ndour, M., Van Mele, P. and Verbeke, W. (2012) Consumer valuation of improved rice parboiling technologies in Benin. *Food Quality and Preference* 23, 63–70.

Demont, M., Rutsaert, P., Ndour, M., Verbeke, W., Seck, P.A. and Tollens, E. (2013a) Experimental auctions, collective induction and choice shift: willingness-to-pay for rice quality in Senegal. *European Review of Agricultural Economics* 40(2), 261–286.

Demont, M., Rutsaert, P., Ndour, M. and Verbeke, W. (2013b) Reversing urban bias in African rice markets: evidence from Senegal. *World Development* 45, 63–74.

Diagana, B. and Reardon, T. (1999) Household consumption responses to the franc CFA devaluation: evidence from urban Senegal. *Food Policy* 24, 495–515.

Diagana, B., Akindes, F., Savadogo, K., Reardon, T. and Staatz, J. (1999) Effects of the CFA franc devaluation on urban food consumption in West Africa: overview and cross-country comparisons. *Food Policy* 24, 465–478.

Engel, J.F., Blackwell, R.D. and Miniard, P.W. (1995) *Consumer Behavior*. Dryden Press, Chicago, Illinois.

Fall, A.A. and Diagne, A. (2008) Etude de la relation qualité-prix du riz produit dans la vallée du fleuve Sénégal. Association pour le développement de la riziculture en Afrique de l'Ouest (ADRAO), Saint-Louis, Senegal.

Fall, A.A., Fall, C.A.K., Gningue, D.R., Ndir, B. and Ndour, M. (2007) Etude sur les critères de qualité et les modes de consommation du riz au Sénégal. Fonds National de Recherches Agricoles et Agro-alimentaires (FNRAA), Dakar, Senegal.

FAO (2012) FAOSTAT. http://faostat3.fao.org (accessed 19 June 2012).

Kelly, V., Reardon, T., Diagana, B. and Fall, A.A. (1995) Impacts of devaluation on Senegalese households: policy implications. *Food Policy* 20, 299–313.

Konkobo, Y.C., Karimou, A.R., Kabore, S. and Diasso, K. (2002) Les pratiques alimentaires à Ouagadougou, Burkina Faso. Centre National pour la Recherche Scientifique et Technique (CNRST), Centre de coopération internationale en recherche agronomique pour le développement (CIRAD), Ouagadougou, Burkina Faso.

Lançon, F. and Benz, H.D. (2007) Rice imports in West Africa: trade regimes and food policy formulation. Paper presented at the 106th seminar of the European Association of Agricultural Economics (EAAE), Pro-poor Development in Low Income Countries: Food, Agriculture, Trade and Environment, Montpellier, France, 25–27 October.

Lançon, F., Touré, A. and Teslim, I. (2001) Rice Consumption Survey in Bouaké: Descriptive analysis. West Africa Rice Development Association, Bouaké, Côte d'Ivoire.

Lançon, F., Erenstein, O., Akande, S.O., Titilola, S.O., Akpokodje, G. and Ogundele, O.O. (2003) Imported Rice Retailing and Purchasing in Nigeria: A survey. West Africa Rice Development Association, Bouaké, Côte d'Ivoire.

Lançon, F., Erenstein, O., Touré, A. and Akpokedje, G. (2004) Quality and competitiveness of local and imported rice on West African urban markets. *Cahiers d'études et de recherches francophones* 13, 110–115.

Lipton, M. (1977) *Why Poor People Stay Poor: Urban bias in world development.* Temple Smith, London.

Moseley, W.G., Carney, J. and Becker, L. (2010) Neoliberal policy, rural livelihoods, and urban food security in West Africa: a comparative study of The Gambia, Côte d'Ivoire, and Mali. *Proceedings of the National Academy of Sciences* 107, 5774–5779.

Nelson, P. (1970) Information and consumer behavior. *Journal of Political Economy* 78, 311–329.

Nelson, P. (1974) Advertising as information. *Journal of Political Economy* 82, 729–754.

Pearson, S.R., Stryker, J.D., Humpreys, C.R., Rader, P.L., Monke, E.A., Spencer, D.S.C., Craven, K., Tuluy, A.H., McIntire, J. and Page, J.M. (1981) *Rice in West Africa: Policy and economics.* Stanford University Press, Stanford, California.

PINORD (2009) *Plan Strategique 2009–2013: Promotion de la Sécurité Alimentaire par la Filière Riz Local dans la Vallée du Fleuve Sénégal.* Programme d'Appui aux Iniatives du Nord, Saint-Louis, Senegal.

PrOpCom (2007) Ofada Rice Demand Characteristics and Measurements. Promoting pro-Poor Opportunities through Commodity and Service Markets, Abuja, Nigeria.

Reardon, T. (1993) Cereals demand in the Sahel and potential impacts of regional cereals protection. *World Development* 21, 17–35.

Reardon, T., Kelly, V., Crawford, E., Diagana, B., Dione, J., Savadogo, K. and Boughton, D. (1997) Promoting sustainable intensification and productivity growth in Sahel agriculture after macroeconomic policy reform. *Food Policy* 22, 317–327.

Ross, C.G. (1983) Review: 'Rice in West Africa: Policy and Economics'. *The Journal of Modern African Studies* 21, 335–336.

Saverimuttu, V. and Rempel, H. (2004) The determinants of cereal grain imports: sub-Saharan Africa, 1970–1997. *African Development Review* 16, 525–548.

Singare, K., Reardon, T., Camara, Y., Wanzala, M., Teme, B. and Sanogo, O. (1999) Household consumption responses to the franc CFA devaluation: evidence from urban Mali. *Food Policy* 24, 517–534.

Tomlins, K.I., Manful, J.T., Larwer, P. and Hammond, L. (2005) Urban consumer preferences and sensory evaluation of locally produced and imported rice in West Africa. *Food Quality and Preference* 16, 79–89.

Tomlins, K.I., Manful, J.T., Gayin, J., Kudjawu, B. and Tamakloe, I. (2007) Study of sensory evaluation, consumer acceptability, affordability and market price of rice. *Journal of the Science of Food and Agriculture* 87, 1564–1575.

USAID (2009a) Global Food Security Response: Senegal Rice Study. United States Agency for International Development, Washington, DC.

USAID (2009b) Global Food Security Response: West Africa Rice Value Chain Analysis. United States Agency for International Development, Washington, DC.

Verbeke, W. and Roosen, J. (2009) Market differentiation potential of country-of-origin, quality and traceability labeling. *Estey Centre Journal of International Law and Trade Policy* 10, 20–35.

Verlegh, P.W.J. and Steenkamp, J.B.E.M. (1999) A review and meta-analysis of country-of-origin research. *Journal of Economic Psychology* 20, 521–546.

24 Tailoring African Rice Value Chains to Consumers

Matty Demont[1*] and David Neven[2]
[1]*Africa Rice Center (AfricaRice), Saint-Louis, Senegal;* [2]*Food and Agriculture Organization of the United Nations (FAO), Rome*

Introduction

The global financial crisis of 2009 onwards and the preceding food price crisis (2007–2008) exacerbated the food insecurity of the rural and urban poor in developing countries. The question of how national, regional and global agri-food systems can effectively respond to these crises and improve food security is once again at the top of the development agenda. This chapter explores potential response strategies for the private and public sectors, and the donor community in West Africa. Merging two new studies on rice in West Africa, this objective is analysed through a value-chain lens from two angles. First, a supply-side perspective based on an extensive, regional value-chain analysis. We present the results of a major study conducted for the Global Food Security Response programme of the United States Agency for International Development (USAID, 2009). Second, a demand-side perspective based on consumer experiments carried out by Africa Rice Center (AfricaRice) in Senegal (Demont *et al.*, 2013a,b).

In most of West Africa, rice production has not been able to meet the increases in demand triggered by population growth, rapid urbanization, increasing incomes, and shifting urban consumer preferences. As a result, the sub-region relies on imports to supply half of its demand for rice. In May 2008, world rice prices tripled in just a few months to reach 30-year, inflation-adjusted highs. With these dramatic price-spikes, combined with an overall import level into West Africa that over time had swelled to 6 million tonnes (Mt) (i.e. 20% of the world's rice imports) (USAID, 2009), governments could no longer afford traditional price interventions to protect domestic rice consumers from the volatility of the global rice market. Although the 2008 rice price crisis may in part have been driven by temporary speculation and a weak US dollar, there are underlying structural trends that have caused an increase in the price of rice (and other staple foods) to a higher plateau.

The comparative advantage of local rice production has been one of the key issues in the food-policy debate in Africa since the early 1980s (Pearson *et al.*, 1981). However, policies for increasing competitiveness have mainly focused on productivity and prices. The 1994 devaluation of the CFA franc – a failed attempt to reverse the historical urban bias and divert African consumers from imported to local grains – clearly illustrated that price policies do not work well in the context of the African rice sector due to a low price response by rice consumers and producers (Diagana *et al.*, 1999). This suggests that

* Corresponding author: m.demont@irri.org

© CAB International 2013. *Realizing Africa's Rice Promise* (eds M.C.S. Wopereis *et al.*)

non-price strategies may be needed to enhance the competitiveness of the sector. Establishing efficient value chains is cited by the World Bank (2008) as a first policy objective in making agriculture more effective in supporting sustainable growth and reducing poverty. Therefore, taking a value-chain perspective, we identify some systemic key constraints and challenges to competitiveness and explore opportunities for upgrading African rice value chains and tailoring them to consumers in end-markets with the aim of simultaneously increasing food security and reducing poverty.

Value Chains

'Value chains' are defined as strategic vertical alliances of non-adversarial relationships between stakeholders within a product's supply chain (Hobbs *et al.*, 2000). Unlike a traditional economic analysis of food staples that assumes a homogeneous product, homogeneous agents, and perfectly competitive markets at various stages along the supply chain, the value-chain approach is a holistic and systemic methodology that takes into account a wide range of factors in assessing agent behaviour and competitiveness (Kaplinsky and Morris, 2001; Neven, 2009). Most notably, the approach takes into account: (i) the heterogeneity of the end-markets and the critical success factors that drive competitiveness (the end-markets are the starting point, i.e. a strategy is developed to take advantage of some identified opportunity in an end-market, which has implications all along the value chain); (ii) the heterogeneity of stakeholders in terms of their capacities and incentives, and how this influences the nature of their interactions vertically (along the chain) and horizontally (at a certain stage of the chain); (iii) the nature of service provision (logistics, extension, finance, etc.); and (iv) the broader environment in which they operate (policies, regulations, infrastructure). Furthermore, value-chain analysis takes a dynamic perspective as it aims to expose the key drivers of change that are present in the system, such as changes in policy, introduction of new technologies, and the behaviour of large firms. In other words, value-chain analysis explicitly recognizes that the value of the end-product to the end-consumer has many components (price, taste, convenience, image, etc.) and is the cumulative outcome of every value-adding activity along the value chain (value can be added or lost at each link in the chain) – this is especially important for staple-food value chains in developing countries that have to become more competitive at every stage.

When domestic production-based staple-food value chains become more competitive they will contribute to food security and income growth. Moreover, by linking producers to consumers through a shared objective, value chains present a more sustainable approach to consumption and production than segmented and adversarial production chains (Demont, 2010). The market demand-driven strategy upon which value chains are based distinguishes them from traditional business relationships. Value chains are built on cooperation rather than adversarial business relationships; their members recognize that participants must create a win–win situation whereby they all benefit financially and all are part of the information-sharing and decision-making process.

A critical challenge in the development of African rice value chains is the governance of quality throughout the supply chain, whereby quality should be tailored to the food preferences of end-market consumers (Rizzotto and Demont, 2011; Demont and Rizzotto, 2012). Sorting by quality is necessary to capture rents in West Africa's highly segmented rice markets, where most consumers want a product of consistent quality. In some cases, however, consumers may prefer to buy mixed quality rice and sort it themselves as this is cheaper and gives them different rice types for different meal types (Demont *et al.*, 2013a,b). In some cases, rice processors or traders may mix different rice varieties to cater to the specific preferences of particular consumers. For example, they may blend a small quantity of aromatic rice into cheaper broken rice to produce a rice blend that is both affordable and aromatic.

Evidence from across the world shows that market forces alone are sub-optimal in achieving governance of quality throughout the supply chain, and other governance mechanisms (long-term contracts, alliances, vertical integration, etc.) are needed to compensate for this market failure and to ensure that suppliers develop the capability to comply with changing consumer demands as rapidly as possible (Swinnen *et al.*, 2010). In high-value markets in developing countries, there has been an evolution towards increasing levels of coordination between different actors in the supply chain and the question is whether similar patterns of supply-chain governance could arise in staple-food supply chains, which typically have lower value added over the chain. Swinnen *et al.* (2010) formally show that the development of chain governance is less obvious in the staple-food sector. Staple crops such as rice are characterized by low value and high storability, and there are a large number of small traders, which increases the possibility of contract breach (Colen *et al.*, 2013). These characteristics make it very unlikely for chain governance to arise spontaneously. They conclude that interlinked contracts or other non-market governance mechanisms might develop, but only if the value of staple crops can be increased and contract enforcement is improved.

The existence of systematic price discounts for local rice relative to imported rice in some African countries suggests that there is underinvestment in upgrading product quality and consumer focus (Demont and Rizzotto, 2012). Evidence from poor Asian countries shows that research on quality has a high payoff (Unnevehr, 1986), which is consistent with the finding that even the very poor have more income elastic demand for food quality than for food quantity (Shah, 1983). Evidence from Africa similarly reveals that quality is an important factor in the demand for rice, even in rural areas (Erhabor and Ojogho, 2011; Demont *et al.*, 2012). The central importance of quality for competitiveness in West Africa's rice markets is demonstrated in several studies (e.g. Boughton and Reardon, 1997; Lançon *et al.*, 2003, 2004; Tomlins *et al.*, 2005; Balasubramanian *et al.*, 2007; Fall *et al.*, 2007; Opoku and Akorli, 2009; USAID, 2009).

The market provision of quality, however, is notoriously fraught with difficulties under asymmetric information: when producers cannot credibly signal the quality of their products, consumers' choices are predicated on the perceived average quality on the market, and this pooling equilibrium leads to market failures (Akerlof, 1970). Therefore, any quality upgrade needs to be accompanied by a credible certification system (Moschini *et al.*, 2008). Quality standards, certification and quality control become more relevant as the value chain moves from being supplier driven to increasingly buyer driven (ACI, 2005). Branding and labelling increase visibility and trust in rice consumption and are an integral part of value-chain upgrading and innovation, as will be illustrated through our case study.

The nature of milling equipment also greatly impacts rice quality. Although large-scale rice milling equipment provides the highest quality of rice most efficiently, it is rare in post-structural adjustment West Africa where it is difficult to aggregate sufficient volumes to operate such facilities economically. Most rice in West Africa is milled using small, sometimes portable, milling machines that largely produce low-quality rice with significant levels of impurities and mixed whole and broken grains. Where development of the value chain for quality rice can generate sufficiently large margins for upgrading, intermediate-scale operations may perhaps provide the best milling option. For example, in Mali, for-hire mini rice mills operate with polishers and graders capable of presenting a clean, polished product and sorting it into homogeneous lots by kernel size.

Methodology

As indicated above, this chapter is derived from two studies on rice in West Africa. The first is a regional rice value-chain study based on three components that build on and inform each other (USAID, 2009). First, a desk study captured and codified a breadth of data on rice value chains in West Africa, comprising a review of over 300 documents as well as key-informant interviews. Second, a field-study component using USAID's (2006) value-chain analysis methodology resulted in detailed value-chain studies on rice in Ghana, Liberia, Mali, Nigeria and Senegal. This component also included a regional market study that entailed further fieldwork in Benin, Côte d'Ivoire, Ghana and Senegal. Third, a 3-day online e-consultation with leading experts on rice in West Africa helped validate the results.

The second study is a Senegal rice-consumer experimental study that used Vickrey second price auctions to elicit consumers' willingness-to-pay (WTP) a price premium, relative to conventional Senegal River valley (SRV) rice as the benchmark, for three rice products of higher quality: labelled and unlabelled high-quality SRV rice and imported Thai 100% broken rice (Demont *et al.*, 2013a,b). The conventional SRV rice type which is commonly available on the market consists of a mix of varieties and has mediocre grain quality. Enhanced-quality SRV broken rice is purified and homogenized through one or two sifting operations and is currently marketed under the Rival® brand name by the Oxfam-funded PINORD (Platforme d'appui aux Initiatives du Nord).[1] Imported Thai 100% broken rice has a grain quality somewhere between the conventional and the enhanced-quality SRV broken rice and contains some impurities. Twenty experimental auctions were conducted in two important urban rice markets (Saint-Louis and Dakar); for each auction, ten women were randomly recruited on the market. Participants were given one kilogram of conventional SRV rice and were offered the opportunity to upgrade it by bidding simultaneously on the alternative rice types during two trials – before and after a sensory test. The experiments thus allowed for a comparison in WTP between imported rice (which dominates urban rice markets) and high-quality local rice.

Findings

The US$10 billion West African rice market is not a single, homogeneous commodity market where competition is driven solely by price (Rutsaert *et al.*, Chapter 23, this volume). Rather there is a wide range of rice products on offer (brown, broken, parboiled, round, aromatic, etc.) and consumer preferences vary greatly among and within countries. Although typically one rice product dominates in a given market, there is often a wide range of rice products available. No matter what the importance of rice in the diet or the income category is, consumers set minimum quality standards and are unlikely to buy lower quality, cheaper rice that does not meet their standards.

West African rice is, in most cases, less competitive than imported rice, but it has the potential to become competitive (USAID, 2009). Local rice is less competitive than imported rice, especially in urban markets, in terms of two main sets of key success factors: (i) local rice in many cases is perceived to be of a lower quality than the comparable imported product that sets the benchmark (The Gambia, Guinea and Mali are notable exceptions); (ii) the absence of trade credit and aggregation in value chains for locally produced rice makes it difficult for the existing urban market distribution system to tap into domestic production. As a result, local rice is largely absent from urban markets and thus not an option for consumers, even if they want to buy it. This largely explains the lack of awareness of the existence of local rice in those markets (Fall *et al.*, 2007; Demont *et al.*, 2013a,b). Nevertheless, in the five country studies conducted in the context of this value-chain analysis, local rice is or can become competitive with imported rice. Margins appear to be sufficiently high to maintain price competitiveness even after taking into account the costs of necessary quality improvements.

A third observation is that rice value chains in West Africa have failed to achieve their considerable potential due to systemic constraints. The latter are related to the **business-enabling environment** (e.g. fickle rice policies[2] that deter private-sector investment, insecure land titles that impede investments by farmers, poor road conditions and corruption that block the linkage between production zones and markets), **vertical linkages** (e.g. the history of government and donor intervention in rice – including input-subsidy programmes – that has generated a mistrust[3] of commercial relationships), **horizontal linkages** (e.g. aggregators or brokers that compete to secure paddy from farmers, who too often engage in side-selling often encouraged by the aggregator/broker), and **support markets** (e.g. years of government and donor interventions that have crowded out private-sector investment in input supply, processing services, extension provision, and finance). Progress through upgrading is needed on many fronts, such as improvements in rice quality, increased efficiency through the uptake of appropriate technology, decreased postharvest losses, and consumer acceptance of local rice as a quality product. For any of these upgrades to take hold, however, supportive government policies and sustainable business-development service provision are needed, as well as strengthened horizontal and vertical partnerships based on trust, transparency and mutual benefits. The widespread absence of these factors represents a key systemic constraint to development.

A fourth observation relates to the critical importance of quality governance. The Senegal experimental auctions show that, while Senegalese consumers were willing to pay an 18% price premium for imported Thai 100% broken rice relative to conventional SRV rice, they were willing to pay a 35% price premium to obtain unlabelled enhanced-quality SRV broken rice, and a 41% price premium to obtain PINORD's Rival-branded enhanced-quality SRV broken rice (Demont *et al.*, 2013b). These findings suggest that Senegalese consumers are willing to pay for intrinsic quality attributes and quality certification and – as the price premiums largely cover the costs of quality improvements and certification – that SRV rice is able to compete against imported rice if its quality is tailored to consumer preferences. However, the experiments also revealed the existence of an important awareness gap: 18% of consumers in Saint-Louis and 47% of those in Dakar were unaware of the existence of quality SRV rice. Quality upgrading of rice in combination with generic promotion programmes that extend the reach of SRV rice might therefore create some opportunities for the development of a certain degree of coordination along the value chain and could have a real impact on urban markets (Rizzotto and Demont, 2011; Demont and Rizzotto, 2012). However, while PINORD's model for the increased commercialization of quality SRV rice is certainly a good preliminary step towards competitiveness, their operational scale is currently too small to significantly impact the market.[4]

The final observation is that the reaction of donors and West African governments to the increased risk of rice price spikes has been to launch programmes and policies to promote local production (AfricaRice, 2011). The impacts of these programmes are still being realized, though forecasts of the results of some of these actions are mixed. While investment in the rice sector is clearly needed, many of the government rice initiatives focus almost exclusively on production to the exclusion of complementary initiatives in processing and marketing, which are critical to match supply to demand. The experimental study findings reported above suggest that supply and demand strategies need to be synchronized in order to tailor the quality of local rice to consumer preferences in urban end-markets (Demont and Rizzotto, 2012). In this example, Rival-labelled rice is produced and marketed in line with the buying behaviour of Senegalese rice consumers. For African policy makers, researchers and donors, this translates into the challenge of finding the optimal mix between both areas of investment (see Tollens *et al.*, Chapter 1, this volume).

Conclusions and Recommendations

Rice production in West Africa can be competitive with imported rice in a far broader range of markets than is currently the case, so it could drastically reduce import dependency and thus food insecurity. The 2008 rice price crisis could provide the jolt needed to unlock this potential, but only if the response strategy is well designed, strikes the right balance between public, private and donor activity, and is implemented with ongoing investment of time and other resources. Governance of quality and clear marketing strategies are crucial for increasing the awareness of urban consumers and strengthening emerging value chains of quality rice, such as the PINORD initiative in Senegal.

A food-security strategy for rice in West Africa needs to satisfy multiple aspects of food security. It must foster the supply of rice to meet the demands of urban and peri-urban populations that currently consume large quantities of imported rice (in most West African countries). At the same time, it must address food access by rural populations, many of whom cultivate rice for subsistence. In addition, trade is essential to the efficient distribution of food to deficit areas from surplus areas that have a competitive advantage to grow large amounts of rice. Although recommended national rice value-chain development plans will differ between countries depending on their unique characteristics,[5] a food-security strategy for rice in West Africa will have three general, distinct but complementary components that need to be balanced in a resource-constrained environment (Fig. 24.1).

First, national value-chain (VC) competitiveness strategies are required to ensure the supply of rice in the quantity and quality needed to effectively compete with imported rice in West Africa's urban markets. While differing in detail by country, national competitiveness strategies will largely be based on the creation of commercial networks characterized by concentrated areas of production (mostly irrigated), market-oriented farmers, and significant investments in storage, processing and marketing. The establishment of these commercial networks implies a time-consuming process of building trust among value-chain stakeholders so that mutually beneficial business models emerge. It also implies a government policy shift to a more market-based approach to food security in which competitive local farms and firms ensure the supply of quality staple foods at the most competitive consumer prices.

Second, national rural rice food-security strategies focused on access to food are needed to improve productivity for the majority of the widely dispersed subsistence rice producers, mainly operating under rainfed production systems. At its core, this strategy takes an incremental and partially subsidized approach to the introduction of basic production and postharvest handling technologies – providing a demonstration effect for replication. It also takes non-distortive approaches to developing links between subsistence farmers and a commercial input-distribution system. Current disincentives to improved rice production will also need to be addressed, including insecure land tenure, dependency on government or donor assistance, and adverse cultural norms such as mistrust of the private sector. A combination of increased sales of cash crops and capital-asset building (savings) will positively affect sustainability and the graduation of farmers from subsistence to market-oriented production. Such rural food-security strategies should focus on a number of different food crops important for nutrition and calorie intake, rather than on rice (or any other staple food) in isolation.

Fig. 24.1. Rice in a regional food-security strategy. VC, value chain.

Third, a regional food-security strategy focused on distribution is needed to facilitate rice flows and learning throughout the sub-region. This facilitation will initially increase flows for imported rice that is already in the market and thus create a more competitive environment for local rice. However, regional improvements will eventually be needed to facilitate trade within West Africa from centres of excellence (characterized by comparative advantages) to the major deficit areas in the region. Moreover, shared learning (rather than just information exchange) will ensure that lessons learned in one country will be applied elsewhere.

Finally, the goals of food security and food self-sufficiency should not be pursued to the point that they undermine economic incentives – they may need to be rationalized within the design of many current government and donor investments. Many post-crisis government interventions have undermined rather than built upon private-sector efforts. Even in countries where liberalization policies have been in place for years, there remains a fundamental lack of trust in markets. At the same time, governments in the sub-region have neither the resources nor the capabilities (as demonstrated by past efforts) to achieve substantial reductions in the need and demand for imported rice. Most donors and researchers would argue that working through the private sector is the most cost-effective strategy for generating the surpluses needed to replace imports. This strategy will have to be demonstrated before many governments in the region will be ready to adopt it.

Notes

[1] Launched in 2006, PINORD introduced and governs the quality of a new enhanced-quality SRV broken rice brand Rival® (Riz de la Vallée) during production (through a quality contract detailing recommended production practices), processing (sifting, cleaning and packaging) and promotion. PINORD also provides microfinancing (PINORD, 2007, 2009).
[2] For example, in 2005, the Nigerian government encouraged large multinational rice companies to invest in rice processing by granting them licences to import brown rice at a preferential tariff rate. Two importers invested in new full-service rice mills, but this investment was undermined when the Nigerian government abandoned the exclusive licensing scheme within 2 years of its introduction (USAID, 2009).
[3] The ubiquitous lack of trust works in all directions. For example, farmers do not trust that processors will pay them a fair price, processors do not trust that farmers will deliver in the promised quantity and quality, and processors do not trust that governments will stick to their policies.
[4] PINORD marketed 500 tonnes of milled Rival rice on the Saint-Louis market in 2007, 5000–6000 t in 2008, 7000–8000 t in 2009, and 12,000 t in 2010. However, the latter still represents only 3% of total SRV rice production in 2010–2011 (Demont *et al.*, 2013b).
[5] This is well illustrated in the country-specific value-chain studies that are attachments to USAID (2009).

References

ACI (2005) Northeast Thailand rice value chain study. Policy Brief Series. Agrifood Consulting International, Bethesda, Maryland.

AfricaRice (2011) *Lessons from the Rice Crisis: Policies for food security in Africa*. Africa Rice Center, Cotonou, Benin.

Akerlof, G.A. (1970) The market for 'lemons': quality uncertainty and the market mechanism. *Quarterly Journal of Economics* 84, 488–500.

Balasubramanian, V., Sié, M., Hijmans, R.J. and Otsuka, K. (2007) Increasing rice production in sub-Saharan Africa: challenges and opportunities. *Advances in Agronomy* 94, 55–133.

Boughton, D. and Reardon, T. (1997) Will promotion of coarse grain processing turn the tide for traditional cereals in the Sahel? Recent empirical evidence from Mali. *Food Policy* 22, 307–316.

Colen, L., Demont, M. and Swinnen, J. (2013) Smallholder participation in value chains: the case of domestic rice in Senegal. In: *Rebuilding West Africa's Food Potential: Policies and private initiatives to promote smallholder-inclusive staple food value chains*. Food and Agriculture Organization of the United Nations and International Fund for Agricultural Development, Rome, Italy (in press).

Demont, M. (2010) Should sustainable consumption and production be a policy priority for developing countries and if so what areas should they focus on? *Natural Resources Forum* 34, 87–88.

Demont, M. and Rizzotto, A.C. (2012) Policy sequencing and the development of rice value chains in Senegal. *Development Policy Review* 30(4), 451–472.

Demont, M., Zossou, E., Rutsaert, P., Ndour, M., Van Mele, P. and Verbeke, W. (2012) Consumer valuation of improved rice parboiling technologies in Benin. *Food Quality and Preference* 23, 63–70.

Demont, M., Rutsaert, P., Ndour, M., Verbeke, W., Seck, P.A. and Tollens, E. (2013a) Experimental auctions, collective induction and choice shift: willingness-to-pay for rice quality in Senegal. *European Review of Agricultural Economics* 40(2), 261–286.

Demont, M., Rutsaert, P., Ndour, M. and Verbeke, W. (2013b) Reversing urban bias in African rice markets: evidence from Senegal. *World Development* 45, 63–74.

Diagana, B., Akindes, F., Savadogo, K., Reardon, T. and Staatz, J. (1999) Effects of the CFA franc devaluation on urban food consumption in West Africa: overview and cross-country comparisons. *Food Policy* 24, 465–478.

Erhabor, P.O.I. and Ojogho, O. (2011) Effect of quality on the demand for rice in Nigeria. *Agricultural Journal* 6, 207–212.

Fall, A.A., Fall, C.A.K., Gningue, D.R., Ndir, B. and Ndour, M. (2007) Etude sur les critères de qualité et les modes de consommation du riz au Sénégal. Rapport Projet FNRAA 010/AP03M010202. Fonds National de Recherches Agricoles et Agro-alimentaires (FNRAA), Dakar, Senegal.

Hobbs, J.E., Cooney, A. and Fulton, M. (2000) *Value Chains in the Agri-Food Sector*. College of Agriculture, University of Saskatchewan, Saskatoon, Canada.

Kaplinsky, R. and Morris, M. (2001) *A Handbook for Value Chain Research.* International Development Research Centre, Ottawa, Canada.

Lançon, F., Erenstein, O., Akande, S.O., Titilola, S.O., Akpokodje, G. and Ogundele, O.O. (2003) Imported Rice Retailing and Purchasing in Nigeria: A survey. West African Rice Development Association, Abidjan, Côte d'Ivoire.

Lançon, F., Erenstein, O., Touré, A. and Akpokedje, G. (2004) Quality and competitiveness of local and imported rice on West African urban markets. *Cahiers d'études et de recherches francophones* 13, 110–115.

Moschini, G., Menapace, L. and Pick, D. (2008) Geographical indications and the competitive provision of quality in agricultural markets. *American Journal of Agricultural Economics* 90, 794–812.

Neven, D. (2009) Three steps in value chain analysis. MicroNote 53. United States Agency for International Development, Washington, DC.

Opoku, R.A. and Akorli, P.A.K. (2009) The preference gap: Ghanaian consumers' attitudes toward local and imported products. *African Journal of Business Management* 3, 350–357.

Pearson, S.R., Stryker, J.D., Humpreys, C.R., Rader, P.L., Monke, E.A., Spencer, D.S.C., Craven, K., Tuluy, A.H., McIntire, J. and Page, J.M. (1981) *Rice in West Africa: Policy and economics.* Stanford University Press, Stanford, California.

PINORD (2007) Charte de qualité du riz de la Vallée du Fleuve Sénégal. Plateforme d'Appui des Initiatives du Nord, Saint-Louis, Senegal.

PINORD (2009) *Plan Stratégique 2009–2013 : Promotion de la sécurité alimentaire par la filière riz local dans la Vallée du Fleuve Sénégal.* Plateforme d'Appui des Initiatives du Nord, Saint-Louis, Senegal.

Rizzotto, A.C. and Demont, M. (2011) Tuning in to consumers: tailoring African rice value chains to urban markets. *Rural 21: The International Journal for Rural Development* 45, 29–31.

Shah, C.H. (1983) Food preferences, poverty and the nutrition gap. *Economic Development and Cultural Change* 32, 121–148.

Swinnen, J.F.M., Vandeplas, A. and Maertens, M. (2010) Governance and surplus distribution in commodity value chains in Africa. In: Sarris, A. and Morrison, J. (eds) *Food Security in Africa: Market and trade policy for staple foods in Eastern and Southern Africa.* Edward Elgar, Cheltenham, UK, pp. 77–98.

Tomlins, K.I., Manful, J.T., Larwer, P. and Hammond, L. (2005) Urban consumer preferences and sensory evaluation of locally produced and imported rice in West Africa. *Food Quality and Preference* 16, 79–89.

Unnevehr, L.J. (1986) Consumer demand for rice grain quality and returns to research for quality improvement in Southeast Asia. *American Journal of Agricultural Economics* 68, 634–641.

USAID (2006) Globalization and the small firm: a value chain approach to economic growth and poverty reduction. MicroReport 42. United States Agency for International Development, Washington, DC.

USAID (2009) Global food security response: West Africa rice value chain analysis. MicroReport 161. United States Agency for International Development, Washington, DC.

World Bank (2008) *World Development Report 2008.* The World Bank, Washington, DC.

25 Improving Grain Quality of Locally Produced Rice in Africa

Koichi Futakuchi,[1]* John Manful[1] and Takeshi Sakurai[2]
[1]*Africa Rice Center (AfricaRice), Cotonou, Benin;*
[2]*Institute of Economic Research, Hitotsubashi University, Tokyo, Japan*

Introduction

Consumption of rice is growing faster than that of any other food commodity in Africa, because it has become a convenience food for the growing urban populations. Imports of rice (close to 10 million tonnes (Mt) per year) cost the African continent more than US$5 billion in 2009 (e.g. Seck *et al.*, 2012). There is an urgent need to substitute imported rice with locally produced rice, but this requires improving the quality of the latter to standards set by the consumer: locally produced rice is often not competitive vis-à-vis imported rice because of a perceived lower quality. In Nigeria, consumers dislike locally produced rice compared to imported rice mostly because of the impurities (stones and other foreign matter) it contains (Tiamiyu *et al.*, 2011). Similarly, 71% of 390 consumers interviewed in Accra, the capital of Ghana, preferred imported over local rice mostly because of impurities (inclusion of foreign matter) and unavailability of local rice in sufficient quantities all year round (Diako *et al.*, 2010).

Grain quality may mean very different things from one country or region to another (see also Rutsaert *et al.*, Chapter 23, this volume). For example, rice with low grain breakage is preferred in most countries, but broken rice is liked in Senegal. Red rice is sold at a higher price than white rice in Kumasi (the second largest city in Ghana), but this is not true for other markets in the country (Sakurai *et al.*, 2006). Nigerians generally prefer cooked rice of a 'harder' texture than consumers in neighbouring countries.

Watanabe *et al.* (2002a,b, 2006) showed that many grain quality characteristics such as grain breakage, grain whiteness after milling, and protein and amylose content were significantly different among 47 rice cultivars cultivated under wet- and dry-season growing conditions in Côte d'Ivoire. In a different trial, where effects of different harvesting dates on milling and related traits were examined in several varieties, interactive effects of harvesting dates and varieties were significant for husking recovery and head-rice ratio, while no significant interactive effects were observed in grain dimensions, milling recovery and chalkiness (Futakuchi *et al.*, 2001 – see definitions of grain quality below). These results illustrate that grain quality is a function of the variety (intrinsic factors) and the production and processing environment (extrinsic factors). It also underlines the importance of conducting grain quality assessments under standardized cultivation and postharvest practices.

Enhancing grain quality of locally produced rice in Africa needs to consider: (i) clear grain

* Corresponding author: k.futakuchi@cgiar.org

© CAB International 2013. *Realizing Africa's Rice Promise* (eds M.C.S. Wopereis *et al.*)

quality targets based on a good understanding of consumer preferences; and (ii) achieving these targets through the choice of the variety and by improving the conditions under which rice is produced, stored and processed.

This chapter starts with an overview of definitions of rice grain quality. Next a case study is presented on consumer preferences for rice as observed in four regions of Benin. Ways to improve grain quality are discussed, ranging from the choice of the variety to management factors before and after harvest. The chapter concludes with recommendations for making African produced rice more competitive vis-à-vis imports.

Defining Rice Quality

We distinguish between rice grain quality indicators in terms of: (i) physical appearance before and after the milling process; and (ii) cooking and eating quality. (For a full description see, e.g., Food Agency, 1995; Manful, 2011.)

Physical appearance before and after milling

- Size and shape – a good rice variety should have a characteristic and stable grain size and shape; consumers generally prefer milled rice of a particular size and shape. The Standard Evaluation System (SES) for rice (IRRI, 1996) provides the following scales: for size, 1 (extra-long, length >7.5 mm), 3 (long, length 6.6–7.5 mm), 5 (medium, length 5.51–6.6 mm), 7 (short, length ≤5.5 mm or less). For shape, 1 (slender, length–width ratio >3.0), 3 (medium, ratio 2.1–3.0), 5 (bold, ratio 1.1–2.0), 9 (round, ratio <1.1). These scores are recorded for brown rice to evaluate these traits as genetic characteristics avoiding the effect of milling on size and shape.
- Impurity – presence of stones and other foreign matter.
- Husking recovery – the percentage of brown rice to paddy on a weight basis after dehusking.
- Milling recovery – the percentage of milled (polished) rice to brown rice on a weight basis after milling.
- Head-rice ratio – the proportion of 'head rice' in milled rice on a weight basis. Any milled grain that is less than three-quarters the size of the whole grain is classified as 'broken'. Milled rice grains after removal of damaged (broken, discolored, etc.), dead and immature grains are referred to as head rice. Breakage is a major damage and high head-rice ratio is considered as an indication of less grain breakage.
- Chalkiness – levels of 'chalkiness', i.e. chalky appearance either in the centres or edges of milled grains due to diffused reflection of light; a chalky grain denotes a grain in which the starch granules are not tightly packed and consequently grain filling can be termed as 'incomplete'. Grains with high chalkiness are generally softer, more likely to break during milling. The scales of chalkiness by SES (IRRI, 1996) are as follows: 0 (none, no chalky area in the kernel), 1 (small, chalky area <10%), 5 (medium, chalky area 11–20%), 9 (large, chalky area >20%).
- Grain colour – colour is an important attribute to the food industry, including for rice. Consumers frequently look at a rice sample and make a judgment decision based largely on overall appearance including colour. Frequently, degree of whiteness is representatively used as a description of colour. The degree of whiteness can be determined by readings (%) of a milling/whiteness meter, with higher values for greater degrees of whiteness.
- Translucency – translucency of milled grains is one of the milled grain traits related to appearance. A milling/whiteness meter also provides a value (%) of translucency and a higher value means that the grain is more transparent.
- Aroma – aromatic rice is sold at a higher price (premium) in most regions. Existence of aroma can be judged with milled grains and is a consumer selection criterion in the market. Aroma is also a factor included in the eating quality.

Cooking and eating qualities

- Apparent amylose content – this is a major factor affecting rice eating texture (usually, the lower the amylose content the softer the rice).

- Gelatinization temperature – the time required to cook rice is determined by the temperature at which the crystalline structures of the starch in the grain begin to melt when heated in the presence of water. This is known as the gelatinization temperature (GT) and it ranges from 55°C to 85°C. Rice with high GT takes a longer time to cook and the cooked rice has a harder texture, while low-GT rice takes a shorter time to cook and has softer to intermediate texture.
- Gel consistency – when rice flour is cooked in excess water it liquefies into a gel. The consistency of the gel can be indicative of the cooked rice texture.
- Viscosity of cooked rice flour – several viscosity parameters (maximum, minimum, breakdown, final and setback viscosities) can be read in the amylograph, which can be drawn by a Brabender viscogram or other similar equipment with a rice flour sample (Food Agency, 1995) and the parameters are associated with eating texture. For those who eat rice after it has cooled, the parameters at low temperatures (final and setback viscosities) are important.
- Protein content – high protein content makes eating texture harder. Amylose content has a much greater effect on texture than protein content (Watanabe *et al.*, 2002a).
- Water-uptake ratio, elongation ratio and swelling ratio – water-uptake ratio is weight of absorbed water during cooking per unit weight of raw rice; elongation ratio is a ratio of the length of a cooked grain to that of a raw grain; swelling ratio is a ratio of the volume of a rice sample after cooking to that of the rice sample before cooking.
- Cooking time.

Comparing Local and Imported Rice: A Case Study in Benin

Consumer preferences

A survey was conducted in Cotonou (the commercial capital of Benin in the southern region, where people are highly exposed to imported rice), Lokossa (a town in southwest region), and Materi and Tangieta (both towns in the northern region) (Study 1). At each site, 125 consumers compared the grain qualities of two imported rice brands (Gino and Sultana), two local – already cultivated in the country – rice varieties (Beris 21 and Tox long) and three newly introduced NERICA varieties (NERICA 1, NERICA 3 and NERICA 4) (Fofana *et al.*, 2010a). The consumers were asked to rate milled and cooked rice samples using the following scale: 1, dislike very much; 2, dislike; 3, neither dislike nor like (neutral); 4, like; 5, like very much. In the test of milled (raw or uncooked) rice, imported rice (Gino and Sultana) and NERICA 1 were appreciated by consumers across Benin; local varieties and NERICA 3 and NERICA 4 were rated lower (Table 25.1). However, cooked rice was appreciated very differently depending on the region (Table 25.2). The imported brands (Gino and Sultana) and NERICA 4 were liked in Cotonou, but they were neutral or disliked and NERICA 4 was disliked in the other areas. NERICA 1 was disliked very much in Cotonou, but received neutral or higher ratings in the other regions. Beris 21 and Tox long were generally disliked and their ratings were especially low in Cotonou. For the three regions other than Cotonou, no rice varieties or brands received a score of 4 or 5. Cotonou stood out as very different from the other regions (Table 25.2).

Preference for imported rice in the raw (uncooked) state seemed to be common across the country, so matching the quality of imported rice can be an excellent target for the improvement of locally produced rice. However, preferences for cooked rice were very site specific. The study also revealed that there can be large variation in consumer preferences even within a relatively small country like Benin. This may suggest a wide range of preference variation in other larger countries and also across Africa, and the necessity to collect more information to expand the study.

Physical appearance

To compare varietal characteristics of local rice with imported rice (Study 2), AfricaRice cultivated

Table 25.1. Consumer acceptability scores[a] for raw (uncooked) rice in the four locations in Benin. (Modified from Fofana *et al.*, 2010a.)

Variety/brand	Cotonou N[b] = 125	Lokossa N = 125	Materi N = 125	Tangieta N = 125	Total N = 500
Gino (imported)	4.4	4.1	4.2	4.3	4.3
Sultana (imported)	4.6	4.4	4.5	4.6	4.6
Beris 21 (local)	2.5	2.8	3.1	2.9	2.9
Tox long (local)	2.5	2.5	2.8	3.1	2.8
NERICA 1 (newly introduced)	4.2	3.8	4.1	4.2	4.1
NERICA 3 (newly introduced)	3.1	2.8	2.5	2.8	2.9
NERICA 4 (newly introduced)	3.4	3.1	2.7	2.7	3.0

[a]1, dislike very much; 2, dislike; 3, neither dislike nor like; 4, like; 5, like very much.
[b]N, number of consumers participating in the test.

Table 25.2. Consumer acceptability scores[a] for cooked rice in the four locations in Benin. (Modified from Fofana *et al.*, 2010a.)

Variety/brand	Cotonou N[b] = 125	Lokossa N = 125	Materi N = 125	Tangieta N = 125	Total N = 500
Gino (imported)	4.2	3.3	3.4	2.6	3.6
Sultana (imported)	4.6	3.1	2.8	2.6	3.5
Beris 21 (local)	2.1	2.9	3.1	2.9	2.8
Tox long (local)	2.2	2.8	3.0	2.7	2.7
NERICA 1 (newly introduced)	1.0	3.0	3.7	3.6	2.6
NERICA 3 (newly introduced)	2.9	1.8	2.1	1.8	2.1
NERICA 4 (newly introduced)	4.2	2.2	1.8	1.9	2.8

[a]1, dislike very much; 2, dislike; 3, neither dislike nor like; 4, like; 5, like very much.
[b]N, number of consumers participating in the test.

five local varieties and five NERICA varieties (three of the five have been introduced to Benin). The harvested samples were processed using a laboratory husker and polisher, and their grain qualities compared to five imported rice brands in 2010 (Fofana *et al.*, 2011a; Table 25.3). The imported rice brands were less chalky, more translucent and had more slender grain shapes than the local rice and the NERICA varieties. According to the standard evaluation system (SES) for rice (IRRI, 1996), the chalkiness scores 0, 1, 5 and 9 indicate non-chalky, small (less than 10% of kernel area), medium (11–20% of kernel area) and large (more than 20%) chalkiness, respectively. As the average of three samples, the imported rice brands had scores of 0 or 1, while the local rice had scores of 4 or 5 (except Adny 11). NERICA varieties also had a high score of 4. For grain hardness, two local varieties (Beris 21 and IDSA 1) and four NERICAs (NERICA 3, 4, 6 and 7) showed low values compared to the imported brands. These varieties also have chalky grains (soft and chalky grains are likely to break). However, the imported brands were not whiter than the other two groups, local and NERICA varieties. In the sensory evaluation of raw (uncooked) rice, imported brands (Gino and Sultana) were highly preferred compared over local varieties (Beris 21 and Tox long) and NERICA varieties (NERICA 3 and 4) (Table 25.1). From these results it follows that efforts to improve local and NERICA varieties in

Table 25.3. Physical characteristics of imported, locally cultivated and NERICA rice. (Modified from Fofana *et al.*, 2011a.)

Rice variety/ brand	Grain size[a]	Grain shape[a]	Chalkiness[a]	Hardness (kgw)[b]	Whiteness (%)	Translucency (%)
Elephant (imported)	Long	Slender	1	7.6	70.8	3.0
Gino (imported)	Long	Slender	0	7.7	66.2	3.4
Savana (imported)	Long	Slender	1	7.2	62.5	3.6
Special rice (imported)	Long	Slender	1	7.0	63.4	3.4
Sultana (imported)	Long	Slender	1	8.3	64.7	3.7
Adny 11 (local)	Long	Intermediate	1	8.9	60.7	2.5
Beris 21 (local)	Long	Intermediate	5	5.5	65.7	2.4
IDSA 1 (local)	Long	Intermediate	5	6.1	68.7	3.4
IR 841 (local)	Long	Intermediate	4	8.1	63.2	2.9
Tox long (local)	Medium	Intermediate	5	7.6	74.6	1.2
NERICA 1 (newly introduced)	Long	Intermediate	4	7.2	73.2	2.7
NERICA 3 (newly introduced)	Long	Intermediate	4	6.5	71.3	2.8
NERICA 4 (newly introduced)	Long	Intermediate	4	5.9	65.3	3.3
NERICA 6	Medium	Intermediate	4	6.1	68.0	3.1
NERICA 7	Long	Intermediate	4	5.9	70.0	3.2

[a]Ranks of a grain size, a grain shape and chalkiness scores were based on Standard Evaluation System (IRRI, 1996). Although brown rice is used for the determinations of the size and shape in SES (IRRI, 1996), milled rice was used here, since the imported rice was all milled rice.
[b]Hardness was measured by grain-hardness tester (Kiya Co Ltd). The units of the meter are kilogram weight (kgw).

terms of appearance should focus on greater grain slenderness, less chalkiness and higher translucency. NERICA 1 was liked as much as the imported brands Gino and Sultana (Table 25.1), although it did not have similar grain slenderness, chalkiness and translucency (Table 25.3). This is most probably due to the wide appreciation of NERICA 1 in Benin for its aroma.

In another survey conducted in 2010 (Study 3), AfricaRice collected 110 milled rice samples (comprising eight varieties) from local millers in both urban and rural areas of eight towns (Cove, Dassa, Glazoue, Lokossa, Malanville, Parakou, Savalou and Save) in Benin. For Cotonou and Porto-Novo (political capital of Benin), no local millers were found in either the cities or the rural areas near the cities. Impurity, head-rice ratio, whiteness and translucency were determined on the samples (Table 25.4). The average whiteness obtained (67.3) is comparable to that of the imported rice (Table 25.3), but the translucency is low compared to that of the local varieties (Table 25.3, it is important to note that the processing in Study 2 was done in the laboratory). This suggests rather poor processing practices of local millers. Impurity was 0.01–5.30% with an average of 1.09% and head-rice ratio was 1.4–71.6% with an average of 28.4%. Although the same measurements were not made on imported rice at the same time, most of the milled rice samples collected from local millers were of poorer quality. For example, in the Philippines, grade 3 (lowest rating for grain quality) rice has 50–60% of head-rice ratio and 0.20–0.50% impurities (Sampang, 1992). Poor quality in appearance may be partly caused by varietal characteristics, but inappropriate postharvest handling can negatively affect the appearance – the

Table 25.4. Physical characteristics of 110 raw (uncooked) rice samples collected from local millers.

	Impurity (%)	Head-rice ratio (%)	Whiteness[a] (%)	Translucency[a] (%)
Average	1.09	28.4	67.3	1.6
SD	0.84	20.2	7.2	0.7
Max	5.30	71.6	83.0	3.6
Min	0.01	1.4	54.0	0.3

[a]One red rice sample found was removed since it was a different type.

current situation requires improvement of post-harvest processing for some traits (impurity, head-rice ratio and translucency).

Cooking and eating quality

In Study 2, amylose content, protein content, gel consistency, water-uptake ratio, elongation ratio, swelling ratio and cooking time were also measured (Table 25.5). The first three factors can affect eating quality. In the sensory test for cooked rice, imported rice (Gino and Sultana), NERICA 1 and NERICA 4 were preferred in some regions of Benin, while local rice (Beris 21 and Tox long) and NERICA 3 were not liked in any region. There could be varietal differences in the varieties and brands tested for characteristics associated with eating quality. All the local varieties, NERICA varieties and imported brands had similar amylose content (21–24%). However, the NERICA varieties had higher protein content than the other two rice categories (higher protein content can make eating texture harder). In gel consistency, the local varieties (except Adny 11) showed lower gel consistencies than the imported and NERICA rices (rice of lower gel consistency can have harder eating texture). These results suggest that the imported rice may have slightly softer eating texture than the others. However, the difference of rating among NERICA 1, NERICA 3 and NARICA 4 in the sensory test of cooked rice (Table 25.2) cannot be explained by these factors – other factors, which were not measured, appear to affect their eating quality. The average scores of the sensory test of cooked rice across all regions were higher for the imported rice brands than the others (Table 25.2), although the imported rice was not liked in the northern regions. The direction of varietal improvement might be to make eating texture a little softer. Since high protein content of the NERICA varieties can be an asset for nutrition supply, improvement of eating texture should be made through the reduction of amylose content, not of protein content.

The local rice (except Adny 11) had higher water-uptake ratio than the other two categories, although one imported brand (Special rice) also had a high value. For elongation rate, most of the varieties and brands showed similar values, but ratios of Tox long (local) and NERICA 1 were relatively high. Swelling ratio was generally higher in the NERICA varieties than in the local varieties. In the imported brands, values of swelling ratios varied greatly, from 3.835 (Savana) to 4.933 (Gino). All five NERICA and three local varieties (Beris 21, IDSA 1 and IR 841) had long cooking times (22.0–25.6 minutes) compared to the imported rices (17.0–19.6 minutes).

From the sensory tests of raw rice and cooked rice (Study 1), we cannot capture the preference for cooking characteristics. However, it is considered that, in general, consumers prefer rice that increases in volume during cooking and short cooking time will be appreciated, especially by urban consumers. The swelling ratio of the existing NERICA varieties (4.202–4.600; average 4.460) is not inferior to that of imported brands (3.835–4.933; average 4.284); however, shortened cooking time will be the point of improvement for the NERICA varieties.

Improving Grain Quality

In this section, various ways to improve grain quality are discussed in more detail.

Table 25.5. Cooking and eating characteristics of imported, locally cultivated and NERICA rices. (Modified from Fofana *et al.*, 2011.)

Rice variety/ brand	Water-uptake ratio	Elongation ratio	Swelling Ratio	Cooking time (min)	Protein content (%)	Amylose content (%)	Gel consistency (mm)
Elephant (imported)	2.074	1.593	4.073	17.3	7.2	23.4	100.0
Gino (imported)	2.221	1.453	4.933	19.6	6.4	23.4	99.5
Savana (imported)	2.191	1.567	3.835	19.3	7.4	22.7	100.0
Special rice (imported)	2.420	1.490	4.364	19.0	7.3	23.0	99.3
Sultana (imported)	2.018	1.539	4.217	17.0	7.0	23.6	99.3
Adny 11 (local)	2.289	1.503	4.117	18.6	7.7	22.9	100.0
Beris 21 (local)	2.516	1.504	4.260	25.6	7.1	22.7	71.3
IDSA 1 (local)	2.386	1.512	4.045	22.3	7.1	21.2	66.8
IR 841 (local)	2.610	1.564	4.132	23.3	7.4	23.1	63.0
Tox long (local)	2.690	1.760	4.365	19.6	6.8	22.6	71.3
NERICA 1 (newly introduced)	2.200	1.603	4.202	26.0	10.6	23.8	99.6
NERICA 3 (newly introduced)	1.895	1.556	4.401	22.0	10.5	23.8	99.0
NERICA 4 (newly introduced)	2.100	1.532	4.600	24.0	10.6	22.8	99.1
NERICA 6	2.004	1.574	4.595	23.0	10.7	22.5	98.8
NERICA 7	1.895	1.542	4.503	24.0	10.8	24.1	95.0

Selecting the right variety

In the case of Benin described above, the breeding target for raw (uncooked) rice appearance is apparently imported rice. It is desirable for new varieties to have less chalkiness, greater translucency and more slender grains than the existing ones. Aromatic rice, NERICA 1, was also preferred in the country, so aroma is also a target. For eating quality, it might be better to aim for a cooked rice texture that is a little softer than that of the existing varieties if consumers in large population cities are targeted (imported rice with a softer eating texture than the existing local rice was rated high in Cotonou). Cooking time of new varieties should also be as short as that of imported rice.

Exact information on consumer preferences is not available in many countries in Africa. In such cases, varietal development for quality is largely dependent on expert knowledge. National agricultural research systems (NARS) may concentrate on developing varieties that match local preferences, as determined by expert knowledge or information from surveys. Given the sensory test results across Benin (preferences related to cooked rice were different across regions, but those related to raw rice were commonly for imported rice) and the fact that a lot of imported rice is consumed in Africa, imported rice can be a model of varietal selection for milled rice appearance. For eating quality, which is more region specific, breeders are best advised to avoid developing 'extreme' varieties.

For instance, all local and NERICA varieties and imported brands evaluated in Study 2 have similar amylose content in the range 21–24%; it will not be necessary to develop varieties with extremely high or low amylose contents. For cooking quality, high swelling ratio and short cooking time are generally desirable breeding directions.

Development of premium-quality rice – non-chalky, translucent, slender and aromatic rice – can be a breeding objective; the 'ORYLUX series' developed by AfricaRice and NARS partners is an example (Futakuchi *et al.*, 2011). However, there are diverse and wide-ranging requirements for varieties and it will be difficult to develop a variety that fulfils all the requirements (high yield potential, resistances to all possible constraints in a target ecosystem, etc.) plus premium quality. In contrast, it will be important for all newly developed varieties – whatever the main breeding objective is – to possess acceptable quality (i.e. not disliked by consumers). For this purpose, the breeders at AfricaRice check all of their materials in terms of basic grain quality characteristics for raw rice appearance from the F_4 generation and unacceptable (poor-quality) progenies (high chalkiness, etc.) are eliminated from further selection.

Most quality traits – such as grain dimensions, grain hardness, whiteness, chalkiness and absence/existence of aroma – are easy to measure on a small sample. Conventional breeding will still be effective for the improvement of these traits. For traits which cost more and take more time to measure, marker-assisted selection (MAS, see Ndjiondjop *et al.*, Chapter 12, this volume) will be a convenient approach. For example, major and minor genes for amylose content occur on chromosomes 6 and 5, respectively, and for gel consistency two QTLs have been identified on chromosomes 2 and 7 (He *et al.*, 1999). Grain breakage (low head-rice ratio) is a highly complicated characteristic associated with various quality traits like grain dimension, chalkiness and hardness, and controlled by a large number of genes. In the comparison of head-rice ratios of 50 varieties between two cropping seasons, a significant correlation was observed ($P < 0.05$). However, its correlation coefficient was low ($r = 0.337$) compared to grain slenderness (length/width ratio) ($r = 0.920$, $P < 0.001$), for example (Watanabe *et al.*, 2002b), suggesting that head-rice ratio is a varietal characteristic but its heritability will be very low compared to grain slenderness. Consequently, selection of head-rice ratio using data from actual measurement will be inefficient. It may be useful to pyramid each characteristic associated with high head-rice ratio by MAS.

Pre-harvest cultivation practices

Sowing seeds that are a mixture of different varieties usually results in asynchronous heading in the same field, which makes harvesting difficult, which will often result in high grain breakage (see below). Farmers usually use their own seeds (harvested the previous season); thus, seed selection and handling should be done carefully. It is recommended that farmers renew their seeds every few years by purchasing certified seeds.

Zhang *et al.* (2012) compared rice grain yield and quality between a conventional lowland irrigation scheme in China and a water-restricted irrigation scheme, where about 16% of the water of the conventional scheme was used. Although yield and 1000-grain weight were lower in the restricted irrigation, quality was higher (higher head-rice ratio and lower chalkiness). Fofana *et al.* (2010b) showed that drought during maturity in upland rice did not affect yield, but decreased grain breakage and amylose content. These results provide some evidence that a water-saving irrigation system may produce rice with a higher head-rice ratio without sacrificing yield.

Fertilizer applications may affect grain quality substantially. Wopereis-Pura *et al.* (2002) showed that late application of 30 kg N/ha at booting stage, in addition to farmers' standard fertilizer management (116 and 127 kg N/ha in the wet and dry seasons, respectively) increased head-rice ratio by 12% and 24% in the wet and dry seasons, respectively, in northern Senegal (Sahel). The late N application also increased yield by 0.4 t/ha and 1.0 t/ha, respectively, in the wet and dry seasons. However, the practice is useful only for rice farmers who can afford to apply additional fertilizer.

Timing of harvest is crucial for grain quality and appropriate timing is often compromised because of competing activities or lack of labour for harvesting. Futakuchi *et al.* (2001) showed that late harvesting (45 days after 50% flowering) decreased head-rice ratio by 7.9% compared to the harvesting at 30 days after 50% flowering in M'bé (in the transition zone between Guinea savannah and forest areas of Côte d'Ivoire). Since lower air humidity during paddy drying decreases head-rice ratio (Aguerre *et al.*, 1986), the negative effect of late harvesting on head-rice ratio will become more severe in the Sahel.

Some insects and diseases may intervene in the field during grain-filling stage. Blast disease reduced head-rice ratio (Hai *et al.*, 2007) and infections of panicle blast increased the number of damaged grains (Shim *et al.*, 2005). Grains attacked by shield bugs (stink bugs) can be damaged with dark spots (Noda and Ishii, 1983) and the spots do not disappear after milling. Management to control pests will be important to produce quality rice. However, little is known about the extent of quality losses caused by these pests in Africa (see also Nwilene *et al.*, Chapter 18, this volume).

Harvesting and threshing

In the harvesting of upland rice, where subsistence farmers are dominant, panicles are harvested with knives; harvested panicles are bundled and kept at home (in humid environments, panicles are stored in the relatively less-humid kitchen). In lowland rice systems, the stems are cut at several centimetres above the base by sickle and manually threshed – for instance, farmers flay harvested plants against a metal barrel or a log. Threshed paddy is dried, sometimes on the ground in direct sunlight. These practices can introduce impurities and induce low head-rice ratio upon milling.

Manual harvesting is a labour-intensive process and proper timing of harvesting may be difficult. As mentioned in the section above, appropriate harvesting time is important for grain quality. For large irrigated lowland plots in the Sahel, a mechanical harvester is a solution to this constraint, and attempts to introduce such a technology have been started by AfricaRice. Introduction of threshing machines, such as the ASI thresher–cleaner (Rickman *et al.*, Chapter 27, this volume), is crucial and has already been carried out in some Sahelian areas. However, this requires an initial investment and backup systems to maintain the machines.

It is important to dry paddy on plastic sheets or tarpaulins to avoid contamination with soil or other foreign materials. It is well known that fast drying of harvested paddy leads to a higher percentage of broken grains (Rhind, 1962). Practices to avoid quick drying of paddy are relatively easy to adopt. One recommendable practice is simply to dry paddy in the shade, in an area with the lowest available relative humidity, not in direct sunshine. In areas where this is difficult, a possible practice is to have a thick layer of paddy on the ground and stir the paddy more frequently during the drying process.

Milling

Most subsistence rice farmers carry out milling by themselves using mortar and pestle and subsequently winnow it in the wind. In this process, almost all grains are broken; meanwhile, professional millers may grade milled rice before it is sold on the market. In Kumasi (Ghana), both Engelberg and so-called 'one-pass' mini rice mills are used (Sakurai *et al.*, 2006). In an Engelberg mill, a steel roller with grooves rotates in a metal casing and paddy grains are husked and then polished by friction between the grooves and grains and among the grains; the machine was developed in 1888 in the USA and is still manufactured and used in South-east Asia (Satake Cooperation, 2006a). The one-pass mini rice mill was developed in Japan in 1956. This machine operates on the same principle of 'friction milling' as the Engelberg, but has a different style of dehusking and polishing with a larger capacity, and the rice husks and bran are removed by aspiration (Satake Cooperation, 2006b). Rice milled by the mini rice mill had significantly higher head-rice ratio and lower levels of impurity than that milled by Engelberg mills in Ghana (Sakurai *et al.*, 2006).

In Benin, all 110 millers from whom AfricaRice collected milled rice samples (Study 3), used Engelberg mills. Their output is of poor quality (see 'Physical appearance' above). To improve the quality of milled rice on the local market, the introduction of improved mini rice mills is one necessary measure. The mini rice mill has greater capacity than the Engelberg type, so a miller who adopts it could process more rice and collect more fees – the fee for milling a unit volume of rice was not affected by technologies of milling, but by location of the mills (Sakurai *et al.*, 2006). The introduction of a new type of rice mill requires a large initial investment and it is important to consider location and effective dissemination modes. Of the 110 millers using Engelberg mills in Benin, 48 used locally produced machines and the remaining 62 used machines imported from Asia. The average levels of impurities and head-rice ratios were 1.02% and 24.4%, respectively, for the locally produced mills and 1.14% and 31.6%, respectively, for the imported mills. Impurity may depend on the condition of the paddy brought to the millers and imported mills produced slightly higher head-rice ratio. Before a shift from Engelberg types to mini rice mills, some improvement could be possible in the local capacity for making mills. Stones that contaminate paddy (through inappropriate threshing and drying) can damage the mills. It is also important to improve local capacity to repair and maintain mills.

More information on rice mills can be found in Rickman *et al.* (Chapter 27, this volume).

Storage conditions

Stored grains may be attacked by weevils or other insects (for more details see Nwilene *et al.*, Chapter 18, this volume). Storage condition and duration may affect grain quality. Bleoussi *et al.* (2011) tested four NERICA varieties and one *Oryza sativa* variety in relation to the effect of storage duration (paddy under room temperature and humidity) on some grain quality traits up to 64 weeks and showed that with the elapse of storage time, husking recovery, milling recovery, head-rice ratio and grain hardness increased, but whiteness decreased; there was no effect on grain chalkiness, but some changes in cooking and eating characteristics caused by extended storage were observed. Little information is available on the effect of existing storage conditions in Africa on grain quality. A study of existing storage practices would be helpful in determining which indigenous practices are appropriate for storing rice and what new practices could be introduced. In Asia, some storage technologies to improve quality and reduce damage to rice are available; for example, IRRI has developed a storage bag for crops (not only for rice but also for other crops such as coffee) called the 'Super Bag'. This is a hermetic storage bag and prevents both oxygen and water from entering from the outside – the following positive effects are expected: approximate doubling of the germination life of seeds; control of insects without using chemicals; improved head-rice recovery of stored grain by about 10% (IRRI, 2005). Such a reputable technology could be directly introduced to Africa relatively easily.

Parboiling

Parboiling is a hydrothermal process in which the crystalline form of starch present in the paddy rice is changed into an amorphous one as a result of the irreversible swelling and fusion of starch. This is accomplished by soaking in cold and then hot water or steam at low pressure, before drying and milling the rice. There are many methods of parboiling rice, but the central processes are essentially the same. The process of parboiling results in physical, chemical and organoleptic changes in the rice, with economic and nutritional advantages (Choudhury, 1991). Parboiled rice is consumed worldwide, except in East Asia, and its consumption is increasing in Africa and Central and South America (Kimura, 1995). Parboiled rice is strongly preferred in some African countries such as Nigeria, where almost all locally produced rice is parboiled. In an assessment of consumer preferences in Benin (Mhlanga, 2010), consumers of all regions – except the south (including Cotonou) – preferred parboiled rice to non-parboiled rice. In countries such as Benin and Nigeria, therefore, there could be a large potential in parboiling technologies to increase incomes, especially for women processors – the main producers of parboiled rice. Fofana *et al.* (2011b) show that a method to steam paddy instead of boiling resulted in a better-quality product with fewer heat-damaged grains, higher head-rice ratio and greater translucency. This method of direct steaming is being disseminated to rice parboilers in Benin by AfricaRice and the Institut national de recherches agricoles du Bénin (INRAB, the Beninese NARS).

Bleoussi *et al.* (2011) report significant interactive effects of paddy storage duration and parboiling on several grain quality traits (grain hardness, husking recovery, head-rice ratio, whiteness, translucency, cooking time, swelling ratio and gel consistency), but not milling recovery. In consideration of the importance of parboiled rice in Africa, further study will be necessary on this topic.

Other factors

Ghanaian rice farmers near Kumasi take paddy directly to millers and pay them on a milled-rice volume basis. Farmers then sell milled rice to brokers or retailers (Sakurai *et al.*, 2006). Sakurai *et al.* (2006) observed that low grain breakage had a significantly positive effect on the rice price in the urban areas of Kumasi but not in the rural areas around the city. For goods to be sold at the right price (i.e. high-value goods sell at high prices and low-value goods at low prices), market information should be available to all actors. In the urban area of Kumasi, there are several clusters of rice millers. In contrast, in the rural areas, millers are much more scattered.

If rice with higher (more appreciated) quality is sold at a higher price, rice value-chain actors will be more motivated to improve the quality of rice; in this example, this is more likely to occur with rice sold in the urban areas than that sold in the rural communities around Kumasi.

Conclusions

Improving the market competitiveness of locally produced rice is crucial for reducing the amount of imported rice and contributing to food security in Africa. The competitiveness of locally produced rice against imported rice in the market depends primarily on its acceptance by the consumer.

Sensory tests were conducted in four regions in Benin using two imported brands, two local (already cultivated) varieties and two newly introduced NERICA varieties with 125 consumers at each site. Preference for raw (uncooked) rice was common across regions with a high preference for imported rice brands and NERICA 1. For cooked rice, preferences differed with region. Cotonou showed very different results (imported rice and NERICA 4 were liked there) from the other regions, most probably due to the fact that its population is more exposed to imported rice compared to the other three regions. The local varieties did not receive high preference scores as either raw (uncooked) rice or cooked rice samples in any region.

Comparing the sensory tests with the study on grain quality traits of imported rice brands, local varieties and NERICA varieties, it was shown that low chalkiness, high translucency, slender grains and aroma are factors that are commonly preferred across Benin for raw (uncooked) rice. The third study showed that locally produced rice was generally of poor quality because of impurities, and a poor head-rice ratio and translucency. These are important indicators to be considered in the improvement of varieties, and with respect to pre- and postharvest practices. Although preference for cooked rice was different among regions, the imported rice brands received relatively high preference scores on average across regions. With respect to cooked-rice characteristics, imported rice possessed a softer texture than the local and NERICA varieties, which was preferred by most consumers.

The diversity of preferences for cooked-rice attributes within a small country like Benin may suggest that there could be huge variations in cooked-rice preferences across the continent. Precise information is not available for most countries in Africa and further consumer-preference surveys will be necessary in those countries. On the other hand, preference attributes for imported raw rice in all regions of Benin was fairly uniform. The physical quality attributes of imported rice could therefore be the benchmark target for breeding, pre- and post-harvest practices in many areas in Africa. It has been reported that despite some superior eating properties of local rice, it is not competitive against imported rice in terms of physical quality in Nigeria (Tiamiyu *et al.*, 2011; Ogundele and Diagne, unpublished data).

In all steps along the value chain where the quality of locally produced rice is affected – i.e.

varietal selection (breeding), cultivation, harvesting and threshing, milling, storage and parboiling – there are some issues to be addressed. Possible improvements in each step are discussed in this chapter based on detailed studies in Ghana and Benin and available knowledge (documents and expert knowledge).

In order to achieve a better physical quality of rice, especially fewer impurities and higher head-rice ratio, the harvesting, threshing and milling processes are crucial with mechanization being important for all these operations. Results from Benin show that locally manufactured rice mills generally produce lower-quality rice than imported mills of the same type. Improvement of the local capacity to construct these mills is needed and this may also contribute to better maintenance of the mills.

Apart from factors such as varietal selection and pre- and postharvest practices discussed above, the socio-economic environments surrounding value-chain actors may have an influence on the quality of locally produced rice. If locally produced rice of better quality is sold at a higher price, the actors will be more motivated to improve rice quality. In our example from Ghana, this was the case in the urban area of Kumasi but not in the surrounding rural areas. This difference between urban and rural areas is because of differences in access to price information (Sakurai *et al.*, 2006). Therefore, better access of value-chain actors to price information is an important prerequisite for rice quality improvement.

References

Aguerre, R., Suarez, C. and Viollaz, P.E. (1986) Effect of drying on the quality of milled rice. *International Journal of Food Science and Technology* 21, 75–80.

Bleoussi, T.M.R., Fofana, M., Innocent, B. and Futakuchi, K. (2011) Effect of parboiling and storage on grain physical and cooking characteristics of the some NERICA rice varieties. In: *Innovation and Partnerships to Realize Africa's Rice Potential.* Proceedings of the Second Africa Rice Congress, Bamako, Mali, 22–26 March 2010. Africa Rice Center, Cotonou, Benin. Available online at: http://www.africarice.org/workshop/ARC/3.3%20Bleoussi%20fin.pdf (accessed May 2012).

Choudhury, N.H. (1991) Parboiling and consumer demand for parboiled rice in South Asia. In: *Rice Grain Marketing and Quality Issues.* Selected papers from the International Rice Research Conference, 27–31 August 1990, Seoul, Korea. International Rice Research Institute, Manila, Philippines, pp. 47–54.

Diako, C., Sakyi-Dawson, E., Bediako-Amoa, B., Saalia, F.K. and Manful, J.T. (2010) Consumer perceptions, knowledge and preferences for aromatic rice types in Ghana. *Nature and Science* 8, 12–19.

Fofana, M., Adhikari, K., Futakuchi, K., Dossou, J., Bleoussi R.T.M. and Dieng, I. (2010a) Urban consumer acceptability of newly introduced rice with local and imported rice in Benin: a comparison. *Indian Food Industry* 29, 37–45.

Fofana, M., Cherif, M., Kone, B., Futakuchi, K. and Audebert, A. (2010b) Effect of water deficit at grain ripening stage on rice grain quality. *Journal of Agricultural Biotechnology and Sustainable Development* 2, 100–107.

Fofana, M., Futakuchi, K., Manful, J.T., Bokossa Yaou, I., Dossou, J. and Bleoussi, R.T.M. (2011a) Rice grain quality: a comparison of imported varieties, local varieties with new varieties adopted in Benin. *Food Control* 2, 1821–1825.

Fofana, M., Wanvoeke, J., Manful, J., Futakuchi, K., Van Mele, P., Zossou, E. and Bléoussi, T.M.R. (2011b) Effect of improved parboiling methods on the physical and cooked grain characteristics of rice varieties in Benin. *International Food Research Journal* 18, 697–703.

Food Agency (1995) *Rice Post-Harvest Technology.* The Food Agency, Ministry of Agriculture, Forestry and Fisheries, Tokyo, Japan.

Futakuchi, K., Watanabe, H., Efisue, A. and Sobambo, B.A. (2001) Milling characteristics of WARDA lowland varieties in different harvesting dates. *Japanese Journal of Crop Science* 70(Extra issue 2), 283–284.

Futakuchi, K., Sié, M. and Wopereis, M.C.S. (2011) Rice breeding strategy at AfricaRice. In: Yanagihara, S. (ed.) Next Challenges in Rice Development for Africa: Workshop for New Collaboration between JIRCAS and AfricaRice. *JIRCAS Working Report* No. 70. Japan International Research Center for Agricultural Sciences (JIRCAS), Tsukuba, Japan, pp. 1–14.

Hai, H.L., Kim, P.V., Du, P.V., Thuy, T.T.T and Thanh, D.N. (2007) Grain yield and grain-milling quality as affected by rice blast disease (*Pyricularia grisea*), at My Thanh Nam, Cai Lay, Tien Giang. *Omonrice* 15, 102–107.

He, P., Li, S.G., Qian, Q., Ma, Y.Q., Li, J.Z., Wang, W.M., Chen, Y. and Zhu, L.H. (1999) Genetic analysis of rice grain quality. *Theoretical and Applied Genetics* 98, 502–508.

IRRI (1996) *Standard Evaluation System for Rice*. International Rice Research Institute, Los Baños, Philippines.

IRRI (2005) Grain storage: the IRRI super bag. *Rice Fact Sheets*. International Rice Research Institute, Los Baños, Philippines.

Kimura, T. (1995) Parboiling. In: *Rice Post-Harvest Technology*. The Food Agency, Ministry of Agriculture, Forestry and Fisheries, Tokyo, Japan, pp. 380–393.

Manful, J. (2011) Developing rice varieties with high grain quality in Africa: challenges and prospects. In: *Innovation and Partnerships to Realize Africa's Rice Potential*. Proceedings of the Second Africa Rice Congress, Bamako, Mali, 22–26 March 2010. Africa Rice Center (AfricaRice), Cotonou, Benin. Available online at: http://www.africarice.org/workshop/ARC/1.20%20Manful%20ed2.pdf (accessed May 2012).

Mhlanga, S. (2010) Economic analysis of consumer based attributes for rice in Benin. MSc thesis, McGill University, Montreal, Canada.

Noda, H. and Ishii, T (1983) Characteristics of peckey rice caused by stink bugs (Heteroptera). *Japanese Journal of Applied Entomology and Zoology* 27, 231–233.

Rhind, D. (1962) The breakage of rice on milling: a review. *Tropical Agriculture* 39, 19–28.

Sakurai, T., Furuya, J. and Futakuchi, K. (2006) Effects of industrial amassment on the improvement of efficiency and quality: a case study for rice millers in Ghana. In: Sonobe, T. and Sawada, Y. (eds) *Market and Economic Development*. Toyo Keizai Shinpou Sya, Tokyo, Japan, pp. 151–179.

Sampang, R.L. (1992) Paddy and milled rice grading. In: Semple, R.L., Hicks, P.A., Lozare, J.V. and Castermans, A. (eds) *Towards Integrated Commodity and Pest Management in Grain Storage*. Food and Agriculture Organization of the United Nations, Rome, Italy. http://www.fao.org/docrep/x5048e/x5048E02.htm (accessed May 2012).

Satake Cooperation (2006a) History of rice mills. *Tasty* 33, 2–4.

Satake Cooperation (2006b) History of Japanese rice mills. *Tasty* 33, 4–6.

Seck, P.A., Diagne, A., Mohanty, S. and Wopereis, M.C.S. (2012) Crops that feed the world 7: rice. *Food Security* 4, 7–24.

Shim, H.S., Hong, S.J., Yeh, W.H., Han, S.S. and Sung, J.M. (2005) Damage analysis of rice panicle blast on disease occurrence time and severity. *Plant Pathology Journal* 21, 87–92.

Tiamiyu, S.A, Usman, A., Gbanguba, A.U., Ukwungu, M.N. and Ochigbo A.A. (2011) Assessment of quality management techniques: toward improving competitiveness of Nigerian rice. In: *Innovation and Partnerships to Realize Africa's Rice Potential*. Proceedings of the Second Africa Rice Congress, Bamako, Mali, 22–26 March 2010. Africa Rice Center, Cotonou, Benin. Available online at: http://www.africarice.org/workshop/ARC/3.4%20Tiamiyu%20fin.pdf (accessed May 2012).

Watanabe, H., Futakuchi, K., Jones, M.P., Teslim, I. and Sobambo B.A. (2002a) Brabender Viscogram characteristics of interspecific progenies of *Oryza glaberrima* Steud. and *O. sativa* L. *Nippon Shokuhin Kagaku Kogaku Kaishi* (Journal of the Japanese Society for Food Science and Technology) 49, 155–165.

Watanabe, H., Futakuchi, K., Teslim, I. and Sobambo, B.A. (2002b) Milling characteristics and grain quality traits of interspecific progenies of Asian rice (*Oryza sativa*) with African rice (*Oryza glaberrima*). *Japanese Journal of Tropical Agriculture* 46, 47–55.

Watanabe, H., Futakuchi, K., Jones, M.P. and Sobambo, B.A. (2006) Grain protein content of interspecific progenies developed from the cross of African rice (*Oryza glaberrima* Steud.) and Asian rice (*O. sativa* L.). *Plant Production Science* 9, 287–293.

Wopereis-Pura, M.M., Watanabe, H., Moreira, J. and Wopereis, M.C.S. (2002) Effect of late nitrogen application on rice yield, grain quality and profitability in the Senegal River valley. *European Journal of Agronomy* 17, 191–198.

Zhang, Y.J., Hua, J.J., Li, Y.C., Chen, Y.Y. and Yang, J.C. (2012) Effects of phosphorus on grain quality of upland and paddy rice under different cultivations. *Rice Science* 19(2). Published online at: www.ricescience.org/EN/abstract/abstract9494.shtml.

26 Developing Competitive Rice Value Chains[1]

J. Dirck Stryker*
Associates for International Resources and Development (AIRD), Arlington, Massachusetts

Introduction

This chapter describes how value-chain analysis can be used to identify key constraints in the rice commodity chain and how these constraints can be overcome through infrastructural investment, institutional change and policy reform. It applies this analysis to the specific case of rice in Rwanda, where a major goal is to slow the growth of rice imports. The analysis reveals that the chief problem is lack of competitiveness of local rice vis-à-vis imports due to the poor quality of milling and a policy environment that discourages investment in mills capable of producing rice that is competitive. This problem of the competitiveness of local rice production is also prevalent in much of West Africa.

The paper is organized as follows. The first section sets out some of the broader dimensions of competition between rice produced and milled in sub-Saharan Africa (SSA) and rice that is imported from outside the continent. Much of this discussion is focused on West Africa and on some of the empirical research that has been undertaken within this sub-region. The second section applies value-chain analysis to this problem in Rwanda. The final section draws some conclusions and looks at the implications for SSA. It raises a number of questions regarding the role of locally grown rice, both as a substitute for imports and as a means of assuring food security.

Can African Rice Compete?

Growth of rice consumption in SSA has been outstripping that of rice production. Between 1961 and 2005, rice consumption in SSA grew at 4.52% annually, compared with growth in production of 3.23% (WARDA, 2007). Imports increased dramatically to fill the gap, as the self-sufficiency ratio (production/consumption) declined from 112% in 1961 to 60% in 2005. The international market thus supplied 40% of SSA's rice needs.

Most growth in production has been due to an expansion of the area under cultivation rather than to an increase in yields, although this situation has changed dramatically in many countries since the 2007–2008 rice crisis (see Seck *et al.*, Chapter 2, this volume). This is partly because of the dominance of rainfed cultivation, both upland and lowland, which accounts for 70% of the area planted to rice in SSA (compared with 37% worldwide; see Diagne *et al.*, Chapter 3, this volume). In fact, substantial increases in yields have been achieved under

* Corresponding author: dstryker@aird.com

© CAB International 2013. *Realizing Africa's Rice Promise* (eds M.C.S. Wopereis *et al.*)

irrigation systems involving total water control, but these have not been sufficiently important in aggregate to offset the relative stagnation of yields under rainfed conditions.

Among the major staple foods in SSA, rice consumption is growing most rapidly (Diagne, 2010). There are a number of reasons for this. Rice is an important convenience food. It requires less time and energy to cook than most of the other staples such as beans, cassava, banana and potato. This is an important attribute given women's increased participation in the labour market and the growing importance of food consumption away from the household. Other desirable features include its ease of storage and handling, and its long shelf life. These make it a highly desirable food in urban areas. Furthermore, when processed, rice gives off several useful by-products, which can be used in the animal feed sub-sector, breweries and other industries. This lowers the price of milled rice to consumers.

If past trends continue (reflecting growth rates seen over the period 2000–2012), consumption in SSA will reach 36 million tonnes (Mt) in 2020 compared with 19 Mt of milled rice production (Seck *et al.*, Chapter 2, this volume). This will imply a self-sufficiency ratio of only about 52%. However, this growing imbalance is based on past trends and does not have to reflect the future. Africa has a number of advantages that may put it in a better position. Production has been increasing, in part spurred on by the food crisis of 2007–2008, and a number of proactive measures have been taken. The fact that most rice is grown under rainfed conditions means that there is substantial potential for investment in irrigation, and a consequent increase in yields up to 7 t/ha or more. Furthermore, the new NERICA varieties developed by the Africa Rice Center for rainfed cultivation mean that gains in yields are not limited to irrigated rice. Finally, a number of studies suggest that SSA has a comparative advantage in rice production in competition with imports, and that comparative advantage appears to be growing, especially given the price increases since 2005 (Lançon and Benz, 2007).

Given this potential to expand production, it would seem that there is a good possibility that the growth in the gap between rice consumption and production could be reversed, and that imports could decline, at least in relative terms. But for this to happen, there needs to be more attention paid to postharvest dimensions of the rice value chain. For it is not at all clear that most of the rice being produced in SSA is competitive with imported rice – not so much in terms of comparative costs as in terms of quality and other dimensions in which these grains compete.

It is well recognized that rice produced locally in SSA suffers a significant price discount in comparison with imported rice (Lançon and Benz, 2007). This appears to be at least partly due to perceived differences in quality. Local African rice generally has more impurities mixed in with it and is not of uniform grain size and colour (Campbell *et al.*, 2009, p 31).[2] Lack of product uniformity leads to longer cooking times and unpredictable preparation. Cleaning and sorting this rice prior to cooking is time-consuming and laborious.

Lack of ready availability of local rice throughout the year may also be a problem. Imported rice tends to be consumed mostly in urban areas, whereas local rice is consumed more in the countryside, close to the production areas. Traders in urban areas are reluctant to carry local rice because of lack of credit from wholesalers, whereas credit is generally available from importers and the wholesalers to whom they sell (Campbell *et al.*, 2009, p 32). In addition, supplies of local rice, which are abundant just after harvest, tend to dwindle thereafter because of lack of adequate incentives for storage.

There is other evidence that local rice and imported rice are not very good substitutes in consumption. In most SSA countries, prices of local rice tend not to follow the movement of prices on the international rice market, at least not very closely. Consumers' price elasticity of response to movements of the prices of imported rice appears to be very low. For example, following the devaluation of the CFA franc in 1994, consumers in Burkina Faso, Côte d'Ivoire, Mali and Senegal responded by reducing their consumption of other foods, such as meat and wheat products, while maintaining their consumption of rice at relatively constant or only slightly reduced levels (Diagana *et al.*, 1999). Consumers tend to set minimum quality standards and are unlikely to shift from imported rice to local rice just because the price of the former has risen. Instead they tend to shift to lower-quality but cheaper imported rice – for example, one with a higher percentage of broken grains (Campbell *et al.*, 2009, p 28).

There is evidence from Senegal that the rise in rice prices on the world market in 2007–2008, which was transmitted into higher retail prices for imported rice in Senegal relatively quickly, did not spill over into a parallel increase in the price of local rice. In fact, prices of local rice actually decreased and there was a substantial widening of the gap between the two (Rutsaert *et al.*, 2010). This has important implications for policy. If there is very little substitutability between imported and local rice, then increased production of local rice, in the face of very inelastic demand, could result in a decline in its price rather than a replacement of imports (Rizzotto and Demont, 2011).

Another question that arises is whether the price discount to which local rice is subject in comparison with imports occurs because of inferior quality, particular consumer preferences, or prejudice in favour of imported rice because it is 'foreign'. This is an important question for policy because, as seen in the case study of Rwanda, the cause will have direct implications for the actions necessary to reduce this discount. Unfortunately, little research has been done that would shed some light on this issue. One exception is a study that was undertaken in Senegal to test the hypothesis that consumers prefer imported rice because it is foreign (Rutsaert *et al.*, 2010). This involved allowing consumers to bid in an auction for four different rices, judged on the basis of their appearance, a sensory taste test, and their own experience.

> Each participant was endowed with one kilogram of the 'benchmark' mediocre-quality SRV (Senegal River Valley) rice and was presented with the option to exchange this kilo into the three alternative rice types. The benchmark rice is a mix of varieties (Sahel 108 and [Sahel] 201) and is commonly available on the market. The imported Thai-25 rice has a grain quality somewhere between the benchmark and the high-quality SRV rice and contains some impurities. The unlabeled and labeled rice types are equal, i.e. high-quality SRV rice (Sahel 108 variety) which is purified, carefully sifted, and branded and marketed as 'Rival'.
>
> (Rutsaert *et al.*, 2010, p 7–8; see also Demont and Neven, Chapter 24, this volume).

Senegalese consumers were willing to pay a premium of CFA 43/kg (€0.07/kg) for quality SRV rice relative to imported rice. They were further willing to add CFA 29/kg (€0.04/kg) for the branded version of the quality SRV rice. Although this suggests that these consumers valued the high-quality SRV rice, especially in relation to the benchmark of mediocre quality, it does not tell us very much about how they would value this rice in comparison with similar high-quality broken rice that was imported. Furthermore, Senegalese consumers may be differentiated to some extent by their preference for broken rice and by the high degree to which they consume imported rice not just in the cities but also in the countryside. Thus, they are more familiar than consumers in many other countries with the differences between imported and local rice.

So what is to be done? Is there any way that African rice can compete on more equal terms with imported rice? If so, what does this require? Better milling? More organized marketing? Improved access to finance? Public promotion? To suggest some answers to these questions, we turn in the next section to a case study of value-chain analysis applied to the issue of the competitiveness of Rwandan rice vis-à-vis imported rice. This analysis involves the following steps.[3]

- Identify where local rice is likely to be most competitive, e.g. as a substitute for rice imports, as a substitute for other locally produced food crops, or for export. This should be based on:
  - projections of local demand and supply;
  - analysis of the economic cost of supplying each of these markets in relation to price; and
  - identification of elements of competitiveness other than cost in each market, e.g. quality, continuity of supply.
- Map different value chains and their spatial relationships and modes of production, collection, milling, transportation and marketing; determining relative quantities of rice passing along each chain.
- Establish the enabling environment, including policies related to production, input taxes and subsidies, investment incentives, trade, regulation, institutional promotion and competition policy.

- Assess costs, value addition, and profitability at each stage of each value chain in both financial and economic prices.[4]
- Identify infrastructural, institutional, financial and policy constraints at each stage of each value chain.
- Establish benchmarks for performance indicators against international competition, e.g. quality in relation to price.
- Prioritize binding constraints.
- Understand political, social, economic and financial issues behind these constraints.
- Formulate actions to overcome these constraints.

Case Study of Rwanda[5]

Background

Rwanda is a relative newcomer to rice. Traditionally, its staple foods have been rainfed crops such as sweetpotato, cassava, beans and maize, which are grown primarily on the steep hillsides using relatively labour-intensive techniques to keep the soil in place and to preserve its fertility. However, beginning in the 1950s, Rwanda began to import rice as a convenience food, especially for urban consumers. These imports accelerated following the ending of the genocide in 1994.

Several years ago, the Government of Rwanda decided to give high priority to the production of rice in the country's marshlands, where, with adequate investment in irrigation infrastructure, the crop is capable of yielding up to 7 t/ha during each of two growing seasons (Ministry of Agriculture and Animal Resources, 2005, pp. 2–3). The government provided this investment and farmers responded by growing rice largely as a cash crop. However, as production increased there was a growing need to expand facilities for processing and marketing. At first, much of this processing was done with small hullers, which produced rice with substantial impurities and little uniformity of grain colour and size. This was similar to much of the rice being produced and processed for home consumption or the local market in West Africa.

In addition to the hullers, the government earlier invested in a number of medium-sized mills of 1–3 t/hour capacity. By the time that the National Rice Program was going into effect, these mills were old and required substantial upgrading or replacement with more modern equipment. The government responded by privatizing these mills, turning them over to cooperatives or selling them to private investors. Often this was done as part of a joint venture arrangement between the cooperatives and the investors. However, even though these mills produced a better quality of rice than the small hullers, they had a hard time competing. Finally, in 2009, the government banned most of the small hullers in an effort to ensure adequate supplies of paddy to the larger mills.

This ban encountered strong opposition from Rwandan farmers, resulting in only about 10% of total production flowing through the mills. This was largely because the prices the mills offered to farmers were substantially lower than those offered by private traders. As a result, the mills continued to operate at less than full capacity, raising their costs and making it even more difficult for them to compete.

A major objective of the study of the rice commodity chain was to make recommendations concerning how the quality of milled rice could be improved so that this rice could compete more favourably with imports, and, at the same time, result in greater efficiency in marketing and processing.

Key findings

The following are some of the key findings from the study that are relevant to these objectives.

- Careful projections of the supply of and demand for rice in Rwanda over 10 years show that, while there is considerable potential for expansion of production, the growth of consumption is such that it is highly unlikely that Rwanda will be able to achieve self-sufficiency in rice – at least as long as production of rice is confined to the marshlands. Given the continuing rice deficit, the major market for locally grown rice is as a substitute for rice imports or to displace other food crops in consumption.
- There are some advantages to competing with imports. In particular, as long as rice is being imported, local production is being

protected by the cost of transporting imported rice from producing countries to Rwanda. In addition, it makes it easy to protect local rice by imposing taxes on rice imports.

- Only about 10% of the paddy being produced is processed in modern mills. The rest is going to seed, feed, losses, hand pounding, or husking by small hullers. This results in a lower-quality product, which commands a lower price than imports. At the same time, several of the modern mills complain about not being able to buy all the paddy they can process. One reason appears to be the low prices that they offer compared with private traders.
- The poor quality of most local rice as currently processed, plus the separation of the market for this rice from that of imported rice, means that it is not a very good substitute for imports. To the extent that it is being sold on local markets, it is competing mostly with other foods. When processed in modern mills, the local rice appears to be of a quality that should be competitive with imported rice, but it does not command the same price as the latter. Frequently, the unexplained discount on local rice is 10–20%. The reasons for this are not well understood and need to be investigated.
- In addition to taxes on imports, the government provides substantial subsidies on production by paying for irrigation infrastructure and subsidizing the transportation of fertilizer. These subsidies amount to close to 30% of total on-farm production costs. Rice production is financially profitable everywhere, partly because of the subsidies involved. It is economically profitable in most, though not all, locations in competition with rice imports.
- Data on the range of small to medium-sized modern mills (0.2–3.5 t/h) that are operating in Rwanda suggests that most of those mills are profitable and would continue to be profitable even if they paid the market price for paddy. There is a wide range of mills of different sizes on the international market that are able to process and grade rice of good quality, with a low percentage of broken grains, uniform appearance and few impurities.

Recommendations to improve competitiveness

The Government of Rwanda has two major objectives regarding the rice sector. The first is to increase supplies of domestic rice to reduce dependence on rice imports. The second is part of the overall objective of assuring food security for the majority of the population. It is clear that the current system for processing most rice in small hullers is not contributing to the first objective because of the poor quality of the hulled rice, as indicated by the substantial discount this rice receives – up to 30% of the price of imported rice. However, it can contribute to food security.

There are some important advantages to having a modest-sized market for this hulled rice. First, the quality is not so bad that no one will eat it. With a little cleaning, it can be quite satisfying, especially for the poor and those living in rural areas. Second, the by-product of such rice is usually a mixture of husks, bran and some broken grains. This is a relatively good animal feed, especially for ruminants, and can contribute to food security via the livestock sector. Third, the small scale of these hullers allows them to be owned and operated by a number of small entrepreneurs, creating income and employment, and – through this – improved food security. Finally, availability of a large number of small rice-processing units creates an environment in which competition flourishes, ensuring that costs to consumers are minimized.[6]

Nevertheless, the poor quality of rice processed in the small hullers does limit its ability to compete with imports. In the longer term, as incomes increase, the consumption of this rice will decline in importance as consumers become more discriminating. It is important, therefore, that a modern milling sector develop alongside the small-scale, poor-quality sector. The analysis of costs and returns in Rwanda suggests that this can be done without the ban on small hullers. There is a ready market for good-quality local rice as a substitute for imported rice, especially if this rice enters the market chain at the same point as the imported rice. Substantial profits can be made all along the value chain even if the larger mills compete freely in the market for paddy.

This does not mean that all rice should be milled in a few larger mills. There are ample markets for and ready sources of supply of modern mills of different sizes – ranging from 250 kg/h to 5 t/h. These mills are capable of cleaning, dehusking, whitening, polishing and grading rice to the specifications that will meet international standards. Small mills are particularly appropriate where rice production occurs in relatively isolated regions. They give to the cooperatives and individual owners the sense that they are able to make productive investments that capture some of the value added in the commodity chain. Furthermore, the existence of a number of mills in this size range, in addition to a few larger mills at strategic locations, helps to spur competition. Such a diversified range of milling activities should therefore be promoted.

There is one caveat. Even when the appearance of imported and local rice is the same, there appears to be a consumer bias towards imported rice, which results in a price discount of 10–20% of the price of imported rice. It is not evident why this is so. There may be characteristics that are not always clearly visible in the market, such as better taste, faster cooking time, more swelling in cooking, greater assurance of availability, and attachment to brand names. These characteristics need to be identified through consumer surveys, blind tastings, and physical testing. Once the sources of bias are understood, efforts should be made to overcome these perceived differences through investment in storage, parboiling and other measures. Finally, to the extent that differences in preferences persist that are unexplained, this should be publicized in order for consumers to learn that their preferences are not based on real differences and that they can gain by buying cheaper local rice.

Conclusions and Implications

The qualitative competitiveness of local rice production and processing in SSA is increasingly being recognized as a major constraint to increased self-sufficiency and reduced dependence on rice imports. The duality of the market for rice is a major factor. On one hand, the market for imported rice is heavily concentrated in urban areas and demands a relatively high standard of quality in terms of lack of impurities and relatively uniform appearance. On the other hand, rice in rural areas is more likely to come from local sources, to have more impurities and to be less uniform with respect to grain size and colour, though it may have certain taste characteristics that are appreciated by local consumers.

One result is that the price of local rice on the domestic market is generally less than that of imported rice. It is not totally evident whether this is because the local rice is of lower quality. Some local rice, processed in modern mills and flowing into similar market channels, would seem to compete very well with imported rice. However, in Rwanda at least, local rice of very similar appearance in terms of lack of impurities, percentage broken grains, uniformity of grain size and colour, etc., sells at a substantial discount over imported rice. Whether this is because of other characteristics that cannot so easily be observed or results from prejudice in favour of imported rice is not known.

This is an important area for research. The effect of the price discount is substantial in terms of foregone profits. Furthermore, lack of substitutability between imported and local rice means that all of the efforts being undertaken to promote local production as a substitute for imports may simply drive down the price of local rice, discouraging investment in expanded production.

It appears from the Rwanda case study and from some other experiences in Africa that it is quite possible to invest in modern mills of various sizes that are capable of profitably milling rice of good quality that should compete very well with imports. Even though the cost of this milling will be higher than using small hullers alone, the price increase resulting from the higher-quality product should be more than enough to offset the higher costs. This is certainly the case in Rwanda.

Given the cleaning, polishing and grading capabilities that these mills have, better milling should take care of the problems of impurities, lack of uniformity, and high percentage of broken grains. With complementary investment in storage, this should also ensure that adequate supplies of local rice are available year round. It will not necessarily solve problems of taste, storability, cooking time, water absorption and other characteristics that are not apparent to the eye. That will require further testing and consumer surveys, including the use of blind sensory tests. It will also require trials to see how improvements can be made. For example, there is some evidence that imports benefit from longer storage during transport, and this improves quality (Bleoussi *et al*., 2010). Parboiling may be used to increase water absorption during cooking. All of this needs to be better understood and tested to see what works best in competition with imports.

To the extent that there do not appear to be any important differences between imported and local rice processed in modern mills – whether visible or not – there may simply be a prejudice in favour of imported rice because it is 'foreign'. The fact that definable differences have not been found and yet imported rice is more expensive than local rice needs to be publicized in order for consumers to learn that their preferences are not based on real differences and that they can gain by buying cheaper local rice.

A final issue to be considered is what happens to food security as progress is made in increasing food self-sufficiency. Here the answer is not so evident. Low prices of rice, even of mediocre quality, help ensure that the poor have adequate supplies of food. If all paddy is channelled towards the modern mills, what happens to this source of food?

The importance of this depends on what other alternatives are available. In Rwanda, there are numerous alternative sources of food upon which the population has traditionally depended. The opportunity to grow rice in the marshlands is an opportunity to make the best use of this scarce resource, which can be most profitably exploited by selling the rice as a cash crop. But this can most profitability occur only if the paddy is processed in modern mills in order to earn the best price in competition with imports.

In other countries, for example Liberia, rice is widely consumed as a staple food throughout the country. Here one wants to be much more careful in moving too rapidly towards the use of modern processing facilities, since this can put rice out of reach of the poor. This does not mean that improvements in processing should not be made, but simply that there are some advantages to the simple hulling of rice to produce a product that is widely consumed throughout the country, yields a by-product that is valued as an input into the livestock sector, and provides income and employment to the rural economy. Thus, a dual approach involving both types of processing might be best until there are reasonable assurances that food security has been attained.

Notes

[1] An earlier version of this chapter was originally presented at the Second Africa Rice Congress, and subsequently published online as part of the proceedings of that Congress (www.africarice.org/workshop/ARC/0P3%20Stryker%20ed2.pdf). It has been updated, restyled and reformatted for this book and some minor errors corrected. (Reproduced with permission from Africa Rice Center.)

[2] Local rice is sometimes preferred over imported rice because of its taste characteristics, even if it has lots of impurities and is poorly graded. However, the preference given to taste is not generally great enough to offset the other negative factors.

[3] The elements of this analysis are drawn from FIAS (2007), Stryker and Salinger (2004) and Pearson *et al.* (1981).

[4] Financial prices are those found in the market; economic prices are financial prices minus taxes and plus subsidies included in them.

[5] The results reported here are from a study undertaken by the author for the Ministry of Agriculture and Animal Resources of Rwanda.

[6] West Africa is filled with examples of how the introduction of small rice hullers contributed to higher prices for producers, lower prices for consumers, and greater market efficiency. For example, in the Office du Niger in Mali, small-scale hullers competed very well with large, inefficient and costly rice mills owned by the state. This led to liberalization of rice marketing and expansion of rice production and processing.

References

Bleoussi, R., Fofana, M., Bokossa, I. and Futakuchi, K. (2011) Effects of storage on grain quality characteristics in NERICA rice. In: *Innovation and Partnerships to Realize Africa's Rice Potential.* Proceedings of the Second Africa Rice Congress, Bamako, Mali, 22–26 March 2010. Africa Rice Center, Cotonou, Benin, pp. 3.3.1–3.3.7. www.africarice.org/workshop/ARC/3.3%20Bleoussi%20fin.pdf.

Campbell, R., Schiff, H., Snodgrass, D., Neven, D., Downing, J. and Sturza, D. (2009) Global Food Security Response: West Africa rice value chain analysis. *USAID microREPORT* no. 161. [United States Agency for International Development, Washington, DC].

Diagana, B., Akindès, F., Savadogo, K., Reardon, T. and Staatz, J. (1999) Effects of the CFA franc devaluation on urban food consumption in West Africa: overview and cross-country comparisons. *Food Policy* 24, 465–478.

Diagne, A. (2010) Rice policy and impact research at AfricaRice. Presentation to CORAF/WECARD Meeting, Cotonou, Benin, 24 May.

FIAS (Foreign Investment Advisory Service) (2007) *Moving Toward Competitiveness Analysis: A value chain approach.* The World Bank, Washington, DC.

Lançon, F. and Benz, H.D. (2007) Rice imports in West Africa: trade regimes and food policy formulation. Poster prepared for the 106th seminar of the European Association of Agricultural Economists, Montpellier, France, 25–27 October.

Ministry of Agriculture and Animal Resources (2005) National Program for the Production of Rice, Program 2006–2016. Ministry of Agriculture and Animal Resources, Kigale, Rwanda.

Pearson, S., Stryker, J.D. and Humphreys, C.H. (1981) *Rice in West Africa: Policy and economics.* Stanford University Press, Palo Alto, California.

Rizzotto, A.C. and Demont, M. (2011) Extending reach to strengthen value chains: increasing consumer awareness of quality Senegal River rice. In: *Innovation and Partnerships to Realize Africa's Rice Potential.* Proceedings of the Second Africa Rice Congress, Bamako, Mali, 22–26 March 2010. Africa Rice Center, Cotonou, Benin, pp. 3.1.1–3.1.7. www.africarice.org/workshop/ARC/3.1%20 Rizzotto%20fin.pdf.

Rutsaert, P., Demont, M., Ndour, M., Verbeke, W., Seck, P.A. and Tollens, E. (2010) Paying for quality: private versus collective valuation of Senegal River Valley rice. Paper presented at the Second Africa Rice Congress, Bamako, Mali, 22–26 March.

Stryker, J.D. and Salinger, B.L. (2004) A guide to commodity-based export diversification and competitiveness strategies for African countries. Paper prepared for UNCTAD, 5 November.

WARDA (2007) *Africa Rice Trends: 2007 Brief – Overview of recent developments in the sub-Saharan African rice sector.* Africa Rice Center, Cotonou, Benin.

27 Mechanizing Africa's Rice Sector

Joseph Rickman,[1]* Jean Moreira,[2] Martin Gummert[3] and Marco C.S. Wopereis[2]

[1]*International Rice Research Institute (IRRI), Maputo, Mozambique;* [2]*Africa Rice Center (AfricaRice), Cotonou, Benin;* [3]*International Rice Research Institute (IRRI), Los Baños, Philippines*

Introduction

Significant increases in rice production can be obtained in sub-Saharan Africa (SSA) by increasing rice grain yield in existing production systems, expanding rice harvested area, and reducing harvest and postharvest losses. However, availability of labour at critical times is often a major constraint, and this situation is aggravated by the effects of the HIV/AIDS pandemic. Delays during harvesting, threshing and drying cause losses in both grain quantity and quality. Rice crops in Africa are often not planted on time due to late and poor land preparation as farmers wait for rain to soften the soil so they can prepare the land using hand implements. This often results in poorly prepared and uneven seed beds and weeds become a major problem. To compensate for this, farmers tend to plant more seeds. Late-planted crops are more susceptible to pest damage, in particular leaf diseases. Unlevelled and uneven fields result in higher water requirements, poor fertilizer-use efficiency and non-uniform crop ripening causing delays at harvest and increased losses to shattering and birds. Rainfed rice crops are often left in the field up to one month longer than necessary because farmers tend to wait until grain moisture content is low enough for easy hand-threshing. At the other end of the scale in some irrigated systems, such as in the Senegal River delta, farmers tend to wait for large combine-harvesters to harvest their fields. These harvesters often break down and can only enter fields when the soil is dry. Such situations increase the risk of shattering and bird damage.

In general, postharvest losses of rice are very high, with estimates ranging from 30% to 50% (Mrema *et al.*, 2008). In some instances, all of the grain is lost, contaminated by mycotoxins or spoiled by rain before harvest and during storage. These losses occur because of poor postharvest management, outdated postharvest technology and poor storage facilities. Losses in quality and quantity combined can reduce the value of the milled rice by 20–50% at the market, thus further reducing farmers' income and providing consumers with poorer-quality rice. In addition to these losses, there is also a loss in potential income by selling the grain at the point of harvest. If farmers are able to safely store their grain, they can often increase the value of the grain by 30% in the 3–4 months after harvest. Rice production in SSA in 2010 was estimated at 18.4 million tonnes (Mt) of paddy (AfricaRice, 2011). Assuming a conservative 20% of postharvest losses, reducing postharvest losses by half would provide an additional 1.84 Mt

* Corresponding author: j.rickman@cgiar.org

© CAB International 2013. *Realizing Africa's Rice Promise* (eds M.C.S. Wopereis *et al.*)

of paddy, equivalent to 10% of regional imports, with a value of about US$550 million per year.

Overcoming labour shortages requires labour-saving technologies and practices, and injecting energy into the farming system through mechanization (Mrema *et al.*, 2008). Agricultural mechanization aims at 'reducing human drudgery, increasing yields through better timeliness of operations because of the availability of more power, bringing more land under cultivation, enabling agriculture-led industrialization and markets for rural economic growth, and ultimately improving the standard of living of farmers' (FAO and UNIDO, 2010). Management practices along the entire rice value chain 'from farm to plate' can be mechanized, i.e. from land preparation, through water management, weeding, pest control and harvesting, to transportation, storage and processing.

Introduction of mechanization can address labour bottlenecks, improve productivity and allow household members to pursue other activities. More importantly, mechanization may allow for intensification and increases in harvested area. For example, introduction of a small-scale combine-harvester may allow timely rice harvesting and increase opportunities for growing a second crop in the same field in one year or opening up new rice fields – creating employment opportunities and increasing farm revenues. Proper land levelling using small hand tractors or four-wheel tractors with laser-assisted equipment will enable farmers to better manage their crops and gain higher returns from inputs such as mineral fertilizer. Development of the mechanization sector itself will also create employment.

Lack of mechanization seriously limits the productivity and competitiveness of rice-based systems in SSA. This is now widely recognized and agricultural mechanization is a key component in all of the national rice development strategies that have been developed under the Coalition for African Rice Development (CARD; 21 African countries; www.riceforafrica.org). There is now a clear commitment at national level to mechanize Africa's rice sector. At the same time, the continent is littered with wrecks of imported agricultural machinery, abandoned because: the technology is not adapted to the field conditions in SSA; it is of inappropriate design; there is a lack of spare parts; or maintenance is costly. Introduction of mechanization, therefore, requires careful analysis of past successes and failures, and discussion of lessons learned.

This chapter starts with a review of the status of agricultural mechanization in SSA. Next, pre-harvest, harvest and postharvest mechanization options that could make a difference in Africa are discussed. Finally, the different roles that partners need to play to enable sustainable mechanization of Africa's rice sector is illustrated, with a discussion of the outcome of a workshop on 'Boosting agricultural mechanization in rice-based systems in sub-Saharan Africa' held at Africa Rice Center, Senegal in June 2011.

Status of Agricultural Mechanization in Africa

The number of four-wheel tractors (with four wheels and two axles) can be used as an indicator of how far a country (or region) has advanced in mechanizing its agriculture. In 1961, SSA had approximately 3.4 times more tractors in use than in Thailand; however, by 2000, Thailand had the same number as the whole of SSA. These tractors were concentrated in a few countries. For example, in 2000, South Africa and Zimbabwe accounted for 50% and 17%, respectively, of the tractors in the Southern African Development Community, while Nigeria accounted for 68% of the tractors in the Economic Community of West African States (Mrema *et al.*, 2008). Few of these tractors worked in rice.

Primary land preparation in SSA relies on human muscle power for about 80% of the cultivated land, with draught animals and tractors being used on only 15% and 5%, respectively. This contrasts strongly with Asia, where land preparation on over 60% of the cultivated land is done by tractors (Mrema *et al.*, 2008). These figures illustrate that farm power is deficient almost everywhere in Africa, and rice-based systems generally are not an exception to this rule.

Pre-harvest mechanization: challenges and opportunities

In Africa, land preparation is done mostly by hand or animals. Power substitution by using either two- or four-wheel tractors is an obvious way to increase labour productivity and to improve the timeliness of operations. Programmes to increase the use of tractors have failed in the past because of a top-down approach to machinery selection, poor maintenance, lack of after-sales service and insufficient operator training. The two-wheel tractor enabled the mechanization of the green revolution in Asia and could also be used by smallholder farmers to do the same in Africa. Initially, the tractor would need to be imported but could eventually be produced locally, for example in joint ventures with Asian companies. The development of contract-service business models for four-wheel tractors for land preparation and other services such as laser levelling could increase the applicability of this technology. Minimum or zero tillage is not a common practice in rice-based systems in Africa.

Due to labour shortage, many farmers have shifted from transplanting to direct seeding or were already direct seeding using manual broadcasting. The drum seeder, which is a simple technology, is available for improved crop establishment in rows, which allows mechanized weeding between rows. More sophisticated direct-seeding technologies, both dry and wet seeded, are being verified in South Asia in conservation farming approaches and could be of interest to Africa at a later stage when successful experiences with mechanization are available.

If mechanical weeders are to be used, crops must be planted in rows. This will require more care and, in some instances, more time or labour at planting. However, if the planting is undertaken using a mechanical seeder, transplanter, or a drum seeder, this will not be a problem. Mechanical weeding using a cone weeder has proved to take one-sixth of the time it takes for traditional hand weeding. The type of weeder will also need to be matched to the soil as, for example, cone weeders are best in heavy clay soils and finger weeders in sandy conditions.

The majority of lowland-rice farmers in SSA level their land by moving soil from higher to lower portions of the field using a hand hoe. In large fields, farmers sub-divide the land into more manageable sizes. This practice tends to reduce the area available for planting because of the space taken up by bunds. Direct-seeding technologies such as the drum seeder, but also more sophisticated mechanized direct-seeding equipment, require improved land levelling for good crop establishment. In areas where four-wheel tractors are being used for contract ploughing, laser-assisted levelling could be added to the service provision for farmers. In India and Vietnam, laser levelling increases yields by 5–10%, reduces irrigation water requirement by 20–40%, reduces herbicide cost, produces better-quality rice and can lead to 4–6% increase of rice area if small rice fields are consolidated into bigger ones. Although the laser-hydraulic control equipment needs to be imported and appears to be expensive, when used in a contract-service business model it can be profitable for the contractor and the farmer.

Engine-driven axial-flow or propeller-type pumps can improve irrigation of individual fields if only a small amount of lift is required. These pumps usually provide flows in the range 150–1500 m^3/h, with heads in the range of 1.5–3 m. These pumps can be driven by the same engines as the hand tractors; these engines can also power stationary threshers.

Harvest and postharvest mechanization: challenges and opportunities

Harvesting

More than 70% of the rice in Africa is harvested by hand using a sickle, knife or machete. This requires a lot of labour, mostly provided by women in rainfed upland and rainfed lowland areas, and by men in irrigated environments. Hand harvesting is fraught with problems, including the time required that could be used in other activities and delays in harvesting, leading to both quantitative and qualitative losses. These losses occur through shattering, low moisture content, attack by rodents, birds and insects, grain germinating in panicle due to rainfall or lodging (panicles touch the soil).

Mechanized harvesting may involve the use of small reapers in combination with mechanical threshers, mini combine-harvesters and large combine-harvesters. Reapers were introduced for use in the rainfed and irrigated lowland environments and have not been really successful in SSA. Constraints in the use of reapers can be attributed to low capacity, less than 1 ha/day; they require an efficient operator, and spare parts may be difficult to access as they need to be imported. Reapers or cutting bars may also be mounted on a power tiller. Reapers are more appropriate for use on medium-sized farms (5–25 ha) and may be affordable for farmer cooperatives. Reapers are commonly used in the Office du Niger (Mali) and northern Senegal.

Since 2000, small combine-harvesters have been introduced from China, India, Japan and Thailand. While they can harvest 2–5 ha/day, their introduction has not been successful because of the higher level of sophistication of the technology, lack of trained operators, poorly prepared and unlevelled field conditions, lack of spare parts and maintenance, and the high initial investment required by small-scale farmers. These challenges are now being addressed by introducing smaller and cheaper 'mini-combines' with less sophisticated technology that may be fabricated and maintained locally. The major parts that need to be imported are the gearbox and drive belts. The local fabrication of these mini-combines will help ensure employment and incomes for the fabricators or local artisans, maintenance-service providers, spare-part fabricators, and others.

Large combine-harvesters capable of harvesting 5–10 ha/day are mostly used in the large irrigation schemes in Egypt, Mauritania and Senegal. In Senegal, only 29 remain in working condition out of the initial 41 introduced (SAED, 2010), They are suitable for large farms but need well-levelled fields to function efficiently. These combines are very expensive, often costing more than $100,000, tend to break down often, and spare parts are not readily available. Consequently, their use is much less now than in the pre-structural adjustment era and often those now being imported are second-hand machines.

Threshing

Threshing is done either manually or mechanically using a pedal or motorized thresher. Manual threshing involves hitting the panicles against a stationary object (e.g. drum, log of wood, wooden box), beating the cut crop with a stick, or running animals or a tractor over the cut panicles to remove the grain. Manual threshing is popular because of its low cost; however, quantitative and qualitative losses can be as high as 20–30%. This is especially a problem with excessively dry or wet panicles. Manual threshing requires the rice straw to be cut long to allow the paddy to be more easily held when hitting against a drum or threshing board to remove the grain from the panicle. Conversely, mechanical threshing requires short straw to avoid clogging the thresher and reducing the machine's threshing efficiency.

In Burkina Faso, Guinea, Liberia, Madagascar and Sierra Leone, hand and pedal threshers have been widely adopted. They are now built locally and are used by small-scale farmers as well as for seed producers. These machines have a threshing capacity of 500 kg/day, and require a lot of physical energy to operate. This has increased the desire within the region for mechanized threshers.

The motorized vortex thresher was introduced in the 1990s and is now widely used in Liberia, Madagascar, Nigeria, Senegal and Sierra Leone. It can thresh 400–800 kg/h and has revolutionized rice threshing in the region. The technology is very simple and the machine is now being fabricated by local companies such as Sismar and Matforce in Senegal. One of the limitations of the vortex thresher is that it does not clean the grain effectively. It also requires up to six people to operate and an additional three or four people are required to clean the grain by winnowing after threshing. This constraint was overcome with the introduction of a thresher–cleaner from Vietnam via the International Rice Research Institute (IRRI). This machine was modified collaboratively by Africa Rice Center (AfricaRice), Société d'aménagement et d'exploitation des terres du Delta du Fleuve Sénégal et des vallées du Fleuve Sénégal et de la Falémé (SAED) and the Institut sénégalais de recherches agricoles (ISRA) to suit local conditions and named 'ASI'. The ASI thresher can thresh and clean 1000–1500 kg grain/h, is operated by four people and the grain does not require winnowing after threshing. This thresher is mounted on two wheels and can be easily transported by draught animals, tractor or other vehicle. It is now fabricated locally and provides employment opportunities in many rural areas.

Crops are ideally harvested when grain moisture content is 20–22% and stored when the moisture level is less than 14%. When crops with high moisture content are threshed, some form of drying will be needed. This can be done by solar drying using the sun, or mechanical drying using some form of hot-air. Sun-drying is practised by most farmers in SSA by drying on mats, roadways or drying pads. Care needs to be taken to ensure that the paddy is not contaminated by soil or other materials when left to dry. With solar drying it is also difficult to control the rate of drying as it is difficult to control ambient temperatures. The ideal temperature is 42°C, but in many instances field temperatures are above 65°C. The depth of the paddy and the duration of drying will also affect grain quality and may result in high levels of grain breakage and low milling yields when processed. This problem can be reduced by turning the grain every hour, and tempering the grain by alternating between sun drying for a short time period and then allowing the grain to cool in the shade. Stacking of moist rice crops for more than 24 h after cutting may cause grain discoloration and spoilage. To avoid this, wet crops need to be threshed quickly after cutting, and then dried as soon as possible. This may be difficult to achieve when trying to sun dry rice harvested in the rainy season.

In these situations hot-air mechanical dryers are an alternative. Small flat-bed batch dryers have been tested in some countries, but they are not popular in Africa because they are expensive to operate and have limited capacity of 2–3 t/day. Rice husk furnaces are now available that can provide hot air for flat-bed and batch dryers. These will reduce the operating cost and use husk (a milling by-product) as the source of energy. Column dryers are used in large-scale mills, but high throughputs are needed and the high cost of fuel is often prohibitive.

Milling

The type of rice mill, the quality of the paddy, postharvest handling, the rice variety, and the miller's skill all influence milling performance. Good-quality paddy processed in a multi-stage rice mill can yield 65–70% of white rice (milling recovery) and 50–60% whole grains (head rice).

The ideal grain moisture level for milling rice is 12–14%. In the Senegal River valley, paddy harvested at the end of the wet season, may have a moisture content (MC) of 10–12%. This moisture content lasts for approximately one month, after which the paddy moisture drops below 10%. This is due to high ambient temperatures during the day, extended drying times and poor storage conditions. Over-dried paddy is more susceptible to breaking during husking and whitening, and this results in reduced white-rice and head-rice yields. Poor postharvest handling of the grain also causes grain breakage (Moreira, 1993; ODI, 2001). When very dry rice is stored it can absorb moisture from the surrounding humid air which may also increase cracking or fissuring in the grain resulting in low head-rice yields. In the Sahel, milled rice often contains 10–20% head rice, 30–40% large broken grains and 30–60% small broken rice (Moreira, 1995; ODI, 2001). Aoki and Seck (2011) also demonstrated the effect of low MC on milling recovery of the irrigated crop harvested in December in Senegal. The reverse occurs with rice harvested during the rainy season in July–August. High grain moisture contents of 15–18%, caused by high humidity and early rains, result in low milling recovery of 55–60%, powdered rice and frequent breakdowns in the mill. These conditions may also act as a constraint to double cropping in the Sahel.

In many rainfed rice areas in West Africa, paddy rice is stored in drums, either metal or plastic, in mud granaries or in the open-air kitchen, and then processed as needed using a pestle and mortar. This type of processing does not remove the bran layer and leaves most of the rice as broken brown rice and husks. For local commercial milling, steel huller mills, often referred to as Engelberg mills, are used. The Engelberg mill often results in yields of 55% white rice and 45% of a mixture of bran and powdered husk, with the latter being used to feed livestock. The cost of milling using this type of mill is CFA 600–1000 ($1.1–1.9) per 80 kg bag of paddy. Many of these steel huller rice mills are now fabricated locally and are widely distributed in SSA.

Surveys from the Office du Niger (Mali; Cruz, 2001a,b), Senegal River valley (SAED, 2010), Guinea (Norsa, 2011) and Cameroon (ODI, 2001) highlighted a large number of Engelberg mills which were originally imported from Europe and China, and also some locally fabricated machines. In 2000, there were more than 700 of these mills in Mali and 350 in Senegal. In Guinea in 1995 there were 200 Engelberg mills and this had risen to 900 by 2003. However, milling rice with an Engelberg mill results in very high percentage of broken rice and low milling recovery. Milling recovery can be less than 55%, which is already 10% below the expected average. This 10% loss is caused by broken rice ending up in the bran and husk.

Two-stage milling, which incorporates a rubber-roll dehusker and a steel polisher for whitening, is also popular in villages and for small-scale commercial milling. These mills can process 250–750 kg of paddy per hour. The rubber rollers remove the husk from the brown rice and then an abrasive or friction polisher removes the brown layer or bran. The brown layer does not contain the powdered husk as in the case with the Engelberg. This type of mill gives a better-quality rice and a higher milling recovery than the Engelberg mills. These mills are not fabricated locally and all spare parts (e.g. the rubber rollers and sieves) must be imported. These components will often need to be replaced every 60–80 tonnes of milled rice if quality is to be maintained.

Rice being processed in small village mills is often a mixture of different varieties and kernel sizes. Because these mills lack paddy-grading facilities they have difficulty competing in quality with imported rice or rice processed by the larger multi-stage commercial mills. Large commercial rice mills pre-grade paddy and milled rice, and in some instances often colour-sort milled rice.

A mechanical grader, with double screens to separate whole and broken grains, was introduced into village mills in Senegal by the Taiwan Agricultural Corporation. This type of grading equipment could be built locally and would make a tremendous improvement to the quality of the final product if adopted in SSA for small village mills which use the Engelberg or two-stage mills.

Multi-stage commercial rice mills, capable of milling 1–2 t/h, are also used in Africa. These mills have separate components for pre-cleaning, husking, whitening and polishing, grading and packaging. In some mills, the milling equipment is not complete, as they lack paddy cleaners, destoners, paddy separators and grading equipment. One way to improve the quality of milled rice has been to grade the rice in two operations. The first operation uses a rotary sifter to separate large grains from small grains, and the second uses an indented cylinder to separate whole grains from large broken grains.

Business models

Mechanization requires equipment to be produced, distributed, serviced, financed and bought by the end-users, who then need to ensure that they make a return on their investment. This means that valid business models are needed for selecting mechanization options, the production and delivery of the machines to the end-users, and provision of support services.

Most schemes for public hire and cooperative ownership of equipment have been unsuccessful worldwide. Business models that have been successfully used include private hire service through contract service providers, private ownership for their own farm and hire-out services for excess capacity, exclusive private owner–user and informal joint ownership (Rijk, 1986). In SSA, particularly in francophone countries, the liberalization of the rice sector and the introduction of agricultural banking led to the establishment of agricultural services providing inputs such as seed, herbicides and fertilizers, land preparation, mechanical harvesting and threshing. Governments completely withdrew from these activities, which were taken over by the private sector, which also took the lead by importing equipment.

Agricultural banks are willing to support the mechanization of the agricultural sector. For example, the West African Development Bank (BOAD) has established a CFA 4 billion ($7.6 million) fund to support the private sector in providing mechanization services to farmers.

Researchers can help to scale out feasible mechanization options by identifying and supporting suitable business models and by providing assistance in business planning. A business plan is required when seeking financing or applying for a loan. Financing institutions are usually risk averse and do not like agricultural credit because of their lack of understanding of the agricultural sector and lack of land tenure and ownership. Business plans can be used to engage with financial service providers with a view to designing custom credit lines for agricultural equipment.

The development of business models for the manufacture and supply of agricultural equipment and service provision is the responsibility of the private sector. However, there is often heavy bureaucracy in granting business permits, high taxes and duties, and other constraints. It is the role of the public sector to create a favourable business environment so that the private-sector players can fulfil their roles. Researchers can assist this by engaging in a policy dialogue and lobbying for a better policy environment.

Can Lessons Learned in Asia Help?

Mechanization of rice production and postharvest processes started in South-east Asia with the promotion of two-wheel tractors, and the development of the axial-flow thresher and low-lift axial-flow pumps in the 1970s. By 2012, land preparation and threshing were fully mechanized in most intensive rice-production systems across Asia. A second wave of mechanization started in the mid-1990s with the development of combine-harvesters in Thailand and Vietnam, and the introduction of laser-assisted land levelling using four-wheel tractors in China, India and Vietnam. The adoption of mechanical dryers started in Thailand and Vietnam in the 1990s and, since the early 2000s, IRRI has facilitated the spread of simple flat-bed dryers, using rice husk as fuel, in Cambodia, Indonesia, Laos, Myanmar, Philippines and Vietnam. More than 80% of the rice in the Central Plains of Thailand is harvested by combines, and there are more than 6000 combines in Vietnam and about 2000 in Cambodia, a country which did not have a single unit in 2007. In India, more than 10,000 contractors provide laser-levelling services. An estimated 6500 flat-bed dryers are used in the Mekong Delta and dryer numbers are rapidly increasing in Cambodia, Indonesia and Myanmar. A 'lessons learned' workshop held in Los Baños (Philippines) in May 2012, with public and private stakeholders and project partners looking at successes and failures in mechanization and postharvest, identified the following factors that led to successful adoption of such technologies:

- Identification or development of appropriate technology options that address end-users' needs, accompanied with adaptive research and development to match the technologies to local needs. This includes proper targeting of end-users – there needs to be a business model for the end-users if the technology is to be applied successfully.
- Private-industry involvement at a very early stage of technology development and dissemination, including local manufacturers, distributors and international machinery companies.
- Private entrepreneurship as the most efficient way to apply and provide mechanization technology, e.g. though machinery contract and hiring services, distribution, and repair services.
- In all successful cases of technology introduction there has been at least one technology champion, who was committed to move the technology forward against all initial hurdles, often supported by a local champion helping to promote the idea to potential users, intermediaries and policy makers.
- Sufficient time horizon for projects that aim to introduce a new technology. The introduction of the axial-flow threshers, combine-harvesters and mechanical dryers in South-east Asia and laser-levelling equipment in India took at least 6 years of support to achieve significant initial adoption that led to sustainable introduction.
- Successful initiatives had some sort of multi-stakeholder platform that embraced the different stakeholders from public and private sectors.
- Policy dialogue for creating an enabling environment for the industry and also for lobbying for support of promising technologies through the government extension service and the provision of support services (e.g. agricultural credit).

Often many well-intended initiatives failed when they included machinery importation or manufacturing by public-sector institutions, top-down decision making on technology choice, subsidized machinery distribution, lack of financing or lack of supporting policy. Machinery use through public hire services, farmer groups or cooperatives instead of service provision by private contract-service providers has been unsuccessful in South-east Asia.

While technologies as such are not necessarily transferrable without modifications, these Asian experiences with approaches and methodologies for fostering mechanization can provide some lessons for new initiatives in Africa.

Towards Sustainable Mechanization in Africa

In June 2011, representatives of national agricultural research systems and local manufacturing and distribution companies from seven countries (Ghana, Mali, Nigeria, Senegal, Sierra Leone, Tanzania and Uganda), farmer organizations, rural credit providers, international research and development organizations (AfricaRice, IRRI, JICA, CIRAD[1]), international agricultural machinery manufacturers (Briggs & Stratton) and CARD discussed opportunities to boost agricultural mechanization in rice-based systems in Africa in Saint-Louis (Senegal). A total of 47 participants attended. The workshop also looked back at successes and failures in terms of agricultural mechanization.

One success story in particular was highlighted: the development of an axial-flow thresher–cleaner in Senegal in the late 1990s based on a design imported from Asia (Donovan *et al.*, 1998; Wopereis *et al.*, 1998). Ten years after its release in 1997, the ASI thresher–cleaner was adapted and in use in six West African countries: Senegal, Mauritania, Mali, Burkina Faso, Ghana and Côte d'Ivoire. The main reason for the success was the establishment of an alliance by AfricaRice between researchers and local agricultural manufacturers. This alliance tested a first prototype imported from Asia via IRRI and adapted it to local conditions. This process meant that the locally built machine could be entirely constructed and manufactured locally, and only the engine needed to be imported.

A similar approach is being used by AfricaRice to develop a local version of a small combine-harvester based on a prototype built by the Philippine Rice Research Institute imported via IRRI. The machine was tested and adapted in the Senegal River valley in northern Senegal during the 2011 growing seasons. In particular, the imported machine became blocked by the tougher Sahel rice straw, so a local manufacturer reinforced the cutting system, and improved the thresher wheel and elevator. After demonstration to farmers, the machine was also adapted to four wheels from the original three. This 'mini-combine' can harvest 1.5–2 ha of rice per day, requiring just three operators to do so. This is at least double the speed of hand-harvesting and ASI threshing, which takes four labourers to complete. Demonstrations to date have shown losses of just 2% and that the machine produces very clean paddy. The payback time on the investment of CFA 4 million ($7620) in such a machine is estimated at 3.5 years on the basis of harvesting 45 ha/year.

Referring to these successful examples, workshop participants stressed the need to avoid massive importation of agricultural equipment without proper testing and evaluation before large-scale release. Governments and national and international research institutes have an important role to play here. Policies also need to be developed with regard to importation and taxation that are supportive of equipment importers, dealerships and local manufacturers. Rules currently vary according to whether it is the whole machine, spare parts or raw materials that are being imported. Tax on new farm equipment can be as high as 35% and on spare parts more than 60%. Where exemptions can be claimed, the tax usually has to be paid up front and then reclaimed – it can take 1–2 years to get the reimbursement. In addition, clearance time at the ports is often very slow resulting in extra demurrage charges being levied.

Governments need to take care that they do not 'over-subsidize'. In Thailand in the 1990s, over-subsidization led to a large number of unviable two-wheel tractor and machinery manufacturers glutting the local market. When subsidies were withdrawn, the number of local manufacturers dropped from more than 150 to just a handful.

Governments also need to be responsible for the certification of equipment and provide support to training institutes. Vocational training institutes need help to develop curricula and provide training for farm-machine operators, mechanics and artisans. This training needs to include technical and business planning and management. Government and private-sector extension officers also require training to support and extend mechanized agriculture at farm level.

Credit institutes require encouragement to structure loans that suit individual farmers and contract-service suppliers. Many micro-credit suppliers work within a restricted radius around their local branch and lend to groups rather than individuals. They also have prepayment rules, often monthly, that make it difficult for farmers to comply. In most localities, interest rates are above 25% and the repayment times less than one year. Many African farmers already work in associations or cooperatives, so collective ownership could be a solution for purchasing equipment in the short term. However, problems always arise in the management and prioritization of communally owned equipment and, eventually, local entrepreneurs will need to be found to offer contract services for activities such as ploughing, harvesting and milling to other farmers.

The workshop formulated the following recommendations for the different key actors to boost agricultural mechanization in rice-based systems in a sustainable manner in Africa.

Government

- Develop coherent strategies to boost agricultural mechanization, particularly in rice-based systems.
- Facilitate enhanced public–private sector collaboration in the development of agricultural mechanization in rice-based systems.
- Reduce taxes and duties on imported machinery, components and raw materials.
- Encourage the establishment of partnerships among local manufacturers to enhance the efficiency and quality of their work.
- Support national research to guide importation of machinery that is well adapted to local rice-growing conditions, effective and durable, and can be serviced and repaired locally.
- Facilitate access to credit for key actors (local manufacturers and end-users) in the mechanization value chain to sustain and increase the supply and demand for agricultural technology.
- Build and support local training centres in agricultural mechanization.

International manufacturers

- Establish direct dealerships in Africa.
- Build local capacity in the use and maintenance of equipment.
- Provide stewardship and quality assurance.
- Develop partnerships with local manufacturers to upgrade their construction capacities.
- Support local training centres in agricultural mechanization.

Local manufacturers

- Construct quality equipment or components that are adapted to local rice-growing conditions and for which local manufacturers have a clear competitive advantage.
- Provide aftersales services for products.
- Create partnerships among local manufacturers to standardize key equipment and to better respond to demands for local manufacture and maintenance of equipment.

National research and extension agencies

- Identify local needs for equipment in partnership with end-users.
- Contribute to the elaboration of policies that strengthen agricultural mechanization of the rice sector.
- Build local capacity through introduction of prototype technology in partnership with local manufacturers.
- Identify and work with key local manufacturers to help adapt prototype machinery that can be fabricated locally.
- Support training of local artisans for constructing and servicing machinery.

- Elaborate technical standards and norms for use, maintenance and manufacturing of equipment based on field testing.
- Provide advice on business planning to mechanize farming in rice-based systems to farmer cooperatives and service providers.
- Provide methods and decision support for extension agents and end-users to guide the use and maintenance of agricultural equipment.
- Provide facilities to train adequate numbers of agricultural engineers.

International research centres such as AfricaRice, IRRI and CIRAD

- Contribute to the elaboration of policies that strengthen agricultural mechanization of the rice sector.
- Contribute to enhanced public–private sector collaboration in the development of agricultural mechanization in rice-based systems.
- Introduce new prototypes for testing under local conditions based on identified needs.
- Help improve local manufacturing technologies.
- Establish a network of local manufacturers and researchers from national and international research (Africa-wide Rice Mechanization Task Force) to enhance and sustain agricultural mechanization in rice-based systems, in particular:
    - Support the development of training curricula on agricultural mechanization in partnership with key actors in Asia and Latin America;
    - Exchange knowledge on agricultural mechanization in rice-based systems in Africa and worldwide;
    - Facilitate South–South cooperation between Africa and Asia and Latin America for local manufacturers and research through exchange visits and training.

Umbrella agricultural research and development forums

- Identify local institutes that can serve as training centres in agricultural mechanization.
- Assess government policies related to agricultural mechanization, in particular with respect to the importation of agricultural machinery and spare parts across Africa.
- Advocate support for mechanization as part of national rice development strategies.

Conclusions

There is an urgent need to inject more energy into Africa's rice-farming systems. Mechanization can provide that extra power, addressing labour bottlenecks, improving productivity per unit of land and labour, and allowing household members to pursue other labour activities.

This chapter reviews the challenges and opportunities related to mechanization in Africa's rice sector, 'from farm to plate', looking at interventions before, during and after harvest. Many possibilities and opportunities exist that can, in principle, be readily exploited, copied and scaled out. However, the African continent is littered with wrecks of imported machinery. As much of the machinery as possible needs to be manufactured locally to create job opportunities, keep costs down and ensure that equipment is adapted to local conditions and can be maintained locally. More thought needs to go into business models that fit African conditions to help scale out feasible mechanization options. Much can be learned from experiences in Asia.

Recommendations formulated during the mechanization workshop in 2011 aim to prevent any repetition of past mistakes and to ensure sustainable and focused mechanization of the rice sector in Africa. Mechanizing Africa's rice sector is a prerequisite to reach the ambitious growth objectives set by African governments for their rice sectors.

Note

[1] JICA, Japan International Cooperation Agency; CIRAD, Centre de coopération internationale en recherche agronomique pour le développement.

References

AfricaRice (2011) *Boosting Africa's Rice Sector: A research for development strategy 2011–2020.* Africa Rice Center, Cotonou, Benin.

Aoki, T. and Seck, Y. (2011) Changement du taux d'humidité et du rendement à l'usinage du riz. Cours de formation des conseillers agricoles de la Société d'aménagement et d'exploitation des terres du Delta du Fleuve Sénégal et des vallées du Fleuve Sénégal et de la Falémé (SAED). (Unpublished).

Cruz, J.F. (2001a) *Transformation du Riz au Mali.* Centre de coopération internationale en recherche agronomique pour le développement (CIRAD), Montpellier, France.

Cruz, J.F. (2001b) Rapport de mission URDOC, Niono du 22 au 26 janvier 2001. Centre de coopération internationale en recherche agronomique pour le développement (CIRAD), Montpellier, France.

Donovan, C., Miézan, K.M., Wopereis, M.C.S., Diack, B.S. and Douthwaite, B. (1998) Technology transfer from Asia to Africa sets the stage for public and private sector collaboration in new technology in Senegal. *International Rice Research Notes* 23(2), 41–42.

FAO and UNIDO (2010) *Investment in Agricultural Mechanization in Africa: Executive summary.* Conclusions and recommendations of a Round Table Meeting of Experts, Arusha, Tanzania, 3–5 June 2009. Food and Agriculture Organization of the United Nations, Rome, Italy.

Moreira, J. (1993) Analyse des caractéristiques de qualité de certaines variétés de riz testées dans la vallée du Fleuve. Société d'aménagement et d'exploitation des terres du Delta du Fleuve Sénégal et des vallées du Fleuve Sénégal et de la Falémé (SAED), Saint-Louis, Senegal.

Moreira, J. (1995) Performances de la rizerie de Ross Bethio : évolution du rendement à l'usinage. Société d'aménagement et d'exploitation des terres du Delta du Fleuve Sénégal et des vallées du Fleuve Sénégal et de la Falémé (SAED), Saint-Louis, Senegal.

Mrema, G.C., Baker, D. and Kahan, D. (2008) Agricultural mechanization in sub-Saharan Africa: time for a new look. *Agricultural Management, Marketing and Finance Occasional Paper* 22. Food and Agriculture Organization of the United Nations, Rome, Italy.

Norsa, J. (2011) Le riz mangrove (Bora Malé) en Guinée : Propositions d'amélioration pour le développement de sa commercialisation. Mémoire de Fin d'Etudes. Ecole supérieure d'Agro-Développement International (ISTOM), Cergy-Pontoise, France.

ODI (2001) La riziculture au Mali. Research reports and studies. Overseas Development Institute, London. http://www.odi.org.uk/resources/details.asp?id=3171&title=riziculture-mali (accessed 27 June 2012).

Rijk, A.G. (1986) The role of farm mechanization in developing countries: experiences in Asian countries. In: *Small Farm Equipment for Developing Countries.* Proceedings of the International Conference on Small Farm Equipment for Developing Countries: Past Experiences and Future Priorities, 2–6 September 1986. International Rice Research Institute, [Los Baños, Philippines], pp. 373–388.

SAED (2010) *Recensement et caractérisation du matériel agricole dans la vallée du Fleuve Sénégal.* Société d'aménagement et d'exploitation des terres du Delta du Fleuve Sénégal et des vallées du Fleuve Sénégal et de la Falémé, [Saint-Louis], Senegal.

Wopereis, M.C.S., Miézan, K.M., Donovan, C., Ndiaye, A.M. and Douthwaite, B. (1998) A new Senegalese thresher/cleaner responds to small-farmer post-harvest needs. *International Rice Research Notes* 23(2), 39–41.

28 Integrating Gender Considerations in Rice Research for Development in Africa

Afiavi Agboh-Noameshie,[1]* Abdoulaye Kabore[1] and Michael Misiko[2]

[1]*Africa Rice Center (AfricaRice), Cotonou, Benin;*
[2]*International Maize and Wheat Improvement Center (CIMMYT), Addis Ababa, Ethiopia*

Introduction

Women's participation in rice production, postharvest and trading operations is well recognized in Africa (Dey, 1984; Nyanteng, 1985; Akande *et al.*, 2007; WARDA *et al.*, 2008; Bunch, 2011). In West Africa, for example, labour supplied by women for rice cultivation varies from 3% for floating rice in Mali, to 80–100% in mangrove-swamp rice in The Gambia and Liberia, where women participate in most of the activities and undertake postharvest processing of the crop (Huvio, 1998). Also a clear gender division of labour exists among crops. In The Gambia, swampland farming is solely women's duty; men cultivate cash crops and their fields are usually larger. In Mali, rice was traditionally grown only by women near rivers and wetlands (Synnevag, 1997, cited by FAO, 2004b). In many African countries, women are responsible for producing subsistence food crops for household consumption on their own plots or in communal household fields. In Côte d'Ivoire, husband and wife farm separate plots and there is some specialization by gender in the crops. Rice is considered a man's crop in some communities, and a woman's crop in others, while in many places, the gender pattern for rice cultivation is complex.

In spite of the active involvement of both men and women in rice farming, processing and marketing, the overall research-for-development agenda has not always fully appreciated or considered the gender perspective (Poats, 1991). Consequently, the technologies and knowledge generated through rice research may not have reached the women end-users. A gender perspective needs to be integrated into agriculture – specifically in rice research for development – as a strategic pathway towards sustainable and effective rice development in Africa.

Gender Actors and Levels in Rice Development

'Gender' is a term used to explain how society constructs the differences between women and men, whereas 'sex' identifies the biological differences between women and men. Therefore, looking at gender does not focus primarily on women or men, but rather on the relationships between their different roles, responsibilities,

* Corresponding author: a.agboh-noameshie@cgiar.org

© CAB International 2013. *Realizing Africa's Rice Promise* (eds M.C.S. Wopereis *et al.*)

opportunities and needs. In sub-Saharan Africa, women, men and youth are key players in rice production, processing and trading; in this chapter, we refer to them as 'gender actors'.

Gendered roles in rice farming in sub-Saharan Africa

In African rice-farming communities, the gender division of activities has been well documented. This division of tasks can be very complex and unbalanced at the expense of women and youth who become the main labour providers. The division of tasks also depends on the rice agroecosystem. In Sierra Leone, women are primarily in charge of planting, weeding and harvesting activities, while men carry out land preparation at the beginning of the cropping season (Kroma, 2002). Similar findings are reported by Fonjong and Mbah (2007) from the rural areas of Ndop (Cameroon), with the difference that some activities such as tilling, transplanting and harvesting were performed by both men and women (Fig. 28.1).

In Yangambi (Democratic Republic of Congo, DRC), women are involved in rice crop establishment and weeding activities along with men (Fig. 28.2); however, some tasks (such as birds scaring) are exclusively carried out by women assisted by children (Kabore and Misiko, 2010). In some farming communities, rice farming is considered as a strictly female activity (e.g. southern Senegal; World Bank, 2008). Generally, rice postharvest activities (threshing, pounding/milling, parboiling, cooking, trading, etc.) are mostly performed by the womenfolk. It has been argued that when some women's operations are mechanized (to save time, reduce the energy burdens or improve the process), they tend to be taken over by males (Stamp, 1990). However, adoption of the 'ASI' thresher–cleaner (see Rickman *et al.*, Chapter 27, this volume) had no adverse effect on the profits of 86% of the sampled women (AfricaRice, 2009).

Gendered access to productive resources for rice farming

Sustainable rice development relies on many factors. Farmers need access to key productive resources such as farmland, labour, agricultural inputs (e.g. quality seed and fertilizer), capital, and complementary rice productivity-enhancing technologies (knowledge, equipment, etc.). Men and women also need equitable control over their farm outputs. Any imbalance in the gendered access to or control of these resources slows rice development. Various studies (e.g. FAO, 2004a) have shown that women have less access than men to critical productive resources and services, including credit, farm inputs (seed, fertilizers, pesticides, etc.), marketing facilities, extension and information. Even when national laws endorse equal rights to own and control land, existing customary laws often prevent women from sustainable access to fertile farmland.

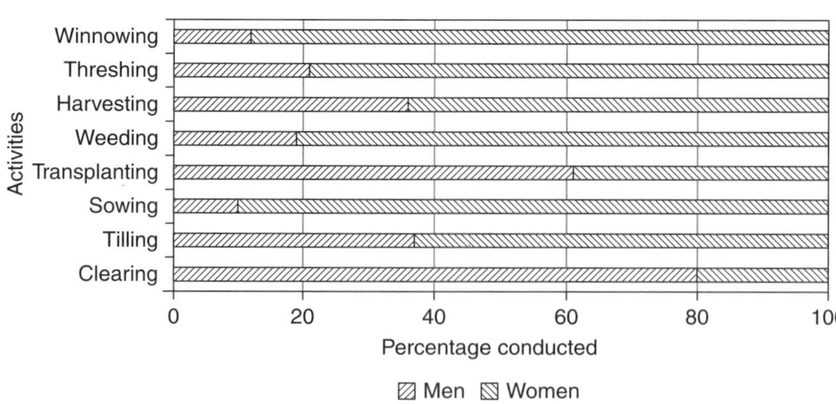

Fig. 28.1. Role of men and women in lowland rice farming in Ndop, Cameroon. (Data from Fonjong and Mbah, 2007.)

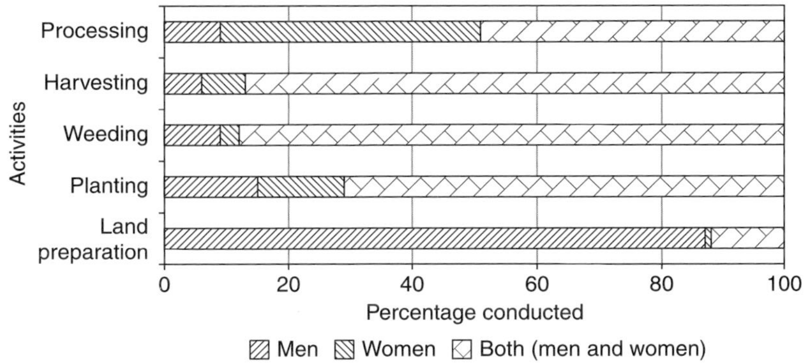

Fig. 28.2. Role of men and women in upland rice farming in the Democratic Republic of Congo. (Data from Kabore and Misiko, 2010.)

In Cameroon, for example, 45% of women are given remote rice fields requiring long trekking time to reach with consequent less time available for working (Fonjong and Mbah, 2007). Organizing themselves into women's groups does not always prove to be an effective solution. IFAD (1998) reports that women's groups were seldom allowed to use the farmland allocated by the community authorities for more than 3 years, because otherwise they would acquire permanent rights to it. Moreover, World Bank (2008) reports that high labour costs and land scarcity concerns are especially important to women farmers with no access to assets and services, and who have specific seasonal labour-use patterns. It appears that gendered access to productivity-enhancing resources and technologies remains a challenge in many African rice-farming communities. Efforts need to be devoted to effectively tackle such gender bias in access to agricultural inputs.

Female rice farmers: skilled in good farming practices

Female rice farmers have proved to possess particular skills in implementing some specific tasks. In-depth community studies conducted in rice-farming communities of Kindia (Guinea) and Yangambi (DRC) revealed that women's small rice plots (as opposed to the main family rice plots that are male dominated) play a critical role in local rice-farming systems. These small female-run rice gardens are used as 'experimental plots' where new varieties are first tested for two or three cropping seasons and the best are earmarked on the basis of female farmers' preferred traits before deciding to scale-up on the main plots (Kabore and Misiko, 2010). These women's plots are also used for managing agro-biodiversity. In fact, several rice varieties (including old treasured ones) are maintained and carefully grown in these women's small plots, which also act as sources of seed for scaling up on the main family plot.

Furthermore, female farmers are known to be better at selecting seed. For example, in Jipalom (Senegal), Linares (2002) observed that rice seed selection is exclusively carried out by female farmers, because they are better in distinguishing different rice varieties. Women's roles as repositories of local knowledge in seed selection, seed storage, genetic conservation and seed health need to be further enhanced through 'hands-on' training.

Gender actors in rice farming in Africa: a socio-cultural and economic perspective

In African rice-farming communities, diverse prevailing practices or wisdoms exist and are generally rooted in the local culture. In such communities, ignoring these socio-cultural practices/ wisdoms could be a barrier for effective scaling-up of gender-sensitive productivity-enhancing technologies. Various issues surrounding gendered access to and control over productive resources are sensitive and need to be tackled with subtlety. Some of these sensitive aspects are related to: (i) access to farmland; (ii) access to and control exerted over some farming technologies (e.g. seed, fertilizers, mechanized equipment); (iii) allocation of labour; (iv) control of the harvested rice – trading their own products; and (v) access to knowledge. For example, in Cameroon, Fonjong and Mbah (2007) report that some men see female income-earning as a threat to their authoritative position as breadwinners. Consequently, they react to women's increased earnings by abandoning their financial responsibilities towards their families and pushing the total burden of household maintenance onto their wives.

Labour exchange is also a common social strategy developed in many rice-farming communities to cope with peak labour periods. The practice cements the social ties in farming communities. Labour exchanges are generally used for assistance to vulnerable farmers, such as elders without other labour resources, widows, sick and disabled farmers, and the poorest farmers. With rice becoming a cash crop and consequent production increases, there is the threat of a shift towards waged or hired labour to cope with peak labour periods, at the expense of labour exchange. The disappearance of such social cement could have detrimental effects on the welfare of resource-poor and vulnerable farmers. Likewise, seed commercialization is not always well accepted. In Gimbi (DRC), seed commercialization is taboo (Misiko, 2009) – seed marketing is disdained by locals. Instead, farmers acquire their rice seed through indigenous reciprocity practices such as barter (seed for seed of different rice varieties or crops), or in-kind seed credit to be paid back at harvest. Female farmers are acknowledged to be the best seed guardians because of their knowledge and abilities in seed management. In such settings, it can be argued that any initiative aiming at developing sustainable seed enterprises should extract valuable insights from the local norms and practices.

Generally, rice farming has been developed in some rural societies following the increasing demand from growing urban centres. The introduction of this 'new' crop has had implications for local cultures. For example, the introduction of rice cultivation in Ndop (Cameroon), in which women's labour constitutes an important component, has dramatically changed the traditional patterns of division of labour and gender roles, which formerly prevented women from engaging in economically profitable activities (Fonjong and Mbah, 2007). This situation has yielded positive outcomes: the introduction of rice farming has brought women not only into food-crop production, but also into the cultivation of cash crops and other income-generating activities, through which some of them have become major breadwinners. The logical outcome is a phenomenal reduction in household poverty among many families, particularly female-headed households.

It can be concluded that in all African societies, diverse socio-economic and cultural practices exist, but are not always known due to the oral nature of the cultures. These practices need to be investigated further to gain important insights for the effectiveness of future rice research and development activities.

Gender Mainstreaming in Rice Research for Development in Africa

New varieties – new gains for women

Gender-blind technology development can obstruct its efficiency and performance. Gender bias has been a problem for agricultural research centres and has been considered, among others, a reason for the non-adoption of some agricultural technologies. Participatory approaches such as participatory varietal selection (PVS) are used by Africa Rice Center (AfricaRice) and national agricultural research and extension (NARES) partners to include both female and male end-users in evaluation and selection of varieties. Research on PVS revealed that men gave importance to short growth duration and plant height, whereas women preferred traits such as good emergence, seedling vigour and droopy leaves that indicate weed competitiveness, since they are mostly involved in the sowing and weeding operations (WARDA *et al.*, 2008).

The PVS approach is complemented by the community-based seed system (CBSS; see Bèye *et al.*, Chapter 14, this volume) to increase the end-users' access to available quality seed of the chosen varieties. Thanks to these two approaches, the new varieties have been introduced in various rice-farming communities in sub-Saharan Africa. The gender-related impacts of improved varieties such as the NERICA varieties have been documented by several authors. For example, in Guinea, Diagne *et al.* (2007) found a higher impact of adoption of NERICA varieties among women (yield increase 1090 kg/ha) than among men (yield increase 442 kg/ha). Similar results are reported from Benin by Agboh-Noameshie *et al.* (2007). Impacts of adoption of NERICA varieties have been noted beyond the farm: Adékambi *et al.* (2008a,b) report improvement in children's schooling in households growing NERICA varieties (i.e. 6% increase in children's school attendance rate and US$20 increase in school expenditure per child) and an increase of about $0.30/day in household consumption expenditure per adult. Various social gains have also been reported following the adoption of NERICA varieties. The development of businesses around postharvest and rice processing technologies has also triggered women's empowerment (built social capital) through their organizations (Zossou *et al.*, 2009).

The reasons underlying these successes are that men and women farmers have been closely associated with the development and adoption of the technologies through assessment of their own needs and the participatory approaches (PVS and CBSS).

Gender-sensitive rice learning and rural innovation

Rice farmers need relevant information and knowledge on productivity-enhancing technologies and opportunities to add value to their production, and to sell their surplus. However, most of them live in rural areas with poor infrastructure (roads, communication facilities, etc.), and they frequently do not have such productivity-enhancing knowledge. Though women are acknowledged as critical players in rice-farming systems, they have limited contact with extension agents; the underlying reasons for this include cultural barriers, heavy workload preventing women from devoting 'spare time' to extension service, and unawareness of the importance of the information to be provided (Chale, 1990; FAO, 1993; Jiggins *et al.*, 1998). As a consequence, only male farmers are generally reached. However, the information directed to male heads of households is not always transferred adequately to their female and youth dependents. For example, in Sierra Leone, agricultural extension messages, when available, often had to be indirectly channelled to women through the men, with little guarantee that the effective transfer of information acquired will occur (Kroma, 2002).

Aware of these shortcomings and hindrances in knowledge dissemination among rice farmers, AfricaRice developed a participatory learning and action-research for integrated rice management (PLAR-IRM) methodology (Defoer *et al.*, 2004; Defoer and Wopereis, Chapter 31, this volume). PLAR-IRM is an intensive methodology that combines ideas from the Farmer Field School approach (such as weekly sessions with groups of about 25 farmers, discussing a specific issue while stimulating learning) with other relevant tools (e.g. cropping calendars or transect walks). Several achievements resulting from the application of PLAR-IRM methodology have been reported. Over 3 years, AfricaRice conducted 183 PLAR training sessions in Mali, Guinea, The Gambia and Ghana, reaching 1248 farmers (61% female, 39% male) (WARDA, 2008). To make the PLAR-IRM more sustainable in Mali, 45 farmer-facilitators (19 female and 26 male) were trained to conduct PLAR in their villages after the 'Participatory Adaptive Research and Dissemination of Rice Technologies in West-Africa' project (WARDA, 2008). Similar PLAR-IRM training activities were conducted in the other project countries (WARDA, 2008). Farmers clearly appreciated the PLAR-IRM training and this approach has helped to break down gender barriers.

Furthermore, a rice seed health video previously produced in Bangladesh was translated into Bambara and used as a video-supported learning tool to reach 3915 farmers (2120 male, 1795 female) in nine villages in Mali. After watching the videos, about 40% of the women changed their seed-drying practices in Bangladesh (Van Mele *et al.*, 2005; Van Mele *et al.*, Chapter 30, this volume). Considering the effectiveness of the seed health video in communicating technologies to a large number of farmers and the success of the PLAR-IRM methodology in farmer learning, AfricaRice developed the zooming-in zooming-out (ZIZO) approach (Van Mele, 2006; Van Mele et al., Chapter 30, this volume) to produce further media-supported learning. In 2009, AfricaRice combined the PLAR-IRM and ZIZO approaches to produce a new series of videos on integrated rice management. Dealing with five modules (land preparation, seedbed, transplanting, weed management, soil fertility management), these videos were translated into local languages in addition to the French or English versions. Through video-supported group learning, 2396 smallholders in DRC (1808 male, 588 female), 920 in Guinea (620 male, 300 female) and 800 in Sierra Leone have been trained (AfricaRice, 2010). These videos were also translated into more than 33 African local languages for a wide dissemination across sub-Saharan Africa, and had already reached 160,000 farmers in 2010 (Wanvoeke and Van Mele, 2010; Van Mele *et al.*, Chapter 30, this volume).

Various studies have shown the effectiveness of these video-based learning tools in communicating agricultural technologies to the poor, women, men and young farmers. For example, in their study on rice parboiling technologies in Benin, Zossou *et al.* (2009) found that 89% of women strongly appreciated the farmer-to-farmer video with two characteristics emerging: (i) the video burns images into the memory (mentioned by 74% of them), and (ii) the video is both educational and entertaining (mentioned by 77% of them). Importantly, the video-supported training has been evaluated as more effective and has reached three times as many women than hands-on training workshops organized by a local NGO. Also more than 90% of the women who watched the video improved the quality of their parboiled rice. These tools provide equitable access to research and extension knowledge by both male and female grassroots actors operating in rice sector.

Towards a Framework for Gender Equity in Rice Research for Development in Africa

Process of gender mainstreaming in research for development

It has become apparent that there is a need to integrate gender as an analytical variable in research for development, especially in rice-based production systems. Innovations should be based on gender analysis,[1] which is a way of looking at a community in its totality to ensure that the interests of all its members – men, women and children – are addressed. Hunt (2004) stresses that gender analysis should be done at the various stages of programme or project design, implementation and impact evaluation, and gives the following definition:

> During program and project design, gender analysis is the process of assessing the impact that a development activity may have on females and males, and on gender relations (the economic and social relationships between males and females which are constructed and reinforced by social institutions). It can be used to ensure that men and women are not disadvantaged by development activities, to enhance the sustainability and effectiveness of activities, or to identify priority areas for action to promote equality between women and men. During implementation, monitoring and evaluation, gender analysis assists to assess differences in participation, benefits and impacts between males and females, including progress towards gender equality and changes in gender relations.

There is a lot of literature on how to integrate gender perspectives into research-for-development programmes. However, this is usually restrained by inadequate capacity to conduct gender-sensitive analysis. Meinzen-Dick *et al.* (2011) report that, despite the evidence that women are being involved in participatory adaptive research and that attention has been focused on gender differences in the impact of agricultural technologies, more work needs to be done in integrating gender in the upstream priority setting and decision making. They recommend more functional linkages among research-for-development actors instead of a unidirectional flow between research and end-users, and present a conceptual framework that will allow a more effective feedback loop (Fig. 28.3).

An important starting point in the implementation of gender mainstreaming is to ensure that the initial definitions of issues and problems are done in a way that allows for the identification of gender differences and disparities.

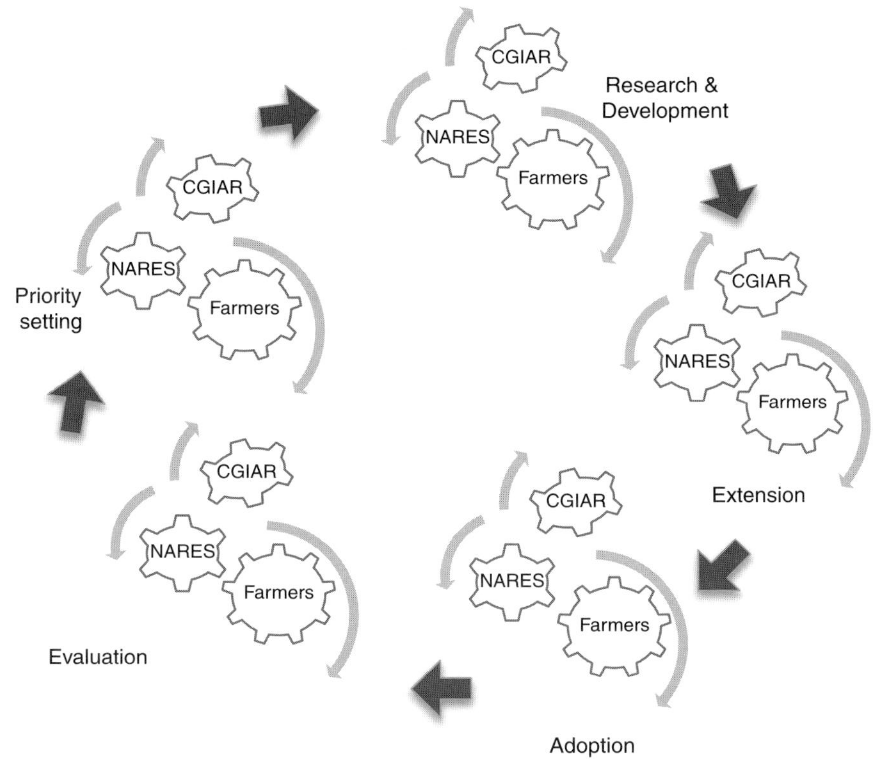

Fig. 28.3. Components of a gender-responsive agricultural research system. (From Meinzen-Dick *et al.*, 2011, reproduced with permission from the International Food Policy Research Institute www.ifpri.org.) NARES: national agricultural research and extension system.

Assumptions that issues and problems are neutral from a gender perspective should be avoided and gender analysis should always be carried out, separately or as part of existing analyses. Meinzen-Dick *et al.* (2011) stress that integrating gender issues into agricultural research and development will require addressing the following critical questions:

- For priority setting: (i) where and how are the differential needs, interests and priorities of women and men reflected? (ii) who makes the decisions regarding the kinds of agricultural R&D that will receive investment? (iii) are there mechanisms to take the needs of women and men as both producers and consumers into account?
- For research and development: (i) who are the researchers and how attuned are they to gender issues?
- During extension phase: (i) who delivers extension services? (ii) who receives the extension services and information? (iii) are women recognized as farmers and clients of the extension services? (iv) how are extension services delivered?
- For adoption of innovations: (i) who can and will adopt agricultural innovations? (ii) who can benefit from them?
- For evaluation and impact assessment: (i) how can both external studies and participatory processes that assess the costs and benefits of agricultural innovations and their related distribution consider gender differences? (ii) how can we use evaluations and assessments that do consider gender differences to inform future research priorities?

These processes can be seen at different scales with different actors in various research-for-development institutions.

Practical example of integration of gender in project cycle

When gender is to be integrated into any project cycle, integration should occur at all stages, from diagnosis and implementation, through to monitoring and evaluation, adoption and impact evaluation.

During the diagnosis phase, constraints and problems, opportunities and needs are determined by reviewing earlier research and undertaking additional studies, especially participatory situation analysis (Cornwall, 2000). Since women and men farmers and processors have shared responsibilities and constraints within the household economic production and household management systems, data disaggregated by sex and age should be collected using gender-responsive tools such as semi-structured questionnaires, checklists and gender-focused discussions (IFAD, 2002). Questions should be structured in such a way that they bring out information on different gender categories in a household. These data are important in monitoring and evaluation (M&E) as they reveal project impacts on men and women, including the young and old, and help in development of gender-responsive technology (Njenga *et al.*, 2008). Once the project design phase is rendered gender-sensitive, all the other phases should automatically take gender issues into account.

Project design: The project's goals and strategies are identified through a joint process of visualizing the desired development and project outcomes, then evaluating alternative strategies that might be applied to realize those outcomes by looking into their viability and effectiveness to produce the required outcomes. In research, some of the strategies identified may include further participatory research into, for example, technical options. Here a gendered assessment of alternative problem-solving strategies can be used as well as a checklist of guiding questions for conducting group discussions as suggested by CIP (2008).

Activity planning: At this stage the goals and strategies are operationalized by: identifying the activities needed to implement the strategies; developing the methods and tools to be applied; dividing responsibilities and tasks among the participating organizations and groups; defining coordination and monitoring mechanisms; and developing a budget and a timeframe. The planning can be done using a participatory gender-responsive planning matrix (Hovorka *et al.*, 2009).

Implementation: Implementation of project activities requires following the designed strategies, goals/objectives and activities, but also requires adaptation to the local situation. Sometimes there is a need to enhance the skills of research teams for effective incorporation of gender in R&D, particularly if there are no partners with gender expertise in the team (Njenga *et al.*, 2008). For instance, the research team should have its capacity for participatory research and gender analysis built.

Monitoring and evaluation: To ensure institutional accountability for gender mainstreaming in the overall rice research-for-development activities and in reporting on gender issues, there is a need to establish relevant gender-sensitive qualitative and quantitative indicators to be regularly monitored and evaluated. This process will ensure gender mainstreaming is supported by gender-sensitive budgeting and relevant time allocation for effective implementation of the planned activities (Njenga *et al.*, 2008).

Impact assessment: Baseline surveys at the outset of each activity need to ensure that relevant benchmark gender-disaggregated data and gendered indicators are captured. Then, at the end of the related activity, the *ex-post* impact assessment generates end-line data and indicators to measure the progress registered and draw the lessons learned from successes or failures for any duplication or future fine-tuning.

It takes political will at the institutional level to operationalize gender mainstreaming – this will ensure that a clear and comprehensive operational gender policy is established and effectively shared.

It should be noted that practical gender mainstreaming is not about running through a checklist of questions to ensure you have not overlooked anything. It is about asking the right questions so that you can see where limited resources should best be used. Gender mainstreaming is a necessary process for achieving gender equality in the most effective and efficient manner. For example, the key questions listed in Box 28.1 could help to ensure that the logical framework of the project includes a gender perspective (UNDCP, 2000).

Box 28.1. Determining the gender sensitivity of a logical framework matrix

Objectives:

- Do the objectives address the problems of both men and women?
- Are the objectives likely to influence relationships between women and men?
- Do the objectives specify **who** is targeted and **who** is expected to benefit, differentiating between women and men?
- Who participated in choosing objectives from the complete set of needs to be addressed?
- What needs of women, and of men, does the project address?

Outputs:

- Are there separate outputs for women and men?
- Are they consistent with the needs of the beneficiary group?

Achievement indicators:

- Are there separate achievement indicators for women and men?

Inputs:

- Are the inputs appropriate for the involvement of both women and men?
- Is there time and is there a budget for gender analysis?
- Are the budgets flexible and reviewable?
- Is the planning flexible enough to enable new activities to be initiated in response to women's and men's constraints?

Risks and assumptions:

- Analysis of risks and assumptions should consider gender-related barriers and constraints that could affect implementation. Where necessary, special project activities should be integrated to counteract barriers and constraints.

Source: Adapted from EC-C (1993).

Challenges related to gender mainstreaming in rice research for development

Despite the increase in awareness and the availability of information on the existing gender disparities in agriculture, integrating a gender perspective in agricultural research for development still faces many challenges. These challenges come from the misconception of gender equality as implying that men and women become equal, while gender equality in fact means that the opportunities and life chances of men and women are equal (Opio, 2003). In agricultural research and development, achieving gender equality will therefore not only require changes in research targeting, system mapping, and diagnosis and intervention, but also in the institutional culture of the research organization to ensure that women are given a strong voice both in shaping research and in shaping the development of their societies (Njenga *et al.*, 2008). Moreover, it is also observed that even though there are well-written gender-mainstreaming strategies at country level, many research and extension institutions have not successfully addressed gender in the design and implementation of their activities. The provision of agricultural services is male dominated and little effort has been made to train men to work with women and be aware of the strategic and practical needs of women within agriculture. Also, despite the fundamental role women play in agriculture very few of them own, control or have guaranteed access to productive resources such as land, credit, technical services, market outlets and information.

Furthermore, very few members of staff have been trained in gender analysis, which therefore limits the scope for promoting equity within most institutions. Members of staff lack experience in mainstreaming gender issues into their programmes. While some are willing to do so, they have no clear guidelines and cannot quite relate the relevance to their working environment.

Conclusion

By analysing and documenting gender-disaggregated data, AfricaRice has shown the impact of some of its research on targeted gender actors. However, more effort needs to be invested to better integrate gender issues in the **whole** rice research-for-development cycle for an effective and sustainable impact on actors' livelihoods. The framework to mainstream gender in rice research-for-development activities presented in this chapter provides important guidelines in this respect. Its application will require substantial awareness-raising and capacity-building efforts among Africa's rice research and extension communities.

Note

[1] A number of different frameworks are used to undertake gender analysis. Some of these have been developed in Northern countries (Overholt *et al.*, 1985; Moser, 1993), and others have been developed and adapted by development practitioners from the South (Longwe, 1991; Parker, 1993; Kabeer, 1994). These articles outline the essential steps that need to be addressed to undertake gender analysis.

References

Adékambi, S.A., Diagne, A., Simtowe, F. and Biaou, G. (2008a) *Does NERICA Rice Cultivation Enhance Child Schooling? A lenses view among rural households from Benin.* Africa Rice Center, Cotonou, Benin.

Adékambi, S.A., Diagne, A., Simtowe, F. and Biaou, G. (2008b) The impact of agricultural technology adoption on poverty: the case of NERICA rice varieties in Benin. Paper prepared for presentation at the International Association of Agricultural Economists' 2009 Conference, Beijing, China, 16–22 August 2009. Africa Rice Center, Cotonou, Benin.

Adesina, A.A. and Djato, K.K. (1997) Relative efficiency of women as farm managers: profit function analysis in Côte d'Ivoire. *Agricultural Economics* 16, 47–53.

AfricaRice (2009) Report of the AfricaRice Research Division, March 2008 – February 2009. Africa Rice Center, Cotonou, Benin.

AfricaRice (2010) Enhancing smallholder access to NERICA for alleviating rural poverty in West and Central Africa (IFAD WCA) Project. Project Report. (Van Mele, P. and Kabore, A., eds). Africa Rice Center, Cotonou, Benin.

Agboh-Noameshie, A.R., Kinkingninhoun-Medagbe, F.M. and Diagne, A. (2007) Gendered impact of NERICA adoption on farmers' production and income in central Benin. Paper presented at the International conference of African Association of Agricultural Economists (AAAE), Accra, Ghana, 20–22 August.

Akande, T., Cisse, Y. and Kormawa, P. (2007) Streamlining policies for enhancing rice production in Africa: past experiences, lessons learnt and the way forward. *African Crop Science Journal* 15(4), 223–235.

Bunch, S. (2011) Women help solve hunger: why is the world still waiting? In: *Our Common Interest: Ending hunger and malnutrition. The 2011 Hunger Report.* Bread for the World Institute, Washington, DC. http://hungerreport.org/2011/report/chapters/one/smallholder-agriculture/196-women-help-solve-hunger-why-is-the-world-still-waiting.

Chale, F. (1990) *Terminal Report 1: Strengthening Agricultural Extension – A study of problems and constraints in extension strategies and methods for reaching rural women.* Food and Agriculture Organization of the United Nations, Rome, Italy.

CIP (2008) Testing Guidelines and Tools for Gender Mainstreaming in Urban Agriculture; Focus City – Building a Sustainable, Cohesive Community through Waste Recycling and Agro-enterprises – Kampala (SNF-K) project. Urban Harvest, International Potato Center, Kampala, Uganda. ftp://ftp.cgiar.org/cip/DIV1/Annex%2025%20Gender%20mainstreaming08.pdf (accessed 25 June 2012).

Cornwall, A. (2000) Making a difference? Gender and participatory development. *IDS Discussion Paper* no. 378. Institute of Development Studies, Brighton, UK.

Defoer, T., Wopereis, M.C.S., Idinoba, P., Kadisha, T.K.L., Diack, S. and Gaye, M. (2004) *Manuel du Facilitateur : Curriculum d'apprentissage participatif et recherche action (APRA) pour la gestion integree de la culture de riz de bas-fonds (GIR) en Afrique Sub-Saharienne*. Centre du Riz pour l'Afrique, Cotonou, Benin.

Dey, J. (1984) Women in rice-farming systems. Focus: sub-Saharan Africa. *Women in Agriculture* vol. 2. Food and Agriculture Organization of the United Nations, Rome, Italy.

Diagne, A., Sogbossi, M.J., Kouame, P., Diawarra, S. and Sidy, M. (2007) Impact of the NERICA in Guinea: productivity and income gains. Africa Rice Center, Cotonou, Benin.

EC-C (1993) *Women and Development: A guide to gender issues in managing European Community co-operation with Latin American, Asian and Mediterranean countries*. Commission of the European Communities, Brussels, Belgium.

FAO (1993) *Agricultural Extension and Farm Women in the 1980s*. Food and Agriculture Organization of the United Nations, Rome, Italy.

FAO (2004a) *Rural Households and Resources: A guide for extension workers*. Food and Agriculture Organization of the United Nations, Rome, Italy.

FAO (2004b) *Building on Gender, Agrobiodiversity and Local Knowledge. Training manual*. Food and Agriculture Organization of the United Nations, Rome, Italy.

Fonjong, L.N. and Mbah, F.A. (2007) The fortunes and misfortunes of women rice producers in Ndop, Cameroon and the implications for gender roles. *Journal of International Women's Studies* 8(4), 133–147.

Hovorka, A., De Zeeuw, H. and Njenga, M. (eds) (2009) *Women Feeding Cities. Mainstreaming gender in urban agriculture and food security*. Practical Action Publishing, Rugby, UK.

Hunt, J. (2004) Introduction to gender analysis concepts and steps. *Development Bulletin* no. 64, pp. 100–106.

Huvio, T. (1998) Women's role in rice farming. http://www.fao.org/sd/wpdirect/wpan0027.htm.

IFAD (1998) *Agricultural Implements Used by Women Farmers in Africa*. International Fund for Agricultural Development, Rome.

IFAD (2002) *Gender and Poverty Targeting in Market Linkage Operations. Toolkit for practitioners. The start of an in-country process*. International Fund for Agricultural Development, Rome, Italy.

Jiggins, J., Samanta, R.K. and Olawoye, J.E. (1998) Improving women farmers' access to extension services. In: Swanson, B.E., Bentz, R.P. and Sofranko, A.J. (eds) *Improving Agricultural Extension: A reference manual*. Food and Agriculture Organization of the United Nations, Rome, Italy.

Kabeer, N. (1994) *Reversed Realities: Gender hierarchies in development thought*. Verso, London.

Kabore, A. and Misiko, M. (2010) Enhancing smallholder access to NERICA for alleviating rural poverty in West and Central Africa Project: Guinea and Democratic Republic of Congo field report XII. Africa Rice Center, Cotonou, Benin.

Kroma, M. (2002) The gender division of labor in rice post-harvest processing in Sierra Leone: implications for extension and technology development. *Journal of Agricultural Education and Extension* 8(4), 181–194.

Linares, O.F. (2002) African rice (*Oryza glaberrima*): history and future potential. *Proceedings of the National Academy of Sciences USA* 99(25), 16,360–16,365.

Longwe, S. (1991) Gender awareness: the missing element in the Third World development project. In: Wallace, T. and March, C. (eds) *Changing Perceptions: Writings on gender and development*. Oxfam, Oxford.

Meinzen-Dick, R., Quisumbing, A., Behrman, J., Biermayr-Jenzano, P., Wilde, V., Noordeloos, M., Ragasa, C. and Beintema, N. (2011) Engendering agricultural research, development, and extension. *IFPRI Research Monograph*. International Food Policy Research Institute, Washington, DC. Available at www.ifpri.org/sites/default/files/publications/rr176.pdf (accessed 28 March 2013).

Misiko, M. (2009) Enhancing smallholder access to NERICA for alleviating rural poverty in West and Central Africa Project: Democratic Republic of Congo field report XII. Africa Rice Center, Cotonou, Benin.

Moser, C.O. (1993) *Gender Planning and Development: Theory, practice and training*. Routledge, London and New York.

Njenga, M., Karanja, N., Kang'ethe, E., Kabiru, C., Munyao, P., Prain, G., Lee-Smith, D., Michael, P., Ngugi, J., Gurung, B. and Gathuru, K. (2008) Enhancing research impacts through mainstreaming gender in organizations and research processes: the case of Urban Harvest and the International Potato Center. Paper presented at the international workshop on Rethinking Impact: Understanding the complexity of poverty and change, Cali, Colombia, 26–28 March.

Nyanteng, V.K. (1985) Women in agriculture in sub-Saharan Africa: implications for agricultural research and technology. In: *Women and Agricultural Technology: Relevance for research*, Vol. 2 – *Experiences*

in International and National Research. Report of a seminar. The Rockefeller Foundation, New York, pp. 153–163.
Opio, F. (2003) Gender mainstreaming in agriculture with special reference to Uganda: challenges and prospects. *African Crop Science Conference Proceedings* Vol. 6, pp. 699–703.
Overholt, C., Cloud, K., Anderson, M. and Austin, J. (1985) Women in development: a framework for project analysis. In: Overholt, C., Cloud, K., Anderson, M. and Austin, J. (eds) *Gender Roles in Development Projects: A case book*. Kumarian Press, West Hartford, Connecticut, pp. 3–15.
Parker, A.R. (1993) *Another Point of View: A manual on gender analysis training for grassroots workers*. United Nations Development Fund for Women, New York.
Poats, S.V. (1991) The role of gender in agricultural development. *Issues in Agriculture* no. 3. Consultative Group on International Agricultural Research, Washington, DC.
Stamp, P. (1990) *Technology, Gender and Power in Africa*. International Development Research Centre, Ottawa, Canada.
UNDCP (2000) Guidelines for Gender Mainstreaming in Alternative Development, Final draft. United Nations International Drug Control Programme, [Vienna, Austria].
Van Mele, P. (2006) Zooming-in zooming-out: a novel method to scale up local innovations and sustainable technologies. *International Journal of Agricultural Sustainability* 4(2), 131–142.
Van Mele, P., Zakaria, A.K.M. and Bentley, J.W. (2005) Watch and learn: video education for appropriate technology. In: Van Mele, P., Salahuddin and Magor, N. (eds) *Innovations in Rural Extension: Case studies from Bangladesh*. CAB International, Wallingford, UK, pp. 77–88.
Wanvoeke, J. and Van Mele, P. (2010) Using videos to improve cross-cultural learning between Asian and African farmers. Presentation at the 3rd International Rice Congress, Hanoi, Vietnam, 8–12 November.
WARDA (2008) Participatory Adaptive Research and Dissemination of Rice Technologies in West-Africa (PADS-Phase II). Final Report. (Van Mele, P., Wanvoeke, J. and Bentley, J., eds). Africa Rice Center, Cotonou, Benin.
WARDA, FAO and SAA (2008) *NERICA: The New Rice for Africa – A compendium*. (Somado, E.A., Guei, R.G. and Keya, S.O., eds). Africa Rice Center, Cotonou, Benin; Food and Agriculture Organization of the United Nations, Rome, Italy; Sassakawa Africa Association, Tokyo, Japan.
World Bank (2008) *Agriculture for Development. World Development Report 2008*. The World Bank, Washington, DC.
Zossou, E., Van Mele, P., Vodouhe, S.D. and Wanvoeke, J. (2009) The power of video to trigger innovation: rice processing in central Benin. *International Journal of Agricultural Sustainability* 7(2), 119–129.

29 Towards a New Approach for Understanding Interactions of Technology with Environment and Society in Small-scale Rice Farming

Edwin Nuijten,[1]* Marina Temudo,[2] Paul Richards,[1] Florent Okry,[1,3] Béla Teeken,[1] Alfred Mokuwa[1] and Paul C. Struik[1]
[1]*Wageningen University, Wageningen, Netherlands;* [2]*Tropical Research Institute (IICT), Lisboa, Portugal;* [3]*Africa Rice Center (AfricaRice), Cotonou, Benin*

Introduction

Modern agricultural technologies have been adopted in many Asian rice farming systems, particularly in areas with good agroecological conditions. But in other rice-growing areas, the adoption of modern technologies has been very slow (Binswanger and Pingali, 1989; Waddington *et al.*, 2010). In West Africa, this has been particularly the case in upland rice cultivation (Richards, 1986; Dalton, 2004; Okry *et al.*, 2011), but also in mangrove-swamp farming systems (Temudo, 2011). Modern technologies are often not adapted to local conditions. Since the 1980s, it has been recognized that farmers need to be more involved in variety development through participatory varietal selection (PVS) and decentralized breeding (Maurya *et al.*, 1988; Sperling and Loevinsohn, 1993; Weltzien *et al.*, 1996; Almekinders and Elings, 2001; Ceccarelli and Grando, 2007). In this vein Africa Rice Center (AfricaRice) has used PVS to improve the adoption of the NERICA varieties (Gridley *et al.*, 2002). However, there is still little attention on the outcomes of farmer technology development and innovation, and ways to build upon these processes or to integrate them with scientific technology development.

Farmer technologies, such as farmer varieties, are the result of long innovation trajectories. These trajectories are shaped by interactions among agroecological, socio-economic and cultural factors and it is therefore essential to understand such interactions when involving farmers in technology development or building upon farmer technology development. The need for a more decentralized technology development approach (for, e.g., varieties) with a larger role for farmers has become even clearer (Dorward *et al.*, 2007; Efisue *et al.*, 2008; Nuijten *et al.*, 2009; Mokuwa *et al.*, 2012). For example, Mokuwa *et al.* (2012) provide evidence that farmer rice varieties can be widely adaptable, as a considerable number performed well under very different ecological conditions and are therefore not restricted to the locality where they were developed. This means that farmers, next to researchers, equally produce technologies that may be relevant for farming over large regions.

* Corresponding author: Edwin.Nuijten@wur.nl

© CAB International 2013. *Realizing Africa's Rice Promise* (eds M.C.S. Wopereis *et al.*)

In many regions, farmers carefully innovate, select and match technologies to local environmental conditions, looking for optimal interactions of technologies with agroecological, socio-economic and cultural factors. The ways farmers organize their farm management and livelihood strategies differ widely, even within small, well-defined regions. Often this variation reflects differences in the agroecological environment, societal organization, and cultural and socio-economic dynamics. We therefore need a better understanding of how technologies interact with the agroecological and socio-cultural environments. Comparisons within and between different areas and countries may yield useful insights into key aspects and mechanisms underlying these complex interactions in rice-farming systems in West Africa. Such comparisons will also provide insights into the way farmer technologies have been developed and how to link farmer technology development with science. For a better understanding of these complex interactions a wide range of scientific disciplines is needed, such as soil science, plant breeding, agronomy, crop physiology, chemistry, economics, anthropology, and communication and innovation studies. As today's farming is also shaped by the past, archaeology and history may also be important to understand the socio-economic and cultural aspects of these complex interactions.

Examples of Complex Interactions

In the examples below, we draw from research conducted on farmer management of rice crop diversity. For example, Teeken *et al.* (2010) show that variety adoption is not the result of 'rational' choice in the narrow sense of optimizing production for given agroecological and socio-economic conditions. Cultural factors (e.g. rituals, food preferences) and socio-political factors (e.g. conflict) play a crucial role in the adoption or rejection of new varieties. For example, in Ghana, Guinea and Sierra Leone people show a cultural preference for rice varieties with a red pericarp over varieties with a white pericarp, whereas in The Gambia, Guinea-Bissau and Senegal people prefer varieties with a white pericarp. Since the late 1950s, the cultivation of African rice (*Oryza glaberrima*) has decreased in Ghana, but its role in rituals has become more important. In The Gambia, a reverse process has occurred: while older women still consider African rice very important (e.g. they like to plant a little bit of it in a field of Asian rice in the belief that it ensures a good harvest) younger women consider African rice as something bad (as being little more than a weed). In southern Guinea-Bissau and Sierra Leone, countries both affected by long periods of armed conflict, the use of African rice has increased. In southern Guinea-Bissau, people appreciate African rice for its medicinal properties, tolerance to salinity, and slow digestibility, while in Sierra Leone farmers appreciate African rice mostly for its ability to grow on poor upland soils, and its short growth cycle in the 'hungry season'. In maritime Guinea, the use of African rice has been maintained as it helps households to achieve food sufficiency in an environment marked by severe economic crises – it is adapted to poor soils and it digests slowly, which reduces daily intake.

Cultural identity can also be associated with tools. A very widely studied identity marker is the fulcrum shovel plough used in mangrove-swamp rice cultivation in Senegal, Guinea-Bissau and Guinea (Linares, 1992; Sarró, 2009). Very small tools like a small harvest knife also function as identity markers, although perhaps less obviously so. Despite the fact that a sickle would increase the speed of harvesting, Mandinka women in The Gambia and Senegal strongly oppose its use. They consider the sickle a Jola tool. Among the Jola the sickle is used by men, although it may also be used by women. Jola women sometimes use a sickle for the harvesting of short-straw (modern) varieties. Some Mandinka women acknowledge that a sickle would make harvesting of short-straw varieties much easier, but still they resist its use. This resistance can be better understood if we realize that in the past sickle harvesting was sometimes violently resisted by some groups in coastal West Africa because it was seen to challenge egalitarian cultural values (Richards, 1996).

The examples given above show how rice technologies (varieties, tools and techniques) are shaped by various factors in very different ways. Technologies such as tools and varieties have different development pathways, but have in common the fact that they are shaped by both environmental and cultural selection pressures. In this chapter, we use variety choice to show how agroecological, socio-cultural and economic factors all influence technology development, use and adoption. Which varieties are chosen not only depends on which varieties best fit a farming system, but also on socio-economic dynamics and the roles different crops and varieties play in larger socio-economic and socio-cultural systems. An example from The Gambia may be used to illustrate this. Before 1970, farmers commonly grew varieties with different crop cycle durations and *O. glaberrima* was the first rice to harvest. With the decrease in rainfall in the early 1970s, varieties with an equally short duration were introduced from Casamance (Senegal). In that same period, labour became less available at household level for rice pounding and bird scaring in the rice fields. Together these factors have affected crop and variety preferences and therefore also the selection and development of crops and varieties by farmers. Farmers explained that more than a generation ago it was not considered difficult to pound (mill) *O. glaberrima*, as pounding fonio (*Digitaria exilis*) was considered even more difficult. Subsequently, most farmers abandoned fonio; moreover *O. glaberrima* is now considered very difficult to pound (Nuijten, 2005). Another example of how social change and access to labour can impact on varietal choice may be given from mangrove-rice farming in southern Guinea-Bissau. In colonial times, Thom and Atanhã were the most extensively grown varieties because of their high productivity and good cooking and eating characteristics. Social change after independence led to a reduction in their cultivation due to the difficulty of finding youth labour-groups willing to do the intensive threshing required for these varieties (Temudo, 2011).

Guinea-Bissau also provides an example of how market conditions may impact on farmer preferences and adoption of new varieties. Since the 1980s, farmers have been adopting cashew (*Anacardium occidentale*) as a cash crop, reducing rice production and relying progressively on the purchase of imported rice during the pre-harvest 'hungry' season. Some farmers with large cashew orchards stopped rice production altogether, other farmers reduced their portfolio, while yet others began to prefer more tasty varieties. However, the decrease in the price of cashew and increase in the price of imported rice has reversed this trend. There is now a renewed interest in non-tasty varieties with long digestion times. Among these varieties attracting renewed interest one can highlight the upland variety Maimuna, which Nuijten *et al.* (2009) identified as belonging to a group of 39 rice accessions representing farmer varieties resulting from hybridization between *O. sativa* and *O. glaberrima*. The exact origin of these materials is beyond current farmer memory. Farmer practices enable such hybridization and spontaneous back-crossing to occur in their fields. Conditions of war and drought have presumably stimulated the spread of these varieties, apparently because they have considerable adaptive plasticity under sub-optimal farming conditions.

Varietal traits also play a complex role in the adoption rate of new varieties. Social processes play a more important role in the adoption of some traits than others (Nuijten, 2005). Crop duration is an example of a trait that is closely linked with the social system. In various upland rice cultivation areas in West Africa, the common rice varieties are chosen in such a way that they are similar in crop cycle duration. One reason for this is the labour calendar. This factor prevails among upland farmers in Guinea. Another important reason in The Gambia, Guinea and Guinea-Bissau is bird scaring. If varieties differ in crop duration, the crops mature over a longer period and hence bird scaring requires more labour. Farmers who sow their rice much earlier or sow earlier-maturing varieties, run the risk of a complete failure due to birds, if they cannot organize proper bird scaring. This implies that if farmers want to adopt a new variety with a shorter duration, farmers will need to test this variety simultaneously as a group, to avoid the risk that birds will destroy the entire crop of a lone pioneer.

The adoption of new varieties often follows a pattern in which a single farmer will test a variety, and then give it to neighbours or family members for further testing. If they also like the new variety it may then be adopted by other farmers in the village, and spread to neighbouring villages. But for some traits, like crop cycle duration, a different way of testing varieties is required, and a critical mass of farmers involved from the outset. This implies that there is a social component in variety choice, and this social component needs to be properly evaluated if varieties are to be introduced effectively. Where adoption of better varieties requires changes in the farming system at village level (as is probably the case for bird scaring of early and long-cycle varieties), there is little to be gained by relying on seed transfer to a handful of (so-called) 'master farmers'. Instead, a group innovation approach is required.

Similar social dynamics apply to traits related to digestibility. In communities where there are strong preferences for 'heavy' varieties, which digest slowly and 'stay long in the stomach', the uptake of 'light' varieties may only be for certain specific purposes, such as to meet the needs of old or sick people. Substantial adoption of light varieties would require re-organizing the food system; in particular, the possibility of increasing the food intake either by the consumption of more rice, and therefore larger fields for which more labour would be needed, or through diversification of the diet, for which a larger range of crops would need to be cultivated with consequent re-organization of tasks and the labour calendar and related rules and rights. This also applies to varieties which cannot be kept long after cooking without becoming either mushy or too hard. In various West African societies, it is important to be able to offer strangers food at any time of the day, or to keep the food till the next morning and eat it as breakfast. These aspects need to be evaluated as part of any rice innovation strategy.

Perhaps paradoxically, a thorough social analysis may be less needed for a trait like taste, which is often seen as a 'cultural' variable. Taste is determined by social dynamics, but is also to an important degree the result of individual preferences. If a variety with a different taste is appreciated by only one farmer, it does not imply any disadvantage for its cultivation. Over time, it may (or may not) be adopted by other farmers. In any case, taste is evaluated differently according to circumstances. A good illustration is the positive evaluation given to certain rices with a poor taste by farmers in southern Guinea-Bissau that might elsewhere be rejected because of this poor taste. The lack of attractive taste is deemed to slow down consumption, and so prevents eating more than necessary in times of food shortage, such as during the pre-harvest season (Temudo, 2011). Equally, farmers in Guinea and northern Guinea-Bissau sometimes mix heavy, poor-tasting rice with very tasty varieties to slow consumption of stock.

These examples show that the preference for traits is not only related to environmental factors, but also to socio-economic, cultural and individual factors. Desired and unwanted traits are conceived in relation to the diversity of available crops and their varieties and thus the concepts of desired and unwanted traits are fluid (Nuijten, 2005). This probably also explains differences in outcomes of studies on the importance to farmers of yield and taste. Taste can only be an important criterion where crop varieties differ greatly in taste and the same can be said for yield (Nuijten, 2005).

Variety preferences and criteria may change over time. In the past, Gambian farmers preferred varieties that matured within the rainy season in order to have as long a growing season as possible. With increasingly erratic rainfall, many farmers now prefer early varieties, although they do not want varieties that mature too early (Nuijten, 2005, 2010). Farmers in a village in Cacheu (Guinea-Bissau) often said that the varieties grown in Casamance and The Gambia today are too early and require intensive bird scaring (Nuijten, unpublished observations, 2007–2008). In the past they grew these early varieties next to varieties with a longer crop cycle duration in the uplands, but were able to do so because children did the bird scaring on early rice while adults worked on other crops. Now that the children go to school, farmers prefer to grow varieties with similar duration in the uplands, so they can do bird scaring and harvesting at the same time.

For other traits, notably yield, genotype-by-environment (G×E) interactions at the local level play a crucial role. For farmers to adopt a new variety for its yield, this new variety needs to perform better than the varieties grown by them over the previous two or three seasons. Farmers do not use multiple replications (or advanced statistical analysis), but only single ones, which implies the yield difference needs to be substantial for a farmer to appreciate it enough to adopt the new variety. Famers may also test the new varieties in the worst part of their plots, and if they perform well there, then they feel they are sure to get good productivity and yield stability elsewhere (Richards, 1986; Temudo, 2011).

In addition to G×E interactions, social processes may play a role when many new varieties are introduced simultaneously in a community. In such a case the adopted variety may not be the best-performing variety, but the variety introduced through the largest or most influential socio-political network. Thus, knowing about these networks and how they are constituted – e.g. as networks of patrons and clients – is important for any successful seed innovation strategy (see Richards, 1986).

Furthermore, variety choice and preference are not the same. Farmers need to choose varieties that do well in the field, even though those varieties do not meet all their preference requirements (Nuijten, 2005). For example, Gambian rice farmers prefer tall rice varieties, but if only short varieties are available with the right crop cycle duration, they will work with short varieties. Women in The Gambia also say they like varieties with big grains, but (particularly in the uplands) the best-performing varieties (both farmer and modern) have small grains. In short, not all preferences count with the same weight. Preferences predict whether (say) a tall, bold-grained rice will do better among Gambian farmers than a rival short, small-grained variety, but cannot indicate whether the tall, bold-grained type will be adopted.

This implies that adoption curves of new varieties are likely to differ depending on particular traits of those varieties in relation to the farming system and the wider socio-cultural and agroecological context. Certain 'permissive' agroecological contexts (in a well-watered country such as Sierra Leone, for example) may allow the adoption of quite a wide range of varieties with different traits. Similarly, socio-cultural and economic conditions may reduce or enlarge the number of desirable varieties, and may narrow or enlarge the variation in traits considerably, for reasons as diverse as ethnic and religious preferences, presence of milling machines, modes of labour organization, and availability of cash income sources and cheap imported rice.

As a result, the uptake of the same new technology (a variety) may show different patterns in different farming systems due to the different agroecological and socio-cultural contexts: the S-curve describing the uptake of a single technology may be quite variable in gradient according to local variations in the configuration of the competing and converging variables. The initial slow growth phase may be slower or faster, and the steepness of the middle part of the curve may also vary. In some areas, the adoption rate of a promising variety may be 100%, while in other areas the same variety may have only a moderate or low uptake. Explanation of what works, where and why, can be quite hard or impossible in the absence of careful analysis of socio-cultural and socio-economic selection factors and their interaction with environmental factors.

Discussion

The implication of the factors described above is that a better understanding of the interactions among technologies, agroecological and socio-cultural factors may allow the development of technologies more suited to farmers' needs. It may also facilitate the distribution of farmer technologies across regions as otherwise diverse as West Africa. Farmer varieties are often considered to be adapted to the local context, whereas modern varieties are developed for wide adaptation.

Research comparing farmer varieties of African and Asian rice in trials in Guinea-Bissau, Guinea, Sierra Leone, Ghana and Togo, shows that many farmer varieties are widely adaptable in agroecological terms (Mokuwa *et al.*, 2012). However, a considerable number of these widely adaptable farmer varieties are not appreciated outside their cultivation zone for socio-economic and cultural reasons. Some farmer varieties are adapted to both the agroecological and the socio-cultural contexts. An example is a variety collected in Guinea-Bissau, called Untufa and belonging to a new interspecific rice type identified by Nuijten *et al.* (2009), that was appreciated by farmers in a case study area in Ghana for both agroecological and socio-cultural reasons (Teeken *et al.*, 2011). In our examples we focused on rice varieties. For other technologies, for example sowing and harvesting tools, or fallow management and postharvest management practices, the same dynamics are also likely to emerge when studied. Some practices may fit various agroecological contexts, but not the socio-cultural contexts. There may be little technology-by-environment interaction, but much technology-by-society interaction, or vice versa. Other practices may fit well across a range of agroecological and socio-cultural contexts (i.e. there is little or no interaction) or be limited to a few localities because there is strong technology-by-environment-by-society interaction.

It may thus be important to test locally developed technologies across the region and to assess whether they are plastic (i.e. have low technology × environment × society interaction effects) or applicable only in specific localities. Given the large diversity in farming systems, and agroecological and socio-cultural variables, conducting experiments together with a wide network of farmers may provide a better understanding of how the various interactions among these variables limit or stimulate the adoption of new (local and modern) technologies (Richards *et al.*, 2009). An important question is how farmer technology development can be better linked to scientific technology development (Offei *et al.*, 2010). An essential first step in achieving this linkage is to realize that farmer technologies need to be valued in a similar way to the technologies developed by scientific research (i.e. without prejudice). If they show potential they deserve region-wide dissemination.

Theoretical perspectives and methodological approaches

At this stage there is no theory that is capable of integrating all perspectives from the various disciplines of soil science, plant breeding, agronomy, crop physiology, chemistry, economics, anthropology, communication and innovation studies, archaeology and history. The examples described above show that both simple and complex interactions may occur. In one case an agroecological factor may play a dominant role, whereas in a seemingly comparable case within a different context cultural factors may be critical. Various researchers have suggested a more systematic or holistic approach in order to fit varieties better to the environment and management practices, and to appreciate complex genotype-by-environment-by-management interactions rather than trying to ignore them (Kronstad, 1996; Kropff and Struik, 2002; Desclaux *et al.*, 2008). The disadvantage of the term 'management' is that it is already the result of interactions among genotypic, environmental and socio-economic factors. Researchers working on participatory plant breeding (PPB) have suggested that the environment should also include socio-economic factors (Ceccarelli and Grando, 2007). To systematically compare and integrate technological, agroecological, socio-cultural and economic factors we suggest the following framework, based on the formula used to describe G×E interactions:

$$P = \mu + T + E + S + T \times E + T \times S + E \times S + T \times E \times S + \varepsilon$$

This framework tangibly brings together the technical, natural and social at the same level.[1] In the context of rice-farming systems, the meaning of the symbols is the following:

- P = outcome (a farming system, resulting from various interactions between a technology and the societal and environment factors)
- μ = a common factor for all farming systems (a common factor shared by all societies in terms of social dynamics and environmental conditions)
- T = technology effects (e.g. differences in performance between varieties)
- E = environment effects (a range of agroecological variables such as climate, soil, landscape, pests)
- S = society effects (socio-cultural and socio-economic; e.g. appreciation of pericarp colour or maturity in relation to the labour calendar, and the societal processes underlying these preferences)
- $T \times E$ = technology-by-environment interaction (e.g. research by Mokuwa *et al.*, 2012, shows strong responses to environment for some farmer varieties and hardly any response for others)
- $T \times S$ = technology-by-society interaction (e.g. varieties that are widely adaptable in agroecological terms are appreciated differently on the basis of how they fit within the farming system in regard to factors such as duration or pericarp colour)
- $E \times S$ = environment-by-society interactions (e.g. the cultivation of rice in different environments [upland or lowland] may be done

by various social groups depending on cultural regulations)

- $T \times E \times S$ = technology-by-environment-by-society interactions (specific outcomes as the result of interactions between all three terms)
- ε = residual variance, not accounted for by any measured factor or their interactions; it includes outcomes that are too complex to explain through reference to any identified mechanism, or behind which there is no particular intentionality other than farmer performance (see Richards, 1993; Batterbury, 1996). Richards (1989, 1993) calls the process of adjusting the farming decisions to ongoing opportunities and constraints 'performance'.[2]

Where appropriate, the factors T, E and S can be further broken down into relevant components. The nature of these factors depends on the research question. The factor ε is often considered unimportant in natural-science research, but in our view it is very important: it not only stands for random noise that is too complex to explain, but also for contexts that at some point in time may yield interesting innovations.

The best methodological approach to describe and/or measure $T \times E \times S$ interactions is a comparative case-study approach, combining experiments (applicable to both social and natural science aspects), questionnaires, various types of interviews, and observation-based and participatory approaches (Nuijten, 2011). Examples of this approach are studies on the development of a new rice type with an interspecific background (Nuijten *et al.*, 2009) and the variation in wide adaptability of farmer varieties (Mokuwa *et al.*, 2012). For a systematic integration of the various social and biological research methods, including qualitative and quantitative research methods (Fig. 29.1), the methodology known as technography (Richards, 2001; Jansen and Vellema, 2011) can be used as a framework. Technography (the description and analysis of technological activity as a systematically related set of material and sociological processes) assumes the existence of real (if deeply embedded) causal mechanisms of both a biological/material and sociological nature. Social facts are real but not necessarily material in constitution. Technography is located within a philosophical framework of critical realism, which assigns an equivalent epistemological status to social and biological variables (see Sayer, 2000). Although technography is in essence a social-science methodology, it can be integrated (because of its attention to causal mechanisms) with various biological research methods that have a strong focus on causal mechanisms and experimentation. The idea is not to generate a single data set of biological and sociological data, but to generate data sets in such a way that a range of hypotheses in relation to the actual functioning of a particular mechanism within a certain context can be validated.

Towards new models for technology development and dissemination

At this stage, there is insufficient data available for a systems approach for rice farming in West Africa. It is also argued that a disadvantage of a systems approach as currently practised is that it largely ignores power issues (inequality, gender, age, etc.) and the openness inherent in farming practices (Jansen, 2001). Another disadvantage is that there is little scope to consider performance, which is an essential element of (both low-input and high-input) farming.

A better understanding of the interactions among technology, environment and society elements is necessary to understand how to build new models for technology development and dissemination in which there is an active role for farmers. In this chapter we have provided some examples of such interactions, mostly in relation to variety choice. We need more examples relating to other technologies, such as soil and water management, and the control of pests and diseases. The local farming system needs to be taken as the starting point, using the suggested methodology to identify key interactions.

A comparison between sub-Saharan Africa and Asia may provide some useful insights in terms of technology development and adoption at a macro level. Certain differences exist in terms of institutional organization at the national level. For example, in China the state plays a dominant role in technology development, which is an important factor

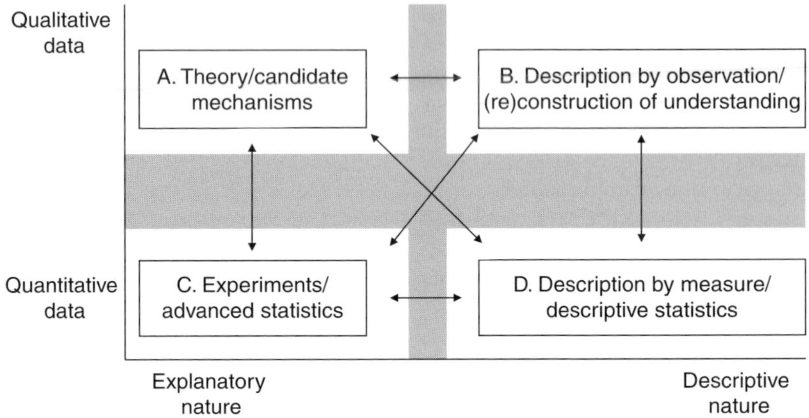

Fig. 29.1. Four basic research styles in the natural and social sciences, and humanities. (After Nuijten, 2011, with permission from Elsevier.)

explaining why China was the first country to develop hybrid rice (Shen, 2010). In India, many NGOs are involved in the promotion of new technologies, such as the System of Rice Intensification (SRI) (Glover, 2011). However, in sub-Saharan Africa the role of the state in agricultural and rural development has declined drastically since the mid-1980s, and many countries have undergone wars and civil conflicts with dramatic impacts on food security, seed security and technological development. NGOs in West Africa have taken up agricultural development only to a limited extent, often only for limited periods of time, and their agendas are influenced by the priorities of international donors. This sometimes results in contrary efforts, such as in The Gambia where some NGOs promoted tree planting in vegetable gardens developed earlier by other NGOs (Schroeder, 1997). More information is needed to better understand the differences in institutions and social contexts between Asia and sub-Saharan Africa.

To improve our understanding of why certain technologies replace others, to what extent and at what speed, the multi-level perspective (Geels and Schot, 2007) may be a useful way of explaining the importance of technographic methodology to technical scientists. The multi-level perspective helps to organize the material and sociological processes related to technological activity and, as such, may help open up the 'black box' of farming systems. In Fig. 29.2, three levels are identified: the socio-technical landscape (exogenous context), the socio-technical regime (the dominant practice) and niche innovations (other practices). The socio-technical regime is shaped by technology, culture, science, industry, market and policy, and adopts innovations developed in niches under certain conditions (Geels and Schot, 2007). The advantage of the multi-level perspective is that it helps to reveal the elements that constitute socio-technical regimes and how they are related to the wider context and to niche innovations. The multi-level perspective may be useful for illuminating strengths and weaknesses of various participatory approaches, and may possibly explain the relative lack of success of some of them. For example, several case studies on participatory approaches and farmer field school approaches in East Africa show that often the old extension approaches were continued using new labels (Isubikalu, 2007; Kamau, 2007). However, a difficulty of the multi-level perspective is where to draw the boundaries between the socio-technical regime, the wider socio-technical landscape, and the niches.

The multi-level perspective agrees with the idea that a broad base is needed for technology development, not only in terms of diversity, but also in terms of technology development pathways/networks. Such networks between farmers and scientists may be based on unsupervised learning, as both practice

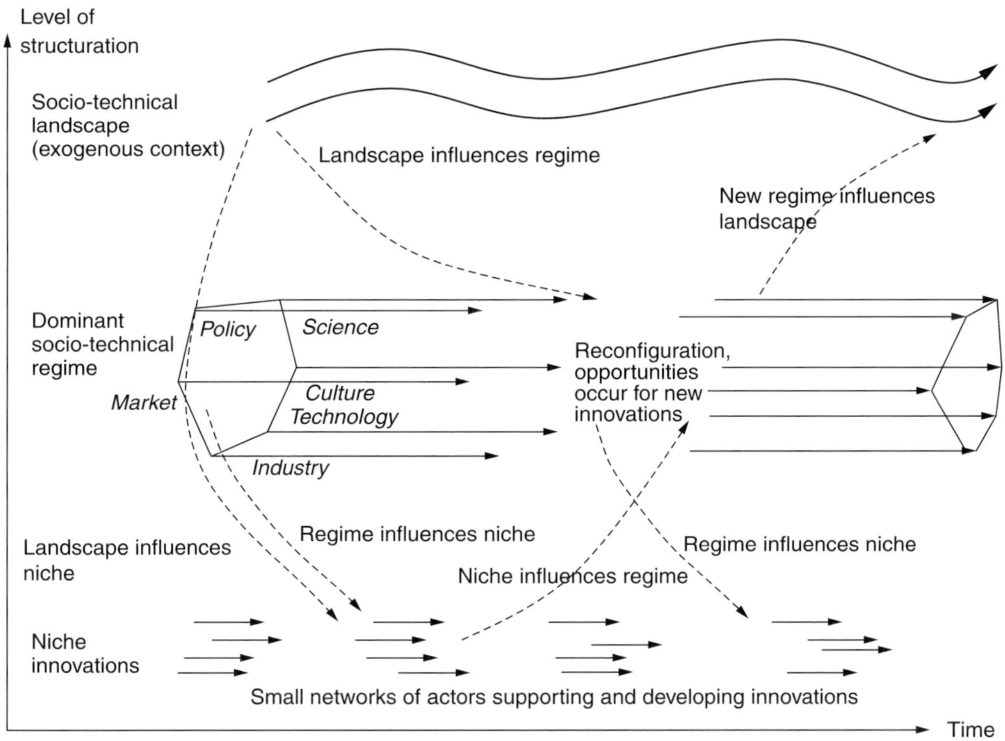

Fig. 29.2. Multi-level perspective. (Adapted from Geels and Schot, 2007.)

observation-based learning and thus no fundamental incompatibility exists between science and farmer innovation (Richards *et al.*, 2009; Offei *et al.*, 2010). However, scientists and farmers use different modes of communication. On-farm trials can be platforms for scientists and farmers to appreciate each other's ways of communication, to share their experiences and to learn from each other, using participatory and/or action-research approaches (Almekinders *et al.*, 2009; Bentley *et al.*, 2010). The trials can have different formats, differing in number of treatments and management, and still be suitable for advanced statistical analysis (Mutsaers *et al.*, 1997; Witcombe, 2002). Research has shown that farmers can handle large numbers of treatments (Ceccarelli *et al.*, 2001) and have a natural interest in experimentation (Richards, 1986; Bentley *et al.*, 2010).

The methodological approach we describe in this chapter facilitates a better understanding of farmer experimentation, and its outcomes, and how it is related to agroecological, sociocultural and economic factors (Nuijten *et al.*, 2009; Mokuwa *et al.*, 2012). With more studies of this kind, the multi-level perspective may more clearly illuminate what institutional adjustments at the level of science and policy are needed to facilitate more interaction and technology exchange between scientists and farmers.

Acknowledgement

Figure 29.1 is reprinted from *NJAS – Wageningen Journal of Life Sciences*, vol. 57, E. Nuijten, 'Combining research styles of the natural and social sciences in agricultural research', pages 197–205, Copyright (2011), with permission from Elsevier and Royal Netherlands Society for Agricultural Sciences.

Notes

[1] At this stage, the framework cannot be used for sophisticated statistical analysis, but its advantage is that it can be used as an analytical framework to better understand how the technical, environmental and societal interact.

[2] Batterbury (1996) argues that planning and performance play different roles in different farming systems, depending on the agroecological and socio-cultural context (including political factors, e.g. conflicts and war). This implies that the factor ε may be smaller in the savannah zone (e.g. in Burkina Faso) where the rainfall period is short and farmers need to plan their farming activities much more, than in the forest zone (e.g. in Sierra Leone) where farmers have the opportunity to adjust their farming activities depending on constraints and opportunities offered by the agroecological and socio-cultural context.

References

Almekinders, C.J.M. and Elings, A. (2001) Collaboration of farmers and breeders: participatory crop improvement in perspective. *Euphytica* 122, 425–438.

Almekinders, C.J.M., Beukema, L. and Tromp, C. (2009) *Research in Action: Theories and practices for innovation and social change.* Wageningen Academic Publishers, Wageningen, Netherlands.

Batterbury, S. (1996) Planners or performers? Reflections on indigenous dryland farming in northern Burkina Faso. *Agriculture and Human Values* 13, 12–22.

Bentley, J.W., Van Mele, P. and Acheampong, G.K. (2010) Experimental by nature: rice farmers in Ghana. *Human Organization* 69, 29–137.

Binswanger, H. and Pingali, P. (1989) Technological priorities for farming in sub-Saharan Africa. *Journal of International Development* 1, 45–65.

Ceccarelli, S. and Grando, S. (2007) Decentralized-participatory plant breeding: an example of demand driven research. *Euphytica* 155, 349–360.

Ceccarelli, S., Grando, S., Bailey, E., Amri, A., El-Felah, M., Nassif, F., Rezgui, S. and Yahyaoui, A. (2001) Farmer participation in barley breeding in Syria, Morocco and Tunisia. *Euphytica* 122, 521–536.

Dalton, T.J. (2004) A household hedonic model of rice traits: economic values from farmers in West Africa. *Agricultural Economics* 31, 149–159.

Desclaux, D., Nolot, J.M., Ciffoleau, Y., Gozé, E. and Leclerc, C. (2008) Changes in the concept of genotype? Environment interactions to fit agriculture diversification and decentralized participatory plant breeding: pluridisciplinary point of view. *Euphytica* 163, 533–546.

Dorward, P., Craufurd, P., Marfo, K., Dogbe, W. and Bam, R. (2007) Improving participatory varietal selection processes: participatory varietal selection and the role of informal seed diffusion mechanisms for upland rice in Ghana. *Euphytica* 155, 315–327.

Efisue, A., Tongoona, P., Derera, J., Langyintuo, A., Laing, M. and Ubi, B. (2008) Farmers' perceptions on rice varieties in Sikasso region of Mali and their implications for rice breeding. *Journal of Agronomy and Crop Science* 194, 393–400.

Geels, F.W. and Schot, J. (2007) Typology of sociotechnical transition pathways. *Research Policy* 36, 399–417.

Glover, D. (2011) Science, practice and the System of Rice Intensification in Indian agriculture. *Food Policy* 36, 749–755.

Gridley, H.E., Jones, M.P. and Wopereis-Pura, M. (2002) Development of New Rice for Africa (NERICA) and participatory varietal selection. In: Witcombe, J.R., Parr, L.B. and Atlin, G.N. (eds) *Breeding Rainfed Rice for Drought-prone Environments: Integrating conventional and participatory plant breeding in South and Southeast Asia.* Proceedings of a DFID Sciences Research Programme/IRRI conference, Los Baños, Philippines, 12–15 March 2002. Centre for Arid Zone Studies, University of Wales, Gwynedd, UK, pp. 23–28.

Isubikalu, P. (2007) Stepping-stones to improve upon functioning of participatory agricultural extension programmes: farmer field schools in Uganda. PhD thesis, Wageningen University, Wageningen, Netherlands.

Jansen, K. (2001) Implicit sociology, interdisciplinarity and systems theory in agricultural science. *Sociologica Ruralis* 49, 172–188.

Jansen, K. and Vellema, S. (2011) What is technography? *NJAS – Wageningen Journal of Life Sciences* 57, 169–177.

Kamau, G. (2007) Researching with farmers: a study of KARI participatory research practices in context. PhD thesis, Wageningen University, Wageningen, Netherlands.

Kronstad, W.E. (1996) Genetic diversity and the free exchange of germplasm in breaking yield barriers. In: Reynolds, M.P., Rajaram, S. and McNab, A. (eds) *Increasing Yield Potential in Wheat: Breaking the barriers.* International Center for Maize and Wheat Improvement, Mexico, DF, Mexico, pp. 19–28.

Kropff, M.J. and Struik, P.C. (2002) Developments in crop ecology. *NJAS – Wageningen Journal of Life Sciences* 50(2), 223–237.

Linares, O. (1992) *Power, Prayer and Production: The Jola of Casamance, Senegal.* Cambridge University Press, New York.

Maurya, D.M., Bottrall, A.A. and Farrington, J. (1988) Improved livelihoods, genetic diversity and farmer participation: a strategy for rice breeding in rain-fed areas of India. *Experimental Agriculture* 24, 311–320.

Mokuwa, A., Nuijten, E., Okry, F., Teeken, B., Maat, H., Richards, P. and Struik, P.C. (2012) Robustness and strategies of adaptation within farmer varieties of African rice (*Oryza glaberrima*) and Asian rice (*Oryza sativa*) across West Africa. *PLoS ONE* 8(3), e34801. doi:10.1371/journal.pone.0034801.

Mutsaers, H.J.W., Weber, G.K. and Walker, P. (1997) *Field Guide for On-farm Experimentation.* International Institute of Tropical Agriculture, Ibadan, Nigeria.

Nuijten, E. (2005) Farmer management of gene flow: the impact of gender and breeding system on genetic diversity and crop improvement in The Gambia. PhD thesis, Wageningen University, Wageningen, Netherlands.

Nuijten, E. (2010) Gender and management of crop diversity in The Gambia. *Journal of Political Ecology* 17, 42–58.

Nuijten, E. (2011) Combining research styles of the natural and social sciences in agricultural research. *NJAS – Wageningen Journal of Life Sciences* 57, 197–205.

Nuijten, E., Van Treuren, R., Struik, P.C., Mokuwa, A., Okry, F., Teeken, B. and Richards, P. (2009) Evidence for the emergence of new rice types of interspecific hybrid origin in West African farmers' fields. *PLoS ONE* 4(10), e7335. doi: 10.1371/journal.pone.0007335.

Offei, S.K., Almekinders, C.J.M., Crane, T.A., Hughes, S.G., Mokuwa, A., Nuijten, E., Okry, F., Struik, P.C., Teeken, B. and Richards, P. (2010) Making better seeds for African food security – a new approach to scientist–farmer partnerships. In: Bruce, T., Foyer, C., Halford, N., Keys, A., Kunert, K., Lawlor, D., Parry, M. and Russell, G. (eds) *Aspects of Applied Biology 96: Africa's 'Engine for Growth' – Plant science and biotechnology hold the key.* The Association of Applied Biologists, Warwick, UK, pp. 141–148.

Okry, F., Van Mele, P., Nuijten, E., Struik, P.C. and Mongbo, R. (2011) Organisation of the seed sector of rice in Guinea: stakeholders, perception and institutional linkages. *Experimental Agriculture* 47, 137–157.

Richards, P. (1986) *Coping with Hunger: Hazard and experiment in an African rice-farming system.* Allen & Unwin, London.

Richards, P. (1989) Agriculture as a performance. In: Chambers, R., Pacey, A. and Thrupp, L.A. (eds) *Farmer First: Farmer innovation and agricultural research.* Intermediate Technology Publications, London, pp. 39–43.

Richards, P. (1993) Cultivation: knowledge or performance? In: Hobart, M. (ed.) *An Anthropological Critique of Development: The growth of ignorance.* Routledge, London, pp. 61–78.

Richards, P. (1996) Agrarian Creolization: the ethnobiology, history, culture and politics of West African rice. In: Ellen, R. and Fukui, K. (eds) *Redefining Nature: Ecology, culture and domestication.* Berg, Oxford, pp. 291–318.

Richards, P. (2001) Technography: Notes on method. Unpublished project document. Convergence of Sciences programme, International Research and Education Fund (INREF), Wageningen University, Wageningen, Netherlands.

Richards, P., De Bruin-Hoekzema, M., Hughes, S.G., Kudadjie-Freeman, C., Offei, S.K., Struik, P.C. and Zannou, A. (2009) Seed systems for African food security: linking molecular genetic analysis and cultivator knowledge in West Africa. *International Journal of Technology Management* 45, 196–214.

Sarró, R. (2009) *The Politics of Religious Change on the Upper Guinea Coast: Iconoclasm done and undone.* Edinburgh University Press, Edinburgh, UK.

Sayer, A. (2000) *Realism and Social Science.* Sage, London.

Schroeder, R.A. (1997) *Shady Practices: Agroforestry and gender politics in The Gambia*. University of California Press, Berkeley, California.
Shen, X. (2010) Understanding the evolution of rice technology in China – from traditional agriculture to GM rice today. *Journal of Development Studies* 46, 1026–1046.
Sperling, L. and Loevinsohn, M.E. (1993) The dynamics of adoption: distribution and mortality of bean varieties among small farmers in Rwanda. *Agricultural Systems* 41, 441–453.
Teeken, B., Nuijten, E., Okry, F., Mokuwa, A., Temudo, M., Struik, P.C. and Richards, R. (2010) Advocating the integration of socio-cultural factors in models for variety dissemination. Presentation at the 28th International Rice Research Conference, Hanoi, Vietnam, 8–12 November.
Teeken, B., Okry, F. and Nuijten, E. (2011) Improving rice-based technology development and dissemination through a better understanding of local innovation systems. In: *Innovation and Partnerships to Realize Africa's Rice Potential.* Proceedings of the Second Africa Rice Congress, Bamako, Mali, 22–26 March 2010. Africa Rice Center, Cotonou, Benin, pp. 4.2.1–4.2.7. http://www.africarice.org/workshop/ARC/4.2%20Teeken%20fin.pdf.
Temudo, M.P. (2011) Planting knowledge, harvesting agro-biodiversity: a case study on southern Guinea-Bissau rice farming. *Human Ecology* 39, 309–321.
Waddington, S.R., Li, X., Dixon, J., Hyman, G. and De Vicente, M.C. (2010) Getting the focus right: production constraints for six major crops in Asian and African farming systems. *Food Security* 2, 27–48.
Weltzien, E., Whitaker, M.L. and Anders, M.M. (1996) Farmer participation in pearl millet breeding for marginal environments. In: Eyzaguirre, P. and Iwanaga, M. (eds) *Participatory Plant Breeding.* Proceedings of a workshop on participatory plant breeding, 26–29 July 1995, Wageningen, Netherlands. International Plant Genetic Resources Institute, Rome, Italy.
Witcombe, J.R. (2002) Mother and baby trial system. In: Witcombe, J.R., Parr, L.B. and Atlin, G.N. (eds) *Breeding Rainfed Rice for Drought-prone Environments: Integrating conventional and participatory plant breeding in South and Southeast Asia.* Proceedings of a DFID Plant Sciences Research Programme/IRRI Conference, Los Baños, Philippines, 12–15 March 2002, pp. 79–89.

30 Innovative and Effective Ways to Enhance Rural Learning in Africa

Paul Van Mele,[1]* Jonas Wanvoeke,[1]
Josephine Rodgers[2] and Blythe McKay[3]
[1]*Africa Rice Center (AfricaRice), Cotonou, Benin;*
[2]*Countrywise Communication, Wilby, UK;*
[3]*Farm Radio International (FRI), Ottawa, Canada*

Introduction

The complexity of language

'Who shall we ask as local translator?' we asked ourselves when going on our first field visit to central Benin in early 2005. The senior author's local colleagues were not sure. None of them actually spoke all the languages of that region, even those who were born there. But they assured him that all would be fine. Upon arrival in the first village, at least five distinct local languages were spoken and farmers seemed often versatile in several of them. One language was generally understood by all (although not spoken by all). The anticipated problem had resolved itself: the community had come up with its own local translators to facilitate group discussion. Still, the challenges seemed daunting. If a country as small as Benin, with only seven million people, had as many as 70 local languages, how were we to strengthen rural learning for rice across Africa?

Observation adds confidence

More than in any other profession, farmers constantly adjust their strategies, responding to emerging needs and opportunities. 'Looking over the fence' is common practice. For instance, when asked how they decided on what seed to use, rice farmers in northern Ghana said they always assessed their neighbours' fields throughout the growing season. If a crop outperformed their own or had an attribute of particular interest, farmers would often attempt to acquire some of their successful neighbour's crop seed to try it out the next season. By that time they are already fairly familiar with its characteristics although they may further test it for yield stability, adaptability and processing traits (in the case of a new variety). Observing the variety regularly in the field and the fact that it has worked for her or his neighbour has given the farmer confidence to test the technology.

Visibility of a technology

Some technologies are easier to observe and hence to assess than others. Most villagers will know when somebody is trying out a distinct new variety, or when a tractor is tilling somebody's land with a new type of plough. Farmers are also quick in calculating how many

* Corresponding author: paul@agroinsight.com

© CAB International 2013. *Realizing Africa's Rice Promise*
(eds M.C.S. Wopereis *et al.*)

work-days can be saved by spraying a field with herbicides instead of manually weeding it. But assessing soil fertility and deciding on ways to maintain or improve it is more complicated. While sophisticated, computerized equipment can help Western farmers to fine-tune fertilizer application based on the soil fertility in a particular part of the field, African farmers have to rely on their life-long experience of working the land. Their in-depth knowledge (which often takes into account soil-living organisms, smell, colour and softness of the soil) may be further strengthened through participatory learning with researchers (Defoer, 2002; Ramisch *et al.*, 2006).

Complexity of a technology

The more complex a technology, the more difficult it becomes to share information about it orally. For instance, integrated pest management skills learned during farmer field schools are not readily passed on to non-participants. For the same reason, there is little evidence that the information required for low external input technology (such as soil-conservation techniques) is transmitted very effectively between farmers (Tripp, 2007). Oral communication has its limitations in conveying complex issues. As sustainable agriculture requires an in-depth understanding of complex relations between farming, nature and society, farmers require additional support to learn.

Changing contexts

Learning tools and methods also need to consider the dynamism in the system and the shelf-life of the information provided. For instance, farmers may adjust their practices based on a weather forecast (that can be easily communicated through mobile phones or radio), or because of changes in climate. Also, the demographic boom in Africa has resulted in rural–rural migration and contributed to an increasingly complex mosaic of local cultures that may require outside facilitation (Saïdou, 2006). With rural electrification being high on the agenda of many African governments, possibilities for wider use of video and television in agricultural development open up.

Lasting impressions

In rural areas with a strong oral culture, unusual events are highly debated and leave a lasting impression in people's minds. When Espérance Zossou visited villages in central Benin more than a year after a video on rice parboiling had been publicly screened, women perfectly recalled most of the details of the video. They had even observed secondary details (like the improved stoves) and sought out more information about them (Zossou *et al.*, 2009a). Although the content of the video was merely technical, women strongly appreciated the event as rural entertainment (Zossou *et al.*, 2009b). While filming with rice farmers in Zianso in southern Mali (Fig. 30.1), one of the elders told Van Mele that he was really excited about contributing. It reminded him of his childhood, when outsiders came to his village to show a black-and-white film on a large screen.

Media and extension revisited

In the 1970s, when the Food and Agriculture Organization of the United Nations (FAO) started to use video as a tool to recover, preserve and reproduce farmers' knowledge, the organization was criticized for using an over-sophisticated medium for a rural setting (Ramírez, 1998). As it turned out, the project paved the way for the use of video as a cost-effective tool to support group training and rural development

Fig. 30.1. Lasting impressions are passed on from generation to generation through storytelling. (Photo: P. Van Mele.)

(Coldevin and FAO, 2001). Rather puzzling is that those promoting information and communications technologies (ICTs) often portray video as inappropriate for use in Africa.

Since the 1990s, communication for development has become more decentralized and gained ground on the agenda of international agencies. This coincided with an explosion in the number of private radio stations across the developing world. Projects emphasize closer interactions with farmers and strengthening research–extension–radio linkages (Hambly Odame *et al.*, 2002; Chapman *et al.*, 2003). Although numerous projects have tried to wean researchers and extension staff away from the linear technology-transfer mind set, most radio broadcasters have not yet been exposed to participatory approaches.

Since the early 1990s, many international agencies have turned their aspirations to new ICTs, but results have not met the expectations. The need to re-adjust strategies was acknowledged during an international meeting of media professionals in Brussels that concluded that 'ICTs are not always the answer to improved information and learning in all circumstances' (CTA, 2009). The Research into Use programme (Lenné, 2008), and the establishment of the Global Forum for Rural Advisory Services (GFRAS) are but two of the signs that the international community is waking up to address the over-due neglect of rural learning.

In what follows, we address four key challenges in rural learning, namely social inclusion, scaling up, collecting farmer feedback and assessing impact. Experiences from Africa Rice Center (AfricaRice) and partners are presented alongside other relevant experiences.

Social Inclusion

Gender inequality in natural-resource use

In Africa, gender division in rice production is intricately interwoven with land use rights. While men often dominate in irrigated rice systems, women more often cultivate rainfed lowlands. As lowlands are playing an increasingly important role in food security, income generation, land regulation and sustainable management of natural resources, researching the contexts, mechanisms and outcomes of interventions is crucial (Van Mele *et al.*, 2011b).

Interventions that add value to the land (e.g. water management structures) often result in reallocation of use rights from women to men, as has been the case in The Gambia (Carney, 1998). External facilitation may prove crucial. For instance, wetland improvement schemes in Burkina Faso only succeeded when reallocation policies changed and women's initial use rights were respected (van Koppen, 2009). Yet, even two decades after the promotion of gender analysis many interventions still fail to address this. One of the reasons is that such experiences are either written up in project reports or in academic literature, both of which are text-based and not easily accessible. Having video documentaries on these topics could help development agencies and communities to anticipate and mediate conflicts over resource use before actual interventions take place.

Gender inequality in access to information

As with any value-adding intervention, differences in access to information may increase the gap between social groups in a community. A heavy emphasis on ICTs without giving proper attention to power relations and marginalized groups is risky and may not be conducive to rural development (Gurumurthy, 2006).

In many cases, development interventions are male-biased because women farmers are restricted by social norms from communicating with men outside their families (Katungi *et al.*, 2008). In Benin, public video screenings created an equal chance for all community members to learn (Zossou *et al.*, 2009b). They strengthened the social capital among women and improved the trust between actors in the rice value chain, a key weakness of markets in Africa (Fafchamps, 2004). Relations between women rice processors, intermediaries and input and output markets improved, and women changed their information-seeking behaviour (Zossou *et al.*, 2010).

Poverty targeting

Since the early 1990s, farmer training has increasingly targeted groups rather than individuals, mainly driven by donor requirements. Numerous groups mushroomed, often opportunistic and prone to local power plays. Extension efforts were rarely based on a good understanding of local contexts or poverty assessments, which continue to be considered only in impact assessments (Kassam, 2006).

The demand-side and the supply-side of service provision, as well as the organizations and donors supporting this, can benefit from better insights into social inclusion issues in rural learning, especially when scaling up and out.

Scaling-up and Scaling-out

Participatory learning and action-research

Although farmers are experimental by nature (Bentley *et al.*, 2010), their opportunities to learn about new ideas and trends are often limited and localized. Few farmers are blessed with the regular visit of an extension agent and even fewer meet a sympathetic researcher willing to listen to their needs or work alongside them over an extended period. The scarce public resources attributed to participatory research must indeed be well targeted and impact pathways carefully assessed from the outset (one of the positive evolutions in allocations of research funds since the late 2000s).

Since 2000, AfricaRice has developed and used participatory learning and action-research (PLAR) with groups of farmers at multiple sites across West Africa. PLAR adopted much of the philosophy of participatory rural appraisal (PRA) (e.g. respect for local agendas and spending long periods of time in villages), as well as several of its tools (e.g. cropping calendars and transect walks). PLAR also adopted weekly meetings and a seed-to-harvest approach from farmer field schools (FFS), but encouraged farmers more to conduct experiments in their own fields on whatever topic they thought relevant for them. At weekly sessions, extension agents encouraged farmers to set up small-scale experiments with various technical options, including timing of land preparation, mineral fertilizer doses and times of application, water management, new rice varieties, and various ways to control weeds. The manual had 28 modules, chapters corresponding to weekly sessions with farmers (Defoer *et al.*, 2004).

Although groups were often formed for the sake of the PLAR, at times groups themselves articulated demand and adjusted the method to suit their needs, as was the case in Mali (Box 30.1).

However, no matter how well the intentions of a project are made or how well a training manual is written, the attitude of those outsiders facilitating the sessions strongly influences the level of participation and learning.

Attitude counts

In 2008, an external evaluation of a project funded by the International Fund for Agricultural Research (PADS) related differences in outcomes between Ghana and Mali to differences in attitude of project staff (Van Mele *et al.*, 2011b). In Ghana, the staff thought of PLAR as an extension method for teaching rice technologies to farmers; in Mali, the staff understood that PLAR was an approach for mutual learning and that it was meant to develop and test technical and institutional innovations with farmers. In Ghana, the staff prompted farmers to say they had adopted project recommendations without change. In Mali, the project staff were proud of farmer innovations and asked farmers to describe them to the project evaluators.

PLAR did stimulate farmers to experiment with new ideas and technologies, especially when the facilitators themselves valued these local experiments (Fig. 30.2). It is little surprise that a positive attitude towards farmers' knowledge and practices helped to nurture mutual learning. It was striking, however, that some staff could go through the entire PLAR manual without gaining a respect for farmer experiments.

The complexity of African farming systems, the costs involved in face-to-face extension and the differences in attitudes of service

Box 30.1. A group for women

Like many of Mali's villages, Zamblara is semi-arid, with rolling hills. During the brief rainy season of about 5 months, men grow maize, sorghum, groundnuts and other crops on the higher ground. Women grow rice in low-lying, seasonally flooded areas near the villages. During the long dry season, men and women grow vegetables in the low areas after harvesting the rice.

Because rice is grown mainly by women in Zamblara, they formed an association of rice producers in 1997. The women created the association to help themselves develop agricultural practices and to increase their income. The name of the group is 'Kotognogontala', which means 'mutual respect'. The group aims to exchange knowledge and good agricultural practices within the community, and in 2002 they requested the participatory learning and action-research (PLAR) training.

From an original group of 27 people, the association has grown to four groups with 115 women and two men. In Mali, most women's groups have at least some men in them. In this one, the village chief is the honorary president and another man attends to monitor the women's activities. The women say the group has helped improve relations between men and women. The group gives the women a place where they can talk about their problems with men, and give each other advice.

Although the women of Zamblara each have their own small plots of rice land, the group works one collective field of 1.5 hectares. They grow rice in the rainy season and vegetables in the dry season. When the women harvest the rice from this plot they sell some of it and keep the money as a group fund. They divide some of the rice among themselves, and keep the rest to use for their meals during group activities.

PLAR has helped increase rice production in the village, and many of their neighbours are now interested in the new techniques. The four PLAR groups each have a farmer-facilitator. Although the PLAR modules were written in French, they have been (verbally) translated into Bambara (the local language). The women have adapted the content, by composing songs and poems about the rice-farming modules.

Source: Wanvoeke *et al.* (2008). Reproduced with permission from the Centre for Information on Low External Input and Sustainable Agriculture (ILEIA).

Fig. 30.2. Learning from farmers requires an open, enquiring and positive attitude towards their knowledge and practices. (Photo: P. Van Mele.)

providers further justify the development of farmer-oriented videos to stimulate learning across organizations and across cultures (Van Mele *et al.*, 2010b). The zooming-in zooming-out (ZIZO) approach provides insights as to how to achieve this.

Zooming-in zooming-out

The ZIZO approach (Fig. 30.3) results in videos that are of regional relevance and locally appropriate (Van Mele, 2006). ZIZO basically revolves around five key steps that are not strictly in chronological order. In particular, the first two may be reversed, or even integrated, depending on the situation.

ZIZO is not a blueprint. No matter how well the video production is planned, the content will change during filming, because of interaction and feedback from farmers. Flexibility and eagerness to learn from people on the ground are crucial.

Since the mid-1980s, there has been a boom of FFS (van den Berg and Jiggins, 2007; FAO, 2008). Their wealth of experiences and regional insights are a gold mine for the development of videos along the ZIZO approach, whereby a number of selected FFS graduates could share their learning in front of the camera.

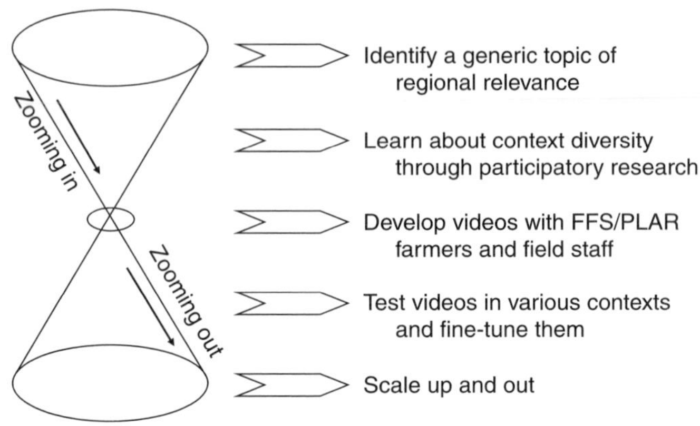

Fig. 30.3. The zooming-in, zooming-out approach. (Modified from Van Mele, 2006.)

An interesting and recurring question from public servants (research or extension) when disseminating the videos to other countries is whether they can change some of the images and the music into local ones, as they (erroneously) believe that this would make the videos more acceptable to their farmers. In fact, civil servants are more likely than farmers to complain about 'cultural barriers' in video-mediated learning (Van Mele *et al.*, 2010b). Nigerian farmers, for instance, never complained about the images of Bangladeshi rice farmers or of the music. After all, these same farmers watch Brazilian telenovelas and Bollywood movies.

Videos made according to the ZIZO approach more easily bridge cultural barriers. Irrespective of the country in which the videos are made, African farmers pay attention to the subject covered. In well-produced videos the pictures tell a story, even if the language is not understood.

Table 30.1. African languages into which rice videos have been translated (2010).

Country	Language
Benin	Bariba, Dendi, Fon, Mina
Burkina Faso	Dioula, Mooré
Central African Republic	Sango
Chad	Arabic
DR Congo	Lingala
Ethiopia	Amharec
Ghana	Buli, Dagaari, Dagbani, Ewe, Gonja, Kusaal, Kasem, Sisaala, Twi
Guinea	Guerze, Susu, Pular
Kenya	Swahili
Madagascar	Malagash
Mali	Bambara
Nigeria	Hausa, Igbo, Yoruba
Senegal	Peulh, Wolof
Sierra Leone	Creole, Mende
The Gambia	Mandinka
Uganda	Ateso, Luganda, Lugbara, Luo, Runyakitara

Tackling the dissemination bottleneck

Appealing to many organizations, the videos were quickly translated into Mandinka. Local language versions boosted local dissemination and use of the videos. Many NGOs, development agencies, farmer organizations, national research and extension staff, as well as radio journalists and TV broadcasters became involved in the translation and national dissemination of the rice videos. By 2010, the rice videos had been translated into 38 African languages (Table 30.1).

Using either the English, French or local language versions, TV stations started to broadcast the rice videos in The Gambia (GRTV), followed by Uganda (UBC), Guinea (RTG), Nigeria (the federal Nigerian Television Authority as well as the state-owned Broadcasting Service of Ekiti State), Burundi (Télévision Nationale du Burundi), Niger (Canal3 in Malanville), the Democratic Republic of Congo (community television of Kinzau-Mvuete) and Central African Republic (Télévision Centrafricaine).

By 2010, AfricaRice had distributed the videos to over 200 organizations, which in turn multiplied and shared them with over 800 organizations (Table 30.2). Development agencies, networks and projects were most active in disseminating the videos, followed by national research institutes and international NGOs. While universities, schools, networks, rural radio and TV surely contributed to making the videos more widely known, so far we have little evidence of them multiplying and further distributing them.

Rural radio stations made good use of the videos to build the capacities of their own staff, and by promoting them to their audience through regular announcements, showing them in villages or in their station during market days. Others creatively broadcast (parts of) the audio track. Some of the stations sold copies to farmers and some were afraid to make additional copies as they thought the videos were copyright protected.

AfricaRice works closely with the national agricultural research systems (NARS), so most of these received copies direct from AfricaRice. However, extension services and farmers' associations received copies mainly via projects and NGOs, indicating how effective and attractive farmer training materials find their way into the system.

Many still believe that videos cannot be readily viewed by farmers, an attitude more prevalent among senior research staff. However, when farmers are asked what they would do if they were given a video but not the equipment to play it on, most will say that they will 'find a way'. Indeed, from farmers' feedback we learned that once farmers watched the rice videos they were eager to obtain a copy and were ready to pay for it. As hardly any of the intermediaries responded to this request, AfricaRice decided to adjust its strategy to ensure that the videos reached the intended audience, namely the farmers.

In 2008, AfricaRice partnered with the Canada-based NGO, Farm Radio International (FRI). At first, the rice videos were used as a resource from which radio scripts were developed and shared through its network. Also, radio broadcasters were provided with contact addresses of people at national research institutes and NGOs who had copies of the videos. We hoped that by doing so, new linkages would be established between rural radio stations and agricultural organizations. Again, we struggled to collect feedback and, apart from some anecdotal evidence, we have no idea whether we succeeded in linking organizations in this way.

In 2009, we then agreed for FRI to insert in their newsletter an English or French DVD of *Rice Advice* (containing the rice video programmes) for those members working in rice-growing countries. The network of more than 350 radio organizations that FRI has established since the early 1980s was a great asset to reach (mainly) rural radio stations and local NGOs directly. Out of the 61 respondents to a survey sent out by FRI to all its members in 2010, fourteen said they had never received the DVD, and 22 said that they had used it to strengthen their own capacities. Some radio stations made copies

Table 30.2. Number of organizations that received the videos through AfricaRice (first level), or indirectly through any of the recipient organizations, 2010.

Type of organization	First-level distribution	Second-level distribution	Third-level distribution	Total
Development agency	26	25	0	51
International NGO	11	9	0	20
Local NGO	13	73	1	87
Research institute	44	43	1	88
Extension service	29	92	3	124
Farmers' association	17	151	46	214
Project	19	36	4	59
University & school	17	24	0	41
Training centre	0	9	1	10
Rural radio station	32	253	4	289
TV	4	6	0	10
Network	4	20	1	25
Total	**216**	**741**	**61**	**1018**

of the *Rice Advice* DVD for farmer groups or members of a cooperative credit union. Others used the videos creatively, e.g. by using the audio tracks of the videos, which they had translated into their local language.

At the same time, Countrywise Communication was contracted to establish public–private partnerships for mass multiplication and dissemination of local language videos, using Ghana and Uganda as test cases. This was no easy task. Ensuring the quality of the translated programmes was the first step, as there was a disconnect between the national scientists and the media people doing the translation work, i.e. they do not 'speak' the same language. Local media people expected to be told what to do, while scientists did not know the process and work needed for a quality product. As local media companies tend to go for the cheapest option and lowest quality (often not taking agriculture seriously), the voice-over recording and editing in many cases had to be done again and again to improve the standard.

The next issue was getting companies and organizations to understand how the DVD would look, feel and work. Many did not understand what was 'on offer' until they saw the finished DVD in multiple languages all on one disc (Fig. 30.4) – at which point the question was, 'Do you also have this for other crops?'

To support the dissemination, private companies were initially reluctant to contribute resources as it was not scheduled in their annual budget plan, or because they had no idea what the DVD would look like, or because they lacked the vision that supporting the dissemination to farmers was a route to reach out to potential customers. This may change as more and more companies realize that farming can be an area of growth for their business.

Most publicly funded organizations (including NGOs) and private companies offered to use their networks to distribute the DVDs, as they could see the economic benefits to their partners – once they could see what the end product was.

Integrating media

Exposure to new ideas drives change and farmers rely on multiple sources of information. A recent inventory by the Forum for Agricultural Research in Africa (FARA) shows how the majority of the initiatives around rural ICTs in agriculture, including the use of mobile telephony, is donor or at least externally driven. Moreover, web and text-based information platforms are often in English. As the African farmer is faced with poor infrastructure, low literacy and limited colonial language use, such models of information delivery have proved to be largely ineffective (Gakuru *et al.*, 2009). Audio-visual media can significantly support rural entrepreneurship, as shown by the numerous cases of successful African seed enterprises (Van Mele *et al.*, 2011a), but media is still missing in national and regional agricultural policies.

Fig. 30.4. Disseminating learner-centred videos poses specific challenges – many companies and organizations did not know what was 'on offer' until they saw the finished DVD with all programmes in multiple languages all on one disc. (Reproduced with permission from Africa Rice Center.)

Farmer Feedback Mechanisms

Monitoring the dissemination and use of the rice videos has been very time-consuming and has relied mainly on good will, as many of the intermediaries are not formal partners of AfricaRice and hence are not accountable to it or required to report back (Van Mele *et al.*, 2010a). Some donors seem to revert to assessing achievements in terms of numbers of farmers reached or additional income generated. These may be easier to communicate to the tax payer, but assessing impacts of projects that aim at strengthening learning across the system (in which the actors are neither predefined nor often identified after the event, as a result of inter-actor linkages) is much harder.

Farmer organizations offer some potential to provide feedback, but none of them have actually been trained in it. And although modern ICT applications open new opportunities to collect, store and (to some extent) respond to farmers' feedback, problems of synthesis and interpretation of often cryptic messages are likely to limit their scope and applicability.

Timely field visits by professionals with a readiness to listen and learn from farmers will remain crucial to understand, document and, in turn, inspire learning-oriented interventions.

Impacts

Household impact assessments of the rice videos are ongoing in Africa, although two studies already provide a clear indication. In Bangladesh, video-mediated group learning about improving the quality of farm-saved seed resulted in farmers' rice yields increasing by an average of 15% (Table 30.3; Chowdhury *et al.*, 2011). After video exposure, marginal and rice subsistence households decreased by 5% and 19%, respectively.

In central Benin, about 69% of the women interviewed were illiterate and nearly all of the women who watched the parboiling video improved their techniques, and therefore the quality of their rice (Zossou *et al.*, 2009a), leading to a 35% increase in price per kg rice sold (Table 30.3).

Women in Benin who watched the videos became motivated to parboil more and improve the quality of rice. The NGOs, impressed by their efforts, helped the women to improve the packaging of parboiled rice and link them to traders. Improved marketing led to increased business. Their rice attracted more buyers and fetched a higher price, which increased the women's profits and strengthened their social cohesion (Zossou *et al.*, 2010).

The video motivated women to start parboiling as a group and to make group-based requests for credit and training (Table 30.4). However, when local NGOs responded by facilitating access to microfinance institutions, these were often unwilling to provide credit to the groups because of past bad experiences in the cotton sector. Instead, informal credit suppliers proved more responsive. Rice producers who attended the open-air video shows together with the women rice processors became more willing to sell them rice on credit (Zossou *et al.*, 2010).

Table 30.3. Changes in rice yield in Bangladesh[a] and price per kg of parboiled rice in Benin[b] after watching different rice videos.

	Video villages		Control villages	
Rice video modules	Before	After	Before	After
Seed management and seedling production[a]	4593 kg/ha	5265 kg/ha	4667 kg/ha	4678 kg/ha
	Yield increase of 15% ($P < 0.001$)		Non-significant change	
Rice quality and parboiling[b]	US$0.55/kg	$0.74/kg	$0.63/kg	$0.64/kg
	Price increase of 35% ($P < 0.001$)		Non-significant change	

[a]Data from Chowdhury *et al.*, 2011.
[b]Data from Espérance Zossou, 2009, unpublished.

Table 30.4. Behavioural and institutional changes triggered by the parboiling video. (From Zossou *et al.*, 2010, reproduced with permission from Taylor & Francis Ltd, www.informaworld.com.)

Types of change	Description of the change	Indicator	Factors that triggered change
Behavioural change	Women had increased motivation to parboil rice	After viewing video, 72% of women became highly motivated to parboil rice	Women realized after watching the video that parboiling was a widespread technology and there was a much larger market for it than they initially realized
	Women developed rice parboiling activity for profit	88% of women who viewed the video parboil rice for profit. Viewers parboiled 70% more rice than those who didn't watch the video	Improved rice quality by using improved parboiling technologies led to an increase in consumer demand
	Women increasingly organized themselves in groups to parboil rice	81% of women who viewed the video parboiled rice in groups after the video show	In some cases, the gift of an improved parboiler for a whole village improved women grouping around rice parboiling Being organized in groups increased trust and facilitated access to informal credit
	Women formulated group-based requests for new training	188 women were trained on the construction and use of improved stoves (that they discovered in the video) and more rice parboiling training was carried out upon women's requests	The discovery of improved stove during video shows led to women's interest in ecological problems during rice parboiling The increasing interest of women led them to seek more information about parboiling
Behavioural and institutional changes	Intermediaries improved their training methods	NGOs strengthened their role as facilitators NGOs increasingly used pictures in their training to capture attention NGO facilitators helped women to better organize themselves	The increasing interest of women in rice parboiling The increasing demands to support rice parboiling The increasing trust of women in NGOs led NGOs to improve their methods and to help women to better organize themselves
Institutional changes	Collaboration strengthened between rural women and input and output markets	Relations between women and credit institutions were improved Relations between women and private-sector traders and sellers were improved	Trust between various actors in the value chain is strengthened by: video shows; women's interest in rice parboiling; and increased demand for parboiled rice

Conclusion

The rice videos discussed in this chapter have demonstrated their power to trigger livelihood changes in multiple countries way beyond improved knowledge, yields and incomes.

As Tripp (2007) says, more robust rural institutions are required to help generate and share relevant information in ways that are sensitive to farmers' busy schedules. Agro-dealers, information centres, plant clinics and radio stations that are near rural markets where farmers commonly gather may be well-suited as rural learning hubs and as places where local language videos can be sold to farmers. Allowing farmers to watch a video or listen to an audio programme at their own convenience helps to reduce farmers' dependency on outside organizations, and at the same time triggers their interest to learn more. Rather than making service providers superfluous, we believe video-mediated rural learning will rather create demand for new services and products.

Rural learning will increasingly need to enhance farmers' and intermediaries' search behaviour (both for information and for potential partners). Apart from technical aspects, institutional innovations related to collective action in natural-resources management, saving and credit cooperatives, marketing and value-chain innovations need to be considered. Here as well, video-mediated learning can pave the way to enhance learning across frontiers. Optimal use should be made of available resources and networks, as too much audio and video material is being 'lost' because it is not shared. To address this, the NGO Access Agriculture was established and created a unique video- and audio-sharing platform.

And finally, more research is needed on the relations between context, mechanisms and outcomes of rural learning. Insights gained will strengthen socially inclusive mechanisms to reach communities with quality audio and video programmes. After all, a good programme is like a good technology: useless if it remains on the shelf.

References

Bentley, J.W., Van Mele, P. and Acheampong, G.K. (2010) Experimental by nature: rice farmers in Ghana. *Human Organization* 69(2), 129–137.

Carney, J.A. (1998) Women's land rights in Gambian irrigated rice schemes: constraints and opportunities. *Agriculture and Human Values* 15(4), 325–336.

Chapman, R., Blench, R., Kranjac-Berisavljevic', G. and Zakariah, A.B.T. (2003) Rural radio in agricultural extension: the example of vernacular radio programmes on soil and water conservation in N. Ghana. *Agricultural Research & Extension Network Paper* 127. AgREN, The Overseas Development Institute, London, 12 pp.

Chowdhury, A.H., Van Mele, P. and Hauser, M. (2011) Contribution of farmer-to-farmer video to capital assets building: evidence from Bangladesh. *Journal of Sustainable Agriculture* 35(4), 408–435.

Coldevin, G. and FAO (2001) Participatory communication and adult learning for rural development. *Journal of International Communication* 7(2), 51–69.

CTA (2009) *The Role of Media in the Agricultural and Rural Development of ACP Countries. Synthesis Report.* CTA Annual Seminar, Brussels, Belgium, 12–16 October 2009. Technical Centre for Agricultural and Rural Development, Wageningen, Netherlands.

Defoer, T. (2002) Learning about methodology development for integrated soil fertility management. *Agricultural Systems* 73, 57–81.

Defoer, T., Wopereis, M.C.S., Idinoba, P., Kadisha, T., Diack, S. and Gaye, M. (2004) *Curriculum d'Apprentissage Participatif et Recherche Action (APRA) pour la Gestion Intégrée de la Culture de Riz de Bas-Fonds (GIR) en Afrique Sub-Saharienne : Manuel du Facilitateur.* Association pour le développement de la riziculture en Afrique de l'Ouest, Bouaké, Côte d'Ivoire.

Fafchamps, M. (2004) Social Capital and Development. *Department of Economics Discussion Paper Series* 214. University of Oxford, Oxford, 25 pp.

FAO (2008) Farmer field schools on land and water management in Africa. In: *Proceedings of an international workshop, Jinja, Uganda, 24–29 April 2006.* Food and Agriculture Organization of the United Nations, Rome, Italy.

Gakuru, M., Winters, K. and Stepman, F. (2009) *Inventory of Innovative Farmer Advisory Services using Information Communications Technologies.* Forum for Agricultural Research in Africa, Accra, Ghana.

Gurumurthy, A. (2006) Promoting gender equality? Some development-related uses of ICTs by women. *Development in Practice* 16(6), 611–616.
Hambly Odame, H., Hafkin, N., Wesseler, G. and Boto, I. (2002) Gender and agriculture in the information society. *ISNAR Paper* No. 55, 8. International Service for National Agricultural Research, The Hague, Netherlands.
Kassam, A.H. (2006) Agricultural institutions and receptivity to social research: the case of the CGIAR. In: Cernea, M.M. and Kassam, A.H. (eds) *Researching the Culture in Agri-Culture. Social Research for International Development.* CAB International, Wallingford, UK, pp. 32–50.
Katungi, E., Edmeades, S. and Smale, M. (2008) Gender, social capital and information exchange in rural Uganda. *Journal of International Development* 20(1), 35–52.
Lenné, J.M. (2008) Research into Use: managing achievements for impact. *Outlook on Agriculture* 37(1), 23–30.
Ramírez, R. (1998) Communication: a meeting ground for sustainable development. In: Richardson, D. and Paisley L. (eds) *The First Mile of Connectivity.* Food and Agriculture Organization of the United Nations, Rome, Italy.
Ramisch, J.J., Misiko, M.T., Ekise, I.E. and Mukalama, J.B. (2006) Strengthening 'folk ecology': community-based learning for integrated soil fertility management, western Kenya. *International Journal of Agricultural Sustainability* 4(2), 154–168.
Saïdou, A. (2006) Converging strategies by farmers and scientists to improve soil fertility and enhance crop production in Benin. PhD thesis, Wageningen University, Netherlands.
Tripp, R. (2007) Crop management innovation and the economics of attention. Paper presented at the Conference on Farmer First Revisited, Institute of Development Studies, Brighton, UK, 12–14 December.
van den Berg, H. and Jiggins, J. (2007) Investing in farmers. The impacts of farmer field schools in relation to integrated pest management. *World Development* 35(4), 663–686.
van Koppen, B. (2009) Gender, resource rights, and wetland rice productivity in Burkina Faso. In: Kirsten, J.F., Dorward, A.R., Poulton C. and Vink, N. (eds) *Institutional Economics Perspectives on African Agricultural Development.* International Food Policy Research Institute, Washington, DC, pp. 389–407.
Van Mele, P. (2006) Zooming-in, zooming-out: a novel method to scale up local innovations and sustainable technologies. *International Journal of Agricultural Sustainability* 4(2), 131–142.
Van Mele, P., Wanvoeke, J. and Zossou, E. (2010a) Enhancing rural learning, linkages and institutions: the rice videos in Africa. *Development in Practice* 20(3), 414–421.
Van Mele, P., Wanvoeke, J., Akakpo, C., Dacko, R.M., Ceesay, M., Béavogui, L., Soumah, M. and Anyang, R. (2010b) Videos bridging Asia and Africa: overcoming cultural and institutional barriers in technology-mediated rural learning. *The Journal of Agricultural Education and Extension* 16(1), 75–87.
Van Mele, P., Bentley, J.W. and Guéi, R.G. (2011a) *African Seed Enterprises: Sowing the seeds of food security.* CAB International, Wallingford, UK.
Van Mele, P., Bentley, J.W., Dacko, R.M., Yattara, K. and Acheampong, G.K. (2011b) Attitude counts: engaging with rice farmers in West Africa. *Development in Practice* 21(6), 806–821.
Wanvoeke, J., Dacko, R.M., Yattara, K. and Van Mele, P. (2008) Women break down barriers in Mali. *LEISA Magazine* 24(3), 14–15.
Zossou, E., Van Mele, P., Vodouhe, S.D. and Wanvoeke, J. (2009a) The power of video to trigger innovation: rice processing in central Benin. *International Journal of Agricultural Sustainability* 7(2), 119–129.
Zossou, E., Van Mele, P., Vodouhe, S.D. and Wanvoeke, J. (2009b) Comparing video and workshops to train rural women about improved rice parboiling in central Benin. *The Journal of Agricultural Education and Extension* 15(4), 329–340.
Zossou, E., Van Mele, P., Vodouhe, S.D. and Wanvoeke, J. (2010) Women groups formed in response to public video screenings on rice processing in Benin. *International Journal of Agricultural Sustainability* 8(4), 270–277.

31 Raising Rice Yields and Beyond: An Experience of Collective Learning and Innovation in Lowland Rice Systems in Madagascar

Toon Defoer[1]* and Marco C.S. Wopereis[2]
[1]*International Centre for development-oriented Research in Agriculture (ICRA), Wageningen, Netherlands;*
[2]*Africe Rice Center (AfricaRice), Cotonou, Benin*

Introduction

In the late 1990s, the Africa Rice Center (AfricaRice), then called the West Africa Rice Development Association (WARDA), and partners conducted yield-gap surveys in the Sahel region of four West African countries (Burkina Faso, Mali, Mauritania, Senegal) to identify agronomic constraints to irrigated rice production. On the basis of these surveys, AfricaRice facilitated the formation of coalitions between research, extension, farmers and the private sector to develop 'integrated rice management' (IRM) options for the region (Wopereis and Defoer, 2007). Coalitions were technology specific, with teams working on mechanization, soil-fertility management and weed management. In this way, a basket of IRM options was gradually developed to address key constraints in the rice cropping cycle from land preparation to harvest. Implementation of the IRM options, particularly for soil fertility and weed management, raised rice yields by an average of almost 2 t/ha, and net benefits for participating farmers in Senegal and Mauritania by 80% (Häfele *et al.*, 2000, 2001). Similar results were obtained in Burkina Faso (Segda *et al.*, 2004, 2005). (See also Tollens *et al.*, Chapter 1, this volume.)

Irrigated systems in these areas are relatively uniform, with good infrastructure and farmers are reasonably well organized. Under these circumstances, relatively 'fixed' options and recommendations were communicated to farmers through training of extension staff and promotional campaigns (field days, rural radio).

Encouraged by the results in the Sahel, AfricaRice moved the work on IRM to rainfed lowland systems in Côte d'Ivoire. After one year of testing various soil-fertility and weed management options at different inland valley sites, however, it was realized that the diversity and variability in the conditions was too great to adopt the same approach as for the Sahelian irrigated systems. The variability of inland valleys means that farmers require flexible technologies that can be adapted to a range of growth conditions. Development of technologies needs to involve farmers at a very early stage; to ensure farmer involvement in technology development, a participatory learning and action-research

* Corresponding author: toondefoer1@gmail.com

© CAB International 2013. *Realizing Africa's Rice Promise* (eds M.C.S. Wopereis *et al.*)

(PLAR) approach to IRM in inland valleys (PLAR-IRM) was adopted. PLAR was originally developed for rainfed agriculture from field to village level in Mali (Defoer and Budelman, 2000).

Due to the different experiences in the irrigated systems in the Sahel and the rainfed inland valley systems in Côte d'Ivoire, AfricaRice concluded that the need to use PLAR-type approaches increases when moving from high- to low-precision systems and from relatively uniform to more diverse production systems. In Sahelian irrigated systems, technology development may be more advanced prior to evaluation by farmers, and recommendations are relatively 'fixed', which allows them to be scaled out more easily. Farmers in 'poorly developed' and variable inland valley systems, however, require more flexible technologies. Further, these farmers require greater insight into the principles of 'good agronomy' to allow them to adjust crop management to highly variable local settings. Experience in Madagascar (discussed below) demonstrates that PLAR approaches to IRM can have high pay-offs in rainfed inland valley systems.

The objectives of this chapter are to: (i) introduce principles and processes of PLAR-IRM; (ii) discuss application of PLAR-IRM in a rainfed lowland rice system in Madagascar; and (iii) describe some of the lessons learned and provide perspectives for wider use of PLAR-IRM in Africa.

PLAR-IRM Principles and Processes

PLAR is a capacity-building process based on collective learning by farmers and other rice stakeholders. PLAR starts from local knowledge and stimulates farmers to identify, try out and learn about alternative management practices. Capacity building through adult learning is acquired through careful and systematic observations, and analysis of the results obtained, leading to improved decision making.

The PLAR approach involves groups of 25 to 35 volunteer farmers who meet on a regular basis (once a week, or once every 2 weeks). The process is facilitated by staff from extension services, research institutes or NGOs. The facilitators facilitate the group learning sessions that take place in the field or in some other convenient location. The sessions generally start before the growing season to deal with aspects related to the planning of the cropping season. During and after the rice-growing season, PLAR groups meet to discuss key crop management decisions, including postharvest issues related to storage, processing and marketing.

Facilitators play an important role as they assist PLAR groups in learning from experiences, observing, analysing, developing new ideas, trying out new options, and adapting and integrating new ways of working into current management practices. PLAR applied to IRM (PLAR-IRM) aims to develop rice management options that enhance rice productivity or quality (or both) in a sustainable manner and to integrate these step-by-step into existing cropping calendars. The options for change not only relate to technological improvements, but also deal with improved organizational and institutional practice and processes.

A PLAR curriculum for IRM was developed that is composed of learning modules, which form the facilitators' basic instruments to run learning sessions with farmer groups. The original PLAR-IRM curriculum (Defoer *et al.*, 2004) is composed of 28 modules, but – as PLAR is a flexible and evolving approach – the modules used with a specific farmer group will depend on needs expressed by the farmers themselves. The approach is open for the development and integration of new modules. The modules are supported by technical and facilitators' manuals (Defoer *et al.*, 2004; Wopereis *et al.*, 2009).

At the heart of the learning modules are PLAR-IRM learning tools that include maps, diagrams, calendars, and observation and monitoring forms. These address crop management practices and agroecological principles, but also provide basic knowledge of soils, pests and diseases, weeds, fertilizers, water management, etc. Many of these tools provide visualization of phenomena to help farmers discover and 'see' things that were previously 'invisible'. Such tools enable farmers to understand causal relationships and to identify more rational management practices, and ways of operating and organizing farming. These learning tools also provide knowledge in an easily accessible form for farmers, and can be used in a flexible way to cater for different situations. This approach stimulates interactive learning among farmers and facilitators.

During PLAR-IRM sessions, participating farmers are encouraged to implement new ideas and technologies in part of their fields ('innovation space'). Farmers discover and learn by doing and, if needed, adapt according to their own local conditions. During the next season, farmers may apply the technique to a larger area, effectively integrating the technology into their production system and cropping calendar. Some innovations may be difficult to implement on only part of the field, e.g. improved land levelling and water management. Farmers' 'innovation spaces' are visited by other farmers during the field observations that are part of PLAR-IRM sessions. Such visits, which are followed by analysis and identification of possible ways of improvement, often generate new ideas that can be tried out by other farmers. Apart from farmers trying out innovations in their individual innovation spaces, PLAR-IRM sessions can generate ideas for 'collective' innovations, calling for the involvement of several farmers, the whole PLAR-IRM group or even the entire village community. This is often the case for innovations related to water management, or improving the general organization among farmers with respect to their position within the value chain, attempting to capture more added product value through collective storage, processing or marketing.

PLAR-IRM can be compared with the widely known farmer field school (FFS) approach, which arose from the same school of thought and paradigm (Braun and Duveskog, 2008). A major difference with the FFS approach is the layout of so-called 'best practices' in a commonly managed field in the FFS. In PLAR-IRM there is no 'commonly managed best-practices field'. Demonstrating 'best practice' may direct farmers' thinking and innovation in a direction that is being pre-established by the facilitator and that is not necessarily adapted to the local settings of the farmer because of the diversity of growth conditions and differences in crop management precision in inland valley systems. With PLAR-IRM, possible options for improvements to the rice production practices are discussed and farmers are encouraged to identify solutions for their own settings.

Madagascar Case Study

Background

In 2005, an integrated rural development project in the Sofia region of northern Madagascar (Projet de soutien de la région Sofia pour le développement rural intégré, PSSDRI) was started by the Aga Khan Foundation (AKF) to improve the revenues of the rural poor by supporting rice producers. The Sofia region is a major rice-producing area representing about 10% (300,000 tonnes) of the national production (Minagri, 2003). With an estimated population of 1.5 million inhabitants, the region has large potential to export to regions with insufficient rice production. The average yield of 1.7 t/ha is substantially below the national average of 2.1 t/ha (FAO, 2001; Bockel, 2003).

Inland valleys represent the largest part of the cultivated rice area of the region, totalling more than 80,000 ha, with only 5% under full irrigation (which enables double cropping). The major rice-growing season, from December to May, coincides with the rainy season, called *Asara*. Rice can be grown during the off-season, called *Jéby*, but only where irrigation facilities are functional.

Approach

PSSDRI activities began in May 2005 with a 2-week workshop to introduce the PLAR-IRM approach and recruit programme staff – six facilitators, a social organizer, technical supervisor, and programme director. The workshop was also attended by staff from government research and extension services and from NGOs from within and outside the region.

Field activities started in the 2005 *Jéby* season, and covered six PLAR-IRM groups in six villages in the districts of Befandriana-Nord and Mandritsara. The PLAR-IRM group facilitators in each district worked closely together and were supported by the programme director and technical supervisor. This intensive way of working and exchange among programme staff allowed for capacity building and reflection on possible adaptations of the PLAR-IRM approach to better fit the local context. In 2007, PSSDRI adapted and republished the PLAR-IRM facilitators' manual, in collaboration with AfricaRice, to make it more relevant to the Sofia region (Defoer *et al.*, 2008).

The project's coverage increased rapidly over 4 years (Table 31.1): by the end of the *Asara* 2008/09 season, PLAR groups had been established in four districts within the Sofia region (Mandritsara, Befandriana-Nord, Port Bergé and Bealanana).

Technological innovations

Out of the wide range of technological and managerial options covering all major issues related to rice growing discussed in the PLAR-IRM groups, farmers selected those that were most relevant for them. They initially tried out these options in their individual innovation spaces, adapted them where and when needed, and then extended them beyond the initial area (Fig. 31.1).

The number of practices or technologies applied in innovation spaces and the rate at which these were extended to larger areas depended essentially on the nature of the land and other resources available. From the second year, farmers declared that they extended the innovations 'to all areas where this would be feasible' (Rowley and Dugué, 2010). The actual area depended upon their access to labour and capital, farm/field characteristics and, most importantly, water control. Access to water differed substantially from one field to another: in some cases resulting in drought, in other cases in flooding. Improving water management often required collective action.

The four most important innovations according to farmers were (PSSDRI, 2009): land preparation, transplanting of young seedlings, transplanting in lines, and weeding using a rotary weeder. The IRM options are discussed below.

Improved land preparation

Traditionally farmers let bullocks 'trample' the field for a number of hours, creating an uneven field that is difficult to level, before flooding for one to two days and harrowing with a wooden frame. As alternatives, three innovations were introduced: 2–3 weeks of flooding (between ploughing and harrowing); harrowing with a traditional wooden frame or with a rotary harrow; and levelling of the field before transplanting. More than 90% of the PLAR farmers adopted one or more of these innovations. Flooding for a longer period was seen as an important preventive measure to combat weeds and insects, and also for making levelling easier. Farmers also reinforced field bunds to maintain water in the field. The rotary harrow was not adopted on a large scale, because it was not available or too expensive. However, locally made versions were starting to appear on the market in 2010.

New varieties and maintaining seed quality

Several new rice varieties were introduced, some of which originated from the national research system (FOFIFA), to 'complement' farmers' local varieties. PLAR farmers received small quantities of seed (200 g) during the first year when they joined the PLAR groups. More than 70% of the PLAR-IRM farmers tried one or more of the new varieties. Half of the farmers adopted variety 'Tox V' because of its high yield potential, resistance to *Rice yellow mottle virus* (RYMV) and adaptability to various water regimes. The adoption of a new variety did not lead to the old or local varieties being abandoned, as farmers

Table 31.1. Coverage of the PSSDRI project in the Sofia region, 2005–2009.

	Jéby 05	*Asara* 05/06	*Jéby* 06	*Asara* 06/07	*Jéby* 07	*Asara* 07/08	*Jéby* 08	*Asara* 08/09
No. facilitators	6	10	17	19	21	22	22	22
No. villages	6	17	31	38	38	58	–	99
No. PLAR-IRM groups	6	21	31	40	41	63	59	102
No. farming families	148	511	734	1274	1195	2658	1608	3782

–, no data.

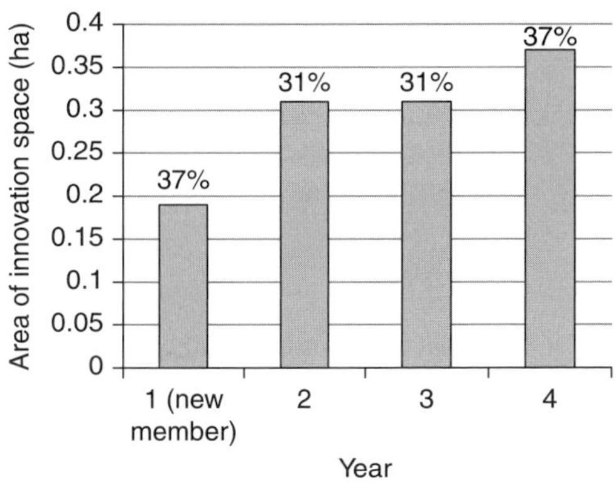

Fig. 31.1. Average area (and percentage of total field) of innovation space as a function of number of years of membership of a PLAR group; *Asara* 2008/09 data. Average area under rice per farmer was 0.56 ha.

generally preferred growing several different varieties suited to their needs. Farmers also improved seed multiplication and conservation practices by selecting an area within their field with vigorous and homogeneous plants of the desired variety, removing off-types, harvesting separately and storing under appropriate conditions.

Nursery preparation

Traditionally, farmers do not pre-germinate seed and tend to use large amounts of seed, which results in high plant density in the seedbed and thin and fragile seedlings. Farmers also establish 'dry' nurseries, outside the lowland area. Two major innovations tested were: (i) pre-germination of seed and (ii) lower seed rates in the nursery (2 kg/10 m^2). These were adopted by 95% of the PLAR-IRM farmers, which led to enhanced seedling vigour and enabled earlier seedling transplanting. Creating 'moist' seedbeds in the lowland area near a water source was adopted by some farmers, but most preferred to keep the nursery outside the lowlands due to the risk of flooding.

Transplanting young seedlings in lines

Usual practice was for farmers to transplant rice seedlings that are about 4–6 weeks old, transplanting these randomly at about five or more seedlings per hill with hills about 10–15 cm apart. Innovations introduced comprised: (i) use of younger seedlings, (ii) fewer plants per hill, and (iii) row planning and reduced hill density. About 80% of the PLAR-IRM farmers started transplanting younger seedlings of 2–3 weeks old, which gave improved tillering and increased yield. Farmers also reduced the number of seedlings per hill to two or three. More than 95% of the farmers moved to transplanting in lines on a 20 cm grid. Transplanting is often done by hired labour and, during the first seasons, transplanting in rows was sometimes hindered by lack of labourers' skill in line transplanting. However, farmers became more convinced about this option because of the ease of weeding and use of the rotary weeder. Transplanting young seedlings in lines (combined with the rotary weeder) spread rapidly beyond the PLAR-IRM farmers and in some lowland valleys almost all the rice is now being transplanted in lines by groups of women specialized in the technique.

Farmers developed an alternative to the use of a rope for marking rows, with the so-called *baobao* (or *fomby*) made of a vein of a raffia leaf with small sticks to mark where to transplant. On each extremity of the vein there is a larger stick (of 20 cm) placed horizontally to indicate the distance between the lines. The *baobao* enables transplanting in lines by one person. Moreover, the *baobao* also allows for some additional levelling.

Weed management

Weed management received much attention during the PLAR-IRM sessions as weeds are a major constraint and because their management is so labour intensive. Most farmers weeded only once each season, by hand or using a hoe, leading to large yield losses due to weed competition. As an innovation, most PLAR-IRM farmers started to flood their fields for 1–2 weeks after ploughing, allowing weeds to germinate, so they could be uprooted by harrowing. Rotary weeders can be manufactured easily by local craftsmen and were adopted by about 95% of the PLAR-IRM farmers, and by all farmers who transplanted in lines.

Soil fertility management

Fertilizer use is very low in Madagascar and particularly so in the Sofia region. The project was initially advised by some national research and extension experts to place major attention on soil fertility as it was generally considered that low soil fertility was one of the major factors responsible for low yields. PLAR-IRM group members received a small quantity of urea for free in the first season. All PLAR-IRM farmers applied half of the dose (at a rate of 50 kg/ha) about one week after transplanting and most of them applied a second dose (at a rate of 50 kg/ha) at panicle-initiation stage. In subsequent seasons, however, when the urea was no longer provided for free, almost none of the PLAR-IRM farmers used urea. Although urea was relatively easily available, farmers were not willing to invest in it, as its benefits appeared negligible under farmers' conditions.

Water management

Measures introduced for field water management included field levelling and bunding. Collective measures were also taken, such as digging and maintaining irrigation and drainage canals. In several villages, important structural improvements to the water distribution system were made, including the rehabilitation of small dams, regular cleaning and rehabilitation of canals.

Rice yields and financial analysis

Rice yields obtained on farmers' innovation spaces averaged 4.6 t/ha over the six seasons between May 2005 and June 2009 – more than double the average yields obtained on the rest of the field (2 t/ha) and three times the average yield obtained by farmers before joining the PLAR groups (1.5 t/ha, the average yield for Sofia region as a whole). Figure 31.2 illustrates a trend of increasing average rice yields on the innovation spaces of PLAR-IRM farmers from *Asara* 2005/06 to *Asara* 2008/09, compared to the rest of their field and average yields of farmers yet to join PLAR groups.

Substantial yield increases were observed from the first season onwards. In subsequent years, increases were more modest than those of the first season (see also Table 31.2).

High yields were maintained in the innovation space and as the total area under innovation substantially increased from year 2 onwards (see Fig. 31.1), the total production increased likewise. On average, farmers were not able to apply the IRM innovations on their entire field over the duration of the project because of the heterogeneous nature of their fields, which prevented the use of IRM innovations in some areas.

Table 31.3 presents a financial analysis of the most popular PLAR-IRM innovations introduced in the Sofia region taking into account additional costs and additional yield obtained as compared to traditional rice practices.

The net return (additional benefit/additional costs) was 10.2, i.e. for each additional dollar invested, farmers gain an average of US$10.2. Taking into account the area under PLAR-IRM options from the second year onwards (0.31 ha), adoption of PLAR-IRM innovations resulted in a net increase in income of more than $200 per year per farmer.

The following section discusses the methodological innovations in the Madagascar PLAR-IRM case study.

Methodological innovations

Modules

Frequent discussions and reviews among project facilitators and with farmers led to a number of important methodological innovations in the PLAR approach. The PLAR curriculum of the first implementation season (*Jéby* 2005) covered the 28 modules of the original manual (Defoer *et al.*, 2004). In the second season (*Asara* 05/06), this was reduced to 22, covering topics that were most relevant to the local situation. Some technological options were more interesting to farmers than others, in particular those related to land preparation, water management, varieties and seed, transplanting, and weed management. Not all farmers and groups had the same level of dedication to participating in learning sessions. The PLAR team therefore adapted the approach to accommodate a **high-intensity level**, covering the complete curriculum in about 25 sessions, and a **low-intensity level** with only 11 sessions, covering topics of greatest interest to farmers, plus the modules on 'planning' and 'making good observations' (Defoer *et al.*, 2008).

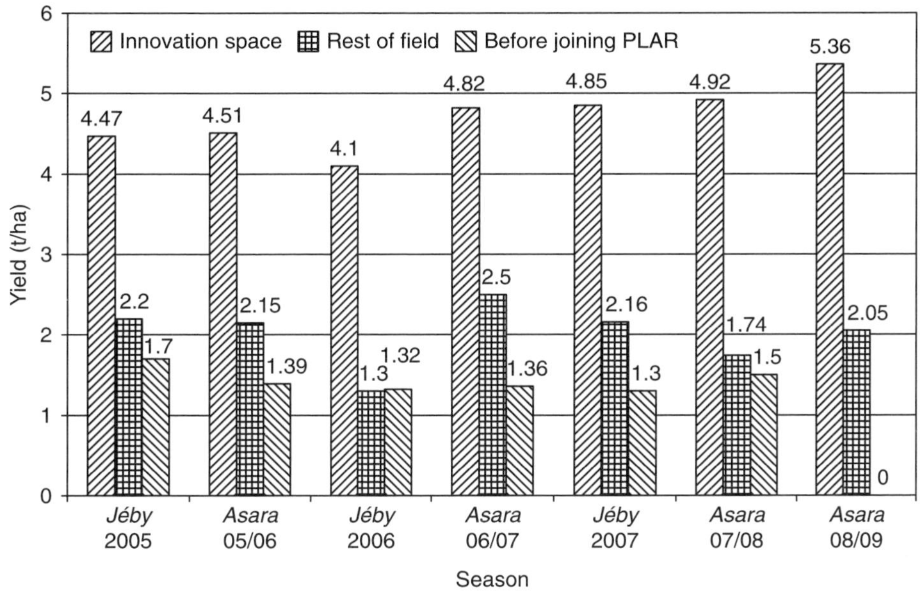

Fig. 31.2. Average rice yields obtained by PLAR-IRM farmers in the Sofia region per season.

Table 31.2. Average rice yields according to the number of years of membership of a PLAR-IRM Group, *Asara* 2008/09.

Location of yield measurement	Rice yield (t/ha)		
	1 year	2 years	3 years
Innovation space	5.26	5.41	5.35
Rest of the field	1.96	2.12	2.03

Farmer-facilitators

It is important to continue the group process that leads to locally adapted innovations, as many more innovations are still to be developed and fine-tuned for further increasing yields and moving beyond yield to a value-chain approach. It is, however, also clear that there is a need to disseminate knowledge to larger numbers of farmers.

Involvement of project-appointed facilitators in PLAR-IRM groups was, therefore, gradually reduced because of the substantial time commitment and the need to start scaling out the results obtained and to reach farmers not directly involved in the PLAR-IRM groups (Table 31.4). Continuous presence of project-appointed facilitators at PLAR-IRM sessions was limited to three consecutive seasons: season 1 covering 11 sessions; season 2 covering the same topics and potentially some new ones; and season 3: development of a new learning programme by the group.

During season 2, 'candidate farmer-facilitators' were identified and asked to lead some of the discussions. During season 3, the project-appointed facilitator handed over some sessions to these farmer-facilitators, thereby coaching and providing support on the job as needed. For the fourth and subsequent seasons, the project-appointed facilitator assisted the group in developing its own learning programme, but no longer participated in the sessions, which were ran entirely by the farmer-facilitators, who could bring in expertise for certain topics as they saw fit. The project-appointed facilitator assisted in the closing session to help the group evaluate the programme and develop ideas for the subsequent season.

Scaling out

Working with PLAR groups is time consuming and intensive and the team therefore developed alternative ways of involving larger numbers of

Table 31.3. Financial analysis of performance of PLAR-IRM innovations (per hectare).[a]

Items	Cost or income ($)
A. Additional costs	
A1 Seed	13.89
A2 Land preparation	
A2.1 Labour (harrowing and levelling)	5.00
A2.2 Depreciation of harrow	22.22
A3 Line transplanting	8.33
A4 Weed management	
A4.1 Labour	8.33
A4.2 Depreciation of weeder	11.11
A. Total	68.89
B. Additional produce value	
3100 kg (= 4600 – 1500)	775.00
B. Total	775.00
C. Net benefit (B–A)	706.11
D. Net additional benefit/additional cost	10.2

[a]Basis for calculations: A1: 40 kg of improved seed at $0.83/kg, to be renewed every 5 years ($8.33) + additional cost for on-farm seed production ($5.56); A2.1: 3 days labour @ $1.67/day; A2.2: cost of harrow = $222.22, depreciated over 10 years; A3: 5 extra person days @ $1.67/day; A4.1: 5 extra person days @ $1.67/day; A4.2: cost of weeder $56.56, depreciated over 4 years; B: value of additional produce = average yield obtained through farmer innovations over the six season/years (4600 kg/ha) minus the average yield of the region under traditional practices (1500 kg/ha) × area under innovation (ha) × $0.25/kg ($1 = MGA1800).

Table 31.4. Involvement of facilitators in PLAR-IRM groups.

Season	Modules	Role of project-appointed facilitators and farmer-facilitators
1	11 'fixed' modules	Project-appointed facilitator leads
2	11 'fixed' modules + other modules of groups' choice	Project-appointed facilitator leads and hands over every now and then to 'candidate farmer-facilitators'
3	Group develops own programme	Project-appointed facilitator starts Farmer-facilitators identified Handing over regularly to farmer-facilitators
4, 5, etc.	Group develops own programme	Project-appointed facilitator assists in planning and evaluating programme and occasionally assists in learning sessions Farmer-facilitators lead

farmers. Rice growing is knowledge intensive and simple extension methods may not be very efficient. The project team opted to develop a 'condensed' training programme for farmers. This training programme was organized for farmers in villages where no PLAR groups existed and where there was no plan to set up PLAR groups, including zones with limited accessibility during the rainy *Asara* season, because of floods. The programmes took 4 or 5 days and brought together about 40 farmers from villages neighbouring a village where a PLAR group operated. Training was organized during the off-season (*Jéby*), when the project-appointed facilitators, who led this process, were in principle not directly involved with the PLAR groups.

This 4–5 day training (called *formation-diffusion*, F-D) included interactive workshops and fieldwork related to those topics that the farmers of the PLAR groups identified as most relevant. The curriculum used during the training week aligned with the PLAR philosophy in terms of adult experiential learning principles. It is important to organize such training weeks in villages and sites with neighbouring PLAR groups, as this allows for effective interaction between the trainees and the PLAR members. Although the project-appointed facilitators took the lead during the training week, in practice farmer-facilitators from the nearby PLAR group contributed to the training sessions. The project developed visual aids (animated cartoon-type brochures and videos) that helped the farmer-facilitators in their work, and also served the trained farmers when they returned to their villages. On returning to their villages, trained farmers organized video sessions for their colleagues and encouraged them to implement some of the innovations shown. The farmers also received a copy of the brochures for further reading. The trained farmers formed a kind of relay between the project and the rest of the villagers; project-appointed facilitators visited at least once a year, together with a member of the monitoring and evaluation (M&E) section of the project to investigate the effectiveness of the use of the extension materials and its effect in terms of application of new technologies and of yield increases. The F-D programme started in *Jéby* 2009 and involved about 4500 farmers. Similar numbers were involved in *Jéby* 2010 and 2011. Preliminary results of a survey indicated that, overall, 31% of F-D trained farmers had adopted at least one IRM practice. The adoption rates were higher in zones and districts where PLAR groups were actively involved during and after the training sessions. Highest adoption rates, among those farmers who had practised IRM, were for a combination of line transplanting of young plants and weed management techniques (AKF, 2011).

Conclusions and Perspectives

The Madagascar case study shows that PLAR-IRM in rainfed rice systems can have a fast and lasting impact on rice yields. The innovations that farmers eventually adopted required in some cases an increase in labour, but this was not seen as a major hurdle.

PLAR-IRM is a time-consuming approach, but for highly diverse systems with low management precision it will deliver concrete results as illustrated in this case study. It is important to think about scaling out of results to other farmers from an early stage, and to develop 'condensed' curricula and easy-to-use learning tools for farmer trainers who will train their colleagues. Video and other tools may be used to scale out essential messages (such as in-line transplanting and the use of a rotary weeder) to farmers outside the project zone.[1]

In the fourth year, PLAR-IRM groups moved from production issues to postharvest and marketing constraints and credit provision, while project facilitators handed over to farmer-facilitators. PLAR groups continue to function even several years after being 'weaned' (i.e. after the project). In most cases, the weaned PLAR groups formed part of an 'operational centre'. These centres are now being organized into formally registered associations that have access to credit services for their members and organize seed and equipment services for their members.

At project level, the benefits of the additional rice produced as a result of project work vastly outweigh the costs of running the project. A simple calculation based on the modest estimation of $200 net income increase per year per farmer and the number of farmers involved in the groups (3780 in *Asara* 2008/09), provided a net figure of $747,400 value added. Compared to the local project costs of about $300,000/year (Rowley and Dugué, 2010),[2] a benefit–cost ratio of more than 2 is obtained. These modest estimations do not include the extra rice produced by farmers who were involved in the F-D programme and who have adopted IRM practices, or the value added through activities related to storage and post-harvest handling of rice. They also do not take into account the considerable enhancement of human and social capital in the communities involved in the project.

The impressive results that can be obtained, as shown in the case study of Madagascar, clearly invites more widespread use and adoption of PLAR-IRM. However, there are important conditions for making things happen. PLAR is quite demanding in terms of time and facilitators' investment. Also any advisory service interested in implementing PLAR should be dedicated to capacity strengthening, above the interest of promoting any technological option for improvement. As we have outlined, PSSDRI adopted a flexible PLAR approach in which farmers themselves discovered what was best for them, what areas needed innovation, and how to innovate. This way of working is unique, however, and many advisory services might face constraints in putting so much effort on strengthening farmers' capacities, in learning together and having confidence in this learning process and in the validity of its outcomes. What is central in PLAR is not what the advisory service believes is good to be promoted, but what the farmers and their groups bring out and innovate. This demands a real paradigm shift and to make it happen organizations should be ready to 'cultivate' real farmer empowerment and refrain from promoting and pushing technologies.

Notes

[1] A video series of six episodes was made on learning about improved rice management practices in a typical PLAR context, where farmers learn from each other and through exchange start putting in place innovations: *The Rice Growers Dream* (http://www.akdn.org/videos_detail.asp?VideoId=103).

[2] Note that this figure does not include cost of AKF Geneva staff supporting the project, or the costs of ad-hoc consultants.

References

AKF (2011) Assessment of *Formation-Diffusion* Initiative for Integrated Rice Farming in 2009 – June 2010. Aga Khan Foundation, Antananarivo, Madagascar.

Bockel, L. (2003) Filière riz et pauvreté rurale à Madagascar : déterminants d'une stratégie de réduction de la pauvreté rurale et politiques publiques de développement. Thèse de doctorat, Université de Metz, France.

Braun, A. and Duveskog, D. (2008) The Farmer Field School Approach – History, global assessment and success stories. Background Paper for the IFAD *Rural Poverty Report 2010*. International Fund for Agricultural Development, Rome.

Defoer, T. (2009) PSSDRI (Programme de soutien de la région de Sofia pour le développement rural intégré) – General presentation (internal document). Aga Khan Foundation, Geneva.

Defoer, T. and Budelman, A. (eds) (2000) *Managing Soil Fertility in the Tropics: A resource guide for participatory learning and action research.* Royal Tropical Institute (KIT), Amsterdam.

Defoer, T., Wopereis, M.C.S., Idinoba, P.A., Kadisha, T.K.L., Diack, S. and Gaye, M. (2004) *Participatory Learning and Action Research (PLAR) for Integrated Rice Management (IRM) in Inland Valleys of Sub-Saharan Africa: Facilitator's manual.* Africa Rice Center (WARDA), Bouaké, Côte d'Ivoire.

Defoer, T., Wopereis, M.C.S., Diack, S., Idinoba, P. and l'équipe du PSSDRI / AKF (2008) *Curriculum d'apprentissage participatif et recherche action (APRA) pour la gestion intégrée de la culture de riz de bas-fonds (GIR) à Madagascar : Manuel du facilitateur.* Le centre du riz pour l'Afrique (ADRAO), Cotonou, Benin and Aga Khan Foundation, Geneva.

FAO (2001) Diagnostic et perspectives de développement de la filière riz à Madagascar. Ministère de l'agriculture, Secrétariat géneral, Unité politique de développement rural, Projet FAO TCP/MAG/8821, Antananarivo, Madagascar.

Häfele, S., Johnson, D.E., Diallo, S., Wopereis, M.C.S. and Janin, I. (2000) Improved soil fertility and weed management is profitable for irrigated rice farmers in Sahelian Africa. *Field Crops Research* 66, 101–113.

Häfele, S., Wopereis, M.C.S., Donovan, C. and Maubuisson, J. (2001) Improving productivity and profitability of irrigated rice production in Mauritania. *European Journal of Agronomy* 14(3), 181–196.

Minagri (2003) *Monographie de la Région de Sofia.* Ministère de l'agriculture, de l'élevage et de la pêche, Antananarivo, Madagascar.

PSSDRI (2009) Analyse de tendance sur le rendement. Programme de soutien de la région de Sofia pour le développement rural intégré. Document interne. PSSDRI, Antsohihy, Madagascar.

Rowley, J. and Dugué, M.-J. (2010) PSSDRI (Programme de soutien de la region de Sofia pour le développement rural intégré) – A review. Aga Khan Foundation, Geneva.

Segda, Z., Häfele, S.M., Wopereis, M.C.S., Sedogo, M.P. and Guinko, S. (2004) Agro-economic characterization of rice production in a typical irrigation scheme in Burkina Faso. *Agronomy Journal* 96, 1314–1322.

Segda, Z., Häfele, S.M., Wopereis, M.C.S., Sedogo, M.P. and Guinko, S. (2005) Combining field and simulation studies to improve fertilizer recommendations for irrigated rice in Burkina Faso. *Agronomy Journal* 97, 1429–1437.

Wopereis, M.C.S. and Defoer, T. (2007) Moving methodologies to enhance agricultural productivity of rice-based lowland systems in sub-Saharan Africa. In: Bationo, A., Waswa, B., Kihari, J. and Kimetu, J. (eds) *Advances in Integrated Soil Fertility Management in Sub-Saharan Africa: Challenges and opportunities.* Springer, Dordrecht, Netherlands, pp. 1077–1091.

Wopereis, M.C.S., Defoer, T., Idinoba, P., Diack, S. and Dugué, M.-J. (2009) *Curriculum for Participatory Learning and Action Research (PLAR) for Integrated Rice Management (IRM) in Inland Valleys of Sub-Saharan Africa: Technical manual.* Africa Rice Center (WARDA), Cotonou, Benin.

32 Impact of Rice Research on Income, Poverty and Food Security in Africa: An *Ex-ante* Analysis[1]

Aliou Diagne,* Didier Y. Alia, Marco C.S. Wopereis, Kazuki Saito, Tebila Nakelse and Papa Abdoulaye Seck
Africa Rice Center (AfricaRice), Cotonou, Benin

Introduction

Rice has always been an important staple in many African countries. Since the 1960s, it has also become the most rapidly growing food source across the continent. However, local production is largely insufficient to meet consumption needs. In 2009, Africa imported 10 million tonnes (Mt) of milled rice, at a cost of US$5 billion. With high food and fuel prices predicted to last well into the 2010s, relying on imports is no longer a sustainable strategy for Africa. Rice-sector development can become an engine for economic growth across the continent, contribute to eliminating extreme poverty and food insecurity within Africa, and improve the social well-being of millions of poor people. Development of the rice sector and related sectors will have considerable impact on the competitiveness of African economies and reduce the need to divert valuable foreign currency exchange to imports. Enhanced local production, processing and marketing will also mean that Africa's cities will have access to affordable food. Rice production will create employment along the value chain and in related sectors, and lead to improved nutritional and health status of the rural agricultural poor. It will allow financing of better education that will give the next generation greater opportunities to break the remaining shackles of under-development (AfricaRice, 2011).

Africa's rice sector faces a large number of biophysical and socio-economic constraints. These generally translate into low productivity of rice produced in Africa. Research areas to address these are numerous and need to be prioritized because resources are scarce – basic economics and general knowledge recognize that this scarcity of resources causes many needs to remain unsatisfied.

Ex-ante assessment of the impact of rice research in Africa is important to: (i) adequately identify priority research themes and target populations; (ii) efficiently allocate scarce resources to priority research themes; (iii) better target research outputs to where they will have the maximum impact; (iv) enhance research relevance and positive impact on the livelihoods and well-being of the target population; and (v) enhance the efficiency of public research organizations (Diagne *et al.*, 2009).

Several assessments have been made of the potential benefit of rice research in Africa. Many of these were conducted by the Africa Rice Center (AfricaRice), one of the 15 members of the CGIAR Consortium and an association of

* Corresponding author: a.diagne@cgiar.org

© CAB International 2013. *Realizing Africa's Rice Promise* (eds M.C.S. Wopereis *et al.*)

24 African member states. The Center has a long tradition of priority-setting dating back to the 1990s (see Diagne *et al.*, 2009, for a detailed review of previous priority-setting exercises). The Center's research priority-setting since 1990 has involved a broad base of rice research and development stakeholder constituencies at different policy- and decision-making levels. Priority-setting is a continuing process and the methodology used has changed over time.

In early 1990, AfricaRice (then the West Africa Rice Development Association, WARDA), conducted a systematic priority-setting[2] exercise that served for both the Center's Strategic Plan for the period 1990–2000 and its Medium Term Plan 1994–1998 (WARDA, 1993). The exercise was implemented through a three-step process. The first step consisted of data-gathering on the relative importance of rice-growing environments (area, production, etc.), and the relative importance of constraints per environment. The second step comprised an analysis of rice-growing environments and main stresses to determine the priority environments and stresses that needed to be addressed, assessment of countries' research capacities for each constraint, and AfricaRice's comparative advantages. The third and final step consisted of validation of the methodology used and the major findings by a task force.

In 2000, a new priority-setting exercise was initiated for the Strategic Plan 2003–2012 (WARDA, 2004a,b). This exercise was essentially based on the outcomes of the previous priority-setting. The earlier constraints analysis was updated through task forces, working groups, surveys and at the meetings of the AfricaRice National Experts Committee (NEC).[3] On the basis of the results of this exercise, revised strategic and medium-term plans were drafted and validated by the NEC, Board of Trustees and Council of Ministers[4] (WARDA, 2004a,b).

Another priority-setting exercise was conducted for the 2005–2007 Medium-Term Plan by a commissioned internal task force (WARDA, 2004b; Diagne *et al.*, 2009). It included a number of innovative features in its methodology – namely, a review of priority-setting methods and approaches adopted by other agricultural research centres (both those inside and outside of the CGIAR), the sub-regional agricultural research organizations (SROs) and national agricultural research systems (NARS). It used a scoring method based on criteria established by the task force, discussed and approved through consultation with a wide range of major stakeholders.

This chapter presents the approach used to assess the potential impact of rice research in sub-Saharan Africa for 2011–2020 and discusses the major findings. The priority-setting exercise for which this *ex-ante* analysis was conducted has borrowed much from the previous one in terms of processes.[5] But in terms of methodology, a number of innovative features were introduced. First, an in-depth farm-household survey was conducted in 21 countries in sub-Saharan Africa to gather data on rice-growing environments, constraints to rice production, and adoption of improved varieties. Second, the research to address the identified production constraints were elicited from scientists through consultation during a 2-day workshop during the 2010 AfricaRice Science Week. Third, econometric models derived from the agricultural household conceptual framework were used to assess expected productivity, poverty and environmental impacts of the proposed research.

The rest of the chapter is organized as follows. We first review the theoretical and conceptual framework that underlies the evaluation of the impact of agricultural research on farmers. Next, we present the methods used to assess the impact on other actors such as processors, traders and consumers, followed by an overview of the different data sets used for the analysis. We then present and discuss the main findings of the analyses and their implications in terms of thematic and geographic priorities for research on rice-based systems in Africa for the period 2011–2020, before concluding the chapter.

Estimation of the Impact of Rice Research on Farmers

This section presents the theoretical and conceptual framework that underlies the models used to assess the potential impact of rice research in Africa. We describe the general framework and then derive the model that explains household decision-making to adopt rice technologies and their impact on household outcomes, and the model that explains how a change in household production environment can affect its outcomes.

The agricultural household model

To identify the impact of rice research on poverty and income at the farmer level, we follow the agricultural household framework (see Deaton and Muelbauer, 1980; Singh *et al.*, 1986; Taylor and Adelman, 2002, for a review of the literature). This framework is summarized in Fig. 32.1.

An agricultural household makes decisions to maximize its utility in the face of some constraints. The decision set, represented by the variables d^* and x^* in Fig. 32.1, includes investment, crop and varietal choices, and resource allocation (seed, land, labour, fertilizer and other inputs), which the agricultural household chooses to maximize the satisfaction it derives from the consumption of food and non-food items represented by the variables c in the figure. The optimal consumption bundle that maximizes the household's satisfaction is determined by a utility function that embodies the household decision-maker's preferences, beliefs and expectations, and by a budget constraint that balances income from all sources as represented

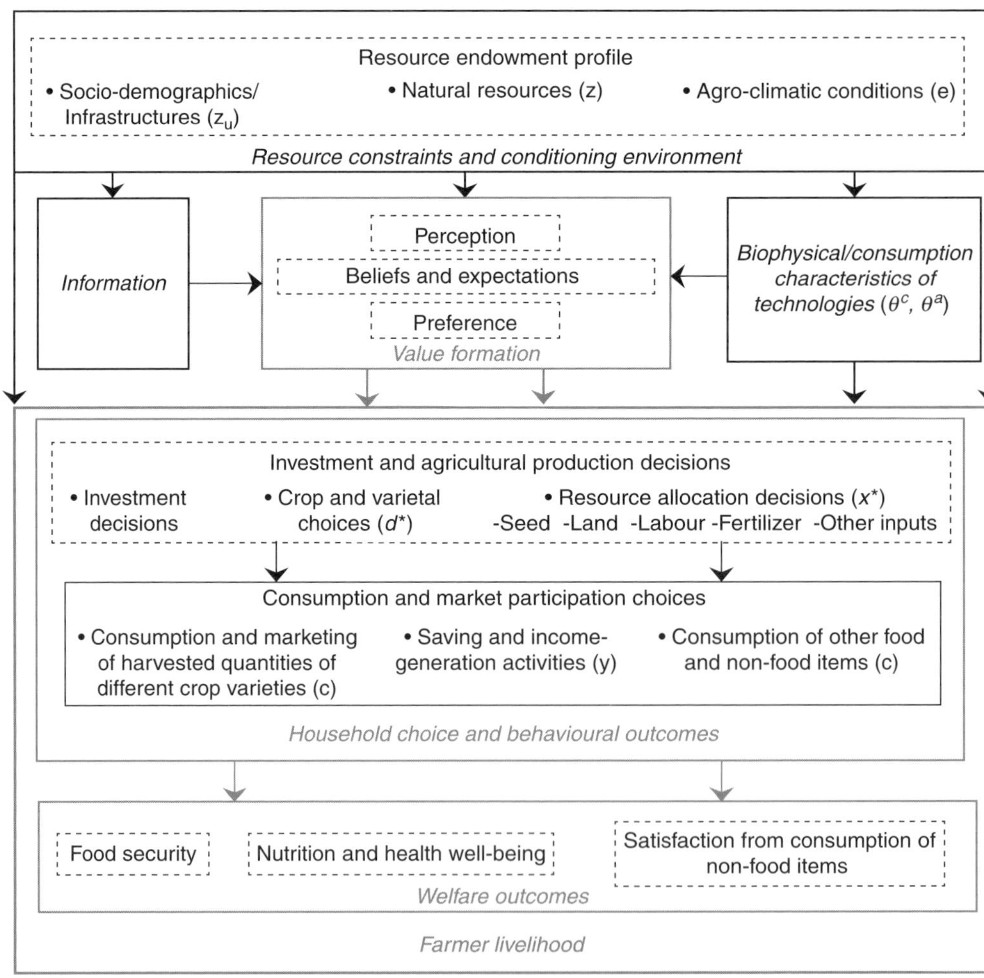

Fig. 32.1. Agricultural household framework (symbols explained in text).

by the variable y in the figure and the cost of acquiring the optimal quantities of the production inputs in d^* and x^* and the non-produced consumption items in c. The maximum utility of consumption attained by the household depends on many factors that are outside of its control. These include the characteristics of the technologies chosen as represented by the variable θ in the figure; the agro-climatic conditions (represented by e in the figure); the household socio-demographics and resources endowment profile, which include the public and private infrastructures and natural resources; the prices of commodities and the information available to the household on all these factors (all represented by z_u in Fig. 32.1).

From the graphical illustration of the agricultural household model in Fig. 32.1 one can see how household decisions (adoption of crop or varietal and agronomic technologies) and the change in the agro-climatic conditions (reduction of yield loss caused by stresses) can affect the household outcomes of interest and permit the identification and estimation of the causal effects. Here we limit the analysis to household total income and village poverty headcount. Note that the impact on income and production is assessed at the household level, while the impact on poverty and food security is assessed at the community level.

Two sets of technologies are assessed. The first relates to adoption of new varieties and is assessed through structural models of farmer demand for varietal characteristics and its relationship to total household income, production or village poverty headcount. The second set relates to adoption of agronomic (including integrated pest management (IPM)) and postharvest technologies assessed through a reduced form model of effect of reduction in yield loss or increase in yield due to farmer adoption of the agronomic or postharvest technologies on total household income or production or village poverty headcount.

Impact of adoption of new varieties

Let us start from the general setting of the agricultural household model as illustrated in Fig. 32.1 and described in the section above and assume that rice-farming households choose among J rice varieties (that include traditional and improved varieties) to produce rice and maximize the utility of consumption of food and non-food items. Thus, the agricultural household's maximization problem can be formally written as:

$$\max_{x \in S(z)}\{U(x, z_u) : subject\ to : p_c . c = p_r . \sum_{j=1}^{J} f(x_j, z_j) - \sum_{k=1}^{K} p_k \sum_{j=1}^{J} x_{jk}\} \quad (32.1)$$

where c is the consumption vector of food and non-food items, with p_c the corresponding price vector, $x_j = (x_{jk})_{k=1,\ldots,K}$, with x_{jk} being the quantity of input k used in producing rice with variety j (one of the inputs being seed); p_k is the price of input k; z_u is a vector of household socio-demographic variables that affect utility; z_j is a vector of exogenous technological and environmental variables conditioning the production of rice using variety j (variety characteristics, plot soil characteristics, weather, etc.); f is a production function; and p_r is the price of rice. There is no loss of generality by assuming a common production function for all varieties because any variety-specific technological parameter can be included in the z_j vector.

There are several rice varieties distinguished by their respective contents of agronomic and consumption characteristics. Farmers make their choice of which varieties to grow and consume on the basis of their preference for these characteristics. The agronomic characteristics are those that affect the rice yield in the field, during harvest and during postharvest grain processing. The most important of these are yield potential, levels of resistance to various biotic and abiotic stresses and other characteristics such as plant height and tillering ability. The consumption characteristics of varieties include grain quality, shape, colour, aroma, taste, and various cooking and eating characteristics (cooking time, swelling capacity, degree of stickiness, storability after cooking, etc.).

Each variety has a fixed constant vector of consumption characteristics $\theta_j^c = (\theta_{jk}^c)_{k=1,\ldots,K}$, with K_c the number of consumption characteristics, and a fixed constant vector of agronomic characteristics $\theta_j^a = (\theta_{jk}^a)_{k=1,\ldots,K}$ with K_a the number of agronomic characteristics. Let also $\theta^c = (\theta_j^c)_{j=1,\ldots,J}$, $\theta^a = (\theta_j^a)_{j=1,\ldots,J}$, $\theta_j = (\theta_j^c, \theta_j^a)_{j=1,\ldots,J}$ and $\theta = (\theta^c, \theta^a)$. Hence, θ_i is the combined vector of agronomic and consumption characteristics of variety j with dimension $K = K_a + K_c$ and θ is the $J \times S$ matrix of agronomic and consumption characteristics of all varieties. Varieties are distinguished and identified uniquely by their full vector of observed and unobserved characteristics θ_j and not by their names or labels (e.g. WABxx, NERICAxx, IRxx, 'traditional', 'improved').

Reformulating the utility function and the production function to take into account variety characteristics, we have: $U(c, z_u) = U(c, \theta^c, z_{u(\theta)})$ and $f(x_j, z_j) = f(x_j, \theta^a, z_{j(\theta)}, e)$. If the focus of the analysis is on the incidence of adoption (i.e. where there is adoption or not) instead of the intensity of adoption, then we define the variable $d^* = I_{\left[x^*_{j0}>0\right]}$ expressing whether the farmer chooses to adopt a variety j by using its seed. As a direct implication from the model described in Fig. 32.1, the farmer optimal choice depends on the conditioning variables e, θ_j and z_u. Taking the expectation conditioning to these variables, we have:

$$P(d_j^* = 1 \mid \theta_j, z_u, e) = g(\beta, \theta_j, z_u, e) \qquad (32.2)$$

with $P(d_j^{**} = 1 \mid \theta_j, z_u, e)$ being the probability of a randomly selected rice farmer adopting variety j when the vector of characteristics of the variety is θ_j, the farmer socio-economic characteristics are z_u, and the vector of agro-climatic conditions is e; β stands for the vector of parameters to be estimated; and g is a function taking value in the interval [0, 1].

Farmers in a village are not often universally exposed to agricultural technologies. Equally, therefore, a variety as a technology is not often universally exposed to all farmers (Diagne and Demont, 2007; Diagne, 2009). This introduces an important bias if one estimates adoption rates and adoption determinants by using traditional methods (Diagne and Demont, 2007). Therefore, to consistently estimate the village variety adoption rate and its determinants, we follow Diagne and Demont (2007) and use the Average Treatment Effect (ATE) estimation framework (see Imbens and Wooldridge, 2009, for a review).

For each characteristic k, we define $\hat{\theta}_k = max\left\{\theta_{jk}\right\}$, the maximum performance of this characteristic across all known varieties. The maximum characteristic value represents for the farmer the 'known technological frontier' for the characteristic. Also, let us define $\bar{\theta}_k = \frac{1}{J_a}\sum_{j=1}^{J}\theta_{jk} \times d_j^*$, the average performance of characteristic k across all varieties adopted by the farmer, with J_a the number of varieties adopted. This is the farmer's 'observed demand' of the characteristic k.

Taking the expectation conditional to e, θ_j and z_u, we obtain the 'predicted demand':

$$E(\bar{\theta}_k \mid e, \theta, z_u) = \tilde{\theta}_k = \frac{1}{J_a}\sum_{j=1}^{J} P\left(d_j^* = 1 \mid \theta_j, z_u, e\right)\theta_{jk} \qquad (32.3)$$

Because varieties are uniquely identified by their varietal characteristics vectors, the introduction of a new variety is equivalent to a change in the set of varietal characteristics vectors of all known varieties and the improvement in a varietal characteristic k is equivalent to the introduction of a new variety that pushes the farmer's known frontier for that characteristic higher. Equation (32.2) shows how this change affects the probability of adoption of a village variety, and equation (32.3) shows theoretically how the demand for characteristic k changes taking into account the adoption.

The structural causal relationships among varietal characteristics, varietal adoption decision, varietal demand and outcome of interest are described in Fig. 32.2, which shows how a change in a given varietal characteristic due to breeding research affects household total income and production and village poverty headcount through adoption and demand.

Impact of adoption of good agricultural practices

The impact of adoption good agricultural practices is assessed through changes in the effects of rice production stresses. These stresses in some cases affect 100% of the harvested area and cause high yield losses. The magnitude and effect of the stresses could be attenuated when the farmer adopts varieties that have the attribute of resistance to the stress. But another important way to reduce losses is for the farmer to adopt good agricultural practices. Such practices can help to overcome

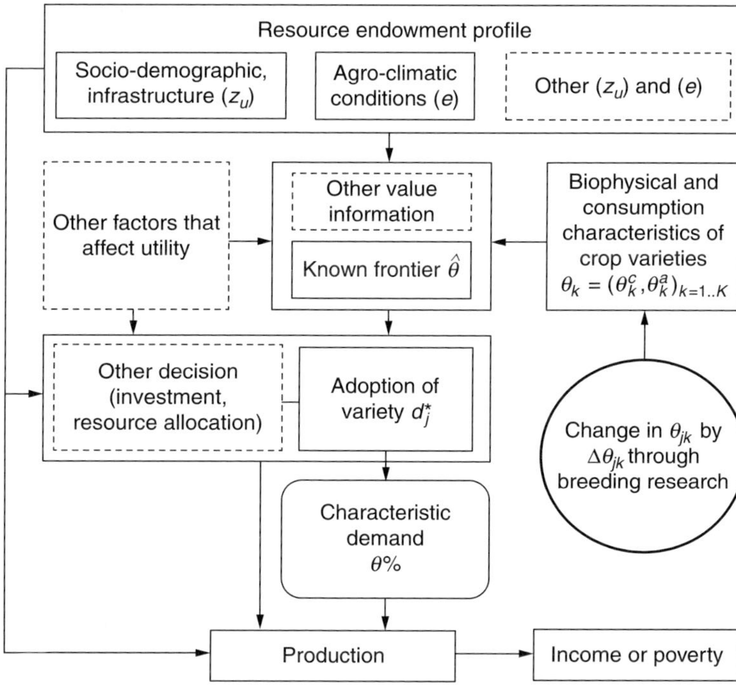

Fig. 32.2. Causal structural relation for the analysis of the impact of adoption of varieties.

stresses during the growing season or tackle postharvest losses. We have not modelled the adoption rate and determinants for this kind of 'agronomic technology'. Using experts' opinions after consultation with AfricaRice agronomists and postharvest specialists, we assume the adoption of non-varietal technologies follows a logistic diffusion curve with a 35% peak adoption rate. The logistic distribution is the most common for technology adoption variables.

The structural relation that shows how agronomic research has impact on income or poverty is described in Fig. 32.3 (the conventions are the same as above). The reader is referred to Diagne *et al.* (Chapter 4, this volume) for a formal derivation of the relationships. The channel linking technologies to impact on farmer outcomes is straightforward compared to adoption of new varieties. When the farmer takes the decision to adopt a good agricultural practice, it is expected that the negative effects that will be faced when the stress hits will be less than if he or she did not have this, or any other, suitable technology.

Specification, identification and estimation of the impact model

To be able to project impact of adoption of a technology derived from research over time, we use an autoregressive (AR) model of the outcomes of interest. The AR model is reasonably realistic in most economic settings. Because of the limited data, we use AR model of order one (AR1). Thus, the general impact model equation that will be estimated is:

$$Y_t^h = \alpha Y_{t-1}^h + \beta D_t^h + \gamma X + \varepsilon_t^h \qquad (32.4)$$

where Y is the outcome variable, D stands for the proxy for the technology that is assessed and x is a vector of variables other than D and which affect Y^t. In such regression, the main econometric concern is the possible biases that would be generated by the possibility that D may not be exogenous.

Breeding research

For the breeding research impact model, D is the demand for a given varietal characteristics k. The impact models are estimated separately for each varietal characteristic. Each equation is

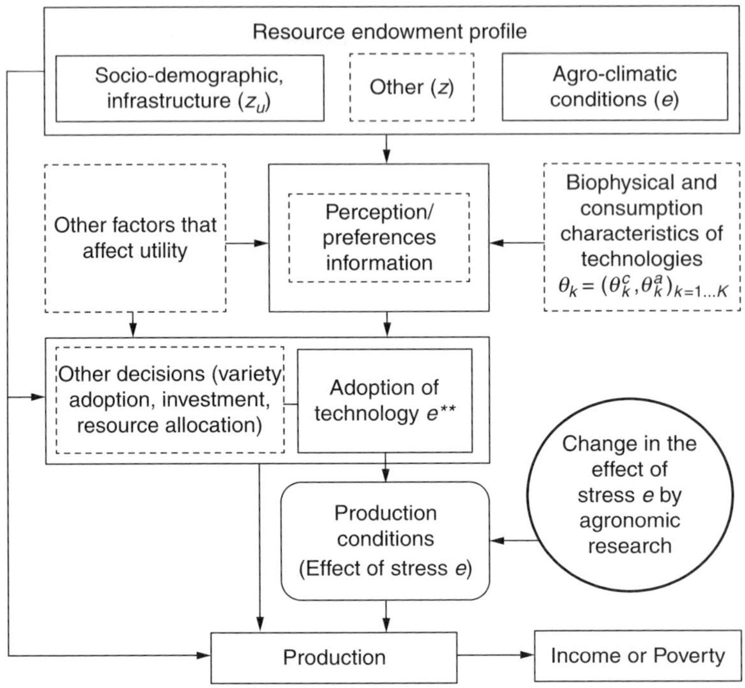

Fig. 32.3. Causal relation for the analysis of the adoption of good agricultural practices.

estimated by the Instrumental Variable method (e.g. Wooldridge, 2002). The varietal technology demand and the outcome (income, production or poverty) can be confounded as shown in the structural causal relation presented in Fig. 32.3. Thus, $D = \widetilde{\theta}_k$ is endogenous in equation (32.4) and Ordinary Least Squares will yield inconsistent parameters. To address this problem, we use the Instrumental Variable estimation method with known frontier ($\hat{\theta}_k$) as instrumental variable for the demand $\widetilde{\theta}_k$. The validity of the frontier as instrument is tested by weak instrument test (see Wooldridge, 2002). The estimated parameter is the total effect of the demand for varietal characteristics on the outcome. Chalak and White (2011) show that in such a regression, the first-stage regression coefficient need not be identified. Each technology is associated with a corresponding varietal characteristic (see Appendix, Table 32.A.1).

Agronomic research

For the agronomic research impact model, D is the yield loss caused by a given stress, when it is experienced by the farmer. We assume that the occurrence of stresses and the yield losses they cause to farm households are exogenous to their decisions. Thus, $D = e$ is exogenous in equation (32.4). So, estimation of the AR1 model by Ordinary Least Squares yields consistent estimates of the parameters. The other variables included in the model are socio-demographic variables, environment and country fixed effect to capture the heterogeneity among environments and countries of stress effect (see Appendix, Table 32.A.1, for complete details of variables used in each model).

From research to impact: methodology

The two diagrams of the structural impact model (Figs 32.2 and 32.3) show how technological options are linked to household- and village-level models. This section explains how the linkage was done. As discussed above, there is a slight difference between how technologies derived from breeding research and agronomic research are assessed. Hence, we distinguish them in the linkage procedure.

Breeding research

STEP 1: TRANSLATING THE SCIENTIST'S ESTIMATE OF PERCENTAGE YIELD GAIN INTO PERCENTAGE INCREASE IN $\hat{\theta}_K$. The average expected yield loss reduction (%) for each proposed research option was converted into an increase in the performance of the characteristic corresponding to the stress that the research option addresses (see Appendix, Table 32.A.1, for correspondence between stress and varietal attribute).[6]

STEP 2: USING PERCENTAGE INCREASE IN $\hat{\theta}_K$ IN THE ADOPTION MODEL TO GET THE PERCENTAGE CHANGE IN VARIETAL DEMAND $\bar{\theta}_K$. By changing the known frontier of characteristic k, the new technology from breeding research (i.e. variety) will induce a change in the farmer's portfolio of adopted varieties. The change $\Delta\bar{\theta}_k$ is captured through the adoption model estimated and the demand function derived from this model (equation 32.3) by plugging in $\Delta\hat{\theta}_k$. This change in demand (as population parameter) already accounts for the adoption of the technology. To account for the uncertainty, the change is multiplied by the probability of success of the research φ_k, and the true impact of the research on demand is $\varphi_k\Delta\bar{\theta}_k$.

STEP 3: PLUGGING $\varphi_K\Delta\bar{\theta}_K$ IN THE IMPACT MODEL TO GET IMPACT AT STARTING YEAR. The induced change in the farmer pool of adopted varieties (demand) is plugged into the impact model corresponding to characteristic k. The impact $\overline{\Delta y} = \beta\,\varphi_k\Delta\bar{\theta}_k$ is then the average impact on farmer income or production or village poverty headcount in the starting year of availability of the technology across adopting and non-adopting farmers.

Agronomic research

An appropriate impact model is estimated for each stress. If the technology is derived from agronomic research, we use the yield loss reduction impact model with the corresponding loss due to the stress that the technology addresses as explanatory variable. If the technology is derived from postharvest research, we use the impact model that has loss caused by postharvest as explanatory variable. Where the technology does not have a specific nature (and is labelled 'other'), we used the impact model that has average yield loss as explanatory variable.

The linkage of technologies derived from agronomic research to impact models is straightforward by plugging the yield loss reduction expected (r_e, %) directly into the impact model and multiplying the result by the probability of success (φ_e) to account for uncertainty. Thus, the impact is $\overline{\Delta y} = \beta\,\varphi_e r_e$ and corresponds to the average impact at farmer or village level in the first year of availability of the technology across adopting and non-adopting farmers.

Projection over time and aggregation across technologies

Projection of impact over time

At the end of the process described above, we had the impact for the first year of availability of the technology. This year corresponds to $t_0 = 2010 + t_d$ with t_d the estimated number of years needed to develop the technology (from 2010) as given by scientists. However, we are interested in projecting impact over time to 2020.

The AR1 models estimated are enough to allow us to forecast the mean impact starting in a given year t_0 and any subsequent year $t_0 + \rho$ as:

$$E\Delta y_{t_0+\rho} = \beta\sum_{j=0}^{\rho-1}\alpha^j E(\Delta y_{t_0+\rho-j})$$
$$= \overline{\Delta y}\frac{1-\alpha^\rho}{1-\alpha} \quad \rho = 1,2,\ldots \qquad (32.5)$$

where $\overline{\Delta y}$ is the impact in the starting year ($\beta\varphi_k\Delta\bar{\theta}_k$ for breeding research and $\beta\varphi_e r_e$ for agronomic research). This formula enables us to forecast the impact at ρ – period ahead forecasted value for the outcome y. Finally, the annual nominal income gained was discounted at the rate of 5% and cumulated to get gross benefit at farmer level.

Aggregation across technologies

The impact parameters calculated, as described above, are for each technology that addresses a specific stress or varietal attribute. These technologies were grouped in major research themes (shown in Appendix, Table 32.A.1):

- Alleviate biotic stresses.
- Alleviate soil-related constraints.
- Alleviate climate-related constraints.
- Alleviate postharvest-related constraints.
- Raise the genetic yield potential.

For each major research theme, the impact parameters were aggregated across all stresses. For the income or production parameters, the aggregation function used was the mean across technology in a given major research theme and was interpreted as the average impact of any technology within that group. For poverty parameters, the mean does not make sense because one individual cannot be lifted out poverty at the same time by two different technologies. Thus, we used a maximum across technologies in a given major research theme as aggregation function so that the result can be interpreted as the minimum number of people that will be lifted out of poverty by any technology within the particular research theme.

Extrapolation from farmer or village level to country level

Number of rice farmers and rice-farming population

The extrapolation from farmer and village level to country level is based on the estimation of the total number of rice farmers in each country. Because of the lack of national estimates of the total number of rice farmers per country, we combined household-survey and secondary data to provide estimates.[7]

The total number of rice-farming households N_h in each of the countries included in our analysis was estimated by taking the ratio of the country's total rice harvested area S (obtained from FAOSTAT, 2010) and the average rice area per household s_h (estimated from the farm-household surveys) and projected over time assuming constant population growth of g = 2.5% (average rural population growth in sub-Saharan Africa from the World Development Indicators [World Bank, 2010]).[8] The formula used is $N_h = \frac{S}{s_h} \times (1+g)^{\rho}$, where ρ stands for time.

To get the distribution across rice-growing environments, we multiplied the proportion of rice farmers by environment (estimated from the household survey) by the total number of rice farmers in the country (see Diagne *et al.*, Chapter 3, this volume, Appendix, for details). This assumes that the structure of the rice farmers by environment will remain stable over time. The total rice-farming population size in the country was estimated by multiplying the total number of rice-farming households (as estimated above) by the average household size. Missing values for average area and household size were estimated by taking the average across neighbours.

From farmer to country level

The potential benefit of rice research and its expected poverty impacts for rice-producing farmers in sub-Saharan Africa were assessed for 38 rice-producing countries. The household- and village-level impacts estimated from data on 16 countries were used to extrapolate impact at national level for the 38 countries. The 38 countries included are: Angola, Benin, Burkina Faso, Burundi, Cameroon, Central African Republic, Chad, Comoros, Congo Republic, Côte d'Ivoire, Democratic Republic of Congo (DRC), Ethiopia, Gabon, The Gambia, Ghana, Guinea, Guinea-Bissau, Kenya, Liberia, Madagascar, Malawi, Mali, Mauritania, Mozambique, Niger, Nigeria, Rwanda, Senegal, Sierra Leone, Somalia, South Africa, Sudan (including South Sudan), Swaziland, Tanzania, Togo, Uganda, Zambia and Zimbabwe. The total rice area harvested in these countries in 2009 was about 9.9 million hectares, which represents 99.3% of the total harvested area of sub-Saharan Africa. Their total paddy rice production for the same year was 19.1 Mt (99.1% of the sub-Saharan African total). Thus, the results of this analysis can be considered to be applicable for all of sub-Saharan Africa.

For each country included in the analysis, the impact on individual farmers' income or production was extrapolated to country level by multiplying the average impact estimates by the estimates of total number of adopting farmers in the country. To see that this provided a consistent estimate of the total benefit to rice farmers at the national level, we let $\bar{y}_h$ be the increase in the average household income, N the average population size of rice farmers in the country and A the adoption rate, so that the total benefit at country level is $T = A \times N \times \bar{y}_h$. Now, the total number of adopting rice farmers, $N^a = A \times N$ and $T = N^a \times \bar{y}_h$.

From village to country level

The poverty impact estimated at the village level was multiplied by the total rice-farming population size in the country. To show that this provides a consistent estimate of the reduction in the total number of poor rice farmers at the national level, let P_v be the reduction in the average village poverty headcount and N_v the average population size of rice farmers in a village, so that the average number of rice farmers in a village lifted out of poverty is $Q_v = N_v \times P_v$. Now, if Q is the total number of people living in rice-farming households lifted out of poverty in the whole country, N the total number of people living in rice-farming household in the country, N^a the total number of adopting rice farmers, K^a the total number of rice-farming villages with adopters, N_v^a the average population size of adopting rice farmers in a village, then we have $N^a = K^a \times N_v^a$, so that $Q = K^a \times Q_v = \frac{N^a}{N_v^a} \times Q_v = \frac{N^a}{N_v^a} \times N_v \times P_v$. If A is the adoption rate (assumed to be the same at both village and national levels), then we have $N^a = A \times N$ and $N_v^a = A \times N_v$. So that $Q = \frac{N^a}{N_v^a} \times N_v \times P_v = N \times P_v$. This extrapolation to country level was done for each proposed research options using the calculated estimate of individual farmer- or village-level impact.

Aggregation

For each proposed research options all income gain and poverty reductions were summed to get the gross income benefit and total poverty reduction for all countries. The income benefit by region, rice environment and research discipline was obtained by summing the benefit across all proposed research options and restricting the sum to the level of disaggregation of interest. The poverty levels by region, rice environment and research discipline were obtained by calculating new parameters at the level of interest using a max aggregation function. The new parameters were then used to obtain impact by country or rice environment and summing the result at the disaggregation level of interest.

Estimation of the Impact of Rice Research on Rice Processors, Traders and Consumers

Rice processors and traders

Rice processing involves several activities. At each step significant numbers of farmers and processors experience significant losses (see Futakuchi *et al.*, Chapter 25, this volume). Postharvest losses in rice can be divided into quantitative and qualitative losses due to the rudimentary handling methods used in many sub-Saharan African countries. It is estimated that, on average, quantitative postharvest losses of rice at farm level in Africa are in the order of 15–20% (AfricaRice, 2010).

In addition, if rice paddy is not harvested and stored on time, or is dried too quickly, the proportion of rice grains that break during the milling process is usually high. Suboptimal drying and storage practices by farmers often result in good-quality paddy rice being mixed with damaged paddy, weed seeds, insect residues, sand and stones. Separation of broken from whole grains and removal of impurities is possible, but only with equipment available in large-scale rice milling operations. Farmers also tend to mix paddy from different varieties when harvesting, drying or storing rice. But, different rice varieties have different milling characteristics.

As a consequence, there is a significant quality gap between locally produced and imported rice. The locally produced rice that is sold on African markets tends to be made up of a mixture of broken and whole-grain rice of different varieties, sizes and colours. Rice grains are also often mixed with weed seeds, stones, sand and insect residues (Lançon *et al.*, 2003). Thus, qualitative losses, estimated by the price differential between imported and locally produced rice, range from 15% to 50%, with an average of 30% in many countries.

To address these constraints, scientists have proposed a set of research options that would generale technologies of good harvest and postharvest handling and timing practices that will be made available to farmers and processors to increase milling performance and the overall quality of locally produced rice. Improved handling practices and technologies can significantly reduce rice paddy and grain losses due to poor harvesting and rice processing technologies (Wadsworth, 1991; Wang and Luh, 1991; Hosokawa, 1995).

We assumed that with adoption of improved processing practices and technologies by rice processors, the milling rate will significantly increase from its current average value of 60% (Totté, 1995) to almost 67% starting from 2013. We further assumed that the percentage of head rice after milling will increase from 11% to 20% and the percentage of broken rice will significantly decrease from 59% to 50%. These estimates are conservative. For the traders, the assessment is done by assuming that the quality of local rice will increase. The increase is measured as the reduction of the price gap between local and imported rice.

In the absence of survey data on rice processors and traders, we made the assumption that the percentage of rice paddy that will be processed using the improved technologies proposed by scientists will follow a logistic curve with an initial adoption rate of 0.5% in starting year 2013 and a peak adoption rate of 35% in 2020. For each year, we forecast the total paddy production and then applied the new technical parameters to compute the increase in milled rice resulting from the adoption of these technologies. The increase in milled rice is valued at processor margins to estimate the income benefit for rice processors. The price increase attributed to the increase in rice quality is used to estimate the benefit for rice traders.

Rice consumers

For rice consumers, it is assumed that the expected increase in production and quality of local rice through adoption of the technologies generated by the proposed research options will be translated into a decrease in rice price compared to the level it may reach in the absence of research.

Using an econometric model with the price of milled rice at each village level as dependent variable and the proxy for the technologies generated by each proposed research option and other village characteristics as covariates, we estimated the level of this price effect. This price effect combined with poverty data from the World Development Indicators (World Bank, 2010), rice expenditure shares provided by the African Development Bank (Koffi Marc Kouakou, AfDB, Tunis, Tunisia, 20 August 2010, personal communication) and estimated population of non-rice farmers in sub-Saharan Africa (obtained by subtracting the number of rice farmers estimated above from the total country population from World Bank, 2010) projected over time, we calculated the expenditure savings on rice by poor consumers. This aggregated expenditure saving has been redistributed to estimate the number of poor consumers that will be lifted out of poverty. Translation into additional rice that can be bought enabled us to calculate the amount of additional caloric consumption and the number of people that could be lifted out of hunger among poor consumers.

Presentation of the Data

The data used for the priority-setting exercise come from various sources, both primary and secondary: (i) household- and community-level surveys conducted in 21 countries in 2009 (AfricaRice, 2010); (ii) a rice experts survey conducted during the AfricaRice Science Week in 2010; and (iii) secondary sources such as FAOSTAT (2010) and World Development Indicators (World Bank, 2010). This section describes the data used and the various transformations made.

Household and community data

Rice data system for sub-Saharan Africa: overview

The rice data system for sub-Saharan Africa was a project funded by the Government of Japan to address the need for better-quality rice data, research and development (R&D) priority-setting and monitoring. Household and community surveys were conducted in 21 countries, members of the Coalition for African Rice Development (CARD)[9] by AfricaRice in collaboration with NARS and national agricultural statistics services. This work aimed to collect household- and community-level data on the biotic, abiotic and socio-economic constraints to rice production. In addition, other data were gathered – farmer knowledge of varieties and their adoption, the farmers' perception of the characteristics of varieties, household demographics, access to seed, area harvested and total production, assets, access to infrastructure, etc. The sample sizes ranged from 370 (The Gambia) to 10,500 (Nigeria) rice-farming households per country in the household surveys.

A wide range of constraints were identified across growth environments. For each constraint, farmers were asked whether they knew it or not. They were also asked to rate the constraint when it occurred in terms of intensity on a three-point scale (high, medium and low). After rating all constraints, they identified the five major ones and were then asked three questions: (i) Have you experienced the constraint in the past 5 years? (ii) What was the proportion of area affected by the constraint in the past 3 years? (iii) What was the yield loss (%) on a whole-field basis when the constraint was experienced in the past 3 years? (See Diagne *et al.*, Chapter 4, this volume, for more details.)

A single list of known traditional and modern varieties was compiled for each village via focus groups. In a given village, each surveyed farmer was asked whether he or she knew each village variety; whether he or she had grown it during the past 5 years; and, if grown, what was the area allocated, the quantity of seed used, the quantity of paddy produced, and other pertinent questions.

A pool of varietal attributes was identified by AfricaRice scientists. Because of the relatively large number of traditional varieties known and cultivated in many of the villages surveyed, the variety performance was evaluated at community level. Measuring the characteristics intrinsic to a variety is complex. Instead of having the exact measure, we used a ranking method to assess varietal characteristics. Each variety's performance for all attributes identified was assessed on a three- or five-point scale by a focus group of rice farmers in the village. The score was then used as a proxy for the varietal attribute. The scores given by the farmers were normalized by dividing by the maximum possible score (3 or 5) to give an index of between 0 and 1. Some of the grouped varietal characteristics scores (e.g. for diseases, insects) were obtained by aggregating the scores of the individual components making up the group using geometric mean (see Appendix, Table 32.A.1). These scores, from village level, were integrated into the household-level file by matching by variety and village. Missing values were corrected using averages by country, village, variety type and rice environment.

Data were collected for 21 countries, but completely processed for only 16 countries.[10] Thus, the models are estimated using only 16 countries' data. The estimated parameters were then used for all the countries included in the analysis. The survey questionnaire was almost the same for the 21 countries, except a few aspects that varied across countries. The data for the 16 countries were pooled.

To assess the probability of adoption of a given variety in a village, we pooled the data as an unbalanced panel. One observation is defined by a pair (h, j), where h is the index for household and j for variety of the village. Thus, the data were balanced at village level in each country.

The income variable used in the impact model was total household income. The survey captured household income from various sources and for the previous 3 years. A household's income comes from rice production, other crop production, livestock production and non-agricultural activities (handicrafts, commerce, work as labourers, formal employment, extraction, processing, etc.). The total household income was obtained as the sum of income from the sources identified for all household members.[11] For uniformity, the incomes were converted from local currency to US dollars using the exchange rate of each currency (from World Bank, 2010). The household total paddy production was obtained by summing the household paddy production across varieties and plots. We calculated each village poverty headcount by using household per-capita income and the poverty line used was the $1.25 per day poverty line multiplied by each country's purchasing power parity (PPP) value (World Bank, 2010).

Rice data system for sub-Saharan Africa: results

A detailed descriptive analysis of the household and community survey data is available in AfricaRice (2010). Here we present some of the key descriptive findings related to the farmers' knowledge and adoption of varieties and the demand for varietal characteristics.

VARIETAL ADOPTION. For a given variety in a village, about 47% of farmers knew it and 29% cultivated it. Among the 'exposed' population (i.e. those who knew it), the adoption rate was estimated at 62.4%. The average treatment effect of adoption of a village variety in the overall population was 61.4%, while it was 60.5% in the population of farmers not exposed to the village variety (see Appendix, Table 32.A.2). The population selection bias was only 1% (see Appendix, Table 32.A.2). This low selection bias may be due to the fact that we focused on any village variety, not on a particular variety. The gap between the potential probability of village variety adoption and the actual adoption rate was estimated at 32.3% (Table 32.A.2).

VARIETAL CHARACTERISTICS DEMAND. Because of the relatively large number of varietal attributes evaluated (27), we grouped some of them using their geometric mean (see Appendix, Table 32.A.1). This geometric mean is the most suitable aggregation function due to the nature of the varietal characteristics to be grouped, the values of which are in the interval 0–1. For all the varietal attributes, the known frontiers, the maximum value of the attribute for the known varieties, are low. The demands, measured as the average value for the adopted varieties, are on average medium and close to the frontier (see Appendix, Table 32.A.3). Thus, there is a need to improve the varietal characteristics by developing improved varieties with higher performance than the existing ones (see also the estimates of the characteristics demand elasticities in the Appendix, Table 32.A.4).

IMPORTANCE OF MAIN RICE PRODUCTION CONSTRAINTS. Constraint analysis focused only on biophysical constraints. The socio-economic constraints are not considered here and will be analysed in later work. Some grouping was made to reduce the number of constraints assessed (see 'sub-category' in Appendix, Table 32.A.1). The yield loss of a given group is the average across the constraints in that group. Also, a farmer experiencing at least one constraint in a group is assumed to have experienced this group of constraints (see Diagne *et al.*, Chapter 4, this volume, for more details on the methodology and findings).

On average, more than 30% of the harvested area of farmers who experienced at least one major constraint was affected. Soil-related constraints, climate-related constraints and weeds were the most common, with 37%, 36% and 33% of the area affected, respectively. The yield losses caused by the constraints when experienced were high and depended on the environment. Climate-related constraints caused on average 28% yield loss in irrigated and upland environments (see Appendix, Table 32.A.5).

Scientist survey

Priority-setting workshop: overview

A 2-day priority-setting workshop was organized during the 2010 AfricaRice Science Week. During this workshop, a questionnaire-based survey developed by the AfricaRice Priority-setting Task Force was conducted. The survey was addressed to AfricaRice experts to elicit research options to address rice production constraints. Scientists proposed research options by rice environment and type of research. The expected impact in terms of yield loss reduction, narrowing of the yield gap or increase in the yield potential under researcher-managed conditions was provided by the experts for each proposed research option. The yield loss reduction R given in tonnes per hectare was turned into a percentage using the formula:

$$r = 100 \times R \times \left(1 - \frac{l}{100}\right) \times \frac{1}{y} \quad (32.6)$$

in which y is the actual on-farm yield, l the actual average yield loss as perceived by the farmer, r the yield loss reduction (%) expected from the technology. This implies that with the adoption of the technology, resulting from the research option the yield loss perceived by the farmer will be reduced from l% to $(l - r)$%. Experts were also asked to indicate associated research costs, probability of success, and the expected year of delivery of the technologies from the proposed research option.

Priority-setting workshop: results

The analysis of the data from the survey of rice experts revealed a wide variety of proposed research options across research disciplines (breeding, agronomy, postharvest) and rice environments. The average yield loss reductions expected from research options mitigating the effects of various biotic and abiotic stresses are 0.6 t/ha in irrigated, 0.5 t/ha in rainfed lowland and 0.45 t/ha in upland environments. The average yield potential increases for research options raising yield potential are 1.5 t/ha for irrigated, 1 t/ha for upland and 1.4 t/ha for rainfed lowland environments. Estimates of yield gains from research options refer to conditions in researcher-managed trials. The average time to delivery of the technologies from the proposed research was slightly more than 4 years and the average probability of success was slightly above 60% (see Appendix, Tables 32.A.6 and 32.A.7 for details of scientists' estimates).

Patterns of Expected Income Benefits and Poverty Reduction

This section presents the main findings of the *ex-ante* analysis of the impact of rice research in sub-Saharan Africa. The results focus on income, poverty reduction and food security. Disaggregation across rice value-chain actors, research option, rice environment, research nature and region are made. The first sub-section focuses on gross income benefit, while the second sub-section presents impact on poverty reduction and food security. The parameter estimates of the model used to project impact are presented in the Appendix, Table 32.A.8. The results disaggregated by country are also presented in the Appendix for reference (Tables 32.A.10 to 32.A.12).

Income benefits

Aggregated income benefits by actors and region

The estimation of the potential impact of research targeted to reduce yield loss due to the major production constraints identified by farmers, to raising the yield potential and to adding quality to rice is a global cumulative 5%-discounted benefit of $10.6 billion over the 7-year period 2014–2020 for the 38 countries included in the analysis, with an average annual benefit of $1.8 billion.

The disaggregation of the income gain across the value-chain actors and per region is presented in Table 32.1.

The estimated potential impact of research targeted to reduce the yield gap and increase grain quality through better crop management and postharvest practices, and to raising the yield potential through higher-yielding varieties is an annual income benefit of $1.1 billion for rice farmers, corresponding to a global cumulative 5%-discounted benefit of $6.9 billion over the 7-year period 2014–2020.

As a result of increased rice supply, domestic prices in major rice-producing countries in Africa are expected to be on average 7.2% lower than the baseline level (i.e. without research). Translating this price effect, it is expected that annual expenditure on rice by non-rice-farming consumers under the $1.25 poverty line will be reduced by $650.6 million (PPP) by 2020 (holding consumption constant), corresponding to a global cumulative 5%-discounted benefit of $3.3 billion.

By improving rice processing technologies and reducing losses, it is expected that the quality of locally produced rice will be increased, generating more revenue for rice processors and rice traders. These benefits are estimated at $64.2 million annually (cumulative 5%-discounted, $323.7 million) for rice processors and $30.8 million annually (cumulative 5%-discounted, $155.3 million) for rice traders.

In terms of regional distribution on income gain, West Africa would have the highest impact, followed by East Africa, Central Africa and finally Southern Africa. This pattern is almost the same for all rice value-chain actors. Country-specific detail is provided in the Appendix, Table 32.A.9.

Income benefits for farmers by type of research

The impact on farmers' income by type of research grouped into major research themes for all sub-Saharan African countries and across rice environments is presented in Table 32.2.

For all types of research and for all sub-Saharan African farmers, the expected gross discounted benefit on farmer income is estimated at \$851.7 million in 2014 (starting year of adoption of the research-generated technology by farmers), reaching \$6.9 billion in 2020. These benefits correspond to an annual discounted benefit of \$1.1 billion from 2014 to 2020. The main assumption underlying these figures is that the impact of the types of research is additive (as explained in the methodology section). Consequently, the gross benefits for all types of research were obtained by summing the individual benefits for each types of research.

Disaggregation across types of research shows that the share of gross discounted cumulated benefits attributable to research that addresses major biophysical production constraints are the most important, with \$336.5 million annually for research that addresses major biotic constraints, \$320.5 million annually for research to alleviate climate-related constraints, and \$220.1 million annually for research addressing soil-related constraints (Fig. 32.4).

Research that raises genetic yield potential is expected to generate an annual income gain of \$126.6 million, representing 11.5% of the total gross benefit. Research that alleviates postharvest loss at farmer level will generate \$94.8 million annually, about 8.6% of the total annual income gained due to successful research in sub-Saharan Africa till 2020.[12]

The corresponding gross cumulative discounted benefits in 2020 are \$2.2 billion for alleviating biotic constraints, \$2.0 billion for alleviating climate-related constraints, \$1.3 billion for alleviating soil-related constraints, \$0.8 billion for raising yield potential and \$0.7 billion for postharvest loss reduction.

Table 32.1. Benefit of research on value-chain actors' income by region (\$ million discounted at 5%) for 2014–2020.

Region	Farmers	Consumers	Processors	Traders	All actors
Gross annual benefit					
Central Africa	109.4	66.6	7.7	3.75	187.5
East Africa	274.3	183.1	18.7	8.97	485.0
Southern Africa	2.6	16.8	1.5	0.77	21.7
West Africa	712.1	384.1	36.3	17.34	1,150.0
sub-Saharan Africa	1,098.5	650.6	64.2	30.80	1,844.2
Gross cumulative benefit					
Central Africa	697.5	333.0	38.6	18.90	1,088.1
East Africa	1,714.0	915.4	94.3	45.19	2,768.8
Southern Africa	16.5	84.1	7.8	3.86	112.3
West Africa	4,458.5	1,920.7	183.0	87.39	49.6
sub-Saharan Africa	6,886.6	3,253.2	323.7	155.30	10,618.7

Table 32.2. Gross discounted income benefits of research options as grouped by major research theme (\$ million).

	Starting year (2014)	End year (2020)	Annual
Biotic constraints	253.6	2,161.9	336.5
Soil-related constraints	159.9	1,349.8	220.1
Climate-related constraints	260.2	2,004.9	320.5
Postharvest-related constraints	87.9	649.0	94.8
Yield potential	90.0	720.8	126.6
All research options	851.7	6,886.6	1,098.5

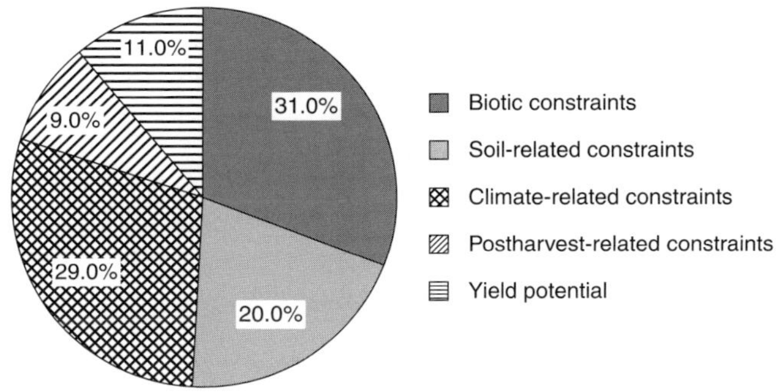

Fig. 32.4. Share of gross annual benefit attributable to technologies as grouped by major research themes.

Income benefits for farmers by rice environment

In terms of rice-growing environment, the rainfed lowland environment comes in first position for all research options with annual income benefit of $497.8 million, upland comes in second place with annual benefit of $430.1 million, irrigated follows with annual benefit of $145.4 million and then other environments with annual benefit of $24.8 million (Fig. 32.5). The high potential impact observed in rainfed lowlands is mainly driven by Nigeria, where this is the major rice environment (70% of rice farmers) (see also Appendix, Table 32.A.10).

Gross cumulated discounted benefit in 2020 will reach $3.0 billion for the rainfed lowland environment, $2.8 billion for upland environment, $0.9 billion for irrigated environment and $0.1 billion for the other environments (Table 32.3).

The impact varies across rice environments for each region (Fig. 32.6).

In general, rainfed lowland is the major rice environment in West Africa, mainly in Nigeria. The annual income benefit for lowland in this region represents 49.8% of the total annual income benefit, while the upland environment has a share of 42.0%, the irrigated environment 6.2% and the other environments 2.0%. In East Africa, the trend is different, with rainfed lowland annual income benefit share equal to 47.3%, irrigated second with 30.7%, upland third with 18.6% and the other rice environments 3.4%. In Central Africa, the dominant environment in terms of annual income benefit is upland (71.6%), followed by irrigated (15.4%), rainfed lowland (12.0%) and the others (1.0%). Southern Africa income benefits come mainly from the upland environment (92.0%) and to some extent from the irrigated environment (8.0%).

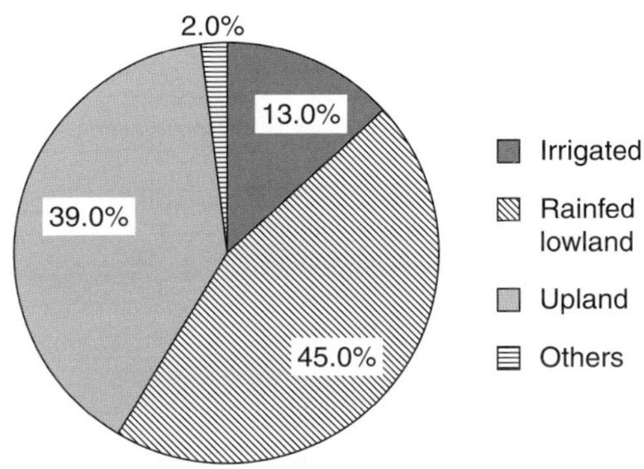

Fig. 32.5. Share of gross annual benefit attributable to each rice-growing environment.

The benefit in terms of value by region and environment is presented in Table 32.4.

Income benefits for farmers by research discipline

In terms of research discipline, breeding research comes in first position with annual income benefits of $426.8 million, followed by agronomic research with annual income benefits of $303.4 million. Postharvest research will have an annual income benefit of $166.0 million, and all other types of research are expected to have an annual income benefit of $202.2 million (Fig. 32.7).

Table 32.3. Income benefits ($ million) of research options as grouped by major research themes and rice environments.

	Irrigated	Rainfall Lowland	Upland	Others	All rice environments
Impact on income in first year of adoption					
Biotic stresses	36.1	114.8	97.3	5.5	253.6
Soil-related constraints	19.7	79.2	56.7	4.3	159.9
Climate-related constraints	36.3	115.0	99.6	9.2	260.2
Postharvest-related constraints	8.3	41.7	36.7	1.3	87.9
Yield potential	11.6	43.8	34.6	0.0	90.0
All research options	112.0	394.5	324.8	20.3	851.7
Aggregate (gross cumulated) discounted impact on income in 2020					
Biotic stresses	303.2	958.7	851.0	49.1	2161.9
Soil-related constraints	185.4	609.2	522.2	33.0	1349.8
Climate-related constraints	263.8	860.1	824.9	56.2	2004.9
Postharvest-related constraints	61.7	305.3	272.5	9.5	649.0
Yield potential	112.3	314.2	294.3	0.0	720.8
All research options	926.3	3047.6	2764.9	147.8	6886.6
Annual discounted impact on income in 2020					
Biotic stresses	47.6	149.4	131.7	7.7	336.5
Soil-related constraints	28.6	104.3	81.5	5.7	220.1
Climate-related constraints	42.6	140.8	127.1	10.0	320.5
Postharvest-related constraints	9.1	44.6	39.8	1.4	94.8
Yield potential	17.5	58.7	50.5	0.0	126.6
All research options	145.4	497.8	430.5	24.8	1098.5

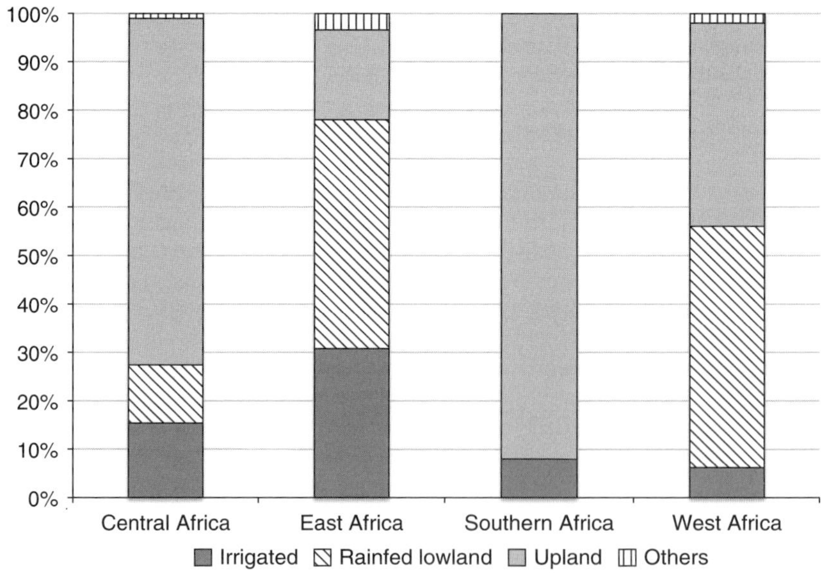

Fig. 32.6. Share of gross annual benefit attributable by rice environment and region.

These annual income gains, aggregated over time to 2020, would yield a gross cumulative discounted benefit of $2.3 billion for breeding research, $2.0 billion for agronomic research, $1.2 billion for postharvest research and $1.4 billion for other research.

Comparison across major research themes shows that the contribution of breeding to research alleviating soil-related constraints will be slightly greater than its contribution to research alleviating biotic stresses and research alleviating climate-related constraints

Table 32.4. Income benefits ($ million) by region and rice environment.

	Irrigated	Rainfed lowland	Upland	Others	All rice environments
Impact on income in first year of adoption					
Central Africa	13.0	10.4	59.1	0.9	83.4
East Africa	64.9	102.8	38.4	7.7	213.8
Southern Africa	0.2	0.0	1.8	0.0	1.9
West Africa	34.0	281.3	225.6	11.7	552.4
sub-Saharan Africa	112.0	394.5	324.8	20.3	851.7
Aggregate (gross cumulated) discounted impact on income in 2020					
Central Africa	107.3	80.4	502.9	6.8	697.5
East Africa	536.9	794.3	326.8	56.0	1714.0
Southern Africa	1.3	0.0	15.2	0.0	16.5
West Africa	280.8	2172.9	1919.9	84.9	4458.5
sub-Saharan Africa	926.3	3047.6	2764.9	147.8	6886.6
Annual discounted impact on income in 2020					
Central Africa	16.8	13.1	78.3	1.1	109.4
East Africa	84.3	129.7	50.9	9.4	274.3
Southern Africa	0.2	0.0	2.4	0.0	2.6
West Africa	44.1	354.9	299.0	14.2	712.1
sub-Saharan Africa	145.4	497.8	430.5	24.8	1098.5

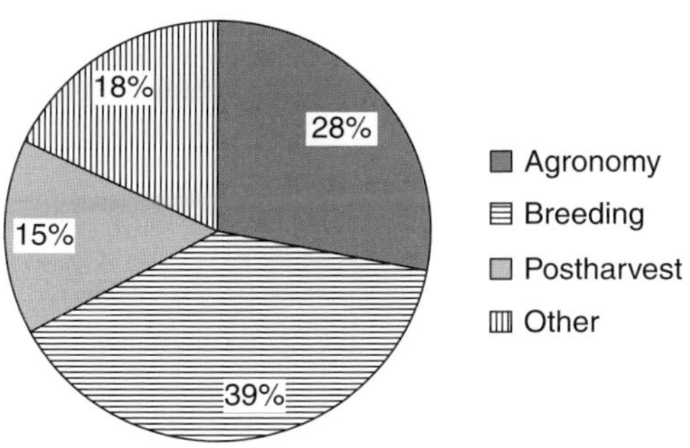

Fig. 32.7. Share of gross annual benefit attributable by research discipline.

(Fig. 32.8). Only breeding research can help in raising yield potential. Table 32.5 provides more details on impact by research theme and research discipline.

Poverty reduction and food security

Giving these income gains, corresponding (i) poverty reduction in terms of number of people lifted out of poverty and (ii) improvement in food security in terms of number of people that can afford to reach caloric sufficiency were also estimated (only for rice-farming households and non-rice-farming consumers).[13] As explained in the methodology, aggregation of poverty reduction across research options was done by using 'max' as aggregation function (to count for the fact that being lifted out of poverty is a one-time event – excluding the possibility of drop back into poverty). So, the results presented here are the minimum yearly poverty reduction.

Aggregated poverty and food-insecurity reduction by actor and region

The results in terms of poverty reduction and food-insecurity reduction are presented in Table 32.6 (see Appendix, Table 32.A.11 for the results disaggregated by country).

As a result of the rice research in sub-Saharan Africa, at least 4.2 million people in rice-farming households will be lifted above the $1.25 poverty line (in 2005 PPP) in 2020. Also the expenditure saving realized by non-rice-farming consumers will equate to 6.8 million urban and rural rice consumers (excluding rice-producing farmers) being lifted above the $1.25 poverty line in 2020. In total, at least 11 million people in the 38 sub-Saharan African rice-producing countries will be lifted out of poverty in 2020, reducing the overall number of poor by 4%.

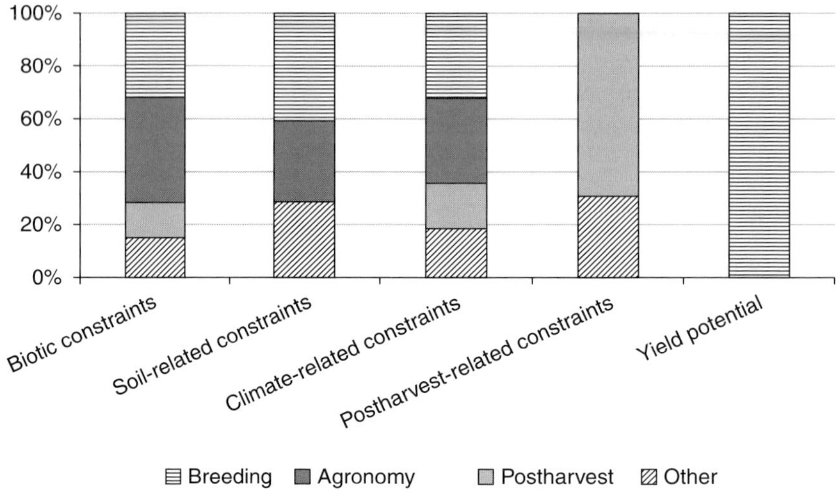

Fig. 32.8. Share of gross annual benefit attributable to research disciplines as grouped per major research theme.

Table 32.5. Income benefits ($ million) of research options as grouped by major research themes and disciplines.

	Agronomy	Breeding	Postharvest	Other	All types
Impact on income in first year of adoption					
Biotic stresses	93.2	74.1	46.5	39.9	253.6
Soil-related constraints	48.5	73.7	0.0	37.7	159.9
Climate-related constraints	86.3	74.2	56.4	43.2	260.2
Postharvest-related constraints	0.0	0.0	63.7	24.3	87.9
Yield potential	0.0	90.0	0.0	0.0	90.0
All research options	228.0	312.0	166.6	145.0	851.7
Aggregate discounted impact on income in 2020					
Biotic stresses	927.1	604.9	308.9	321.0	2161.9
Soil-related constraints	470.5	436.5	0.0	442.8	1349.8
Climate-related constraints	627.3	584.6	388.7	404.4	2004.9
Postharvest-related constraints	0.0	0.0	459.3	189.8	649.0
Yield potential	0.0	720.8	0.0	0.0	720.8
All research options	2024.9	2346.7	1156.9	1358.1	6886.6
Annual discounted impact on income in 2020					
Biotic stresses	133.6	107.4	44.9	50.5	336.5
Soil-related constraints	67.2	89.6	0.0	63.3	220.1
Climate-related constraints	102.6	103.2	55.5	59.2	320.5
Postharvest-related constraints	0.0	0.0	65.6	29.2	94.8
Yield potential	0.0	126.6	0.0	0.0	126.6
All research options	303.4	426.8	166.0	202.2	1098.5

Table 32.6. Poverty and food-insecurity reduction for rice farmers and consumers by region in 2020 (millions of people).

Region	Farmers	Consumers	Total
Number of people lifted above the PPP $1.25 poverty line			
Central Africa	0.3	0.7	1.0
East Africa	1.0	1.6	2.7
Southern Africa	0.0	0.5	0.5
West Africa	2.8	4.0	6.8
sub-Saharan Africa	4.2	6.8	11.0
Number of people no longer undernourished			
Central Africa	0.1	0.4	0.5
East Africa	0.4	1.0	1.4
Southern Africa	0.0	0.1	0.1
West Africa	0.7	2.9	3.6
sub-Saharan Africa	1.2	4.4	5.6

It is anticipated that the improved purchasing power generated by the uptake of improved rice technologies will help undernourished people in Africa to be able to afford to reach caloric sufficiency and more balanced diets. As a result of increased availability and reduced prices, 5.6 million undernourished people will reach caloric sufficiency in Africa, reducing the number of food-insecure people by 6%.

In terms of regional distribution of poverty reduction, it is expected that by 2020 some 6.8 million people will be lifted out of poverty in West Africa, 2.7 million in East Africa, 1.0 million in Central Africa and 0.5 million in Southern Africa.

In terms of regional distribution of reduction of undernourished people, it is expected that by 2020 some 3.6 million undernourished people will be able to afford to reach caloric sufficiency in West Africa, 1.4 million in East Africa, 0.5 million in Central Africa and 0.1 million in Southern Africa (see Appendix, Table 32.A.12 for the results disaggregated by country).

Poverty reduction for farmers by research theme

Analysis for the major research themes shows a wide divergence in their impact upon farmer poverty (Table 32.7).

Poverty reduction will be greatest for research that addresses biotic constraints, followed by research that addresses climate-related constraints, research to alleviate soil-related constraints, research to reduce postharvest losses and finally research to raise yield potential (Table 32.7). Only the most effective research theme contributes to the overall value to avoid double counting.

Poverty reduction for farmers by environment

As noted earlier, the major environment in terms of number of farmers is rainfed lowland. It is also in this environment that the expected poverty reduction will be highest. The general picture in poverty reduction is almost the same as the distribution of farmers across major rice-growing environments. The number of people living in rice-farming households that will annually be lifted out of poverty will be 2.0 million in the rainfed lowland environment; 1.4 million for upland, 0.7 million for irrigated and 0.1 million for other environments (Fig. 32.9).

Poverty reduction for farmers by research discipline

Agronomic research will yield the highest poverty reduction in the early years (2.5 million people), followed by breeding (0.83 million), postharvest research (0.31 million) and other research (0.24 million). By 2020, this trend will change as the starting year and growth of impact significantly differs from one research discipline to another. Thus, breeding will come in first position, followed by agronomy, postharvest and finally other research types (Fig. 32.10).

Table 32.7. Poverty reduction for farmers by major research theme (millions of people).

	First year (2014)	End year (2020)
Biotic stresses	2.47	3.51
Soil-related constraints	0.66	3.34
Climate- and water-related constraints	1.66	3.85
Postharvest-related constraints	0.31	1.43
Yield potential	0.82	4.37
All research themes	2.47	4.37

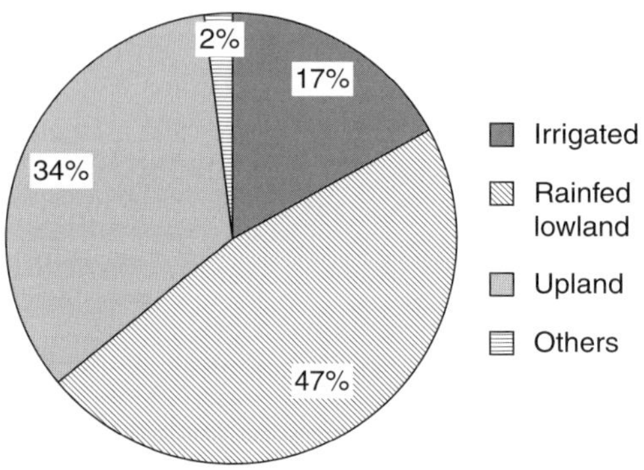

Fig. 32.9. Share of poverty reduction for farmers by rice environment.

As with income impact, there is a significant difference among research types when one differentiates by major research options. Poverty impacts of research to alleviate major stresses (biotic, climate-related, soil-related) derive mainly from agronomic research and to some significant extent from breeding. On the other hand, poverty reduction due to raising yield potential derives solely from breeding. Impact of research alleviating postharvest losses is due to postharvest research and 'other' types of research that could not be clearly classified (Table 32.8).

Research, economics and financial results and impact on production

This section presents the estimation of the direct and indirect research costs. It also describes the methods used to calculate economic and financial indicators.

Estimation of research costs and economic and financial indicators

This research and development (R&D) will be conducted mainly within the framework of the Global Rice Science Partnership (GRiSP), a CGIAR Research Programme on rice, led globally by the International Rice Research Institute (IRRI) in the Philippines and by AfricaRice for the African continent. Research costs include the GRiSP budget for Africa for the period 2011–2015 and a forecasted value for 2016–2020 – a total of about $420 million. They also include indirect costs of dissemination of the technologies (estimated from various past projects at about $1.2 billion).

The benefits and costs were aggregated and discounted to derive the rate of return and the benefit–cost ratio indicators. The financial rate of return for all research activities within the period 2011–2020 is estimated at 84% and the economic rate of return (assuming 20% price distortion) is 61%, showing that rice research in Africa within GRiSP is financially and economically very profitable.

SUMMARY IMPACT ON PRODUCTION AND RELATED INDICATORS.

Impact on rice production. By 2020, Africa's rice paddy production will have increased from 18.4 Mt in 2010 to 46.3 Mt. Without the R&D proposed and assessed in this chapter and projecting each country's production on the basis of 1980–2010 growth rate, the baseline level of paddy production would be 31.7 Mt (20.6 Mt of milled rice) in 2020. Thus, the research and its associated technology dissemination activities will result in a rice production increase of 14.6 Mt of paddy (9.5 Mt of milled rice), corresponding to a 46% increase over the baseline scenario.

Impact on rice imports. Simple projection of rice consumption under the baseline scenario using each country's rice consumption for the period 1980–2010 shows that rice consumption will rise from 19.8 Mt in 2010 to 35.0 Mt by 2020. Under the baseline scenario of no R&D,

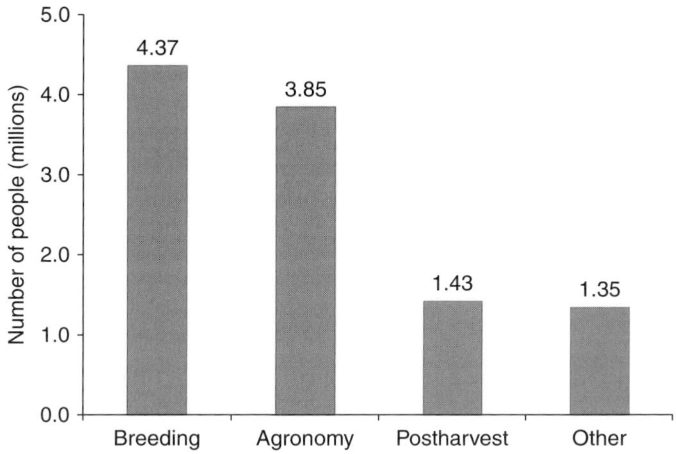

Fig. 32.10. Poverty reduction for farmers (numbers of farmers lifted above the $1.25 PPP poverty line) by research discipline by 2020.

Table 32.8. Number of people (millions) lifted out of poverty through adoption of technologies as grouped by major research themes and disciplines.

	Research discipline				
Research theme	Breeding	Agronomy	Postharvest	Others	All types
Impact on poverty in first year					
Biotic stresses	0.32	2.47	0.20	0.24	2.47
Soil-related constraints	0.38	0.66	0.00	0.20	0.66
Climate-related constraints	0.35	1.66	0.21	0.22	1.66
Postharvest-related constraints	0.00	0.00	0.31	0.22	0.31
Yield potential	0.82	0.00	0.00	0.00	0.82
All technologies	0.82	2.47	0.31	0.24	2.47
Impact on poverty in 2020					
Biotic stresses	1.48	3.51	0.86	1.26	3.51
Soil-related constraints	1.63	3.34	0.00	1.35	3.34
Climate-related constraints	1.86	3.85	0.96	1.24	3.85
Postharvest-related constraints	0.00	0.00	1.43	1.14	1.43
Yield potential	4.37	0.00	0.00	0.00	4.37
All technologies	4.37	3.85	1.43	1.35	4.37

Africa would have to import about 14.0 Mt of milled rice in 2020. But, with this proposed R&D and the production increase it will generate, imports will no longer reach this level, but rather 4.9 Mt in 2020 – corresponding to a reduction of 67%.

The production increase and the increase in quality of local rice attributed to the technologies generated by this research should lead to an increase in the continental rice self-sufficiency ratio from the current level of 60% to at least 83% in 2020. Under the baseline scenario, this ratio would remain close to 60%. In 2011, only five countries had a self-sufficiency ratio greater than 70% (Tanzania, 90%; Madagascar, 89%; Mali, 84%; DRC, 84%; Guinea, 74%); with the production increases predicted, at least nine more countries should reach this level, and all countries will increase their self-sufficiency ratios by 2020.

Contribution to agricultural GDP. The share of rice in agricultural gross domestic product (GDP) of sub-Saharan African countries should increase from the 2010 level of 3.82% to 5.19% in 2020. This corresponds to a 26.5% increase from the baseline scenario, which assumes that the agricultural GDP will maintain its current trend. Thus, R&D on rice in Africa will contribute to achieving the Comprehensive Africa Agriculture Development Programme target of 6% per year agricultural growth.

Conclusion

This chapter presents the AfricaRice research priority-setting exercise for 2011–2020. It covers methodology and presents the projected impact on income and poverty reduction from 2011 to 2020. The methodology used for this priority-setting borrows a lot from the methodologies used in the past, but also includes a number of innovative features.

We used a systematic approach and various data and econometric methods. The data on rice-growing environments, constraints to rice production, varietal characteristics and adoption of improved varieties were collected from household and community surveys. Secondary data were also collected from FAOSTAT, the World Bank and the African Development Bank. Research-based technologies were elicited from scientists, together with their expected efficacy (yield loss reduction), projected costs, probability of success and year of delivery. Econometric models were developed to assess: (i) farmer demand for rice traits and impact on adoption; (ii) impact of varietal technology demand on household total income and village-level poverty headcount; and (iii) impact of reduction of negative effects of production constraints on household total income and village-level poverty headcount. The model parameters were used to estimate impact at country level and for countries for which survey data were not available. Estimations were projected over 10 years, but taking into account the projected year of delivery and the probability of success. In total, the results of the exercise covered 38 major rice-producing countries, which represent more than 99% of the total sub-Saharan African rice area and production.

The priority-setting showed that the total cumulative discounted income benefit expected for all research and all sub-Saharan African countries will be $0.9 billion in 2014 and $10.6 billion in 2020, corresponding to an annual income gain of $1.8 billion. As a consequence of these income gains, 2.3 million people will be lifted above the $1.25 PPP poverty line in 2014 and 11.0 million in 2020. These figures hide important differences across research options and disciplines, regions and rice environments.

In terms of type of research, the impact of research to alleviate major biophysical constraints is most important. Thus, the main focus should be given to this area of research. However, the significant share of research that addresses postharvest constraints in the total benefit suggests that there is a need to consider this area in the future research agenda. Also, research that raises yield potential needs to continue to be undertaken. In terms of research discipline, breeding is the most important, followed by agronomy. Postharvest research, even though coming in last position, provides a significant share of the total benefit.

In terms of geographical area, the main rice-producing region in sub-Saharan Africa is West Africa, which accordingly will receive the highest research benefit. Research efforts need to continue to be focused in this region, with a specific focus on Nigeria, Guinea, Sierra Leone and Côte d'Ivoire. East Africa will be the second major beneficiary region and Central Africa third.

In general, the rainfed rice-growing environments predominate on the continent. This understanding was reinforced by the priority-setting results that showed that the rainfed lowlands will receive the greatest benefit from research, closely followed by the uplands. Irrigated environments, whose importance is increasing, will be the third major environment. The picture is slightly different across the regions. Rainfed lowlands and uplands are the major rice environments in West Africa and of almost equal importance. In East Africa, the two major rice environments are rainfed lowlands and irrigated. In Central Africa, upland is the major environment, followed by irrigated and rainfed lowlands in almost the same proportion.

Notes

[1] An earlier version of this chapter was presented at the International Association of Agricultural Economists (IAAE) Triennial Conference, Foz do Iguaçu, Brazil, 18–24 August 2012. The results of this study have also been used in the AfricaRice Strategic Plan 2011–2020 (AfricaRice, 2011).
[2] See WARDA (1993, 1997, 2001a,b) for more details on priority-setting at WARDA during the 1990s.
[3] The National Experts Committee (NEC) is composed of the directors general of the national agricultural research systems (NARS) of AfricaRice's member states; the NEC meets once a year at AfricaRice headquarters to discuss research progress and new directions (i.e. strategic decision-making).
[4] Like other CGIAR-supported centres, AfricaRice has a Board of Trustees composed of nominees from member states and from non-member states. The Council of Ministers (COM), composed of Ministers of Agriculture or Scientific Research of member states is the highest governing body of the Center with statutory meetings being held once every 2 years.
[5] As in the priority-setting exercises conducted in 1990 and 2000, this one also involved a broad range of rice research and development stakeholder constituencies at different policy- and decision-making levels.
[6] Example: a breeding research solution addressing weed competitiveness by reducing the losses due to weeds by x% will result in an improved variety with a weed-competiveness performance $\frac{x}{100}$ higher than the average weed-competiveness of existing varieties.
[7] The extrapolation weight in the rice statistics data that is needed to estimate the total number of rice farmers is available for only a few countries (Guinea, Nigeria, Senegal and Sierra Leone).
[8] This assumes that the rice cropping intensity is once per year. Dawe *et al.* (2010) use the same method for some African and Asian countries.
[9] The 21 member countries of CARD in which household and community surveys were conducted were: Benin, Burkina Faso, Cameroon, Central African Republic, Côte d'Ivoire, DRC, The Gambia, Ghana, Guinea, Kenya, Liberia, Madagascar, Mali, Mozambique, Nigeria, Rwanda, Senegal, Sierra Leone, Tanzania, Togo and Uganda (see www.riceforafrica.org).
[10] Data for Guinea, Liberia and Mozambique were not in the right format and had not been aggregated with the others. Data for The Gambia and Tanzania were not completely processed.
[11] The survey does not directly measure income at household-member level. During the interviews, the enumerators evaluated the income of each member and summed these to get the income of the household.
[12] These do not include postharvest-research benefits at processor level.
[13] Processors and traders are considered as non-rice-farming consumers.

References

AfricaRice (2010) Rice data systems for sub-Saharan Africa: contribution to the Japan–AfricaRice Emergency Rice Project Synthesis Report. Africa Rice Center, Cotonou, Benin.

AfricaRice (2011) *Boosting Africa's Rice Sector: A research for development strategy 2011–2020.* Africa Rice Center, Cotonou, Benin.

Chalak, K. and White, H. (2011) An extended class of instrumental variables for the estimation of causal effects. *Canadian Journal of Economics* 44, 1–51.

Dawe, D., Pandey, S. and Nelson, A. (2010) Emerging trends and spatial patterns of rice production. In: Pandey, S., Byerlee, D., Dawe, D., Dobermann, A., Mohanty, S., Rozelle, S. and Hardy, B. (eds) *Rice in the Global Economy: Strategic research and policy issues for food security.* International Rice Research Institute, Los Baños, Philippines, pp. 015–025.

Deaton, A. and Muelbauer, J. (1980) *Economics and Consumer Behavior.* Cambridge University Press, Cambridge.

Diagne, A. (2009) Technological Change in Smallholder Agriculture: Bridging the adoption gap by understanding its source. *CEGA Working Paper* No. AfD-0905. Center of Evaluation for Global Action, University of California, Berkeley, California.

Diagne, A. and Demont, M. (2007) Taking a new look at empirical models of adoption: average treatment effect estimation of adoption rates and their determinants. *Agricultural Economics* 37, 201–210.

Diagne. A., Kormawa, K., Youm, O., Keya, S. and N'cho, S. (2009) Priority assessment for rice research in sub-Saharan Africa. In: Raitzer, D.A. and Norton, G.W. (eds) *Prioritizing Agricultural Research for Development.* CAB International, Wallingford, UK, pp. 116–135.

FAOSTAT (2010) http://faostat.fao.org/site/452/default.aspx (accessed November 2010).
Hosokawa, A. (1995) Introduction. In: Hosokawa, A. (ed.) *Rice Post Harvest Technology.* The Food Agency, Ministry of Agriculture, Forestry and Fisheries, Japan.
Imbens, G.M. and Woodridge, J.M. (2009) Recent developments in the econometrics of program evaluation. *Journal of Economic Literature* 47, 5–86.
Lançon, F., Erenstein, O., Akande, S.O., Titilola, S.O., Akpokodje, G. and Ogundele, O.O. (2003) Imported rice retailing and purchasing in Nigeria: a survey. Project report. The Nigerian rice economy in a competitive world: constraints, opportunities and strategic choices. West Africa Rice Development Association, Bouaké, Côte d'Ivoire.
Singh, I., Squire, L. and Strauss, J. (1986) The basic model: theory, empirical results, and policy considerations. In: Singh, I., Squire, L. and Strauss, J. (eds) *Agricultural Household Models.* The Johns Hopkins University Press, Baltimore, Maryland, pp. 17–47.
Taylor, E.J. and Adelman, I. (2003) Agricultural household models: genesis, evolution, and extensions. *Review of Economics of the Household* 1(1–2), 33–58.
Totté, A. (1995) *Les étapes de l'après-récolte du riz et le suivi de la qualité.* Food and Agriculture Organization of the United Nations, Rome.
Wadsworth, J.I. (1991) Milling. In: Luh, B.S. (ed.) *Rice Production,* 2nd Edn, Vol. 1. Van Nostrand Reinhold, New York.
Wang, C.Y. and Luh, B.S. (1991) Harvest, drying, and storage of rough rice. In: Luh, B.S. (ed.) *Rice Production*, 2nd Edn, Vol. 1. Van Nostrand Reinhold, New York, pp. 311–346.
WARDA (1993) Medium Term Plan 1994–1998 for Presentation to TAC on the 25th of March 1993. West Africa Rice Development Association, Bouaké, Côte d'Ivoire.
WARDA (1997) WARDA Medium Term Plan 1998–2000 for Presentation to the Mid-term Meeting of the Consultative Group on International Agricultural Research, Cairo, Egypt, 26–30 May 1997. West Africa Rice Development Association, Bouaké, Côte d'Ivoire.
WARDA (2001a) *Potential for Green Revolution in Rice in West and Central Africa. Report of Second Biennial WARDA/National Committee Meeting, 20–21 March 2000, M'bé, Côte d'Ivoire.* West Africa Rice Development Association, Bouaké, Côte d'Ivoire.
WARDA (2001b) *Summary of WARDA/NARS Task Forces Activities 1991–1997.* West Africa Rice Development Association, Bouaké, Côte d'Ivoire.
WARDA (2004a) *Strategic Plan 2003–2012.* West Africa Rice Development Association, Bouaké, Côte d'Ivoire.
WARDA (2004b) The Medium Term Plan 2005–2007: Charting the future of rice in Africa. West Africa Rice Development Association, Bouaké, Côte d'Ivoire.
Wooldridge, J.M. (2002) *Econometric Analysis of Cross Section and Panel Data.* MIT Press, Cambridge, Massachusetts.
World Bank (2010) World Development Indicators. Available at: http://data.worldbank.org/data-catalog/world-development-indicators (accessed 2 December 2010).

Appendix

Table 32.A.1. Stress grouping and correspondence between stress and varietal characteristics.

Major research theme	Sub-category	Varietal characteristics	Stress
Alleviate biotic stresses	Weeds	Weed competitiveness	Weeds
	Insects	Resistance	Insects (termites, African rice gall midge, stem borers)
	Birds	Resistance	Birds
	Diseases	Resistance	Diseases (bacterial leaf blight, blast, *Rice yellow mottle virus*)
			Nematodes
Alleviate soil-related constraints	Soil and nutrients	Resistance	Zn deficiency
			Salinity/Alkalinity
			Deficiency/low use efficiency of N, P, K
			Iron (Fe) toxicity
			Acidity
Alleviate climate-related constraints	Climate and water	Resistance	Drought
			Flooding
			Heat stress
			Cold stress
Alleviate postharvest-related constraints	Postharvest	Threshability, milling recovery	Postharvest losses
Raise the genetic yield potential	–	Yield potential	–

Table 32.A.2. Village variety exposure and adoption rate.

	Parameter	SE	z-statistic
Mean population potential adoption rate (ATE)	0.644	0.004	185.52***
Potential adoption rate among exposed (ATT)	0.645	0.004	203.48***
Potential adoption rate among exposed (ATU)	0.643	0.004	154.02***
JEA rate	0.399	0.002	203.48***
GAP	–0.246	0.001	–154.08***
PSB	0.001	0.001	7.3
Observed			
Exposure rate	0.467	0.002	188.98***
Observed adoption rate	0.291	0.002	129.51***
Observed adoption rate among exposed	0.624	0.005	129.51***

ATE, Average Treatment Effect; ATT, average treatment effect on the treated; ATU, average treatment effect on the untreated; JEA, joint exposure and adoption; GAP, adoption gap; PSB, population selection bias; SE, standard error.

Table 32.A.3. Varietal characteristics frontier, demand and relative demand.

Characteristics	Frontier $\max_{j=1,\dots J}\{\theta_{jk}\}$	Demand $\frac{1}{J}\sum_{j=1}^{J}\theta_{js}\times d_j^{**}$	Relative demand $\frac{\bar{\theta}}{\hat{\theta}_s}$
Weed competitiveness	0.78	0.70	0.91
Resistance to birds	0.71	0.64	0.92
Resistance to insects	0.76	0.69	0.92
Resistance to diseases	0.78	0.71	0.92
Resistance to soil stresses	0.70	0.62	0.90
Resistance to drought	0.73	0.66	0.91
Threshability, milling recovery	0.69	0.62	0.91
High yield potential	0.66	0.58	0.90

Table 32.A.4. Varietal characteristics demand elasticities.

Varietal characteristic	Demand elasticities (adoption parameter)
Weed competitiveness	0.62
Resistance to birds	0.66
Resistance to insects	0.63
Resistance to diseases	0.60
Resistance to soil stresses	0.58
Resistance to drought	0.61
Threshability, milling recovery	0.57
High yield potential	0.60

Table 32.A.5. Yield loss and area affected by major constraints.

Constraint	Actual yield loss in 2008 (%)				Area affected in 2008 (%)
	Irrigated	Rainfed lowland	Upland	Others	
Weeds	17.6	22.5	22.6	20.7	33.4
Birds	21	17.6	18.7	13.7	26.4
Insects	19.5	21.3	22.8	16.8	29.4
Diseases	21.4	19.8	20.8	21.7	27.3
Soil and nutrient	30.2	28.7	24.9	26.3	36.5
Climate	28.2	25.6	27.6	17.7	35.9
Postharvest	28.8	23.8	13.6	34.0	26.9

Table 32.A.6. Scientist survey results grouped by major research themes and rice-growing environment.

Research theme	Environment	Yield loss reduction (t/ha)	Fixed cost ($)	Annual cost ($)	Time to delivery (years from 2010)	Probability of success (%)
Alleviate biotic stresses	Irrigated	0.53	151,981	171,708	3.95	75
	Rainfed lowland	0.38	149,715	179,391	4.12	73
	Upland	0.37	112,370	186,000	3.81	76
	Others	0.27	53,483	80,225	3.66	69
	All environments	0.39	120,889	157,313	3.88	73
Alleviate soil-related constraints	Irrigated	0.55	130,714	151,607	4.10	71
	Rainfed lowland	0.55	106,357	133,357	4.81	75
	Upland	0.40	114,077	135,857	4.09	75
	Others	0.31	37,273	55,273	4.77	72
	All environments	0.46	100,231	122,632	4.43	73
Alleviate climate-related constraints	Irrigated	0.61	157,867	203,200	4.77	73
	Rainfed lowland	0.47	113,912	120,735	4.45	73
	Upland	0.48	162,143	196,643	4.37	72
	Others	0.33	58,600	42,500	4.84	75
	All environments	0.49	127,866	147,830	4.59	73
Alleviate postharvest-related constraints	Irrigated	0.33	45,333	88,667	3.43	77
	Rainfed lowland	0.41	28,625	62,250	3.61	78
	Upland	0.39	32,875	60,000	3.55	78
	Others	0.20	25,600	41,000	2.69	84
	All environments	0.35	34,267	66,033	3.39	79
Raise the genetic yield potential	Irrigated	1.32	103,143	97,571	4.12	71
	Rainfed lowland	1.23	99,000	124,000	5.68	67
	Upland	1.02	83,333	145,000	4.67	72
	Others	0.0	0	0	0.0	0
	All environments	1.20	155,000	120,722	4.74	70
All options	Irrigated	0.60	130,080	156,123	4.11	73
	Rainfed lowland	0.51	113,328	136,769	4.4	74
	Upland	0.45	110,945	158,621	4.03	75
	Others	0.28	47,425	61,793	4.06	72
	All environments	0.48	105,556	134,063	4.15	74

Table 32.A.7. Scientist survey results grouped by major research theme and discipline.

Research to	Type	Yield loss reduction (t/ha)	Fixed cost ($)	Annual cost ($)	Time to delivery (years from 2010)	Probability of success (%)
Alleviate biotic stresses	Breeding	0.42	194,550	237,300	4.55	78
	Agronomy	0.38	65,367	97,244	2.93	63
	Postharvest	0.25	21,000	23,500	3.75	96
	Other	0.35	46,100	98,400	4.58	81
	All types	0.39	120,889	157,313	3.88	73
Alleviate soil-related constraints	Breeding	0.45	90,929	96,857	5.96	74
	Agronomy	0.51	111,083	151,500	2.89	75
	Postharvest	0.00	0	0	0.00	0
	Other	0.39	0	0	0.00	66
	All types	0.46	100,231	122,632	4.43	73
Alleviate climate-related constraints	Breeding	0.52	212,875	177,083	5.11	68
	Agronomy	0.47	82,775	173,500	4.82	74
	Postharvest	0.45	43,250	76,750	3.26	74
	Other	0.45	27,875	31,438	3.09	85
	All types	0.49	127,866	147,830	4.59	73
Alleviate postharvest-related constraints	Breeding	0.00	0	0	0.00	0
	Agronomy	0.00	0	0	0.00	0
	Postharvest	0.40	42,150	82,050	2.84	79
	Other	0.24	18,500	34,000	4.48	78
	All types	0.35	34,267	66,033	3.39	79
Raise the genetic yield potential	Breeding	1.20	95,389	120,722	4.74	70
	Agronomy	0.00	0	0	0.00	0
	Postharvest	0.00	0	0	0.00	0
	Other	0.00	0	0	0.00	0
	All types	1.20	95,389	120,722	4.74	70
All options	Breeding	0.58	155,945	169,336	5.06	74
	Agronomy	0.44	83,434	131,548	3.37	69
	Postharvest	0.38	40,692	76,731	3.03	81
	Other	0.35	31,036	56,268	4.08	78
	All types	1.20	95,389	120,722	4.74	70

Table 32.A.8. Parameters for impact models.

		Impact on income		Impact on poverty	
	Type of research	β	α	β	α
Birds	Agronomy	–4.81	0.61	0.72	0.14
	Breeding	161.16	0.74	–4.23	0.85
	Postharvest	–2.01	0.42	0.07	0.77
	Other	–4.96	0.61	0.21	0.83
Climate-related constraints	Agronomy	–3.17	0.63	0.26	0.72
	Breeding	138.54	0.74	–4.13	0.85
	Postharvest	–2.01	0.42	0.07	0.77
	Other	–4.96	0.61	0.21	0.83
Diseases	Agronomy	–4.62	0.63	0.20	0.73
	Breeding	143.63	0.73	–3.91	0.85
	Postharvest	–2.01	0.42	0.07	0.77
	Other	–4.96	0.61	0.21	0.83
Insects	Agronomy	–1.03	0.61	0.07	0.73
	Breeding	134.80	0.73	–4.49	0.85
	Postharvest	–2.01	0.42	0.07	0.77
	Other	–4.96	0.61	0.21	0.83
Postharvest-related constraints	Agronomy	–2.01	0.42	0.07	0.77
	Breeding	135.10	0.74	–4.51	0.85
	Postharvest	–2.01	0.42	0.07	0.77
	Other	–4.96	0.61	0.21	0.83
Soil-related constraints	Agronomy	–1.47	0.61	0.14	0.77
	Breeding	132.48	0.73	–4.36	0.85
	Postharvest	–2.01	0.42	0.07	0.77
	Other	–4.96	0.61	0.21	0.83
Weeds	Agronomy	–1.40	0.61	0.20	0.72
	Breeding	124.42	0.73	–4.03	0.85
	Postharvest	–2.01	0.42	0.07	0.77
	Other	–4.96	0.61	0.21	0.83
Yield potential	Agronomy	0.00	0.00	0.00	0.00
	Breeding	73.81	0.73	–4.84	0.85
	Postharvest	0.00	0.00	0.00	0.00
	Other	0.00	0.00	0.00	0.00

β, α, as per Eqn 32.4.

Table 32.A.9. Annual income benefit by actors and country ($ million).

Country	Farmers	Consumers	Processors	Traders	Total
Angola	5.3	9.6	0.0	0.0	15.0
Benin	6.1	6.2	0.3	0.1	12.8
Burkina Faso	11.8	32.4	0.6	0.3	45.0
Burundi	4.6	8.6	0.4	0.2	13.7
Cameroon	12.7	3.5	0.3	0.1	16.6
Central African Republic	6.0	3.2	0.2	0.1	9.5
Chad	25.6	6.3	0.5	0.2	32.6
Comoros	3.5	0.9	0.3	0.1	4.8
DRC	49.2	28.7	0.9	0.4	79.2
Congo, Republic	0.4	2.6	0.0	0.0	3.0
Côte d'Ivoire	68.8	5.5	2.2	1.1	77.6
Ethiopia	3.1	0.5	0.9	0.4	4.9
Gabon	0.2	0.1	0.0	0.0	0.3
The Gambia	9.1	1.7	0.1	0.0	11.0
Ghana	19.9	10.9	0.9	0.4	32.1
Guinea	80.8	1.2	4.5	2.1	88.6
Guinea-Bissau	16.8	19.2	2.3	1.1	39.4
Kenya	0.7	4.1	0.4	0.2	5.3
Liberia	42.4	4.4	0.9	0.4	48.1
Madagascar	121.6	91.7	8.5	4.0	225.8
Malawi	17.1	0.2	4.0	2.0	23.3
Mali	51.7	15.4	4.5	2.1	73.7
Mauritania	3.4	0.7	2.2	1.1	7.4
Mozambique	11.1	19.8	0.4	0.2	31.5
Niger	4.7	20.6	0.2	0.1	25.6
Nigeria	308.9	247.2	11.3	5.4	572.9
Rwanda	5.4	4.0	5.4	2.7	17.5
Senegal	8.1	16.6	1.2	0.6	26.4
Sierra Leone	75.2	0.5	3.2	1.5	80.4
Somalia	0.0	0.0	0.0	0.0	0.0
South Africa	0.2	9.6	1.2	0.6	11.7
Sudan[a]	0.7	11.3	0.1	0.0	12.1
Swaziland	0.0	0.0	0.0	0.0	0.0
Tanzania	103.5	43.8	3.7	1.7	152.7
Togo	4.6	1.7	1.8	0.9	9.0
Uganda	13.1	10.8	0.5	0.3	24.7
Zambia	2.4	4.2	0.3	0.1	7.0
Zimbabwe	0.0	3.0	0.0	0.0	3.0
All sub-Saharan Africa	**1098.5**	**650.6**	**64.2**	**30.8**	**1844.2**

[a]Including South Sudan.

Table 32.A.10. Annual income benefit for farmers by rice-growing environment ($ million).

Country	Irrigated	Upland	Rainfed lowland	Others	Total for farmers
Angola	0.7	3.9	0.6	0.1	5.3
Benin	0.1	1.7	3.9	0.4	6.1
Burkina Faso	0.1	2.1	8.8	0.7	11.8
Burundi	2.2	2.0	0.3	0.1	4.6
Cameroon	1.8	8.0	2.9	0.0	12.7
Central African Republic	0.0	5.5	0.4	0.1	6.0
Chad	0.3	22.0	3.2	0.0	25.6
Comoros	0.3	0.6	2.6	0.1	3.5
DRC	6.2	36.4	5.7	0.9	49.2
Congo, Republic	0.1	0.3	0.0	0.0	0.4
Côte d'Ivoire	1.6	45.1	21.7	0.4	68.8
Ethiopia	2.5	0.0	0.6	0.0	3.1
Gabon	0.0	0.1	0.0	0.0	0.2
The Gambia	4.1	2.1	2.4	0.4	9.1
Ghana	2.3	1.5	15.9	0.2	19.9
Guinea	2.2	48.9	27.3	2.4	80.8
Guinea-Bissau	7.4	1.8	3.6	3.9	16.8
Kenya	0.3	0.4	0.0	0.0	0.7
Liberia	0.7	28.9	11.6	1.2	42.4
Madagascar	52.1	14.5	46.1	8.8	121.6
Malawi	1.4	15.7	0.0	0.0	17.1
Mali	10.6	8.1	32.6	0.4	51.7
Mauritania	3.4	0.0	0.0	0.0	3.4
Mozambique	3.6	1.4	5.7	0.4	11.1
Niger	0.1	2.2	2.4	0.1	4.7
Nigeria	4.0	95.3	209.7	0.0	308.9
Rwanda	5.4	0.0	0.0	0.0	5.4
Senegal	7.2	0.6	0.0	0.3	8.1
Sierra Leone	0.0	58.8	12.6	3.7	75.2
Somalia	0.0	0.0	0.0	0.0	0.0
South Africa	0.0	0.2	0.0	0.0	0.2
Sudan[a]	0.5	0.1	0.1	0.0	0.7
Swaziland	0.0	0.0	0.0	0.0	0.0
Tanzania	23.7	12.1	67.6	0.1	103.5
Togo	0.3	1.8	2.5	0.0	4.6
Uganda	0.0	6.1	7.0	0.0	13.1
Zambia	0.2	2.2	0.0	0.0	2.4
Zimbabwe	0.0	0.0	0.0	0.0	0.0
All sub-Saharan Africa	**145.4**	**430.5**	**497.8**	**24.8**	**1098.5**

[a]Including South Sudan.

Table 32.A.11. Number of people lifted above the PPP $1.25 poverty in 2020 by country.

Country	For farmers by rice environment					
	Irrigated	Upland	Rainfed lowland	Others	Total	Total for consumers
Angola	2,704.1	10,936.9	1,992.3	329.5	15,962.9	101,180.2
Benin	549.9	5,006.4	13,106.8	1,430.0	20,093.0	85,623.7
Burkina Faso	952.0	9,502.9	45,249.2	3,924.6	59,628.6	391,033.2
Burundi	8,908.8	5,550.7	1,011.2	167.2	15,637.9	55,704.9
Cameroon	7,377.4	22,205.8	9,361.6	0.0	38,944.8	101,715.2
Central African Republic	85.1	9,784.7	774.9	147.5	10,792.1	24,240.7
Chad	1,299.6	61,071.0	10,311.0	23.0	72,704.5	62,648.1
Comoros	1,016.8	1,555.4	8,541.0	176.3	11,289.5	9,852.9
DRC	26,165.4	105,828.8	19,278.3	3,188.7	154,461.2	265,059.6
Congo, Republic	206.9	837.0	152.5	25.2	1,221.7	26,462.9
Côte d'Ivoire	7,756.5	155,469.9	87,342.2	1,566.9	252,135.5	216,689.9
Ethiopia	9,914.0	0.0	1,985.9	0.0	11,899.9	11,011.6
Gabon	92.1	277.1	116.8	0.0	485.9	28,681.4
The Gambia	16,573.1	5,882.6	7,787.3	1,465.0	31,708.0	37,734.1
Ghana	2,935.0	1,320.6	16,385.0	240.2	20,881.6	238,026.9
Guinea	8,743.1	135,346.7	88,421.1	7,941.0	240,452.0	20,450.5
Guinea-Bissau	29,468.1	5,077.0	11,769.6	13,081.1	59,395.8	139,768.6
Kenya	1,280.6	1,374.9	50.4	0.0	2,706.0	153,298.3
Liberia	2,791.9	80,051.1	37,465.2	3,970.2	124,278.3	22,784.6
Madagascar	259,396.9	50,044.4	185,358.8	36,518.4	531,318.4	802,251.4
Malawi	5,436.9	43,562.7	0.0	0.0	48,999.6	1,513.9
Mali	100,480.5	52,859.5	249,231.8	3,226.1	405,797.8	190,761.9
Mauritania	13,637.3	0.0	0.0	0.0	13,637.3	57,329.8
Mozambique	14,471.7	3,793.7	18,587.9	1,361.1	38,214.3	150,887.1
Niger	296.5	5,972.0	7,657.8	198.8	14,125.0	172,633.5
Nigeria	19,365.7	319,761.5	821,680.3	0.0	1,160,807.5	2,013,711.1
Rwanda	16,543.2	0.0	0.0	0.0	16,543.2	25,095.4
Senegal	35,693.5	2,038.0	0.0	1,339.0	39,070.4	353,501.4
Sierra Leone	0.0	275,379.6	68,970.6	20,818.1	365,168.3	6,078.8
Somalia	0.0	0.0	0.0	0.0	0.0	0.0
South Africa	60.3	483.1	0.0	0.0	543.4	419,157.8
Sudan[a]	1,809.3	395.4	273.8	0.0	2,478.5	120,607.1
Swaziland	0.0	0.0	0.0	0.0	0.0	0.0
Tanzania	94,945.8	33,512.1	218,582.5	276.1	347,316.5	255,127.7
Togo	1,407.0	6,126.6	10,084.2	0.0	17,617.8	33,947.1
Uganda	67.3	15,989.5	21,525.7	0.0	37,582.6	133,725.5
Zambia	747.9	5,992.6	0.0	0.0	6,740.5	29,225.3
Zimbabwe	10.8	86.3	0.0	0.0	97.0	65,623.6
All sub-Saharan Africa	**693,191.4**	**1,433,076.3**	**1,963,055.8**	**101,413.9**	**4,190,737.4**	**6,823,145.9**

[a]Including South Sudan.

Table 32.A.12. Number of people no longer undernourished in 2020 by country.

Country	Farmers by environment					
	Irrigated	Upland	Rainfed lowland	Others	Total	Total for consumers
Angola	1,356.4	5,486.1	999.4	165.3	8,007.2	173,945.0
Benin	149.3	1,359.3	3,558.6	388.2	5,455.4	32,009.5
Burkina Faso	95.5	953.6	4,540.7	393.8	5,983.6	151,401.7
Burundi	5,166.1	3,218.8	586.4	97.0	9,068.3	34,480.2
Cameroon	2,764.8	8,321.9	3,508.4	0.0	14,595.1	23,557.3
Central African Republic	72.5	8,338.1	660.3	125.7	9,196.6	18,297.8
Chad	667.5	31,369.4	5,296.3	11.8	37,345.0	50,181.1
Comoros	638.0	976.0	5,359.4	110.6	7,084.0	7,145.3
DRC	4,756.8	19,239.3	3,504.7	579.7	28,080.6	38,882.0
Congo, Republic	70.5	285.0	51.9	8.6	415.9	36,412.2
Côte d'Ivoire	1,544.3	30,952.9	17,389.2	312.0	50,198.3	28,394.4
Ethiopia	4,222.6	0.0	845.8	0.0	5,068.4	1,384.6
Gabon	56.8	170.9	72.0	0.0	299.7	2,334.4
The Gambia	4,663.8	1,655.4	2,191.4	412.3	8,922.8	3,429.3
Ghana	1,520.8	684.1	8,487.6	124.4	10,816.8	27,969.5
Guinea	2,166.5	33,537.9	21,910.1	1,967.7	59,582.2	1,768.6
Guinea-Bissau	8,693.1	1,497.7	3,472.0	3,859.0	17,521.8	62,128.4
Kenya	497.3	533.9	19.6	0.0	1,050.7	37,421.7
Liberia	1,061.6	30,437.9	14,245.4	1,509.6	47,254.5	36,021.2
Madagascar	74,257.2	14,326.1	53,062.4	10,454.1	152,099.9	429,113.7
Malawi	2,094.2	16,779.2	0.0	0.0	18,873.4	781.8
Mali	8,964.2	4,715.8	22,234.7	287.8	36,202.4	51,061.8
Mauritania	3,701.1	0.0	0.0	0.0	3,701.1	2,662.0
Mozambique	6,176.1	1,619.0	7,932.8	580.9	16,308.9	96,691.9
Niger	77.2	1,554.9	1,993.9	51.8	3,677.7	66,638.9
Nigeria	5,816.1	96,034.4	246,776.4	0.0	348,627.0	2,314,629.5
Rwanda	7,128.6	0.0	0.0	0.0	7,128.6	12,743.2
Senegal	7,939.9	453.3	0.0	297.8	8,691.1	89,394.6
Sierra Leone	0.0	67,566.8	16,922.5	5,107.9	89,597.2	2,103.8
Somalia	0.0	0.0	0.0	0.0	0.0	0.0
South Africa	9.6	77.2	0.0	0.0	86.8	28,833.5
Sudan[a]	567.1	123.9	85.8	0.0	776.9	58,256.6
Swaziland	0.0	0.0	0.0	0.0	0.0	0.0
Tanzania	44,407.9	15,674.2	102,235.1	129.1	162,446.3	354,425.6
Togo	488.9	2,129.0	3,504.2	0.0	6,122.2	6,769.0
Uganda	26.6	6,315.1	8,501.7	0.0	14,843.4	33,546.6
Zambia	410.8	3,291.3	0.0	0.0	3,702.1	55,139.1
Zimbabwe	4.3	34.5	0.0	0.0	38.8	5,739.2
All sub-Saharan Africa	**198,305.6**	**409,969.2**	**561,583.7**	**29,012.1**	**1,198,870.7**	**4,375,695.1**

[a]Including South Sudan.

33 Realizing Africa's Rice Promise: Priorities for Action

Marco C.S. Wopereis,[1]* Aliou Diagne,[1] David E. Johnson[2] and Papa Abdoulaye Seck[1]
[1]*Africa Rice Center (AfricaRice), Cotonou, Benin;*
[2]*International Rice Research Institute (IRRI), Los Baños, Philippines*

Introduction

Rice is the most rapidly growing food source in Africa. In 2010–2011, the rice self-sufficiency ratio in sub-Saharan Africa was about 60% with imports close to 10 million tonnes (Mt) per year or about one-third of that available on the world market, and costing almost US$5 billion per year. As witnessed during the food crisis of 2007–2008, this is a risky, expensive and unsustainable situation that may lead to severe food insecurity and civil instability in some countries. Soaring and highly volatile rice prices and relatively low levels of global rice stocks are predicted to remain the norm until at least 2020, and predicted demand for rice remains strong. Total rice consumption in sub-Saharan Africa is projected to rise from 24.0 Mt in 2012 to 36.0 Mt by 2020 (Seck *et al.*, Chapter 2, this volume).

The critical challenge facing the African rice sector is to sustainably enhance production, processing and marketing, and to turn this major concern into an opportunity as rice becomes a preferred staple. Africa's rice sector has the potential to be a powerful engine of economic growth as the huge demand for rice is currently filled by imports. Rice production concerns millions of people and not just rice farmers. There are the producers and manufacturers of inputs (seeds, fertilizers, pesticides) and machinery, and the traders who sell these, while on the post-production side, there are processors, traders, wholesalers, retailers and consumers.

This book deals with a diverse range of topics that are all of relevance to realizing 'Africa's rice promise', defined in the Introduction: 'Africa has sufficient land and water resources to produce enough rice to feed its own population and, in the long term, generate export revenues'. This concluding chapter brings together the main ideas presented in this book and traces a way forward to develop Africa's rice sector in a sustainable and equitable manner. We will discuss a number of priorities that are grouped in four main action areas:

- sustainably increasing rice production and rice productivity;
- enhancing rice quality and marketing;
- promoting conducive policies for smallholder and agribusiness development; and
- strengthening impact-oriented rice research, extension and knowledge management.

Many priorities will need to be addressed simultaneously to ensure effective and sustainable connections among value-chain actors (often women, see Agboh-Noameshie *et al.*, Chapter 28,

* Corresponding author: m.wopereis@cgiar.org

© CAB International 2013. *Realizing Africa's Rice Promise* (eds M.C.S. Wopereis *et al.*)

this volume) – from production to processing to marketing – to avoid fragmentation of efforts, and to promote equitable development.

Sustainably Increasing Rice Production and Rice Productivity

Enhancing rice production in Africa will necessitate concerted efforts to increase the productivity of rice per unit of land, water, labour and nutrients applied, and through the development of new land and water resources in a responsible and sustainable manner. Since the rice crisis of 2007–2008, good progress has been made in sub-Saharan Africa. According to USDA data (USDA, 2013), rice production in sub-Saharan Africa increased by a spectacular 8.4% per year over the period 2007–2012 (see Seck *et al.*, Chapter 2, this volume). Approximately 71% of this production increase can be attributed to yield increase, and 29% to harvested area expansion. Despite two relatively bad years, average rice yield in sub-Saharan Africa increased over that period by 108 kg/ha per year. In comparison, rice yield worldwide, driven by the Green Revolution in Asia, increased by 52 kg/ha per year over the period 1960–2010. Cereal growth rates after the Second World War amounted to 78 kg/ha per year in the UK and to 50 kg/ha per year in the USA. The rice yield growth rate in sub-Saharan Africa (as a response to renewed commitments to boosting Africa's rice sector after the rice crisis of 2007–2008) is, therefore, similar to growth rates witnessed on other continents after the introduction of technological innovation and institutional change. These trends are visible in all regions, except Central Africa (virtually no change in yield in the period 2007–2012) and North Africa (decline in average yield, driven by Egypt). In comparison, rice production in Asia over the period 2007–2011 grew by only 1.6% (Seck *et al.*, 2012). It is essential that this remarkable production growth rate of 8–9% per year is maintained to keep rice imports at about 10–12 Mt/year (Seck *et al.*, Chapter 2, this volume). In fact, this growth rate needs to be even greater (12% per year) to reach the ambitious goal of lifting the rice self-sufficiency ratio in sub-Saharan Africa to 87% by 2020 and reducing rice imports to about 5 Mt/year, as called for in the newly approved 2011–2020 strategic plan for rice research for development in Africa (AfricaRice, 2011b), which is discussed in more detail below.

It is, therefore, paramount that more rice is harvested on the African continent and this needs to be done in the most effective, efficient and sustainable manner. To achieve this, we distinguish five priorities for action:

1. Raising labour productivity through mechanization.
2. Establishing effective and efficient seed systems.
3. Closing yield and productivity gaps, and sustainably intensifying and diversifying rice-based farming systems.
4. Expanding rice harvested area.
5. Adapting to climate change.

Raising labour productivity through mechanization

As shown in this book (e.g. Rickman *et al.*, Chapter 27, this volume; Stryker, Chapter 26, this volume), lack of appropriate mechanization is one of the most important bottlenecks to the development of Africa's rice sector. Mechanization is particularly needed for land preparation, harvest and postharvest operations, which are still mostly manual and are extremely time consuming, causing severe delays which impact directly on rice yield and rice quality. Rice crops in Africa are often not planted on time because of late and poor land preparation as farmers wait for rain to soften the soil so that they can prepare the land using hand implements. This often results in poorly prepared and uneven seed beds and weed infestation. Small- to medium-scale equipment for land levelling, seeding, transplanting and weeding can make a huge difference in the time investment for crop establishment and free up time for farmers (often women) to pursue other activities.

Harvesting and threshing are also serious bottlenecks for farmers and these operations are also still mostly done manually. Paddy may sit in the field for weeks or even months waiting to be harvested or threshed, during which time its quality deteriorates. The introduction of locally adapted small combine-harvesters (as is being done in Senegal, Mali, Nigeria and several other countries) should go a long way towards enabling timely harvesting and threshing. This could provide the incentive for farmers to sell their paddy quickly and focus on producing a second crop. The early removal of paddy from the farm would not only enable farmers to focus on their core farming business (i.e. crop production), but would also open up the prospect for greater aggregation of the marketable surplus of paddy (AfricaRice, 2011a).

Africa is, however, littered with abandoned machines that rust away because they were not adapted to the growing environment, there is a lack of spare parts and inadequate local knowledge. Research organizations must be involved in the testing of machines, and massive introduction of purchased machinery must be avoided. Instead, local manufacturing and maintenance of machinery needs to be stimulated and governments need to facilitate the import of machine parts and raw materials. Much can be learned from experiences with mechanization in Asia and Latin America, and active knowledge exchange in the field of mechanization for rice-systems needs to be promoted.

Establishing effective and efficient seed systems

Farmers' access to quality seeds and rice varieties that are adapted to their rice-growing environments, is the backbone of rice-sector development in Africa. However, many farmers in Africa do not have access to improved rice varieties that could make a huge difference to their lives. Bèye *et al.* (Chapter 14, this volume) show that the informal sector is the dominant source of seed for African rice farmers. Farmers retain seed from their previous harvest or buy, exchange or receive seed from other farmers within their own village or from neighbouring villages. There is no blueprint solution for seed-system development in Africa, and the best possible approaches are likely to be specific to rice agroecosystems and the degree of market access. Seed systems will also evolve – for example, with emerging formal systems catering to specific market niches. Governments need to strengthen both commercial and development-oriented rice seed systems. Quality breeder or foundation seed must form the basis for seed production–supply systems. Governments must also support farmer seed systems and strengthen farmer knowledge and capacity to select and conserve quality seed for the next season. The emergence of a commercial seed sector will occur where farmer seed systems are strong, where farmers are keenly aware of what varieties are available, are engaged in seed and information exchange, and are knowledgeable consumers (Louwaars and De Boef, 2012). The introduction of hybrid rice (see El-Namaky and Demont, Chapter 13, this volume) may further catalyse rice seed-sector development in Africa.

At the regional level, seed legislation needs to be harmonized or complemented by guidelines to facilitate efficient seed flows across borders. Adequate rice seed security stocks must also be maintained to respond to emergency situations. At both national and regional levels, there is a need to adopt an integrated rice seed-sector development approach, aiming to promote diversified seed systems, meeting the seed needs of all of Africa's rice farmers.

Closing yield and productivity gaps and sustainably intensifying and diversifying rice-based farming systems

As shown by (e.g.) Saito *et al.* (Chapter 15, this volume), yields in farmers' fields are often below what would be possible with improved management ('potential yield'). A good understanding of such 'yield gaps' enables us to identify progress in farmers' fields and also helps us identify the extent to which increased costs can be justified to raise yields or reduce losses. Identifying the yield gaps also enables the major yield-limiting factors (e.g. drought, excess water, nutrient deficiencies, extreme temperature) and yield-reducing factors (e.g. pests, diseases) to be identified. Interventions may solve more than one constraint – for example, improving water control through building bunds and land levelling in lowland fields may at the same time improve weed management (see Rodenburg and Johnson, Chapter 16, this volume). Characterization of the key constraints also raises questions, such as: what is the variability in management among farmers, what are farmers doing who obtain relatively good yields, and what are others doing who obtain relatively poor yields?

Yield gaps occur in low-input systems with poor water control and relatively low-input management, but often also in high-input systems with good water control that allows for more precise management. The degree of crop management precision that is possible is of great importance to rice productivity. Lowlands in the inland valleys with scarce infrastructure to retain water or drain excess water do not allow precise management (e.g. timing of rice transplanting, weed control or nitrogen application). Rice growth and development can be severely disrupted by drought or floods. Absence or late availability of critical inputs may also undermine farmers' ability to manage precisely. With increasing control over water and other resources, improved crop management becomes possible. As a result, systems can be intensified (greater use of inputs, higher cropping intensity) or diversified (e.g. a legume crop after rice).

We propose here to identify two types of yield gap on the basis of what is presented in this book. The first yield gap (type I) is defined as the difference between actual farmers' yields and what would be possible through the introduction of technological innovations (e.g. a new variety) given the level of management precision under which farmers operate. The second yield gap (type II) is defined as the difference between actual farmers' yields and what would be possible if the system could be moved to a higher level of crop management precision with less production risk (e.g. by introducing bunding, supplementary irrigation or digging a drainage canal).

Defoer and Wopereis (Chapter 31, this volume) illustrate how type I yield gaps can be closed by working with farmers in an inland-valley setting in Madagascar, who all work under similar agroecological conditions and with similar crop-management precision and yet still obtain vastly different yields. The most important innovations identified for increasing and stabilizing yields were improved land preparation, transplanting of young seedlings, transplanting in lines, and weeding using a rotary weeder. Some rice technologies may work like an 'insurance policy' and help farmers to reduce risk. This is the case with varieties that have been 'upgraded' to include tolerance to certain biotic or abiotic stresses. Introduction of a variety with the *Sub1* gene in flood-prone areas (see Dramé *et al.*, Chapter 11, this volume) helps reduce risk for farmers with fields at risk of flooding. In years when flooding occurs, farmers with *Sub1* varieties would have much higher yields or be able to harvest much earlier than farmers without such varieties, effectively helping to close a type I yield gap.

Investing in bunding in drought-prone lowlands may radically change the production environment in which farmers operate, reducing risk and enabling greater farming precision, this would then open avenues for intensification and diversification and lift the yield ceiling to another, higher level – resulting in a greater (type II) yield gap between actual and potential yields that can be exploited. Investment in key production infrastructure (bunding, levelling, irrigation) and providing access to key resources (credit, fertilizer, etc.) will allow Africa's farmers to get out of high-risk, low-input, unsustainable agriculture and open up new production horizons.

As we have seen in this book, the degree of precision that can be applied to crop management is important for the type of technologies that can be used by the farmer (Wopereis and Defoer, 2007). Technologies suited to well-defined and precise conditions (e.g. good water control, high fertilizer input, and 'narrow windows' for field operations) will be of little use to a farmer in an inland valley without any water control. In low-precision or 'fuzzy' systems, farmers need flexible technologies that give reasonable results under a range of biophysical conditions. Especially in such low-precision systems, farmers can contribute to technology development at an early stage, which gives them the opportunity to evaluate a range of prototype options. Precision management systems will be better served with a smaller range of well-defined technological options. Technology development for these systems can be done under more controlled research conditions with less intensive farmer involvement, resulting perhaps in a few well-defined final products. In other words, the degree of flexibility of a technology required is related to the degree of crop management precision of a system. A lower degree of crop management precision increases the need for flexible technologies with large application domains. To ensure adequate farmer involvement in the technology development, methodological approaches are needed to strengthen farmers' capacity in technology innovation, experimentation, evaluation and adaptation, making optimal use of the available resources and best choices of alternative ways of managing resources.

In the past, the introduction of new technologies has had limited impact, because attention was often focused on only one aspect of the cropping calendar (e.g. fertilizer management or varietal improvement). As shown in this book (e.g. Defoer and Wopereis, Chapter 31, this volume), much better results are obtained if a more holistic approach is used, where a new technological option is not so much introduced but rather integrated into the prevailing production system, taking into account interactions with other production factors and management practices. As shown in irrigated rice-systems in the Sahel (Tollens *et al.*, Chapter 1, this volume), a new soil-fertility management strategy may require new options for weed management. Gradually, other technological options may be integrated, eventually leading to a range of technological options that encompass the entire growth cycle, from the initial planning phase to harvest and postharvest stages. This process is called integrated crop management, indicating the step-wise integration of new technological options into production systems with full farmer participation, thereby raising production levels in a sustainable way. When applied to rice, this process ultimately results in 'baskets of integrated rice management options' or 'baskets of good agricultural practices' for different rice production systems. As these systems are dynamic, baskets of options will evolve over time (Wopereis and Defoer, 2007).

Expanding rice harvested area

Across the continent, the most fertile and productive lands for rice are found in the flood plains and inland valleys, and the potential to expand rice harvested area in sub-Saharan Africa is huge. There is large untapped irrigation potential in sub-Saharan Africa, estimated at about 21 million ha (Zwart, Chapter 21, this volume). There is scope to increase the area under irrigation in many countries through expansion or rehabilitation of irrigation structures. This is especially important in countries like Mali and Senegal, where farm size per household in irrigated systems has been declining since the 1970s because of population growth and lack of new land that has been developed for irrigation and is reaching critically low levels (SWAC/OECD, 2011).

With irrigation, farmers will reduce rice production risk and will be able to lift their rice farms to a higher production level through intensification. It will also open up possibilities to grow two or even three rice crops per year depending on the prevailing climate. With an estimated land area of 130 million ha, inland valleys are common landscapes in many parts of Africa that have traditionally not been used for agriculture. Only a fraction (3–4%) of these inland-valley lowlands is used for rice. This is partly because inland-valley lowlands are difficult to manage and are often associated with water-borne disease. Moreover, inland-valley exploitation is often complicated by unfavourable land-tenure arrangements. However, inland valleys are increasingly being used for rice cultivation in the wet season and for other crops, including vegetables, in the dry season near large urban centres because of high population density and proximity of markets (Rodenburg, Chapter 22, this volume). For upland rice, too, there are possibilities for area increase if accompanied with appropriate crop management practices, such as proper soil-fertility management, through balanced use of organic inputs and mineral fertilizers. For example, the growth of Uganda's rice sector in recent years has depended mostly on upland rice systems (Ministry of Agriculture, Animal Industry and Fisheries, 2009). One of the most striking examples is the rapid expansion of rice cultivated areas in Ethiopia, where upland NERICA varieties are grown with supplementary irrigation in hydromorphic zones (AfricaRice, 2012a). On the other hand, experience from Brazil has shown that the area under upland rice may eventually decrease upon intensification and increased yield per unit of harvested area (Pinheiro *et al.*, 2005).

While the rice harvest area expands, it is important to consider the value of environmental services of land and water resources, and how these will be affected by developments in rice-based systems. Inland valleys in particular are important for local flood and erosion control, water storage, nutrient retention, stabilization of the micro-climate, as well as for recreation and tourism, and as sources of water, clay and sand for crafts and construction. While the main crop is often rice, inland valleys and their fringes are used to grow a variety of other crops (e.g. maize, vegetables, fruit trees), and are also often used for cattle grazing – particularly during the dry season when the water table recedes below the soil surface of the valley bottoms, but there is sufficient residual moisture to support crop growth. Furthermore, these environments provide important forest, wildlife and fisheries resources, and contribute to biological diversity as well as local cultural heritage. The water resources available in inland valleys are often used by rural communities to fulfil a variety of daily household needs. Besides the water resources, biological diversity of inland valleys is probably among the most important functions for the local communities – inland valleys are important locations for the collection of non-agricultural plant resources, and local communities have considerable knowledge of the useful plant species, their use, abundance and collection places (Rodenburg, Chapter 22, this volume).

Such environmental services are often neither well understood nor quantified. Such understanding of who would potentially benefit or lose from land and water developments is, however, extremely important, particularly in the light of a changing climate and the growing interest in Africa's natural resources. Policies are needed (e.g. on land tenure) to facilitate socially acceptable and environmentally sound expansion of rice-producing areas.

Adapting to climate change

Changes in the climate of Africa are expected to include major changes in rainfall distribution, increased frequency of extreme weather events, and generally rising temperatures and CO_2 levels. Farmers have great experience in dealing with climate risk, but the expected pace of change may mean that local knowledge and technologies may be insufficient to cope with new conditions. Anticipating likely climate changes will help in providing alternatives or measures to enable farmers to adapt (e.g. to lower and erratic rainfall, changing river discharges). New climate-resilient varieties, crop- and resource-management technologies, and institutional innovations such as insurance against crop failure may help farmers adapt to rapidly changing environments. Mitigation opportunities are also important. The impact of the predicted enhanced use of Africa's lowlands for rice, land clearance and burning in upland environments, and increased use of nitrogen fertilizer needs more study to develop ways to limit additional release of greenhouse gases into the atmosphere. A global effort is needed to develop targeted technological options to help African farmers to adapt to and mitigate the effects of climate change.

Enhancing Rice Quality and Marketing

Locally produced rice needs to find a market, and it must be of sufficient quality to compete with imports. Quality of locally produced rice is often poor because of sub-optimal harvest and postharvest practices, leading African rice consumers to prefer imported rice. Further, varieties grown may not be to the liking of urban consumers. Hence, to attract Africa's rice consumers and lift the demand for locally produced rice, marketing campaigns are required to make African consumers aware of quality rice produced in Africa. In short, locally produced rice must respond to the preferences of Africa's rice consumers and be promoted to find a market. We distinguish three priorities in this action area:

1. Promoting investment in improved processing technologies.
2. Enhancing access to appropriate financial products for the local rice value chain.
3. Improving market knowledge, branding and policy sequencing.

Promoting investment in improved processing technologies

Rice processing in Africa is dominated by small-scale rice millers, the majority of which produce an end-product that fails to meet the quality requirements of urban consumers. Moreover, most operators only provide milling services – they do not buy paddy and sell on milled rice. This practice contributes to the fragmentation of the market for milled local rice and discourages private-sector investment. The promotion of private-sector investment in efficient rice processing technologies, such as 'mini-rice mills' with built-in capacity for de-stoning, polishing and sorting homogeneous high-quality rice, will go a long way towards stimulating the local rice value chain (see Stryker, Chapter 26, this volume). Government support of modern rice processing should include mechanisms to provide incentives for processors to upgrade their technologies, such as duty-free imports on processing equipment, tax concessions or access to finance. As paddy production expands, there is an urgent need to process good-quality milled rice that matches the quality benchmark of imported rice (see Futakuchi *et al.*, Chapter 25, this volume). The objective, however, should not be to repeat past misguided policies of government-run large-scale industrial rice mills, but rather to promote modern processing technologies capable of producing high-quality rice. Such mills should include medium-scale operations to deliver large quantities for urban centres and 'mini-rice mills' for rural locations (see Stryker, Chapter 26, this volume).

Enhancing access to appropriate financial products for the local rice value chain

Access to agricultural finance is critical for all rice value-chain actors and to stimulate development in the supply and competitiveness of local rice. For instance, limited access to production credit to purchase productivity-enhancing inputs can counter otherwise profitable production decisions. Without access to adequate financial products, rice farmers often end up selling paddy on credit to traders who may then delay payment to farmers because of their own lack of access to adequate finance.

Failures in credit provision constitute major bottlenecks in the development of a well-integrated value chain for locally produced rice, which adversely affects the overall competitiveness of the chain. Fixed investments in improved processing technologies, warehouses and farm machinery require long-term financing rather than the short-term financing needs of paddy-production credit.

In many African countries, little locally produced rice is available in urban areas, and the development of storage capacity (warehouses) requires appropriate financial products to guarantee rice supply. As paddy production expands, it will be necessary to promote year-round availability and marketing of local rice. Greater investments are therefore needed for the development of effective warehouse systems. Successful experiences – for example, with warehouse-receipt systems – could be promoted. Incentives should be given to rice millers to facilitate access to finance for investment and working capital necessary to procure and hold paddy and milled rice in storage. Adequate financing mechanisms and facilities for the marketing of rice could be extended to local rice wholesalers.

Improving market knowledge, branding and policy sequencing

Africa Rice Center (AfricaRice) research using experimental auctions (Demont and Neven, Chapter 24, this volume) has shown that women consumers are willing to pay a premium for locally produced quality rice compared to imported rice of the same quality. Some consumers, however, prefer the lower price of local mediocre-quality rice. The implication is that different market segments exist for different rice consumers, and development of the rice sector should take such differences into account. Looking at the effect of branding, Costello *et al.* (2013) found that the majority of rice consumers in Dakar, for instance, preferred 'local' sounding brands over 'foreign' sounding brands. Rice value-chain development should, therefore, ensure that rice of different quality remains available to cater for consumer preferences and differences in purchasing power.

The key lessons from the value-chain work conducted in Senegal by AfricaRice and partners are that the availability of quality local rice needs to be promoted among the population, and that production of quality rice requires investment and conducive policies. Several steps may be required starting with increasing the quality of local rice to the level of imported rice, thus adding value to the local product, then scaling up local rice production while running promotional programmes to market the surplus, with the goal of eventually replacing imported rice in urban markets. A branding exercise in Saint-Louis in 2006 failed to impact the market because of lack of promotion (M. Demont, Saint-Louis, Senegal, 2012, personal communication). Fragmentation of the local paddy contributed to this as producers act alone in processing and selling their surplus paddy, which is a major disincentive to private-sector investment in the domestic rice value chain.

Promoting Conducive Policies for Smallholder and Agri-business Development

Development of Africa's rice sector will require coherent, evidence-based policy making at both national and regional levels. There is greatly increased awareness of rice as a strategic commodity capable of fuelling economic growth and contributing to hunger and poverty reduction across the continent. Many African countries have embarked on ambitious programmes to boost their rice production capacity as a response to the 2007–2008 rice crisis. With the upward spikes in food prices, many policy tools have been implemented that were out of vogue following the implementation of market-oriented reforms in the 1990s. These include price controls, export bans, subsidies on retail prices, the release of food security stock, subsidies on inputs such as seed and fertilizer, farm machinery and postharvest equipment, and the establishment of a minimum producer price. Some of the policy measures taken by African governments to alleviate the impact of soaring staple prices on consumer welfare are well founded given the strategic importance of rice. Many untargeted subsidies and outlays, however, will make it more difficult to balance the public budget. Governments should avoid undermining incentives for domestic rice production by misjudged policy measures such as price controls and reduced import taxes introduced in an unpredictable manner – which increases market uncertainty – or maintaining these for unnecessarily long periods. In short, policy intervention needs to be rules-based and predictable so as not to discourage investment or undercut the emergence of a dynamic private rice sector (see Seck *et al.*, Chapter 2, this volume).

As a result of the global food crisis in 2008, Africa has also become a target for direct foreign investment in agriculture. Major companies are acquiring large tracts of land for food production and biofuel plantations. It is clear that African countries need to move cautiously with respect to this new situation because of the complexity, political sensitivity and context specificity of the land issue within and across countries. They need to ensure that these investments lead to win–win situations for all involved, not least the resource-poor local farmers (AfricaRice, 2011a).

Regional economic communities should be strengthened to contribute in such areas as harmonizing seed legislation, import tariffs and regulating rice imports, in line with the Comprehensive Africa Agriculture Development Programme Framework for African Agricultural Productivity (FARA, 2006). National governments need to take the lead in promoting public–private partnerships across the rice value chain for production, storage, processing and distribution infrastructure for quality rice in the African market. To boost Africa's rice sector, policies conducive to developing investment in domestic production capacities are required. Further, investments to rehabilitate and expand areas under irrigation should continue, along with support provided to raise the productivity of smallholder rice producers through access to improved varieties, good-quality seed and fertilizer. We have learned, however, that targeting investment efforts uniquely on production can create gluts at harvest time because of insufficient processing and marketing capacity in the value chain. Therefore, it is vital to simultaneously invest in the harvesting, processing and marketing nodes of the local rice value chain by a combination of public support to farmers and privileging private-sector investment. Also, as proposed by the Economic Community of West African States (ECOWAS; CEDEAO, 2012), it is important to establish local (regional) rice stocks that are well managed (not necessarily by the public sector), to protect local markets from short-term price shocks on the global market.

Strengthening Impact-oriented Rice Research, Extension and Knowledge Management

Rice research and development efforts need to be strengthened in Africa and become more interlinked and impact-oriented. The following priorities will need to be addressed in this action area:

1. Capacity strengthening among research and extension communities and rice value-chain actors.
2. Improving rice knowledge management.
3. Implementing a new Africa-wide rice research agenda.

Capacity strengthening among research and extension communities and rice value-chain actors

Lack of investment in agriculture in the 1990s led to a desperate lack of capacity at all levels in the rice value chain and gross neglect of Africa's agricultural research and extension capacity, which jeopardizes progress in developing Africa's rice sector. A survey conducted among AfricaRice's then 22 member states in 2008 showed that approximately 250–275 researchers (including about 15 women) were involved to some extent in rice research. Most of these worked on many other crops and spent only a fraction of their time on rice, and the average age of researchers was 47. Egypt alone took the lion's share of this research pool, with 50 highly qualified researchers working full time on rice, including 12 breeders. In comparison, a country the size of Nigeria had only two rice breeders (AfricaRice, 2011b).

Extension services in Africa are largely understaffed and starved of access to consistent and relevant rice information and improved extension tools. Public and private extension services often focus on high-value export commodities and less on staple food crops such as rice. This is seriously hampering rice-sector development, which will depend to a large extent on the development of rice technologies adapted to local settings and their dissemination to actors in the rice value chain. There is a clear need for improved accessibility to rice-related information and functional infrastructure for research and development, and the development of a critical mass of trained scientists and public, NGO and private-sector extension agents. There is also a clear lack of appropriate technology-delivery and information-exchange mechanisms. There is, therefore, an urgent need to rebuild Africa's research and extension capacity for rice. The Second Africa Rice Congress held in Mali in 2010 called for a 'Marshall plan' by African governments and their development partners to substantially strengthen the training and retention of new staff, while updating agricultural curricula in vocational training schools and universities, ensuring efficient spillover to actors in the rice value chain and strengthened information exchange. Conducive working environments and appropriate budgetary provisions are needed to strengthen and retain an effective capacity in agricultural research and extension.

Improving rice knowledge management

Knowledge management and creating rural learning opportunities at the regional, national and local levels presents challenges. Public- and private-sector agents continually renegotiate their roles and build new sets of skills and expertise, either in-house or by partnering with others. New actors in the rice sector, however, often work with 'top-down mindsets', lacking awareness of gender, poverty and sustainability issues. Strategic partnerships, and learning alliances and methods may help to increase awareness and increase rice-sector performance in a sustainable and equitable manner. Promising opportunities to strengthen learning and innovation systems include: the establishment of partnerships with NGOs and producer organizations; the emergence of brokers between the supply and demand sides of innovation; the use of information and communications technology (ICT) to enhance rural learning and 'self-diagnosis' of problems and testable solutions; the development of 'new professionals' through institutional change in higher education; and new funding sources and mechanisms to better address resource-poor and women farmers' needs.

Rapid development of ICT in sub-Saharan Africa (mobile phones, rural radio, internet, etc.) offers exciting new opportunities for AfricaRice and partners to facilitate exchange of information and learning modalities that are demand-driven and tailored to specific needs. Given the high illiteracy rates in sub-Saharan Africa, there is a need for quality audio- and video-based learning content, and local capacities for content creation and adaptation. This will be of particular importance to rural women.

Implementing a new Africa-wide rice research agenda

After a series of workshops and meetings that started in 2008, the AfricaRice Council of Ministers approved, in September 2011, a new Strategic Plan to boost Africa's rice sector for the period 2011–2020 (AfricaRice, 2011b) – seven Priority Areas are considered:

1. Conserving rice genetic resources and providing smallholder farmers with climate-resilient rice varieties that are better adapted to production environments and consumer preferences.
2. Improving rural livelihoods by closing yield gaps and through sustainable intensification and diversification of rice-based systems.
3. Achieving socially acceptable expansion of rice-producing areas, while addressing environmental concerns.
4. Creating market opportunities for smallholder farmers and processors by improving the quality and the competitiveness of locally produced rice and rice products.
5. Facilitating the development of the rice value chain through improved technology targeting and evidence-based policy-making.
6. Mobilizing co-investments and linking with development partners and the private sector to stimulate uptake of rice knowledge and technologies.
7. Strengthening the capacities of national rice research and extension agents and rice value-chain actors.

Priority Areas 1–5 will result in new rice technologies that will make a positive, sustainable and lasting difference in the livelihoods of farmers and other rice value-chain actors. Through Priority Area 6, links will be established with large rice-sector development initiatives and the private sector to obtain co-investments to stimulate uptake of appropriate rice knowledge and technologies and to obtain feedback on technology performance. Priority Area 7 addresses the desperate lack of trained capacity across the rice value chain and in rice research and development in Africa (discussed above). Across Priority Areas, there is a need for working closely with women farmers, researchers, extension agents and agribusiness women in order to maximize efficiency, effectiveness and impact (AfricaRice, 2011b).

As an Association of currently 24 African member states (January 2013), and recognized by the African Union as the Center of Excellence for Rice Research in Africa, AfricaRice is well placed to coordinate these rice research-for-development efforts across the continent over the next decade. AfricaRice will act as both a developer and broker of rice knowledge, and will tap sources from within and outside the African continent, with each partner contributing to the rice research-for-development agenda according to its comparative advantage.

Links between research and development investments from both public and private sector are often very weak in Africa. As a result, opportunities for large-scale exposure of farming and agribusiness communities to new rice technologies and crop and natural-resource management principles are often lacking. Rice-sector development will require better linkages (feed-forward and feedback loops) between research networks and development initiatives in the public sector, civil society and private sector. This rice research-for-development agenda will, therefore, be implemented through a range of partnerships from strategic upstream research to linking with development partners to achieve impact on the ground, with the national agricultural research systems (NARS) as the key entry point in each country.

To achieve impact and boost Africa's rice sector, it is essential to: (i) focus efforts; (ii) build critical mass; (iii) connect actors in the research and development communities; and (iv) communicate results. To follow these four principles, AfricaRice will implement its Strategic Plan through three mechanisms.

The **first mechanism** is through AfricaRice's participation in CGIAR Research Programmes (CRPs), in particular the Global Rice Science Partnership (GRiSP), led globally by the International Rice Research Institute (IRRI) (see www.cgiar.org/our-research/cgiar-research-programs/rice-grisp/). AfricaRice is leading the implementation of GRiSP in Africa; it will play a broker role to mobilize rice knowledge from outside Africa and will also ensure that knowledge from Africa will benefit other continents.

The **second mechanism** consists of the Rice Task Force mechanism: an Africa-wide systematic collaborative research effort on critical thematic areas in the rice sector (e.g. rice breeding, rice agronomy), based on the principles of sustainability, build-up of critical mass, and ownership by the NARS. The Task Force mechanism will contribute to the development of a new generation of rice scientists across the continent.

The **third mechanism** consists of 'Rice Sector Development Hubs' – zones where rice research products from the CRPs and the Task Forces will be integrated across the rice value chain to achieve development outcomes and impact. These Hubs represent key rice-growing environments and different market opportunities across African countries, and are linked to major national and regional rice-development efforts to facilitate broader uptake of rice knowledge and technologies. The geographic positioning of each Hub is determined in national workshops, convened by the NARS (AfricaRice, 2011b, 2012b).

Activities in the Hubs focus on producing sufficient quantities of the right quality of rice and rice-based products of interest to the national or regional markets in a sustainable manner. Hubs are regions strategic for rice development, where local innovations and research products and services are tested, adapted and integrated in 'baskets of good agricultural practices' (integrated rice management options) with feedback provided to researchers on technology performance. Hubs are built around large groups of farmers and involve other value-chain actors, such as rice millers, input dealers and rice marketers. Change agents from research, NGOs and extension agencies work with these actors to evaluate technological and institutional innovations, facilitate diffusion of knowledge and establish linkages along the rice value chain. This type of interaction is stimulated through the establishment of multi-stakeholder platforms (Tollens *et al.*, Chapter 1, this volume). Care is taken that women and youth are not marginalized, but rather strengthened in the process of rice value-chain development. By January 2013, there were already 59 Hubs identified in 20 sub-Saharan African countries; and in support of these, Task Force activities focus on the Hubs to avoid dispersion of activities.

Diagne *et al.* (Chapter 32, this volume) assessed the potential impact of this Africa-wide rice research agenda for 2011–2020. They show that the total cumulative discounted income benefit expected for all the research-based technologies and all sub-Saharan African countries will be \$0.9 billion in 2014 and \$10.6 billion in 2020, corresponding to an annual income gain of \$1.8 billion. As a consequence of these income gains, 2.3 million people will be lifted above the \$1.25 purchasing power parity (PPP) poverty line in 2014 and 11.0 million in 2020. In terms of research, rice breeding is expected to yield the greatest benefits, closely followed by agronomic research. Postharvest research, even though coming in last position, provides a significant share of the total benefit. In terms of geographical area, research efforts need to continue to be focused on West Africa, with a special focus on Nigeria, Guinea, Sierra Leone and Côte d'Ivoire. East Africa will be the second major beneficiary region and Central Africa third. Rainfed rice-growing environments predominate on the continent. Priority-setting results show that the rainfed lowlands will receive the greatest benefit from research, closely followed by the uplands. Irrigated systems, whose importance is increasing, will be the third major environment. These figures hide important differences across research options, rice environments, research types and regions.

Conclusions

Rice is critical for food security and political stability throughout Africa, and it has great potential to fuel economic growth. For many decades, rice has had the fastest-growing consumption rate among the staple crops, in large part driven by huge growth in urban demand; however, about 40% of the rice consumed on the continent is imported. Despite being a global crop, only 7% of world production is traded internationally. Africa's reliance on the international market is therefore a very risky strategy – any rupture in the global market supply would likely have major political implications due to the numbers of (especially urban) people who rely on rice for their daily food, as seen during the 2007–2008 rice crisis in Africa.

Research has shown that the potential for rice production on the African continent exceeds anticipated consumption levels in the distant future by far and that domestic rice can be competitive. Even though aggregate yields are lower for Africa than for Asia, closer examination of yield by growing environment and season suggests that rice yields in Africa are often on a par with those in Asia, especially for irrigated systems (AfricaRice, 2011a). Moreover, Africa has huge untapped natural resources in the form of land and water – resources that are now scarce in other parts of the world, such as Europe, Asia and North America. Although local rice has for a long time suffered the stigma of poor quality, its taste is preferred by many consumers over imported varieties, and when quality concerns are met consumers are prepared to pay a premium for local varieties.

The critical challenge facing the African rice sector is to enhance production, processing and marketing to enable a major concern to be turned into an opportunity – the growing demand for rice as a preferred staple. To respond to this challenge, Africa will have to become a global powerhouse of rice production, avoiding the trap of fragmentation of production, processing and marketing factors, to produce sufficient quantities of the right quality of rice to compete against imports on the African markets.

In terms of enhancing production, we need to understand factors limiting and reducing rice productivity per unit of land, water and labour, and work together with farmers to develop 'baskets of good agricultural practices' that can gradually close the current large yield gaps in farmers' fields. There is also a need to mobilize investments to raise the yield ceiling in farmers' fields, reduce risk and enable greater farming precision. One way of achieving the latter is to enable greater control over water resources in a collective and sustainable manner. For inland valleys in Africa, this will often entail the construction of main and secondary drainage channels and the identification, bunding and levelling of individual fields with minimal soil disturbance. There is also a need to rehabilitate and expand irrigation structures, because rice in Africa is still mainly grown under rainfed conditions. Improved access to credit, quality seed of improved varieties (tailored to specific growth environments and market demands) and mineral fertilizer will continue to play a key role in boosting rice productivity.

With the prospect of increased production, there is a need for investment in the harvesting, processing and marketing described above, so that the whole value chain works together to ensure more and good-quality local rice reaches consumers' tables. Financial products (e.g. credit) need to be adapted to the target borrower – there is no 'one size fits all' credit mechanism for everyone involved in the value chain. Moreover, access to good storage facilities along the rice value chain (including warehouses) should be expanded to improve the storage and marketing of quality rice (AfricaRice, 2011a).

Past experiences have shown that encouraging farmers to do their own processing is not helping in the drive to improve and maintain quality. For this reason, we need to move towards systems where farmers focus on production, using appropriate machinery to maximize their output in terms of both quantity and quality. Processing may then be carried out with medium- and large-scale machinery owned by producer associations or private entrepreneurs. These could contract farmers to grow specific varieties with quality seed and specified inputs and other management practices – some form of outgrowers' scheme. In this way, the processors will collect and aggregate rice of a single variety of a similar quality, which will enable them to produce grain of uniform quality ready for the market (AfricaRice, 2011a).

Marketing is another important aspect. There is strong evidence that Africans prefer local rice varieties and will pay a premium for them if the quality is right. Contractual arrangements for delivery of quality grain between processors and wholesalers, and appropriate branding are logical steps to improve the value and consumption of local rice on the continent.

Thus, the 'final' step in the value chain is that wholesalers will buy and brand (package and label) quality local rice for onward sale to retailers and thence consumers (AfricaRice, 2011a).

In our vision, Africa's rice farmers will operate modernized family farms, most of which will be mechanized and in many cases farmers will grow a second crop (either rice or some other crop). Farmers, many of them women, will be well informed, using quality rice seed and good agricultural practices that are environmentally sustainable from land preparation to harvest. Farmers' prices will be differentiated to reflect quality grain and farmers' associations or millers will aggregate quality paddy; rice will be mainly milled by dedicated quality millers using good storage and processing practices; credit will be available to all stakeholders in the rice value chain; contractual arrangements will be the norm, between farmers or farmers' associations and processors, and between processors and wholesalers or importers; wholesalers will bulk-buy quality rice for branding and onward sale to retailers; and the commercial rice product will carry a label indicating not only its origin, but also its quality.

Working together towards that vision will allow Africa as whole, and the various subregions, to realize 'Africa's rice promise'.

Acknowledgements

Some of the Conclusions in this chapter are reproduced from AfricaRice (2011a) with permission from Africa Rice Center.

References

AfricaRice (2001a) *Lessons from the Rice Crisis: Policies for food security in Africa.* Africa Rice Center, Cotonou, Benin.

AfricaRice (2011b) *Boosting Africa's Rice Sector: A research for development strategy 2011–2020.* Africa Rice Center, Cotonou, Benin.

AfricaRice (2012) *Boosting Africa's Rice Sector: Implementing the AfricaRice 2011–2012 strategic plan.* Africa Rice Center, Cotonou, Benin.

CEDEAO *et al.* (2012) Réserve Régionale de Sécurité Alimentaire : Etude de faisabilité. Rapport provisoire. Communauté économique des États de l'Afrique de l'Ouest, Abuja, Nigeria.

FARA (2006) *Framework for African Agricultural Productivity.* Forum for Agricultural Research in Africa, Accra, Ghana.

Louwaars, N. and De Boef, W. (2012) Integrated seed sector development in Africa: a conceptual framework for creating coherence between practice, programs and policies. *Journal of Crop Improvement* 26, 39–59.

Ministry of Agriculture, Animal Industry and Fisheries (2009) Uganda National Rice Development Strategy (UNRDS), 2nd draft. Ministry of Agriculture, Animal Industry and Fisheries, Government of the Republic of Uganda, Entebbe, Uganda.

Pinheiro, B. da S., de Castro, E. da M., de Moraes, O.P. and Breseghello, F. (2005) Developing aerobic rice in Brazil. In: Toriyama, K., Heong, K.L. and Hardy, B. (eds) *Rice is Life: Scientific perspectives for the 21st century.* Proceedings of the World Rice Research Conference held in Tokyo and Tsukuba, Japan, 4–7 November 2004. CD-ROM. International Rice Research Institute, Los Baños, Philippines, and Japan International Research Center for Agricultural Sciences, Tsukuba, Japan, pp. 154–156.

Seck, P.A., Diagne, A., Mohanty, S. and Wopereis, M.C.S. (2012) Crops that feed the world 7: rice. *Food Security* 4(1), 7–24.

SWAC/OECD (2011) *West African Challenges* no. 2. Sahel and West Africa Club Secretariat, Paris, France.

USDA (2013) USDA database. www.fas.usda.gov/psdonline/psdQuery.aspx (accessed on 31 January 2013).

Wopereis, M.C.S. and Defoer, T. (2007) Moving methodologies to enhance agricultural productivity of rice-based lowland systems in sub-Saharan Africa. In: Bationo, A., Waswa, B., Kihari, J. and Kimetu, J. (eds) *Advances in Integrated Soil Fertility Management in sub-Saharan Africa: Challenges and opportunities.* Springer, Netherlands, pp. 1077–1091.

Index

Page numbers in **bold** refer to illustrations and tables

accreditation 80, 181
    see also certification; Truthfully Labelled Seed
action priorities 424–436
Advanced Spaceborn Thermal Emission and Reflection Radiometer (ASTER) instrument 198–200
AfRGM (African rice gall midge)
    biological control 10, 233–234, 237
    damage type 230
    host plants 236
    outbreaks, cropping and fertilizer association 233
    resistance 71, 174, 176, 232
    susceptability 175
African rice 95, 161–162, 356
    see also Oryza glaberrima
African rice gall midge *see* AfRGM
AfricaRice Genebank Information System (ARGIS) 94
AfricaRice Strategic Plan 2, 15, 391, 425, 433
agri-business development 431–432
agriculture
    conservation 257–258
    good practices adoption 394–395
    practices 394–395, 436
    water use competition 265–266
    see also cropping-systems; farmers
agroecological-zones 190–**192**, **195**, 197, 359
agroecosystems 405
agronomy
    research
    benefit 405, **407**, **408**, 434
    discipline importance 412, **418**
    impact model parameter 397, **419**
    poverty reduction 409–**410**, **411**
    task force 3, 198
alleles
    evaluation need 117, 123
    introgression 165–167, 168
    marker 122
    resistance 72, **153**, 162, 168
alternate wetting and drying (AWD) 267, 268–269, 273
amylose content 312–313, 316, **317**, 318
analysis
    ex-ante 12, 241, 390–413
    financial 384, **386**
    impact **395**
    molecular diversity 89
    spatial 191, 273
    structural 135
approaches
    ecogeographic 91
    genomic 91
    methodological 360–361, 363
    participatory 150, 346–347
    systematic 412
    value-chain 11
area
    average **383**
    disaggregated 39
    distribution **38**, **252**
    estimated 35–45, **191**, **192**, **195**
    expansion **26**, 188–200, 433
    improved varieties percentage 12
    major constraints **416**

aromatic-rice 174, 299, 312, 317, 321
Asara season 382
ASI thresher-cleaner 339
Asian rice 95, 130, 161–162, 356
    see also Oryza sativa
attributes
    consumer perspective 297
    cooked rice 321
    credence attributes 300, 301
    defined 297
    experience attributes 299–300
    search 297, 299, 301
    varietal, pool 401
    see also characteristics; QTL; traits
auctions 301, 306, 430–431
auto-regressive (AR) model 50, 395
Azolla spp. 256

backcrossing 100–103, **114**, **137**, **139**, 152, 175, 357
    see also breeding; genetics; hybrids
bacterial blight (BB) 162, 215–216, 217–218, 221–222
bacterial leaf blight
    bacterium responsible 11, 214
    importance ranking 55
    perception index 47
    resistance 70, 71, 112, 174, 221
    symptoms 215
bacterium 10, 214, 217
Bamè 282
Bateson-Dobzhansky-Muller (BDM) model 133, **134**
Benin-China Cooperation 282
best practices 381, 394–395, 436
biases 49, 63–65, 294, 301, 329
birds 50, 55, 61, 241–247, 357, 358
blast 10, 72, 175, 214–215, 216, 218–220, 319
Blétou valley 281, 287
branding 300, 305, 306, **314**, 430–431, 436
    see also packaging
breeding
    adaptive 121–122
    gene pools use 112–114
    genomics-assisted 108, **113**, 124
    goals **71**
    interspecific 71–72
    lines 4, 117
    methods 69–76, 152–153, 161
    molecular 152
    new schemes application 138–139
    objective 318
    open source collaborative 122
    populations 123
    programmes 4–**5**, 79, 150, 154, 174
    research **395**–396, 397, 434
    RYMV resistance use 163–164
    strategies 113, 174
    streams 114
    task forces 69–70, 74–**75**, 150–151
    technology adoption 393–394
    see also backcrossing; genetics; genomics
bridge materials 114
broken-rice 297, 318, 321
Bt-rice 232
bunding 8, 209, 259, 280, **281**, 289, 427
business models 337–338, 341
business-enabling environment 306

C_4 metabolic pathway 154
catalogues 80, 82
    see also varieties, lists
cereals 242, **245–246**, 425
certification 80, 181, 305, 339
chalkiness 312, 314, 315, 317, 321
characteristics 89–90, 110, 200, 205, **315**, 426–427
    see also attributes; QTL; traits
chromosomal regions, comparison 133
chromosome segment substitution lines (CSSLs) 135–**137**
climate
    change
        adapting 70, 96, 211, 425, 429
        effects anticipation 153–154, 223
        impacts 210
        pest movement effect 234–236
        threat 278
        water stress effect 266, 270, 272
    constraints **54**, 58, 60, 402
    see also rainfall; temperature
cloning, positional 167–168
co-flowering 98–99
co-investment mobilizing 433
CO_2 increase 210
cold tolerance 152
    see also temperature
collaboration 71–72, 123, 139, 198
    see also partnerships
commercialization 186, 308, 346
Common External Tariff (CET) 31–32
communication 260, 363, 369, 374, 432, 433
    see also media; videos
communities 286, 400–402
    see also participation
community-based seed system (CBSS) 185–186, 346–347
competitiveness 31, 303–304, 306, 308, 321, 324–330
compounds, plant-specific, detection 235
conservation 88–**89**, 257–258
constraints
    abiotic 47–49, 50–56, 60, 144, **145**

see also climate; drought; flooding; iron (Fe) toxicity; nutrients, deficiencies; salinity; soil; temperature; water
agronomic 7
biophysical 56–61, **59**, 65–68, **283**, 390, 412
see also disease; insect pests; iron (Fe) toxicity; nutrients, deficiencies; weeds
biotic 47–49, 50–56, 61, 63–65
see also birds; diseases; insects; rodents; weeds
characterization 236, 426–427
importance 402
indicators 50
perceptions 48, 63–65
production 2, 6, 282–284, 289, 402
ranking **283**
real effects 48
research impacts 402
see also climate; soils; stresses; water; weeds
consumer acceptability scores **314**
consumers
attributes perspective 297
behaviour, *defined* 294
bias, imported rice 329
preferences 294–301, 317
quality standards 325–326
research impacts 399–400
value chain tailoring 303–308
consumption
annual evolution **295–296**
attributes 294
characteristics 393–394
levels 294
rate 24, 28, **29**–30, 324, 325, 434–435
cooked-rice 299, 312–313, **316–317**, 318, 321
cooperation 122
see also partnership, Task Force Mechanism, Rice Sector Development Hubs
core collection concept 90–91
costs
estimations **31**
genotyping 119, 123
imported rice 311
phenotyping 116
projects 387
reduction R&D 15
research 410–412
country-of-origin labelling (COOL) 300
credence 300, 301
credit 299, 340, 375, 436
crisis
food 1, 11–12
price 213, 304, 307
rice 24–33, 179, 213, 425
crop
development 103–104
establishment method 243
growth duration 154
integrated management 428
losses assessment techniques 244
management 6–11, 144, 154, 254, 258–259, 427
simulation models 7, 196–197, 200
see also cropping-systems
cropping-systems
adoption 209
alternative 257
calendar 7–8, 428
double cropping 198
intensification-induced yield loss 9–10, 256
irrigated rice 7
operation timing 242
rotation 9–**10**, 200, 208, 209, 257, 259
synchronized 242
upland 209
crosses 4, 102, 166
see also backcrosses; hybrids
CSSL (chromosome segment substitution lines) 136–**137**
cultivars 206–207, 289
see also varieties
cultivated-areas 35–45, 154, 322, 324–325
cultural identity 345–346, 356, 359
cyber-seed concept 185
cytoplasmic male sterile (CMS) lines multiplication 173–174

data
management 76
presentation 400–402
sources 37, 39, **40**, 47, 80, 104, 197, 412
survey 196, 379, 402–403, **417**, **418**
databases 25, 72, 88, 92, 161, 162
decision-support 7, 9, 121, 122, 341
deep-water systems 190, 251–252
delay-payment mechanisms 299
DEM (digital elevation model) 198–199
demand
increasing 24
lifting 15, 32, 301
perspective 303
prediction 31, 394, 424
urban 346
varietal characteristics 402, **416**
denaturing gradient gel electrophoresis (DGGE) technique 236
deregulation, formal seed systems 186
DGGE (denaturing gradient gel electrophoresis) technique 236
DIARPA (Diagnostic Appraisal System for Water management) 287, **288**, 289
digestibility trait 358
digital elevation model (DEM) 198–199

disease
 control 218–222
 during grain-filling stage 319
 genes, controlling 121
 hotspot screening 72
 incidence 195–196
 incomes impact 58
 major 213–**216**
 management 213–223
 mapping potential epidemics 195–196
 model 195–196
 research 58
 resistance 111, 175, **176**
 spatial variability 11
 technology 56, 60–61
 types 284
 water-borne 277
dissemination bottleneck 372–374
distinctness, uniformity and stability (DUS) criteria 79, 80, **82**, 83
diversification 426–428, 433
diversity
 analysis 89, 110–112
 biological 284–285, 429
 environment 188
 farm-produced seed 183–184
 gene pool 112–114
 genetic 87, 130–131, 135–138, **166**
 reduction 130
 rice and related wild species 87–92
DNA markers 72
domestic resources cost (DRC) estimations **31**
domestication 103–104, 111, 130–131, 133–135
donors 72, **137**, 151
drainage infrastructure 251
drought
 constraints 62, 281, **283**
 grain effects 318
 management 56
 research 72
 resistance 6
 yield gap determinant 196, 200
dryers 336, 338
DUS (distinctness, uniformity and stability) criteria 79, 80, **82**, 83
DVDs 373–374
 see also media; videos

eating qualities 312–313, **316**, **317**–318
ecology 277–279
econometric methods 412
economy 11, 24–33, 204, 277–279, 359
ecosystems 35, **41**, **145**, 208, **283**, 285–286
 see also inland-valleys; irrigated-systems; rainfed-systems; upland-systems
elite lines 121–122
elongation rate 316
enemies, natural 237
environment
 characterization 35
 classification **36–37**
 diversity 188
 division 69
 genotype interactions 175, 358, 360–361
 research 399, 405–**406**, **407**, **410**, **417**, **421**
 services 428–429
 technology interactions 355–363
 see also lowland-systems; rainfed-systems; upland-systems
EPIRICE diseases model 195–196
equality 351
 see also gender issues;men; women
Equatorial forest agroecological zone 190, 276
equitable development 424
evapotranspiration 266, 267, **272**
Experimental auctions 301, 430–431
extension 340–341, 368–369, 424, 432–434

facilitators **386**, 387
fallow-systems **10**, 200, 209, 257
farm-household *see* household
farmer field schools (FFS) 370, 381
farmers
 adoption model **67**
 characteristics 402
 confidence 367
 decision making 7
 facilitators 385
 feedback mechanisms 375
 female 281
 meeting the needs 179–186
 modern 436
 number estimation 67, 398
 perceptions 46–62
 practices 253–254, 256, 258
 research impacts 391–398, **421**
 seed involvement 182, **183**, 375
 varieties 359
FCC (Fertility Capability Classification) system 193, 194, 252
FERRIZ model 255
fertilizer
 application 8, 318
 management 10, 259
 mineral 8, 251–253, 255–256, 258, 260, 269
 needs 260
 price reduction policy 11
 rates **8**
 recommendations 255, 256, 259, 260
 subsidies 32
 timing 154

use 253
see also nitrogen; nutrients; phosphorus; potassium
finance 80, 384, 386, 430
flooding 56, 62, 153, 209, 267, 269
flowering 98–99, 175, **176**
food baskets 24, 276–290
food security 173, 303, 304, 307–**308**, 330, 390–413
formation-diffusion: F-D training 386–387

gall midge 10, 162, 284
see also African rice gall midge
gametes 131, 133, 134, 135, 151
GEBV (genomic estimated breeding value) 120, 121, 123
gender issues
access 344–345, 346, 369
actors and levels 343–346
analysis 3, 348–349, 351
considerations, project integration 343–352
equity framework 348
implications 280
improved varieties 347
inequality 369
mainstreaming 346–351
see also women
gene-bank, AfricaRice 69, 88
gene-flow 95–104, 210
see also interspecifics
genes
candidate 134–135
characterization 110
identified 71, 72, 136–138, 152
mapping 136–138
pools 103–104, 112–114, 145, 151–152
resistance 71, 137, 162, **164**, 168–169, 382
transfer 136–138
genetic improvements **12**, **150**, 162–169
see also breeding; genetics; genomics; quality
genetics 4–5, 114, 130–131, 135–137, 144–156
see also breeding; genes; genomics
Genome-wide association studies (GWAS) **113**, 115, 116, 117, 118–119
see also QTL, mapping
genomic estimated breeding value (GEBV) 120, 121, 123
genomic selection (GS) 114, 120–121, 122
genomics 5, 108–124
see also breeding; marker-assisted selection
genotype-phenotype relationships 114–122
genotypes 114–122, 175, 358, 360–361
genotyping 115–121, 123, 139
geographic information system (GIS) 91, 116, 287, 289
see also remote-sensing
germplasm
collection 88–**89**, 90–91, 156
elite resources 121–122
evaluation 10, **71**, 231
exchange 152
multiple streams 114
preservation need 156
resources 112, 118
streams 114
see also breeding; genes; phenotyping; resources, genetic; seed
GIS (geographic information system) 91, 116, 287, 289
see also remote-sensing
Giza lines **5**, 175
government strategies 339–340
see also policies
grader, mechanical 337
grain
appearance 312, 313–316, 317, 329
breakage 297, 318, 321
characteristics 206–207, 299, 312, 313–316, 394
quality 13, 74, 121, **176**, 206–207, 311–32
hardness 314
sensory evaluation 300, 314–315, 317, 321, 435
taste 435
whiteness 311, 312, **315**, 316, 318, 320, 321
grasses 209–210, 284
growing environments **37**, **38**, 412
GS (genomic selection) 114, 120–121, 122
guidelines
efficient seed flows facilitation 426
manuals 7, 370
partial water control structures 288
see also recommendations
Guinea savannah agroecological zone 190
GWAS (genome-wide association studies) **113**, 115, 116, 117, 118–119
see also QTL, mapping

haplotypes 117, **134**
harvested-area
estimates 36, 37
expansion 25, **27**, 198–200, 425, 428–429
growth rates **28**
under rice **26**
harvesters 335, 339, 425–426
harvesting
hand-harvesting problems 334, 356
investment 431
mechanization 322, 334–335, 425–426
practices 399–400
timing 7, **8**, 154, 318–319, 399–400
head-rice ratio 312, 315, 318, 319, 321
heat 151, 154
see also temperature

herbicides **8**–9, 206, **208**, 209, 210
heterosis 173, 177
homogeneity 80, 301, 303, 304, 306
host plants 221–222, 236
host-pathogen relationships 218
hullers 327
humid forest agroecological zone 197, 276
husking recovery 312
hybridization 135–136, 357
hybrids
- breeding strategy 174
- development 100–103, 173–177
- evaluation 74, 174–175
- fertility 72, 95–96, 102, 131–132, **139**, 151
- sterility 132, 133, 151
- technology 173–177
- *see also* breeding; genetics

hydrology **277**
- *see also* water-management systems

iBridges development 72, **139**, 151
ICT (information and communication technology) 369, 374, 375, 432, 433
impact
- assessment 350, 375–**376**
- boosting 433
- extrapolation 398–399
- model parameters **419**
- parameters 50
- projection over time 66–67, 395, 397
- research 46–62, 391–413
- scientific options 403
- structural, model **396**–397
- studies 12–13

implementation 350, 432, 433–434
imported-rice
- attributes 298, 299
- consumption preference 294, 297, **298**, 301
- cost 69, 311, 390
- flows increase 308
- increase 324
- percentage 24, 279, 303, 435
- quality 299, 306, 307
- reduction 321
- regulating 33, 431
- reliance on 25, 303
- research impact 410–412
- surge **30**
- tariffs 32, 431
- tonnes 69, 311, 390

impurities 311, 312, 315, 321
- *see also* quality

incentives 273, 308
- *see also* subsidies

inclusion, social 369–374
income
- gain, regional distribution 403
- production constraints effect **58**–60
- projections 31
- research benefits 403–407, **420**, 434
- research impacts 58–60, 67–**68**, 390, 398–399, **419**
- variable, household 401
- yield loss impact **68**

incompatability
- interspecific **134**
- *see also* reproductive barrier

information
- access 124, 210, 322
- gender issues 347, 369
- sharing 122
- sources 297, 374
- *see also* databases; knowledge; learning; training

information and communications technology (ICT) 369, 374, 375, 432, 433
infrastructure 184, 251, 281–**282**, 287
inland valley systems 9, 195, 198, 276–290, 381, 429
innovations 347–348, 355, 362, 367–377, 379–388, 427
- *see also* land, preparation; weeders, mechanical

inputs, organic 256–258, 260
- *see also* fertilizer

insect pests
- area affected 55
- grain-filling stage intervention 319
- management 56, 229–237
- pre-and postharvest 230–231
- research 60
- resistance 175, **176**
- species 200, 284
- technology adoption 57
- *see also* integrated pest management

Instrumental Variable method 396
integrated pest management (IPM)
- approach 214
- challenges 237
- field-level 232–233
- genetic control component 218
- landscape-level 233–234
- post-harvest 234
- research solutions 393
- strategies 236

integrated rice management (IRM) 7, **8**, 9, 379–380, 428, 434
Integrated Transect Method (ITM) 287, 289
integrated-systems 179–186, 381–382, 428
intellectual property rights (IPR) 177
intensification 426–428, 433
International Rice Information System (IRIS) 92
interspecific bridge lines (iBridges) development 72, **139**, 151

interspecifics 3, 96, **101**, 104, 236
introgression 136, 138, 151, 162, 168
investment
  commercial, seed systems 184
  domestic rice value chain upgrade 32, 431
  foreign 431
  genotyping 123
  pay-back time 339
  promotion 429–430
  rate of return **12**
IPM *see* integrated pest management
IRM (integrated rice management) 7, **8**, 9, 379–380, 427, 434
iron (Fe) toxicity
  avoidance 8, 259
  cause 284
  distribution 148
  effects 9, 258
  reducing, nutrients 6
  severity 154
  tolerance 150, 151
irrigated-systems
  area 191, 266
  defined 190
  description 379
  ecosystems 208
  evapotranspiration 267
  infrastructures 251
  large scale 270–272
  management 6–8
  nutrient uptake 251
  performance 270–271
  potential 265, 428
  research **7**, 412, 434
  salinity problem 146–147
  soil constraints 193
  untapped potential 198
  water productivity 265–274
  yields estimates 39
  zones 270
isolines, RYMV-resistant, creation **167**
isotopes, stable 235
ITM (Integrated Transect Method) 287, 289

Jéby (off-season) 381–382, 384, **385**, 386, 387
  knowledge
  communication 260
  dissemination 347
  gaps, crop and weed interactions 211
  improved 9
  lack 254
  management 424, 432–433
  plant species 285–286
  uptake stimulating 433
  see also communication; information; learning; training; videos

labelling 186, 300, 305, 436
  see also packaging
labour 332, 333, 346, 357, 425
land
  issue 431
  levelling 259, 289, 333, 334, 338
  preparation **8**, 333–334, 338, 382
  tenure 200, 280
  use 286, 287, 289, 428–429
landraces 88
landscapes 194, 277
languages 367, **372**
  see also videos, translated
learning
  audio-based 433
  collective 379–388
  contexts changing 368
  facilitation 308
  gender-sensitive 347–348
  methods 368
  modules 384
  participatory 370
  requirements **371**
  rural, enhancement 367–377, 432
  tools 7, 368, 380–381, 387
  unsupervised 362–363
  see also knowledge; media; videos
legumes 9–10, 257, 259
lines 4, 72, 117, **137**, 151
livelihoods 277, 356, 377, **392**, 433
local-rice
  availability 325, 329
  characteristics 313–316
  comparative advantage 303
  competitiveness 13, 31–33, 306, 308, 324–330
  discounted 13, 305
  imported rice comparison 313–316
  preference 294, **298**, 299, 300, 312
  price 300, 329
  production 11
  quality 307, 311–322, 400, 431, 435
  research impact 400, 410
  unavailability 311
  value chain 430, 436
  varieties 96, 145
loci **132**, **166**
losses
  areas **416**
  assessment techniques 244
  causes 144, 204, 284
  impact **68**
  parasitic infection 284
  postharvest 242–243, 273, 332–333, 395, 399
  qualitative 399
  quantification 236
  reduction 50–56, **57**

lowland-systems
biotic stresses 200
collective learning and innovation experience 379–388
fertilizer use 253
intensified 35
irrigated 6–7, 189, 196–197, 253–256
rainfed 8–9, 189–190, 197, 251, 258–259, 412, 434

MABC (marker-assisted backcrossing) 114, 152
machinery 7, 8, 11, 206, 340
see also mechanization
mangrove systems 251–252
mangrove-swamp production system 190
manure 256, 257
see also fertilizer; nutrients
mapping 117–118, 132–133, 136–138, 167–168
marker-assisted backcrossing (MABC) 114, 152
marker-assisted recurrent selection (MARS) 73, 153
marker-assisted selection (MAS)
approach 115, 122, 318
breeding 2, 72–73, **113**, 119, 152
capacity-building 169
iBridges development **139**
QTL mapping 118
rice improvement **167**
marker-phenotype information 73
markers 72, 121, 138–139, 161
see also marker-assisted selection; molecular-markers
market
access importance 9
conditions 357
duality 329
enhancing 424
knowledge improving 430–431
opportunities creating 433
regulation policies 32–33
marketing 13, 15, 174, 429, 431, 435–436
MARS (marker-assisted recurrent selection) 73, 153
mechanization 8, 322, 332–342, 425–426, 436
see also machinery
media 260, 368–369, 374
see also videos
mega-environments 121–122
megagametophytes **134**
meiosis **134**
Meloidogyne incognita, resistance gene 137–138
MET (multi-environmental testing) **71**, 74–**77**
micro-credit 340
milled-rice 305, 315, 327
see also millers; milling; mills
millers 321, 436
milling
competitiveness factor 324
effects 312, **316**, 399
equipment, rice quality impacts 305
harvesting date effects 311
methods 319–320, 336–337
rate 175, **176**, 400
recovery 29, 312
rice quality impacts 305
mills 322, 327, 328, 329
mineral deficiencies 154
see also fertilizer; nutrients, deficiencies
molecular-markers 72–73, 89, 90, 161–169, 235–236
morphology 89–90, 136–137, 276, **278**
mulching 10
multi-environmental testing (MET) **71**, 74–**75**
Multi-Lateral Systems 88
multi-stakeholder platforms (MSPs) 14, 287, 338, 434

national performance trials (NPTs) 83
national research and extension agencies 340–341
national varietal release committee (NVRC) 79, 80–81, 83
natural-resources 6–10, 194, 285, 369
see also land; water
NDVI (Normalized Difference Vegetation Index) 271
near-isogenic lines (NILS) 135–**137**, 152
nematodes 10, 137–138
NERICA rice varieties
adoption 347, 355
amylose content 318
area 13
backcrosses 175
characteristics 290, **315**, 394
cooking and eating qualities 316–**317**, 321
development 3, 5, 162
growing conditions adapted 131
Oryza spp. content 139, 151
preference 316, 321
promotion 4, 11
resistance trait 6
storage duration 320
New Rices for Africa *see* NERICA rice varieties
Niger River 270
NILs (near-isogenic lines) 135–**137**, 152
nitrogen (N)
application 8, 259, 269
Azolla replacement 256
deficiencies 194, 258
legumes accumulation 257
levels 284
losses 146
recommended rates 255
recovery rates 197
see also fertilizer; mineral deficiencies; nutrients

Normalized Difference Vegetation Index (NDVI) 271
nucleotides 115
nursery preparation 383
nutrients
balances 250–251
deficiencies 147–148, 194, 284
inputs 194
iron toxicity reducing 6
management 251, 253–254, 260
recommendations 259
recovery rate improving 197
use 250–260, 289
see also fertilizer; nitrogen
NVRC (national varietal release committee) 79, 80–81, 83

observation 367
Ofada rice 300
off-types 97, 100, **102**–103, 104, 135, **139**
Office du Niger **268**, 270–272, 273, 335, 336
Orseolia oryzivora 10
'ORYLUX series' 318
Oryza genus 87
ORYZA model 6–7, 196, 255
Oryza spp.
diversity 110–112
donors **137**
genome types **111**
O. brachyantha (FF genome) 112
O. eichinngeri CC genome species 112
O. glaberrima
characteristics 89, 161–162
domestication 87, 95, 130–131
genetic diversity 130–131, 135–137
population structure 89
resistance 71, **138**
trait identificaton 90
unlocking 130–139
O. longistaminata 111
O. punctata weed 111–112
O. sativa 15, 87, 131–132, **139**, 161–162
sequencing 110

Pack-Mutator-like (Pack-MULE) transposable element 134–135
packaging 32, 184, 185, 234, 297, 337, 375, 436
see also branding; labelling
parasitoids 10, 174, 233–234, 237
parboiling 299, 320–321, 322, 330, 375, **376**
parents 72, **137**
participation 281, 282, 286, 287, 289, 290, 434
see also communities
Participatory Learning and Action Research for Integrated Rice Management (PLAR-IRM) 288–289, 347, 379–381, 382
Participatory Learning and Action Research (PLAR) 9, 289, 370, 379–380
participatory rural appraisal 370
participatory varietal selection (PVS) 155, 346–347, 355
partnerships 2–4, 431, 432, 433
see also collaboration; Task Force mechanism; Rice Sector Development Hubs
paspalum gall midge (PGM) 10
pathogens 10, 214, 216–218
see also blast; rice yellow mottle virus
pests 195–196, 200, 231–232, 234–236, 319
see also birds; insect pests; integrated pest management; nematodes; parasitoids; rodents
phenotypes 73, 89, 91, 114–122
phenotyping 72, 115–121, 155, 156
phosphorus (P)
deficiency 10, 147–148, 255, 257, 258
deficiency tolerance 72, 152, 154
photoperiod, varietal responses 5
photothermal responses 5
physiology research 5–6
plant material development **167**
see also breeding; genetics
planting density, bird damage association 243
PLAR (Participatory Learning and Action Research) 9, 289, 370, 379–380
PLAR-IRM (Participatory Learning and Action Research for Integrated Rice Management) 288–289, 347, 379–381, 382
policies
agricultural equipment importation 339
conducive, promoting 431–432
dialogue 338
instruments 33
R&D 15
sequencing 430–431
value-chain approach 11–12
Policy Analysis Matrix (PAM) 11
pollination, supplementary, rope use **177**
polymorphism analysis 236
population 31, 49, 63–64, 65, 67, 266
post-zygotic barrier 135
postharvest handling and management
facilities 290
see also dryers; milling; storage; threshing; transport
insect pests 230–**231**
integrated pest management 234
losses 242–243, 273, 332–333, 395, 399
practices 7, **8**
research 412, 434
timing 399–400
potassium (K) 154, 255, 259
pounding *see* milling

poverty
  reduction 60–61, **409**, **411**
  research impacts 67, 390, 398–399, 407–409, **419**, **422**
  targeting 370
  yield loss impact **68**
pre-breeding activities **71**
pre-harvest cultivation practices 318–319
precision management systems 427–428
prediction models 154
  see also RIDEV (Rice Development) simulation model
preferences 294–301, 317, 321, 358, 359
price
  attribute 299
  crisis 213, 304, 307
  differentiated 436
  discount 305, 325–326
  local rice 329
  rise 326
  spikes 12, 25, 307
  stabilization 33
priority-setting 391–412, 432, 433, 434
processing
  activities 399
  enhancement 424, 435
  investment 431
  practices 315, 436
  technologies promotion 429–430
processors, research impacts 399–400
production systems
  see irrigated-systems; rainfed-systems; deep-water systems
protection 24–25, 328
  see also subsidies; taxes
protein content 135, 313
puddling 266, 289
pumps 334, 338
purchase, aggregation 33
purchasing power parity (PPP) poverty line 434
PVS (participatory varietal selection) 155, 346–347, 355

QDS (Quality Declared Seed) 185
QTL (quantitative trait locus)
  abiotic stresses identified 156
  identifying 115, 136–137
  introgressing 113
  mapping 108, 116, 117–118
  moving across species 114
  Pup1 152, 155
  traits identification 72, **138**
  yield component 73
quality
  acceptable 183
  attributes, physical 321
  benchmark 306
  characteristics 311
  see also amylose content; grain, breakage; grain, whiteness; protein content
  comparison 314
  control 180, 186
  demand 305
  enhancing 15, 429–431
  gap 399
  governance 304, 307
  improvement 13, 311–322
  standards, consumer-set 325–326
  upgrading 183–184, 306–307
  value chain role 435, 436
  see also standards
Quality Declared Seed (QDS) 185
quantitative trait locus *see* QTL

R&D *see* research & development
R-BIP (Retrotransposon-based insertion polyorphism) markers 135
radio 369, 373
  see also media
rainfall
  bird damage association 243
  distribution knowledge 258
  fertilizer application relationship 258
  knowledge 259–260
  regimes 276, **278**
  relative yield reduction 196
  temperature 153
  variability 210
  see also climate; rainfed-systems
rainfed-systems
  area 37, 191
  cultivation 324–325
  drought risk 200
  herbicides use 209
  high-input/commercial systems 190
  inland valley 380
  low-input systems 191, 200
  lowland 8–9, 189–190, 197, 251, 258–259, 412, 434
  subsistence farming 191
  upland 190, 197–198, 251, 256–258, 412, 434
rats 284
  see also rodents
reapers 334–335
recombinant inbred line (RIL) 136–137
recommendations 255, 256, 259, 260, 269, 341
  see also guidelines
Red-billed Quelea 242, 243, 247
RefSeqs 110, 111
remote-sensing 270–272, 273, 287, 289
reproductive barrier **132**, 133, 135, 139
  see also S_1 locus

research
agenda 432, 433–435
benefits 144–145, 390–391, **405–407**, **408**
disciplines **407**, **408**, **411**, **418**
gender-responsive components **349**
impact-orientated 144, 424, 432–434
impacts 46–62, 65–70, 144, 390–413, 415–423
priority-setting 391–412, 433
themes **404**, **405**, **406**, **408**, **410**, **411**, **417**, **418**
see also research & development
research & development (R&D)
agricultural categories 15
costs 409–412
demand-lifting 15, 32, 301
forums, agricultural 341, 374
gender considerations integration 343–352
partnerships 3–4, 7
rate of returns analysis **13**
strategic plan 425
see also research; Task Forces
resistance
alleles 72, **153**, 162, 168
biotic stresses 71, **73**, 221–222
breeding 73
evaluation **168**
genes 137, 162
high 164
hybrids **176**
management 218
partial 163–164
sources identification 10
study 162–169
traits **112**, **138**
resources
access 260, 344–345
genetic 87, 90–92, 433
natural 6–10, 194, 285, 369
see also land; water
restriction fragment length polymorphisms (RFLPs) 115
Retrotransposon-based insertion polyorphism (R-BIP) markers 135
RFLPs (restriction fragment length polymorphisms) 115
Rhamphicarpa fistulosa parasitic weed 284
Rice Advice DVD 373–**374**
Rice Sector Development Hubs 13–14, 15, 434
rice stripe necrosis virus (RSNV) 137, **138**
rice types 297
rice yellow mottle virus (RYMV)
description 215
genetic control 162–169, 220–221
near-isogenic lines **137**
perception 51, 52, 55
resistance 90, 137, 163–**164**, 168–169, 382
resistant and tolerant rice species 71, **73**, **112**, 221–222, 382
studies 10
susceptibility 175
symptom development delay 163
variability 216–217
rice-sector development 13–14, 15, 390, 434
RICEPEST simulation model 236
RIDEV (Rice Development) simulation model 5, 6–7, 8, 154, 255–256
rodents 50, 55, 60, 61
roguing 102
round-grain (japonica) rice 297, 298, 299
RYMV *see* rice yellow mottle virus

S_1 locus 132–135, 139, 151
see also reproductive barrier
Sahel 6–8, 175, 190, 197, 273, 379
salinity
climate change effects 153–154
irrigation problem 200
locus 151
modifiers 193, 194
resistant germplasm identified 71
tolerance 6, 71, 146–147, 151, 152, 174
Saltol 73, 151, 152, 155
sawah systems 280, **281**, 288
scaling 8, 123, 337, 370–374, 385–387
SEBAL (Surface Energy Balance Algorithm for Land) model 271
seed
breeder 182
collection 88
farmer-saved **182**, 375
foundation-seed 183
health video 347
hybrid 174, 177
legislation 184, 186
quality 180, 375, 382–383, 436
release systems 180
sources **181**, 182
value chain 183, 184–186
seed-systems
access 32, 86, **181**
boards 184
commercialization 186, 308, 346
control infrastructures 184
development 80, 186
emerging formal systems 186
enterprises, performance 184–185
establishing 425
informal/farmer 179, **181**–182, 183–184
integrated sector 179–186
legislation 11, 426, 431
regulation 79, 184
sector development, key issues **183**

seed-systems (*continued*)
- security stocks 186
- transaction involvement, farmers 182

seeding **8**, 209, 334
- *see also* sowing

seedlings 89–90, 382, 384
selection
- criteria 155
- genomics-facilitated 108–109, 114, 120–121, 122
- genotypic 161
- marker assisted 72–73, **139**
- pressures 104
- recurrent 73–74
- *see also* breeding; genetics; marker-assisted recurrent selection; marker-assisted selection

Senegal River valley (SRV) rice 7, **8**, 306, 307, 309, 326
sensory evaluation 300, 314–315, 317, 321, 435
sequencing 109–112, 116
shield bugs, grain damage 319
simple sequence repeat (SSR) markers 89, 90, 115, **139**, 162, 167
single nucleotide polymorphisms (SNPs)
- arrays 108, 115–116, 117
- chips 155
- density 117
- genome sites 109–110
- markers **113**, 114, 116, 138–**139**

site-specific nutrient management (SSNM) research 260
small-scale systems 355–363
smallholders 250, 355–363, 431–432
SNPs *see* single nucleotide polymorphisms
society, technology interactions 355–363
socio-cultural conditions 359
socio-cultural practices 345–346
socio-economic factors 280, 283, 322, 356, 358, 390
socio-technical landscape 362
soil
- classification system 193, 194, 252
- constraints **53**, 56, 60, 61, 144, 192–195
- degradation 250, 273
- fertility 7, 193, **194**, 258, 278, 284, 383
- problem 145, 154, 200
- quality **252**, 273
- resources 251–253
- *see also* iron (Fe) toxicity; salinity

sowing **8**, 154, 242
- *see also* seeding

soybean, dual-purpose 257
spatial analysis 43–45, 191, 270, 273
- *see also* remote-sensing

species
- cultivated rice 161–162
- *see also* varieties
- insect pests 200, 284
- pest birds 242
- weeds 111–112, **205**, 285–286
- wild 87, 88, 95, 98–99, 110

SRI (System of Rice Intensification) 269, 270, 273
SRV (Senegal River valley) rice 7, **8**, 306, 307, 309, 326
SSNM (site-specific nutrient management) research 260
SSR (simple sequence repeat) markers 89, 90, 115, **139**, 162, 167
standards 79, 80, 181, 306, 325–326
- *see also* quality

staple food crops contribution **25**
starch content 300
sterility
- barrier 71–72, 104, 114
- interspecific 131–132, 133, 151
- restorer lines 175
- spikelet 196
- *see also* reproductive barrier

stocks, local, establishment 431–432
storage 32–33, 300, 320, 322, 330, 436
strategies
- best-bet 286, 287, 289
- food security 307–**308**
- government 339–340
- locally adjusted 288–289
- non-price 304
- supply and demand 307
- supply-shifting 32

stresses
- abiotic 144–156
- biophysical 56–61, 65–68, **283**, 390, 412
- combinations 155
- environmental side-effects 394–395
- grouping **415**
- resistance 72, **112**
- tolerance 71, **112**, 144–156
- *see also* constraints; soil; temperature; water

Striga hermonthica resistance 6
sub-humid agroecological zone 191
submergence tolerance 152–**153**
subsidies 32, 180–181, 328, 339
Sudan savannah agroecological zone 190
supply and demand 69, 307, 327
supply-shifting 15, 32
supply-side perspective 303
support markets 306
Surface Energy Balance Algorithm for Land (SEBAL) model 271
sustainability
- action priority 426–428
- development 3, 344–345, 424
- genomics-assisted breeding potential 123
- harvested areas expansion potential 198–200
- inland valley strategies 286–289

integrated pest management role 229
issues 254
management options 229
mechanization 339–341
production/productivity 425
System of Rice Intensification (SRI) 269, 270, 273

tariffs 31–32, 431
Task Force mechanism 2–3, 69–70, 74–**75**, 150–151, 198
see also partnerships
taxes 328, 339
technography 361, 362
technologies
access 116, 150
adoption **57**, 62, 150, 338–339, 395
complexity 368
development approaches 9, 338
development and dissemination models 361–363
flexible 427–428
innovation 382–384
interactions 355–363
uptake 359, 433
visibility 367–368
see also tools; varieties
temperature
extremes 56, 148–150, 154, 196, 200
increases 210
knowledge 259–260
variations **149**, 153
varietal responses 5, 151, 154
threshers 8, 338, 339
threshing 290, 319, 322, 335–336, 425–426
tillers, power 288
tolerance
breeding 72, 73, 145, 150, 152, 155–156, 175–176
differential varieties 72
gene discovery 71
improvement strategies 150–153, 174
mechanisms 154
screening 154–155
traits **112**, 150–153, 154
upgraded varieties 427
tools 7, 272–273, 348, 368, 380–381, 387
see also technologies
toposequence 194–196
toxicity *see* Iron (Fe) toxicity
traceability 186
tractors 333, 334, 338
see also mechanization
traders 399–400
training
facilities 260
institutes, government support 339–340
manual 370
materials 373
populations 123
programme 386
sets 123
videos 8, 387
see also knowledge; learning
traits
adoption rate role 357
composite 116
genes/QTLs localization 71
heritability 116, 117, 145, 161, 218, 318
identified 72, 90, 112
list 83
morphological 89
taste 358
yield, genotype-by-environment (GxE) interactions 358–359
see also genetics; genomics
translucency 312, 314, 315, 317, 321
transmission ratio distortion (TRD) 131–132, 133
transplanting 209, 382
transport 290
TRD (transmission ratio distortion) 131–132, 133
Truthfully Labelled Seed 185

undernourishment **423**
upland-systems
biotic stresses 200
cropping 9–10, 209
rainfed 190, 197–198, 251, 256–258, 412, 434
urbanization 24, 294, 301
urea super granules (USGs) technology 255

value for cultivation and use (VCU) requirements for varietal release 80, **82**, 83
value-chains
analysis 303, 326–327
competitiveness 308, 324–330
defined 304
development 13–14, 15, 307–308, 324–330, 433
development-orientated 185–186
finance access enhancement 430
income, research benefits **404**
systemic constraints 306
tailoring to consumers 303–308
upgrading 32
variation 114–122
varieties
acceptability scores **314**
access lack 70
adaptable 355
adoption

varieties (*continued*)
cultural and socioeconomic factors 356
curves 359
data collection 412
demand elasticities **416**
history 83, **84–85**
impact 12
probability 401
rate 394, **415**
register 79
research impacts 402, 404, **406**, **407**, **408**, **411**
role 357–358
catalogue 80
characteristics **392**, 402, **415**, **416**
choice **8**, 243, 359
comparing 359
criteria change 358
development 69, 70, **71**, 101, 152–153
duration 357
introducing, testing, disseminating 4–5
lists 5, 401
new 12, 260, 346–347, 382–383
rejection 356
release systems 79–86
selection 317–318, 322
testing and disseminating 4–6
see also breeding; genetics; marker-assisted selection
VCU (value for cultivation and use) requirements for varietal release 80, **82**, 83
vegetables 259
see also legumes
Vickrey second price auctions 306
videos
based learning 348, 433
impact assessment 375
translated 8, 347, 348, 372
viscosity, cooked rice flour 313
visual aids 387
see also videos

water
applied by irrigation 260
availability 145, 278
consumption 271, **272**
control 8, 258–259, 287, 427, 435
excess 147
see also flooding
intake reduction 267
problem, farmer perception 56
productivity 265–274
re-use 266
regimes 6, 382
related abiotic constraints **54**
see also drought; flooding
resources 284, **285**, 428–429
saving 268, 269, 273, 318
scarcity 200, 260, 266, 272–273
see also drought
sources 35
standing, maintaining 208
stress 266
table continuum **36**
uptake ratio 313, 316
water-management systems
continuum **36**
design 287
facilitation 259
fertilizers/herbicides efficiency maximizing **8**–9
harvesting 251
improved 144, 288
infrastructures 281–**282**, 287
measures 384
structures 279–282
watershed 194
weather-forecasting 258
weeders, mechanical 334, 382, 383, 427
weeds
bird-damage association 243
competition 6, 206, 210, 284, 289
control **208**, 243, 269–270
herbicide resistance 210
impact 61
importance perception 50, **51**, **52**, 53–54
management 7, **8**, 204–211, 254, 383
parasitic 284
problems 200
species 111–112, **205**, 285–286
technology options 60
yield loss factor 57
wetlands 198, 276–277
wild species 87–92, 95, 98–99, 110–112
willingness-to-pay (WTP) 306, 326
wisdoms 345–346
women 281, 343, **344**, **345**, **371**
see also gender issues
WP_I (water productivity) 266, 267, 268, 269, 271, 273

yield
actual **145**
attainable 46
average **41**
defining factors 46
estimation 35–42, 271, 282
gains 288
gap 188–200, 379, 425, 426–428, 433
growth rate **27**, 28
high, breeding for 73–74
increase 15, 25, **26**, **27**, 33, 254

integrated management effects 289
limiting factors 46, 189
see also drought; nutrients, deficiencies; temperature, extremes; water, excess
potential 46, 72, 73, **145**, 188–189, 196–198, 382
reducing factors 46, 144, 189, 426
see also birds; constraints; disease; insect pests; rodents; weed
variability 7, **175**
videos impact **375**

zinc (Zn) deficiency 61
zones, agroecological 190–**192**, **195**, 197, 270, 359
zooming-in zooming-out (ZIZO) approach 347–348, 371–**372**